GRAPHITE, GRAPHENE, AND THEIR POLYMER NANOCOMPOSITES

Edited by

Prithu Mukhopadhyay
Rakesh K. Gupta

CRC Press
Taylor & Francis Group
Boca Raton London New York

CRC Press is an imprint of the
Taylor & Francis Group, an **informa** business

CRC Press
Taylor & Francis Group
6000 Broken Sound Parkway NW, Suite 300
Boca Raton, FL 33487-2742

Printed in the United States of America on acid-free paper
Version Date: 20120627

International Standard Book Number: 978-1-4398-2779-6 (Hardback)

Library of Congress Cataloging-in-Publication Data

Graphite, graphene, and their polymer nanocomposites / [edited by] Prithu Mukhopadhyay and Rakesh K. Gupta.
p. cm.
Summary: "Here we go again. After intercalated compounds of graphite (1974), fullerenes (1985), and carbon nanotubes (1991), it is time for another allotrope of elemental carbon to be at the forefront of scientific curiosity. The allotrope is: "graphene". By graphene, we mean the basal plane of graphite, a one atom thick two dimensional honeycomb layer of sp2 bonded carbon. Conversely, when many graphene layers are stacked regularly in three dimensions, graphite is created"-- Provided by publisher.
Includes bibliographical references and index.
ISBN 978-1-4398-2779-6 (hardback : acid-free paper)
1. Graphite. 2. Graphene. 3. Polymeric composites. 4. Nanocomposites (Materials) I. Mukhopadhyay, P. (Prithu) II. Gupta, Rakesh K. (Rakesh Kumar), 1954-

TA455.G7G73 2013
620.1'18--dc23 2012018248

Visit the Taylor & Francis Web site at
http://www.taylorandfrancis.com

and the CRC Press Web site at
http://www.crcpress.com

To the memory of my father who taught me the value of true education.

Prithu Mukhopadhyay

To my mother who helped me become what I am today.

Rakesh K. Gupta

Contents

Preface

Here we go again. After intercalated compounds of graphite (1974), fullerenes (1985), and carbon nanotubes (CNTs) (1991), it is time for another allotrope of elemental carbon to be at the forefront of scientific curiosity (Boehm 2010). The allotrope is "graphene." By graphene, we mean the basal plane of graphite, a one-atom-thick two-dimensional honeycomb layer of sp^2-bonded carbon. Conversely, when many graphene layers are stacked regularly in three dimensions, graphite is created.

Why such an interest in graphene? It is all about the digital world, and the search for materials that will make integrated circuits smaller, faster, and cheaper! Graphene is a semiconductor with a zero band gap and an exceptionally high charge mobility. In fact, electron mobilities in graphene could reach values that are more than an order of magnitude higher than those encountered in an Si transistor. This opens up the tantalizing possibility that one day graphene might replace silicon as the building block of the electronic industry and revolutionize nanoelectronics. Although the existence of graphene had been known for a long time, the material had never actually been synthesized. This had to await the work of Andre K. Geim and Konstantin S. Novoselov of the University of Manchester, U.K., who were awarded the 2010 Nobel Prize in Physics for their ability to isolate a defect-free, single sheet of carbon atoms through micromechanical cleavage of graphite, whereby monolayers are peeled from graphite crystals (Novoselov et al. 2004). This method, however, produces a very small amount of pristine graphene, which makes it unsuitable for efficient and scalable high-volume manufacturing. Nevertheless, this pioneering work paved the way to the rise of intense graphene research.

Important characteristics of graphene are that it is nanoscale in dimension, and it is derived from graphite, an inexpensive precursor. Consequently, a key goal of worldwide research has been to produce a large enough volume of pristine graphene safely and in a cost-efficient manner. Other researchers are seeking practical applications of graphene that will benefit society, especially in the electronics area.

In the technical literature, a number of methods have been described for producing graphene, which can broadly be classified as (1) micromechanical exfoliation, (2) epitaxial growth of graphene films, (3) chemical vapor deposition (CVD), (4) unzipping of CNTs, and (5) reduction of graphene oxides. Each method has its own benefits and related drawbacks. A "bottom-up" approach using chemical synthesis is an interesting strategy (Cai et al. 2010; Choucair et al. 2009). However, scaling-up to produce large quantities remains a formidable challenge. Research efforts have, so far, established that an easier route to manufacture large amounts of graphene is via the chemical exfoliation strategy. During chemical exfoliation, such as oxidation and subsequent reduction of graphite oxides, one produces partially or highly reduced graphene oxides. From a chemical viewpoint, these graphene sheets contain various types of residual oxygen-containing species. From a physical viewpoint, graphene sheets become corrugated, and the graphene platelets can contain a variety of defects such as topological, adatoms, edges/cracks, vacancies, loops, adsorbed impurities, and so on within the graphene-like structures (Terrones et al. 2010). When the dimensions of these platelets fall in the nanoscale range, they are commonly termed "graphene nanoplatelets" (GNPs). One of the most technologically promising applications of nanographene materials is in polymer reinforcement. Studies have shown that stress transfer takes place from the polymer matrix to monolayer graphene, indicating that graphene acts as a reinforcing phase (Gong et al. 2010).

Polymers have been combined with other polymers to form blends and copolymers, toughened by the addition of elastomers, mixed with talc, calcium carbonates, and clays to give filled systems and extruded and molded with fibers and other anisotropic reinforcements to yield composites and hybrid materials. This simple "mix-and-match" approach has allowed polymer scientists and engineers to utilize a small library of polymers to produce a bewildering array of useful products capable of possessing extremes of property values. Traditional filled-polymer composites typically utilize high loadings of micron-sized filler particles to obtain desired properties. If the filler particle size is reduced to its nanoscale dimension from its micron size, similar properties are achieved but with a drastically reduced filler loading level to achieve percolation. Primarily, this is due to the surface area to volume ratio of the nanoparticles, which is several orders of magnitude larger than that of micron-sized fillers. To qualify to be called a "nanoparticle," the particle has to have at least one dimension in the nanometer range. Therefore, when nanometer particles are dispersed in a polymer matrix, the result is termed as "polymer nanocomposite" (PNC); the matrix itself can be single or multiphase. The critical reinforcing effects of nanosized particles come from their aspect ratio, very large specific surface area, and the particle–matrix interactions. The original concepts for PNCs owe their origin to the invention of polyamide–clay composites at the Toyota Research Corporation in 1985 (Okada et al. 1988). At that time, the objective was to make plastics used in under-the-hood applications heat resistant and lighter than metal. Since then, the list of nanoparticles has grown and PNCs have seen numerous commercial applications ranging from autoparts to packaging to coatings (Ashton 2010).

The latest addition to this palette of nanomaterials is graphene. Graphene layers can be stacked, functionalized, and modified to provide numerous types of graphene-based nanoscale materials. Rolled-up graphene, known as CNTs also has some structural flexibility. However, the performance and cost advantages of graphene challenge CNTs in nanocomposites, coatings, sensors, and energy storage device applications. For instance, the quality of graphene's crystal and band structures yields uniquely low noise levels, increasing the sensitivity of the sensors (Yang et al. 2010). Although incorporation of CNTs in large-scale integrated electronic architectures is a daunting task, graphene is highly amenable to microfabrication (de Heer et al. 2007). On the other hand, the cost advantage of graphene, graphene oxides, or its nanoplatelets over CNTs stems from easy access to the graphitic precursor material, the cost, and the scalable method. In addition, due to its structure, graphene raises fewer toxicity issues as compared to CNTs.

This book attempts to compile, unify, and present the emerging research trends in graphene-based polymer nanocomposites (GPNC). Researchers from several disciplines across the continents share their expertise and research knowledge about graphene, its properties, and the behavior of graphene-based composites. To the best of our knowledge, there is no other published book that provides this kind of a comprehensive snapshot of graphite, graphene, and their PNCs. Without a broad perspective of the underlying physics and the chemistry of graphene, the full story of GPNC remains untold. That is indeed our premise, and this is where the story begins.

Organization of the Chapters

The preface provides the foundation for this book and describes how readers can benefit from the developing story of the most sought-after carbonaceous nanomaterial, *graphene*. In Chapter 1, John Zondlo from West Virginia University (USA) introduces us to natural and synthetic graphite, their properties, and characterization techniques; graphite, after all, is the precursor to graphene. The chapter lucidly describes where natural graphite is found and how synthetic graphite is manufactured. The author lists the prominent commercial applications of graphite in this chapter. It is evident that societal growth would be impeded without the use of graphite. Applications range from graphite electrodes in the electric arc furnace to graphite refractories in the steel and aluminum industries to nuclear reactors as both a moderator and a reflector. A description of the importance of the different forms of graphite provides the link to the next few chapters.

Chapters 2 and 3 prepare us about graphene and its unique characteristics. Pierre Carmier of CEA–INAC/Université Joseph Fourier, Grenoble (France), elaborates in Chapter 2 as to what has made graphene to climb to the top of the materials research chart. The author enumerates the electronic transport properties of graphene and explains how graphene's bipartite Bravais lattice and its gapless band structure make graphene so exotic! While delving into the theoretical issues of graphene, the author reminds us about the obstacles that have to be overcome for real carbon-based electronics. Most of us will agree that characterization of graphene is one of the most critical tasks in graphene research. Once graphene is synthesized, there are numerous techniques for its characterization. However, some of these are still evolving as technology advances. In Chapter 3, Viera Skákalová and Dong Su Lee of Max-Planck Institute, Stuttgart (Germany), and Alan B. Kaiser of Victoria University of Wellington (New Zealand) address this critical task. Not only do the authors summarize different characterization techniques available in microscopy (AFM, SEM, TEM, STM) and in spectroscopy (Raman, Augur, ARUPS, XPS), but they also discuss unique features of electrical conduction properties depending on the degree and the nature of defects in graphene. The authors conclude this chapter by highlighting thermal conduction properties of graphene and how the thermoelectric power of graphene could aid in the thermoelectric conversion of heat to electrical energy.

In the "top-down" approach, exfoliation of graphite is the key to the preparation of single or multiple graphene nanosheets. The quality of graphene materials can vary depending on the preparative methods. In reality, only cost-effective approaches to produce graphene sheets in large quantities would have commercial importance. In Chapter 4, Martin Matis, Urszula Kosidlo, Friedemann Tonner, Carsten Glanz, and Ivica Kolaric of Fraunhofer Institute for Manufacturing Engineering and Automation, Stuttgart (Germany), detail electrochemical exfoliation of graphite and recent advances in the process. They show that graphite can be exfoliated using different nontoxic electrolytes such as polyelectrolytes, surfactants, and ionic liquids. In addition to the low cost, this exfoliation process has the advantage of producing functionalized GNPs in bulk quantities, which is crucial to graphene-based nanocomposites manufacturing. In Chapter 5, Weifeng Zhao and Guohua Chen of Huaqiao University (China) discuss different exfoliation routes to producing graphene and graphite nanoplatelets for their use in polymer composites. Numerous studies on polymer-based graphene nanocomposites are discussed. Ways to bring graphene nanomaterials into the real world of polymer processing are critically examined. The authors of this chapter propose the use of the wet ball-milling method to further exfoliate graphite nanoplatelets into graphenes. Chapters 4 and 5 describe possible routes to bringing graphene in bulk quantities to the market.

Chapters 6 through 9 showcase how graphene as the newest nanomaterials can be used in numerous applications of commercial interest.

Most of us will agree that clean energy is essential for securing the future of our planet. Chapter 6 deals with emerging clean energy applications of graphene-based materials in solar cells, lithium ion batteries, supercapacitors, and catalysts. Bin Luo, Minghui Liang, Michael Giersig, and Linjie Zhi of the National Center for Nanoscience and Technology, Beijing (China), discuss each of the clean energy applications in depth and provide a glimpse of graphene's applications in the domain of clean energy technology. These authors cite an impressive number of studies available in the current literature to capture the progress and future directions of each of these technologies. Because of its versatile application potential, the authors surmise that graphene could be used as a base material for various optoelectronic devices in the future. Martin Pumera of Nanyang Technological University (Singapore) delves into the electrochemistry of graphene-based nanomaterials in Chapter 7. It is known that chemical activity drastically changes at the edges of graphene depending on their carbon termination. However, the author argues that there is no significant difference between the electrochemical response of single-, few-, and multilayer graphene sheets. He discusses the importance of electrochemical performance of graphene nanostructures in applications such as sensing and biosensing, supercapacitors, and batteries.

Chapter 8 examines the fabrication of graphene-PNCs and their applications as saturable absorbers for pulse lasers. Both graphene-based Q-switched lasers and mode-locked lasers are examined. In this chapter, Kian Ping Loh, Qiaoliang Bao, Dingyuan Tang, and Han Zhang of Nanyang Technological University (Singapore) show that the functionalization of graphene via covalent linking of a dye to the basal plane, and noncovalent attachment of aromatic molecules, aids in tuning the optical properties. The authors believe that electrospun graphene-PNCs are promising candidates for practical and efficient photonic materials in the generation of high energy or ultrashort pulses.

Epoxies are a class of thermoset polymers utilized extensively in products ranging from floor coatings to aircraft fuselages. Chapter 9 written by Iti Srivastava, Mohammad A. Rafiee, Fazel Yavari, Javed Rafiee, and Nikhil Koratkar of Rensselaer Polytechnic Institute, New York (USA) discusses the potential of graphene as a nanofiller in epoxy-based composite materials technology. The practical relevance of the nanocomposites' mechanical properties such as tensile strength, Young's modulus, buckling resistance, and ductility to material properties including fracture toughness and fatigue resistance are examined. The authors demonstrate that the graphene content required to significantly boost the mechanical properties of epoxy systems is one to two orders of magnitude lower than with the use of other competing nanofillers such as CNTs, nanoclays, as well as silica/aluminum/titania nanoparticles. The process of making hierarchical graphene/epoxy/E-glass fiber composites and their properties is also discussed. Finally, the authors have summarized technical issues that require attention in order to realize the full potential of graphene-based epoxy nanocomposites.

Nanoparticles come in different shapes and sizes. Chapter 10 provides an overview of the various types of nanofillers (ranging from metallic nanoparticles to silicates to biosource nanoparticles to CNTs and graphenes) and reviews their utilization in a variety of PNCs. This chapter authored by Musa R. Kamal and Jorge Uribe-Calderon of McGill University, Montreal (Canada), provides information on the enhancement of a range of polymer properties and also discusses challenges during the preparation of the nanocomposites. Chapters 11 and 12 address two distinctly different methods of GPNC preparation. Instead of employing the conventional route to PNCs preparation, such as solution or melt mixing, Chapter 11 reveals how a robust and yet a simple technique for controlled polymerization, namely, atom transfer radical polymerization (ATRP), is utilized to produce graphene-based PNCs. Chapter 11 written by Arun K. Nandi, Rama K. Layek, Sanjoy Samanta, and Dhruba P. Chatterjee of the Indian Institute for the Cultivation of Science, Kolkata (India), discusses the work of these authors and suggests that the ATRP method has the potential to modify the surface of graphene by grafting polymeric moieties. This improves filler dispersion and allows one to fine-tune various properties of graphene-based PNCs. Chapter 12 deals with the synthesis of GPNCs in a biodegradable polymer matrix and utilizes a solution mixing procedure. In this chapter, Gui Lin and James E. Mark of the University of Cincinnati (USA) and Isao Noda of the Procter & Gamble Company, Ohio (USA) report on the structure–mechanical properties relationships of PHBHx reinforced by expanded graphite, graphene oxide, and reduced graphene oxide.

Chapters 13 through 15 are devoted to the specialized properties of graphene-based PNCs.

In Chapter 13, Olga Shepelev and Samuel Kenig of Shenkar College of Engineering & Design, Ramat Gan (Israel), explore the opportunities to modify lower-cost graphene nanoplatelets for further compounding and processing in conventional plastics processing equipment. The authors compounded treated graphene into a polypropylene matrix to prepare PNCs and showed that strategies such as surface treatments, ball milling, and ultrasonification improved mechanical, thermal, and electrical properties. Chapter 14 deals with barrier properties of GPNCs. In this chapter, Matthew S. Thompson, Sushant Agarwal, and Rakesh K. Gupta of West Virginia University (USA), and Prithu Mukhopadhyay of IPEX Technologies, Inc., Verdun (Canada), describe the process of molecular diffusion through polymers and show that graphene-based PNCs have similar mass transfer barrier behavior to clay-based PNCs, but the loading level of graphene needed is much lower. By conducting sorption experiments, these authors predict significant permeability reductions in the nanocomposites despite enhanced solubility of the diffusing molecule with increasing graphene content! Chapter 15, authored by J. S. van der Wal of Composite Agency, Amsterdam (the Netherlands), addresses the potential reduction in service life of fiber-reinforced composites and multilayer polymer structures due to ingress of small molecules into the polymer. These permeants can plasticize the polymer matrix and also cause delamination at the fiber–matrix interface. Using methanol as a diffusing molecule, the author shows how the incorporation of a small amount of functionalized graphene sheets in epoxy can counteract the tendency of chemically driven filler-matrix failure.

In Chapter 16, a review of graphene/PNCs is reproduced. Here, Hyunwoo Kim and Christopher W. Macosko of the University of Minnesota, Minneapolis (USA), and Ahmed A. Abdala of the Petroleum Institute, Abu Dhabi (UAE), provide a lucid perspective of graphene-based PNCs research.

Chapter 17 is concerned with the preparation, properties, and limitations of highly filled graphite–polymer composites. The superior electrical conductivity of graphite is attractive for the preparation of bipolar plates of proton exchange membrane (PEM) fuel cells. This chapter coauthored by Sadhan C. Jana and Ling Du of the University of Akron, Ohio (USA), employed a synergistic combination of expanded graphite and electrically conductive carbon black in epoxies to examine both in-plane and through-plane electrical conductivity.

The Opportunities and the Challenges

Graphene research has caught the world's attention. Start-up companies that supply graphene materials are emerging in different parts of the world, and nanomaterial providers are adding graphene to their product portfolio. Market research entities are computing graphene sales numbers and projecting encouraging future sales volumes. Even governments are allocating money for funding graphene research. From a commercial standpoint, this is indeed good news for graphene-based PNCs. This new material is entering a crucial segment in its product life cycle from innovation to applications. But challenges still need to be overcome in order to bring about synergy between graphene research and its myriad anticipated applications.

The chemistry part of graphene and its derivatives has, however, begun to unfold. Graphene sheets are individually very strong, but, in graphite, sheets slide past one another making the material soft as is the case with pencil lead. Likewise, the thermal conductivity of suspended graphene differs considerably from graphene grown via CVD. Although composed of identical atoms, only differently arranged, the material properties are drastically different. Can the basal surface of graphene be made reactive? A knowledge of the interactions at the liquid–graphene interfaces is crucial to the application of graphene in electrochemical energy storage systems. So are the understanding and control of hydrophobic interactions in the field of protein folding and self-assembly. Studies are being conducted to examine the controllable interaction of water with epitaxial graphene films of different thickness values (Zhou et al. 2012). Indeed, graphene's wetting transparency comes from its extreme thinness. How can one exploit the wetting response in the design of conducting, conformal, and impermeable surface coatings (Rafiee et al. 2012)? Only future studies will tell! Curiosity will reveal one day what happens to graphene when it is subjected to high compression. These new pieces of information will allow researchers to continue to address fundamental questions that go to the very core of our understanding of chemical interactions in materials while simultaneously opening the doors to innovative applications.

The synthesis and properties of graphene nanoplatelets, its oxides, and the control of type and quantity of oxygen-containing species have been the research foci of the graphene-based nanocomposite community (Mukhopadhyay and Gupta 2011; Potts et al. 2011; Terrones et al. 2011). Strategies such as electrolytic exfoliation, ball milling, and microwave heating are being advanced as the means of obtaining single, few and multilayered graphene flakes in bulk quantities. Spectroscopic techniques (Raman, SEM, TEM, AFM, XPS, Auger, ARUPS) are being used to understand the surface and edge chemistry of graphene. Various techniques are needed since the morphology of graphene sheets changes when they are derived from different synthetic routes. A proper characterization of graphene is critical to understanding and exploiting GPNC properties. To tailor polymer chain length and molecular weight, modern synthetic tools (ATRP, RAFT, NMRP) could be used to develop application-specific GPNCs.

Graphene-based composites could be used in well-established technologies (cars, aircrafts, fuel cells, etc.) as well as in emerging green technologies (solar cells, batteries, catalysis). For instance, transparent and conducting GPNCs could replace indium tin oxide one day. Although not necessarily transparent, highly conducting GPNCs could find applications in nanoelectromechanical systems. In structural applications, the buckling resistance of a composite material is of immense practical interest. So are the load transfer and understanding of fatigue life. Studies are piling up to develop engineering data and design guidelines for GPNCs. Then there are applications that can only be dreamed about.

One of the major obstacles in the path to progress is to understand the detailed evolution of chemical structures during oxidation/reduction and controlled functionalization of graphene. A range of defects in a graphene-like structure can influence the physicochemical properties of graphene. The path forward is then to accurately identify the defects and to systematically quantify them. Essentially, these defects pose challenges, but also afford opportunities to anchor polymer chains to the surface, and thereby maximize the application potential of synthesized GPNCs. Although CNTs and graphene appear to provide comparable mechanical and electrical properties, graphene-based composites potentially provide larger thermal conductivity enhancements and superior barrier properties than CNT-based composites. Graphene nanocomposites, when used as thermal interface materials, outperform those containing CNT or metal nanoparticles due to graphene's aspect ratio and lower Kapitza resistance at the composites' interface (Khan and Balandin 2012). The problem with barrier studies, however, is the lack of comprehensive data and a suitable model to better understand barrier properties of GPNCs.

Nonetheless, the main challenge that remains is how to produce a large enough volume of graphene safely and in a cost-efficient manner. That is where the stage is set and the race is on.

The study of graphene-based PNCs is a multidisciplinary research field. Latest breakthroughs can emerge only when convergent thinking of various fields meets and learns from one another's work. The tree of knowledge of several branches such as chemistry, physics, and biology to chemical, mechanical, electrical, and civil engineering can allow the rise of graphene to attain its true potential. Meaningful advancements to bridge the gap between GPNC research and its applications are likely to occur only if a broader scientific and engineering perspective is in view. This is what each chapter author, who is a specialist in his or her own field, brings to the fore in the story of *Graphite, Graphene, and Their Polymer Nanocomposites*. We hope that this book will contribute to the advancement of both science and technology in this exciting area.

Prithu Mukhopadhyay
Montreal, Québec, Canada

Rakesh K. Gupta
Morgantown, West Virginia

July 22, 2012

Acknowledgment

One of the editors, Prithu Mukhopadhyay, would like to thank IPEX Technologies, Inc. for allowing him to dabble in polymer science and technology with a keen eye for commercial development of products.

References

Ashton, H. C. 2010. "The incorporation of nanomaterials into polymer media." In *Polymer Nanocomposites Handbook*, eds. R. K. Gupta, E. Kennel, and K.-J. Kim, 21–43. Boca Raton, CRC Press.

Boehm, Hanns-Peter. 2010. "Graphene—how a laboratory curiosity suddenly became extremely interesting." *Angew. Chem. Int. Ed.* 49: 9332–9335.

Cai, J., P. Ruffieux, R. Jaafar et al. 2010. "Atomically precise bottom-up fabrication of graphene nanoribbons." *Nature* 466: 470–473.

Choucair, M., P. Thordarson, and J. A. Stride. 2009. "Gram-scale production of graphene based on solvothermal synthesis and sonication." *Nature Nanotechnol.* 4: 30–33.

de Heer, W. A., C. Berger, X.S. Wu et al. 2007. "Epitaxial graphenes." *Solid State Commun.* 143: 92–100.

Gong, L., I. A. Kinloch, R. J. Young, K. S. Novoselov et al. 2010. "Interfacial stress transfer in a graphene monolayer nanocomposites." *Advanced Mater.* 22: 1–4.

Khan, M. F. S., and A. A. Balandin. 2012. "Graphene based nanocomposites as highly efficient thermal interface materials." *Nano Letters* 12: 861–867.

Mukhopadhyay, P., and R. K. Gupta. 2011. "Trends and frontiers in graphene-based polymer nanocomposites." *Plast. Eng.* 67: 32–42.

Novoselov, K. S., A. K. Geim, S. V. Morozov et al. 2004. "Electric field effect in atomically thin carbon films." *Science* 306: 666–669.

Okada, A., Y. Fukushima, M. Kawasumi et al. 1988. "Composite material and process for manufacturing same." US Patent 4,739,007.

Potts, J. R., D. R. Dreyer, C. W. Bielawski, and R. S. Ruoff. 2011. "Graphene-based polymer nanocomposites." *Polymer* 52: 5–25.

Rafiee, J., X. Mi, H. Gullapalli et al. January 22, 2012. "Wetting transparency of grapheme." *Nature Materials* 11: 217–222.

Terrones, M., A. R. Botello-Méndez, J. Campos-Delgado et al. 2010. "Graphene and graphite nanoribbons: Morphology, properties, synthesis, defects and applications." *Nano Today* 5: 351–372.

Terrones, M., O. Martin, M. González et al. 2011. "Interphases in graphene polymer-based nanocomposites: Achievements and challenges." *Adv. Mat.* 23: 5302–5310.

Yang, W., K. R. Ratinac, F. Braet et al. 2010. "Carbon nanomaterials in biosensors: Should you use nanotubes or graphene?" *Angew. Chem. Int. Ed.* 49: 2114–2138.

Zhou, H., P. Ganesh, V. Presser et al. 2012. "Understanding controls on interfacial wetting at epitaxial graphene: Experiment and theory." *Physical Review B* 85: 035406.

Editors

Prithu Mukhopadhyay is a scientist with IPEX Technologies, Inc. in Montreal, Canada. Since 1997, he has been engaged in R&D as well as manufacturing and quality management for injection molding and extrusion. He earned his master's degree in organic chemistry and his PhD in polymer chemistry from the Indian Institute of Technology, Kharagpur. He came to Montreal, Canada, in 1991 as a post-doctoral fellow in chemical engineering at Ecole Polytechnique after having worked as a chemist for the Oil and Natural Gas Commission of India. Prithu is passionate about new plastics technologies, and he has been a member of the New Technology Committee of the Society of Plastics Engineers since 1998. He has chaired the committee in the past, and he has been active in developing New Technology Forums at the Annual Technical Conference of the Society since 2000. He is an expert on plastics piping materials, and he has published and spoken extensively on this and a variety of polymer topics. He is the founding editor of Plasticstrends, an educational website that was established in 2000.

Rakesh K. Gupta is a George B. and Carolyn A. Berry professor and department chairman of chemical engineering at West Virginia University where he has been teaching since 1992. He holds B. Tech. and PhD degrees in chemical engineering from the Indian Institute of Technology, Kanpur, and the University of Delaware, respectively. Before coming to WVU, he taught at the State University of New York at Buffalo for 11 years. He also worked briefly for the Monsanto and DuPont Companies. His research focuses on polymer rheology, polymer processing, and polymer composites. He has published more than 100 journal papers, 65 conference papers, and 12 book chapters on these topics. He also holds three U.S. patents. He is the author of *Polymer and Composite Rheology*, the coauthor of *Fundamentals of Polymer Engineering* and the coeditor of *Polymer Nanocomposites Handbook*.

Contributors

Ahmed A. Abdala
Chemical Engineering Program
The Petroleum Institute
Abu Dhabi, United Arab Emirates

Sushant Agarwal
Department of Chemical Engineering
West Virginia University
Morgantown, West Virginia

Qiaoliang Bao
Department of Chemistry
National University of Singapore
Singapore

Pierre Carmier
CEA–INAC (Commissariat à l'Energie Atomique et aux Energies Alternatives–Institut Nanosciences et Cryogénie)
Université Joseph Fourier
Grenoble, France

Dhruba P. Chatterjee
Polymer Science Unit
Indian Association for the Cultivation of Science
Jadavpur, Kolkata, India

Guohua Chen
Institute of Polymer and Nanomaterials
Huaqiao University
Quanzhou, Fujian, China

Ling Du
Department of Polymer Engineering
University of Akron
Akron, Ohio

Michael Giersig
National Center for Nanoscience and Technology
Zhongguancun, Beijing, China

Carsten Glanz
Process Engineering of Functional Materials
Fraunhofer Institute for Manufacturing Engineering and Automation (Fraunhofer IPA)
Stuttgart, Germany

Rakesh K. Gupta
Department of Chemical Engineering
West Virginia University
Morgantown, West Virginia

Sadhan C. Jana
Department of Polymer Engineering
University of Akron
Akron, Ohio

Alan B. Kaiser
MacDiarmid Institute for Advanced Materials and Nanotechnology
and
School of Chemical and Physical Sciences
Victoria University of Wellington
Wellington, New Zealand

Musa R. Kamal
Department of Chemical Engineering
McGill University
Montreal, Québec, Canada

Samuel Kenig
Shenkar College of Engineering and Design
Israel Plastics and Rubber Center
Ramat Gan, Israel

Hyunwoo Kim
Department of Chemical Engineering and Materials Science
University of Minnesota
Minneapolis, Minnesota

Ivica Kolaric
Process Engineering of Functional Materials
Fraunhofer Institute for Manufacturing Engineering and Automation (Fraunhofer IPA)
Stuttgart, Germany

Nikhil Koratkar
Department of Mechanical, Aerospace and Nuclear Engineering
Rensselaer Polytechnic Institute
Troy, New York

Urszula Kosidlo
Process Engineering of Functional Materials
Fraunhofer Institute for Manufacturing Engineering and Automation (Fraunhofer IPA)
Stuttgart, Germany

Rama K. Layek
Polymer Science Unit
Indian Association for the Cultivation of Science
Jadavpur, Kolkata, India

Dong Su Lee
Korea Institute of Science and Technology of Jeonbuk
Seoul, Republic of Korea

Minghui Liang
National Center for Nanoscience and Technology
Zhongguancun, Beijing, China

Gui Lin
Department of Chemistry and Polymer Research Center
University of Cincinnati
Cincinnati, Ohio

Kian Ping Loh
Department of Chemistry
National University of Singapore
Singapore

Bin Luo
National Center for Nanoscience and Technology
Zhongguancun, Beijing, China

Christopher W. Macosko
Department of Chemical Engineering and Materials Science
University of Minnesota
Minneapolis, Minnesota

James E. Mark
Department of Chemistry and Polymer Research Center
University of Cincinnati
Cincinnati, Ohio

Martin Matis
Process Engineering of Functional Materials
Fraunhofer Institute for Manufacturing Engineering and Automation (Fraunhofer IPA)
Stuttgart, Germany

Prithu Mukhopadhyay
IPEX Technologies, Inc.
Verdun, Québec, Canada

Arun K. Nandi
Polymer Science Unit
Indian Association for the Cultivation of Science
Jadavpur, Kolkata, India

Isao Noda
Beckett Ridge Technical Center
The Procter & Gamble Company
West Chester, Ohio

Martin Pumera
Division of Chemistry and Biological Chemistry
School of Physical and Mathematical Sciences
Nanyang Technological University
Singapore

Javad Rafiee
Department of Mechanical, Aerospace and Nuclear Engineering
Rensselaer Polytechnic Institute
Troy, New York

Mohammad A. Rafiee
Department of Mechanical, Aerospace and Nuclear Engineering
Rensselaer Polytechnic Institute
Troy, New York

Sanjoy Samanta
Polymer Science Unit
Indian Association for the Cultivation of Science
Jadavpur, Kolkata, India

Olga Shepelev
Shenkar College of Engineering and Design
Israel Plastics and Rubber Center
Ramat Gan, Israel

Viera Skákalová
Max-Planck Institute for Solid State Research
Stuttgart, Germany

Iti Srivastava
Department of Materials Science and Engineering
Rensselaer Polytechnic Institute
Troy, New York

Dingyuan Tang
School of Electrical and Electronic Engineering
Nanyang Technological University
Singapore

Matthew S. Thompson
Department of Chemical Engineering
West Virginia University
Morgantown, West Virginia

Friedemann Tonner
Process Engineering of Functional Materials
Fraunhofer Institute for Manufacturing Engineering and Automation (Fraunhofer IPA)
Stuttgart, Germany

Jorge Uribe-Calderon
Materials Unit
Yucatan Center for Scientific Research
Merida, Yucatan, Mexico

J. S. van der Wal
Composite Agency
Amsterdam, the Netherlands

Fazel Yavari
Department of Mechanical, Aerospace and Nuclear Engineering
Rensselaer Polytechnic Institute
Troy, New York

Han Zhang
School of Electrical and Electronic Engineering
Nanyang Technological University
Singapore

Weifeng Zhao
Institute of Polymer and Nanomaterials
Huaqiao University
Quanzhou, Fujian, China

Linjie Zhi
National Center for Nanoscience and Technology,
Zhongguancun, Beijing, China

John W. Zondlo
Department of Chemical Engineering
West Virginia University
Morgantown, West Virginia

1

Graphite: Structure, Properties, and Applications

John W. Zondlo*

CONTENTS

* Correspondence to: John W. Zondlo (John.Zondlo@mail.wvu.edu)

1.1 Introduction

The element carbon (element 6 on the periodic table; atomic weight of 12.010) is unique among all the elements. Carbon has the ability to form different types of bonds not only with itself but also with a variety of other elements such as hydrogen, oxygen, nitrogen, and sulfur. In fact, the entire subject of organic chemistry is based on these types of compounds. Carbon can bond to itself by single, double, or triple bonds and is capable of forming a nearly infinite variety of chains, rings, and interconnected three-dimensional (3-D) networks. Crystalline forms of carbon that are normally encountered at ambient conditions are diamond and graphite (Marsh 1989). Figure 1.1 shows these two structures. In diamond, carbon atoms are in the sp^3 configuration, all bonds are sigma bonds, and all bonds are arranged in a tetrahedral lattice with all the covalent bond lengths equal to 1.54 Å. This rigid stable structure results in diamond being the hardest of all natural materials. Indeed the name diamond comes from the Greek word *adamus*, which means invincible.

In contrast, graphite, which is naturally a very soft material, is found in layers of sp^2 hybridized carbons in the form of planar hexagonal rings. These "graphene" layers are 3.35 Å apart and the carbon–carbon covalent sigma-type bond length within the layer is 1.42 Å. In addition, the remaining p orbitals are

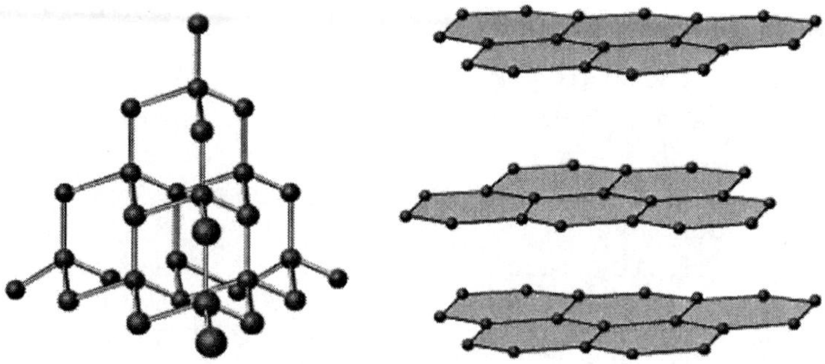

FIGURE 1.1
Molecular structures of (a) graphite and (b) diamond showing distinct differences in bonding.

delocalized π-type bonds that extend perpendicular to the planes and give rise to weak van der Waals attractions between the planes. The difference between the bonding forces parallel and perpendicular to the planes is responsible for the diverse, anisotropic, and industrially relevant properties of graphite such as thermal conductivity, hardness, porosity, electrical conductivity, surface effects, and chemical reactivity. Indeed the graphite industry manipulates the crystal structure of graphite to produce the wide variety of products in use today. It should be noted that an additional form of carbon, fullerene, is recognized by workers in the field as the third form of crystalline carbon. This type of carbon has been popularized by the discovery of buckyballs and single-walled carbon nanotubes. Finally, there is a huge range of noncrystalline or amorphous carbons in which a small number of layered planes are randomly oriented with respect to each other. Such carbons have wide industrial applications. Although all these forms of carbon are extremely interesting, the main focus of this chapter is on graphite. The chapter is organized as follows: First, a discussion on the occurrence of natural graphite and the production of synthetic graphite is presented. This is followed by sections on the structure and properties of graphite, techniques for its structural and mechanical characterization, several of the more important industrial applications of graphite and, finally, some of its newly discovered and emerging forms.

1.2 Graphite: Occurrence and Manufacture

1.2.1 Natural Graphite

Natural graphite is considered a mineral (Mantell 1968). It has been used since prehistoric times as a decorative pigment and its presence has been confirmed in several burial sites in Europe (Marsh et al. 1997). It has been used for centuries as a tool for writing and drawing. The name graphite was coined by the mineralogist Abraham Werner in 1789 from the Greek word *graphein*, which means "to write." The early chemist Scheele (1742–1786) identified the mineral as carbon by burning it in an oxygen atmosphere. Until then, graphite had been considered a derivative of lead because of its appearance. In the Middle Ages, natural graphite was fashioned into graphite/clay crucibles and lubricants.

Natural graphite is commonly divided into three classes based on the differing geological environments in which they occur. The three classes are as follows: (1) Amorphous, (2) vein or plumbago, and (3) flake graphite. The calculated crystallographic density of graphite is 2.26 g/cc, whereas densities of its natural counterparts can range from 2.20 to 2.30 g/cc depending on the amount of ash constituents. The natural formation of graphite can occur in either metamorphic or igneous conditions, but the former is the most common. The conversion of amorphous carbon to highly crystalline graphite requires both high pressure and high temperature; however, the form of the carbon precursor is not always clear. For example, some deposits of amorphous and flake graphite are believed to originate from organic matter derived from plants, much in the same way that coal deposits are formed. In other cases, the conversion of carbonate to carbon in limestone deposits is suspected. Still other theories purport that some "proto-carbon" species are actually carried by geologic fluids and deposited in locations far removed from the original source material. This may be especially true for vein graphite.

Amorphous graphite is actually a misnomer since by definition all graphite is crystalline. It is found either in beds or distributed within metamorphic rocks like slate or shale. This graphite occurs in the form of minute particles of crystals, which are so small that they cannot be resolved by the naked eye or the optical microscope. It is sometimes referred to as "microcrystalline graphite." Seams of amorphous graphite originate from metamorphosed anthracite coal deposits. The graphite content of the seam could vary from 20% to 85% and depends on the original carbon content of the coal seam, the remainder being trapped mineral matter in the form of quartz and clays. Amorphous graphite is the least "graphitic" of the naturally occurring graphite due to its microcrystalline structure and lack of significant long-range order. Due to the intimate contact of graphite and mineral matter, this type of graphite is difficult to refine. Its uses vary widely and depend on the nature of impurities present in it. Typical applications are in greases, lubricants, and coating material such as paints and polishes. It is currently mined in Mexico and China, and deposits have been found in Korea, Italy, and Rhode Island, to name a few locations.

As the name implies, flake graphite occurs in the form of flakes or lamellar structures. It is formed primarily in a metamorphic environment and is most often found as isolated hexagonal plates. It is the most common form of naturally occurring graphite with large deposits found in North America,

Central and South America, Europe, Africa, Russia, and China. It can be formed from either organic or inorganic carbon sources with the former being the source of most deposits. The as-mined raw material can be refined by conventional mechanical ore-processing techniques involving crushing, milling, and flotation operations. Depending on ore source and the efficacy of the mechanical and chemical cleaning processes, flake graphite is capable of yielding a product containing over 99% carbon. Flake graphite finds application in the manufacture of pencil "lead," powdered lubricants, coatings, electronics, refractories, and crucibles.

Vein graphite is believed to be of igneous or metamorphic origin with the carbonaceous precursor being transported by a high-temperature, high-pressure geologic fluid and precipitated as graphite into fractures or fissures of the surrounding host rock. The subterranean fluid known as "pegmatite" originates at high pressures from sources deep within the earth. It is believed that these fluids penetrate a deposit of organic or inorganic carbon, or a previously existing flake graphite deposit, and carry the mobilized carbon through rock fractures, where under proper conditions the graphite precipitates as veins. Since solid graphite is deposited directly from the fluid phase, its purity is extremely high. It is not uncommon for the as-mined graphite to be as high as 99% carbon. Vein graphite is mined underground only in Sri Lanka. Interestingly, most vein graphite is cleaned by laborers handpicking the mined material to separate the product from the admixed rock. (The story goes that over the years, attempts were made to mechanize the cleaning process only to have the equipment sabotaged by native laborers who feared losing their jobs.) The veins of graphite can be as much as 6 ft. thick in some deposits and are mined by conventional underground mining techniques. Since vein graphite is deposited directly from a high-temperature fluid phase, it has a very high degree of crystallinity making it ideal for thermal and electrical applications.

1.2.1.1 Natural Graphite: Sources and Production

The worldwide demand for natural graphite was flat until 2008, but since then it has increased slightly to just over 1 million tons per year. China leads the production of natural graphite and is followed by India, Brazil, and North Korea. Table 1.1 gives a breakdown of graphite production for the top producing countries in 2009 and 2010 (Olson 2011).

China and Canada were the leading producers of flake graphite. Sri Lanka continued to be the sole supplier of lump or vein graphite and Mexico supplied mainly amorphous graphite. It is estimated that the worldwide recoverable reserve of all types of natural graphite is over 800 million tons. Estimated prices for imported natural graphite in the United States for 2010 were US$667 per ton for flake graphite, US$1910 per ton for lump and chip graphite, and US$251 per ton for amorphous graphite. The main applications of natural graphite are in foundry and steelmaking operations, refractory

TABLE 1.1

Worldwide Production of Natural Graphite (Thousand Tons)

Country	2009	2010
Brazil	76	76
Canada	25	25
China	800	800
India	130	130
North Korea	30	30
Madagascar	5	5
Mexico	5	5
Norway	2	2
Sri Lanka	11	11
Ukraine	6	6
United States	0	0
Others	3	3
Total production	1093	1093

and crucible manufacture, and brake linings and lubricants manufacture. There is also a growing demand for graphite recycled from spent refractory bricks. Such material is finding increased application in brake linings and thermal insulation.

1.2.2 Synthetic Graphite

Synthetic or artificial graphite (Mantell 1968) is a manufactured material; it is made by heating a carbonaceous precursor in an inert atmosphere to temperatures in excess of 2400°C. A high temperature is required to effect the solid-state phase transition, which leads to the formation of graphite crystallites. The raw materials from which graphite is fashioned are numerous and can originate from petroleum, coal, and natural or synthetic organic chemicals. However, it is important to note that not all carbon-containing compounds can be converted to graphite. These so-called nongraphitizable carbons do not form graphite even if they are heated to high temperatures. Wood is an example of this type of carbon. Thus, the source of raw material for graphite has been, and continues to be, very selective. The discovery of electricity and methods to generate and control the extremely high temperatures required for the formation of graphite were key elements in the development of graphite manufacture.

Perhaps the first record of "manufactured carbon" is from the experiments of Sir Humphrey Davy who, in the early 1800s, was searching for better electrode materials to use in his electric arc studies. He blended powdered charcoal and thick tar, molded the mixture under pressure, and baked the mixture, thus developing a process that is still in use today. During the remainder of the nineteenth century, most of the work in synthetic carbon was focused on the evaluation of a variety of naturally occurring feedstocks. In addition, many novel processing concepts were developed, such as impregnating the baked electrode with resin and rebaking it to increase the density. Techniques for the purification of carbon artifacts were also studied around this time. In 1876, Carre patented a mixture of coke, lampblack, and sugar syrup that could be hydraulically pressed through a "drawplate" to form the first extruded rod-shaped electrodes. In 1877, Charles Brush discovered the use of calcined petroleum coke as an excellent feedstock for electrodes. The decade between 1890 and 1900 saw many developments including incorporation of the electric furnace to achieve the high temperatures necessary for graphitization. Perhaps the most significant discovery occurred in 1896 when E. G. Acheson was granted a patent for the manufacture of graphite in his electric resistance furnace. Here, a series of prebaked electrodes are tightly packed in the furnace between two electrical contacts. When a high electric current is passed through the system, the resistance of the packed rods is such that the so-called "$I \times R$" or resistive-type heating can reach temperatures in excess of 3000°C. The first commercial trial of the furnace occurred in June 1897 when 2905 rectangular rods were packed between the contacts 16 ft apart, covered with a 2 in. layer of calcined coke, and subjected to an electric current of 7800 A. The latter part of 1897 saw the production of over 142,000 lb of such electrodes. Spurred on by the need for carbon anodes in the aluminum industry and electrodes in the steel industry, the graphite industry experienced tremendous growth in the first part of the twentieth century. More recently, the nuclear industry has found applications for graphite in reactors. Markets in the aerospace, defense, marine, and automobile sectors have also skyrocketed the potential applications of graphite.

1.2.2.1 Raw Materials of Synthetic Graphite

The raw materials needed for the manufacture of synthetic graphite fall into three main categories: (1) The "body" or filler material, which supplies the bulk of the carbon found in the final product; (2) the "binder" material, which holds the fine particulate body material together; and (3) a variety of other materials for processing and refining the finished products. As noted in Section 1.2.2, not all carbon-containing feedstocks are suitable for graphite production. The conversion from an amorphous or unordered state to the highly ordered crystalline graphite structure places constraints on the selection of viable feedstock candidates. In the process of graphitization, the carbon atoms must undergo movement and rearrangement, which is necessary for the formation of the graphitic structure, during the high-temperature treatment. Graphitizing carbons must first pass through a molten or fluid stage during the process of pyrolysis. This liquid stage is crucial in that it allows the planar aromatic molecules to align with each other and form a pre-graphite lattice. The subsequent high-temperature heat treatment completes the solid-state transformation into the final crystal lattice of graphite.

Diverse graphitizable precursors are available. Such carbons include vitrinite and solvent extracts from bituminous coal, coal tars and coal-tar pitches, petroleum pitches, polymers, and model polyaromatic species such as naphthalene. The graphite industry relies almost exclusively on pitch-based precursors from both petroleum and coal sources. Interestingly, most untreated pitches are isotropic, structureless solids at room temperature. However, when the pitches are heated to temperatures around 350°C a new phase of small anisotropic spheres is formed and appears as suspended particles in the isotropic pitch matrix (Brooks and Taylor 1968; Taylor 1961). As the process continues, the spheres continue to grow and eventually coalesce and solidify into the structures required for subsequent graphite formation. This highly ordered liquid-crystalline phase is known as "mesophase" and all forms of carbon must pass through this state on their way to becoming graphite. It is a result of the distillation of smaller pitch components, allowing the remaining larger polyaromatic molecules to associate with each other. These disklike associations are free to move in the fluid phase and due to cohesive forces coalesce with each other and form extended regions of planarity and similar orientation. If the process is allowed to continue, pitch is converted to 100% bulk mesophase. Figure 1.2a shows a polarized-light micrograph of 100% mesophase. The highly ordered flow domains are evident in the figure. Figure 1.2b shows the spheres growing in the melt (top) and coalescing into the bulk mesophase (bottom).

In the United States, the primary body material for graphite production is petroleum coke. Such coke is formed in the process of delayed coking, which, at temperatures around 450°C, converts heavy distillation residues from crude oil into the last remaining light fuel fractions and a solid by-product known as "green delayed coke." Originally, this solid product was considered a waste stream and was either landfilled or burned for heat. However, the petroleum companies soon realized the value of this coke and optimized the delayed coker to produce a graphitizable, highly ordered "needle coke," which is suitable as a bulk material for the graphite industry. At this stage, the green coke still contains about 10%–20% volatile matter, and it must be subjected to further heating to drive these species off and make it suitable for graphite manufacture. This final process known as "calcination" results in the production of calcined petroleum coke. Typically, the calcining operation is performed in a rotating kiln where the green coke moves through the kiln and is subjected to temperatures between 1250°C and 1350°C for approximately 1 hour. Figure 1.3 shows a typical commercial rotating kiln. Finally, it should be noted that the quality of petroleum coke depends intimately on the composition and source of the crude oil entering the refinery. Thus, the graphite industry must maintain rigid standards on the characterization and specification of its coke supplies. Some of the most important properties of delayed coke with respect to graphite production are sulfur levels, residual volatile matter content, ash content, hardness, and the coefficient of thermal expansion. As an aside, anthracite coal has high carbon content

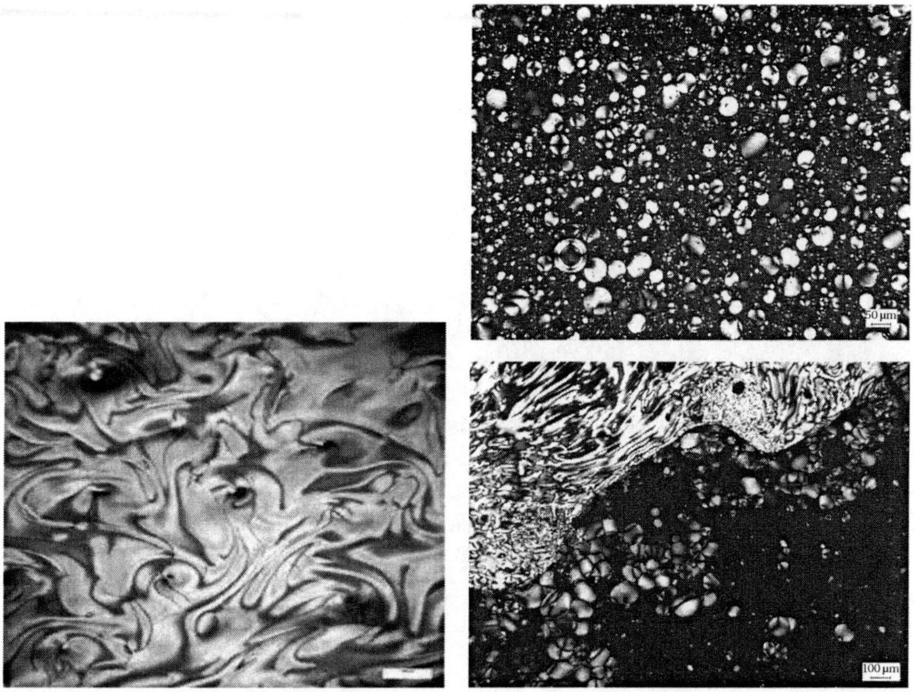

(a) (b)

FIGURE 1.2
Polarized-light photomicrographs. (a) Micrograph of 100% mesophase and (b) micrograph of mesophase spheres growing (top) and coalescing into bulk mesophase (bottom). (From Miller, B. G., and S. Winton. An Industrial-Based Consortium to Develop Premium Carbon Products from Coal Final Report, Prepared for U.S. Department of Energy, National Energy Technology Laboratory, Cooperative Agreement No. DE-FC26-03NT42874, 1822, January, 6, 2011. With permission.)

FIGURE 1.3
A commercial rotating kiln for the calcination of green petroleum coke.

and can serve as a feedstock in the production of graphite electrodes. Thus, some low-ash anthracites can be used as the bulk material; but high-quality graphite is generally formed from calcined petroleum needle coke.

The second most important ingredient for making synthetic graphite is the binder. The binder is mixed with the coke filler particles and serves several functions. The binder holds the coke particles together during processing. Hence, it must demonstrate suitable adhesion to coke. It must also allow the coke/binder mix to be fluid, which not only facilitates the forming steps such as molding and extrusion but also enables coke particles to move into an ordered alignment under the shear forces generated during the forming process. During the subsequent heating operations, the binder is itself pyrolyzed to form permanent coke bridges between filler particles, adding structural strength to the body. Moreover, the binder coke should be graphitizable. The main binder material in use today is derived from the distillation of coal tar and is known as coal-tar pitch. Coal tar originates as a condensable by-product in the production of metallurgical coke in the recovery coke oven. Binder pitch is characterized by a broad range of properties; two of the most important properties are the softening point and the viscosity/temperature behavior. Table 1.2 shows some properties of a typical coal-tar pitch. It should also be noted that a suitable binder pitch should contain a small fraction of tiny particulate matter (in the range of 0.25–2 μm) known as "quinoline insolubles" or QIs (Marsh et al. 1997). In coal-tar pitch, QI comes mainly from the cracking of the volatile coal species in the coke oven as they move along the hot surfaces on their way out of the oven. In addition, some coal fines are carried over to the coal tar during the charging of coal to the coke oven. The

TABLE 1.2

Some Properties of Typical Coal-Tar and Petroleum Pitches

Property	Coal-Tar Pitch	Petroleum Pitch
Softening point (°C)	110	120
Sp. Gr. (specific gravity referenced to water) at 25°C (g/mL)	1.33	1.24
Coking value (wt%)	60	51
Ash (wt%)	0.10	0.16
QIs (wt%)	14	0.2
Sulfur (wt%)	0.8	3.0
Viscosity (Pa·s)		
150°C	3.6	15.0
160°C	1.5	4.5
170°C	0.8	1.7
Average molecular weight	350	500

Source: Hupp, T. R., I. C. Lewis, J. M. Criscione, R. L. Reddy, C. F. Fulgenzi, D. J. Page, F. F. Fisher, A. J. Dzermejko, and J. B. Hedge. Kirk-Othmer Encyclopedia of Chemical Technology, published online: August 15, 2003, DOI: 10.1002/0471238961.0201110512052309.a01.pub2.

former QI is known as "normal QI" and the latter is referred to as "carryover QI." Together, they are termed "primary QI" and make up about 80%–90% of all the QI found in coal-tar pitch. Coal-tar pitch typically contains about 2–12 wt% total QI depending on the properties of the coal and the coking conditions. The QI plays several important roles in the pitch during the baking process and is very beneficial to the performance of the final electrodes. The QI allows the pitch to coat the coke particles much better than low QI pitches. Also, QI blocks the pores of the coke and thereby prevents the pitch from penetrating too deeply into the coke, resulting in more efficient pitch utilization and the formation of well-anchored pitch/coke bridges between the coke particles. During baking, QI provides escape paths for the pitch volatiles, reducing cracking and increasing strength. Finally, normal QI has an inherently high coking value and facilitates the conversion of pitch to coke.

Finally, numerous other materials are added at various steps in the manufacturing process of graphite. The most important among these materials is impregnation pitch. Since the coal tar binder pitch has about 40% volatile matter, significant porosity results after the binder/coke mixture is pyrolyzed. Depending on the application of the final graphite product, such porosity may be objectionable. To overcome this problem, the once-baked article is treated with a pitch of lower viscosity, which is forced into the voids and rebaked, thus depositing additional carbon into the pores. This impregnation step may be repeated several times depending on the desired density of the final graphite. Typical impregnation pitches are generally made from the thermal processing of petroleum distillates, and they normally have higher softening points and lower viscosities than binder pitch. These impregnation pitches must also have very low QI so as not to inhibit the flow of the pitch into the pores. This is normally not a problem since petroleum pitch is a distillation product. Table 1.2 lists some properties of petroleum pitch. Other materials such as light oils are added to the green mix as lubricants, which migrate to the surface of the electrode and aid in the extrusion process. Chemical inhibitors are sometimes included to prevent the deleterious effects caused by heteroatoms such as sulfur or nitrogen present in the coke, which vaporize during heat treatment and cause voids or even cracking of the graphite article, a phenomenon known as "puffing." Ultrapure graphite may require some final treatment with halogen compounds to remove trace metal contaminants.

1.2.2.2 Manufacturing Processes for Synthetic Graphite

Specific processes for the manufacture of graphite are varied and dependent on the type and final properties of the desired graphite product. What follows is a description of the typical processing steps for the manufacture of electrically conductive graphite electrodes such as the ones used in the steel industry (Marsh et al. 1997). These electrodes must be of high purity and exhibit low electrical resistance especially in the axial direction. The flow diagram of the overall process is shown in Figure 1.4.

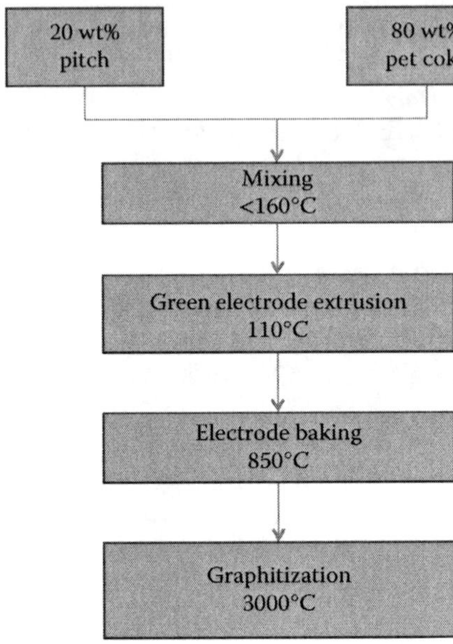

FIGURE 1.4
Process flowchart for electrode production.

The process begins by crushing and sizing the calcined petroleum coke to specific size fractions and subsequently blending them to a prescribed particle-size distribution. The coke particles are then mixed with the binder pitch to distribute the pitch thoroughly so that it evenly coats each coke particle. The proportion of coke to binder is usually around 80/20 by weight. This step is performed at elevated temperatures (in the range of 160°C) so that the binder is kept soft and fluid. Once mixed, the hot mixture is allowed to cool to a temperature just above the softening point of the binder. It is then extruded by forcing it through a die in the shape of a cylinder to a dimension that is as close as possible to the finished dimensions of the graphite electrode. Using this fabrication method, manufacture of electrodes with diameters as large as 48 inches is possible. After extrusion, the hot, plastic article is cooled by rolling it into a water bath, which causes the binder pitch to solidify. The resultant article is now referred to as a "green" (i.e., unbaked) body. At this stage, the article comprises calcined petroleum coke particles held together in a solid coal-tar pitch matrix. In addition to imparting the final shape to the electrode, the shear forces generated during extrusion cause the anisotropic layered planes of needle coke to orient along the axis of the cylinder, which imparts higher axial strength and electrical conductivity to the electrode.

The next step in the fabrication process is the baking operation. Here the green artifact is heated to 800°C–1000°C to drive off the volatile species from the binder and convert the remaining binder to a solid coke after it passes through the mesophase state. The green body must be physically supported during this operation; otherwise, it would remelt and loose its shape (i.e., it would slump or sag). The support is usually provided by surrounding the green article with calcined coke particles as the electrodes are packed in the baking furnace. The baking step must be performed slowly and carefully so as to avoid cracking and promote a smooth transition of binder pitch to pitch coke. The baking step is one of the most critical operations in the manufacture of synthetic graphite, and for larger electrodes it can take as long as 12–14 days for heating and 3–5 days for cooling. The baked body can be described as a carbon article comprising calcined petroleum coke particles held together in a solid, coked pitch matrix. After baking, the electrode is ready to be sent to the Acheson graphitization furnace (described in Section 1.2.2), where it is subjected to temperatures as high as 3000°C. A diagram of the Acheson furnace is shown in Figure 1.5.

During this final heat treatment, the carbon atoms in the coke particles and the binder coke undergo transformation to become the highly crystalline graphite structure. Heating rates are on the order of 40°C–60°C/h in furnaces rated at 6,000 kW with a direct current (DC) of 60,000 A. The graphitization heat-treatment step takes about 3 days to complete with another 8–10 days required for cooling and unloading the charge. After graphitization heat treatment, the resulting article can be referred to as a graphite shape

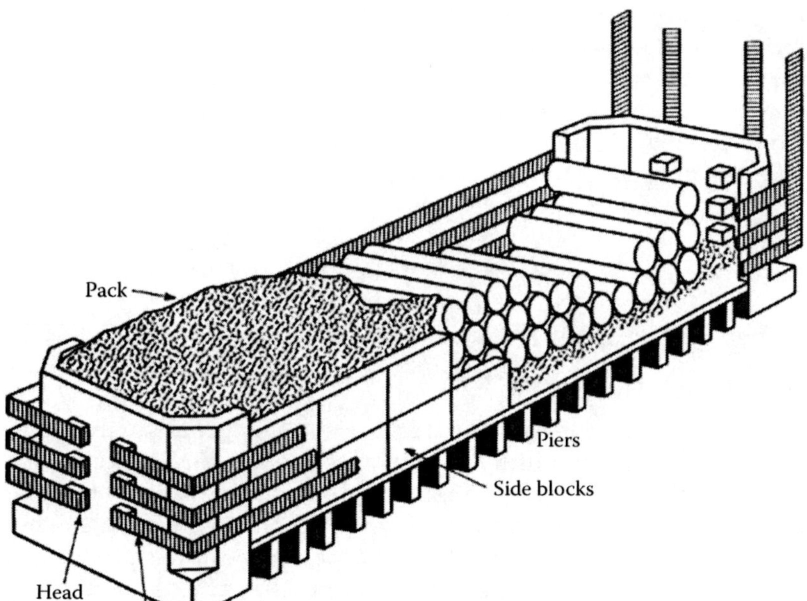

FIGURE 1.5
The Acheson electric furnace showing the electrodes packed and ready for graphitization. (From Hupp, T. R. et al. Kirk-Othmer Encyclopedia of Chemical Technology, published online: August 15, 2003, DOI: 10.1002/0471238961.0201110512052309.a01.pub2. With permission.)

composed of graphitized petroleum coke particles held together by a graphitized coal-tar pitch matrix. Following graphitization, some machining may be required to meet the dimensional tolerances required by the final graphite article. After final inspection and packaging, the finished product is ready for shipment to the customer.

It should be noted here that the aforementioned procedure is followed for most grades of synthetic graphite, but there are a few exceptions. When very dense graphite is required, a pitch impregnation step is performed after the first bake. This is then followed by an additional baking step and repeated impregnations until the desired density is achieved. The resultant graphite is known as "high-density" graphite and is used for structural applications in which high strength and low porosity are desired.

In the case of isotropic graphite production, an alternate shaping operation is conducted. Recall that during extrusion of the green mix, the shear stresses generated in the die cause the needle-like coke particles to align along the axis of the electrode and thereby impart a large degree of anisotropy to the piece. Isotropic graphite comprises graphitic domains randomly oriented in the solid with respect to each other. Hence, no preferential alignment can occur. To achieve this isotropic structure, the green article is molded instead of extruded. This process, called "isostatic pressing," densifies the green mix by applying uniform pressure to the green mix, thereby preventing any alignment of the particles due to shear. In addition to this modified method of fabrication, most isotropic graphite artifacts are formulated using "isotropic coke" fillers, which are milled to very fine particle size. These isotropic types of coke are generated from less aromatic petrochemical feeds, and they contain less needle or anisotropic macro- and microstructures than the needle coke fillers used in conventional electrographite. The resulting isotropic graphite can exhibit bulk densities as high as 2.0 g/cc.

1.3 Structure and Properties of Synthetic Graphite

Carbon has several crystalline allotropes, that is, graphite, diamond, and fullerenes, of which graphite is the thermodynamically stable form under standard conditions. The graphite structure consists of carbon atoms bonded trigonally and arranged in the form of stacked planar sheets, also known as graphene layers. As shown in Figure 1.6, the planes are normally stacked in the hexagonal ABAB sequence, although the ABCABC rhombohedral sequence appears in a small fraction (around 10%) of natural graphites and graphite materials subjected to high shear rates that result from milling or other industrial mechanical processes.

In the ideal case, there are no stacking faults or cross-linking between the layers in the crystallographic *c* direction, giving the ideal interlayer spacing

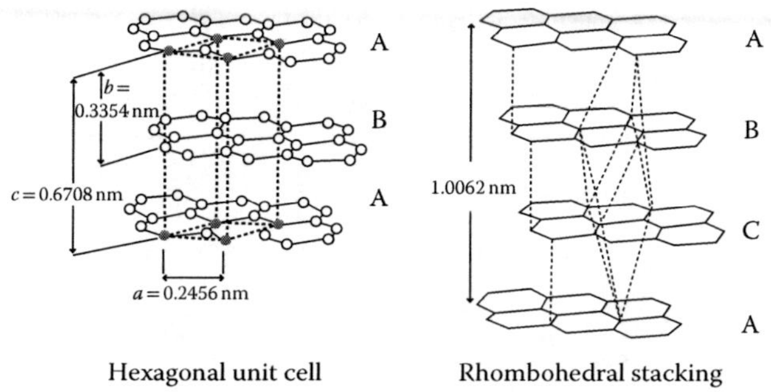

FIGURE 1.6
Hexagonal and rhombohedral packing of graphene layers in graphite.

of 3.354 Å and a density of 2.268 g/cc. Covalent sigma-type bonds exist between the carbon atoms within the graphene sheets (1.42 Å), whereas only weak van der Waals–like attractions are observed between the layers. In fact, it is this difference in forces parallel and perpendicular to the layers that is responsible for the unique anisotropic properties of graphite. For example, the conjugated π-orbitals of the graphene layers delocalize π-electrons, resulting in electrical conduction similar to metallic conductors. However, across the layered structure there is no such electron movement and, hence, there is very poor electrical conductivity in that direction.

When highly crystalline particles of graphite are combined to form a bulk material, the macroscopic properties of the material may differ significantly from those of pure graphite. This is due to the nature of the arrangement of the crystallites in the bulk structure. It is the relative orientation of one crystallite to another as well as the purity and presence of defects within the bulk structure that allow the manufacturer to create synthetic graphite with a wide range of structures and properties. Moreover, synthetic graphite typically comprises a highly ordered material intertwined with some less-ordered material. The greater the long-range order of the crystallites the more anisotropic the bulk properties. Hence, it is possible to have isotropic graphite comprising microscopic highly graphitic structures arranged in a random pattern in the bulk. In addition, the presence of heteroatoms such as sulfur, oxygen, and nitrogen; occlusions of minerals; and species intercalated between the graphene layers contributes to the degree of disorder.

1.3.1 Mechanical Properties of Synthetic Graphite

As noted in Section 1.3, the selection of raw materials (e.g., the source and properties of pet coke) and processing conditions (such as size of coke particles, method of formation of particles, and final graphitization temperature) can influence the yield of manufactured graphites having a wide range of properties. These properties are also dependent on the degree of anisotropy

TABLE 1.3
Properties of Graphite Crystals

Property	In the Basal Plane	Across the Basal Plane
Resistivity (μΩ·m)	0.40	60
Elastic modulus (GPa)	1020	36
Tensile strength (GPa)	96	34
Thermal conductivity (W/m·K)	2000	10
Thermal expansion (1/°C)	-0.5×10^{-6}	27×10^{-6}

Source: Hupp, T. R., I. C. Lewis, J. M. Criscione, R. L. Reddy, C. F. Fulgenzi, D. J. Page, F. F. Fisher, A. J. Dzermejko, and J. B. Hedge. Kirk-Othmer Encyclopedia of Chemical Technology, published online: August 15, 2003, DOI: 10.1002/0471238961.0201110512052309.a01.pub2.

of the finished graphite. Hence, there can be a difference in properties relative to the direction of the molding force applied during the extrusion of the green mix. The properties parallel to the extrusion axis are termed "with-grain" and those perpendicular to the extrusion axis are termed "cross-grain." Table 1.3 illustrates the variation in some properties for highly ordered graphite crystals as the extreme case. Manufactured graphite reflects this anisotropy, but it is not nearly as pronounced as in highly ordered graphite, and specific grades of graphite have been established on this basis.

For many applications, the following mechanical properties of graphite are important: Bulk density, tensile strength, flexural strength, and compressive strength. These properties are discussed briefly in Sections 1.3.1.1 through 1.3.1.4. It should be noted however that due to the great variety of industrial grades of graphite, only average properties are reported here. The reader is referred to manufacturers' data sheets for more specific information.

1.3.1.1 Bulk Density

Bulk density of a material is defined as its weight per unit volume. Average values for some of the most commonly used grades of graphite are between 1.40 and 1.78 g/cc. Bulk density can be useful for inferring trends in other properties. For example, a higher bulk density usually implies improved mechanical strength and thermal conductivity, whereas electrical resistance and permeability decrease with increase in bulk density.

1.3.1.2 Tensile Strength

Tensile strength of a material is defined as the force per unit area needed to break the material under pure tension. Average room-temperature values run from around 3 to 30 MPa in the with-grain direction and from 2.5 to 20 MPa in the cross-grain direction. In general, tensile strength increases with temperature to a maximum at about 2500°C; on reaching that limit it drops sharply.

The maximum value is roughly twice that at room temperature. Also, a nearly linear trend has been found between bulk density and tensile strength. See *The Industrial Graphite Engineering Handbook* (1963) for more details.

1.3.1.3 Flexural Strength

Flexural strength of a material is the maximum stress on the material that causes it to break in the bending mode. Average room-temperature flexural strength ranges from 5 to 30 MPa (with-grain) and from 5.5 to 25 MPa (cross-grain). As expected, flexural strength increases with temperature as do the other strength parameters.

1.3.1.4 Compressive Strength

Compressive strength of a material is the maximum compressive load referenced to the original cross-sectional area, which causes the material to crush. Average room-temperature values for graphite range from 10 to 55 MPa (with-grain) and from 13 to 60 MPa (cross-grain). The compressive strength increases with increasing temperature.

1.3.2 Thermal Properties of Synthetic Graphite

The main thermal properties considered here are thermal conductivity, coefficient of thermal expansion (CTE), specific heat, and thermal shock resistance. Graphite can have a very high value of thermal conductivity and as such is an ideal material for use in high-temperature thermal management systems.

1.3.2.1 Thermal Conductivity

Compared with other refractory materials, graphite can have very high thermal conductivity. The dominant method for heat transfer in graphite is through lattice vibrations with little contribution from the electronic component. The thermal conductivity for most conventional synthetic graphite varies from 50 to 150 W/m·K. Above room temperature, the thermal conductivity decreases in an exponential fashion and tends to level off around 3000°C. For anisotropic graphite, the thermal conductivity is generally higher in the with-grain direction as expected. Interestingly, the ratio of thermal and electrical conductivities at room temperature is a constant; so the value of thermal conductivity can be calculated from the more easily measured electrical conductivity to within about 5% error (*The Industrial Graphite Engineering Handbook* 1963).

1.3.2.2 CTE

The CTE reflects the change in length per unit length of a material with temperature. Typical CTE values range from 0.5×10^{-6} to 5×10^{-6} per degree Celsius and with- and cross-grain values are very strongly dependent on the anisotropy of the graphite. Note from Table 1.3 that for highly crystalline graphite the ratio of these two quantities can be on the order of 25 or greater, but most commercial graphite has values much lower than this. Generally, a higher CTE indicates a more isotropic graphite and a lower value of CTE suggests significant anisotropy in the graphite. The CTE increases with temperature and it has been observed that most graphite follows the same linear trend above about 200°C. Thus, a generalized series of correction factors have been derived and high-temperature CTE can be easily predicted from the value measured near room temperature.

1.3.2.3 Specific Heat

Specific heat of a material reflects the amount of heat added to a unit mass of the material to increase its temperature by 1°C. At temperatures above about 300 K, the specific heat of all types of natural and synthetic graphite is the same within the errors of the measurements. The value of c_p ranges from 0.1723 cal/g·K at 300 K to 0.527 cal/g·K at 3000 K. Correlations exist to predict the specific heat to within 1.5% over the entire temperature range from 300 to 3200 K.

1.3.2.4 Thermal Shock Resistance

Resistance to thermal shock is one of the important properties unique to graphite. No standard test exists to evaluate thermal shock, so it is difficult to quantify. A "thermal shock index" has been defined based on some mechanical and thermal properties, and this parameter has been used to assess the relative values of shock resistance between different grades of graphite. The reader is referred to manufacturer's data for more detailed information. Table 1.4 is a compilation of most of these properties for two typical commercial graphite grades: (1) Coarse-grained and (2) fine-grained graphite.

1.3.3 Electrical Properties

The most important electrical characteristic of graphite is its electrical resistivity. The resistivity is dominated by conduction along the basal planes. Thus, resistivity of synthetic graphite is determined by the orientation and size of the crystallites along the extrusion axis or in the with-grain direction, as well as the bonding between the crystallites from the graphitized binder. Obviously, it is highly dependent on the selection of raw materials. For high-quality petroleum coke-based graphite, typical room-temperature values range from 5 to 15 μΩ·m with values decreasing with temperature until about 500°C and slowly increasing thereafter.

TABLE 1.4

Representative Data for Fine-Grained and Coarse-Grained Graphite

Temperature (°C)	Thermal Conductivity (W/m·K)		CTE[b] (cm/cm × 10^6/°C)		Specific Heat[c] (kJ/kg·K)	Tensile[d] Modulus (GPa)		Tensile[d] Strength (MPa)		Compression[d] Modulus (GPa)		Compression[d] Strength (MPa)	
	wg[a]	ag[a]	wg	Ag		wg	ag	wg	ag	wg	ag	wg	ag
Fine-grained graphite, 180 μm maximum grain size													
21	150	114	2.15	3.10	0.63	11.5	7.9	17.4	15.0	9.7	7.2	26.6	20.1
260	117	93	2.50	3.46	1.30	11.6	8.0	19.3	17.2	10.0	7.4	27.9	21.7
538	91	72	2.82	3.84	1.63	11.7	8.1	21.7	19.7	10.3	7.6	29.3	23.4
816	73	57	3.16	4.12	1.80	11.9	8.3	24.1	22.1	10.6	7.9	30.9	25.2
1093	60	46	3.45	4.45	1.95	12.1	8.6	26.0	24.3	11.4	8.3	32.4	26.9
1371	52	40	3.70	4.69	2.03	12.5	9.0	28.3	26.2	12.4	9.0	35.2	29.3
1649	46	35	3.95	4.91	2.11	13.2	9.6	29.9	27.9	13.4	9.7	38.1	31.6
1927	42	32	4.17	5.16	2.16	13.7	10.5	31.0	29.3	13.4	9.7	32.2	37.2
2204	40	29	4.35	5.39	2.18	11.5	8.4	31.7	30.1	12.1	9.0	37.9	32.6
2482	38	28	4.58	5.71	2.20	8.0	5.9	31.0	29.3	10.0	7.9	32.4	26.6
2760	36	28	4.83	6.04	2.20	5.2	4.3	26.9	24.8	7.9	6.2	26.6	19.1

(Continued)

TABLE 1.4 ***(Continued)***

Representative Data for Fine-Grained and Coarse-Grained Graphite

Temperature (°C)	Thermal Conductivity (W/m·K)		CTE[b] (cm/cm × 10^6/°C)			Tensile[d]				Compression[d]			
					Specific Heat[c]	Modulus (GPa)		Strength (MPa)		Modulus (GPa)		Strength (MPa)	
	wg[a]	ag[a]	wg	Ag	(kJ/kg·K)	wg	ag	wg	ag	wg	ag	wg	ag
Coarse-grained graphite, 6400 μm maximum grain size													
21	156	108	0.46	1.03		4.2	2.6	3.75	2.91	3.0	2.6	9.3	12.1
1371	30	22	2.4	3.2		5.8	2.9	5.34	4.54	3.4	2.8	12.0	14.7
1927	24	19	2.7	2.85		6.5	3.7	5.39	4.36	4.3	3.3	14.1	17.4
2427	24	20	3.0	4.2		5.6	3.0	7.32	5.17				

Source: Hupp, T. R., I. C. Lewis, J. M. Criscione, R. L. Reddy, C. F. Fulgenzi, D. J. Page, F. F. Fisher, A. J. Dzermejko, and J. B. Hedge, Kirk-Othmer Encyclopedia of Chemical Technology, published online: August 15, 2003, DOI: 10.1002/0471238961.0201110512052309.a01.pub2.

[a] wg = With-grain; ag = across-grain.

[b] CTE = coefficient of thermal expansion.

[c] To convert kilojoules to kilocalories, divide by 4.184.

[d] To convert megapascals to pounds per square inch, multiply by 145.

1.3.4 Chemical Properties

The amount of impurities in graphite is assessed as ash, that is, the residue remaining after complete combustion of graphite in oxygen. In general, ash levels in synthetic graphite are low since most of the contaminant species entering with the coke and pitch are driven off during the high-temperature thermal processing steps. Typical ash levels can run as high as 1% for graphite specimens of very large size to as low as 20 ppm for highly purified grades. The primary impurities tend to be iron, vanadium, sulfur, and calcium.

If exposed to air at high temperatures, graphite can react with oxygen to form CO and CO_2. Since graphite is most often used at elevated temperatures, this oxidation reaction can have a significant effect on performance. As expected, the oxidation rate of graphite is a function of raw materials, porosity, extent of graphitization, and the amount of impurities. In fact, certain metals can catalyze the reaction and greatly increase the oxidation rate especially in the 400°C–800°C range. Graphite can also react with CO_2 and steam through the traditional "gasification" reactions. Nonetheless, in most applications graphite can be used even at high temperatures with carbon losses that are permissible. In very severe oxidizing environments or for operation at extreme temperatures in air, some sort of oxidation protection may be applied to the surface of graphite.

1.4 Structural and Mechanical Characterization of Graphite

A plethora of instrumental methods of analysis are available today, and it is safe to say that nearly all have been applied to the study of graphite. Chemical analyses can determine the elemental composition of both feedstock and finished product. Mechanical testing evaluates strength and other physical properties. In this section, the focus is on methods commonly used to elucidate the structural characteristics of graphite on a microscopic scale. These characteristics ultimately determine bulk behavior and hence applications of various grades of graphite. They also provide insight into the relationship between structure and properties. It should be noted at the start that all synthetic graphite is, in general, a combination of highly ordered graphitic domains in contact with amorphous or less-ordered material. In the ideal situation, the graphene layers occur with no stacking faults in the stacking direction perpendicular to the layers. This direction is known as the crystallographic *c* direction. As stated in Section 1.3, the ideal interlayer distance is 3.354 Å, which is commonly termed the "*d*" spacing. Disorder in graphite can be attributed to other lattice defects such as vacancies, dislocations, and the presence of interstitial heteroatoms. Two additional lattice parameters commonly referenced are the "effective crystallite sizes," L_c and L_a. It is noted that L_c is a measure of the height of the stacked graphene layers and L_a is a measure of the width of the stack in a graphite sample. The dimensions of L_c and L_a are usually given in angstroms. As L_c and L_a increase and the d spacing decreases (i.e., it approaches 3.354 Å), the structure of synthetic graphite approaches the structure of ideal graphite and, consequently, the anisotropy exhibited in physical properties increases. Finally, another parameter often quoted is the "degree of graphitization," g, which is an indicator of how much a synthetic graphite structure deviates from the highly crystalline theoretical graphite structure. The parameter g is defined as follows:

$$g = \left(3.440 - d_{(002)}\right) / (3.440 - 3.345)$$

where $d_{(002)}$ represents the actual d spacing of the specimen, which is determined by x-ray diffraction (XRD), as explained in Section 1.4.1 (Seehra and Pavlovic 1993).

The physical and chemical properties of graphite are to a large extent determined by variations in the d spacing, the crystallite dimensions, and the relative orientation of the crystallites.

1.4.1 XRD

This section is a brief introduction to XRD and the discussion is not intended to be rigorous or comprehensive. Sufficient background on the topic is presented so that the application of XRD to the determination of structural parameters of graphite can be explained. The crystallographic parameters and the degree of graphitization of synthetic graphite are commonly determined by XRD.

All solid matter can be classified as either amorphous or crystalline. Amorphous materials have their constituent atoms arranged in a random or disordered fashion. Crystalline materials are characterized by highly ordered and repeated arrangements of atoms in a 3-D array or lattice. The unit cell is the smallest structural element that, if repeated, describes a crystal. All crystalline materials have a unique chemical composition and crystal structure. The position of atoms in the unit cell is given by the 3-D coordinate axes, x, y, and z, and the unit cell has dimensions a, b, and c, respectively, in these directions. Graphite has a hexagonal unit cell with $a = b = 2.456$ Å and $c = 6.708$ Å. Within the crystal structure, there exist faces or lattice planes of atoms and their orientation is described by the intersections that each plane makes with the three coordinate axes. The common convention for the identification of these planes is by means of the Miller indices, h, k, and l, which is notated as (hkl). For example, the (100) plane is a plane of atoms parallel to the y and z axes and intersecting the x axis at the distance $1a$. It is the collection of these crystallographic planes that interact with incoming x-rays and give rise to an XRD pattern, which is unique to each and every crystalline material. The x-rays have wavelengths on the order of atomic dimensions, around 0.1–10 Å, and hence are ideally suited for probing the structure of crystalline matter.

When a parallel beam of x-rays is directed to a surface of a crystal, it interacts with the crystallographic planes in the crystal resulting in a scattering of the x-rays by the atoms in the crystal. These "diffracted or scattered waves" can interfere with each other either constructively or destructively. If the atoms are arranged in periodic layers as in the parallel faces of a crystal, it can be shown that the diffracted waves can add constructively so as to satisfy the Bragg's equation (Bragg 1913):

$$n\lambda = 2d \sin\theta$$

where λ is the wavelength of the incident x-ray; d is the spacing between parallel lattice planes in the crystal; θ is the angle between the lattice planes and the incident x-ray beam; and n is the so-called order of reflection, which is an integer (1, 2, 3, …). Normally for x-rays, n is set to 1 since first-order reflections are usually the strongest. The wavelength of the incident x-rays is determined by the particular x-ray source on the XRD system. So by searching for peaks over a range of θ values, it is possible to calculate the d spacing for various planes in the crystal lattice. Figure 1.7 shows this effect schematically. Incidentally, if the parameters a, c, and λ are known a priori, the θ location for each peak can be predicted for any combination of the Miller indices (hkl). For graphite, the equation resulting from such an analysis is as follows (Seehra et al. 1992):

$$\sin^2\theta = 0.1314(h^2 + hk + k^2) + 0.0132\, l^2$$

Materials such as graphite have numerous crystallographic planes from which x-rays can scatter. In order to view all possible planes simultaneously, powdered samples are used (Jenkins and Snyder 1996). In this technique, graphite is ground to a fine powder and placed as a pellet onto the XRD stage. Each particle is a crystal, and having a large number of such particles randomly oriented on the stage ensures that every possible set of crystallographic planes is likely to be illuminated by the incident x-ray beam and

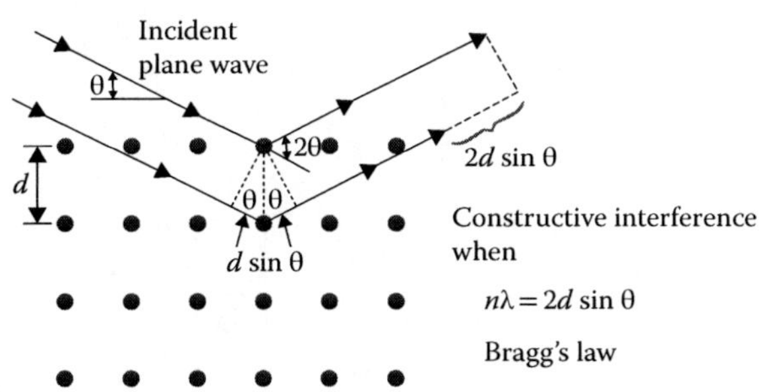

FIGURE 1.7
Scattered x-rays reflected off parallel crystal planes and constructively interfering with each other. (From Nave, R. HyperPhysics Project, Department of Physics and Astronomy, Georgia State University, Atlanta, Georgia, 30302–4106, 2011. With permission.)

hence detected. For very precise work, an internal standard such as silicon is added to the powder so that peak positions and widths can be calibrated.

A much-simplified diagram of a typical XRD system is shown in Figure 1.8. The system consists of an x-ray source, a series of filters and collimators, and a detector that is capable of being moved in the 2θ direction to scan all possible angles. The x-ray source is usually the *K*α emission from a target metal, such as copper, being bombarded by a high-energy stream of electrons. The wavelength for the copper *K*α line, for example, is 1.54056 Å. The filter and the collimator ensure a monochromatic and focused x-ray beam strikes the sample. As the detector moves at a constant angular velocity, the intensity of the diffracted beam is recorded as a function of 2θ, the so-called diffraction angle. The resultant diffraction pattern is a plot of Bragg-peak intensity versus the diffraction angle. An example of the diffraction pattern for a sample of highly crystalline commercial-grade graphite is shown in Figure 1.9, in which the

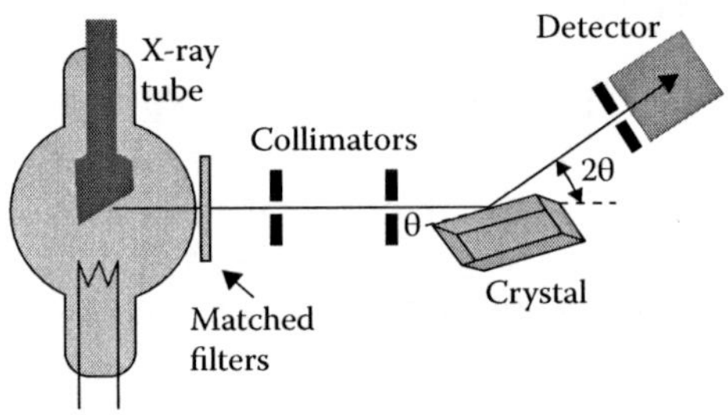

FIGURE 1.8
Schematic representation of a typical XRD system. (From Nave, R. HyperPhysics Project, Department of Physics and Astronomy, Georgia State University, Atlanta, Georgia, 30302–4106, 2011. With permission.)

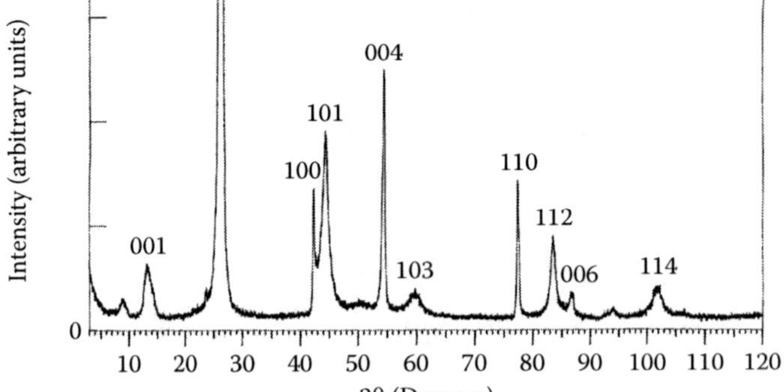

FIGURE 1.9
The XRD pattern of a commercial graphite sample with the Miller indices of major peaks identified. (From Seehra, M. S. et al. Is it graphite yet? Interim Report, West Virginia University, Morgantown, West Virginia, 1992. With permission.)

TABLE 1.5

Representative Values of Lattice Parameters for Two Commercial Graphite Products and Purified Flake Graphite

Sample	$d_{(002)}$ (Å)	a (Å)	L_c (Å)	L_a (Å)	g
HP	3.3565	2.4635	310	545	0.97
IN	3.3634	2.463	360	220	0.89
FL	3.3547	2.4617	∞	∞	0.99

Source: Seehra, M. S. and A. S. Pavlovic, Carbon, 31, 557–564, 1993.

Note: In the table, HP = high-purity commercial graphite; IN = isotropic nuclear graphite; FL = natural flake graphite.

various crystallographic planes have been identified by means of their corresponding Miller indices (Seehra et al. 1992). The parameters usually quoted for graphite as a result of XRD analysis are the interlayer spacing, $d_{(002)}$; lattice parameters, a and c; degree of graphitization, g; and stack height and width lengths of the crystallites, L_c and L_a, respectively. Table 1.5 lists some representative values of these quantities for high-purity graphite, isotropic nuclear-grade synthetic graphite, and highly crystalline flake graphite.

Finally, the cardinal rule of XRD is that every crystalline material has a unique XRD pattern and the same material always has the same pattern. If more than one crystalline substance is present in a sample, each behaves independently of the other and the resultant pattern is the sum of individual patterns. The analyst can then easily confirm the presence of each substance in the sample by comparing the XRD pattern with thousands of patterns in a computerized database. With proper calibration, quantitative assessments of each species can also be made. The reader is referred to some studies by researchers for more detailed treatments of XRD (Ewing 1969; Jenkins and Snyder 1996; Skoog et al. 2007; Stanjek and Hausler 2004).

1.4.2 Microscopic Examination

The two main microscopic techniques applied routinely to the study of carbon and graphite are optical and electron microscopy. The main impetus for microscopic examination of any material is to assess its surface morphology and optical texture. In addition, quantitative measurements such as measuring the characteristic domain size and other pertinent distances can be performed. All this information eventually reveals a more complete picture of the structure of graphite as it relates to precursor materials and processing conditions.

1.4.2.1 Polarized-Light Optical Microscopy

The optical microscope is a device containing one or more lenses that produces an enlarged image of the object placed in the focal plane of the lens. The resolution of the optical microscope is limited by diffraction effects and approaches the wavelengths of transmitted light, in the region of about 200 nm. There are basically two options available in optical microscopy: (1) Transmission microscopy, in which the light beam passes through a transparent object or thin section, and (2) reflection microscopy, in which the incident light is reflected off the surface of a solid sample. An additional enhancement involves the use of plane-polarized light. This technique is particularly suited to the study of anisotropic materials such as graphite (Forrest and Marsh 1977). Normal white light comprises all wavelengths of visible light with light waves vibrating in all directions in a plane perpendicular to the direction of propagation. When such light is passed through a polarizing filter, a wave with only a single vibration direction parallel to the polarizer, or plane-polarized light, emerges. When two such polarizers are placed back-to-back no light is passed if the polarizers are at 90° to each other (crossed polarization), and the original plane-polarized light is passed if both polarizers are oriented in the same direction.

Anisotropic or crystalline materials possess a highly ordered arrangement of atoms in the crystal lattice. This arrangement of atoms determines the optical properties of the crystal, which are generally different in different directions. When such a material is illuminated with plane-polarized light, the orientation of the incident light is varied due to the crystallographic orientation of the impinged surface. For hexagonal crystals such as graphite, the various orientations of the anisotropic directions respond differently to polarized light and produce two different reflected beams that, when recombined, can form a beam with different colors and intensities by constructive or destructive interference. Such effects can reveal information on the character of the sample along with qualitative and quantitative data regarding the anisotropic nature of the material. This is the basis for the reflected light–polarized microscope.

A diagram of a typical polarized-light microscope is shown in Figure 1.10. White light is generated in the epi-illuminator and directed through a polarizer where it emerges as plane-polarized light. It is sent to a half-mirror, which directs the light down the tube to the sample. The light strikes the sample and is reflected back through the objective where the image is magnified. The anisotropic or birefringent sample produces two wave components, which are combined in a second polarizer known as the "analyzer." The analyzer is located between the objective and the eyepiece. The light passes back through the half-mirror to the eyepiece where it is viewed by the operator. The stage can be rotated so that orientation of the crystallites with respect to the plane of the light can be varied. As this occurs, dramatic color changes are observed in the sample. In practice, the solid sample is embedded in an epoxy and the surface of the sample is highly polished to optical flatness. When graphite is observed, interference color patterns are generated due to the graphitic layers at the surface of the sample. The colors range from purple to blue and yellow depending on the orientation of the layers (Marsh 1989). Figure 1.2a and b comprises such images showing the ordered structure of mesophase pitch. Polarized-light photomicrographs of some green cokes are shown in Figure 1.11. Even in black-and-white, the highly ordered structures are evident.

Over the years, methods for quantifying the appearance of carbon surfaces have been developed. The anisotropic features are referred to as optical

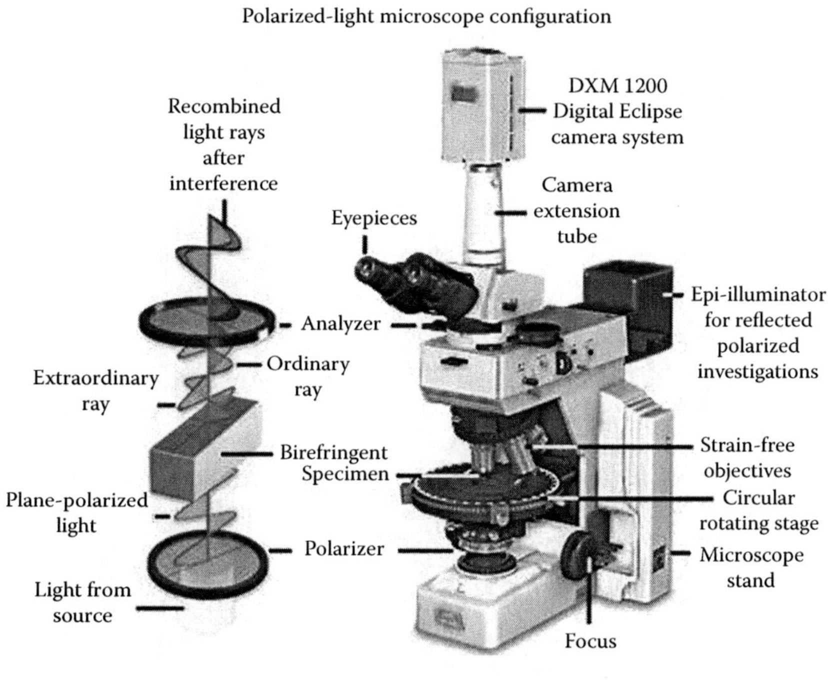

FIGURE 1.10
A polarized-light microscope configured for both transmission and reflected modes.

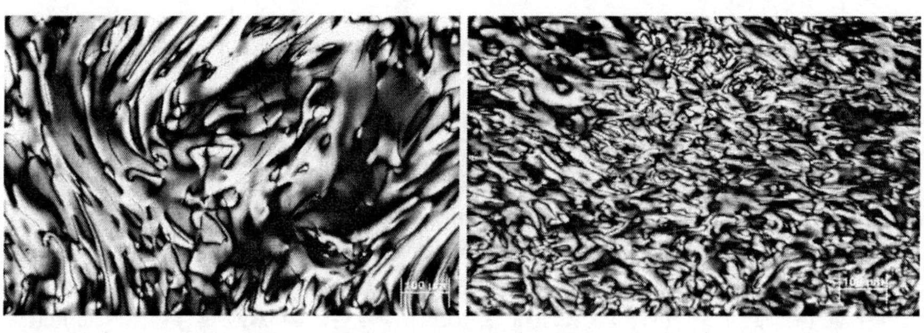

FIGURE 1.11
Polarized-light micrographs showing highly anisotropic (coarse mosaic) green coke (left) and more isotropic (medium mosaic) green coke (right). (From Miller, B. G., and S. Winton. An Industrial-Based Consortium to Develop Premium Carbon Products from Coal Final Report, Prepared for U.S. Department of Energy, National Energy Technology Laboratory, Cooperative Agreement No. DE-FC26-03NT42874, 1822, January 6, 2011. With permission.)

texture and are broken down into arbitrarily defined classes, which are defined by the optical texture index or OTI. (Marsh 1989; Moreland et al. 1988). Briefly, an OTI of 0 indicates no optical texture or isotropy. Lamellar structures over 20 μm in length and 10 μm in width are given an OTI of 30. For an overall image, individual factors are assigned by point-counting the specific features, assigning them each an OTI and averaging to get the OTI for the entire specimen. More recently, computer pattern recognition software has been used to automate the entire process. Such a technique has been utilized, for example, to calculate the percentage of mesophase formed as a function of heat-treatment time (Khandare 1995).

The polarized-light microscope finds significant application in research areas such as geology, chemistry, material science, biology, and medicine. An entire branch of science known as "petrography" is focused specifically on the use of polarized-light microscopy for the study of minerals and coal.

1.4.2.2 Electron Microscopy

The resolution of optical microscopy (around 1 μm) is limited by the wavelength of visible light. Much higher resolution (on the order of 1 nm) can be obtained when a beam of electrons scans a surface. The two most important electron-microscopic methods are scanning electron microscopy (SEM) and transmission electron microscopy (TEM). SEM reveals images of the external morphology and topology of a surface by looking at signals from the electrons scattered off the surface. TEM, on the other hand, reveals details of the internal structure of a solid by passing the electron beam through a thin section of the sample. With both techniques, magnification on the order of 10× to over 500,000× is possible.

1.4.2.2.1 SEM

In SEM, electrons are generated by an electron gun and focused into a high-energy beam on the order of 2–10 nm in width when it strikes the sample. The beam is controlled electronically by a series of lenses and made to scan a desired area of a surface. The electron beam can range in energy from 1 to 40 keV. To reduce the mean free path of the electrons and to prevent arcing of high voltages, SEM is performed under ultrahigh vacuum. The electrons interact with the atoms in the sample and are scattered from the surface of the sample producing a variety of signals. The secondary and backscattered signals are usually captured and converted to an image of the surface. The very narrow electron beam produces images with a large depth of field so that 3-D aspects of the surface may be viewed. In addition, other signals such as x-rays and Auger electrons are produced from this scattering. Using proper detectors, these signals can be used for performing chemical analysis of the surface allowing both the qualitative and quantitative identification of

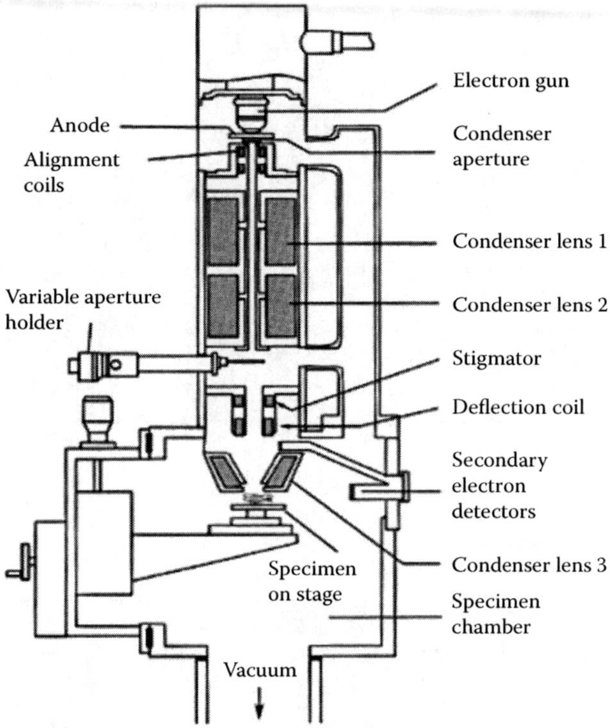

FIGURE 1.12
Schematic of a scanning electron microscope showing its main components. (Courtesy of Hitachi, Tokyo, Japan.)

elements in the sample. One such technique is known as "energy-dispersive x-ray spectroscopy" or EDX and it is commonly used for elemental analysis. Figure 1.12 shows the main features of a typical SEM.

For conventional SEM imaging, samples must be electrically conducting to prevent the accumulation of charge. Nonconducting samples are coated with gold, which makes their surfaces conductive and prevents image-distorting charge accumulation. SEM is routinely applied for viewing carbon and graphitic materials. The samples could be highly polished, but sometimes etching the surface either chemically or by ion bombardment can enhance the surface texture of the images. Representative SEM micrographs of some carbon materials are shown in Figures 1.13 and 1.14.

1.4.2.2.2 TEM

In TEM, the electron beam is formed and focused in much the same way as it is in SEM. Once formed, the beam is made to pass through an extremely thin section of the sample. Three main types of electronic lenses are used in TEM. The condenser lenses focus the beam prior to entering the sample. After the beam passes through the sample, it is once again focused by the objective

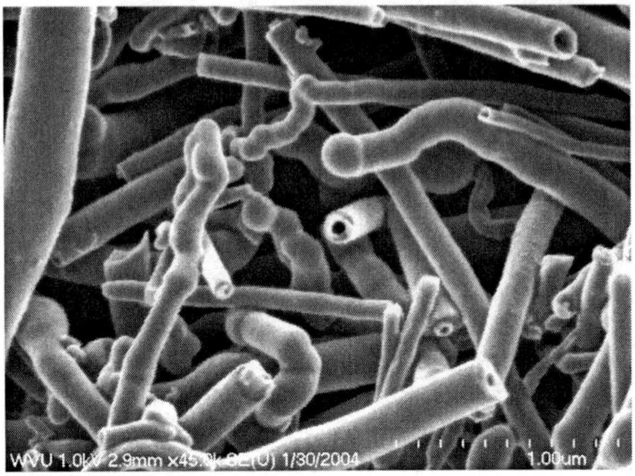

FIGURE 1.13
A SEM micrograph of vapor-grown carbon nanotubes. (Courtesy of Magean, L. Private Communication, Department of Chemical Engineering, West Virginia University, Morgantown, West Virginia. 2011.)

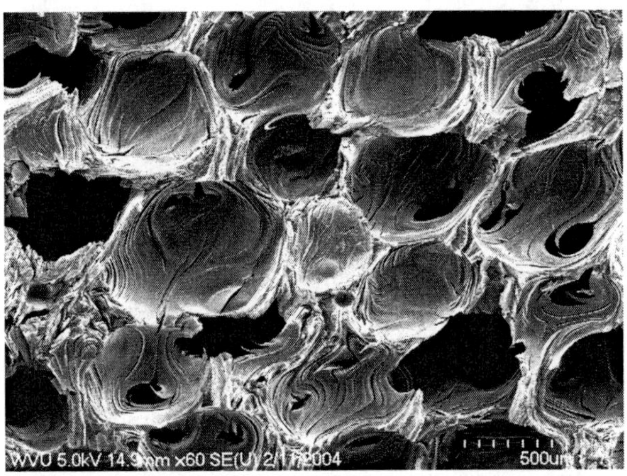

FIGURE 1.14
A SEM micrograph of anisotropic carbon foam. (Courtesy of Magean, L. Private Communication, Department of Chemical Engineering, West Virginia University, Morgantown, West Virginia. 2011.)

lenses and, finally, the beam is expanded using the projector lenses to the viewing screen or some other data acquisition device. As in SEM, to reduce the mean free path of electrons and to prevent arcing of high voltages, TEM is performed under ultrahigh vacuum. A schematic diagram showing the main features of a TEM is given in Figure 1.15.

Sample preparation for TEM is quite demanding. Very thin sections of sample with uniform thickness are required. For a specimen to be electron

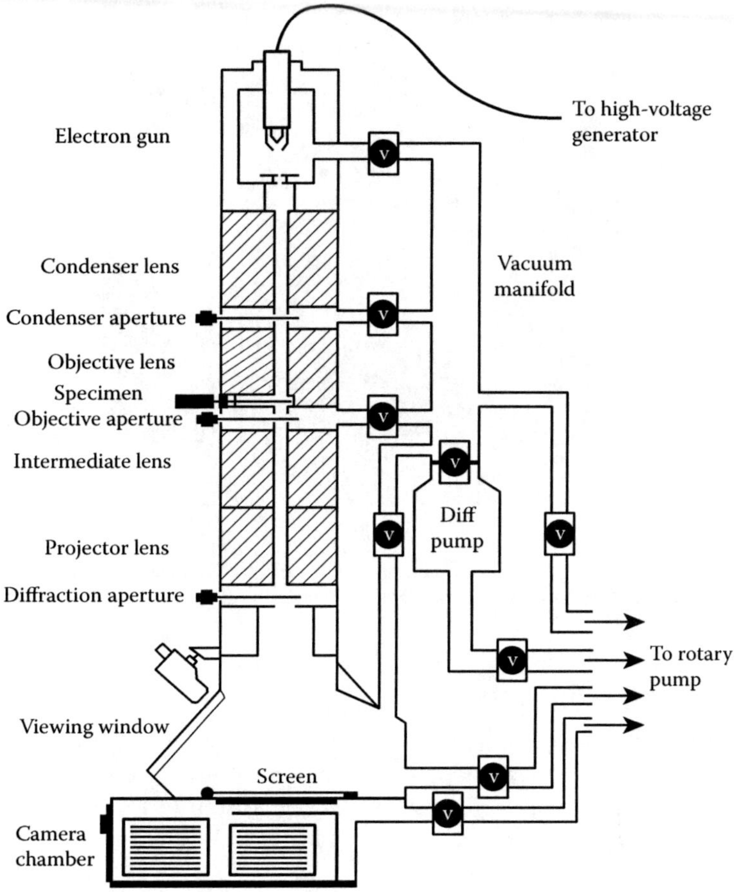

FIGURE 1.15
Schematic of a transmission electron microscope showing its main components. (Courtesy of Hitachi, Tokyo, Japan.)

transparent, the thickness of the specimen must be around 100 nm, which necessitates great care in handling. The sample is mounted on a grid and placed in a transmission electron microscope using a sample holder or stage where it is imaged. Interestingly, by manipulating the lenses it is possible to obtain a diffraction pattern of the sample, thus revealing additional structural information. "Staining" the sample with elements of high atomic numbers such as lead, osmium, or gold is sometimes done to enhance the TEM image. This is especially useful for biological specimens. Finally, samples may be thinned by the incorporation of either chemical or ion etching. An excellent TEM photomicrograph of a graphene flake embedded in a polymer matrix is shown in Figure 1.16. Note that the actual graphene lamellae are visible.

FIGURE 1.16
A TEM micrograph showing actual graphene lamellae. (Courtesy of Gupta, R. Private Communication, Department of Chemical Engineering, West Virginia University, Morgantown, West Virginia. 2011.)

1.4.3 Mechanical Analyses

The mechanical characteristics of graphite are assessed by a variety of mechanical testing procedures, some of which are described in Sections 1.4.3.1 through 1.4.3.4. Since these tests are commonly found in the laboratory, details are not presented.

1.4.3.1 Tensile, Compressive, and Torsion Analysis

Tensile, compressive, and torsion data can be gathered by means of a device having a precisely controlled moveable head equipped with a suitable load cell for force measurement. The sample is placed between the moveable heads and stretched, compressed, or twisted while simultaneously recording the displacement and applied force. Typical stress–strain curves are generated from these data allowing the strength and modulus of the sample to be calculated. Samples of graphite, coke, fibers, foams, and composites are all amenable to this technique.

1.4.3.2 CTE

The CTE is found by means of a dilatometer whereby the actual displacement caused when the sample is heated can be measured. Some devices measure the CTE near room temperature, whereas others can operate at temperatures as high as 1000°C. By placing the same sample in the instrument in different directions, with-grain and cross-grain CTE can be determined, thus revealing the anisotropic nature of the specimen.

1.4.3.3 True Density and Bulk Density

The density of solid, foam, or composite graphite is usually determined using the helium pycnometer. Helium is allowed to expand into the specimen, and from measurements of the volume and pressure before and after expansion both intrinsic density and bulk density can be calculated.

1.4.3.4 Electrical Resistivity

Electrical resistivity is measured using the so-called four-point technique in which a voltage is imposed over a precisely known length of sample. Measurement of the corresponding current along with the cross-sectional area allows the calculation of resistivity. Since electrical resistivity is directly related to thermal conductivity, the latter can be found from the former.

1.5 Applications of Graphite

Because of its unique thermal and structural properties, graphite is utilized in nearly all commercial and industrial sectors. With new uses being discovered daily, the versatility of graphite is evident. In Sections 1.5.1 through 1.5.7, several of the more common applications of graphite are discussed (Criscione et al. 2000).

1.5.1 Electrode Applications

The use of synthetic graphite as electrodes in the electric arc furnace represents the largest consumption of the material today. In fact, it was the need to prepare better electrodes for electric arc lights and furnaces that led Davy and others in the nineteenth century to many of the discoveries that now form the basis of the carbon and graphite industries. A modern electric arc furnace such as the furnaces used in the steel industry consists of a refractory-lined, bowl-shaped hearth into which the charge, usually scrap iron, is placed. A refractory-lined, spherically shaped roof extends over the bowl and allows the graphite electrodes to pass through and enter the furnace (see Figure 1.17).

The electrode support structure and power supply occupy the area above the furnace. During operation, an electric current, usually alternating current (AC), is passed through the electrodes causing an arc to form between the electrode tip and the charge in the furnace. Heating of the charge occurs due to both the current passing through the charge and the radiation from the arc. The temperature reached in the arc can be in excess of 3400 K, and at this temperature the scrap metal quickly melts. The large amount of electrical energy fed to a typical electric furnace ranges from 400 to 600 V at

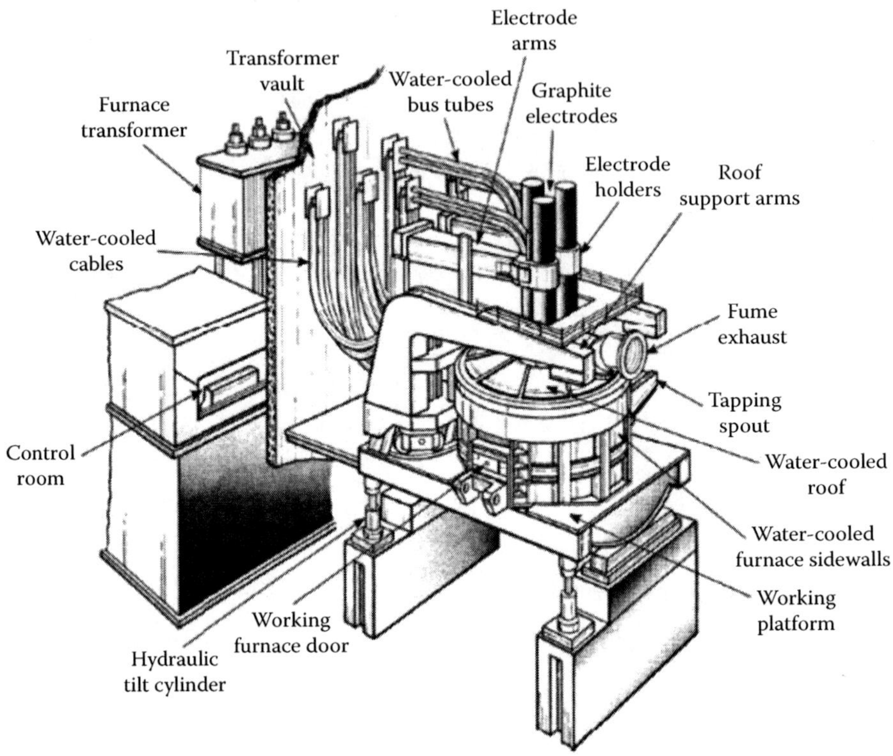

FIGURE 1.17
An electric arc furnace showing the graphite electrodes and associated electrical connections at the top. (From Criscione, J. M. et al. Kirk-Othmer Encyclopedia of Chemical Technology, published online: December 4, 2000, DOI: 10.1002/0471238961.01161612.a01. With permission.)

electric current in excess of 40,000 A, and results in a throughput of about 80 t of metal per hour. Obviously, this operation places severe electrical and mechanical demands on the electrodes.

Commercial graphite electrodes range in size from 32 to 700 mm in diameter and are up to 2800 mm in length. They are marketed in two broad classifications: (1) Regular grade and (2) premium grade. Some typical properties of these electrodes are given in Table 1.6.

The ability to operate under these conditions makes graphite an excellent material for this application. Its low electrical resistivity keeps power losses small, whereas its high strength and low CTE promote excellent resistance to thermal shock and other mechanical stresses. Unfortunately (or fortunately for the graphite producer), graphite electrodes are consumed in the furnace operation. The consumption mainly occurs at the electrode tip and is caused by graphite vaporization and physical abrasion from contact with the charge. Also, some air oxidation transpires at the hot sidewall of the electrode and further graphite is lost due to breakage resulting from thermal or mechanical

TABLE 1.6

Typical Room-Temperature Properties of Regular and Premium-Grade Graphite Electrodes

Property[a]	Regular-Grade Electrodes	Premium-Grade Electrodes
Bulk density (g/cc)	1.60	1.70
Resistivity (μΩ·m)	7.3	5.5
Flexural strength (kPa)		
wg	6900	9100
cg	5800	7000
Elastic modulus (GPa)		
wg	5.3	7.6
cg	3.5	5.0
CTE (10^{-6}/°C)		
wg	0.60	0.40
cg	1.40	1.10
Thermal conductivity (W/m·K)		
wg	134	168
cg	67	101
Thermal shock parameter[b]		
wg	290×10^3	490×10^3
cg	80×10^3	130×10^3

Source: Hupp, T. R., I. C. Lewis, J. M. Criscione, R. L. Reddy, C. F. Fulgenzi, D. J. Page, F. F. Fisher, A. J. Dzermejko, and J. B. Hedge, Kirk-Othmer Encyclopedia of Chemical Technology, published online: August 15, 2003, DOI: 10.1002/0471238961.0201110512052309.a01.pub2.

[a] wg = With-grain; cg = cross-grain.

[b] Thermal shock parameter = (thermal conductivity × strength) × (CTE × elastic modulus)$^{-1}$

shock. Thus, as the electrode is consumed, fresh electrode must be continually lowered into the furnace. Once an electrode is completely spent, an ingenious system has been developed to attach a new electrode to the spent one. This operation is performed at the top of the furnace where the two electrodes are joined by a "double-tapered" threaded connecting pin, which is screwed into the ends of the electrodes. Over the years, significant improvements have been made, both in the overall operation and in the electrodes themselves, so that the electrode consumption is minimized. Currently, the average consumption rate is approximately 2–5 kg carbon per ton of steel. Figure 1.18 shows typical graphite electrodes as they are used in the electric arc furnace. Note the threaded connecting pins at the ends of the electrodes.

1.5.2 Metallurgical Applications

Metallurgical applications are applications in which graphite comes into direct contact with the molten metal being processed. Such operations generally involve high temperatures; so the physical and thermal properties

FIGURE 1.18
A pair of graphite electrodes along with their threaded connecting pins in place.

of graphite make it an ideal choice for processing metals, ceramics, and glasses. Some of the more important characteristics of graphite in relation to this application are as follows: Graphite is unreactive with many molten metals, it does not adhere to molded or cast metal artifacts, its strength increases with temperature, it has excellent dimensional stability due to its low CTE, it has good resistance to thermal shock, and it does not melt or soften at the processing temperatures of most metals. Crucibles, which are used widely for heating and refining metals, are frequently made of graphite. They can be heated by a variety of mechanisms, such as conduction, induction, radiation, and resistance heating, and do not react with or dissolve in the metal they contain. Crucibles can range in size from a few centimeters in diameter to sizes over 1 m in diameter for commercial applications, such as in the production of beryllium and its compounds. Due to its ease of machining, graphite promises significant use as molds for metal casting or as dies for continuous metal forming. Some examples are the use of graphite molds for casting steel wheels for railroad cars and for the continuous extrusion of aluminum components. Purification of metals sometimes requires the injection of trace reactants into the melt, and graphite tubes and nozzles are widely used for this application. Due to its strength and dimensional stability, graphite is used in the aerospace industry as tooling for the high-temperature forming of aircraft parts and wings. Complex machined-graphite shapes see wide application as resistive heating elements and thermal insulation where they are utilized at temperatures in excess of 3000 K. Figure 1.19 shows such a heating element. One important example of this application is the furnace for the production of single-crystal silicon from its melt.

FIGURE 1.19
An electric heating element machined from graphite.

1.5.3 Refractory Applications

The use of graphite as a refractory material dates back to 1850, and since then graphite refractories have found wide use in the steel and aluminum industries. The blast furnace is used for the production of iron from its ore. The first reported use of carbon refractory in the blast furnace was in France in 1872 and later in the United States in 1892. Considerable development occurred in Europe over the next 50 years or so and interest was renewed in the United States in the 1940s. Originally, graphite refractory served as the lining of the lower portion of the blast-furnace hearth; but today, nearly all the temperature-sensitive parts can be made of graphite. Of critical importance in this application is the ability of the lining to contain the molten metal and slag over a service life, which can be as long as 15 years. Carbon and graphite materials have many of the important characteristics needed by refractory linings in the blast furnace. They maintain their strength in the high-temperature environment of the furnace, which is typically around 1200°C. They have high thermal conductivities so that with active cooling on the outside, the surface temperature of the working face is low enough to allow solidification of the metal or slag, which in turn prevents erosion of the refractory. They are chemically resistant to attack by the molten charge in the furnace. Finally, their resistance to thermal shock and low CTE maintains dimensional stability and prevents cracking. Individual blocks of graphite refractory can be as long as 6 m and 1 m thick. Due to different wear mechanisms acting on the hearth bottom and sidewalls, different formulations of refractory are used along with specific installation techniques.

Carbon materials find extensive application in the production of aluminum from its ore in the Hall–Heroult process. This electrolytic process uses carbon anodes, which act as the reducing agent for the alumina. Since the carbon here is a reactant, it is consumed in the smelter. Every pound of aluminum produced requires approximately 0.4 lb of carbon. The aluminum smelter not only operates at high temperatures (around 1000°C) but also contains corrosive sodium and fluoride species from the added cryolite, which serves as a fluxing agent. Graphite blocks, sometimes made from calcined anthracite coal, are installed at the bottom of the cell as the cathode. Obviously, the cathode must have low electrical resistance and be resistant to chemical attack from the molten aluminum. The sidewalls of the cell are also lined with graphite blocks and these require high thermal conductivity and strength, along with thermal shock resistance and immunity to chemical attack.

1.5.4 Chemical Applications

The many unique properties offered by graphite make it an ideal material of construction for the chemical process industry. Its chemical inertness allows it to come in contact with corrosive acids, bases, and organic and inorganic species without degradation. This property of graphite combined with its strength and thermal shock resistance enables operation at processing conditions that are unattainable by other materials. Vessels lined with graphite can store extremely corrosive fluids such as hydrofluoric acid. Entire reactor systems have been constructed from graphite materials. Its high thermal conductivity permits efficient heat removal from highly exothermic reactions. Graphite tubes have been used in the construction of corrosion-resistant heat exchangers. Figure 1.20 shows a heat exchanger constructed from graphite.

FIGURE 1.20
A heat exchanger constructed from graphite.

In the field of separations, carbon and graphite Raschig rings and packing support plates are available for packed-tower separators and carbon bubble-cap trays exist for distillation columns. For high-pressure applications, certain grades of highly impregnated graphite have been manufactured. Moreover, graphite pumps, pipes, fittings and valves, and safety devices are available for the safe transport of toxic and corrosive fluids. Finally, a new class of porous graphite (foam) has been developed for the filtration of corrosive and high-temperature fluids.

1.5.5 Mechanical Applications

In the present context, the mechanical properties of graphite are defined in relation to sliding surfaces in contact and under load. For this application, the primary attributes of graphite are its strength, low friction, lubricity, and compatibility with other materials. Specialized formulations and processing conditions are used to create graphite having these specific properties. Unlike most metals, graphite is unique in its ability to function under high speeds and loads without galling, its capability of high-temperature operation, and its resistance to chemical attack. Typical applications in this area include all types of seals for jet engines, high-temperature bearings and bushings, packing glands, pistons and piston rings, flat-plate slider bearings for machinery support, turbine and rotor blades, and a variety of gaskets especially useful at high temperatures. Special additives such as metallic sulfides and other impregnants can be mixed to the formulation to improve oxidation resistance and performance with high-pressure gases and liquids, and to allow operation at ultrahigh vacuums such as that encountered in deep space. In the medical field, graphite has been used successfully for the fabrication of prosthetics and artificial heart valves. In the aerospace industry, nose cones, rocket nozzles, leading edges on wings and turbines, and insulating components are commonly fabricated from graphite. Figure 1.21 shows examples of a graphite heart valve and high-performance graphite racing pistons.

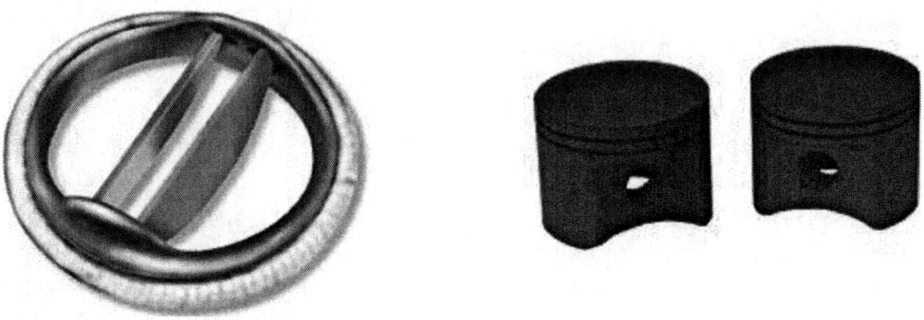

FIGURE 1.21
A graphite replacement heart valve (left), and graphite pistons for a high-performance racing engine (right).

1.5.6 Nuclear Applications

Nuclear-grade graphite is used extensively as both a moderator and a reflector in nuclear reactors primarily due to its excellent thermal stability and strength at high temperatures and its ultrahigh purity. Other especially attractive properties are its thermal shock resistance and its stability under nuclear radiation. In the nuclear reactor, the moderator acts to decrease the velocity and, hence, the kinetic energy of fast-moving free neutrons created by fission from energies around 2 MeV (fast neutrons) down to 1 eV (thermal neutrons) so that the fissile nuclei can capture the neutron and thus sustain the fission chain reaction. The purpose of the reflector is to reflect neutrons back into the active core region and thereby further aid in sustaining the fission reaction. This helps in reducing the critical size of the reacting pile. A graphite reflector can also serve as a moderator. Due to a variety of characteristics, it can be shown that graphite has high "figures of merit" when used as both a moderator and a reflector. It is also less expensive and safer than other materials such as beryllium, and it can simultaneously serve as a structural material, unlike moderators such as heavy water.

The first successful application of graphite as a moderator material was in 1942 by Enrico Fermi in his pioneering work at the University of Chicago. Similar positive results were found by workers in the then USSR. Interestingly, workers in Germany at the same time determined graphite to be unsuitable as a moderator, but this false conclusion was attributed to their very poor grade of graphite and its corresponding low level of purity. For this application, most graphite has sufficient strength, density, and other relevant properties, but the most severe constraint is placed on purity. To be a suitable moderator, the candidate material must be free from trace impurities (notably boron) that possess large neutron-capture cross sections. This in turn requires modification of the fabrication process either to graphitize at much higher temperatures for longer periods of time or to include chemical purification following graphitization. Post-treatment of graphite with halogens such as chlorine and fluorine converts the trace contaminants to volatile species, which leave the graphite by diffusion. Obviously, these procedures add to the cost and level of complexity of production. Table 1.7 lists some of the more important properties of nuclear-grade graphite.

In 1942, Eugene Wigner theorized that under the influence of the very fast and energetic neutron flux, graphite could over time undergo serious changes to its structure and properties. This behavior has been confirmed by a significant experimental program. The dimensional changes of graphite are a function of both the neutron flux and the temperature at the time of exposure and can be influenced somewhat by the starting raw materials and processing conditions. In general, the results indicate that during exposure the volume first decreases and then later increases as a function of neutron flux. The effect is accelerated as temperature rises (Kasten 1969). Other properties such as CTE, thermal conductivity, and specific heat also change as

TABLE 1.7

Properties of Nuclear Graphite

Property	Anisotropic Graphite	Isotropic Graphite
Density (g/cc)	1.71	1.86
Resistance (μΩ·cm)	735	1,000
Tensile strength (kPa)	9,930	46,172
CTE (10^{-6}/°C)		
With-grain	2.2	5.3
Against-grain	3.8	5.3
Anisotropy ratio (CTE ratio)	1.73	1.0
Total ash (ppm)	740	400
Boron content (ppm)	0.4	0.3

Source: Criscione, J. M., R. L. Reddy, C. F. Fulgenzi, D. J. Page, F. F. Fisher, A. J. Dzermejko, and J. B. Hedge, Kirk-Othmer Encyclopedia of Chemical Technology, published online: December 4, 2000, DOI: 10.1002/0471238961.01161612.a01.

a result of the applied neutron flux. The designer must be aware of these effects and compensate for them. Nevertheless, more than 100 graphite reactors have operated successfully over the last 60 years.

Finally, graphite can be purposely doped with good neutron absorbers such as boron and can subsequently be used as shielding material in reactors, thus adding to their safety. New reactor designs such as the advanced gas-cooled reactor (which uses carbon dioxide as the coolant) and the high-temperature gas-cooled reactor (which uses helium as the coolant) use large quantities of bulk graphite as the moderator and the structural material around the core. The advantage of these reactors is that they can operate at much higher temperatures (up to 950°C) and hence have higher efficiencies. The core structures are machined into complex shapes, known as "prismatic blocks" (see Figure 1.22), and are fabricated from isotropic graphite. As shown in Table 1.7, the anisotropy ratio of this graphite is 1, indicating complete isotropic properties. This is particularly important for CTE since these structures are subjected to very high thermal gradients and must maintain their dimensional stability to prevent cracking and structural failure.

1.5.7 Graphite in Drilling Mud

A "drilling fluid" is used during a drilling operation into the earth's surface. The current application of drilling fluids is almost exclusively limited to rotary drilling technology, frequently used for oil, gas, and water wells. Drilling fluids are classified by the nature of their principal ingredients, that is, water, oil, or gas, which are sometimes present in combinations. If the principal component of a drilling fluid is a liquid (i.e., water or oil), the term "mud" is given to the mixture of the liquid vehicle and the suspended solids. Presently, aqueous suspensions are used in most drilling fluids. Originally,

FIGURE 1.22
Prismatic graphite structures in the core of a high-temperature nuclear reactor.

the main function of the drilling fluid was as an aid in removing the cuttings formed at the drill tip in the borehole and transporting them up to the surface. The cuttings were then separated from the fluid and the liquid was recycled to the borehole. As the technology developed, additives were incorporated into the fluid, each having a specific technical purpose along with ensuring improved safety, more efficient use of raw materials, and lower cost. Graphite is one such additive.

In the rotary drilling operation, the drilling fluid serves many functions. Some of the more important ones include carrying cuttings from the drill tip up and out of the borehole, keeping the drill head clean and cool, reducing friction at the drill tip and along the shaft of the drill, exerting a hydrostatic pressure on the borehole, and forming a thin cake that seals fissures in the rock as drilling proceeds. The main use of the numerous additives in the mud is to prevent the flow of drilling fluid into the subterranean pores and fractures exposed during drilling so as to minimize both the loss of drilling fluid and potential adverse environmental impacts. It is of utmost importance that the mud penetrates and seals the exposed formations and forms a low-permeability "filter cake." For the proper formation of the cake, a range of particle sizes is used. The larger, "bridging" particles become trapped in the pores and initially retard the influx of fluid. Subsequently, smaller particles flow into the pores and fill the voids between the bridging particles. The density and the viscosity response (usually shear-thinning) of the mud under the various flow regimes encountered during drilling are also critical parameters.

Graphite is an ideal additive to drilling mud and has particular advantages as a stop-loss material (i.e., it is effective in sealing subterranean fissures) and a lubricant. For most applications, graphite is sized in the range of –20 to +200 U.S. standard mesh. Graphite is inert, is not soluble in groundwater, does not soften in the presence of oil or alkali, and can withstand extremes of temperature and pressure. It also has a relatively high density (about 2 g/cc), which facilitates the generation of high hydrostatic pressures in a borehole. There is some evidence to indicate that graphite is resilient so that once it is forced into a fissure under pressure, the graphite particle actually increases in volume on pressure reduction, thereby creating a more effective filter cake (Zaleski et al. 1998). Graphite has been shown to be an excellent lubricant for the drill bit as well as the drill shaft and casing. Thus, the torque and drag of the drilling process can be reduced, significantly decreasing the amount of power needed to drive the bit and the shaft. It also aids in reducing friction and wear in the shaft and casing and in ancillary equipment such as pumps, pipes, and filters.

All types of graphite are used in these "down-hole" applications with synthetic and amorphous graphite being the most common. Approximately 50,000 tons of graphite are incorporated into drilling muds each year with a large fraction of this going to the North Sea and other areas where active exploration is taking place. More recently, graphite has seen considerable application in the area of geothermal wells where it is used as a highly thermally conductive grout, thereby enhancing heat transfer of the thermal well.

1.6 Novel Forms of Graphite

Carbon materials are unique in that the basic graphene building block can be "assembled" in so many different ways to create a wide variety of structural forms. A few of the many novel forms of carbon that are distinct from the bulk graphite considered thus far in this chapter are described in Sections 1.6.1 through 1.6.5. Some of these novel forms have been known for many decades, whereas others are newly emerging structures whose applications are evolving almost daily.

1.6.1 Intercalation Compounds

Graphite intercalation compounds (GICs) are formed when an inorganic species incorporates itself between the graphene layers of the graphite lattice. In this structure, the intercalated species reside in the interplanar interstitial sites between the layers, whereas the layered structure of the graphite lattice remains largely intact (Chung 2002). Figure 1.23 is a representation of such a structure. Hundreds of these compounds have been produced over the years since they were first reported by Schaffautl in 1841. A very comprehensive review of the intercalation compounds of graphite is contained in the study by Dresselhaus and Dresselhaus (2002).

FIGURE 1.23
Foreign atoms intercalated in the graphite lattice: Here, the foreign compound is the potassium compound KC_8.

Graphite exhibits the so-called staging phenomenon in which the stage index, *n*, indicates the number of graphene layers between the intercalated layers. For example, stage 1 GICs are those in which a layer of intercalated species resides between every graphene layer, whereas in stage 2 GICs there are two graphene layers between the intercalated layer. Obviously the interlayer distance of graphite increases on intercalation. The nature of interaction of the guest atom or molecule with graphite can provide a means of controlling the physical properties of a GIC. The most profound effects have been seen and characterized on electrical conductivity, which can range from insulating to nearly superconducting. These effects are largely due to the nature of the interaction between graphite and the intercalant. Charge-transfer or donor-type ionic bonds lead to increase in electrical conductivity, whereas covalent bonding disrupts the graphite structure and destroys conductivity.

The preparation techniques for GICs differ from each other in the manner in which graphite is exposed to the intercalant. The intercalant could be in the vapor or liquid phase or present in solution. Electrochemical methods have also been used. The significant process variables include temperature, vapor pressure, graphite characteristics, and properties of the intercalant.

A few covalently bonded intercalation compounds are known. Among these are graphite oxide and the graphite fluorides. Graphite monofluoride (CF) is formed by reacting graphite and fluorine at high temperature and pressure, whereas tetracarbon monofluoride (C_4F) is formed at temperatures below 80°C. Both these GICs are electrical insulators.

The majority of intercalation compounds of graphite are those in which there is some degree of ionic bonding between the graphite layer and the intercalant. These types of GICs are formed spontaneously when graphite is exposed to the intercalant in either the vapor form or the liquid form at a prescribed minimum temperature for hours or days. The graphite/bromine (Br_2) compound has been extensively studied and exits in several stages (2, 3, 4, and 5).

Alkali-metal GICs are known to exist for most alkali metals including Li, Na, K, Rb, and Cs. Graphite/alkali-metal intercalation compounds are prepared by contacting the graphite and the alkali metal and heating them either together or in separate interconnected bulbs using the two-zone vapor transport technique. Since these forms of GICs represent n-type materials, the electrical conductivity of these structures increases and the anisotropy ratio of conductivity decreases.

Finally, intercalation compounds with metal halide species have been synthesized. Graphite–ferric chloride is perhaps the oldest and the most studied of these materials. In fact, it has been shown that once intercalated, $FeCl_3$ can be reduced with hydrogen to $FeCl_2$ while remaining in the graphite lattice. These compounds are of interest for their magnetic properties. Also, some mixed metal species such as $FeCl_3$–K have been shown to have potential as catalysts for ammonia synthesis (Ichikawa et al. 1972).

There are several characterization techniques useful for investigating GICs. Some of these include visual inspection, chemical analysis, dilatation assessment, optical and electron microscopy, and XRD. The simple analytical techniques give information on chemical formula and stage number. XRD yields information on stage number as well as the spacing distance between layers in the lattice. Electron microscopy provides insight into the microstructure of GICs and TEM images enable individual graphite and intercalant lattice planes to be seen directly.

Besides being noteworthy for their fundamental physics and chemistry, GICs have many potential applications. They are attractive for use in battery and electrode applications because of their low weight, high energy density, and high discharge capacity. The modern lithium-ion battery contains such carbon structures. GICs are used as catalysts in a wide range of chemical reactions. Catalysis has been reported for the synthesis of ammonia from nitrogen and hydrogen; for organic reactions such as halogenation, esterification, and polymerization; as well as the synthesis of hydrocarbon fuels from H_2 and CO, to name just a few applications. Intercalation compounds have been formed with graphite fibers to effect electrical "wires" of very high conductivity. Some studies of graphite fibers intercalated with K have shown an increase in conductivity by over a factor of 20. Finally, GICs show a very large expansion in volume when heated. This is due to vaporization of the intercalant in the graphite lattice. The volume expansion can be on the order of many hundred times. The process is called "exfoliation" and it is generally an irreversible process (Chung 2002). When exfoliated graphite is subsequently compressed, it forms a flexible sheet that can be easily cut to any desired shape. Such graphite is used extensively as gasket material and is especially suited for high-temperature and severe chemical applications. Such compounds also serve as precursors in the production of graphene (see Section 1.6.4.3).

1.6.2 Carbon and Graphite Fibers

A traditional carbon fiber is by definition a filament comprising 92 wt% carbon made by the pyrolysis of some precursor fiber. More recently, short nanofibers or carbon whiskers have been grown from the vapor phase with the aid of metallic catalysts. Graphite fibers are obtained by the alignment of graphite planes along the axis of the fiber upon heat treatment. The graphite formation can be confirmed by XRD. In general, these fibers are characterized by high tensile modulus and high thermal and electrical conductivity in the axial direction. The first carbon fibers were made from a cellulose precursor by Edison in 1880 in his search for a suitable filament for the light bulb. (His search also led him to pyrolyze a length of his wife's hair!) Modern fiber technology was first developed in the United States by Bacon in 1964 (Bacon and Tang 1964) using rayon precursor, by Shindo in Japan in 1961 (Shindo 1961) using polyacrylonitrile (PAN) precursor, and in the United Kingdom by Watt in 1963 (Watt et al. 1966) with PAN fiber. The need for such fibers arose from the aerospace industry, which sought a strong, lightweight fiber as reinforcement in advanced composites. Carbon fibers are generally classified into three grades: (1) General-purpose (GP), (2) high-performance (HP), and (3) activated carbon fibers (ACP). The GP fibers are amorphous, low-cost fibers with relatively low strength. The HP fiber is more graphitic and has high strength and modulus. Activated fibers consist of carbon with a high degree of microporosity, and exhibit large adsorption capacity and rapid adsorption/desorption.

Carbon fibers have been prepared from a wide variety of natural and organic precursors, but only two, PAN- and pitch-based fibers, are commercial today (Marsh et al. 1997). The PAN is a polymer produced from the polymerization of acrylonitrile. The PAN is first spun into a fiber typically by a process known as "wet-spinning" to a fiber diameter of about 15 μm. Before the fiber can be processed at a high temperature, it must first be stabilized through an oxidation step. Oxidation is usually carried out around 220°C in air with the fiber kept under tension. During this process, the nitrile groups undergo cross-linking, forming what is termed a "ladderlike" structure. Some oxygen is also incorporated into the fiber at this time and this further stabilizes the structure. Stabilization time is on the order of hours. Next, the fiber is carbonized in nitrogen at a temperature between 1000°C and 1700°C. This treatment drives away most foreign atoms leaving behind a fiber having a carbon content of over 90%. The carbonization sets up the 3-D carbon network having some microcrystalline structure. Further heat treatment of

2000°C–3000°C leads to the formation of the final graphitic structure in the fiber. Maximum tensile strength (as high as 7 GPa) usually occurs at a temperature of about 1500°C, whereas the modulus continues to increase with temperature reaching values as high as 500 GPa at 2500°C. The final diameter of the fiber is between 5 and 10 μm. Such fibers have a thermal conductivity greater than 50 W/m·K. It should be noted that even after the carbonization step, fibers with reasonable mechanical properties are produced. The PAN-based fibers have a carbon yield of around 50% based on the weight of starting material.

Carbon and graphite fibers may also be spun from pitch precursors. Such pitches are generally petroleum-based, although some are coal-based. The green fiber can be spun from either isotropic or 100% mesophase pitch, and the latter is produced from the former by heat treatment. If spun from isotropic pitch the resultant fiber has a relatively low modulus (100 GPa), whereas fibers spun from mesophase pitch are capable of a very high modulus of over 900 GPa. Thus, the resultant properties of the mesophase pitch–based fibers can exceed the properties of typical PAN-based fibers. These fibers are melt-spun under tension at a temperature above the softening point of the pitch. Isotropic pitch has a softening point around 100°C, whereas mesophase softens at about 300°C. Prior to spinning, the pitches must be purified to remove any trace particulate matter, which could lead to fiber breakage during spinning. Typical pitch fibers leave the spinneret at about 100 μm in diameter and are drawn down under tension to about 10 μm. Since the pitch is essentially a thermoplastic, the fiber must be stabilized prior to carbonization or graphitization; otherwise, it would melt. The stabilization process occurs slowly at a temperature between 200°C and 300°C in air. The process is essentially an oxidation reaction whereby oxygen is incorporated into the structure through cross-linking and renders it infusible. The stabilization step is critical and final properties of the fiber depend on it; so it must be done carefully over a precisely prescribed period of time. Heating rates on the order of tenths of a degree Celsius per minute are required. Once stabilized, the fiber can be carbonized and graphitized in a manner similar to that noted earlier in this section for PAN fibers. Pitch-based fibers have a carbon yield of around 75% based on the weight of starting material.

Carbon fibers have also been grown from the gas phase by vaporizing hydrocarbons and promoting fiber formation on metallic catalysts such as iron and nickel (Edie and Diefendorf 1992). Such fibers are known as vapor-grown carbon fibers (VGCFs). Typical hydrocarbons such as methane and benzene have been used as precursors. Fiber diameter can be between 0.1 and 100 μm, and the product usually leaves the reactor in the form of intertwined mats.

Finally, the Japanese pioneered fibers created from a pure synthetic mesophase pitch, which is made from the polymerization of naphthalene (Ono and Oya 2006). This pitch has been commercialized by the Mitsubishi Gas Chemical Company, Tokyo, Japan, under the name "AR pitch."

Carbon fibers have been characterized by many techniques including XRD and electron microscopy. XRD elucidates structural characteristics and various dimensional parameters of the fiber, whereas microscopy gives visual indications of the structure, surface, and internal texture and is also capable of performing elemental analysis. Correlations of structure with processing conditions, die geometry, and fiber fracture can easily be made through microscopic observation. Figure 1.24 indicates some of the cross-sectional features found in carbon fibers.

The applications of carbon fibers are numerous. They are employed in such diverse fields as aerospace, defense, the automotive industry, sporting goods, and the biomedical industry. They are also used for heat and electromagnetic shielding. Carbon-fiber composites are made from two-dimensional (2-D) or 3-D woven fabrics of carbon fiber set in a carbon, polymer, or ceramic matrix (Edie 1990). A variety of surface treatments have been proposed to allow better incorporation of fibers into the composite matrix and improve adhesion of fibers to the matrix. These composite structures have found extensive use in aerospace and aircraft industries as they represent strong, lightweight structural materials. The Stealth bomber is one such high-profile example. The incorporation of these composites in aircraft and automotive applications allows structures of lighter weight and results in corresponding fuel efficiency with increased safety. Carbon–carbon composites are finding increased utility in aircraft brakes particularly due to their ability to absorb energy while operating at temperatures well beyond the melting point of most metals. Recently, the U.S. Navy constructed above-deck components of ships using carbon–carbon composites. High-performance racing cars use composite clutch discs for improved gear shifting, wear, weight savings, and acceleration.

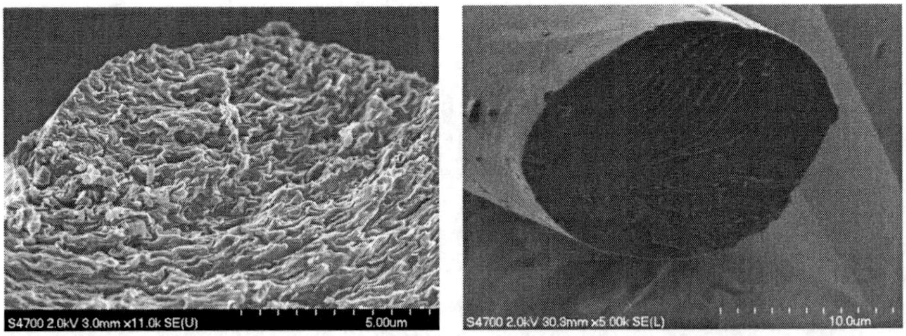

(a) (b)

FIGURE 1.24
Cross-sectional views of carbon fibers showing some highly ordered structures (a) and a more isotropic cross-section (b). (Courtesy of Magean, L. Private Communication, Department of Chemical Engineering, West Virginia University, Morgantown, West Virginia. 2011.)

1.6.3 Mesocarbon Microbeads

Mesophase is formed during the thermal pyrolysis of selected petroleum- and coal-based pitches. Mesophase is a liquid-crystalline state represented by highly ordered organic molecules in a layered or lamellar structure. The transition from isotropic pitch to mesophase occurs through the formation of tiny spheres from the disklike associations of the planar molecules present in the melt. As the transition proceeds, eventually the spheres contact each other and coalesce ultimately into 100% mesophase pitch (see Figure 1.2). The spheres can grow to about 25 μm before they start touching each other. A process has been developed in which spheres are grown, collected, and subsequently used in the fabrication of small-grained, high-density, highly isotropic graphite (see, e.g., Noguchi et al. 1983; Yamada et al. 1974). Unfortunately, the controlled growth and collection of these tiny structures is fraught with many technical challenges. Simple filtration has been attempted, but the small size of particles renders the filter cake impermeable. Some processes involving pyrolysis of the pitch at supercritical conditions or coprocessing with an insoluble silicon oil have been recorded (Cheng et al. 2009). Atomization of the pitch into small droplets by means of an injector has also been proposed (Pennington 1996).

Since the mesophase microbeads represent tiny, highly anisotropic graphite-like domains, they can be molded and isostatically pressed to yield highly isotropic graphite. The incorporation of other trace materials such as graphite and coke powders has been observed to further enhance the final properties of graphite (e.g., Oshima et al. 1991). This technology is capable of producing fine-grained graphite, which is ideally suited for the nuclear industry. Finally, there has been much effort in incorporating microbeads in battery electrodes and fuel-cell components. They have also been utilized as catalyst supports for electrochemical oxidation (Bai et al. 2002).

1.6.4 Fullerenes, Nanotubes, and Graphene

1.6.4.1 Fullerenes

Broadly, fullerenes are spherical structures in which carbon atoms are arranged in the form of "closed-cage" hexagons or pentagons. They are not simply a new geometry of a graphene sheet since they contain both 6- and 5-membered rings. Hence, their properties are different from those of graphite. Indeed, they are deemed the "third form" of carbon. They are formed by laser ablation of graphite and are found in the soot resulting from electrical discharge between carbon electrodes. They must be synthesized in an inert atmosphere since they readily oxidize in air during their synthesis. Interestingly, they always appear with an even number of carbon atoms and are thus given the formula C_{2n}, where n is greater than 10. The most common and widely publicized fullerene is the spherical isomer of C60, which is known as the buckyball. It is named after Buckminster Fuller who popularized the geodesic dome, which has a structure similar to C60. Figure 1.25 shows the typical fullerene structure of C60. The discovery of the C60 molecule is credited to Kroto (University of Sussex, Brighton, East Sussex, United Kingdom), Smalley, and Curl (Rice University, Houston, Texas) in 1985 (Kroto et al. 1985), although evidence for its existence was published a year earlier by a team from Exxon Research (Annandale, New Jersey) (Rohlfing et al. 1984). In 1996, Kroto, Smalley, and Curl were awarded Nobel Prize in Chemistry for their discovery. Since then, C60 has been found to exist in nature and recently its infrared signature was observed in a cloud of cosmic dust 6500 light years away. Larger quantities can be fabricated in an electric arc surrounded by helium or argon, the soot from which can be extracted and the various fullerenes isolated.

The C60 molecule comprises 20 hexagons and 12 pentagons with no two pentagons being adjacent. It has an outer diameter of 10.1 Å with an average bond length of 1.44 Å. The carbon atoms are in sp^2 hybridization, but the hexagon to hexagon bonds (6-6) are slightly shorter than those between hexagons and pentagons (6-5). There are 1812 possible isomers of C60, but only one is stable. Its molecular weight is 720.66 and it appears as a stable black crystalline powder. It melts above 280°C and has a density of 1.72 g/cc. The C60 has a thermal conductivity of 0.4 W/m·K and a resistivity of 1014 Ω/m. Because of its uniquely high symmetry, the various spectra of C60 are very simple. For example, the infrared spectrum has only four bands and the ^{13}C NMR has only one peak since all carbons are identical. Fullerenes have inspired research for theoretical calculations as well.

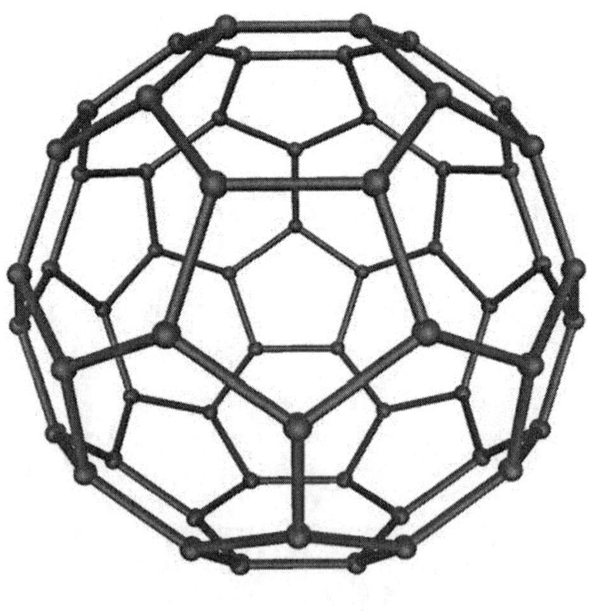

FIGURE 1.25
The structure of C60 fullerene.

Most of the common chemical reactions of C60 occur across the 6-6 bond as 1-2 additions. The C60 is slightly soluble in some organic solvents, including benzene, toluene, carbon disulfide, 1,2-dichlorobenzene, and 1-methylnaphthalene. An incredible variety of derivatives have been synthesized from C60. Incar-fullerenes have species such as La,Y, Ce, Sc, Sc_3N, and atomic nitrogen trapped within the sphere. Some of these products are stable, whereas others are not. The stable ones have extremely interesting properties and are the focus of much research. Multiwall or nested fullerenes have been created under special processing conditions. Heterofullerenes are made by replacing a carbon atom with a foreign atom in the sphere. Nitrogen and phosphorous substitutions are examples of these compounds. Finally, two C60 structures are joined together to form dimers (a.k.a. dumbbells) and fused species. The reader is referred to the excellent review article by Taylor (2002) for more details.

To date, over 1000 compounds related to fullerenes have been documented. Intense research is underway on possible applications of these species. Perhaps the most exciting study is in the area of medicine. Specific molecules have been attached to or encased in C60 as a means of developing new drug therapies or diagnostics. When attached to a metal, fullerenes have been used as catalysts and have even shown some superconducting properties. Because of their spherical shape, they may act as lubricants in the form of "nano ball-bearings."

1.6.4.2 Carbon Nanotubes

Carbon nanotubes are thin hollow cylinders comprising carbon. They can be thought of as a graphene layer rolled up to form a cylinder. They can be single-walled nanotubes (SWNTs) or multiwalled nanotubes (MWNTs). Their ends can be open or capped with a fullerene-type structure. Carbon nanotubes were first discovered by S. Iijima in 1991 (see Iijima 1991). Their diameter can range from 0.4 to 3 nm and they can be hundreds of microns in length, giving them an L/D ratio in the millions. Depending on how the graphene layer is rolled up, they can have several possible configurations. The three main types are armchair, zigzag, and chiral, depending on the shape formed by adjacent carbon atoms in the cylinder (see Figure 1.26). This orientation is important since it determines the properties of the resultant tube. For example, tensile strength and electrical conductivity are both strongly influenced by the configuration.

When cylinders are formed by multiple layers and coaxially arranged, the structure is referred to as an MWNT. In these structures, the spacing between layers is similar to that found in graphite, that is, about 3.4 Å.

Carbon nanotubes are grown by a variety of methods. The earliest effort was based on techniques used to produce fullerenes, that is, growth in the arc discharge from a carbon electrode. This method continues to be the most widely practiced one today. The next technique is laser ablation where a laser vaporizes graphite at a high temperature and the nanotubes are condensed

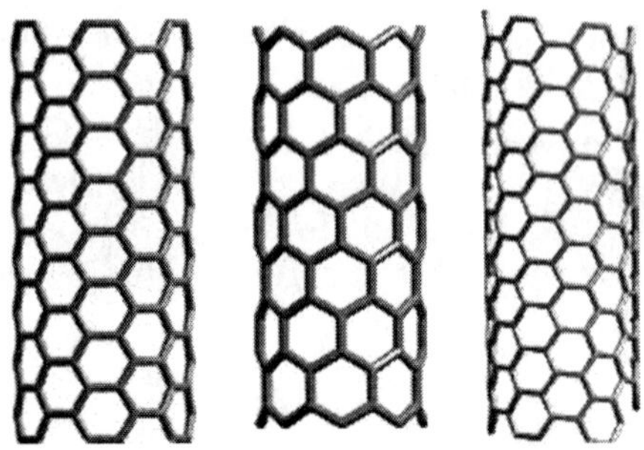

FIGURE 1.26
The three main configurations possible for carbon nanotubes.

from the vapor. The final method is chemical vapor deposition (CVD) in which the nanotubes grow from the vapor on a metallic substrate of fine catalyst particles. Typical metal catalysts include iron, cobalt, and nickel. There are many sources of carbon for CVD such as carbon monoxide, methane, ethylene, acetylene, and benzene. Although there is no clear winner, each technique has its own advantages and disadvantages. In all these synthesis methods, the challenge is to gain control of the fiber diameter, length, and electronic properties while simultaneously harvesting the product in an easy and efficient manner.

The unique properties of carbon nanotubes are derived from the strength of the carbon–carbon bond, the orientation of graphene sheets in the axial direction, and the electronic behavior of the delocalized electrons in the p orbitals. A Young's modulus of over 1 TPa has been reported for both SWNT and MWNT (Krishnan et al. 1998; Wong et al. 1997). Tensile strengths of over 200 GPa have been estimated. Nanotubes have the highest strength-to-weight ratio for any material. They are excellent for applications in reinforced composites and nanoelectromechanical systems. Wind turbine blades containing carbon nanotubes are much lighter and stronger than conventional blades. Thus, larger blades can be incorporated allowing a unit to generate more electrical power. Depending on the configuration of the nanotubes, they can behave as either a metallic conductor or a semiconductor, which makes them ideal building blocks for nanoelectronics and circuits. Thus, nanodevices completely constructed from nanotubes can be envisioned in which metallic nanotubes carry signals into and out of such devices while semiconductive nanotubes behave as active elements. The high flexibility of nanotubes has led to their use as tips for scanning-tunneling and atomic force microscopes. The surface of nanotubes can be functionalized permitting improved performance and processing in composites. Functionalized nanotubes can have medicinal applications related to selective cancer treatments and drug delivery carriers. Finally, it has been postulated that nanotubes can serve as hydrogen storage devices; however, there is still debate over the reported results. These applications are but a few currently being pursued. The future of nanotechnology is perhaps one of the brightest in all of material science, and fullerenes and nanotubes will figure prominently in its development.

1.6.4.3 Graphene

Graphene is essentially a single layer of graphite in the form of a hexagonal lattice of carbon atoms bonded in the sp^2 configuration. All the bond lengths in graphene are equal to 1.42 Å. Flake graphite is made up of stacks of graphene sheets as discussed in Section 1.2.1. It also constitutes the basic structure of SWNTs and MWNTs. Graphene was first isolated in 2004 by peeling off a layer of graphite on a piece of Scotch tape and transferring it to a stable SiO_2 substrate. In fact, in 2010 the Nobel Prize in Physics was awarded to Geim and Novoselov for their pioneering research on the structure and properties of graphene.

There are several synthesis routes for the production of graphene (Zhu et al. 2010). The mechanical exfoliation of graphite involves the transfer of graphene layers to tape, which is subsequently dissolved allowing the single layers to be deposited on a substrate. The next method produces epitaxial graphene by the evaporation of silicon from silicon carbide at temperatures above 1100°C. Graphene has also been synthesized by CVD onto thin metal films of copper and nickel. Such sheets, when successfully transferred to other substrates, show much promise in electronic applications. Finally, the chemical exfoliation of graphite can produce derivative species such as graphene oxide and graphene fluoride, which convert to graphene following chemical reduction (Compton and Nguyen 2010). Currently this last method offers the most promising route for the manufacture of large bulk graphene sheets capable of being fabricated into a variety of graphene-based devices.

The unique physical, chemical, and electrical properties of graphene allow many new and exciting applications (Huang et al. 2011). Because of its high strength, good conductivity, and high thermal stability, graphene is being incorporated as a filler in graphene–polymer composites. Graphene has been used as a template for the synthesis of metal nanoparticles to form graphene–metal nanoparticle composites, which have application in catalysis and electrochemical sensors. In addition, graphene–semiconductor nanomaterial composites are promising increased potential in various energy-related areas such as Li-ion batteries, solar cells, and supercapacitors. Graphene, particularly in the form of nanostrips and nanomesh, has been incorporated into various electronic components such as field-effect transistors, integrated circuits, and memory devices. Due to its 2-D planar structure, functionalized graphene makes an excellent platform for the detection of chemical and biological molecules whereby the absorption of just a few molecules can lead to a measurable change in resistance. A more detailed discussion of the synthesis, properties, and applications of graphene follows in subsequent chapters of this book.

1.6.5 Carbon and Graphite Foam

Carbon foam has recently ignited considerable interest in the field of carbon science because of its unique properties. It is a cellular, highly porous material in which the cells are formed by ligaments comprising either graphitized or nongraphitized carbon. A photomicrograph of a representative glassy carbon foam is shown in Figure 1.27.

Foams can have open or closed cells, although the majority is open cell. Since it is interconnected by a network of ligaments making up the cell walls, its bulk properties are isotropic. It is lightweight with density ranging from 0.1 to 0.8 g/cc and can have a compressive strength between 0.1 and 20 MPa. Depending on the precursor material, it can exhibit very high or very low values of electrical and thermal conductivities. Thermal conductivities can range from 0.03 to over 150 W/m·K. Once graphitized, it is mechanically stable at temperatures up to 3000°C and chemically inert up to around 500°C. The porosity is 80%–90%. The CTE can be varied from 2×10^{-6} to 7×10^{-6} per degree Celsius.

Carbon foams are produced by carefully carbonizing resinous polymeric foams made from phenolics or polyurethanes. They have also been made from isotropic and mesophase pitches (Klett et al. 2000), raw coal, and solvent extracts from bituminous coal (Stiller et al. 1999). If the precursor is graphitizable, then graphitic foam can result if the material undergoes heat treatment to a sufficiently high temperature. Thus, for example, raw coal and isotropic pitch precursors can yield insulating foams, whereas anisotropic feedstocks such as mesophase pitch can yield foams that are highly electrically and thermally conductive (Gallego and Klett 2002). Most coal- and pitch-based foams are prepared in a manner similar to baking a cake. The precursor is

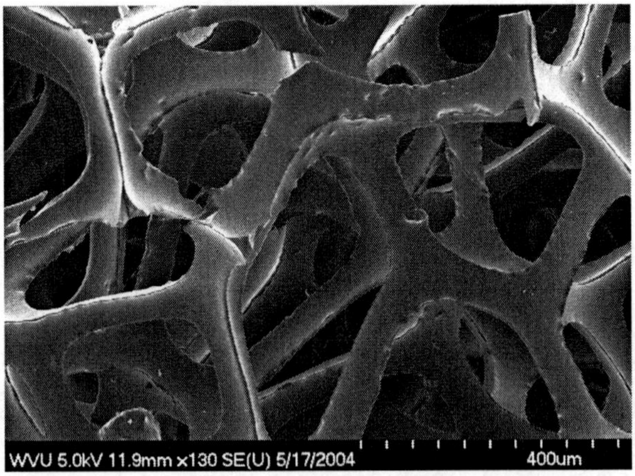

FIGURE 1.27
Photomicrograph of a reticulated glassy carbon foam. (Courtesy of Magean, L. Private Communication, Department of Chemical Engineering, West Virginia University, Morgantown, West Virginia. 2011.)

slowly heated until it begins to devolatilize forming gas bubbles throughout the structure. If the temperature is just right, the gas evolution is made coincident with the point where the larger precursor molecules begin to cross-link, thus solidifying the mass while setting and stabilizing the foam structure. Some techniques heat the precursor under pressure of an inert gas and, at the proper temperature, quickly decompress the system causing rapid foaming and solidification. Still other methods use a foaming agent (e.g., dissolved CO_2 or some chemical agent) to aid the creation of gas bubbles. Certain pitch-based foams may require thermal stabilization similar to that found in carbon-fiber manufacture. Some workers have correlated properties of the pitch precursor, such as fluidity, dilatation, and weight-loss characteristics, to define the foaming conditions and, ultimately, determine the final properties of the foam (Chen et al. 2006). Thus, there is a complex interrelationship among the precursor material, its chemical and physical characteristics, and the foaming conditions with the final properties of the resultant foam. Finally, researchers have augmented the properties of the foam by mixing carbon and graphite fibers with the pitch prior to foaming (Miller and Winton 2011). The shear forces generated during foaming cause alignment of the fibers along the ligaments resulting in remarkable variation in the characteristics of the foam. Much work continues to be done in this area.

The applications of carbon and graphite foam are limited only by the imagination. The low thermal conductivity of some carbon foam makes it an ideal thermal insulation/protection material, which is capable of operating at very high temperatures such as those found in high-temperature furnaces and during reentry of space vehicles. It has been used successfully as a core material for strong, lightweight, fire-resistant laminated structural panels such as those found on decks of ships and in the walls of mine refuge chambers. It is impregnated with resin to form a carbon–carbon composite where the foam is a low-cost replacement for carbon fibers. Its open cell structure and high porosity make it an ideal candidate for making high-temperature, corrosion-resistant filters and catalyst supports. If the foam is made electrically conducting, it can function as an element in an electrostatic precipitator. It has been used as electrodes and grids in batteries resulting in lower weight and higher efficiency of these devices. One particularly novel application of foam is as an energy-absorbing material; on impact, the cellular structure of foam collapses, absorbing the energy of the impact, which means the force is not transmitted through to the other side. Pitch-based graphitized foam with high thermal conductivity has been fashioned into heat sinks and radiator bodies (Klett et al. 2000). Finally, and by no means lastly, foams can be easily machined into complex shapes such as rocket nose cones, rocket nozzles, heat-exchanger cores, and tooling for aircraft wings and other parts.

The phenomenal growth of the carbon foam industry is evidenced by the fact that 10 years ago there were perhaps two or three foam producers in the United States, whereas today there are dozens all over the world. Thus, carbon and graphite foams could develop into the real superstars of the carbon industry.

1.7 Conclusions and Future Outlook

Although it is difficult to predict, the future seems bright for carbon and graphite. Certainly bulk products such as electrodes and specialty graphite will be in demand since many heavy industries (such as iron, silicon, and aluminum producers) depend on them. With increasing pressure being placed on the burning of fossil fuels, nuclear power will play an important role in our energy future. Reactors that incorporate graphite components have emerged as the safest, most reliable, and most efficient of all candidate reactor systems. Due to their high strength and low weight, carbon-fiber composites and carbon foams will play an ever-increasing role in the aerospace, defense, and automotive markets as strong, heat-resistant, and lightweight structural materials. Finally, novel nanostructures embodied in fullerenes, nanotubes, and graphenes will see phenomenal growth in the areas of microelectronics, sensors, and medical devices for drug delivery and diagnostics. The following chapters in this book give the reader a glimpse into a small part of this new and exciting area.

References

Bacon, R., and M. M. Tang. 1964. *Carbon* 2: 211.

Bai, Y., J. Li, X. Qiu, and J. Wu. 2002. Mesocarbon microbeads supported Pt-Ru catalysts for electrochemical oxidation of methanol. *Journal of Power Sources* 111: 160–164.

Bragg W. L. 1913. The Diffraction of Short Electromagnetic Waves by a Crystal. *Proceedings of the Cambridge Philosophical Society* 17: 43–57.

Brooks, J. D., and G. H. Taylor. 1968. The formation of some graphitizing carbons. *Chemistry and Physics of Carbon* 4: 243–286.

Chen, C., E. B. Kennel, A. H. Stiller, P. G. Stansberry, and J. W. Zondlo. 2006. Carbon foam derived from various precursors. *Carbon* 44: 1535–1543.

Cheng, Y., T. Li, F. Li, Q. Zhuang, and D. Jing. 2009. Preparation of mesocarbon microbeads and subsequent heating treatment. *The International Journal of Chemical Reaction Engineering* 7: A27.

Chung, D. D. L. 2002. Review graphite. *Journal of Material Science* 37: 1475–1489.

Compton, O. C., and S. T. Nguyen. 2010. *Small* 6 (6): 711–723.

Criscione, J. M., R. L. Reddy, C. F. Fulgenzi, D. J. Page, F. F. Fisher, A. J. Dzermejko, and J. B. Hedge. 2000. Kirk-Othmer Encyclopedia of Chemical Technology, Published Online: 4 December 2000, DOI: 10.1002/0471238961.01161612.a01.

Dresselhaus, M. S., and G. Dresselhuas. 2002. Intercalation compounds of graphite. *Advances in Physics* 51 (1): 1–186.

Edie, D. D. 1990. In *Carbon Fibers, Filaments and Composites*, J. L. Figueiredo (Ed.). Dordrecht: Kluwer Academic.

Edie, D. D., and R. J. Diefendorf. 1992. Carbon fiber manufacturing. In *Carbon-Carbon Materials and Composites*. NASA 1254: 19.

Ewing, G. W. 1969. *Instrumental Methods of Chemical Analysis*. New York: McGraw-Hill.

Forrest, R. A., and H. Marsh. 1977. Reflectance interference colours in optical microscopy of carbon. *Carbon* 15: 348.

Gallego, N. C., and J. W. Klett. 2002. Carbon foams for thermal management. International Seminar on advanced Applications for Carbon Materials, Jeju Island, Korea.

Gupta, R. 2011. Private Communication, Department of Chemical Engineering, West Virginia University, Morgantown, WV.

Huang, X., Z. Yin, S. Wu, X. Qi, Q. He, Q. Zhang, Q. Yan, F. Boey, and H. Zhang. 2011. *Small* 7 (14): 1876–1902.

Hupp, T. R., I. C. Lewis, J. M. Criscione, R. L. Reddy, C. F. Fulgenzi, D. J. Page, F. F. Fisher, A. J. Dzermejko, and J. B. Hedge. 2003. Kirk-Othmer Encyclopedia of Chemical Technology, Published Online: 15 August 2003, DOI: 10.1002/0471238961.0201110512052309.a01.pub2.

Ichikawa, M., T. Kondo, and K. Tamaru. 1972. Ger. Offen., DE 2114769 A, 19720316.

Iijima, S. 1991. *Nature* 354: 56.

The Industrial Graphite Engineering Handbook. 1963. New York: The National Carbon Company.

Jenkins, R., and R. Snyder. 1996. *Introduction to X-ray Powder Diffractometery*. New York: Wiley.

Kasten, P. R. 1969. U.S. Oak Ridge National Laboratory, ORNL-TM-2136.

Khandare, P. 1995. Characterization of mesophase pitch materials from petroleum and coal-derived precursors: Kinetics and rheology at elevated temperatures. Ph.D. Thesis, Department of Chemical Engineering, West Virginia University, Morgantown, WV.

Klett, J., R. Hardy, E. Romine, C. Walls, and T. Burchell. 2000. High-thermal-conductivity, mesophase-pitch-derived carbon foams: effect of precursor on structure and properties. *Carbon* 38: 953–973.

Krishnan, A., E. Dujardin, T. W. Ebbesen, P. N. Yianilos, and M. M. J. Treacy. 1998. *Physical Review B. Condensed Matter and Materials Physics* 58: 14013.

Kroto, H. W., J. R. Heath, S. C. O'Brien, R. F. Curl, and R. E. Smalley. 1985. *Nature* 318: 162.

Magean, L. 2011. Private Communication, Department of Chemical Engineering, West Virginia University, Morgantown, WV.

Mantell, C. L. 1968. *Carbon and Graphite Handbook*. New York: John Wiley.

Marsh, H. 1989. *Introduction to Carbon Science*. London: Butterworths.

Marsh, H., E. A. Heintz, and F. Rodriguez-Reinoso. 1997. *Introduction to Carbon Technologies*. Alicante: Universidad.

Miller, B. G., and S. Winton. 6 January, 2011. An Industrial-Based Consortium to Develop Premium Carbon Products from Coal Final Report, Prepared for U.S. Department of Energy, National Energy Technology Laboratory, Cooperative Agreement No. DE-FC26-03NT42874 1822 pages.

Moreland, A., J. W. Patrick, and A. Walker. 1988. Optical anisotropy in cokes from high-rank coals. *Fuel* 67: 730.

Nave, R. 2011. HyperPhysics Project, Department of Physics and Astronomy, Georgia State University, Atlanta, GA 30302–4106.

Noguchi, K., H. Tanaka, Y. Kumura, E. Kitajima, N. Tsuchiya, and T. Sunada. 1983. Process of producing mesocarbon microbeads of uniform particle size distribution. US Patent No. 4381990.

Olson, D. W. January, 2011. U.S. Geological Survey, Mineral Commodity Summaries.
Ono, H., and A. Oya. 2006. Preparation of highly crystalline carbon nanofibers from pitch/polymer blend. *Carbon* 44: 682–686.
Oshima, H., T. Okamoto, S. Tanaka, and T. Kawakami. 1991. Japanese Patent Publication No. 37,108.
Pennington, W. 1996. Assessment of mesophase rheology. M.S. Thesis, Department of Chemical Engineering, West Virginia University, Morgantown, WV.
Rohlfing, E. A., D. M. Cox, and A. Kaldor. 1984. *Journal of Chemical Physics* 81: 3322.
Schaffautl, P. 1841. *Journal of Practical Chemistry* 21: 155.
Seehra, M. S., and A. S. Pavlovic. 1993. X-ray diffraction, thermal expansion, electrical conductivity and optical microscopy studies of coal-based graphites. *Carbon* 31: 557–564.
Seehra, M. S., J. Zhao, and A. S. Pavlovic. 1992. Is it graphite yet? Interim Report, West Virginia University, Morgantown, WV.
Shindo, A. 1961. Studies of graphite fiber. Govt. Indust. Res. Inst., Osaka, GIRIO Report 317.
Skoog, D. A., F. J. Holler, and S. R. Crouch. 2007. *Principles of Instrumental Analysis.* Belmont: Brooks/Cole.
Stanjek, H., and W. Hausler. 2004. Basics of X-ray diffraction. *Hyperfine Interactions* 154: 107–119.
Stiller, A. H., P. G. Stansberry, and J. W. Zondlo. 1999. Method of making carbon foam material and resultant product. US Patent 5,888,469.
Taylor, G. H. 1961. Development of optical properties of coke during carbonization. *Fuel* 40: 465–472.
Taylor, R. 2002. Fullerenes. In Kirk-Othmer Encyclopedia of Chemical Technology, Published Online: 20 September 2002, DOI: 10.1002/0471238961.062112122001 2512.a01.
Watt, W., L. N. Phillips, and W. Johnson. 1966. High-strength, high-modulus carbon fibers. *The Engineer* 221: 815–816.
Wong, E. W., P. E. Sheehan, and C. M. Lieber. 1997. *Science* (Washington, DC) 277: 1971.
Yamada, Y., T. Imamura, H. Kakiyama, H. Honda, S. Oi, and K. Fukuda. 1974. Characteristics of meso-carbon microbeads separated from pitch. *Carbon* 12: 307–319.
Zaleski, P. L., D. J. Derwin, D. J. Weintritt, and G. W. Russell. 1998. Drilling fluid loss prevention and lubrication additive. US Patent 5,826,669.
Zhu, Y., S. Murali, W. Cai, X. Li, J. W. Suk, J. R. Potts, and R. S. Ruoff. 2010. *Advanced Materials* 22: 3906–3924.

2

Electronic Transport in Graphene: Theory and Applications

Pierre Carmier*

CONTENTS

* Correspondence to: Pierre Carmier (pierre.carmier@cea.fr)

2.1 Introduction

Until recently, purely two-dimensional (2D) materials were thought perhaps not to exist in nature. Physicists who wanted to study transport or any other physical manifestation of quantum mechanics in two dimensions had no other choice but to fabricate 2D systems artificially by confining 3D ones in a plane (such that the energy spectrum in the confined dimension is effectively reduced to a single quantum level). Two-dimensional electron gases (2DEGs) were thus obtained by forming 2D quantum wells at the interface of semiconducting heterostructures such as GaAs–AlGaAs using elaborate electrostatic confinement techniques. Since their discovery, these systems have been the source of meticulous studies and shown to exhibit novel quantum phenomena stemming from the influence of confinement on quantum particles: the quantum Hall effect, Coulomb blockade, and conductance quantization in ballistic quantum point contacts, to name but a few (Beenakker and van Houten 1991). However, these systems suffer from the experimental challenges that have to be met to efficiently control them since they are buried deep in the aforementioned heterostructures. The synthesis of graphene, a truly 2D crystal both stable and controllable at room temperature, was nothing shy of a revolution in this respect. What is this amazing material?

2.1.1 Carbon Once More

Graphene is a single plane (or monolayer) of graphitic carbon. Graphite is hence made of many layers of graphene weakly coupled with each other by the van der Waals interaction. The idea of obtaining graphene by peeling it off from graphite is very old and dates back to the early twentieth century. It is the microscopic counterpart of pencil writing on a sheet of paper. Applied to graphite using the Scotch-tape technique popularized by Geim and Novoselov, which is more formally referred to as *mechanical exfoliation* and basically consists in repeatedly pulling off few-layer graphite films from graphite crystals using adhesive tape, it was the first step that led to the experimental isolation and detection of a monolayer of graphite in 2004 (Novoselov et al. 2004). This experiment has initiated an arguably unprecedented wave of interest in the condensed matter community and beyond. The reasons for this situation are diverse as will be illustrated in this book, but one must admit that the exceptional character of the material's electronic properties stands out and certainly provided the initial motivation. Graphene had indeed long been studied theoretically and predicted to feature very special properties (as compared to other low-dimensional materials) arising from its band structure (Wallace 1947; McClure 1956). Most significantly, low-energy excitations behave like massless pseudorelativistic particles (Slonczewski and Weiss 1958; Semenoff 1984), so-called Dirac fermions, obeying the Dirac equation instead of the quadratic dispersion common to most of the 2DEGs formed in AlGaAs heterostructures or silicon metal-oxide-semiconductor field-effect transistors (Si-MOSFETs).

The field is moving at such a pace that it is almost impossible to keep track of the current status of our understanding and grasp of graphene science (Figure 2.1). Nevertheless, I shall attempt to review some of the developments in the field, which have set graphene at the center of the stage for the past few years. This chapter has naturally no claim for exhaustiveness and also reflects in part some of the author's personal tastes. Many other reviews have

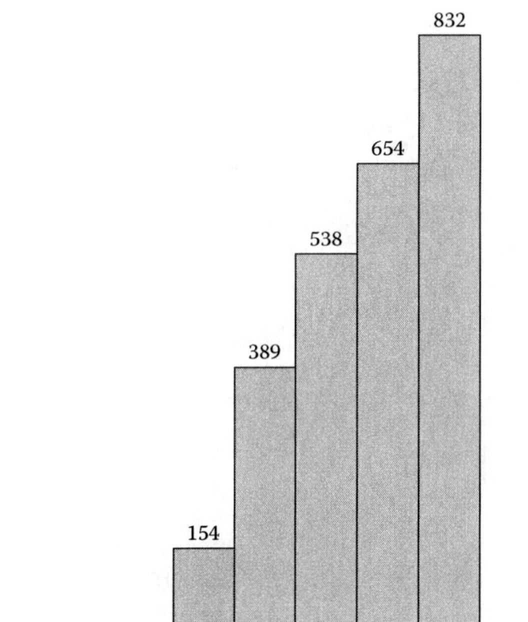

FIGURE 2.1
A histogram representing the number of papers posted on the web site arXiv containing the word "graphene" in their titles. Both the sudden burst of activity in 2006 and the steady increase in the amount of posted material since then are remarkable.

already been written on this very active subject, and the interested reader is referred to them for more detailed information (Geim and Novoselov 2007; Castro Neto et al. 2009; Geim 2009; Abergel et al. 2010; Das Sarma et al. 2011). This chapter will henceforth be split into three parts. In Section 2.2, I will discuss some of the electronic transport experiments that have launched the field of graphene science. Basic information concerning graphene's lattice and band structures will be introduced, and its connection with other carbon-based nanomaterials—in particular, carbon nanotubes (CNTs)—will be mentioned. In Section 2.3, I will address some of the essential electronic properties of graphene from a fundamental perspective. Theoretical notions in condensed matter physics will be introduced along the way when necessary in a pedestrian fashion. Some of the implications of controlling and manipulating these exotic electronic properties for other fields of physics will be briefly touched upon. Finally, in Section 2.4, I will discuss the promises held by graphene for applications on a short-term and long-term basis. The possibility of a "graphene revolution" in the transistor industry will be analyzed and the challenges left to overcome will be described.

2.2 New Material

In this section, I will review some of the experimental developments that have accompanied the rise of graphene in the condensed matter community. Graphene's place in the family of carbon nanomaterials will be mentioned, with a particular emphasis as to why graphene has arguably attracted more attention than CNTs had at the time. Finally, a detailed presentation of the material's lattice and band structures will be provided and related elementary electronic properties will be briefly discussed.

2.2.1 "Groundbreaking Experiments" Lead to a Nobel Prize

In the early 2000s, several groups were exploring the possibility of synthesizing atomically thin crystals, with high-level research being done in particular in the groups led by W. de Heer in Georgia Tech and P. Kim in Columbia. At the time, the sole possibility of isolating strictly 2D crystals was still a matter of debate. A number of experiments in the past had indeed repeatedly failed to obtain single-layer crystals, due to the inherent strong thermal fluctuations at the high temperatures which were required to grow such crystals (Venables et al. 1984): the target material inevitably ended up breaking or curving into 3D space before the monolayer limit could be attained. Pessimistic expectations were somewhat wrongfully nurtured by a theorem proved in the 1960s by Mermin and Wagner, which forbids the existence of long-ranged order in 2D crystals *living in 2D space*. Importantly, however, "real" crystals such as graphene live in 3D space and nothing prevents them from relaxing through this transverse degree of freedom: graphene was indeed later shown to be stabilized by out-of-the-plane mechanical deformations, so-called ripples (Meyer et al. 2007), which remain small enough that the 2D character of the material is preserved.

An important step forward was made in 2004 in the first of a series of "groundbreaking experiments"* conducted by Geim and Novoselov. By repeatedly cleaving high-quality graphite using Scotch tape and depositing the obtained multilayer samples on a substrate, the two future Nobel laureates and their collaborators from the Manchester Centre for Mesoscience and Nanotechnology were able to isolate monolayers of graphite by identifying them optically using a standard optical microscope (Novoselov et al. 2004). The optical detection process turned out to be efficient, thanks to a clever (and somewhat fortunate) choice of substrate thickness, which significantly increased the visibility of those multilayers that were only one atom thick (Abergel et al. 2007; Blake et al. 2007), actually optimizing the optical contrast of carbon monolayers in the wavelength range of the human eye sensitivity. A more accurate way of detecting single layers of graphite was later found in Raman spectroscopy, which basically consists in probing a material's crystallographic structure with a laser to excite phonon modes (vibrational degrees of freedom). Graphene's phonon spectrum indeed features a characteristic peak that can be readily identified (Ferrari et al. 2006). More information on experimental characterization of graphene can be found in Soldano et al. (2010).

Once electrically contacted using standard electron-beam lithography techniques, graphite monolayers already showed exceptional transport characteristics, including a tunable electric-field effect (Novoselov et al. 2004), that is, the ability to continuously tune the density of charge carriers n_s from electron to "hole"-like by applying a gate voltage, which was a testimony to the predicted gapless band structure. Very high electrical mobilities μ were also measured, indicating that charge carriers could travel through the material unscattered over large distances: μ outreached 10^4 cm^2/V/s for graphene deposited on a SiO_2 substrate and later 10^5 cm^2/V/s for suspended graphene (Bolotin et al. 2008), the highest ever reported value for a semiconductor at room temperature. Recent developments (Dean et al. 2010; Mayorov et al. 2011; Xue et al. 2011) have seen mobilities on substrate-supported graphene increase by an order of magnitude when hexagonal boron nitride (hBN) is sandwiched between the sample and the SiO_2 substrate,† while mobilities reaching 10^6 cm^2/V/s on suspended graphene have been reported (Elias et al. 2011).

* The experiments carried out by Geim and Novoselov and their team were praised in those terms by the 2010 Nobel committee.

† This is due to the smaller density of trapped charges in hBN (as compared to SiO_2) and the smaller induced corrugations in the graphene sheet, thanks to hBN's identical lattice structure.

2.2.1.1 Quantum Hall Measurements

Soon after the inaugural 2004 experiment, Hall measurements were carried (Novoselov et al. 2005; Zhang et al. 2005), which truly launched the field by uncovering novel physics and attesting the purely 2D character of the material. In this experimental setup, a strong perpendicular magnetic field B (typically of the order of a few Teslas) is applied to the graphene sheet and a current I propagates along the boundaries of the sample in response to a small voltage bias V applied between source and drain metal contacts (also referred to as incoming and outgoing leads). In the quantum Hall regime (see Section 2.3.2), the conductance $G = I/V$ is quantized and varies in a steplike fashion as a function of the magnetic field. In graphene, this quantization of the Hall plateaus in the conductance is actually anomalous, which can be traced back to the existence of a Landau level of zero energy equally shared by electron and "hole"-like excitations (Gusynin and Sharapov 2005). The observation of this characteristic signature in 2005 was thus another fingerprint of the gapless band structure and unambiguously confirmed the 2D character of the sample. Additionally, the anomalous scaling of the Landau levels in graphene (see Section 2.2.3) led to the observation of the quantum Hall effect at room temperature (Novoselov et al. 2007), whereas the latter previously required low temperatures to be observed in 2DEGs. This resilience to ambient chemical conditions actually holds true for many other graphene's properties and represents a crucial aspect in the material's revolution, along with the fact that graphene can be easily manipulated and functionalized (as opposed to typical 2DEGs formed at buried semiconductor interfaces).

2.2.1.2 Wonder Material

Later, other amazing properties reaching far outside the field of electronic transport attracted the attention of scientists. Graphene behaves indeed remarkably, whether it be from optical, mechanical, or thermal perspectives. It is optically quasi-transparent in the entire spectrum of visible light (Nair et al. 2008), which alongside carbon's inert chemical nature makes graphene an ideal candidate for acting as a support over which to manipulate biological molecules such as DNA (Nair et al. 2010). It is both exceptionally strong—more than 100 times stronger than steel with a breaking strength of the order of 40 N/m—and flexible (Lee et al. 2008). Its (phonon-mediated) thermal conductivity is equally impressive (of the order of 1000 W/m/K), outplaying even that of copper at room temperature (Balandin 2011), and could enable graphene electrodes to serve as efficient heat removers for evacuating the ever-larger amounts of heat dissipated by modern electronic devices. Graphene is also impermeable to gas molecules as small as helium, paving the way for gas sensor applications (Schedin et al. 2007).

All in all, the pioneering achievement of Geim and Novoselov is to have both elaborated a complete methodology for isolating graphene (which may not be viable industrially but is in fact very convenient for experimentalists) and succeeded in making electrical transport measurements demonstrating the material's exceptional properties. Alongside graphite (3D), CNTs (1D), and buckyballs (C_{60} molecules wrapped in a, so to say, 0D object, discovered in the 1980s shortly before CNTs), graphene (2D) has therefore now filled the gap in the carbon family and is made routinely available for conducting fairly intricate experiments. This breakthrough also successfully launched the synthesis of other monolayer materials such as hBN or MoS_2.

2.2.2 Link with Carbon Nanotubes

Before proceeding with our introduction on graphene's lattice and band structures, let us make a short detour and clarify where graphene stands with respect to other carbon nanomaterials. The reader might wonder why this latest addition to the carbon family has attracted so much attention, when fullerenes had already been synthesized experimentally long ago and also shown to feature very interesting and related properties. Before answering this question, let us begin by noting that graphene and CNTs are closely related systems. Indeed, single-wall CNTs, which were obtained a couple of years after the initial discovery of multiwall CNTs by Iijima (1991), are nothing else than graphene sheets rolled in tubes. The boundary conditions applied when rolling the sheet strongly influence the properties of the obtained CNT, even affecting the resulting band structure, which can be semimetallic or insulating. A CNT can in fact be fully characterized by a set of integers (n, m), which measure the distance in the primitive vector basis of the honeycomb lattice between a given lattice site and the one to which it is mapped when rolling the sheet into a tube (Saito et al. 1998). For a given value of n, a CNT with any edge boundary condition can be obtained, from zigzag termination for $m = 0$ to armchair termination for $m = n$, and any mixed situation for intermediate values of m (zigzag and armchair boundary conditions are defined in Figure 2.2). Control and growth of CNTs is now a mature branch of experimental physics (Charlier et al. 2007), with the notable exception of electronic properties that are inherently affected by our current inability to tailor nanotubes with given values of (n, m). This issue of atomic-scale tailoring is also crucial for the prospects of engineering graphene-based transistors (see Section 2.4.2).

Even though graphene and CNTs have much in common, they fundamentally differ in their dimensionality. The main reason for the scientific surge that has followed graphene's experimental detection is indeed its 2D aspect, making graphene amenable to the full armada of experimental techniques, which have been thoroughly studied and optimized in the semiconductor industry for the past decades. On the contrary, CNTs are 1D systems (effectively confined in the radial direction) and thus conceptually less simple objects than graphene sheets in terms of experimental manipulation.

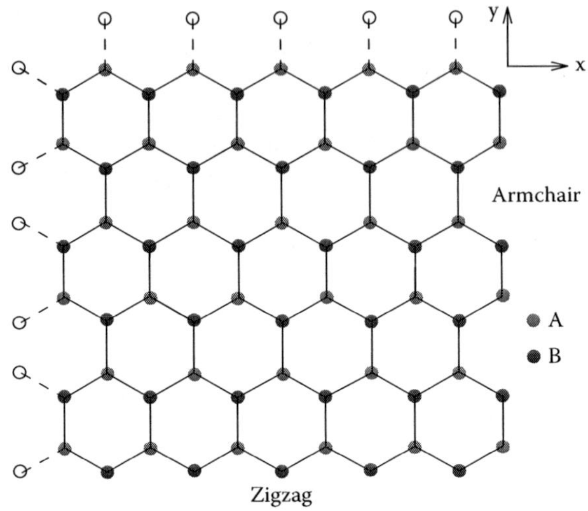

FIGURE 2.2
Graphene's hexagonal lattice. It is bipartite: atoms from sublattice A (light gray) cannot be mapped on those from sublattice B (dark gray) using the primitive vector basis. Also shown are the two main types of edge terminations considered for this type of lattice: zigzag and armchair. The former affects atoms from one sublattice only, whereas the latter affects both the sublattices.

Additionally, 2D electronic systems are known to be challenging for theorists to tackle analytically, as they stand midway between higher-dimensional systems whose properties are often efficiently captured by a mean-field description and 1D systems for which exact approaches (such as the Bethe ansatz, bosonization, and quantum Monte Carlo) exist. Phase space of 2D systems is both nontrivial topologically (particles can be exchanged without crossing each other as in 1D systems) and has reduced degrees of freedom (which decreases the odds of seeing mean-field behaviors as in 3D systems). The combination of these features largely explains the huge interest that has been devoted to graphene in the past years, even though manifestations of nontrivial many-body phenomena have remained scarce until now. This interest has been rewarded by providing the electronic transport community with fascinating phenomena unobserved in CNTs, such as Klein tunneling, weak antilocalization, or the relativistic quantum Hall effect, all of which will be discussed in Section 2.3.

2.2.3 Lattice and Band Structures

To conclude this introductory section, I will now present graphene's lattice and band structures, which are the sources of the material's singularity. Most of the electronic properties discussed in Section 2.3 directly follow from graphene's bipartite Bravais lattice and its gapless band structure.

2.2.3.1 Lattice Structure

Graphene is made of carbon atoms arranged in a honeycomb lattice, as is illustrated in Figure 2.2. Each carbon atom possesses four valence electrons, three of which chemically interact with the atom's nearest neighbors and form strong in-plane σ bonds, resulting in sp^2 hybridization of the atomic orbitals. The remaining free electron forms a weak π bond, which points out of the plane, resulting in a p_z orbital. It can hop from one atom to another and lead to the propagation of a charge current through the crystal. Actually, a simple tight-binding model, which considers only hopping (with amplitude $\gamma_0 \approx 3$ eV) of conduction electrons between nearest neighbors (separated by a distance $a \approx 0.1$ nm), is enough to explain most of graphene's electronic properties.

The elementary unit cell of the lattice (from which the entire plane of graphene can be generated) is made of two atoms. This implies the existence of two corresponding sublattices and an associated degree of freedom for the electron; it can be either on sublattice A or on sublattice B. This degree of freedom acts as a *pseudospin* and is responsible for the internal structure of electrons in graphene: they are described by two-component wave functions, which are the analogue of spinors in relativistic quantum field theory. This internal structure turns out to be a crucial aspect in the specificity of electrons in graphene and grants them an internal phase degree of freedom that is closely connected with Berry's (1984) phase. This phase manifests itself in many physically observable phenomena such as the anomalous quantization of the Hall conductance, the absence of backscattering for smooth potentials, and a quantum-related increase of the classical conductivity known as weak antilocalization.

2.2.3.2 Band Structure

Let us now take a look at the band structure of graphene. This can be done by going to reciprocal space and by investigating for which values of the wave vector $\boldsymbol{k}$ a wave can propagate through the crystal. The spatial periodicity of the crystal restricts these values to a unit cell in $\boldsymbol{k}$-space (the first Brillouin zone), and the energy-momentum dispersion relation in this cell can be established using Bloch's theorem (Ashcroft and Mermin 1976). It is shown in Figure 2.3a for graphene.

As one can see, the first Brillouin zone is hexagonal, reflecting the symmetry of real space. The corners of this hexagon stand out among other points, as the dispersion relation vanishes there. Note, however, that only two of these corners (which we will denote K and K′) are actually inequivalent, the other four being related to them by the symmetry of reciprocal space. The existence of these two points constitutes an additional degree of freedom, which is traditionally referred to as the valley degree of freedom (due to the usually valley-shaped quadratic dispersion relation in 2DEGs).

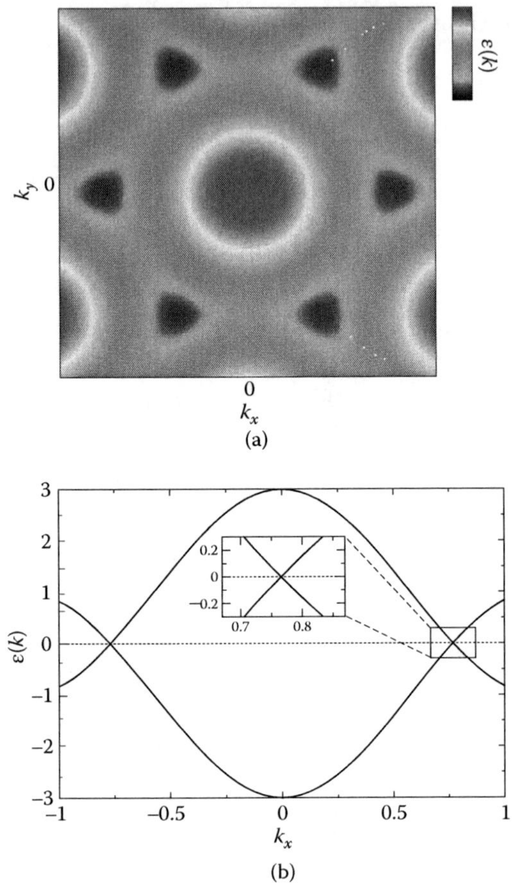

FIGURE 2.3
(a) Color plot of the energy-momentum dispersion relation in reciprocal space (for the conduction band $\varepsilon > 0$ only). The hexagonal nature of the first Brillouin zone is apparent. The corners of this unit cell are special in that the dispersion relation vanishes at these so-called Dirac points. (b) Energy-momentum dispersion along the line $k_y = 0$ (for both conduction and valence bands). Notice that, in the vicinity of the Dirac points, the band structure is linear in momentum (inset). The energy is expressed in unit of γ_0 and the momentum in unit of π/a.

Since the dispersion relation vanishes at points K and K′, it can be expanded in the vicinity of these points to describe low-energy excitations (Figure 2.3b). One can show this leads to a linear dependence in momentum $\varepsilon = \pm v_F|\boldsymbol{p}|$, which is that of massless pseudorelativistic particles, with an effective speed of light v_F that is about 300 times smaller than the speed of light in vacuum.*

* Contrary to usual 2DEGs, the Fermi velocity is here a constant independent of energy; at higher energies, however (of the order of 1 eV), this is no longer true due to k-anisotropic terms in the dispersion relation known as trigonal warping (Plochocka et al. 2008): this is visible in the triangular shape of the energy contour around the Dirac points shown in Figure 2.3a.

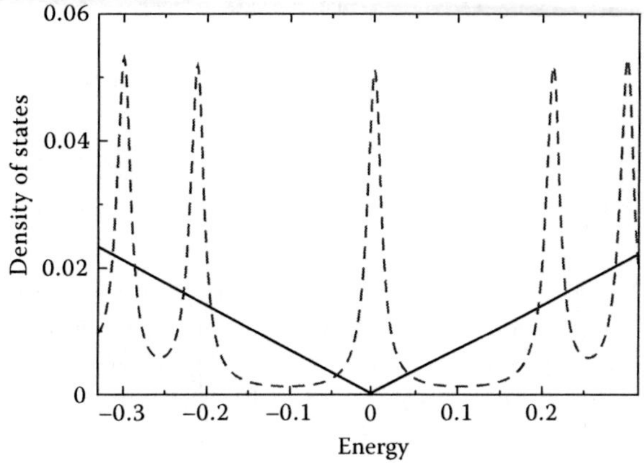

FIGURE 2.4
Density of states for low-energy excitations in graphene with (dashed light gray) and without (full dark black) magnetic field. In the former case, a small amount of disorder has been included to smoothen the Landau level peaks into Lorentzians of width $\delta_{dis} = 0.01\gamma_0$. Notice the peak at exactly zero energy originating from the B-independent $n = 0$ Landau level. The energy is expressed in unit of γ_0 and the density of states in unit of $(\gamma_0 a^2)^{-1}$. The magnetic length here is $l_B = 10a$.

Charge carriers can thus be positive or negative energy electrons, the latter being occasionally referred to as "holes" in analogy with semiconductors. One ends up with a very unusual band structure, which is that of a semiconductor with zero gap. The absence of a gap is a direct consequence of the sublattice symmetry (the fact that the system is indistinguishable upon exchanging A and B sites) and is the reason why low-energy excitations in graphene are effectively massless. Conduction and valence bands meet at the Dirac point (the K/K′ point introduced earlier), where the density of states vanishes. Indeed, contrary to 2DEGs where the density of states is constant, the latter depends linearly on energy in graphene, as can be seen in Figure 2.4. This enables the carrier density n_s (the energy-integrated density of states) in graphene to be conveniently tuned by applying a gate voltage V_g: $n_s = \pm k_F^2/\pi$, with the Fermi wave vector $k_F \approx eV_g/(\hbar v_F)$ at low carrier densities ($\hbar$ is the reduced Planck's constant).

The vanishing of the carrier density at the Dirac point is a highly unusual feat that is expected to give rise to many exotic phenomena, including features of quantum criticality (Müller 2009). Screening of electrostatic charges is also predicted to be significantly affected at the Dirac point (where a naive Thomas–Fermi calculation yields a diverging screening length $l_{TF} \sim k_F^{-1}$), thereby allowing the long-range Coulomb interaction to play an increasingly important role. This could lead to a small departure from the celebrated Landau–Fermi liquid paradigm, according to which the effect of moderate electron–electron interactions can be phenomenologically taken into account by simply renormalizing some of the parameters of the system such as its

Fermi velocity and mass (Vozmediano 2011). The Dirac point has, however, proved until now very difficult to probe experimentally because of the inevitable presence of residual doping (impurities, ripples, defects, etc.) in the samples.

2.2.3.3 Landau Levels

The linear energy dependence of the density of states is drastically modified when a strong perpendicular magnetic field is applied to a 2D electron system such as graphene. This is due to the concept of Landau level quantization: the energy spectrum transforms from a quasi-continuum to discrete Landau levels (Figure 2.4) whose macroscopic degeneracy is roughly given by the number of magnetic flux quanta threading the sample. Because of the linear band structure of graphene, these Landau levels, instead of being equidistant as $\varepsilon_n \sim B(n + 1/2)$ in 2DEGs, scale as $\varepsilon_n = \pm(\hbar v_F/l_B)\sqrt{(2|n|)}$ (McClure 1956), where $l_B = \sqrt{(\hbar/(eB))}$ is the magnetic length and n is the quantum number. This leads to the previously mentioned possibility of room-temperature quantum Hall effect since $k_B T_{\text{room}} << \hbar v_F/l_B$, that is, thermal activation is too weak to blur Landau level quantization. In addition, the absence of a gap in the band structure is responsible for the existence of a zero-energy Landau level (at $n = 0$), which shows up as a shift in the quantization of the Hall plateaus in the conductance, as will be discussed in more detail in Section 2.3.2.

2.3 Electronic Properties of Massless Dirac Fermions

Let us now discuss in more detail some of the electronic properties that make graphene so special. Transport in generic quantum disordered systems such as graphene (see Mucciolo and Lewenkopf 2010 and Peres 2010 for an introduction to this topic) can be characterized by several important length scales: the Fermi wavelength λ_F, the elastic scattering length (or mean free path) l_e, and the phase-coherence length l_Φ. Additionally, the typical size of the system L (for mesoscopic systems) is relevant. Unless otherwise stated, I will ignore inelastic events and take the limit $l_\Phi = \infty$, and I will consider the diffusive regime $\lambda_F \le l_e << L$ in which electrons diffuse from one scattering center to another. Various kinds of disorder can be expected in graphene. They are generally split into two categories, according to their ability to scatter electrons between valleys K and K′: disorder acting on length scales shorter than the lattice spacing (e.g., vacancies and dislocations) is referred to as being "short range" and couples the valleys, while disorder acting on length scales larger with respect to the lattice spacing (e.g., charged impurities and ripples) is referred to as being "long range" and gives rise to no intervalley scattering. In the following, I will almost always consider the latter category.

This section will be divided into three parts. In Section 2.3.1, I will start by dealing with transport in graphene in the absence of a magnetic field and illustrate the robustness of the electrical conductivity through various examples. Then, in Section 2.3.2, transport in the presence of a magnetic field will be described with a particular emphasis on the quantum Hall setup. Connections with the rising field of topological insulators (TIs) will be mentioned. Finally, in Section 2.3.3, the effect of correlations on the single-particle picture described until then will be briefly discussed.

2.3.1 Noninteracting Transport

One of the trademarks of graphene's low-energy excitations is their ability to transport a charge current in situations where this would not be the case in most other 2DEGs, and in particular, in presence of disorder, which in most materials strongly hinders electrical transport. This remarkable current-carrying capability will be illustrated by several examples: Klein tunneling, minimum conductivity, weak antilocalization, and zero-energy modes.

2.3.1.1 Klein Tunneling

Whereas classical particles are perfectly reflected by potential barriers of higher energy, quantum charge carriers such as those present in most 2DEGs can tunnel through potential barriers with a probability $T \sim \exp(-2k_F L)$, which is exponentially suppressed by the width L of the barrier (dashed light gray in Figure 2.5a). This is the famous quantum tunneling effect, which nevertheless remains strongly limiting in terms of charge transport.

This is no longer the case when one considers Dirac fermions such as in graphene instead of Schrödinger-like excitations. The reason for this situation is that graphene is a gapless system:* conduction and valence bands touch and interband transitions are no longer "classically" forbidden as it is the case in semiconductors, where charge carriers can only go from the conduction to the valence band by "Zener tunneling" through the gap. When electrons in graphene approach a potential barrier of higher energy, they can easily travel through it by populating one of the "hole"-like energy states in the valence band (Katsnelson et al. 2006). Therefore, there is no need in this situation to resort to the creation of particle–antiparticle pairs at the barrier as proceeded by Klein in 1929 when he considered Dirac particles in the vacuum. This so-called Klein tunneling enables electrons in graphene to tunnel through potential barriers with much higher probability than expected for Schrödinger-like excitations. Instead of being exponentially suppressed

* Opening a gap can be achieved by breaking sublattice symmetry, for example, applying a sublattice-dependent chemical potential; another possibility is the famous Kane–Mele transition to the topological insulating phase induced by strong spin–orbit coupling (see Section 2.3.2).

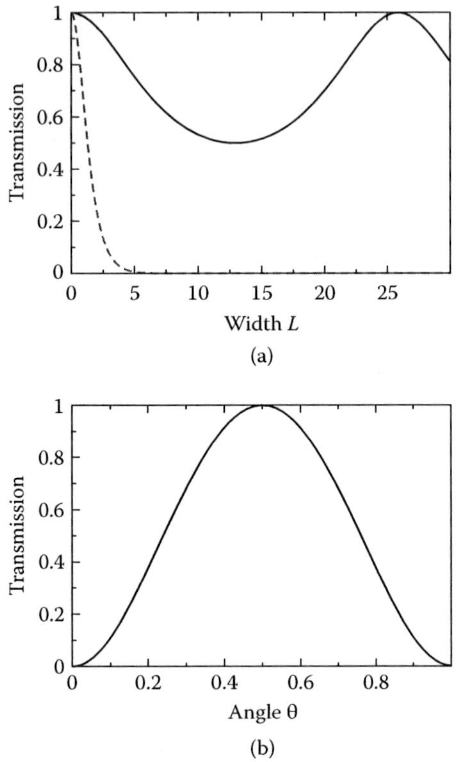

FIGURE 2.5
(a) Transmission probability T of an electron with incoming angle $\theta = \pi/4$ through a potential barrier of width L (in unit of a). The black solid line is for an electron in graphene and the gray dashed line is for an electron with a quadratic Schrödinger-like dispersion. Unlike in the latter case, T in graphene is not a monotonously decreasing function of L: There are resonant values of this width at which transmission is perfect. (b) Transmission probability of an electron in graphene through a potential step as a function of the incoming angle θ (in unit of π). At normal incidence, transmission is perfect. In both the figures, the height of the potential barrier/step is taken to be twice that of the energy of the incoming electron.

by the width of the barrier, the transmission probability T oscillates smoothly in graphene (Figure 2.5a).

It can even become perfect when electrons approach barriers at normal angular incidence (Figure 2.5b). This counterintuitive phenomenon, an experimental evidence of which was only recently and conclusively obtained for the first time (Stander et al. 2009; Young and Kim 2009), stems from the absence of backscattering first predicted by Ando et al. (1998) in CNTs. It can be classically understood by recalling that the velocity of charge carriers is constant in graphene: under normal incidence on a potential barrier, an electron has no other choice—to conserve its velocity—but to keep going. On a more fundamental level, this absence of backscattering can be related to the existence of a topological Berry phase, which is responsible for the cancellation of the backscattering amplitude to all orders in a perturbation expansion for the disordered potential.

The ability of charge carriers in graphene to easily tunnel through potential barriers can make it difficult to confine them. However, there are ways to do so (DeMartino et al. 2007; Silvestrov and Efetov 2007; Bardarson et al. 2009). In particular, one should notice that the earlier discussion was valid only for potentials abrupt on the scale of λ_F. For smoother potentials, it is possible to efficiently filter electrons around the normal angle (Cheianov and Falko 2006) (at which transmission remains perfect since this is an intrinsic and robust property stemming from the band structure). In the context of *n-p* junctions, which can be experimentally achieved in graphene by using two electrostatic gates to independently tune the carrier densities on both sides of the junction (Ozyilmaz et al. 2007; Williams et al. 2007), this angular filtering could serve as a basis for an electron lens (Cheianov 2007a).

2.3.1.2 Minimum Conductivity

One of the specificities of graphene is that its electrical conductivity experimentally does not scale to zero as the Dirac point is approached, despite the fact that the semiclassical dependence ($\sigma \sim |n_s|^\alpha$, with $\alpha = 1/2$ for ballistic systems, and $\alpha = 1$ for diffusive ones due to the disorder-induced constant density of states) should naively imply the opposite (Figure 2.6). Naturally, this semiclassical Boltzmann approximation breaks down at the Dirac point since the Fermi wavelength of charge carriers diverges and quantum corrections

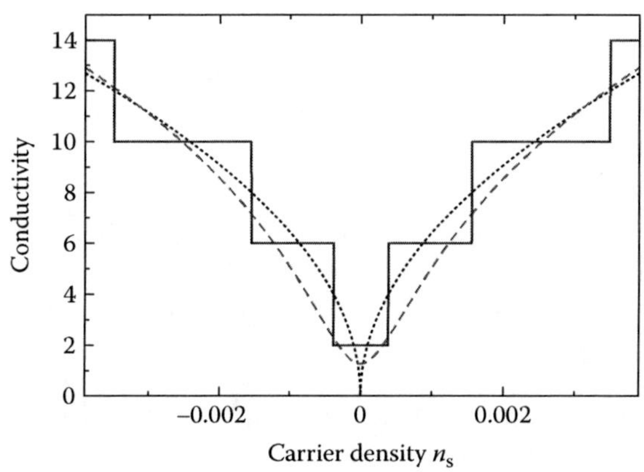

FIGURE 2.6
Graphene electrical conductivity σ (in unit of the conductance quantum e^2/h) as a function of the (gate-voltage induced) carrier density n_s (in unit of a^{-2}). The dotted black line indicates the semiclassical scaling $\sigma \sim \sqrt{(|n_s|)}$, which is valid for clean graphene. Experimentally, the vicinity of the Dirac point is characterized by a minimum in conductivity, yielding a curve that is closer in spirit to the dashed gray one. Finally, for a clean mesoscopic system, the conductance becomes quantized and features plateaus as in the thick black curve, often including a zero-energy channel as is the case here.

$O(\lambda_F/l_e)$ become important. Experimentally, however, the existence of a minimum in the conductivity is due to the fact that the *global* vanishing of the carrier density does not forbid the existence of *local* density inhomogeneities. This has been most clearly observed for graphene samples deposited on SiO_2 substrates, where local density fluctuations possibly originating from charged impurities trapped in the underlying substrate have been shown to turn the sample into a network of electron and "hole" puddles (Martin et al. 2008). These local density inhomogeneities have also been observed in suspended samples—though less pronounced—and could be in this case due to the existence of ripples (Meyer et al. 2007). In any case, whereas these puddles would tend to localize charge carriers in usual materials, the ability of massless Dirac fermions to Klein-tunnel through potential barriers allows a percolating current to flow through the puddle network (Cheianov 2007b),* which could explain the existence of a minimum conductivity of the order of the conductance quantum e^2/h.

Remarkably, even in the absence of disorder (taking $l_e = \infty$, which corresponds to a perfectly clean or ballistic system), the conductivity in graphene is predicted to remain finite at zero energy and of the order of the conductance quantum because of the ability of charge carriers at the Dirac point to easily undergo virtual interband transitions. This is a manifestation of Heisenberg's uncertainty principle for massless relativistic particles: uncertainty in momentum translates into uncertainty in energy, which at the Dirac point manifests itself in the ambiguous electron/"hole" nature of charge carriers (Katsnelson et al. 2006). This jittery motion of charge carriers, so-called Zitterbewegung, acts as a kind of intrinsic disorder that has led transport in undoped graphene to be coined "pseudodiffusive" due to the similarities it shares with that in disordered metals (Tworzydlo et al. 2006). These features have remained until now difficult to observe experimentally, however, due to the previously mentioned disorder-induced density inhomogeneities at the Dirac point.

In both ballistic and diffusive cases, the fact that the conductivity does not scale to zero as the carrier density vanishes is problematic for transistor switching applications, as we will see in Section 2.4.

2.3.1.3 Weak Antilocalization

The picture of transport I have described until now, which essentially corresponds to the quasi-classical diffusive regime of $\lambda_F << l_e$ in which $\sigma \sim k_F l_e$, can break down in two important ways. The first case is when the typical scale L of the system is reduced such that the limit $\lambda_F << L$ ceases to be valid. This will be addressed in the next paragraph. The second case is when the elastic mean free path l_e becomes of the same order of magnitude as the Fermi

* This theoretical model, however, relies on a classical (phase-incoherent) description of transport, whose validity at such low energies is debated.

wavelength λ_F. In this quantum limit, the effect of disorder is generally expected to induce a localization of the charge carriers and therefore suppress the electrical conductivity. This is a profound quantum manifestation of disordered systems discovered by Anderson (1958), which deals with the destructive interference of waves (not necessarily electrons) in a scattering medium. To get a first idea, one can look at it from a semiclassical perspective in which the conductivity through a disordered region can be shown to depend on the interference of phases accumulated along the classical trajectories connecting the initial and the final points: $\sigma \sim \sum_{(i \neq j)} A_i A_j^*$, where A_i is the (complex-valued) probability amplitude accumulated along classical trajectory *i*. In the presence of disorder, these phases are randomized and interfere destructively with each other, leading to an exponential suppression of the conductivity on a characteristic length scale called the localization length. In other words, the spatial extent of the wave function of an electron is exponentially localized on this scale. It is almost a theorem that, in two spatial dimensions (and less), this localization length is nonzero for any finite amount of disorder and is responsible for a metal–insulator transition. However, from a practical point of view, this localization length must be compared to the typical size of the system, and the Anderson (strongly) localized regime is usually reached only when $l_e \leq \lambda_F$.

In graphene, Anderson localization is altogether absent as long as disorder does not couple the valleys (Bardarson et al. 2007; Nomura et al. 2007). The unusual robustness of electrical conductivity in graphene is intimately connected with this remarkable result, which can be traced back to considerations of symmetry. To illustrate this, let us consider the simpler case of the first quantum correction $O(\lambda_F/l_e)$ to the classical conductivity, which is known as weak localization. This correction is obtained by coupling time-reversed pairs of trajectories in the above formula for the conductivity. It is generally negative and is therefore often considered to be a precursor of a disorder-induced suppression of transport. In graphene, however, coupling time-reversed pairs of trajectories for charge carriers located in a single valley leads to the appearance of a topological Berry phase that makes this correction positive (Suzuura and Ando 2002): it improves the conductivity and is therefore referred to as weak antilocalization. As in any other 2DEG, this correction vanishes when time-reversal symmetry is broken, for example, by applying a magnetic field *B*. The experimental signature of weak antilocalization therefore shows up in the negative sign of the magnetoconductivity $\Delta\sigma = \sigma(B) - \sigma(0)$, as was observed in graphene by Tikhonenko et al. (2009).

2.3.1.4 Zero-Energy Modes

Until now, I have restricted the discussion of electronic properties to bulk-like systems, that is, systems sufficiently large that their boundaries can be considered irrelevant. When structural confinement is taken into account, transport properties can be altered in important ways. Indeed, quantum confinement in clean mesoscopic systems causes transport of current to be quantized. The number of current propagating modes (or channels) is roughly given by the ratio of the sample confining dimension W and half of the Fermi wavelength $\lambda_F/2$. When a small voltage bias is applied between incoming and outgoing leads, a current can flow through the sample and is carried by the aforementioned channels. The conductance* of the system as a function of the chemical potential then increases in a steplike fashion each time a new channel opens, which can be quantitatively expressed using the Landauer formula as $G = G_0 \sum_n T_n$, where T_n is the transmission probability of channel n (which equals 1 for a ballistic system) and $G_0 = 2e^2/h$ is the spin-degenerate quantum of conductance. In the semiclassical limit of a large number of channels $k_F W >> 1$, this yields a conductance $G/G_0 \approx k_F W/\pi$.

Among the various finite-size objects made of graphene sheets that have been considered as toy models in the literature, nanoribbons have attracted considerable attention because they are conceptually simple and experimentally accessible, despite the fact that atomic-scale tailoring remains an important challenge in the field, a problem referred to as edge roughness as will be further discussed in Section 2.4 of this chapter. Nanoribbons are semi-infinite graphene sheets, with one direction (the transverse one) much smaller than the other (the longitudinal one) but not so small that it would become a quasi-1D system (a quantum wire). It has actually been shown that the finite width W of the transverse direction can cause the opening of a gap $\Delta \sim \hbar v_F/W$ in the band structure of the ribbon (Han et al. 2007), which could be paramount in the advent of fabricating graphene transistors (see Section 2.4).

In clean nanoribbons, the possibility of opening a gap, however, relies on the nature of the edge termination (Brey and Fertig 2006). Like for CNTs, structural termination of graphitic carbon at the edges of a sample is indeed crucial in determining the nanoribbon's electronic properties. Edges can be of zigzag or armchair type (Figure 2.2) or any combination of both.† In most cases, the corresponding ribbon will remain gapless (metallic) and therefore feature a potentially very robust zero-energy mode (see Wakabayashi et al. 2009 for a comprehensive discussion of low-energy modes in graphene nanoribbons). This holds true for any zigzag-terminated ribbon, whatever its width W, and leads to a longitudinal electrical conductance with plateaus at values $(2n + 1)G_0$. On the contrary, armchair-terminated ribbons can be metallic or semiconducting depending on their widths W (Brey and Fertig 2006). The band structure of an armchair nanoribbon can therefore be dramatically affected by changing the number of transverse rows of carbon hexagons by a single unit. Additionally, armchair edges couple valleys and

* In mesoscopic systems, one generally trades conductivity for conductance as the former is a scale-dependent quantity.

† Despite the fact that edges of experimental nanoribbons are certainly far from being clear-cut one or the other, in part due to the tailoring issues mentioned earlier, which create vacancies and dislocations, theoretical models often assume as simplification edges to be either zigzag or armchair.

therefore lift the twofold valley degeneracy present in zigzag ribbons. This is manifest in the corresponding expression for the longitudinal conductance, which features plateaus at values $(n + \alpha_W)G_0$, with $\alpha_W \neq 0$ when $W/(\sqrt{3}\, a) = 2$ (modulo 3).

Graphene's current-carrying ability remains therefore relatively robust when structural confinement is taken into account. The presence of disorder in such systems (and in particular, edge disorder, which is "short range") can nevertheless lead to a potentially strong suppression of charge transport (Mucciolo et al. 2009).

2.3.2 Magnetotransport

In this subsection, we will investigate what happens to electrical transport in graphene when a magnetic field is applied to the system. I will begin by discussing how electric and magnetic fields actually play symmetric roles in systems such as graphene, which are provided with a (pseudo)relativistic structure. Then, specificities of transport in the quantum Hall setup will be presented and magnetically induced spin effects will be briefly touched upon.

2.3.2.1 Lorentz Invariance

The pseudorelativistic structure describing quasi-particle excitations in graphene provides the latter with Lorentz invariance, that is, the existence of a symmetry that guarantees that certain quantities are independent of the reference frame in which they are considered. For an infinite graphene sheet with applied constant electric $\boldsymbol{E}$ and magnetic $\boldsymbol{B}$ fields, the Lorentz-invariant quantities of interest are $\boldsymbol{E}\cdot\boldsymbol{B}$ and $\boldsymbol{E}^2 - v_F{}^2\boldsymbol{B}^2$. The most significant information is contained in the latter, as the former is identically zero for crossed fields (in-plane $\boldsymbol{E}$ and perpendicular $\boldsymbol{B}$), which will be our main focus here. Its Lorentz invariance implies that there are essentially two distinct transport regimes characterized by $\beta = v_F|\boldsymbol{B}|/|\boldsymbol{E}|$: an electric regime $\beta < 1$, in which transport is dominated by the electric field (since one can make a boost to a reference frame where the magnetic field vanishes and the electric field is reduced to $\boldsymbol{E}\,\sqrt{(1-\beta^2)}$), and a magnetic regime $\beta > 1$, in which transport is correspondingly dominated by the magnetic field.

This theoretical argument allows making physical predictions. In the case where $\beta \ll 1$, one is indeed back to the setup of Klein tunneling but with a weak magnetic field. Transmission through the potential step can then be shown to remain qualitatively similar as to what was described in Section 2.3.1.1, but with perfect transmission shifted at angle $\theta_\beta = \cos^{-1}(\beta)$ (Shytov et al. 2009). On the contrary, when $\beta \gg 1$, the electric field acts as a perturbation on the well-formed Landau levels and can eventually lead to their breakdown as the critical value $\beta = 1$ is approached (Lukose 2007). Additionally, in the quantum Hall regime (see discussion in Section 2.3.2.2), the formation of a *n-p* junction with $\beta \gg 1$ can lead to a suppression of charge transport through the sample for all edge channels but the lowest one (Carmier et al. 2011), in analogy with quantum adiabatic transport (Beenakker and van Houten 1991).

2.3.2.2 Relativistic Quantum Hall Effect

Let us now address the issue of how charge transport is modified when a strong perpendicular magnetic field leads to the formation of Landau levels. This situation was already mentioned on several occasions in Section 2.2 and claimed to host a fundamental mesoscopic manifestation known as the quantum Hall effect. The latter was discovered in the early 1980s (Klitzing et al. 1980) and consists in the quantization (with unprecedented accuracy) of the current traversing a 2D sample threaded by a strong perpendicular magnetic field. In a semiclassical picture, current is carried by edge states since bulk electrons are confined by the magnetic field (Figure 2.7) as soon as the latter is large enough that the associated magnetic length l_B is smaller than the width of the sample.

Because their direction of propagation is imposed by the magnetic field at a given edge, these edge states cannot be backscattered by disorder (Figure 2.7). The only backscattering channel is on the opposite edge, which means that if edges are sufficiently far from each other with respect to the scale set by the magnetic length (or the radius of the cyclotron orbit), then current is essentially perfectly transmitted through the sample. The presence of weak to moderately strong disorder, as well as the precise nature of the boundary conditions, is therefore irrelevant. These edge states are not immune to Anderson

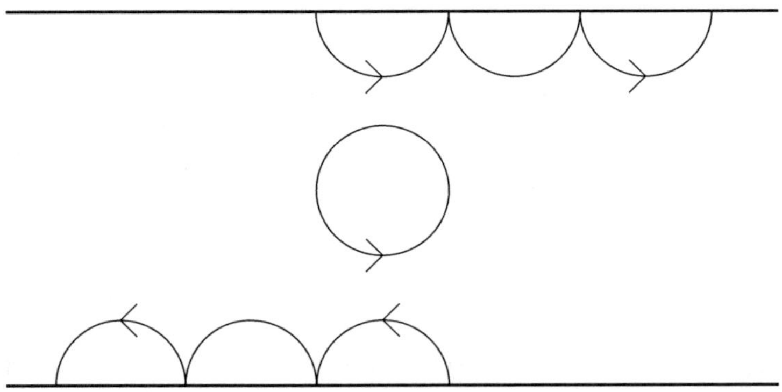

FIGURE 2.7
In the quantum Hall regime, current transport in the sample can be semiclassically described by the trajectories depicted in the figure. In the bulk, the strong magnetic field confines the charge carriers along closed cyclotron orbits rendering this part of the material insulating. Close to the edges, however, the curving effect of the magnetic field enables charge carriers to propagate along the edges of the sample and therefore to transport current from one side to the other.

localization, however, which roughly takes place when $l_e \leq l_B$ (where l_B plays the role of the Fermi wavelength in this case).

In terms of transport, to each bulk Landau level can be associated an edge channel that carries a quantum of conductance. As the carrier density is raised, the number of edge channels remains unchanged until the next Landau level is crossed. Therefore, the conductance as a function of the carrier density (which is referred to as the filling factor in this context) is a steplike function with integer plateaus at quantized values $G_n = nG_0$. In graphene, this expression is modified as $G_n = (2n + 1)G_0$. The factor of 2 takes into account the additional valley degree of freedom, and the constant shift is due to the zero-energy Landau level, which is equally shared between electron-like and "hole"-like charge carriers (Novoselov et al. 2005; Gusynin and Sharapov 2005; Zhang et al. 2005). This is another example, by the way, of a situation where charge transport in graphene is nonzero close to the Dirac point.

2.3.2.3 Connection with Topological Insulators

Upon closer inspection, the quantum Hall phase appears rather peculiar as it is not governed by any symmetry-breaking order parameter such as other familiar phases like crystals or magnets (Moore 2010). The ordering in the quantum Hall phase is in fact "topological," and a 2D electron system in the quantum Hall regime is indeed the simplest example of a so-called TI. Unlike band insulators in which no current can flow due to a large energy gap, this latest addition to the family of band structures is characterized by the existence of gapless edge states while the bulk electrons remain insulating. These edge states are topologically protected from certain perturbations such as moderate disorder and interactions. In the case of the quantum Hall insulator, this topological protection is rooted in the Chern integer n_C, which is equal to the number of edge channels and yields the topological relation $G = n_C G_0$.

The development of the field of TIs (Hasan and Kane 2010) has recently accompanied that of graphene and climbed the ladder of important preoccupations in the condensed matter community after Kane and Mele showed that graphene could in principle be tuned into a new type of TI, characterized by a Z_2 integer, if the spin–orbit coupling were sufficiently large (Kane and Mele 2005a, b). This Z_2 integer distinguishes trivial insulators from topological ones by counting the number of pairs of Dirac cones and yields for TIs a topologically robust conductance $G = G_0$. Although the intrinsic spin–orbit coupling ultimately proved too weak in graphene to observe the transition, Z_2 TIs were effectively discovered in certain band insulators with a strong spin–orbit coupling, such as 2D HgTe/CdTe quantum wells (König et al. 2007) or 3D binary compounds containing bismuth (Hsieh et al. 2008). In 2D Z_2 TIs, the edge states are no longer skipping orbits like in quantum Hall systems but counterpropagating spin-polarized channels protected by time-reversal symmetry. These gapless states can carry quantized values of spin giving rise to a quantum spin Hall effect that could revolutionize the field of spin transport, as the latter is hindered by inherent difficulties associated to the fact that spin is not a conserved quantity in solids, as opposed to charge, due to the existence of spin decoherence processes.

2.3.2.4 Magnetically Induced Spin Polarization

Let us conclude our discussion of magnetic field effects by mentioning its influence on spin transport. Most of the time, due to a weak intrinsic spin–orbit coupling, the spin degree of freedom in graphene is merely a degeneracy factor. At very strong perpendicular magnetic fields ($B > 10$ Teslas), however, the spin degeneracy of the lowest Landau level can be lifted, opening a bulk Zeeman gap in the band structure around the Dirac point (Zhang et al. 2006). This degeneracy lifting is predicted to give rise to spin-polarized counterpropagating edge states akin to those in the quantum spin Hall phase (Abanin et al. 2006). Applying an in-plane magnetic field can also lift this degeneracy and polarize the spin of charge carriers. Typical distances over which the spin polarization remains well defined are promising: due to the weak spin–orbit coupling and electron–phonon interactions (which stand among the main spin decoherence processes), spin-coherence lengths reaching the micrometer range have been reported (Tombros et al. 2007).

As a closing remark, one should point out that alongside spin manipulation, the possibility of controlling other internal degrees of freedom such as pseudospin and valley indices is also being explored to offer alternative routes from conventional Si-MOSFET logic devices and pave the way for quantum information processing. A valley-polarized state, for example, would have the advantage of being robust to the presence of smooth disorder. The hottest proposal that is currently studied to make a two-state quantum information unit (or qubit), that is, the building block of a quantum computer, is to use Majorana fermions. These elusive particles have been notably predicted to appear as bound states in the core of superconducting vortices in Z_2 TIs (Fu and Kane 2008). They should exhibit non-Abelian exchange statistics and offer the advantage that, due to topological protection, their quantum state is robust to measurement operations; that is, it is protected from decoherence and allows fault-tolerant quantum computation (Hasan and Kane 2010). Observing a glimpse of these particles is, however, a challenging endeavor, as they are both massless and charge neutral, and experimental shortcomings have so far prevented checking these exciting predictions.

2.3.3 Many-Body Effects

In this final subsection, I will say a few words on what kind of correlated phenomena can be expected in graphene, considering the material's unusual band structure. A broad survey of this advanced topic is provided in Kotov (2011). After discussing the possibly nontrivial nature of the ground state of undoped graphene, I will briefly mention the effect of electron–electron (and electron–phonon) interactions on transport. In this subsection, I will consider clean graphene, as the simultaneous treatment of disorder and many-body interactions is an extremely challenging endeavor, which is beyond the scope of this chapter.

2.3.3.1 Correlated Ground State

Before discussing the effect of correlations on electronic transport, let us make a short detour and discuss what kind of thermodynamical quantum phase graphene should lie in at the Dirac point. We have seen that, from a single-particle perspective, graphene is a semimetal that, as far as transport is concerned, remains gapless at the Dirac point due to Zitterbewegung. However, the absence of a transport gap does not mean an interaction-induced spectral gap cannot exist. Undoped graphene is indeed a very peculiar system, as the electron–electron (Coulomb) interaction is unscreened there and can potentially lead to a Mott insulator transition in which the single-particle wave functions are localized by the repulsive interaction. Actually, various models have predicted that graphene, depending on the value of the effective fine-structure constant $\alpha_g = e^2/(4\pi\hbar v_F)$ that measures the strength of Coulomb interactions, could become insulating at the Dirac point due to the opening of a gap originating from the formation of tightly bound electron–"hole" pairs (excitons). This has been confirmed by lattice Monte Carlo numerical simulations (Drut and Lähde 2009), which have estimated the required critical value of α_g to be lower than that of graphene in vacuum (and thus experimentally relevant). However, no trace of this phase transition has been observed so far, perhaps due to the electrostatic screening induced by disorder in the samples. A transition to the more familiar antiferromagnetic insulating phase (opposite spin polarizations on sublattices A and B) has also been considered in the literature, but the required strength of the Coulomb interaction in graphene seems unrealistic.

The possibility of intrinsic superconductivity in graphene has also been given some thought but remains highly debated. The critical temperature for the standard phonon-mediated BCS (Bardeen-Cooper-Schrieffer) transition is indeed extremely low, and other competing orders (such as charge-density and spin-density waves) should be taken into account.* However, like any metal, graphene can be made (weakly) superconducting when brought in close contact with a superconductor: this is the proximity effect. When this is the case and graphene is sandwiched between two superconductors, a superconducting Josephson current (which is nonzero even at the Dirac point) can

* Recent proposals suggest to investigate highly doped graphene, where van Hove singularities in the density of states along with Fermi surface nesting provide more favorable conditions for a superconducting order to develop (see Einenkel and Efetov 2011 and Nandkishore et al. 2012 and references therein).

propagate through the created junction. Additionally, due to graphene's gapless band structure, Andreev reflection* at a graphene–superconductor interface has been predicted to be specular at low energies. An introduction to this active ongoing research field (and corresponding references) can be found in Beenakker (2008).

2.3.3.2 Inelastic Scattering

Returning once and for all to the physics of transport in mesoscopic systems, let us consider what is the effect of electron–electron interactions on charge transport. When a finite value of the phase-coherence length l_Φ is restored such that it is smaller than the typical size of the system, inelastic events such as electron–phonon and electron–electron interactions have to be taken into account. These processes tend to suppress the quantum nature of the electrons by inducing phase-breaking events. Indeed, quantum particles, as opposed to classical ones, possess a phase degree of freedom as they are represented by complex wave functions. The phase-coherence length l_Φ is the typical length scale over which the particle has a memory of its phase. For distances greater than this length, the particle's phase is randomized by so-called dephasing processes such as the previously mentioned electron–phonon and electron–electron scattering events.

So far, a single-particle description has proved sufficient to capture most of graphene's observed electrical transport characteristics. Despite the absence of electrical screening at the Dirac point (due to the vanishing carrier density), which is therefore expected to reinforce the long-range nature of the Coulomb interaction, very few signatures of electron–electron interactions have been observed experimentally, with the notable exception of the fractional quantum Hall effect (see discussion in Section 2.3.3.3). A possible explanation of this situation could be due to the predicted (and recently observed, see Elias et al. 2011) logarithmic increase in the renormalized Fermi velocity as the Dirac point is approached, causing the effective fine-structure constant $\alpha_g \sim v_F^{-1}$ to actually decrease and paradoxically set graphene in a noninteracting regime (Vozmediano 2011). As for electron–phonon interactions, they are expected to be rather weak in graphene (Chen et al. 2008; Morozov et al. 2008), especially at low temperature.

2.3.3.3 Fractional Quantum Hall Effect

The most striking signature of electron–electron interactions in graphene so far is the fractional quantum Hall effect, which has been successfully observed in graphene using very clean suspended samples (with mobilities

* When an electron from a normal metal arrives at an interface with a superconductor, it can be reflected as a hole in the metal, provided a Cooper pair is added to the superconductor. This process is known as Andreev reflection.

$\mu > 10^5$ cm^2/V/s) in the quantum Hall regime (Bolotin et al. 2009; Du et al. 2009). The fractional quantum Hall effect is the manifestation of a correlated quantum state of matter for fractionally filled Landau levels, which shows up in transport measurements through the quantization of the conductance at values equal to certain rational multiples of the conductance quantum. Its observation in graphene is a strong indication that electron–electron interactions surely cannot totally be ignored in this system, even though their increased strength in this particular situation is expected as a consequence of the quenching of the kinetic energy of electrons when Landau levels are formed and electrons start following classical cyclotron trajectories. A detailed account of our current understanding of the effect of strong perpendicular magnetic fields on graphene is given in Görbig (2011).

2.4 Toward Carbon-Based Electronics?

Now that I hope to have established graphene's worthiness for fundamental science, let us move on and analyze how relevant graphene could be for industrial applications. Potential applications in the field of electronics will be described and the difficulties lying ahead to compete with current field-effect transistors will be presented. I will then conclude by discussing how graphene-based structures such as multilayers, ribbons, and other typical mesoscopic objects could provide crucial help in this respect.

2.4.1 Graphene Transistors

Semiconductor electronics is generally divided into two main fields of application: digital logic devices, which are dominated by Si-MOSFET technology, and radio frequency devices, which have become increasingly relevant with the advent of wireless communications (Schwierz 2010). If graphene holds promise in both fields, its prospects in the latter are more encouraging since the importance of a band gap is less crucial. Indeed, radio frequency performance is less sensitive to on/off ratios, as it is concerned with amplification of signals. The cut-off frequency is its primary figure of merit, and graphene transistors have already been shown to operate with cut-off frequencies reaching 100 GHz (Lin et al. 2010), which are comparable to the highest values achievable for standard GaAs devices.

Before going on, let me remind the reader how a field-effect transistor actually operates. A typical example is shown in Figure 2.8. It consists of a semiconductor deposited on a substrate that acts as a gate dielectric insulator. A top gate, separated from the semiconductor by an insulating oxide barrier, is deposited above and enables controlling the carrier density in the sample. When no voltage is applied, the system is said to be in the "off" state: the current

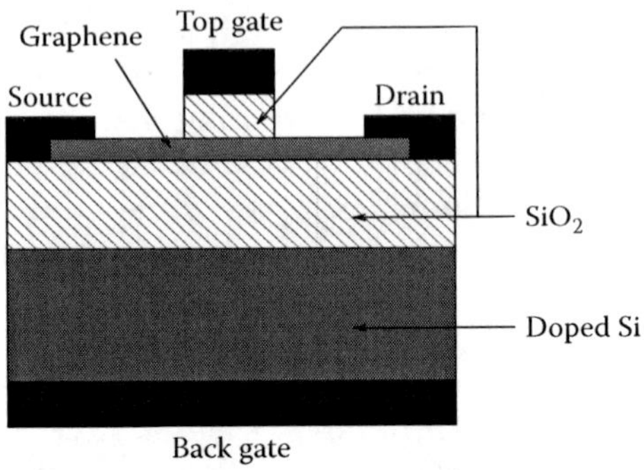

FIGURE 2.8
The figure depicts a graphene field-effect transistor with an insulating oxide barrier separating the sample from a top gate that enables continuously tuning the carrier density in the channel. Depending on the voltage applied, a current of variable magnitude can flow between the source and the drain electrodes.

I_{off} flowing between the source and the drain electrodes should then be as small as possible. When a gate voltage is applied, the current in the system increases and the "on" state is reached when the applied voltage leads to a saturation of the carrier velocity. The current flowing is then correspondingly referred to as I_{on}. The on/off ratio I_{on}/I_{off} is an important measure of how well a field-effect transistor can distinguish both signals. Typical semiconductors nowadays have a band gap $\Delta \approx 0.5$ eV, an on/off ratio $10^4 < I_{on}/I_{off} < 10^7$ and a saturation velocity $v_{sat} \approx 10^7$ cm/s. The saturation velocity has become an increasingly important figure of merit, surpassing mobility in this respect. Indeed, as gate lengths have become ever smaller, the associated electric fields have become so large that they cause the speed of charge carriers to saturate (through optical phonon relaxation).

Since the first micrometer-sized Si-MOSFETs have been developed, the road to improve the computing speed of electronic chips has been size scaling, that is, reduce the size of transistors to incorporate an increasing amount of them in a single chip. As of today, this number has exceeded the billion landmark, and silicon transistors have successfully been scaled down to gate lengths of 20–30 nm. This continuous effort has been shown to lead to the so-called Moore's law, that is, the fact that the computing power of processors doubles every 18 months. There is, however, a general agreement in the transistor community according to which the size of silicon transistors is reaching the critical lower bound at which quantum short-channel effects (Schwierz 2010) are expected to destroy the classical transport phenomenology, which rules the behavior of conventional field-effect transistors.

The need for another material able to overcome these limitations is therefore becoming dire and graphene is considered a serious candidate for this task. Its most significant advantage over silicon is that it can be scaled down to sizes an order of magnitude lower (2–3 nm), even though the specific issue of Klein tunneling will have to be dealt with. Other perks include a much lower transistor on-resistance V_{top}/I_{on} than in Si-MOSFETs (nonetheless counterfeited so far by high contact resistances at the interface of graphene and the leads, despite not having to deal with the issue of Schottky barriers) (Avouris 2010) and a much higher saturation velocity due to the higher frequencies required to activate optical phonon excitations in graphene (however, v_{sat} could also be intrinsically limited by the coupling with surface phonons from the underlying substrate). Naturally, graphene's already stated high electrical mobility is also a good point since it increases switching speed.

2.4.2 Practical Issues

Even though a number of applications based on graphene will very likely be achievable in the near future as will be discussed at length in this book, applications in electronics and in particular graphene transistors still face several important obstacles. Here, I will discuss two of the main ones: the need for a better control on the fabrication process of the samples and the prohibitively small on/off ratios achieved so far.

2.4.2.1 Atomic-Scale Tailoring

A very important obstacle to the fabrication of optimal graphene field-effect transistors is the current impossibility to engineer graphene samples with a given shape and size or, in other words, to tailor graphene at the atomic scale in a fully controlled fashion. A similar issue plagues the development of CNTs. As of today, we are still searching for an experimental method enabling us to act on the shape of materials with an atomic precision, which can evidently not be achieved by a technique such as mechanical exfoliation. Such a method is very much needed considering the influence of edges on electronic transport properties. Currently used is electron-beam lithography, which cannot of yet act efficiently on such small scales. Recent methods investigated to improve our control on nanoribbon synthesis include helium ion beam nanopatterning (Bell et al. 2009), which seems fairly accurate down to 15 nm, and standard focused electron beam sculpting techniques but at unusually high temperatures (800–900 K) where a self-repair mechanism (suppressing the created vacancies by lattice reconstruction) was identified (Song et al. 2011). Successful steps in this direction would enable bridging the gap with molecular electronics.

Among other promising fabrication processes stands epitaxial growth, which consists in the sublimation of silicon from SiC crystals in a furnace (as pioneered by the group led by W. de Heer at Georgia Tech in collaboration with C. Berger at Néel Institute in Grenoble, Berger et al. 2004). This technique still suffers from a proper theoretical understanding of the influence of residual coupling between the surface layer and those below it. In addition, it requires high temperatures and remains an expensive undertaking. On the contrary, and like chemical vapor deposition (Li et al. 2009), it enables the synthesis of much larger samples of graphene but at the cost of (until now) lower electrical mobility. Progress along these lines is, however, going so fast that some of the information may very well already be out of date by the time this book reaches print.

2.4.2.2 Opening a Band Gap

One of the main challenges that must be overcome to make transistor devices out of graphene is to open a gap. Indeed, current on/off ratios ($I_{on}/I_{off} < 5$) are way too small as compared to those of Si-MOSFETs (which are of the order of 10^6) to satisfy switching requirements (Liao et al. 2010). A solution that sounded reasonable was to replace the SiO_2 substrate by hBN, which possesses the same lattice structure as graphene but with chemically different atoms on A and B sublattices. Besides increasing the mobility in the samples (Dean et al. 2010), this was hoped to lift the sublattice symmetry in graphene and therefore open a gap in the band structure. However, recent experiments (Xue et al. 2011) seem to show that this actually does not happen due to a small mismatch between hBN and graphene's lattice constants. Other explored possibilities include reversible hydrogenation, which can turn semimetallic graphene into insulating graphane (Elias et al. 2009), random deposition of magnetic impurities potentially capable of opening a gap by spatially ordering through the electron-mediated RKKY interaction (Cheianov et al. 2009; Abanin et al. 2010), and strong in-plane mechanical strain that would break the isotropy of the lattice and could lead to gap-opening phenomena such as the generation of pseudomagnetic fields (Guinea et al. 2010) or the merging of the Dirac points* (Montambaux et al. 2009). Ideally, however, one should be able to continuously tune the gap. This is actually possible in a Bernal-stacked bilayer of graphite (Novoselov et al. 2006), a system that has correspondingly attracted increasing theoretical and experimental interest in the past couple of years. This system and other graphene-based structures will now be briefly highlighted.

2.4.3 Different Flavors

Let us begin with the misleadingly named bilayer graphene (BG), which is actually a bilayer of graphite. BG is different from both graphene and other graphite multilayers: it is a gapless semiconductor like graphene, but with a parabolic dispersion. Charge carriers are therefore massive chiral fermions carrying a Berry phase 2π, which exhibit a different kind of Klein

* The values of strain required to observe these phenomena may prove unrealistic in single-layer graphene, but could be far more reasonable in bilayer graphene.

tunneling with perfect transmission through potential barriers at certain magic angles and zero transmission at normal incidence (Katsnelson et al. 2006). The Landau level spectrum is also anomalous, but this time proportional to $B\sqrt{(n(n-1))}$, featuring an eightfold degenerate zero-energy Landau level (McCann and Falko 2006) whose signature is visible in quantum Hall conductance measurements. Many-body interactions are also predicted to be stronger than in graphene, and scenarios of pseudospin ferromagnetism or excitonic electron–"hole" pair condensation between top and bottom layers have been proposed. But the biggest advantage over graphene is the possibility to open a tunable gap in the band structure by applying a perpendicular electric field (Castro et al. 2007) that lifts the layer degeneracy (which for pure BG adds up to the spin and valley degrees of freedom). On/off ratios of 100 and gaps of 0.1 eV have been reported (Ohta et al. 2006; Zhang et al. 2009). These values remain too low to fulfill requirements of modern digital logic devices, but BG, along with few-layer graphene systems, could nevertheless turn out to play an important part in the development of carbon-based electronics.

Other graphene-based mesoscopic systems studied include the more familiar quantum point contact—a very narrow constriction in which conductance is quantized—and quantum dot. The latter is a cavity of small dimensions which electrons can enter and exit through quantum point contacts, which act as tunnel barriers (effectively due to the width-induced gap in the constriction). It is the 0D member in the family of quantum mesoscopic systems and is known to exhibit spectacular manifestations as a consequence of confining quantum particles, such as Coulomb blockade; that is, the fact that the quantum dot can only contain so many electrons that the Coulomb repulsion between them can tolerate. Quantum dots can therefore be tuned (by reducing the width of the constriction) such that they effectively possess a small quantized amount of electric charge. These widespread systems in 2DEGs have only begun to be studied in graphene (Miao et al. 2007; Ponomarenko et al. 2008; Tombros et al. 2011) and could serve as a basis for fascinating applications such as single-electron transistors and spin filters (Recher and Trauzettel 2010).

In many respects, however, the ideal candidate to follow in silicon's footsteps would consist of clean and properly tailored narrow graphene nanoribbons. Ribbons 10 nm wide with band gaps of several tens of millielectron volts were achieved, additionally featuring on/off ratios larger than 10^3 and even reaching those of Si-MOSFETs at low temperatures (Wang et al. 2008). Transfer to room temperature was met with variable success but progress is under way (Lin et al. 2011). To open a required band gap of typically 0.5 eV, one should be able to make clean ribbons of width 2–3 nm, which is impossible with today's lithographic techniques. Improvements in this direction are highly sought after, illustrating a collateral effect of the graphene surge, that is, the driving force it generates to improve our knowledge in highly advanced engineering techniques.

2.5 Summary and Outlook

We have seen that graphene in many respects deserves the attention that it has been granted over the past few years in the condensed matter physics community. Its outstanding characteristics (very high electrical mobility at room temperature, residual conductivity at zero carrier density, optical transparency, unprecedented tensile strength and thermal conductivity, etc.) alongside with its relatively straightforward experimental synthesis make it an ideal landscape for both theorists and experimentalists eager to revisit solid-state physics from a different perspective.

The lack of space has prevented me from discussing some exciting possibilities that combine graphene's electronic spectrum with its other remarkable properties. For example, transparent conducting electrodes made of graphene are considered serious competitors for indium tin oxide (ITO) in the field of optoelectronics, being among other things both cheaper and less fragile. Envisioned applications include bendable touch screens made of liquid crystals coated by graphene, which would be almost impossible to break or scratch, "smart" windows whose transparency could be continuously tuned by an electric field, and new generations of solar cells (Bonaccorso et al. 2010). The swift recombination rate of photoexcited electron–"hole" pairs in graphene (due to the material's gapless band structure) could also allow it to serve as an efficient photodetector. Graphene's lightness and elasticity are ideally suited to nanoelectromechanical systems (NEMSs) and open the prospect of individual atom sensing (Geim 2009). Finally, graphene's potential as a good thermoelectric material (in which an electrical current is induced by applying a gradient of temperature) is also attracting attention lately. Covering graphene with specific types of adatoms could indeed induce a transition to a topological insulating phase (Weeks et al. 2011) characterized by a very high thermoelectric efficiency $ZT \approx 1$ at low temperatures (Shevtsov et al. 2011), provided the phonon-mediated thermal conductivity is significantly lowered by the presence of short-range disorder (such as edge roughness or vacancies).

Even if industrial applications are still scarce as of today, which is not too surprising considering the short time span that has gone by since the surge launched by the experimental breakthrough of Geim and Novoselov, many of them are foreseen to reach maturity in the upcoming years, numerous examples of which will be presented elsewhere in this book. Carbon-based transistor electronics will nonetheless probably have to wait a little longer. Tailoring with an atomic-scale precision and opening a large enough band gap (to achieve decent switching performances) remain important challenges. However, I would not be surprised if our current expectations were proved too shortsighted, considering the activity going on in this rich and stimulating field.

Acknowledgments

It is my pleasure to thank Jean-Noël Fuchs (Univ Paris-Sud, LPS, Orsay), Caio H. Lewenkopf (Univ. Fed. Fluminense, Instituto de Fisica, Niteroi RJ, Brazil), Denis Ullmo (CNRS/Univ Paris-Sud, LPTMS, Orsay), and Xavier Waintal (CEA-INAC/UJF, SPSMS, Grenoble) for their support and for helpful discussions and comments regarding the contents of this chapter. I also acknowledge financial support from STREP ConceptGraphene.

References

Abanin D. A., Lee P. A., and Levitov L. S. 2006. Spin-filtered edge states and quantum Hall effect in graphene. *Phys. Rev. Lett.* 96: 176803.

Abanin D. A., Shytov A. V., and Levitov L. S. 2010. Peierls-type instability and tunable band gap in functionalized graphene. *Phys. Rev. Lett.* 105: 086802.

Abergel D. S. L. et al. 2010. Properties of graphene: A theoretical perspective. *Adv. Phys.* 59: 261–482.

Abergel D. S. L., Russell A., and Falko V. I. 2007. Visibility of graphene flakes on a dielectric substrate. *Appl. Phys. Lett.* 91: 063125.

Anderson P. W. 1958. Absence of diffusion in certain random lattices. *Phys. Rev.* 109: 1492–1505.

Ando T., Nakanishi T., and Saito R. 1998. Berry's phase and absence of back scattering in carbon nanotubes. *J. Phys. Soc. Jap.* 67: 2857–2862.

Ashcroft N. W. and Mermin N. D. 1976. *Solid state physics*. Fort Worth, TX: Saunders College Publishing.

Avouris P. 2010. Graphene: Electronic and photonic properties and devices. *Nano Lett.* 10: 4285–4294.

Balandin A. A. 2011. Thermal properties of graphene and nanostructured carbon materials. *Nat. Mater.* 10: 569–581.

Bardarson J. H. et al. 2007. One-parameter scaling at the Dirac point in graphene. *Phys. Rev. Lett.* 99: 106801.

Bardarson J. H., Titov M., and Brouwer P. W. 2009. Electrostatic confinement of electrons in an integrable graphene quantum dot. *Phys. Rev. Lett.* 102: 226803.

Beenakker C. W. J. 2008. Andreev reflection and Klein tunneling in graphene. *Rev. Mod. Phys.* 80: 1337–1354.

Beenakker C. W. J. and van Houten H. 1991. Quantum transport in semiconductor nanostructures. *Sol. Stat. Phys.* 44: 1–228.

Bell D. C. et al. 2009. Precision cutting and patterning of graphene with helium ions. *Nanotechnology* 20: 455301.

Berger C. et al. 2004. Ultrathin epitaxial graphite: 2D electron gas properties and a route toward graphene-based nanoelectronics. *J. Phys. Chem.* 108: 19912–19916.

Berry M. V. 1984. Quantal phase factors accompanying adiabatic changes. *Proc. R. Soc. Lond. A* 392: 45–57.

Blake P. et al. 2007. Making graphene visible. *Appl. Phys. Lett.* 91: 063124.

Bolotin K. I. et al. 2008. Ultrahigh electron mobility in suspended graphene. *Sol. Stat. Comm.* 146: 351–355.

Bolotin K. I. et al. 2009. Observation of the fractional quantum Hall effect in graphene. *Nature* 462: 196–199.

Bonaccorso F. et al. 2010. Graphene photonics and optoelectronics. *Nat. Photonics* 4: 611–622.

Brey L., and Fertig H. A. 2006. Electronic states of graphene nanoribbons studied with the Dirac equation. *Phys. Rev. B* 73: 235411.

Carmier P., Lewenkopf C. H., and Ullmo D. 2011. Semiclassical magnetotransport in graphene n-p junctions. *Phys. Rev. B* 84: 195428.

Castro E. V. et al. 2007. Biased bilayer graphene: Semiconductor with a gap tunable by the electric field effect. *Phys. Rev. Lett.* 99: 216802.

Castro Neto A. H. et al. 2009. The electronic properties of graphene. *Rev. Mod. Phys.* 81: 109–162.

Charlier J.-C., Blase X., and Roche S. 2007. Electronic and transport properties of nanotubes. *Rev. Mod. Phys.* 79: 677–732.

Cheianov V. V. et al. 2007a. Random resistor network model of minimal conductivity in graphene. *Phys. Rev. Lett.* 99: 176801.

Cheianov V. V. et al. 2009. Hidden Kekulé ordering of adatoms on graphene. *Sol. Stat. Comm.* 149: 1499–1501.

Cheianov V. V. and Falko V. I. 2006. Selective transmission of Dirac electrons and ballistic magnetoresistance of n-p junctions in graphene. *Phys. Rev. B* 74: 041403(R).

Cheianov V. V., Falko V. I., and Altshuler B. L. 2007b. The focusing of electron flow and a Veselago lens in graphene p-n junctions. *Science* 315: 1252–1255.

Chen J.-H. et al. 2008. Intrinsic and extrinsic performance limits of graphene devices on SiO_2 *Nat. Nanotechnol.* 3: 206–209.

Das Sarma S. et al. 2011. Electronic transport in two-dimensional graphene. *Rev. Mod. Phys.* 83: 407–470.

Dean C. R. et al. 2010. Boron nitride substrates for high-quality graphene electronics. *Nat. Nanotechnol.* 5: 722–726.

De Martino A., Dell'Anna L., and Egger R. 2007. Magnetic confinement of massless Dirac fermions in graphene. *Phys. Rev. Lett.* 98: 066802.

Drut J. E., and Lähde T. A. 2009. Is graphene in vacuum an insulator? *Phys. Rev. Lett.* 102: 026802.

Du X. et al. 2009. Fractional quantum Hall effect and insulating phase of Dirac electrons in graphene. *Nature* 462: 192–195.

Einenkel M., and Efetov K. B. 2011. Possibility of superconductivity due to electron-phonon interaction in graphene. *Phys. Rev. B* 84: 214508.

Elias D. C. et al. 2009. Control of graphene's properties by reversible hydrogenation: Evidence for graphane. *Science* 323: 610–613.

Elias D. C. et al. 2011. Dirac cones reshaped by interaction effects in suspended graphene. *Nat. Phys.* 7: 701–704.

Ferrari A. C. et al. 2006. Raman spectrum of graphene and graphene layers. *Phys. Rev. Lett.* 97: 187401.

Fu L. and Kane C. L. 2008. Superconducting proximity effect and Majorana fermions at the surface of a topological insulator. *Phys. Rev. Lett.* 100: 096407.

Geim A. K. 2009. Graphene: Status and prospects. *Science* 324: 1530–1534.

Geim A. K., and Novoselov K. S. 2007. The rise of graphene. *Nat. Mater.* 6: 183–191.

Görbig M. O. 2011. Electronic properties of graphene in a strong magnetic field. *Rev. Mod. Phys.* 83: 1193–1243.
Guinea F., Katsnelson M. I., and Geim A. K. 2010. Energy gaps and a zero-field quantum Hall effect in graphene by strain engineering. *Nat. Phys.* 6: 30–33.
Gusynin V. P., and Sharapov S. G. 2005. Unconventional integer quantum Hall effect in graphene. *Phys. Rev. Lett.* 95: 146801.
Han M. Y. et al. 2007. Energy band-gap engineering of graphene nanoribbons. *Phys. Rev. Lett.* 98: 206805.
Hasan M. Z., and Kane C. L. 2010. Topological insulators. *Rev. Mod. Phys.* 82: 3045–3067.
Hsieh D. et al. 2008. A topological Dirac insulator in a quantum spin Hall phase. *Nature* 452: 970–974.
Iijima S. 1991. Helical microtubules of graphitic carbon. *Nature* 354: 56–58.
Kane C. L., and Mele E. J. 2005a. Quantum spin Hall effect in graphene. *Phys. Rev. Lett.* 95: 226801.
Kane C. L., and Mele E. J. 2005b. Z_2 topological order and the quantum spin Hall effect. *Phys. Rev. Lett.* 95: 146802.
Katsnelson M. I., Novoselov K. S., and Geim A. K. 2006. Chiral tunnelling and the Klein paradox in graphene. *Nat. Phys.* 2: 620–625.
Klitzing K. v., Dorda G., and Pepper M. 1980. New method for high-accuracy determination of the fine-structure constant based on quantized Hall resistance. *Phys. Rev. Lett.* 45: 494–497.
König M. et al. 2007. Quantum spin Hall insulator state in HgTe quantum wells. *Science* 318: 766–770.
Kotov V. N. et al. 2011. Electron-electron interactions in graphene: Current status and perspectives. *arXiv:1012.3484* (accepted in *Rev. Mod. Phys.*).
Lee C. et al. 2008. Measurement of the elastic properties and intrinsic strength of monolayer graphene. *Science* 321: 385–388.
Li X. S. et al. 2009. Large-area synthesis of high-quality and uniform graphene films on copper foils. *Science* 324: 1312–1314.
Liao L. et al. 2010. High-speed graphene transistors with a self-aligned nanowire gate. *Nature* 467: 305–308.
Lin M.-W. et al. 2011. Room-temperature high on/off ratio in suspended graphene nanoribbon field-effect transistors. *Nanotechnology* 22: 265201.
Lin Y.-M. et al. 2010. 100-GHz transistors from wafer-scale epitaxial graphene. *Science* 327: 662.
Lukose V., Shankar R., and Baskaran G. 2007. Novel electric field effects on Landau levels in graphene. *Phys. Rev. Lett.* 98: 116802.
Martin J. et al. 2008. Observation of electron–hole puddles in graphene using a scanning single-electron transistor. *Nat. Phys.* 4: 144–148.
Mayorov A. S. et al. 2011. Micrometer-scale ballistic transport in encapsulated graphene at room temperature. *Nano Lett.* 11: 2396–2399.
McCann E., and Falko V. I. 2006. Landau-level degeneracy and quantum Hall effect in a graphite bilayer. *Phys. Rev. Lett.* 96: 086805.
McClure J. W. 1956. Diamagnetism of graphite. *Phys. Rev.* 104: 666–671.
Meyer J. C. et al. 2007. The structure of suspended graphene sheets. *Nature* 446: 60–63.
Miao F. et al. 2007. Phase-coherent transport in graphene quantum billiards. *Science* 317: 1530–1533.
Montambaux G. et al. 2009. Merging of Dirac points in a two-dimensional crystal. *Phys. Rev. B* 80: 153412.

Moore J. E. 2010. The birth of topological insulators. *Nature* 464: 194–198.

Morozov S. V. et al. 2008. Giant intrinsic carrier mobilities in graphene and its bilayer. *Phys. Rev. Lett.* 100: 016602.

Mucciolo E. R., Castro Neto A. H., and Lewenkopf C. H. 2009. Conductance quantization and transport gaps in disordered graphene nanoribbons. *Phys. Rev. B* 79: 075407.

Mucciolo E. R., and Lewenkopf C. H. 2010. Disorder and electronic transport in graphene. *J. Phys. Condens. Matter* 22: 273201.

Müller M., Schmalian J., and Fritz L. 2009. Graphene: A nearly perfect fluid. *Phys. Rev. Lett.* 103: 025301.

Nair R. R. et al. 2008. Fine structure constant defines visual transparency of graphene. *Science* 320: 1308.

Nair R. R. et al. 2010. Graphene as a transparent conductive support for studying biological molecules by transmission electron microscopy. *Appl. Phys. Lett.* 97: 153102.

Nandkishore R., Levitov L., and Chubukov A. 2012. Chiral superconductivity from repulsive interactions in doped graphene. *Nat. Phys.* 8: 158–163.

Nomura K., Koshino M., and Ryu S. 2007. Topological delocalization of two-dimensional massless Dirac fermions. *Phys. Rev. Lett.* 99: 146806.

Novoselov K. S. et al. 2004. Electric field effect in atomically thin carbon films. *Science* 306: 666–669.

Novoselov K. S. et al. 2005. Two-dimensional gas of massless Dirac fermions in graphene. *Nature* 438: 197–200.

Novoselov K. S. et al. 2006. Unconventional quantum Hall effect and Berry's phase of 2π in bilayer graphene. *Nat. Phys.* 2: 177–180.

Novoselov K. S. et al. 2007. Room-temperature quantum Hall effect in graphene. *Science* 315: 1379.

Ohta T. et al. 2006. Controlling the electronic structure of bilayer graphene. *Science* 313: 951–954.

Ozyilmaz B. et al. 2007. Electronic transport and quantum Hall effect in bipolar graphene p-n-p junctions. *Phys. Rev. Lett.* 99: 166804.

Peres N. M. R. 2010. The transport properties of graphene: An introduction. *Rev. Mod. Phys.* 82: 2673–2700.

Plochocka P. et al. 2008. High-energy limit of massless Dirac fermions in multilayer graphene using magneto-optical transmission spectroscopy. *Phys. Rev. Lett.* 100: 087401.

Ponomarenko L. A. et al. 2008. Chaotic Dirac billiard in graphene quantum dots. *Science* 320: 356–358.

Recher P., and Trauzettel B. 2010. Quantum dots and spin qubits in graphene. *Nanotechnology* 21: 302001.

Saito R., Dresselhaus G., and Dresselhaus M. S. 1998. *Physical properties of carbon nanotubes*. London: Imperial College Press.

Schedin F. et al. 2007. Detection of individual gas molecules adsorbed on graphene. *Nat. Mater.* 6: 652–655.

Schwierz F. 2010. Graphene transistors. *Nat. Nanotechnol.* 5: 487–496.

Semenoff G. W. 1984. Condensed-matter simulation of a three-dimensional anomaly. *Phys. Rev. Lett.* 53: 2449–2452.

Shevtsov O. et al. 2011. Tunable thermopower in a graphene-based topological insulator. *Phys. Rev. B* 85: 245441.

Shytov A. et al. 2009. Atomic collapse, Lorentz boosts, Klein scattering, and other quantum-relativistic phenomena in graphene. *Sol. Stat. Comm*. 149: 1087–1093.

Silvestrov P. G., and Efetov K. B. 2007. Quantum dots in graphene. *Phys. Rev. Lett*. 98: 016802.

Slonczewski J. C., and Weiss P. R. 1958. Band structure of graphite. *Phys. Rev*. 109: 272–279.

Soldano C., Mahmood A., and Dujardin E. 2010. Production, properties and potential of graphene. *Carbon* 48: 2127–2150.

Song B. et al. 2011. Atomic-scale electron-beam sculpting of near-defect-free graphene nanostructures. *Nano Lett*. 11: 2247–2250.

Stander N., Huard B., and Goldhaber-Gordon D. 2009. Evidence for Klein tunneling in graphene p-n junctions. *Phys. Rev. Lett*. 102: 026807.

Suzuura H., and Ando T. 2002. Crossover from symplectic to orthogonal class in a two-dimensional honeycomb lattice. *Phys. Rev. Lett*. 89: 266603.

Tikhonenko F. W. et al. 2009. Transition between electron localization and antilocalization in graphene. *Phys. Rev. Lett*. 103: 226801.

Tombros N. et al. 2007. Electronic spin transport and spin precession in single graphene layers at room temperature. *Nature* 448: 571–574.

Tombros N. et al. 2011. Quantized conductance of a suspended graphene nanoconstriction. *Nat. Phys*. 7: 697–700.

Tworzydlo J. et al. 2006. Sub-Poissonian shot noise in graphene. *Phys. Rev. Lett*. 96: 246802.

Venables J. A., Spiller G. D. T., and Hanbucken M. 1984. Nucleation and growth of thin films. *Rep. Prog. Phys*. 47: 399.

Vozmediano M. A. H. 2011. Renormalization group aspects of graphene. *Philos. Transact. A Math Phys Eng Sci* 369: 2625–2642.

Wakabayashi K. et al. 2009. Electronic transport properties of graphene nanoribbons. *New J. Phys*. 11: 095016.

Wallace P. R. 1947. The band theory of graphite. *Phys. Rev*. 71: 622–634.

Wang X. et al. 2008. Room-temperature all-semiconducting sub-10-nm graphene nanoribbon field-effect transistors. *Phys. Rev. Lett*. 100: 206803.

Weeks C. et al. 2011. Engineering a robust quantum spin Hall state in graphene via adatom deposition. *Phys. Rev*. X 1: 021001.

Williams J. R., DiCarlo L., and Marcus C. M. 2007. Quantum Hall effect in a gate-controlled p-n junction of graphene. *Science* 317: 638–641.

Xue J. et al. 2011. Scanning tunnelling microscopy and spectroscopy of ultra-flat graphene on hexagonal boron nitride. *Nat. Mater*. 10: 282–285.

Young A. F., and Kim P. 2009. Quantum interference and Klein tunnelling in graphene heterojunctions. *Nat. Phys*. 5: 222–226.

Zhang Y. et al. 2005. Experimental observation of the quantum Hall effect and Berry's phase in graphene. *Nature* 438: 201–204.

Zhang Y. et al. 2006. Landau-level splitting in graphene in high magnetic fields. *Phys. Rev. Lett*. 96: 136806.

Zhang Y. et al. 2009. Direct observation of a widely tunable bandgap in bilayer graphene. *Nature* 459: 820–823.

3

Different Characterization Techniques to Evaluate Graphene and Its Properties

Viera Skákalová*, Dong Su Lee, and Alan B. Kaiser

CONTENTS

* Correspondence to: Viera Skákalová (V.Skakalova@fkf.mpg.de)

3.1 Introduction

The highlight of this chapter is *graphene*. What is it? There is a clear definition:

Graphene is an allotrope of carbon, whose structure is a one-atom-thick planar sheet of sp^2-bonded carbon atoms that are densely packed in a honeycomb crystal lattice.

However, most of the materials produced as graphene and presented in this book do not fully satisfy that definition. Each of the synthesis routes impresses its fingerprint into the structure, varying the concentration of defects in the crystal lattice, the size of crystal domains, the number of layers, substitutional impurities, and so on. Does it mean that the material properties of each of these "graphenes" will be different? In this chapter, we aim to present methods that help to characterize different aspects of graphene-like materials, their structures, and properties.

We introduce characterization techniques in Sections 3.2 through 3.5.

Microscopy techniques help to visualize samples in vastly different scales, from millimeters down to ångströms. Nowadays, a significant technological development enables imaging with atomic resolution. We can literally see a missing or an additional atom in the crystal lattice. Microscopy, therefore, represents a crucial tool for understanding the relation between the structure and properties. In Section 3.2, we present various microscopy techniques and examples of important results delivered by these techniques, especially, related to graphene of various origins.

On the other hand, spectroscopy provides information on energy and/or momentum distributions of electronic and phonon systems. Section 3.3 introduces various spectroscopic methods that are very powerful in characterizing graphene samples for dopant and defect concentrations, covalent interactions with functional groups, the number of graphitic layers (to a certain extent), and so on.

Section 3.4 provides a basic overview of experimental methods used for characterizing electrical transport and shows how the electrical characteristics differ when the crystal structure of graphene samples varies depending on the synthesis route.

Finally, examples of thermal properties characterization are presented in Section 3.5.

Besides introducing the methods and techniques used for the characterization of graphene samples, this chapter offers a general overview of variations and trends in structural, electronic, and electrical properties determined by the origin of graphene.

3.2 Microscopy

3.2.1 Optical Microscopy

The story of the recent explosion of interest in graphene started in an optical microscope at the Department of Physics of Manchester University. The microscope objective with a magnification of 100× displayed tiny shapes of different contrasts spread over the surface of a Si/SiO_2 specimen. The objects visible through the optical microscope were very thin layers of exfoliated graphite crystal. There were apparently lucky circumstances involved in the discovery of the easy technique of detecting a single layer of graphite—graphene. Later, detailed studies of the optical contrast of graphene on the Si/SiO_2 substrate with respect to the optical properties of the SiO_2 reported in several papers showed that detecting a single atomic layer is not always so trivial (Blake et al. 2007; Gaskell et al. 2009; Roddaro et al. 2007). The thickness of the SiO_2 layer covering the Si substrate as well as the wavelength of the visible light were found to play the most crucial role for the optical resolution, which developed to such a point that the number of graphitic layers can be distinguished with single-layer precision (Figure 3.1).

The relation between the optical image contrast, the wavelength of light entering the optical system, and the thickness of SiO_2 layer is demonstrated in Figure 3.2.

Akcöltekin et al. (2009) showed that it is possible to optically identify ultrathin sheets of graphene not only on an Si substrate covered with a thin layer of SiO_2, but also on other crystalline substrates such as $SrTiO_3$, TiO_2, Al_2O_3, and CaF_2. Even though there are more accurate techniques for the detection of the number of graphitic layers, optical microscopy remains the fastest and

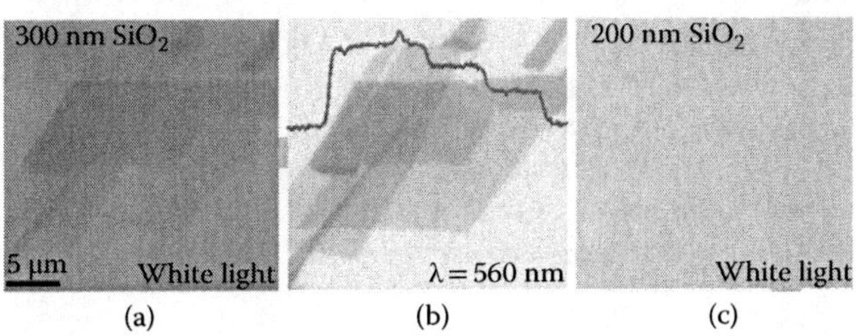

FIGURE 3.1
(a) Graphene crystallites on 300 nm SiO_2 imaged with white light and (b) green light and (c) another graphene sample on 200 nm SiO_2 imaged with white light. Single-layer graphene is clearly visible on the left image (a), but even three layers are indiscernible on the right image (c). Image sizes are 25 × 25 μm². (From Blake, P. et al., *Appl. Phys. Lett.*, 91, 063124, 2007. With permission.)

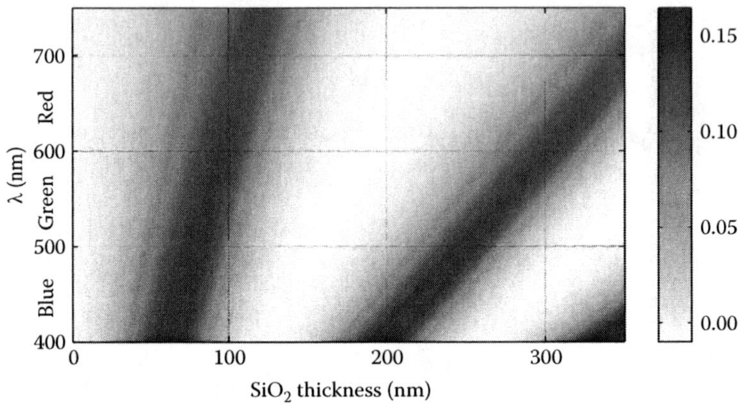

FIGURE 3.2
Plot of the contrast as a function of wavelength and SiO_2 thickness. The scale on the right shows the expected contrast. (From Blake, P. et al., *Appl. Phys. Lett.*, 91, 063124, 2007. With permission.)

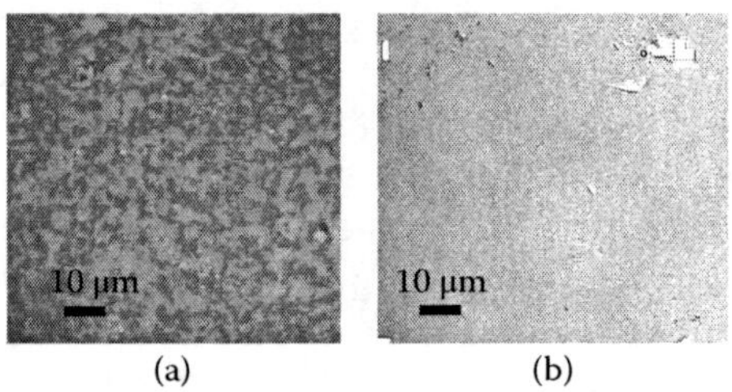

FIGURE 3.3
Comparison of large-area graphene layers prepared by the chemical vapor deposition (CVD) technique with different catalyst substrates, (a) Ni and (b) Cu, imaged by optical microscope with the same magnification.

most convenient tool in basic research laboratories while searching for graphene. Optical microscopy provides reasonable resolution for evaluation of the homogeneity of graphene layers prepared by various techniques. Optical micrographs in Figure 3.3 taken at the same resolution clearly demonstrate a significant difference in morphology between two samples of large-area graphene. The layers were prepared by chemical vapor deposition (CVD) technique, but using different catalyst substrates, namely Ni (Figure 3.3a) or Cu (Figure 3.3b). Although the sample prepared on the Ni substrate in Figure 3.3a is covered with a dense network of multilayer domains, the sample grown on Cu (Figure 3.3b) looks smooth and homogeneous.

3.2.2 Atomic Force Microscopy

Imaging graphene by optical microscopy is applicable only when a dielectric substrate below graphene has specific optical properties. Atomic force microscopy (AFM) imaging still requires a dielectric substrate underneath

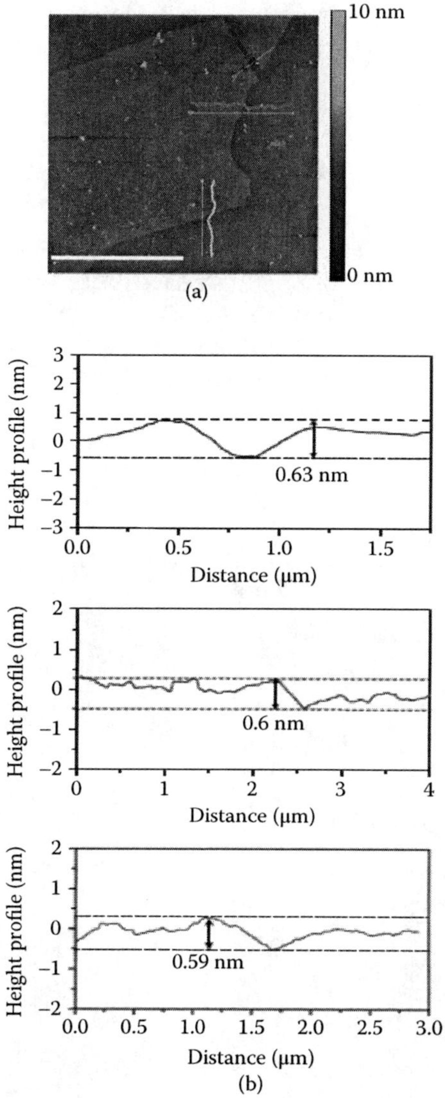

FIGURE 3.4
(a) Image of an individual chemically converted graphene sheet. (b) Height profiles taken along the solid lines show the step heights of about 0.6 nm. (From Tung, V. C. et al., *Nat. Nanotechnol.*, 4, 25–9, 2009. With permission.)

a sample, but without any specific requirements. Moreover, besides imaging the sample, AFM provides information on topography. It is, therefore, an appropriate method for characterizing thickness and surface roughness of graphene sheets. Figure 3.4a taken from Tung et al. (2009) presents an AFM image of a chemically converted graphene sheet from graphene oxide (GO).

In order to investigate the thickness of the sheet, height profiles of the sample were measured at three different positions. The height profiles plotted in Figure 3.4b show steps from the SiO_2 substrate to the surface of a chemically converted graphene sheet as high as 0.6 nm for a given cross section.

3.2.3 Scanning Tunneling Microscopy

Unlike AFM, scanning tunneling microscopy (STM) displays samples on electrically conductive substrates. Current passes between a conductive tip and a conductive substrate through a sample probing its electronic structure and so distinguishing different atoms with particular chemical interactions on the substrate. At special conditions (cryogenic temperature, ultrahigh vacuum, vibration-free fixation, and atomically sharp tip), STM can perform with an accuracy of atomic resolution.

Figure 3.5 is taken from Yu et al. (2011), where the authors demonstrate a controlled growth of single-crystal graphene grains by the CVD method on Cu. Figure 3.5a presents a large-scan-area STM topography image of a

FIGURE 3.5
Scanning tunneling microscopy (STM) of a single-crystal graphene grain on Cu. (a) STM topography image taken near the corner of a graphene grain on Cu. Dashed lines mark the edges of this grain. (b–d) Atomic-resolution STM topography images (filtered to improve contrast) taken from three different areas in the grain as indicated in (a). The gray (b), black (c), and white (d) squares (not to scale) indicate the approximate locations where the images were taken. (From Yu, Q. et al., *Nat. Mater.*, 10, 443–9, 2011. With permission.)

grain with nearly straight edges (dashed lines) meeting at the corner. The angle between the edges was determined as 120°. Figure 3.5b–d show atomically resolved STM topography images taken from three different locations marked in Figure 3.5a. The same lattice orientation in the three images (parallel with the edges shown in Figure 3.4a) is consistent with the single-crystalline nature of the grain.

Gao et al. (2010) synthesized a sample of graphene on a copper single crystal Cu(111) and characterized it by STM. Figure 3.6a shows an STM topography image of graphene monolayer islands and Figure 3.6b, a differential conductance dI/dV image recorded simultaneously with the topography image (Figure 3.6a). The dI/dV signal (Figure 3.6b) clearly differentiates graphene from copper surface showing dark contrast on this image.

Domain boundaries of graphene on a Cu(111) substrate and Moiré patterns resulting from a mismatch between the crystal orientation of Cu(111) substrate and the graphene domains are observed in STM images in Figure 3.7. Figure 3.7a shows a domain boundary between two domains with different

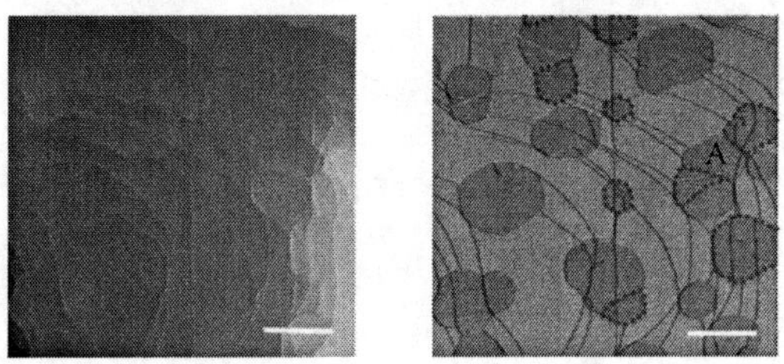

(a) (b)

FIGURE 3.6
(a) STM topography image of monolayer graphene on Cu(111). (b) Differential conductance dI/dV image recorded simultaneously with the topography image. (a) Scale bar: 200 nm. (From Gao, Li. et al., *Nano Lett.*, 10, 3512–6, 2010. With permission.)

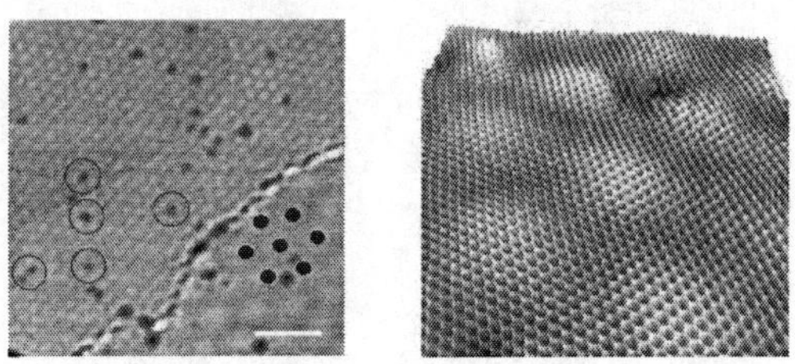

(a) (b)

FIGURE 3.7
(a) STM topography images at the domain boundaries of graphene on Cu(111) and two most observed Moiré patterns. Adsorbates on the graphene surface are marked with circles. Scale bar: 10 nm. (b) Atomic resolution STM topography image of graphene showing the Moiré pattern and the honeycomb structure. (From Gao, Li. et al., *Nano Lett.* 10, 3512–6, 2010. With permission.)

Moiré patterns corresponding to different crystal orientations of graphene domains. The periodicity of the Moiré pattern for the upper left domain is ~2.0 nm, and that for the lower right domain is around 5.8 nm. From the periodicity of the Moiré pattern, the orientation of graphene crystal domains can be resolved. The STM image in Figure 3.7b shows the hexagonal structure of graphene with atomic resolution and a Moiré pattern on top of it. The periodicity of the Moiré pattern is an order of magnitude greater than that of the graphene crystal lattice.

3.2.4 Scanning Electron Microscopy

Scanning electron microscopy (SEM) has been used effectively to visualize graphene products grown through various growth techniques. A comparison of SEM and optical microscopy imaging is clearly shown in Figure 3.8 (Dong and Chen 2010). Few-layer graphene is clearly visible in the SEM image (Figure 3.8b), but not so well in the optical image (Figure 3.8a).

Figure 3.9 displays a typical sample of few-layer graphene synthesized by the CVD technique on a Ni catalyst where numerous multilayer domains decorate the surface (Park et al. 2010), while the growth on Cu substrate (Figure 3.10) leads to more homogenous morphology of predominantly single-layer graphene (Li et al. 2009a).

The SEM image in Figure 3.11 clearly shows that oxidative chemical exfoliation of expanded graphite and the subsequent reduction process result in a disordered structure of GO (Jeong et al. 2008).

In this work, GO samples were prepared by a simplified Brodie method. After the reactions, hydroxyl, epoxide, carboxyl, and some alkyl functional groups were detected to be present in the GO. This caused the interlayer

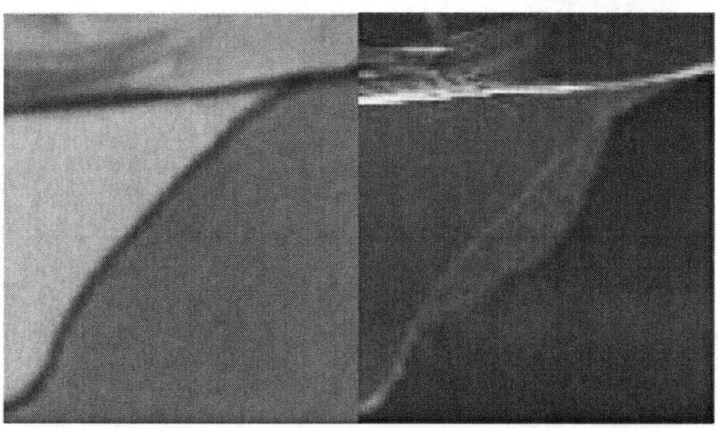
(a) (b)

FIGURE 3.8
Images of a thin graphitic flake in (a) optical and (b) scanning electron microscopes.

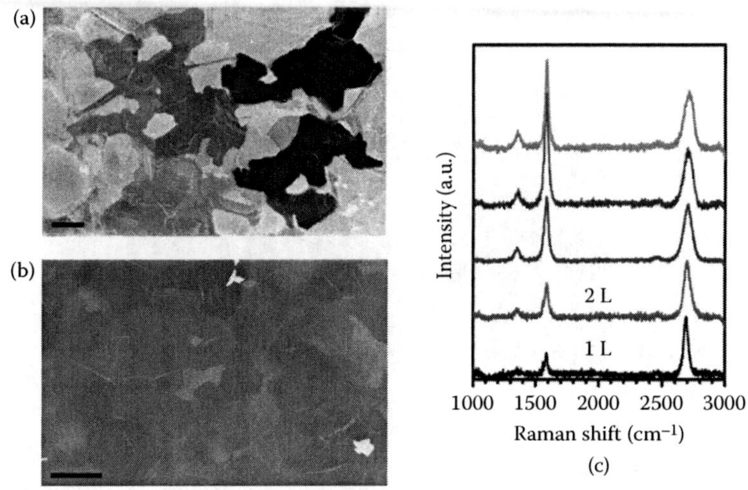

FIGURE 3.9
Scanning electron microscopy (SEM) image of a chemical vapor deposition (CVD)-grown graphene sheet on a nickel-coated SiO_2/Si substrate (a) for 30 seconds and (b) for 7 minutes, and (c) Raman spectra of a CVD-grown graphene sheet on a nickel-coated SiO_2/Si obtained from different positions on the sample (scale bar: 1 μm). (From Park, H. J. et al., *Carbon*, 48, 1088–94, 2010. With permission.)

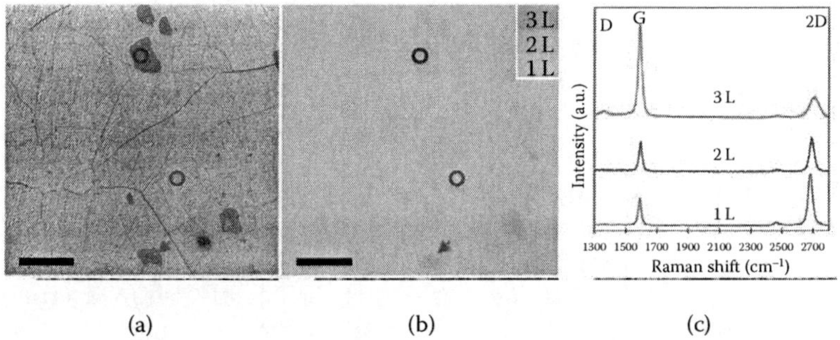

FIGURE 3.10
(a) SEM image of graphene transferred onto an SiO_2/Si (285-nm-thick oxide layer) showing wrinkles, as well as two- and three-layer regions. (b) Optical microscope image of the same regions as in (a). Scale bars: 5 μm. (c) Raman spectra from the marked spots indicated by circles or arrows showing the presence of one, two, and three layers of graphene. (From Li, X. et al., *Science*, 324, 1312–4, 2009a. With permission.)

separation to increase starting from 3.36 Å for pyrolytic graphite up to 7.37 Å after 24-hour treatment. An SEM image of graphite shows "chunky" layered structures (Figure 3.11a). With a treatment time of 24 hours (GO1), these layered structures were exfoliated and transformed into smaller sized flakes (Figure 3.11b). With a treatment time of 48 hours (GO2), the structures were

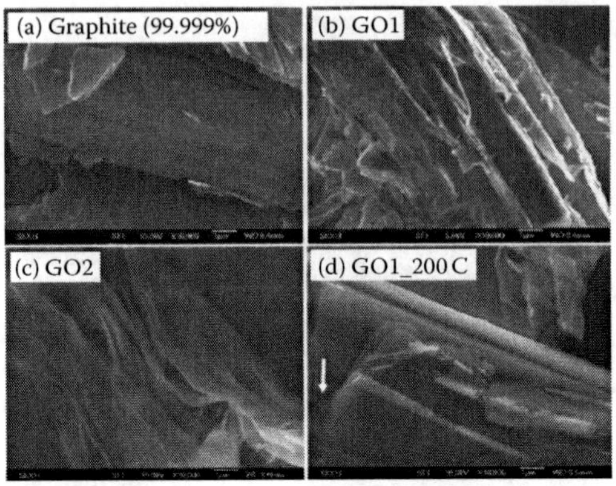

FIGURE 3.11
(a) SEM images of pyrolytic graphite (PG) and graphite oxides (GO) after different treatment: (b) GO1—24 hours reaction time, (c) GO2—48 hours reaction time, and (d) GO1 at 200°C under Ar (g) for 1 h. Accelerating electron energies are 15.0 kV for GO2 and PG, 5.0 kV for GO1, and 10.0 kV for GO1_200C. All images show a scale bar of 1 μm. (From Jeong, H.-K. et al., *J. Am. Chem. Soc.*, 130, 1362–6, 2008. With permission.)

further exfoliated as shown in Figure 3.11c. With the exposure of GO1 at 200°C under Ar(g) for 1 h, ordered layered structures similar to those seen in the graphite sample resulted, but the exfoliated layered structures still remained unchanged as indicated by the arrow (Figure 3.11d).

SEM was also able to illustrate the growth mechanism reported in Li et al. (2010b). The authors report on the effect of growth parameters such as temperature, methane flow rate, and partial pressure on the growth rate, domain size, and surface coverage of graphene as determined by Raman spectroscopy (shown also in Section 3.3.1 on Raman spectroscopy), transmission electron microscopy (TEM), and SEM. A two-step CVD process to synthesize graphene films with domains having an area of hundreds of square micrometers was presented. SEM images in Figure 3.12 clearly show an increase in domain size by changing the growth parameters. The separated domains are crystallographically rotated with respect to each other within a range of angles from about 13° to nearly 30° as shown in Figure 3.12 (Li et al. 2010b).

A study of the mechanism of growth for self-aligned graphene sheets (nanosheets) is reported in Zhu et al. (2007). To synthesize carbon nanosheets (CNSs), mostly the radio frequency (RF) plasma-enhanced chemical vapor deposition (PECVD) system has been used. In contrast to ordinary CVD graphene synthesis, CNS deposition in an RF PECVD system does not require a catalyst to predetermine the structure and/or location. The mechanism of growth suggested in Zhu et al. (2007) is dominated by surface diffusion processes in the presence of energetic ion bombardment realized within inductively coupled plasma at simultaneous etching by atomic hydrogen coupled to a vertical electrical field near the substrate surface (Figure 3.13).

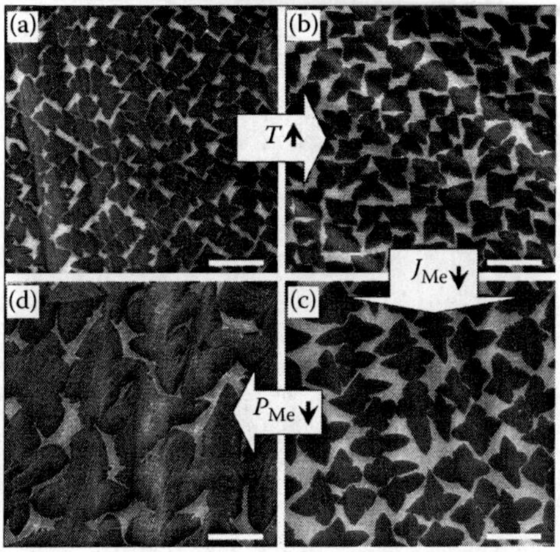

FIGURE 3.12
SEM images of partially grown graphene under different growth conditions: temperature T (°C)/flow of methane J_{Me} (sccm)/pressure of methane P_{Me} (mTorr). (a) 985/35/460; (b) 1035/35/460; (c) 1035/7/460; and (d) 1035/7/160. Scale bars are 10 μm. (From Li, X. et al., *Nano Lett.*, 10:4328–34, 2010. With permission.)

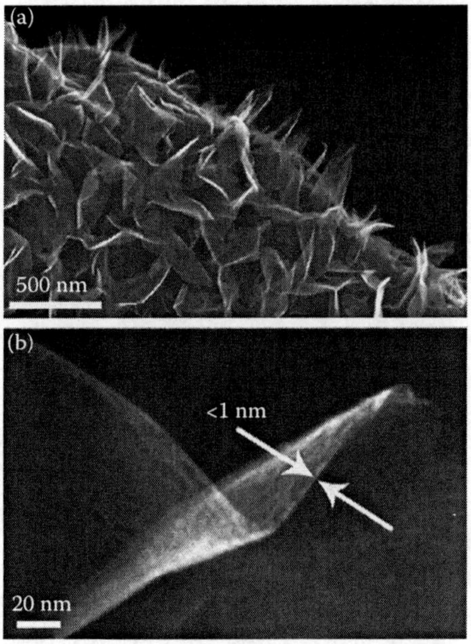

FIGURE 3.13
(a) SEM image of carbon nanosheets directly grown on the curved surface of an Ni wire of a transmission electron microscopy (TEM) grid shows that the nanosheets have a flat surface morphology and are free-standing roughly vertical to the substrate surface. (b) SEM of an enlarged nanosheet edge with a thickness less than 1 nm. (From Zhu, M. et al., *Carbon*, 45, 2229–34, 2007. With permission.)

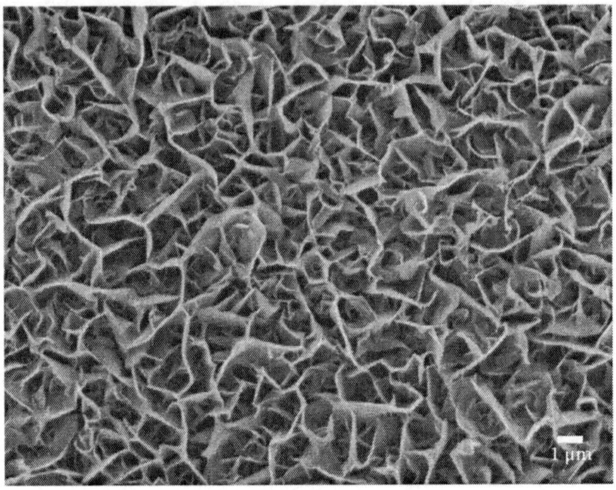

FIGURE 3.14
Top-view SEM micrograph of as-grown carbon nanosheet (CNS). (From Malesevic, A. et al., *J. Appl. Phys.*, 104, 084301, 2008. With permission.)

In Malesevic et al. (2008), CNSs were found to be a good field emitter characterized by turn-on fields as low as 1 V/μm and field amplification factors up to several thousands, as shown in Figure 3.14. The CNS field emission behavior depends on the synthesis parameters as follows: Hydrogen is identified as an efficient etchant to improve field emission, and samples grown on titanium show lower turn-on field values and higher amplification factors when compared to samples grown on silicon.

Hexagonal-shaped single crystal domains of few-layer graphene were synthesized on Cu foils using atmospheric pressure CVD with a high methane flow (Robertson and Warner 2011). SEM in Figure 3.15 reveals that the graphene domains have a hexagonal shape and are randomly orientated on the copper foil. However, the sites of grapheme nucleation exhibit some correlation by forming linear rows. TEM results give evidence that the individual domains contain approximately 5–10 layers in the central region and thinning out toward the edges of the domain.

Similar ambient CVD conditions to ones presented in Robertson and Warner (2011) were used for the synthesis of single-crystal graphene grains on polycrystalline Cu in Yu et al. (2011). The SEM images of the graphene grains presented in Figure 3.16 show that there is no definite epitaxial relationship with the Cu substrate, and the graphene crystals can cross Cu grain boundaries. The edges of these grains are found to be predominantly parallel to zigzag directions. Here the artificially initiated seeds lead to controllable synthesis of single-crystal graphene. The novel approach presented in Yu et al. (2011), using prepatterned growth seeds to control graphene nucleation, opens a route toward scalable fabrication of single-crystal graphene devices without grain boundaries.

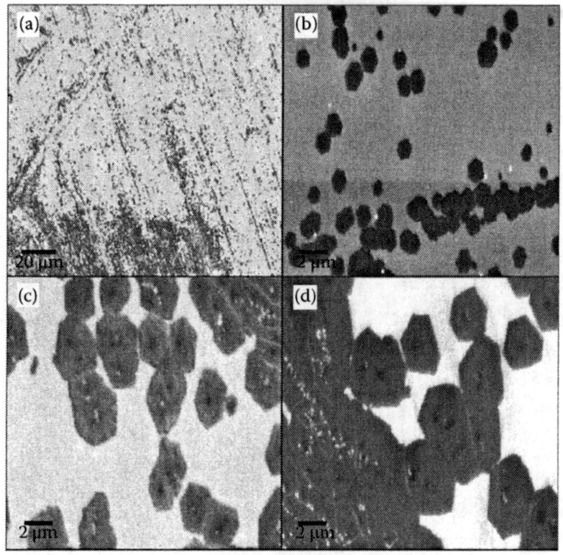

FIGURE 3.15
(a) A low magnification image from the 2-min deposition showing a typical distribution of the graphene domains (dark contrast) on the copper foil (bright contrast). (b–d) High-magnification images of 3, 6, and 10 min growths, respectively. (From Robertson, A. W. and Warner, J. H., *Nano Lett.*, 11, 1182–9, 2011. With permission.)

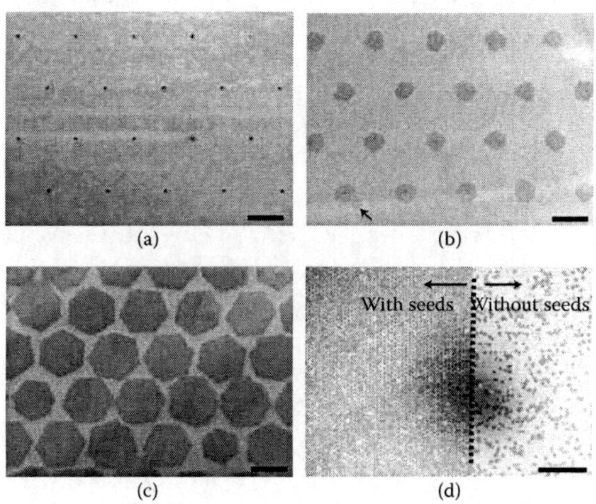

FIGURE 3.16
SEM images of (a) an array of seed crystals patterned from a pregrown multilayer graphene film on Cu foil by e-beam lithography; (b) a graphene grain array grown from an array of seed crystals, with a growth time of 5 min; (c) a graphene grain array from seeded growth following a growth time of 15 min; (d) a seeded array of grapheme grains (to the left of the dotted line), next to a randomly nucleated set of graphene grains in an area without seeds (to the right of the dotted line). Scale bars in (a–c) are 10 μm and the scale bar in (d) is 200 μm, respectively. (From Yu, Q. et al., *Nat. Mater.*, 10, 443–9, 2011. With permission.)

3.2.5 Transmission Electron Microscopy

The discovery of graphene has posed new challenges for TEM characterization. There are various TEM techniques such as classical methods, selected area electron diffraction (SAED), and dark field (DF) imaging involved in the characterization of graphene, but of crucial importance are state-of-the-art aberration (Cs), corrected TEM imaging, and other novel techniques such as nanoprobe diffraction analysis in scanning transmission electron microscope mode (D-STEM'). High-resolution TEM (HRTEM) suffers from a serious flaw as the energy of electrons necessary to obtain atomic resolution (200–300 keV) is far above the kick-off damage threshold for graphitic carbon (about 60–80 keV). Practical implementation of Cs correctors in the last decade opened the possibility to decrease the acceleration voltage to 80 kV, preserving the resolution at the atomic level. For the monatomic layer of graphene, this means that there are hardly any secrets left in understanding all structural details. From the impressive library of experimental data acquired by TEM technique, here we will present only a few examples to demonstrate different aspects of the structural studies.

A suspended graphene was successfully fabricated for TEM studies (Meyer et al. 2007). Electron diffraction analysis performed at various tilted angles led the authors to the conclusion that graphene does not form a simple planar surface but is corrugated with a typical period of a few nanometers. It is expected that a flat sheet is transformed into a set of rods (rods in the inset of Figure 3.17a) in reciprocal space, directed perpendicular to the reciprocal lattice of graphene (black hexagon). For the corrugated sheet, a superposition of the diffracting beams from microscopic flat areas effectively turns the rods into cone-shaped volumes (cones in the inset of Figure 3.17b). The simulated image in Figure 3.17b became a symbol of graphene as a promise for future science and technology. The roughness shown imitates quantitatively the roughness found experimentally.

Large-area graphene prepared by the CVD method relatively enabled easy fabrication of a satisfactory number and size of specimens, which significantly sped up the structural investigations. Images in Figure 3.18 are the result of one of these studies (Li et al. 2009b). Graphene of size tens of microns presented in Figure 3.18a contains areas with variable numbers of layers (Figure 3.18c), but monolayer graphene can be proven by the electron diffraction pattern in Figure 3.18b, where sets of six spots within two rings all have identical intensity as a signature of a single layer of atoms organized in hexagons.

TEM characterization with atomic resolution of crystal domain sizes and domain boundaries in CVD-grown graphene was presented in Kim et al. (2011). Figure 3.19a shows an image of two crystal grains, which are relatively tilted by 26.6°. The dashed rectangle is the field of view for panel (d). The inset shows the Fourier transform of the image. It shows the two sets of hexagonal patterns from two grains. Figure 3.19b and c show atomic-resolution images of the same region after removing the right set (b) or left set (c) of hexagonal

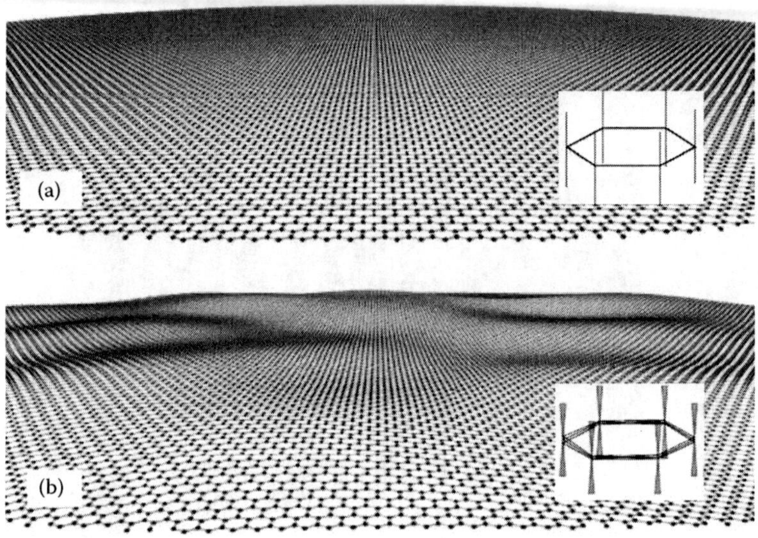

FIGURE 3.17
Microscopically corrugated graphene: (a) Flat graphene crystal in real space (perspective view). Inset: The reciprocal space for a flat sheet represents a set of rods (gray) directed perpendicular to the reciprocal lattice of graphene (black hexagon). (b) The same for corrugated graphene. Inset: The reciprocal space for a corrugated sheet represents a set of cones. (Taken and modified from Meyer, J. C. et al., *Nature*, 446, 60–3, 2007.)

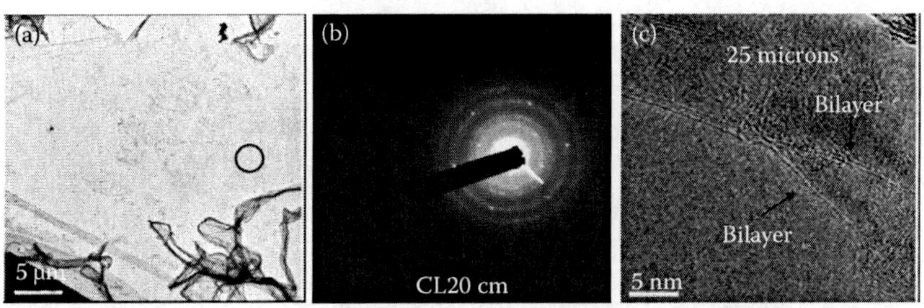

FIGURE 3.18
TEM images of graphene films. (a) Low-magnification image of graphene film on TEM grid and (b) electron diffraction pattern from circled area in (a). (c) Two- and three-layer graphene transferred from 25-μm-thick Cu foil. (From Li, X. et al., *ECS Trans.*, 19, 41–52, 2009b. With permission.)

patterns using the fast Fourier transform. The contrast of the right side (b) or left side (c) of the image completely disappears, indicating that the graphene has no defects inside the right-side grain. A zoom-in image of the high-angle tilt grain boundary of graphene is shown in Figure 3.19d. The pentagon, hexagon, and heptagon are overlaid with light gray, black, and dark gray polygons, respectively, as shown in Figure 3.20 (Meyer, J. C. et al. 2008).

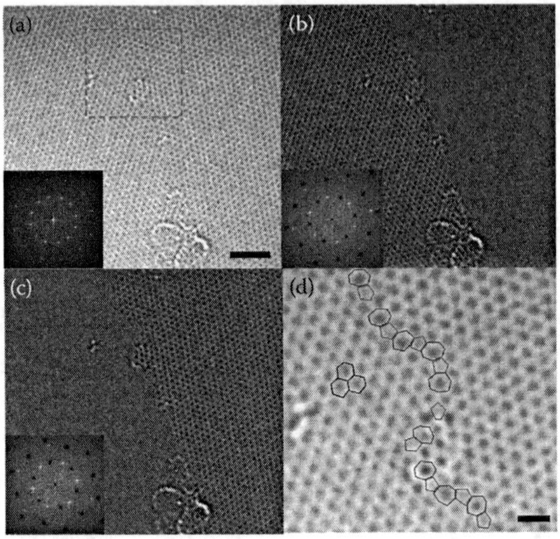

FIGURE 3.19
(a) Atomic-resolution images of a high-angle tilt grain boundary of graphene. The scale bar is 2 nm. (b, c) Atomic-resolution image of the same region after removing right set (b) or left set (c) of hexagonal patterns using the fast Fourier transform. (d) Zoom-in image of the high-angle tilt GB of graphene. The scale bar is 0.5 nm. (Taken from Kim, K. et al., *ACS Nano*, 5, 2142–6, 2011.) In situ observations by high-resolution TEM of defect formation in the graphene structure and the dynamics of the edges were reported by Girit et al. (2009). The TEM image in Figure 3.20 (left) shows the zigzag type of a part of the edge around the hole in the graphene layer. In the right panel, it is shown how, during in situ observation, the zigzag edge exposed to electron irradiation was transformed into a 5–7 defect edge (Girit et al. 2009).

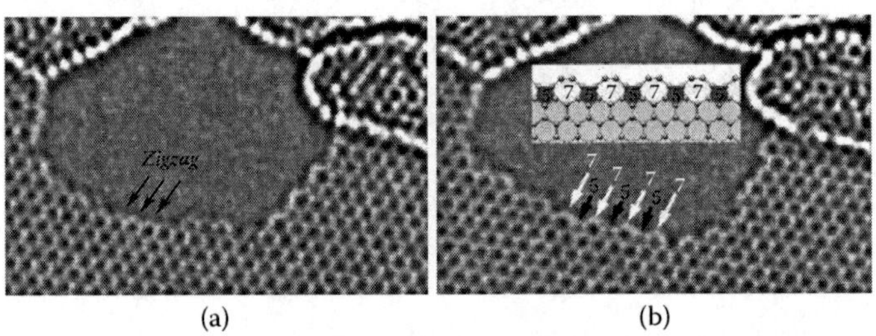

FIGURE 3.20
High-resolution TEM (HRTEM) images showing a zigzag edge near a hole region of a graphene layer (a) transformed into a 5–7 defect edge (b). (Images are adapted from movies in the supplementary material of Meyer, J. C. et al., *Nano Lett.*, 8, 3582–6, 2008.)

Rotational stacking order in bilayer and multilayer graphene was analyzed in Warner et al. (2009). Figure 3.21a shows a bilayer graphene sheet where misorientation of the layers gives rise to a Moiré pattern. By filtering in the frequency domain using a Fourier transform, the graphene lattice of each sheet was reconstructed and the packing structure and relative orientations was determined.

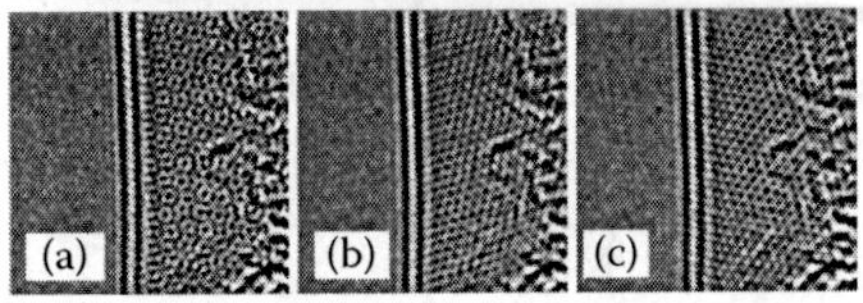

FIGURE 3.21
Bilayer graphene nanostructure imaged using aberration-corrected HRTEM with an electron accelerating voltage of 80 kV. (a) Moiré pattern due to misorientation of the layers, (b) and (c). Reconstruction of the graphene lattice in each of the layers by filtering in the frequency domain using a Fourier transform. (From Warner, J. H. et al., *Nano Lett.*, 9, 102–6, 2009. With permission.)

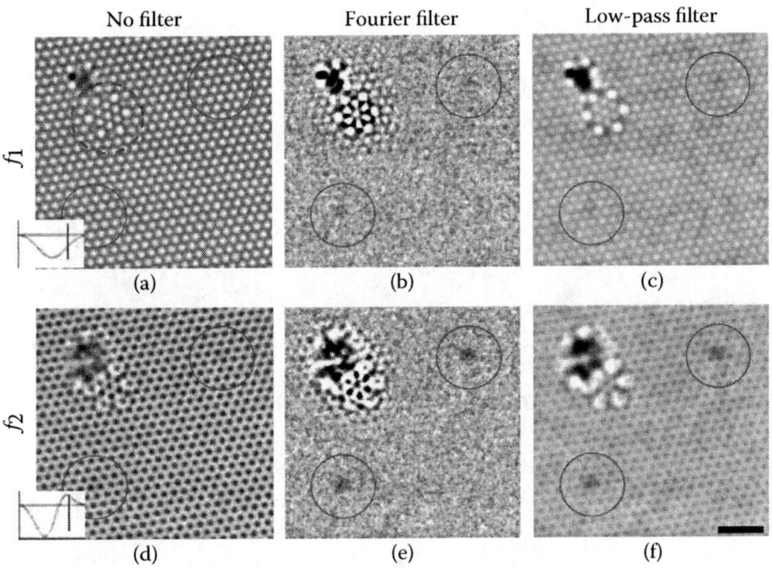

FIGURE 3.22
Nitrogen dopants in graphene (synthesized on Ni). In the Scherzer defocus image (a), dark contrast can be directly interpreted as atomic structure. The flower defect, consisting of seven rotated hexagons, six pentagons, and six heptagons, is frequently found in these samples (indicated by a dashed line in panel (a). However, the nitrogen substitution defects are not significantly above the noise (circles, a–c). In a larger defocus image (*f*2, d–f) with similar noise level, the nitrogen defects are clearly detected as a smooth dark contrast. A filter (e, f) is needed to find the N dopants in the much stronger signal of the single layer graphene lattice. Scale bar is 1 nm.

Detection of nitrogen atoms substituting carbon in the hexagonal structure was reported in Meyer et al. (2011). The electronic charge density distribution or the electrostatic atomic potential of a solid contains information not only on the atomic structure, but also on the electronic properties, such as the nature of the chemical bonds or the degree of ionization of atoms. Due to a very close atomic mass of carbon and nitrogen, it is hardly possible to distinguish them in an HRTEM image (Figure 3.22a). However, taking into account the different charge density distribution for carbon versus nitrogen bound within the hexagonal lattice of graphene (Meyer et al. 2011), the technical feasibility of an experimental analysis of charge redistribution due to chemical bonding by means of HRTEM was demonstrated.

In addition to the TEM analysis of graphene itself, the use of graphene as TEM substrates for the analysis of other species is already established.

The company Electron Microscopy Sciences, which provides supplies and equipment for electron microscopy, has announced the addition of graphene transmission electron microscope support films to its product line and characterized graphene as follows:

"Graphene is a single atomic layer of carbon atoms tightly packed in a two-dimensional honeycomb lattice. This novel material is atomically thin, chemically inert, consists of light atoms, and possesses a highly ordered structure. Graphene is electrically and thermally conductive, and is the strongest material ever measured. These remarkable properties make graphene the ideal support film for electron microscopy."

3.3 Spectroscopy

3.3.1 Raman Spectroscopy

3.3.1.1 Raman Spectra of Graphene Produced by Different Routes

Raman spectroscopy is a powerful tool to characterize graphene (Malard et al. 2009). Many important properties of graphene, such as electronic structure, number of layers, disorder, doping level, and strain, can easily be determined by a short-time measurement in ambient conditions without serious degradation on the sample. Figure 3.23 shows representative Raman spectra measured on graphene samples produced by different routes: mechanical exfoliation from highly oriented pyrolytic graphite (HOPG), CVD growth on Cu substrate, and epitaxial growth on SiC. The most prominent features in the Raman spectra of graphene are the G band around 1580 cm^{-1} and the 2D band around 2700 cm^{-1}. Another feature around 1300 cm^{-1} is the D band usually absent when the Raman spectrum is measured inside a flake exfoliated from HOPG. For graphene on SiO_2, all the features are well separated from the substrate modes. However, the Raman features of CVD graphene on metal substrate and epitaxial graphene on SiC are disturbed by a strong Raman signal from the substrates, which limits the accuracy of analysis of Raman features from the raw data. Baseline subtraction is a way of extracting the graphene signal from the originally measured data. In the case of CVD graphene, the metal substrate usually gives a large background baseline without showing any peak in addition to the graphene peaks. However, the baseline of a pristine metal substrate changes when it is covered with graphene after CVD growth. Therefore,

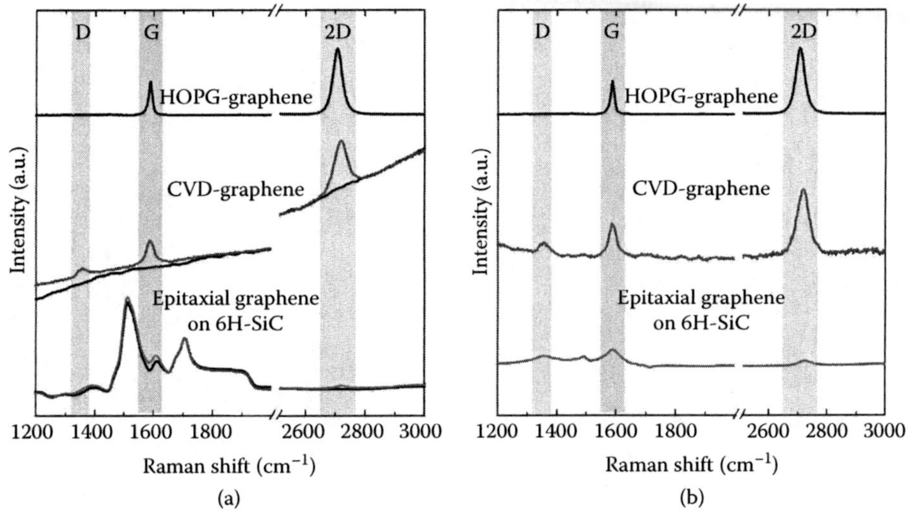

FIGURE 3.23
(a) Raman spectra of various graphene samples exfoliated from highly oriented pyrolytic graphite (HOPG), grown on a Cu substrate by chemical vapor deposition (CVD) and epitaxially grown on the Si-face of SiC. (b) Spectra after the baseline subtraction. (The data for epitaxial graphene are taken from Lee, D. S. et al., *Nano Lett.*, 8, 4320–5, 2008.)

one should note that the baseline subtraction does not work well if one uses the Raman signal of the pristine metal substrate before the growth as a reference signal. For a better fit, the reference signal can be obtained from the metal substrate treated at similar temperature as the graphene growth without a source gas flow. In the case of epitaxial graphene, one can use the SiC substrate spectrum acquired before the graphene growth as a reference for the baseline subtraction because the baseline does not significantly change after the growth.

3.3.1.2 G, D, and 2D Modes

The Raman G mode, which is named after *graphite*, is the only first-order phonon mode in graphene, being mediated by a simple scattering process with the longitudinal vibration mode as shown in Figure 3.24 (Marlard et al. 2009). As a result, the Raman G mode is independent of the energy of incident laser. The peak position of the G mode, P(G), is sensitively tuned by changing the doping level of the sample (see Section 3.3.1.4).

Other modes of D (*diamond*) and 2D (also often referred to G′ or D*) originate from so-called double resonance processes. Here the resonance condition is determined by the angular dependence of electron-phonon scattering and the Raman signal is dominated by scatterings with a singularity of the phonon density of states. Such a resonance condition changes with different

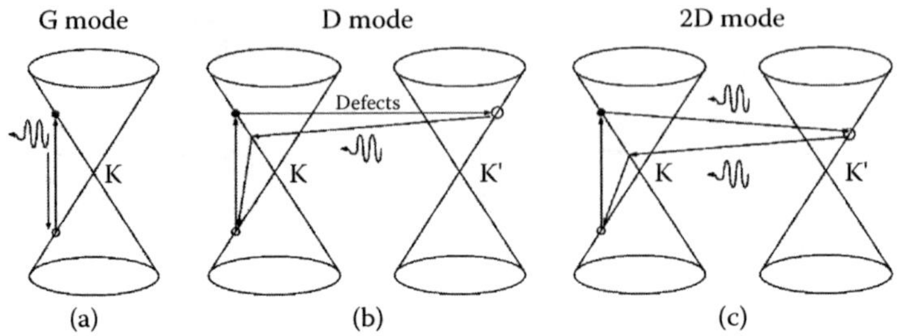

FIGURE 3.24
Schematic illustrations of the Raman scattering process for (a) the G mode, (b) the D mode, and (c) the 2D mode.

excitation energy for the creation of electron-hole pairs and the Raman D and 2D peaks are not identical in position, in intensity, and in width for different incident laser energies. The double resonance process for the D mode is mediated by an elastic scattering with defects and an inelastic scattering with a phonon. The intensity of the D mode with respect to that of the first-order phonon mode, that is, G mode, is therefore widely used as a quantity determining the concentration of defects. The 2D mode mediated by two inelastic scatterings is sensitive to the electronic structure of graphene. One can unambiguously distinguish monolayer graphene from a few-layer graphene from the Raman 2D mode.

3.3.1.3 Determination of Number of Layers

The electronic properties of monolayer graphene are unique due to the relativistic nature of charge carriers resulting from the linear dispersion relation. However, the electronic properties drastically change as the number of layers increases in the Bernal-stacking fashion. Even for bilayer graphene, the linear dispersion converts to parabolic near zero energy. It is, therefore, important to determine the number of layers of the sample.

The Raman spectrum for monolayer graphene exfoliated from HOPG exhibits a sharp symmetric 2D peak, whereas for multilayers the peak is rather broad and asymmetric with a shoulder structure, as shown in Figure 3.25a and b. The 2D peak for a monolayer is fitted by a single Lorentzian. The single component of the 2D band is due to the restricted double resonance condition and can be used as a fingerprint to distinguish monolayer graphene from multilayer (Ferrari et al. 2007). For bilayer, the asymmetric 2D peak can be decomposed into four Lorentzian peaks, as shown in Figure 3.25c. The four components stem from the four different double resonance processes.

The fitting of the 2D peak with a single Lorentzian is widely used to distinguish monolayer graphene. However, one should note that the single 2D peak is *not* a fingerprint of monolayer graphene in the case of CVD graphene

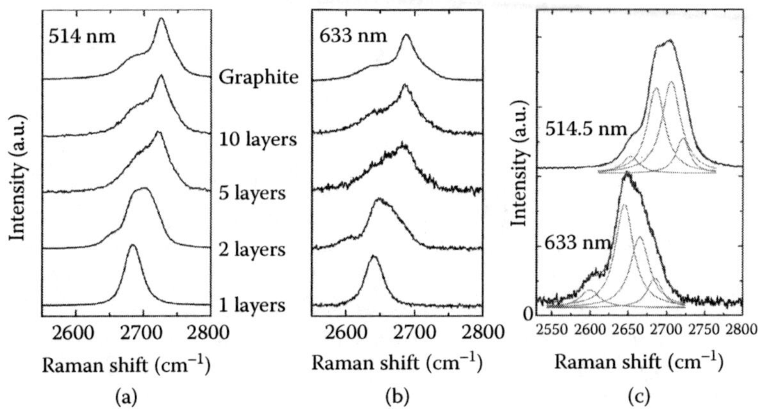

FIGURE 3.25
(a, b) Raman spectra near the 2D peak for exfoliated graphene samples with various numbers of layers. (c) Raman spectra of bilayer graphene and the four components of the 2D band. (From Ferrari, A. C. et al., *Phys. Rev. Lett.*, 97, 189401, 2007. With permission.)

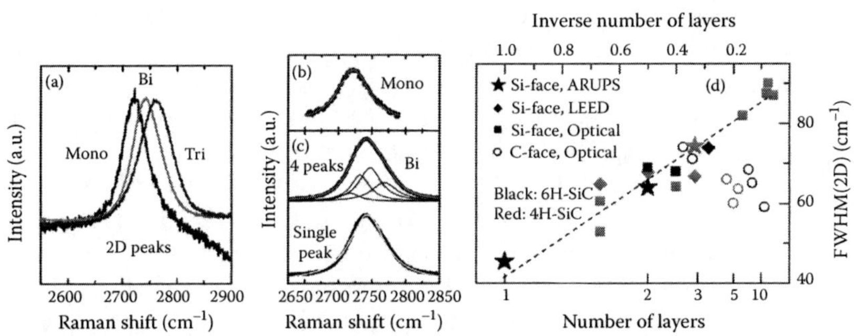

FIGURE 3.26
(a) Raman spectra near the 2D peak of mono-, bi-, and trilayer graphene grown on the Si-face of SiC. (b) The 2D peak of monolayer and the single Lorentzian fit. (c) Four components of the 2D peak of bilayer grown on SiC. (d) FWHM(2D) as a function of inverse number of layers. (From Lee, D. S. et al., *Nano Lett.*, 8, 4320–5, 2008. With permission.)

and epitaxial graphene. Raman spectra of as-grown samples produced by CVD (Figure 3.9) and by thermal decomposition of SiC (Figure 3.26) reveal that the 2D peak is almost symmetric without showing a significant shoulder structure even for multilayers. However, for epitaxial graphene grown on a Si-face, the full width at half maximum of the 2D peak, FWHM(2D), is roughly proportional to the inverse number of layers and can be used to estimate the number of layers.

There is another way of determining the number of layers of epitaxial graphene based on Beer's Law (Ni et al. 2008a; Sharf and Singer 2003). The model assumes that light passing through a film is attenuated exponentially

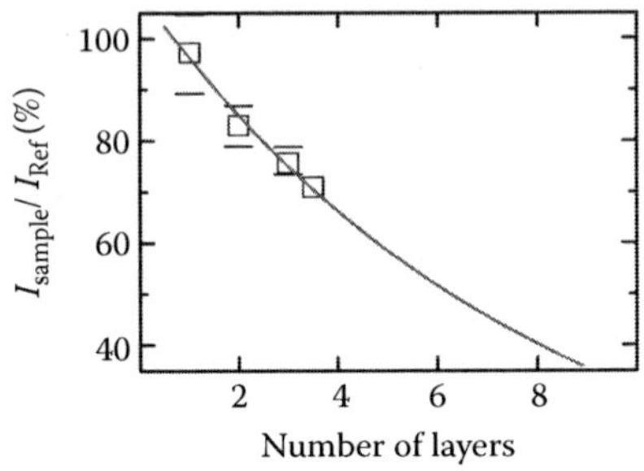

FIGURE 3.27
The ratio between the intensity of the strongest SiC substrate Raman peak obtained on graphitized samples and on a hydrogen-etched reference sample as a function of the graphene layer thickness. Experimental data points are for mono-, bi-, and trilayer samples and a three- to four-layer thick graphene sample. The curve corresponds to an exponential attenuation fit. (From Lee, D. S. et al., *Nano Lett.*, 8, 4320–5, 2008. With permission.)

according to $\exp(-t/\lambda)$, where t is the film thickness and λ is the penetration depth of the laser light. Here the SiC peak at ~1510 cm^{-1}, which has the highest intensity, can be used to estimate the attenuation from the graphene layers. Figure 3.27 shows the attenuated intensity of the SiC peak as a function of the number of graphene layers for graphene samples whose layer thicknesses were predetermined by other methods. The data points are fitted to $I_{Sample}/I_{Ref} = \exp(-2t_0N/\lambda)$, where t_0 is the stacking distance of the graphene layers and N is the number of layers. The factor of 2 is added to take into account the reflection geometry. The fitting curve can be used to determine the number of layers of an epitaxial graphene sample.

One should also note that when the layers are not Bernal stacked but turbostratic (twisted), as for graphene grown on the C-face of SiC (Faugeras et al. 2008; Lee et al. 2008) and some cases of CVD graphene, the Raman 2D peak (Li et al. 2010a) is single even for multilayers.

3.3.1.4 Doping Level of Graphene

The Raman G mode is often used to determine the charge carrier concentration. As shown in Figure 3.28, the G peak position and width change systematically with the change of charge carrier density, which is tuned by gate-voltage while the 2D peak is much less influenced by charge carrier density. The G peak for the sample with charge neutrality appears around 1583 cm^{-1}. The G peak position of graphene exfoliated from HOPG and transferred onto SiO_2 often ranges from 1585 to 1587 cm^{-1} due to compensation of p-type dopants such as adsorbed oxygen or water molecules upon exposure to air. One should note that the G peak is blue shifted both upon hole doping and electron doping from the charge neutral point. Thus, even though one can estimate the doping level, the type of charge carrier should be determined by other ways as shown in Figure 3.29 (Coletti et al. 2010).

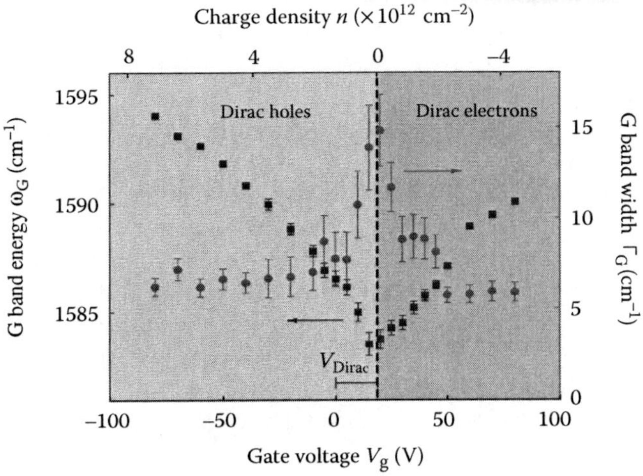

FIGURE 3.28
G band energy (squares) and G band width (circles) as a function of charge carrier density tuned by Si back-gate voltage. (From Yan, J. et al., *Phys. Rev. Lett.*, 98, 166802, 2007. With permission.)

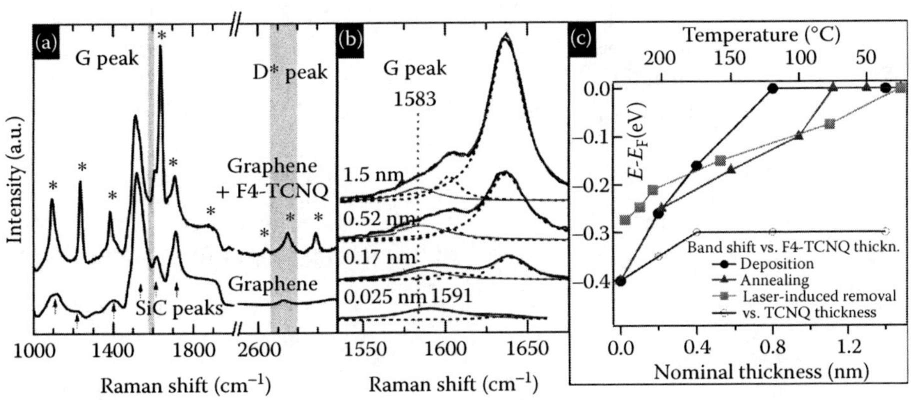

FIGURE 3.29
(a) Raman spectrum of pristine and F4-TCNQ-modified epitaxial graphene. Molecular peaks are marked by the stars and SiC peaks by the arrows. (b) Differential spectra (contribution of graphene and molecule) of as-prepared sample (on the top) and of the sample after extensive laser exposures. Black dashed lines show that the Lorentzian curves of molecular contribution. Light gray solid lines show the graphene contribution (assigned as G peak). The molecular peaks decrease by the laser exposure although the G peak does not. The G peak position shifts from 1583 to 1591 cm^{-1} as the molecules are removed by the laser exposure. (c) Displacement of Dirac point from Fermi energy measured by ARPES as a function of film thicknesses after F4-TCNQ deposition (filled circles), TCNQ deposition (empty circles), and evaluated from Raman measurements after laser-induced F4-TCNQ desorption (squares). Additionally, the band shift is shown as a function of the annealing temperature of a molecular multilayer of F4-TCNQ (filled triangles). (From Coletti, C. et al., *Phys. Rev. B*, 81, 235401, 2010. With permission.)

The estimation of charge carrier concentration using the G peak position can also be applied for CVD graphene and epitaxial graphene. In Lee et al. (2008), the charge carrier concentration estimated by G peak position is consistent with that measured by angle-resolved ultraviolet photoelectron spectroscopy (ARUPS), which unambiguously determines the band structure. In Coletti et al. (2010), where the laser exposures sequentially remove the as-deposited dopant of tetrafluorotetracyanoquinodimethane (F4-TCNQ) molecules, the charge carrier densities estimated by the sequential Raman measurements show reasonable agreement with the ARUPS results. However, one should note that the Raman G peak position is affected not only by the doping level but also by other properties of graphene such as strain. Although the strain effect on the G mode is usually much smaller than the doping effect, the behavior of the Raman 2D mode, which is more affected by strain should be taken into account.

3.3.1.5 Strain of Graphene

The Raman 2D peak is affected very little by the doping level (Yan et al. 2007), but strongly by the electronic structure of graphene. For instance, it is sensitive to Fermi velocity (Ni et al. 2010), stacking order, and strain (Ni et al. 2008b). Figure 3.30 shows the behavior of the 2D and G modes graphene transferred on a polyethylene terephthalate (PET) substrate upon the applied uniaxial tensile strain. Both the 2D peak and the G peak positions are red shifted. In this sense, the blue-shifted 2D peaks in epitaxial graphene and CVD graphene are understood as a result of compressible strain (Lee et al. 2008; Röhrl et al. 2008).

3.3.1.6 Defects and Edge

A schematic illustration of the holes of different edge structure is shown in Figure 3.31a. AFM images of round and hexagonal holes are shown in Figure 3.31b and d.

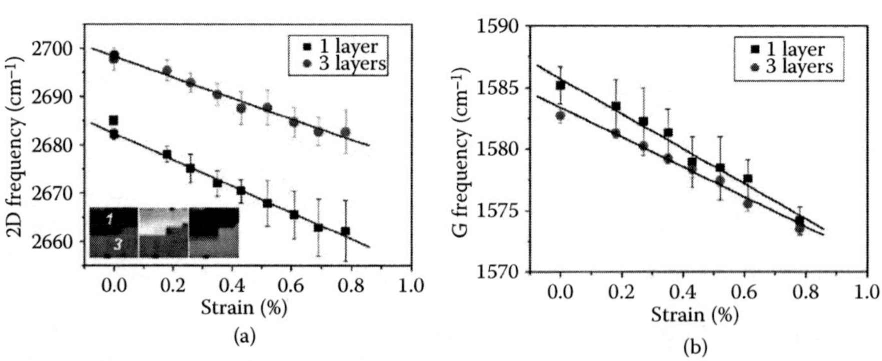

FIGURE 3.30
(a) The 2D peak position and (b) and the G peak position of monolayer graphene as a function of uniaxial strain. (From Ni, Z. H. et al., *ACS Nano*. 2, 2301–5, 2008b. With permission.)

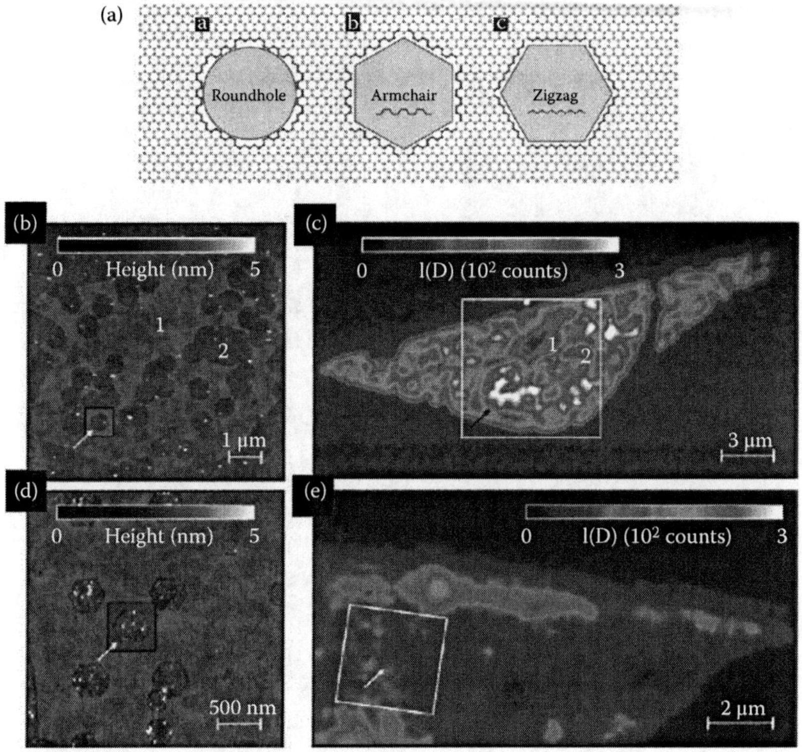

FIGURE 3.31
(a) Schematic illustration of holes of different edge structures. (b–d) Atomic force microscopy (AFM) images of round holes and hexagonal holes developed by Ar annealing at ~700°C (b and d) and intensity maps of the D mode on the same samples (c and e). (From Krauss, B. et al., *Nano Lett.*, 10, 4544–8, 2010. With permission.)

Raman D peak intensity is used to characterize a defect density. Exfoliated graphene from HOPG does not possess the D peak inside the flake showing its high quality without defects, although epitaxial graphene and CVD graphene usually exhibit D peak with a significant intensity. The high D peak for epitaxial graphene and CVD graphene is attributed to inhomogeneous crystal domains, which can be seen in SEM or STM.

However, the D peak can also be found on the edge of the exfoliated flake. It has been known that the D peak is enhanced on a so-called *armchair* edge, but a *zigzag* edge does not give rise to the D mode (Gupta et al. 2009; Krauss et al. 2010).

3.3.2 Auger Spectroscopy

Auger electron spectroscopy (AES) is based on the analysis of energetic electrons emitted from an excited atom after a series of internal relaxation events. The electron beam excitation is limited to the first few atomic layers.

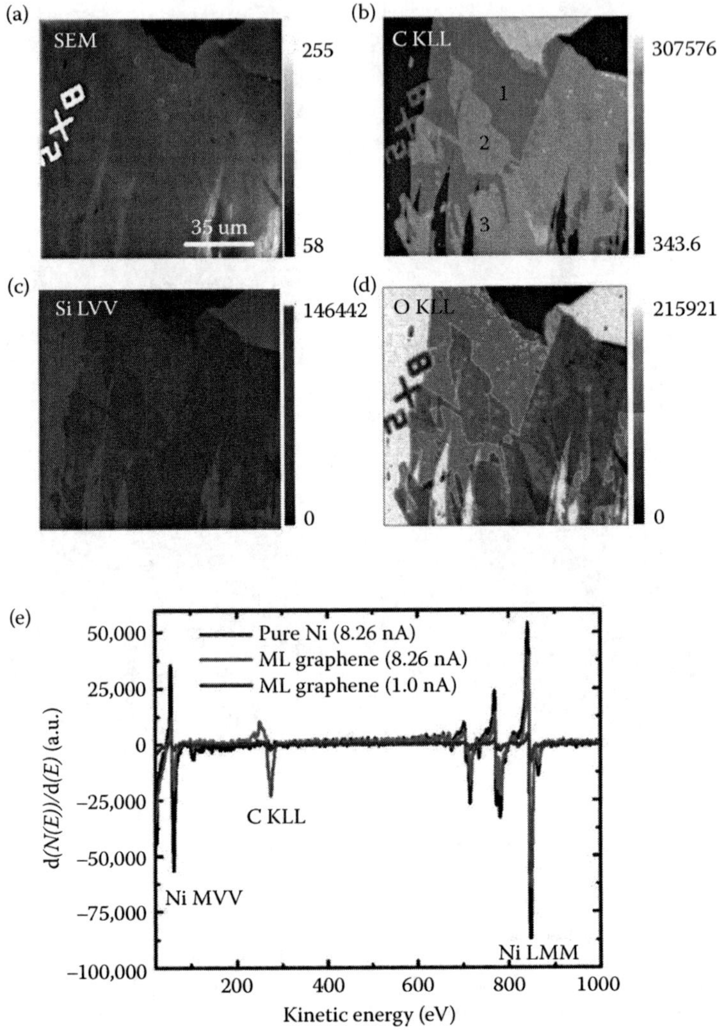

FIGURE 3.32
(a–d) A SEM image (a), Auger spectroscopy maps of C KLL (b), Si LVV (c), and O KLL (d) signals of a graphene sample. (e) Auger spectra of CVD graphene grown on Ni substrate as well as a reference spectrum of Ni substrate. (From Xu, M. et al., *ACS Nano.*, 4, 2937–45, 2010. With permission.)

The AES has been used to determine the elements in the sample and also their content by analyzing the spectral intensity. The Auger intensity map of C KLL peak provides clear contrast for mono-, bi-, and trilayer graphene, as shown in Figure 3.32a. The substrate signals of Si or O in the case of exfoliated graphene on SiO_2 and of Ni in the case of CVD graphene grown on Ni can also be used to determine the layer thickness, where the graphene layers attenuate the substrate signals.

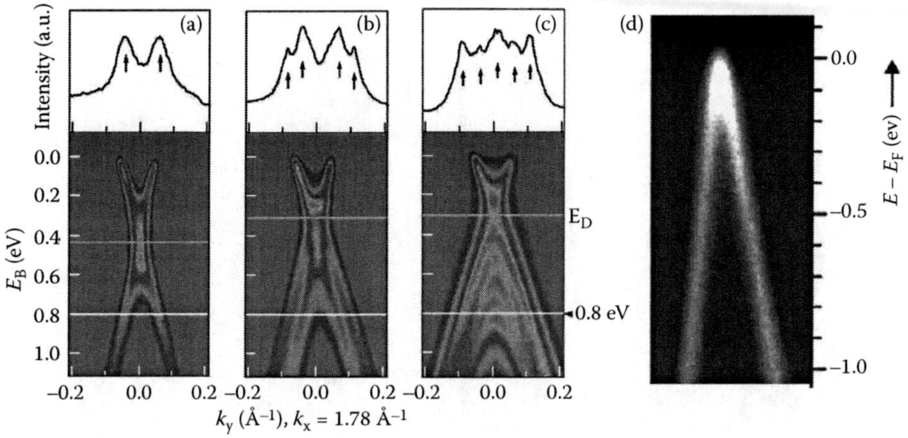

FIGURE 3.33
(a–c) Angle-resolved ultraviolet photoelectron spectroscopy (ARUPS) images and cross section (momentum distribution curve) at E_B = 0.8 eV of (a) mono-, (b) bi-, and (c) trilayer graphene grown on the Si-face of SiC. (From Lee, D. S. et al., *Nano Lett.*, 8, 4320–5, 2008.) (d) ARUPS of the graphene quasiepitaxially grown on a Ni single crystal substrate after Au intercalation between graphene and Ni. (From Iwasaki, T. et al., *Nano Lett.*, 11, 79–84, 2010. With permission.)

3.3.3 Angle-Resolved Ultraviolet Photoelectron Spectroscopy

ARUPS is a direct experimental technique to observe the distribution of the electrons in the reciprocal space as shown in Figure 3.33 (Lee et al. 2008; Iwasaki et al. 2010). The ARUPS measurement on graphene gives its complete electronic band structure. A large-area (at least over millimeter in size) crystalline sample is, however, required for the data resolution. Exfoliated graphene is not suitable for the ARUPS measurement due to its small size. The domains with different crystalline directions in CVD graphene also prevent the angular resolution. The ARUPS measurement is, therefore, only applied to a large-area epitaxial graphene grown on SiC (Lee et al. 2008) or graphene produced by a quasiepitaxial growth on a single crystalline substrate (Iwasaki et al. 2010).

3.3.4 X-Ray Photoemission Spectroscopy

X-ray photoemission spectroscopy (XPS) is used to detect the binding energy of core-level electrons, which is well defined for different elements. It serves to determine the chemical composition of the sample surface. Figure 3.34 shows examples of XPS data on zero and monolayer graphene grown on SiC. The spectra are decomposed by several peaks, the positions of which represent the elements or chemical bondings. The peak intensity area is proportional to the number of atoms in the detected sample area and thus a quantitative analysis is also possible (Riedl et al. 2010).

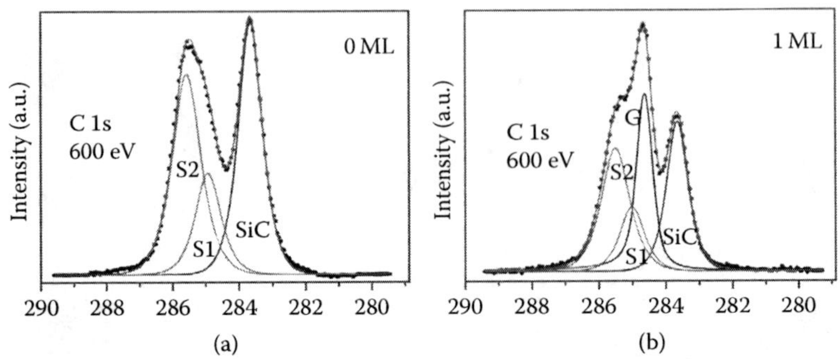

FIGURE 3.34
C 1s spectra of (a) the buffer layer and (b) the monolayer graphene grown on SiC. The X-ray photoemission spectroscopy (XPS) data shows the buffer layer (zero layer) contribution (S1 and S2) in addition to the contribution of graphene and SiC. (From Riedl, C. et al., *J. Phys. D: Appl. Phys.*, 43, 374009, 2010. With permission.)

3.4 Electrical Conduction

The electrical conduction properties of graphene show some remarkable features and are of crucial importance for many applications. To reveal their intrinsic conduction properties as closely as possible, monolayer graphene layers have been suspended above a substrate: extraordinarily large values of the electronic mobility have been measured, nearly 200,000 cm^2/Vs at very low temperatures (Du et al. 2008), reducing to 120,000 cm^2/Vs at a temperature of 240 K (Bolotin et al. 2008). These mobilities far exceed those in any conventional semiconductors (e.g., 1350 cm^2/Vs for electrons in silicon and 8500 cm^2/Vs in GaAs [Kasap 2005] at room temperature). The principal origin of these high mobilities in graphene is the high degree of crystalline perfection that can be attained.

In general, the conductivity of graphene samples shows a wide variation of values depending on the degree and nature of disorder (which depends on the method of preparation of the samples), and can be changed by the application of a gate voltage or a change in temperature. We now look at how the conducting properties of graphene vary with gate voltage, temperature, and type of disorder.

3.4.1 Electrode Configurations for Measurements on Graphene

An important issue in performing electron transport measurements is the contacts between the sample and the measuring apparatus. This is particularly the case for highly conducting materials such as graphene, for which the contact resistances may exceed the resistance of the sample. To investigate the intrinsic behavior of graphene, four-terminal measurements are

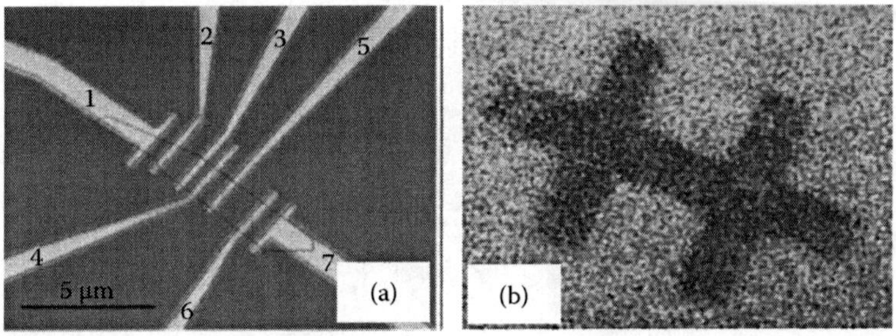

FIGURE 3.35
Electrode configurations for the measurements of the electronic properties of (a) a graphene flake (outline indicated by the thin black line) and (b) a CVD graphene layer of length 0.0.15 μm between the inner edges of the Hall bar electrodes. Both samples are on SiO_2/Si substrates.

desirable. For experiments on exfoliated graphene, a suitable graphene flake on a substrate consisting of a 300 nm SiO_2 layer on doped Si has been often used, with a metal electrode structure fabricated on top of the sample and substrate by electron-beam lithography. An example is shown in Figure 3.35a, for which four-terminal measurements of resistance can be made for adjacent samples of different lengths (Skákalová et al. 2009). Another electrode system is the Hall-bar configuration illustrated in Figure 3.35b, which allows measurements transverse to the main current flow, as required to investigate the quantum Hall effect (Skákalová et al. 2011).

3.4.2 Gate Voltage Dependence of Conduction

One of the most remarkable features of graphene is that the nature of the charge carriers (electrons or holes) and the magnitude of the conductivity can be changed by applying different gate voltages V_g to a conducting gate separated by an insulating layer from the graphene sheet. We can understand this behavior in terms of the unusual band structure of graphene near the Fermi level, which is sketched in Figure 3.36.

The first panel (a) in Figure 3.36 represents the calculated energy E of the electronic states in pure uncharged graphene as a function of wave vector magnitude k (since electron motion can occur in two dimensions in graphene, the energy bands display a conical shape in the E–$\mathbf{k}$ plot with both k_x and k_y directions included) (Wallace 1947). The electronic states are filled (at low temperatures) up to the Fermi level that is located where the electron energy cones meet, so the density of electron states at the Fermi level is nominally zero. Neutral graphene can, therefore, be described as a zero-gap semiconductor or as a semi-metal with zero density of states at the Fermi level. It is, therefore, poised just on the border between semiconductor and metal.

Application of a positive gate voltage V_g attracts electrons and fills electron states in the upper cone as in Figure 3.36b, leading to metal-like conduction

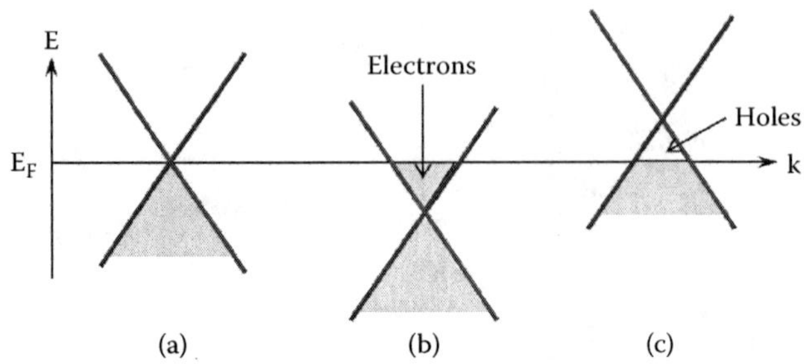

FIGURE 3.36
(a) Sketch of the electron energy E versus wave vector magnitude k for graphene, showing the intersection of the dispersion relations for two bands at the Fermi energy E_F for a neutral graphene sheet. (b) Application of a positive gate voltage attracts electrons leading to mobile electrons at energies above the band crossing. (c) Application of a negative gate voltage yields hole carriers below the band crossing.

by electrons, while a negative gate voltage leads to metal-like conduction by holes, as shown in Figure 3.36c. Because the energy–wave vector relation is linear (unlike the parabolic behavior for free electrons), the density of electrons or holes is expected to be proportional to the magnitude of gate voltage, as observed experimentally in good quality samples (Zhang et al. 2005). The carrier density n can easily be calculated from the formula

$$n = \frac{C_g V_g}{e} \tag{3.1}$$

where e is the electronic charge and C_g is the capacitance per unit area for the plate capacitor formed by the insulating layer between the graphene layer and the gate; for a 300 nm SiO_2 layer, $C_g = 115\ \mu F/m^2$. In this way, carrier densities of electrons or holes of up to more than $10^{12}\ cm^{-2}$ can be obtained. Hence, graphene is an unusually versatile electronic material in which conduction can easily be tuned by a gate voltage and switched from electrons to holes.

Since conductivity increases with carrier density, a minimum conductivity is observed as gate voltage is varied and the Fermi level goes through the energy where the conical electron energy surfaces touch (although the conductivity does not go to zero at the charge neutrality point [CNP], as shown in Figure 3.37a). The minimum often occurs at a nonzero gate voltage owing to the presence of charged impurities between the graphene layer and the substrate or elsewhere, contributing to the electric field.

The sharpness of the increase in resistivity at the CNP, recorded as the width of the resistance peak at half maximum, can be used as an indicator of the "quality" of the graphene sample (the resistivity peak widens as disorder increases). As the charge carrier density decreases near the CNP, inhomogeneous "puddles" of electrons and hole carriers develop (Martin

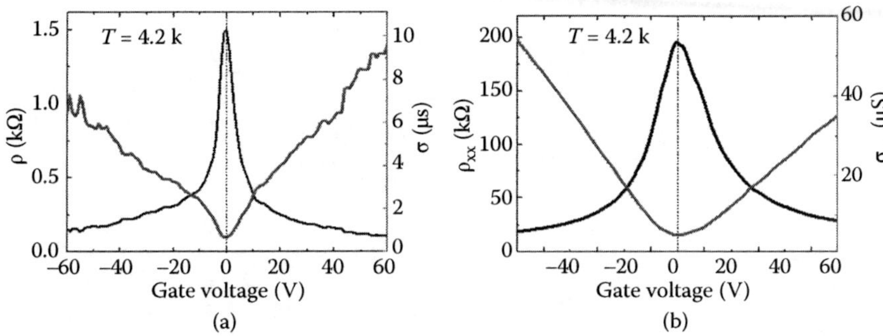

FIGURE 3.37
Variation of the resistivity ρ (left scale) and conductivity σ (right scale) at a temperature of 4.2 K as gate voltage V_g varies for (a) graphene exfoliated from highly oriented pyrolytic graphite (HOPG) and (b) a graphene layer grown on Ni (Skákalová, V. et al., *Phys. Status Solidi C*, 8, 3191–4, 2011). The conductivity increases linearly with the magnitude of gate voltage (and so carrier density) except near the charge neutrality point (CNP). (Note that the universal conductance fluctuations are observed for the more ordered sample (a), as discussed in the text). For 2D graphene, resistivity is defined as resistance per square.

et al. 2008), leading to a change in the nature of charge transport as discussed in Section 3.4.3.

Chemical doping of graphene can change the local carrier density and thereby the conductivity. The dominant effect is the change in the conductivity corresponding to a change in the carrier density, but there is also a small broadening of the resistivity maximum (Schedin et al. 2007). For micrometer-sized graphene crystals, the adsorption of individual gas molecules can be detected by step-like changes in the resistance, indicating the potential of graphene for ultrasensitive chemical detectors (Schedin et al. 2007).

3.4.3 Temperature Dependence of Conductivity for Single-Crystal Graphene

The temperature dependence of the conduction is often a useful indicator of what conduction mechanisms are occurring and/or what scattering mechanisms are limiting the conductivity.

Mechanically exfoliated graphene layers, as used in the experiments of Geim and Novoselov (2007) that won them the 2010 Nobel Prize for Physics, reveal the intrinsic properties of near-perfect crystalline graphene. Considering the regime where there is a substantial concentration of mobile carriers, that is, away from the vicinity of the CNP, the resistance increases with temperature as in conventional metals. For suspended graphene layers, the increase in resistance is linear in temperature (Bolotin et al. 2008), consistent with scattering by acoustic lattice vibrations (phonons), as observed in conventional metals over a wide range of temperatures.

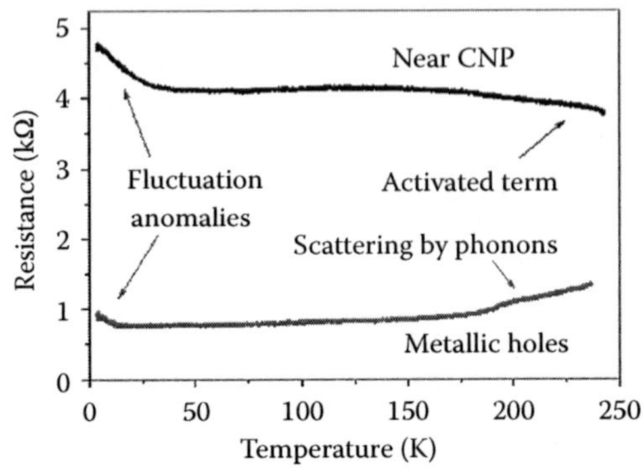

FIGURE 3.38
Resistance of a mechanically exfoliated monolayer graphene crystal in the regime of large carrier density (lower curve) showing the increase of resistivity at higher temperatures (ascribed to scattering by high-energy phonons), and the anomaly at low temperatures due to mesoscopic resistance fluctuations (Skákalová, V. et al., *Phys. Rev. B*, 80, 153404, 2009.). The upper plot is for conduction near the CNP showing the change to nonmetallic temperature dependence at higher temperatures (as well as the fluctuation anomaly at low temperatures).

At higher temperatures (above around 150 K), there is a strong nonlinear increase with temperature, as shown in Figure 3.38, which can be explained by scattering from high-energy phonons (Chen et al. 2008b; Martin et al. 2008; Skákalová et al. 2009). This scattering is proportional to the number of phonons excited; that is, to the Bose–Einstein excitation factors for these high-energy phonons (Chen et al. 2008b). The phonons causing this scattering could include SiO_2 interface phonons (Chen et al. 2008b) and the 160 meV zone-boundary phonons that are known to couple strongly to conduction electrons (Mahelona et al. 2010). Other sources of scattering are atomic-scale defects (Stauber et al. 2007) of various types and charged impurities (Chen et al. 2008a). Further discussion of the temperature-dependent resistivity with expressions for the temperature dependence due to various mechanisms, and a comparison with carbon nanotube networks and conducting polymers, is given in our recent review (Kaiser and Skakálová 2011).

3.4.4 Conduction Near the Charge Neutrality Point

Near the CNP, conduction in graphene tends to lose its metallic temperature dependence as the inhomogeneous "puddles" of electrons and hole carriers (Martin et al. 2008) develop: resistance often decreases as temperature increases (Bolotin et al. 2008; Skákalová et al. 2009), as illustrated in Figure 3.38. This behavior suggests an activated term possibly associated with transport between puddles (an activation energy of order 50 meV was indicated for samples of exfoliated graphene [Skákalová et al. 2009]). However, the resistance does not tend to infinite values as temperature approaches absolute zero temperature as for semiconductors.

3.4.5 Mesoscopic Conductance/Resistance Fluctuations

At low temperature in graphene crystals of mesoscopic size, strong universal conductance fluctuations are often present owing to the long phase coherence lengths and small number of scatterers, as shown in Figure 3.37a (Lee et al. 1987). These fluctuations are observed in samples of mesoscopic size (i.e., of the order of 1 μm) due to interference of electron waves scattered from a small number of fixed scattering centers: as the gate voltage is changed, so is the phase of the scattered electron waves and the resulting interference pattern. The pattern such as that in Figure 3.37 is fully repeatable as gate voltage is cycled up and down, provided that the scattering centers remain at fixed positions.

These conductance fluctuations are not directly seen in the temperature dependence of resistivity since the phase of the wave functions is not greatly changed by a temperature increase, but the resistance below 50 K can be increased or decreased depending on whether the gate voltage is at a value for constructive or destructive interference of scattered electron waves. This low-temperature interference resistance anomaly shows a monotonic exponential decay (or increase if the low-temperature resistance is decreased) as temperature increases from helium temperatures and thermal averaging washes out the interference peaks and troughs (Skákalová et al. 2009). These fluctuations are suppressed in large graphene samples where the fluctuation effects are also averaged out.

Weak localization, representing quantum interference effects in scattered electron waves, is a major source of temperature dependence of resistivity at lower temperatures in disordered metals as scattering increases, but appears to be strongly suppressed in graphene (Geim and Novoselov 2007).

3.4.6 Graphene Prepared by CVD

Conduction in graphene made by CVD shows some features different from those for exfoliated graphene (Bae et al. 2010; Reina et al. 2009; Skákalová et al. 2011). An example is shown in Figure 3.39: the temperature dependence shows nonmetallic behavior (i.e., the resistance decreases as the temperature increases) for the more highly conducting regime as well as near the CNP. In fact, the temperature dependence is qualitatively similar for large and small carrier densities. In this case, the resistivity is much larger than for the exfoliated graphene sample. The reason for this difference is that the CVD graphene layer is not a single crystal but a patchwork of crystalline areas with thin boundaries separating them (as observed by high-resolution TEM [Kaiser and Skakálová 2011; Park et al. 2010]). Although these barriers prevent observation of the characteristic metallic conduction temperature dependence shown by the low-resistivity sample in Figure 3.38, the conduction is still metal-like in the sense that the conductivity remains nonzero in the zero temperature limit where thermally assisted conduction mechanisms

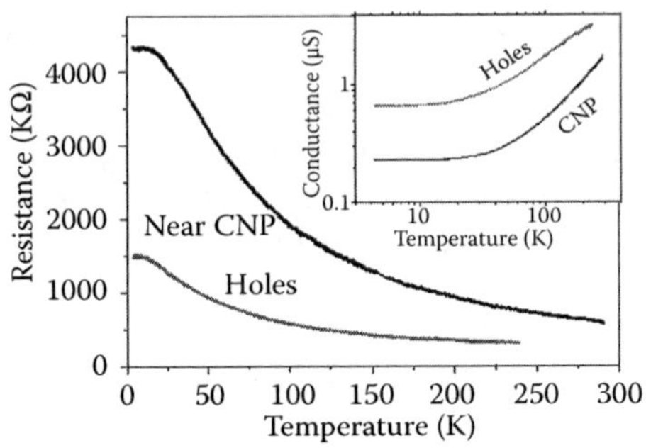

FIGURE 3.39
Resistivity of graphene made by CVD (Skákalová, V. et al., *Phys. Status Solidi C*, 8, 3191–4, 2011), showing the overall nonmetallic sign of the temperature dependence, but also the metal-like nonzero conductance at very low temperatures (shown more clearly in the inset by replotting the data using logarithmic scales).

are frozen out (Skákalová et al. 2011). This is most clearly seen using logarithmic scales as in the inset to Figure 3.39. This behavior is qualitatively similar to that seen in highly conducting polymers where the quantum tunneling between metallic conducting regions is possible through thin barriers without thermal assistance, as reviewed elsewhere (Kaiser and Skakálová 2011).

3.4.7 Reduced Graphene Oxide

Another method of forming extensive graphene layers is via the formation of GO by exposing graphite to oxygen (Eda et al. 2011; Gilje et al. 2007; Gómez-Navarro et al. 2007) and separating out single layers of partial GO in aqueous solution. These GO layers can then be deposited on a substrate and chemically treated to remove the oxygen. This process leaves pure graphene, but with disordered regions where the oxygen has been removed interspersed with more conducting regions that were not oxidized. The presence of these wide barriers means that the variable-range hopping (i.e., phonon-assisted quantum tunneling between localized states) makes the most noticeable contribution to the temperature dependence of conductance, but there is also a temperature-independent metal-like conductivity contribution even in this more disordered graphene, as shown in Figure 3.40.

$$\sigma = \sigma_0 \exp\left[-\left(\frac{T_0}{T}\right)^{1/3}\right] + \text{constant} \tag{3.2}$$

where the first term represents variable-range hopping in 2D (σ_0 is taken as constant and T_0 is a constant depending on the density of localized states and their spatial extent (Kaiser et al. 2009; Mott and Davis 1979). The constant term becomes larger for higher bias voltages. Kaiser et al. (2009) ascribe it to

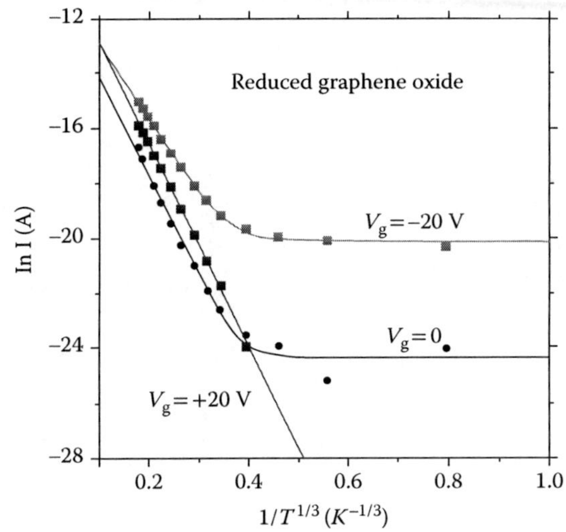

FIGURE 3.40
Natural logarithm of the current (in A) of reduced graphene oxide (at bias voltage 0.5 V) for different gate voltages V_g, showing the variable-range hopping behavior above 40 K, and the existence of a small metal-like term at low temperatures. (Gómez-Navarro, C. et al., *Nano Lett.*, 7, 3499–503, 2007; Kaiser, A. B. et al., *Nano Lett.*, 9, 1787–92, 2009.) The plot of ln*G* (where *G* is the conductivity in siemens) is the same except for a small upward shift of the vertical scale by ln(0.5) = 0.693.

electric field-driven tunneling through barriers between the regions with delocalized metallic states or field-driven hopping between localized states. Interestingly, a somewhat similar behavior of variable-range hopping and temperature-independent metal-like conductivity contribution also occurs in thin networks of single-wall carbon nanotubes (Ravi et al. 2010).

3.4.8 Nanostructured Graphene and Devices

The very high mobilities of charge carriers in graphene implying near-ballistic transport suggest possible applications in high-frequency electronics (Allen et al. 2010; Geim 2009); the increase of conductivity for both negative and positive gate voltages (Figure 3.37) in graphene naturally leads to ambipolar field-effect transistor (FET) devices tunable to electron or hole operation (Avouris et al. 2007). The dominantly metallic nature of intrinsic conduction in graphene means that it is not obviously suitable for use in devices requiring a large on/off ratio. However, it is possible to form nanoribbons of graphene by electron-beam lithography and etching techniques that have a distinct band gap, with the magnitude inversely proportional to the width of the nanoribbon (Han et al. 2007), raising the possibility of wide application of graphene in semiconductor devices. Graphene nanoribbon FETs have been demonstrated with an on/off current ratio of one or two orders of magnitude

(Han et al. 2007). More recently, graphene nanoribbons isolated from bulk graphite have yielded on/off ratios of 10^6 for the thinnest nanoribbons (Li et al. 2008); it appears that the perfection of the edges of the nanoribbons is crucial to such achievements.

The continuing improvements in the preparation of graphene nanostructures with different configurations show considerable promise for future applications in electronics: for example, graphene single-electron transistors (SETs) in which electrons tunnel into a graphene quantum dot one by one have been fabricated by etching trenches in graphene flakes to create the components needed, as reviewed by Ihn et al. (2010).

There is also hope that the cost of synthesis and preparation of graphene sheets at the scale and quality required for applications may be reduced so that graphene may replace indium tin oxide (ITO) in major applications such as solar cell electrodes and touch screens requiring strong, transparent, and flexible conducting sheets (Bae et al. 2010).

3.5 Thermal Transport Properties

In addition to the electrical conduction, the high thermal conductivity of graphene is of considerable interest and is leading to further possibilities for practical applications such as the removal of heat generated by electronic and optoelectronic devices. Carbon in the form of diamond and HOPG has long been known as a material with outstanding heat conduction properties, and more recently carbon nanotubes and graphene have been found to be capable of even higher thermal conductivity (an extensive review of the thermal properties of carbon materials including methods of measurement and theoretical aspects is given by Balandin [2011]). In this section, we highlight the most significant aspects of thermal conduction in graphene and carbon nanotubes (rolled-up graphene). We also discuss the thermoelectric power of graphene, which is larger than that of typical crystalline metals and could be of assistance in thermoelectric conversion of heat to electrical energy.

3.5.1 Thermal Conduction

The process of heat conduction in solids takes place by two main mechanisms: lattice (i.e., atomic) vibrations for which phonons are the energy quanta (present in all solids) and mobile electrons (in solids with free electrons). Both processes occur in metals, but the high thermal conductivities of good metals arise mostly from the free electrons: "hot" electrons with greater kinetic energy diffuse faster down the temperature gradient than "cold" diffuse in the opposite direction, giving an electronic contribution to the thermal conductivity that is typically larger than the phonon contribution.

However, although graphite shows metal-like behavior, it appears that phonons dominate the thermal conduction owing to the strong sp^2 bonds (Balandin 2011), and the even higher thermal conductivity of diamond arises from the strong sp^3 bonds between the carbon atoms, allowing efficient transport of heat energy by small-amplitude lattice vibrations (Ho et al. 1974). The phonons carrying the heat current are scattered by impurities, lattice defects, and crystal boundaries, and also by other phonons. At low temperatures, the thermal conductivity K increases with temperature T as more phonons are thermally excited. However, as the temperature increases beyond around 100 K, the thermal conductivity K decreases as more back-scattering phonon–phonon collisions can occur (Balandin 2011) (as shown by the example in Figure 3.41 for HOPG [Taylor 1966]). A somewhat similar peak is also observed in the thermal conductivity of individual single-wall carbon nanotubes (SWCNTs) near room temperature (Pop et al. 2006; Yu et al. 2005).

Table 3.1 shows representative values of the thermal conductivity of graphene (calculated from the measured conductance by assuming a thickness of one carbon atom). The most significant feature of the results is that, like carbon nanotubes (Pop et al. 2006; Yu et al. 2005), both exfoliated and CVD-grown graphene can have larger room-temperature thermal conductivities than HOPG or diamond. Unfortunately, no data are available for the thermal conductivity of graphene below room temperature, but the conductivity at 600 K is significantly less than near room temperature (Chen et al. 2011), as expected owing to greater phonon–phonon scattering (as for HOPG and diamond). This decrease at higher temperatures is sketched in Figure 3.36 for the thermal conductivity

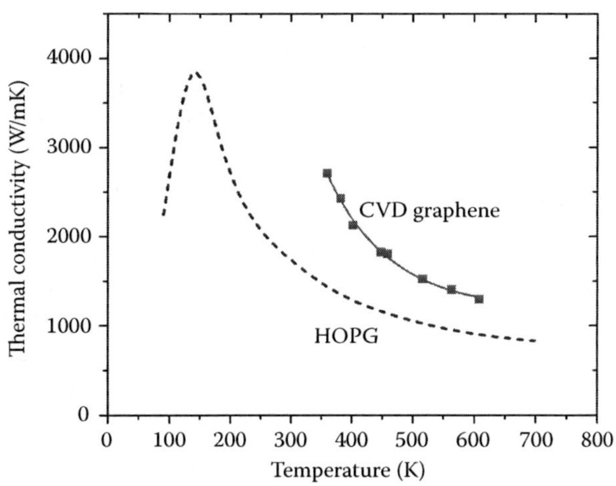

FIGURE 3.41
Example of the measured thermal conductivity of chemical vapor deposition (CVD) graphene suspended over a 9.7 μm hole (Chen et al. 2011), compared to that of highly oriented pyrolytic graphite (HOPG). (From Taylor, R., *Phil. Mag.*, 13, 157–66, 1966. With permission.) For more extensive data and discussion see Chen et al. (2011).

TABLE 3.1

Some Measured Thermal Conductivity Magnitudes K for Different Graphene Samples Near Room Temperature (RT) or at Higher Temperature T, Compared to Those for Other Carbon Materials: Individual Single-Wall Carbon Nanotubes (SWCNTs), Highly Oriented Pyrolitic Graphite (HOPG) and Diamond

	T (K)	K (Wm/K)	Reference
Graphene (suspended, exfoliated)	RT	2000–5000	Balandin et al. (2008) Ghosh et al. (2008)
Graphene (suspended, CVD)	RT 600 K	~2800 ~1200	Chen et al. (2011)
SWCNT (suspended)	RT 600 K	~3500 ~1500	Pop et al. (2006)
Diamond	RT	~2300	Ho et al. (1974)
HOPG	RT	~2000	Ho et al. (1974)

For more extensive data, see Balandin (2011).

of CVD graphene (Chen et al. 2011), which nevertheless remains higher than that in graphite. It should be borne in mind that the measurement of thermal conductivity in a monolayer such as graphene is a difficult task (Balandin 2011; Chen et al. 2011), so the values have quite large uncertainties. (Other authors [Faugeras et al. 2010] suggest smaller thermal conductivity values.)

It is clear that graphene and carbon nanotubes have great potential in heat dissipation applications. Graphene, like SWCNTs, can be used as an efficient filler to greatly enhance the thermal conductivity of composite materials even for low filling factors (e.g., a 1000% increase for 5 vol% graphene) (Balandin 2011). With the relatively small boundary resistance for the interfaces of graphene with other materials, graphene composite materials show great promise as thermal interface materials for use up to high temperatures, for example, in 3D integrated electronics (Ghosh et al. 2008).

3.5.2 Thermoelectric Properties

The direct generation of electrical power from a temperature difference across a sample using the thermoelectric effect is a topic of great current interest (Vineis et al. 2010). The efficacy of a material in thermoelectric energy conversion is determined by the dimensionless thermoelectric figure of merit

$$ZT = \frac{S^2 \sigma T}{K} \tag{3.3}$$

where S is the thermoelectric power of the material, and σ and K are the electrical and thermal conductivities, respectively. It is immediately clear that for efficient thermoelectric energy conversion, a material must have large thermopower and electrical conductivity, but very low thermal conductivity is required. Unfortunately, high electrical conductivity is usually associated with high thermal conductivity. Graphene can have a thermopower of up to 80 μV/K (Zuev et al. 2009), which is higher than for conventional metals, but the high thermal conductivity is a problem. One promising route to useful thermoelectric materials is the composites of disordered polymers with a small fraction of electrically conducting material that provides good electrical conduction and thermopower (e.g., SWCNTs in disordered polymer can significantly raise the thermoelectric figure of merit [Hewitt et al. 2011]). Disordered graphene can achieve greatly reduced thermal conduction relative to electrical conduction (Teweldebrhan and Balandin 2009), as for conventional thermoelectric materials with nanoscale structure (Liu et al. 2011).

3.6 Concluding Remarks and Future Prospects

In this chapter, in the sections covering microscopy, spectroscopy, and electrical conduction, we introduced characterization techniques for different aspects of graphene-like materials, their structures, and properties. We have illustrated the considerable variety of electronic and structural properties of graphene samples prepared in different ways. These differences arise from the morphology of the graphene sheets, for example, the presence of thin grain boundaries between graphene crystallites in graphene prepared by CVD, and the presence of more extended disordered regions in graphene prepared by the reduction of GO. Isolated defects and impurities can also play a key role in affecting the properties given the perfection of structure achievable in the best samples. This in turn means that changes in graphene conductivity, for example, can be used as an ultrasensitive detector of adsorbed molecules.

Recently, there have been rapid advances in developing new techniques for preparing graphene in many forms for a wide variety of applications, with costs reducing and electronic and mechanical properties improving, for example, for the use of graphene layers in touch screens as a possible replacement for ITO. Future developments will undoubtedly include more precise tailoring of graphene samples for use in nanoscale electronic devices such as FETs and sensors.

References

Akcöltekin, S., Kharrazi, M. E., Köhler, B., Lorke, A., and Schleberger, M. 2009. Graphene on insulating crystalline substrates. *Nanotechnol.* 20:155601.

Allen, M. J., Tung, V. C., and Kaner, R. B. 2010. Honeycomb carbon: A review of graphene. *Chem. Rev.* 110:132–45.

Avouris, P., Chen, Z., and Perebeinos, V. 2007. Carbon-based electronics. *Nat. Nanotechnol.* 2:605–15.

Bae, S., Kim, H., and Lee, Y. 2010. Roll-to-roll production of 30-inch graphene films for transparent electrodes. *Nat. Nanotechnol.* 5:574–8.

Balandin, A. A. 2011. Thermal properties of graphene and nanostructured carbon materials. *Nat. Mater.* 10:569–81.

Balandin, A. A., Ghosh, S., Bao, W. et al. 2008. Superior thermal conductivity of single-layer graphene. *Nano Lett.* 8:902–7.

Blake, P., Novoselov, K. S., Castro Neto, A. H. et al. 2007. Making graphene visible. *Appl. Phys. Lett.* 91:063124.

Bolotin, K. I., Sikes, K. J., Hone, J., Stormer, H. L., and Kim, P. 2008. Temperature-dependent transport in suspended graphene. *Phys. Rev. Lett.* 101:096802

Chen, J.-H., Jang, C., Adam, S., Fuhrer, M. S., and Williams, E. D. 2008a. Charged-impurity scattering in graphene. *Nat. Phys.* 4:377–81.

Chen, J.-H., Jang, C., Xiao, S., Ishigama, M., and Fuhrer, M. S. 2008b. Intrinsic and extrinsic performance limits of graphene devices on SiO_2. *Nat. Nanotechnol.* 3:206–9.

Chen, S., Moore, A. L., Cai, W. et al. 2011. Raman measurements of thermal transport in suspended monolayer graphene of variable sizes in vacuum and gaseous environments. *ACS Nano* 5:321–8.

Coletti, C., Riedl, C., Lee, D. S. et al. 2010. Charge neutrality and band-gap tuning of epitaxial graphene on SiC by molecular doping. *Phys. Rev. B.* 81:235401.

Dong, L.-X., and Chen, Q. 2010. Properties, synthesis, and characterization of graphene. *Front. Mater. Sci.* 4:45–51.

Du, X., Skachko, I., Barker, A., and Andrei, E. Y. 2008. Approaching ballistic transport in suspended graphene. *Nat. Nanotechnol.* 3:491–5.

Eda, G., Ball, J., Mattevi, C. et al. 2011. Partially oxidized graphene as a precursor to graphene. *J. Mater. Chem.* 21:11217–23.

Faugeras, C., Faugeras, B., Orlita, M., Potemski, M., Nair, R. R., and Geim, A. K. 2010. Thermal conductivity of graphene in corbino membrane geometry. *ACS Nano* 4:1889–92.

Faugeras, C., Nerrière, A., Potemski, M. et al. 2008. Few layer graphene on SiC, pyrolitic graphite and graphene: A Raman scattering study. *Appl. Phys. Lett.* 92:011914.

Ferrari, A. C., Meyer, J. C., Scardaci, V. et al. 2007. Raman spectrum of graphene and graphene layers. *Phys. Rev. Lett.* 97:189401.

Gao, Li., Guest, J. R., and Guisinger, N. P. 2010. Epitaxial Graphene on Cu(111). *Nano Lett.* 10:3512–6.

Gaskell, P. E., Skulason, H. S., Rodenchuk, C., and Szkopek, T. 2009. Counting graphene layers on glass via optical reflection microscopy. *Appl. Phys. Lett.* 94:143101.

Geim, A. K. 2009. Graphene: Status and prospects. *Science* 324:1530–4.

Geim, A. K., and Novoselov, K. S. 2007. The rise of graphene. *Nat. Mater.* 6:183–91.

Ghosh, S., Calizo, I., Teweldebrhan, D. et al. 2008. Extremely high thermal conductivity of graphene: Prospects for thermal management applications in nanoelectronic circuits. *Appl. Phys. Lett.* 92:151911.

Gilje, S. S., Han, S., Minsheng, W., Kang, L. W., and Kaner, R. B. 2007. A chemical route to graphene for device applications. *Nano Lett.* 7:3394–8.

Girit, Ç. Ö., Meyer, J. C., Erni, R., Rossell, M. D., Kisielowski, C., Yang, L., Park, C.-H., Crommie, Marvin L. Cohen, M. L., Louie, S. G., and Zettl, A. 2009. Graphene at the edge: Stability and dynamics. *Science* 323:1705–8.

Gómez-Navarro, C., Weitz, R. T., Bittner, A. M. et al. 2007. Electronic transport properties of individual chemically reduced graphene oxide sheets. *Nano Lett.* 7:3499–503.

Gupta, A. K., Russin, T. J., Gutiérrez, H. R., and Eklund, P. C. 2009. Probing graphene edges via Raman scattering. *ACS Nano* 3:45–52.

Han, M. Y., Ozyilmaz, B., Zhang, Y., and Kim, P. 2007. Energy band-gap engineering of graphene nanoribbons. *Phys. Rev. Lett.* 98:206805.

Hewitt, C. A., Kaiser, A. B., Roth, S., Craps, M., Czerw, R., and Carroll, D. L. 2011. Varying the concentration of single walled carbon nanotubes in thin film polymer composites, and its effect on thermoelectric power. *Appl. Phys. Lett.* 98:183110.

Ho, C. Y., Powell, R. W., and Liley, P. E. 1974. *Thermal conductivity of the elements: A comprehensive review*. American Chemical Society, Washington.

Ihn, T., Güttinger, J., Molitor, F. et al. 2010. Graphene single-electron transistors. *Mater. Today* 13:44–50.

Iwasaki, T., Park, H. J., Konuma, M., Lee, D. S., Smet, J. H., and Starke, U. 2010. Long-range ordered single crystal graphene on high quality heteroepitaxial Ni thin films grown on MgO(111). *Nano Lett.* 11:79–84.

Jeong, H.-K., Lee, Y. P., Lahaye, R. J. W. E. et al. 2008. Evidence of graphitic AB stacking order of graphite oxides. *J. Am. Chem. Soc.* 130:1362–6.

Kaiser, A. B., Gómez-Navarro, C., Sundaram, R., Burghard, M., and Kern, K. 2009. Electrical conduction mechanism in chemically derived graphene monolayers. *Nano Lett.* 9:1787–92.

Kaiser, A. B., and Skákalová, V. 2011. Electronic conduction in polymers, carbon nanotubes and graphene. *Chem. Soc. Rev.* 40:3786–801.

Kasap, S. O. 2005. *Principles of electronic materials and devices*, 3rd ed. McGraw-Hill, New York (p. 386).

Kim, K., Lee, Z., Regan, W. et al. 2011. Grain boundary mapping in polycrystalline graphene. *ACS Nano* 5:2142–6.

Krauss, B., Nemes-Incze, P., Skákalová, V., Biro, L. P., von Klitzing, K., and Smet, J. H. 2010. Raman scattering at pure graphene zigzag edges. *Nano Lett.* 10:4544–8.

Lee, D. S., Riedl, C., Krauss, B., von Klitzing, K., Starke, U., and Smet, J. H. 2008. Raman spectra of epitaxial graphene on SiC and of epitaxial graphene transferred to SiO_2. *Nano Lett.* 8:4320–5.

Lee, P. A., Stone, A. D., and Fukuyama, H. 1987. Universal conductance fluctuations in metals: Effects of finite temperature, interactions, and magnetic field. *Phys. Rev. B* 35:1039–70.

Li, G., Luican, A., Lopes dos Santos, J. M. B. et al. 2010a. Observation of Van Hove singularities in twisted graphene layers. *Nat. Phys.* 6:109–13.

Li, X., Cai, W., An, J. et al. 2009a. Large-area synthesis of high-quality and uniform graphene films on copper foils. *Science* 324:1312–4.

Li, X., Cai, W., Jung, I. et al. 2009b. Synthesis, characterization, and properties of large-Area graphene films. *ECS Trans.* 19: 41–52.

Li, X., Magnuson, C. W., Venugopa, A. et al. 2010b. Graphene films with large domain size by a two-step chemical vapor deposition Process. *Nano Lett.* 10:4328–34.

Li, X., Wang, X. R., Zhang, L., Lee, S. W., and Dai, H. J. 2008. Chemically derived, ultrasmooth graphene nanoribbon semiconductors. *Science* 319:1229–32.

Liu, C.-J., Liu, G.-J., Liu, Y.-L., Chen, L.-R., and Kaiser, A. B. 2011. Enhanced thermoelectric performance of compacted Bi0.5Sb1.5Te3 nanoplatelets with low thermal conductivity. *J. Mater. Res.* 26:1755–61.

Mahelona, K. K., Kaiser, A. B., and Skákalová, V. 2010. Resistance and mesoscopic fluctuations in graphene. *Phys. Status Solidi B* 247:2984–7.

Malard, L. M., Pimenta, M. A., Dresselhaus, G., and Dresselhaus, M. S. 2009. Raman spectroscopy in graphene. *Phys. Rep.* 473:51–87.

Malesevic, A., Kemps, R., Vanhulsel, A. et al. 2008. Field emission from vertically aligned few-layer graphene. *J. Appl. Phys.* 104:084301.

Martin, J., Akerman, N., Ulbricht, G. et al. 2008. Observation of electron–hole puddles in graphene using a scanning single-electron transistor. *Nat. Phys.* 4:144–8.

Meyer, J. C., Geim, A. K., Katsnelson, M. I. et al. 2007. The structure of suspended graphene sheets. *Nature* 446:60–3.

Meyer, J. C., Kisielowski, C., Erni, R. et al. 2008. Direct imaging of lattice atoms and topological defects in graphene membranes. *Nano Lett.* 8:3582–6.

Meyer, J. C., Kurasch, S., Park, H. J. et al. 2011. Experimental analysis of charge redistribution due to chemical bonding by high-resolution transmission electron microscopy. *Nat. Mater.* 10:209–15.

Mott, N. F., and Davis, E. A. 1979. *Electronic processes in non-crystalline materials*, 2nd ed. Clarendon, Oxford.

Ni, Z. H., Chen, W., Fan, X. F. et al. 2008a. Raman spectroscopy of epitaxial graphene on a SiC substrate. *Phys. Rev. B* 77:115416.

Ni, Z. H., Wang, Y. Y., Yu, T., You, Y. M., and Shen, Z. X. 2010. Reduction of Fermi velocity in folded graphene observed by resonance Raman spectroscopy. *Phys. Rev. B* 77:235403.

Ni, Z. H., Yu, T., Lu, Y. H., Wang, Y. Y., Feng, Y. P., and Shen, Z. X. 2008b. Uniaxial strain on graphene: Raman spectroscopy study and band-gap opening. *ACS Nano* 2:2301–5.

Park, H. J., Meyer, J., Roth, S., and Skákalová, V. 2010. Growth and properties of few-layer graphene prepared by chemical vapor deposition. *Carbon* 48:1088–94.

Pop, E., Mann, D., Wang, Q., Goodson, K., and Dai, H. 2006. Thermal conductance of an individual single-wall carbon nanotube above room temperature. *Nano Lett.* 6:96–100.

Ravi, S., Kaiser, A. B., and Bumby, C. W. 2010. Improved conduction in transparent single walled carbon nanotube networks drop-cast from volatile amine dispersions. *Chem. Phys. Lett.* 496:80–5.

Reina, A., Jia, X., Ho, J. et al. 2009. Large area, few-layer graphene films on arbitrary substrates by chemical vapor deposition. *Nano Lett.* 9:30–5.

Riedl, C., Coletti, C., and Starke, U. 2010. Structural and electronic properties of epitaxial graphene on SiC(0001): A review of growth, characterization, transfer doping and hydrogen intercalation. *J. Phys. D: Appl. Phys.* 43:374009.

Robertson, A. W., and Warner, J. H. 2011. Hexagonal single crystal domains of few-layer graphene on copper foils. *Nano Lett.* 11:1182–9.

Roddaro, S., Pingue, P., Piazza, V., Pellegrini, V., and Beltram, F., 2007. The optical visibility of graphene: Interference colors of ultrathin graphite on SiO_2. *Nano Lett.* 7:2707–10.

Röhrl, J., Hundhausen, M., Emtsev, K. V., Seyller, Th., Graupner, R., and Ley, L. 2008. Raman spectra of epitaxial graphene on SiC(0001). *Appl. Phys. Lett.* 92:201918.

Schedin, F., Geim, A. K., Morozov, S. V. et al. 2007. Detection of individual gas molecules adsorbed on graphene. *Nat. Mater.* 6:652–5.

Sharf, T. W., and Singer, I. L. 2003. Quantification of the thickness of carbon transfer films using Raman tribometry. *Trib. Lett.* 14:137–45.

Skákalová, V., Kaiser, A. B., Yoo, J. S., Obergfell, D., and Roth, S. 2009. Correlation between resistance fluctuations and temperature dependence of conductivity in graphene. *Phys. Rev. B* 80:153404.

Skákalová, V., Lee, D. S., Beringer, T., Kaiser, A. B., Srivastava, A., Park, H. J., and Roth, S. 2011. Variations of electronic transport in graphene of different origins. *Phys. Status Solidi C* 8:3191–4.

Stauber, T., Peres, N. M. R., and Guinea, F. 2007. Electronic transport in graphene: A semiclassical approach including midgap states. *Phys. Rev. B* 76:205423.

Taylor, R. 1966. The thermal conductivity of pyrolytic graphite. *Phil. Mag.* 13:157–66.

Teweldebrhan, D., and Balandin, A. A. 2009. Modification of graphene properties due to electron-beam irradiation. *Appl. Phys. Lett.* 94:013101.

Tung, V. C., Allen, M. J., Yang, Y. et al. 2009. High-throughput solution processing of large-scale graphene. *Nat. Nanotechnol.* 4:25–9.

Vineis, C. J., Shakouri, A., Majumdar, A., and Kanatzidis, M. G. 2010. Nanostructured thermoelectrics: Big efficiency gains from small features. *Adv. Mater.* 22:3970–80.

Wallace, P. R. 1947. The band theory of graphite. *Phys. Rev.* 71:622–34.

Warner, J. H., Rümmeli, M. H., Gemming, T., Büchner, B., and Briggs, G. A. D. 2009. Direct imaging of rotational stacking faults in few layer graphene. *Nano Lett.* 9:102–6.

Xu, M., Fujita, D., Gao, J., and Hanagata, N. 2010. Auger electron spectroscopy: A rational method for determining thickness of graphene films. *ACS Nano* 4:2937–45.

Yan, J., Zhang, Y., Kim, P., and Pinczuk, A. 2007. Electric field effect tuning of electron-phonon coupling in graphene. *Phys. Rev. Lett.* 98:166802.

Yu, C., Shi, L., Yao, Z., Li, D., and Majumdar, A. 2005. Thermal conductance and thermopower of an individual single-wall carbon nanotube. *Nano Lett.* 5:1842–6.

Yu, Q., Jauregui, L. A., Wu, W. et al. 2011. Control and characterization of individual grains and grain boundaries in graphene grown by chemical vapour deposition. *Nat. Mater.* 10:443–9.

Zhang, Y., Tan, Y.-W., Stormer, H. L., and Kim, P. 2005. Experimental observation of quantum Hall effect and Berry's phase in graphene. *Nature* 438:201–4.

Zhu, M., Wang, J., Holloway, B. et al. 2007. A mechanism for carbon nanosheet formation. *Carbon* 45:2229–34.

Zuev, Y. M., Chang, W., and Kim, P. 2009. Thermoelectric and magnetothermoelectric transport measurements of graphene. *Phys. Rev. Lett.* 102:096807.

4

Electrochemical Exfoliation: A Cost-Effective Approach to Produce Graphene Nanoplatelets in Bulk Quantities

Martin Matis, Urszula Kosidlo*, Friedemann Tonner, Carsten Glanz, and Ivica Kolaric

CONTENTS

* Correspondence to: Urszula Kosidlo (urszula.kosidlo@ipa.fraunhofer.de)

4.1 Introduction

Graphene, due to its remarkable properties, has attracted a tremendous amount of attention from both fundamental and applied sciences research groups. So far, studies have focused mainly on the physical properties of graphene, and in particular, its electronic transport behavior such as the quantum Hall effect (Zhang et al. 2005). However, the fascination with graphene is not due to its electrical properties alone. Graphene is interesting also because of its excellent mechanical properties, including its Young's modulus of approximately 1100 GPa (Lee et al. 2008); fracture strength of 125 GPa (Lee et al. 2008); thermal conductivity, which is close to 5000 W/m·K (Balandin et al. 2008); and specific surface area, which is calculated to be as high as 2630 m^2/g (Stoller et al. 2008). These fascinating properties make graphene a potential candidate for applications in electronics (Li et al. 2008), sensor technology (Schedin et al. 2007), and energy storage (Stoller et al. 2008) and as a filler for polymer composites.

The discovery of graphene, its extraordinary properties, and its potential to be applied in a vast amount of products has triggered the intense development of production methods. However, for graphene to be successfully utilized in any engineering application, the production of high-quality graphene and graphene nanoplatelets (GNPs) on a massive scale is required.

In this chapter, we discuss one of the graphene production methods belonging to the group of exfoliation processes. In this approach, electrochemical treatment in the presence of various electrolytes is used to exfoliate the graphene and GNPs from the graphitic electrode material. This method, sharing the benefits of chemical exfoliation such as readiness for scale-up and utilization of economically feasible starting materials, offers the advantage of using nontoxic materials, thus ensuring greater environment and worker friendliness. The as-prepared GNPs are available in the form of stable aqueous suspensions, which are ready to be isolated as monolayer or multilayer graphene sheets. Further, the process gives an advantage of simultaneous functionalization and surface modification of graphene sheets, which results in simplification of the process and later advantageous interaction of graphene with matrix materials when graphene is used as filler for polymer-based nanocomposites.

4.1.1 Electrochemical Process Background

In chemistry and manufacturing, electrolysis is a method that uses most often a direct electric current (DC) to drive an otherwise nonspontaneous chemical reaction. Electrolysis is commercially very important as a stage in the separation of elements from naturally occurring sources such as ores using an "electrolytic cell."

Electrolysis is carried out in a specially designed apparatus called an electrolytic cell. An electrolytic cell consists of a container made of either glass or any suitable plastic material, which is called the "electrolytic tank"; two electrodes, which are placed in the cell; and a solution of an electrolyte. Electric current is passed through the solution by connecting the electrodes to a source of DC. The schematic representation of such a cell is given in Figure 4.1.

For the purpose of investigations and the electrochemical exfoliation of graphene materials, such a standard electrochemical cell as the one shown in Figure 4.1 is used. Electrodes can be made from any sufficiently conductive materials, such as metals, semiconductors, graphite, and even conductive polymers. As explained in Section 4.2, two graphite electrodes (the anode and the cathode) are used in the electrolytic exfoliation process. The anode is defined as the electrode where oxidation occurs and the cathode is the electrode where reduction takes place. Each electrode attracts ions of the opposite charge. Positively charged ions (cations) move toward the electron-providing (negative) cathode, whereas negatively charged ions (anions) move toward the positive anode. In case of graphite exfoliation, anode is the electrode at which graphene materials are produced.

The key process of electrolysis is the interchange of atoms and ions by the removal or addition of electrons from the external circuit. When external voltage is applied to the electrodes, the electrolyte provides ions that move to and from the electrodes, where charge-transferring, Faradaic, or redox reactions can take place. Only at an external electrical voltage of large enough magnitude can an electrolytic cell decompose a normally stable or inert chemical compound in the solution.

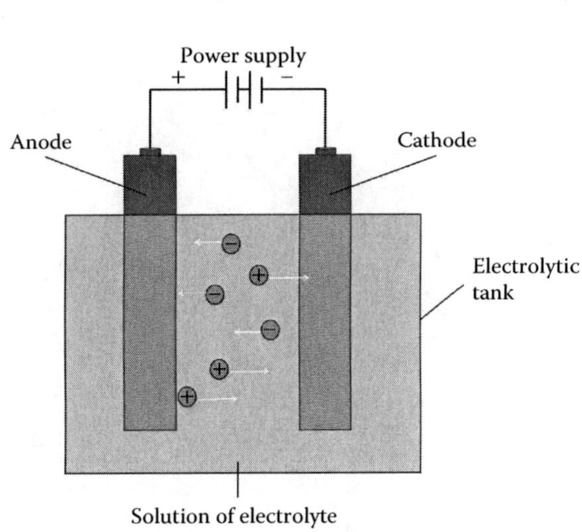

FIGURE 4.1
Schematic representation of an electrochemical cell setup.

4.1.2 Electrolytic Exfoliation of Graphite

Electrochemical treatment of graphite was proposed as an economical and practical approach for the exfoliation of graphite by simple electrochemical oxidation in an aqueous ammonia medium (Weng et al. 2003). For this purpose, two identical graphite electrodes were placed in 2 M aqueous ammonia solution as cathode and anode, and a voltage of 15 V (DC) was applied between the electrodes. The anodic graphite rod was gradually oxidized and exfoliated into the electrolyte, resulting in a colloidal solution. At the same time, some coarse particles were exfoliated from the graphite matrix, which precipitated at the bottom of the container.

The analysis (Weng et al. 2003) of obtained materials revealed that the dispersed particles are sphere-like and form irregular aggregates (Figure 4.2). The diameters of the dispersed particles were found to be around 40 nm, with a remarkable narrow size distribution. Additionally, no characteristic feature of graphite crystalline structures was found, suggesting the deconstruction of graphite structure during the oxidation process.

Further investigations and x-ray diffraction (XRD) patterns indicated that the as-prepared powder contains a significant number of randomly oriented single-layer sheets as well as multilayer ones of small size (Weng et al. 2003).

Additionally, evidence (Weng et al. 2003) indicated that graphite was electrolytically oxidized during the exfoliating process, producing oxygen-containing groups such as carboxyl and phenolic hydroxyl groups, which not only made the nanoparticles highly polar and readily dispersible in water but also enabled them to react with ammonium to form salts. It is predicted that such exfoliated graphite derivatives may be critical to the construction of functional nanometer-scale structures, which may find applications in advanced devices and sensors (Weng et al. 2003).

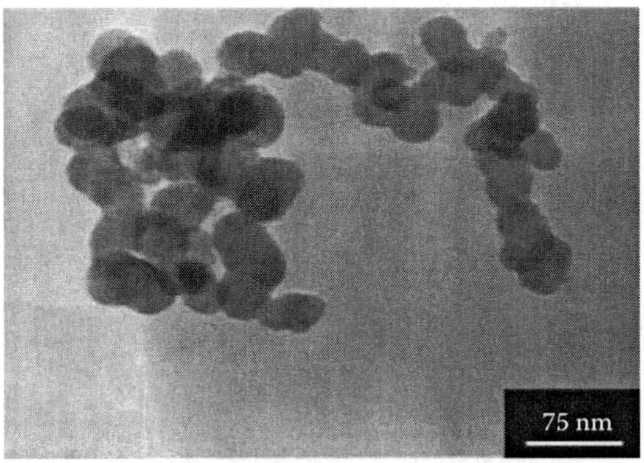

FIGURE 4.2
The TEM image of nanoparticles obtained through electrochemical treatment of graphite. (From Weng, W.-G. et al. Preparation and characterizations of nanoparticles from graphite via an electrochemically oxidizing method, *Syn. Met.* 139: 221–225, 2003. With permission.)

Electrochemical treatment of graphite was further described to generate a colloidal suspension of chemically modified graphene sheets by Liu et al. (2008). In this case, the commercial graphite rods used as anode and cathode were immersed in a phase-separated mixture of water and ionic liquid. Ionic liquid was chosen so as to provide new opportunities for the electrochemical functionalization of graphite, such that a rapid and mild green chemical functionalization process results (Liu et al. 2008).

In a synthesis, 1-octyl-3-methyl-imidazolium hexafluorophosphate ($[C_8mim]^+[PF_6]^-$) and water were used as the electrolyte. A voltage of 15 V was applied between the two graphite electrodes. After corrosion of the anode, the precipitate of GNPs was obtained at the bottom of the reactor.

Transmission electron microscopy (TEM) and field-emission scanning electron microscopy (FESEM) (Figure 4.3) observations showed that the average length of GNPs was about 700 nm and the width was 500 nm and that the graphene sheets were crumpled. The average thickness was confirmed by atomic force microscopy (AFM) to be around 1.1 nm, leading to the conclusion that complete exfoliation of graphite down to individual graphene sheets was achieved. Whereas a pristine graphene sheet is atomically flat with a well-known van der Waals thickness of approximately 0.34 nm, a functionalized graphene sheet is expected to be thicker due to the presence of hydrocarbon chains attached to it slightly above and below the original graphene plane. The ionic liquid group content in the GNPs was determined quantitatively by thermogravimetry, and the presence of 6 wt% of ionic liquid groups was suggested.

In contrast to graphite oxide, GNPs obtained from this process no longer disperse in water but readily form stable and homogenous dispersions in polar aprotic solvents. This characteristic makes GNPs an ideal advanced

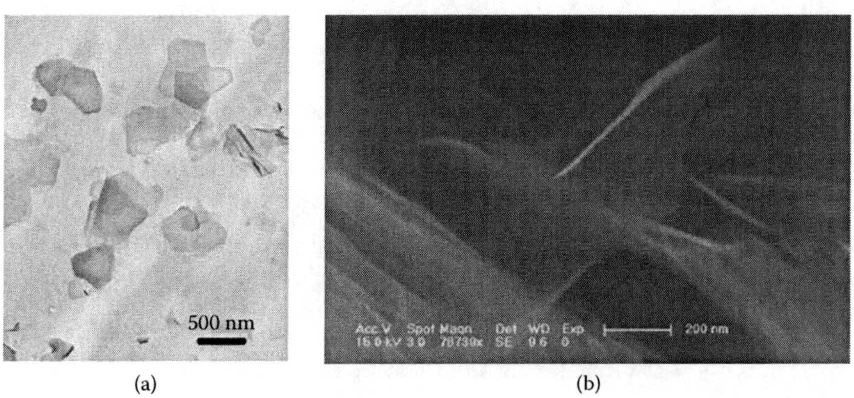

FIGURE 4.3
Images of GNPs obtained through ionic liquid–assisted electrochemical exfoliation of graphite: (a) TEM and (b) FESEM images. (From Liu, N. et al. One-step ionic-liquid-assisted electrochemical synthesis of ionic-liquid-functionalized graphene sheets directly from graphite, *Adv. Funct. Mater.* 18: 1518–1525, 2008. With permission.)

filler material and an ideal candidate for facilitating synthesis of functional graphite–polymer composites (Liu et al. 2008).

Thereafter, Wang et al. (2009) reported what was claimed a highly efficient synthesis of graphene by electrolytic exfoliation from graphite, which could be easily scaled up for large-scale production. Similar to previous experiments, in this case two graphite rods were used as electrodes. In one synthesis, two electrodes were immersed in a water-based solution of poly(sodium-4-styrenesulfonate) or PSS, which was used as the electrolyte, between which a voltage of 5 V (DC) was applied. As a result of electrolysis, a dispersion of graphitic material was obtained as a product. As postprocessing, the dispersion was centrifuged at a low speed in order to remove large agglomerates, after which the top of the dispersion was decanted. This supernatant was directly filtered to form graphene paper, or washed with water and ethanol and dried in a vacuum oven to obtain bulk powders of graphene (Wang et al. 2009).

The characterization of the as-received powders allowed the authors to confirm the graphitic crystal structure of graphene with most of the graphene sheets being stacked, multilayered GNPs. The morphology of individual graphene sheets (or GNPs) revealed crumpling and rippling in accordance with previously reported intrinsic nature of graphene sheets (Meyer et al. 2007). Figure 4.4 shows images of bulk graphene powders and those of graphene sheets.

It is suggested that this graphene preparation method be scaled up for mass and continuous production of graphene materials; however, the current yield of reaction is defined to be 15 wt% (Wang et al. 2009).

Moreover, in a patent application publication in 2009, Zhamu et al. (2009) proposed an electrochemical method of producing GNPs with the formation of a graphite compound intercalated by carboxylic acid. In their study, carboxylic acid was used as both an electrolyte and an intercalate source. In subsequent processing steps, the intercalated graphite compound was exposed to a thermal shock to produce exfoliated graphite and subjected to mechanical shearing treatment to produce GNPs. These nanoplatelets are suggested to be suitable reinforcement filler candidates for polymer nanocomposites.

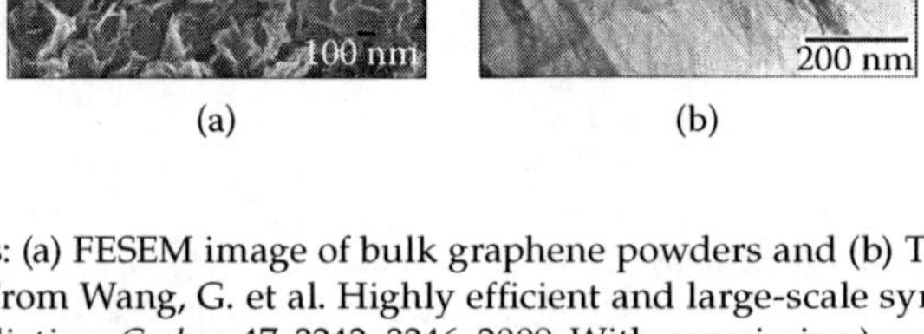

(a) (b)

FIGURE 4.4
Microscopy images: (a) FESEM image of bulk graphene powders and (b) TEM image of many graphene sheets. (From Wang, G. et al. Highly efficient and large-scale synthesis of graphene by electrolytic exfoliation, *Carbon* 47: 3242–3246, 2009. With permission.)

The most recent study in the field of electrochemical synthesis of carbon nanomaterials focused on the molten salt electrolytic method (Schwandt et al. 2010). In this study, it was confirmed that the electrolysis of molten lithium chloride in the presence of a graphite electrode generates a carbonaceous product that contains nanostructure constituents such as particles, fibers, and tubes. Through the process of parameter optimization, it was possible for the authors to achieve a content of nanotubes in the carbonaceous product of approximately 80%. The synthesized nanotubes were of multiwalled type having varying diameters and occurring as aggregates of considerable size.

In Section 4.1, we reviewed the state of the art in electrolytic production of functionalized GNPs and nanocarbons reported in the studies by Liu et al. (2008), Schwandt et al. (2010), Wang et al. (2009), Weng et al. (2003), and Zhamu et al. (2009). The focus of these research activities was mainly on obtaining the proof of principle for this production method, including investigations of various electrolytes, for example, ionic liquids, polyelectrolytes (PSS), and short carboxylic acids, followed by evaluation of product quality. Table 4.1 gives a summary of these activities.

When reviewing Table 4.1, one should keep in mind that in the study by Schwandt et al. (2010) the product of primary interest was carbon nanotubes

TABLE 4.1

Summary of Materials Used in the Literature for the Production of GNPs, and Characteristics of the Resulting Materials

Reference	Graphite	Graphite Supplier	Electrolyte	Graphene Thickness (nm)	Lateral Dimensions (nm)
(Weng et al. 2003)	Graphite model SES	Shanghai Carbon Elements Factory	Aqueous ammonia solution	Not given	40
(Liu et al. 2008)	High-purity graphite rods	China National Medicines; Shenyang Co., Ltd.	Mixture of water and ionic liquid	1,1	700 × 500
(Wang et al. 2009)	High-purity graphite rods	Aldrich	Water solution of PSS	≥0,8	300 × 420
(Schwandt et al. 2010)	Electro carb graphite (EC4, EC5, EC15, EC17)	Tokai Carbon, United Kingdom	Molten lithium chloride	—	—
(Zhamu et al. 2009)	Layered graphite material	—	Carboxylic acid	<30	1,000–20,000

and thus the characteristics of GNPs are missing. As for the reference Zhamu et al. (2009) in Table 4.1, being a patent application publication, a broader range of graphitic materials is covered in general. Among layered graphite materials, authors consider natural graphite, synthetic graphite, highly oriented pyrolytic graphite (HOPG), graphite fiber, graphitic nanofiber, graphite oxide, graphite fluoride, chemically modified graphite, or a combination of these materials.

In Section 4.2, we present our contribution to the field of electrolytic production of functionalized GNPs. The focus is on the mass production of functionalized graphene for use in polymer composites with altered mechanical and electrical properties. As for mechanical reinforcement, the coupling strength of the filler to the matrix is most important. An appropriate functionalization of GNPs is necessary and is ideally achieved at the needed level in the production process already. The proposed electrochemical exfoliation process has a functionalization step that achieves the exfoliation of nanoplatelets from graphite.

To optimize the electrolytic exfoliation process with the aim of mass production, different grades of electrodes (HOPG and normal graphite electrodes), different types of electrolytes (ionic liquids, detergents, and polyelectrolytes), and different posttreatment methods were tested. Liu et al. (2008), Wang et al. (2009), and Zhamu et al. (2009) each used ionic liquid, polyelectrolyte (PSS), or short carboxylic acid, respectively. In this study, both the electrodes and the products of electrolysis and posttreatment procedures were characterized by scanning electron microscopy (SEM), Raman spectroscopy, optical microscopy, and electrical conductivity measurements. Comparison between different electrolytes and electrodes is facilitated. From comparative measurements, general conclusions about the reaction mechanisms can be gained.

Finally, process limitations and potential scale-up are discussed.

4.2 Experimental Setup

The following subsections (4.2.1 and 4.2.2) give an overview of various electrolytes and graphitic materials that have been tested in this work. The information about the experimental setup are completed by the list of the analytical equipment used in the investigations.

4.2.1 Chemical Compounds and Graphite Materials

In this study, the following chemicals were used: PSS, poly(acrylic acid, sodium salt) or PAA, poly(2-acrylamido-2-methyl-1-propanesulfonic acid) or PAMPS, alginic acid sodium salt (AL), sodium dodecyl sulfate (SDS), sodium oleate (SO), sodium palmitate (SP), sodium dodecanoate (SD), dodecyldimethyl(3-sulfonatopropyl) ammonium (DDSA), and 1-dodecyl-3-methylimidazolium iodide ([C_{12}-mim]I). Chemical structures of these electrolytes are illustrated

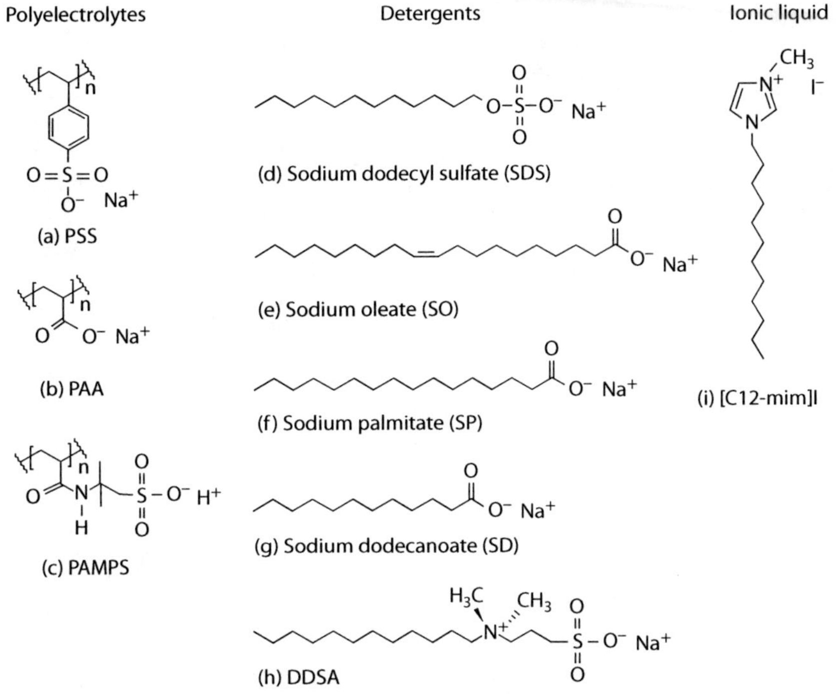

FIGURE 4.5
Chemical structures of tested electrolytes: Chemicals are grouped into polyelectrolytes (a–c), detergents (d–h), and ionic liquids (i). Full names of abbreviated materials are (a) poly(sodium 4-styrenesulfonate), (b) poly(acrylic acid, sodium salt), (c) poly(2-acrylamido-2-methyl-1-propanesulfonic acid), (h) dodecyldimethyl(3-sulfonatopropyl) ammonium, and (i) 1-dodecyl-3-methylimidazolium iodide.

in Figure 4.5. All chemical compounds were purchased from Sigma-Aldrich and were used as received without further purification. Polymers were purchased as water solutions, detergents were purchased as powders, and ionic liquid was purchased in the solid state.

Electrographite materials MREG-E, MR75, and MR190 (proprietary product designators) were obtained from Müller & Rössner GmbH & Co. KG (see Figure 4.6 and Table 4.2), Troisdorf, Germany, in the form of blocks, which were cut to pieces for further usage. The HOPG was purchased from NT-MDT and used without further treatment.

4.2.2 Experimental and Analytical Equipment

All electrolytic processes were conducted using standard laboratory current sources. Sonicator Bandelin Sonopuls HD3200 was used with titanium-tapered tip and glass processing vessel. Eppendorf Centrifuge 5804 was used with glass centrifuge tubes. Filtration towers Sartorius 16249 and 16274 were

TABLE 4.2

List of Suppliers, Prices, Maximum Grain Size, Crystal Size, and Electrical Resistivity for Electrode Materials Used

Material	Supplier and Price	Maximum Grain Size	Crystal Size	Electrical Resistivity
MR190, electrographite	Müller & Rössner; US$26.90 per kilogram	<25 μm (data sheet)	<10 μm (from SEM)	112 mΩ·cm
MR75, electrographite	Müller & Rössner; US$11.70 per kilogram	<0.8 mm (data sheet)	<40 μm (from SEM)	75–85 mΩ·cm
MREG-E	Müller & Rössner; US$10.40 per kilogram	<2.0 mm (data sheet)	<80 μm (from SEM)	65 mΩ·cm
HOPG	NT-MDT Co.; US$55–95 per sample (10 mm × 10 mm)	Not applicable; highly oriented, 1.2 mm thick	Typically 0.5–1 mm, quality ZYB	Typically parallel to layer plane: $3.5–4.5 \times 10^{-2}$ mΩ·cm; perpendicular: 150–250 mΩ·cm

Note: Crystal size was estimated from SEM images.

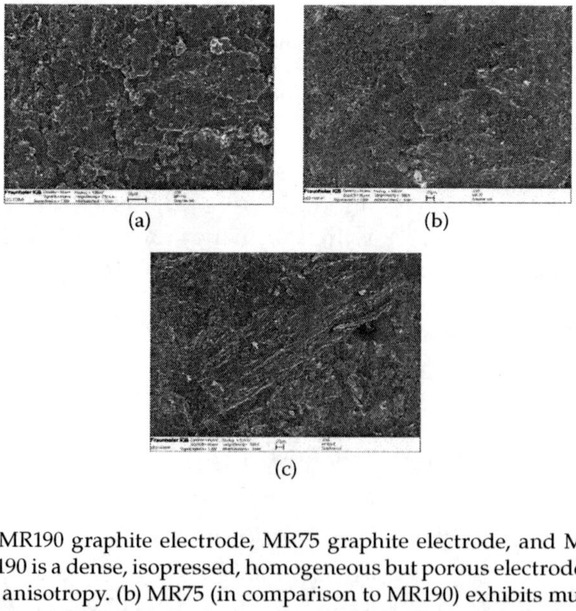

FIGURE 4.6
SEM images for MR190 graphite electrode, MR75 graphite electrode, and MREG-E graphite electrode. (a) MR190 is a dense, isopressed, homogeneous but porous electrode with very small grains and small anisotropy. (b) MR75 (in comparison to MR190) exhibits much larger grains, with their in-plane orientation in the bulk material. (c) MREG-E is a graphite material with the largest grain size (in comparison with MR190 and MR75), which exhibit a high degree of orientation in the bulk.

used with cellulose (as a supporting filter), and polycarbonate (0.4 μm diameter of pores) and polyamide (0.45 μm diameter of pores) filters, respectively.

The SEM images were made by FESEM ZEISS model 1530VP.

Raman spectra were taken using a Jobin-Yvon Type LabRam V 010 single grating spectrometer equipped with a double super razor edge filter and a peltier-cooled charge-coupled device (CCD) camera.

Spectra were taken in quasi-backscattering geometry using the linearly polarized 632.817 nm line of a He/Ne gas laser having power less than 5 mW, which is focused to a 15–20 μm spot through a 20× microscope objective onto the top surface of the sample. The resolution of the spectrometer (grating 1800 L/mm) was 1 wave number (cm^{-1}). The heater used was from the company Linkam (Tadworth, United Kingdom), type THS 600, with a small argon gas flow rate. For electrical measurements, standard two-probe measurement with probes kept approximately 5 mm apart pressed on the film surface was used.

4.3 Results and Discussion

The aim of this study was to test different electrode materials combined with various electrolytes in respect to the kind and quality of resulting materials. To emphasize this approach, all products are abbreviated as a combination of electrolyte–electrode abbreviations. For instance, product of electrolysis of polystyrene sulfonate (PSS) with a HOPG graphitic electrode is abbreviated as PSS-HOPG.

4.3.1 Selection of Electrolytes

Different types of electrolytes have been used for the electrochemical preparation of GNPs. In general, groups of electrolytes of interest in these investigations include polyelectrolytes, detergents (surfactants), and ionic liquids.

Table 4.3 summarizes the results obtained for each specific electrolyte and specifies whether electrochemical exfoliation has been achieved. A negative result means that exfoliation of GNPs was not achieved under described conditions.

Polyelectrolytes are promising candidates for utilization as electrolytes as almost every tested polyelectrolyte results in successful exfoliation, leading to functionalized GNPs with surface characteristics that might be necessary for improved interfacing with a matrix material in a composite. The only exception is alginic acid (AL), which generates an insulating gel covering the anode.

Detergents in the form of ordinary soaps are interesting as electrolytes, because of their availability and price. Unfortunately, typical detergents such as SO, SP, and DDSA lead to exfoliation of large particles and exfoliation occurs only in the presence of sulfuric acid as an inert electrolyte. In addition, SDS does not facilitate the exfoliation process at all.

TABLE 4.3

Table of Electrolytes Tested in This Work for the Exfoliation Process, Giving Their Full Name, Abbreviation, Active Ion, Molar Mass, CAS Number, and Exfoliation Success in This Work

Group of Electrolytes	Electrolyte	Abbreviation	Active Ion	Molar Mass (g/mol)	CAS Number	Successful Exfoliation (Yes—Y/Partial—P/No—N)	Price[a] ($/kg)
Polyelectrolyte	Poly(sodium 4-styrenesulfonate)	PSS	Anion	70,000	25704–18–1	Y	264
Polyelectrolyte	Poly(acrylic acid, sodium salt)	PAA	Anion	15,000	9003–04–7	Y	819
Polyelectrolyte	Poly(2-acrylamido-2-methyl-1-propanesulfonic acid)	PAMPS	Anion	2,000,000	27119–07–9	Y	3,876
Polyelectrolyte	Alginic acid sodium salt	AL	Anion	—	9005–38–3	N	179
Detergent	Sodium dodecyl sulfate	SDS	Anion	288.38	151–21–3	N	78
Detergent	Sodium oleate	SO	Anion	304.45	143–19–1	P	28,360
Detergent	Sodium palmitate	SP	Anion	278.41	408–35–5	P	7,500
Detergent	Sodium dodecanoate	SD	Anion	222.3	629–25–4	P	3,075
Detergent	Dodecyldimethyl (3-sulfonatopropyl) ammonium	DDSA	Inner salt	335.55	14933–08–5	P	2,468
Ionic liquid	1-dodecyl-3-methylimidazolium iodide	$[C_{12}\text{-mim}]I$	Cation	378.34	81995–09–7	N	15,060

Note: The electrolytes are grouped into polyelectrolytes, detergents, and ionic liquids. Partially operational electrolytes denoted by P in the one before last column (surfactants) work only if the ionic strength is high enough; higher conductivity is reached by the addition of sulfuric acid.

[a] Prices are recalculated from Sigma-Aldrich products. Prices were calculated in the case of solutions to 1 kg of pure chemical species. In the case of wholesale products, the prices could drop dramatically.

Although the ionic liquid we used did not lead to exfoliation under the conditions tested, Liu et al. (2008) reported successful exfoliation with the same type of ionic liquid.

From the chemical point of view, the influence of various anions in the electrolyte on exfoliation was tested by exchanging the anion part alone. Carboxylic acids and sulfates (sulfonates) were chosen as the anion constituent of the electrolyte, and for both of them sodium cation was chosen. Polyelectrolytes did not require any additional inert electrolyte to reach required conductivity. In the case of small-molecule electrolytes (SO, SD, SP, SSDA), sulfuric acid was used as the indifferent electrolyte to increase conductivity. Similar alkyl chain lengths were chosen for both ionic liquids and detergents.

Polyelectrolytes are considerably cheaper than ionic liquids or detergents in a pure form (for comparison of prices, see Table 4.2), which makes them suitable for utilization in mass production processes. Additionally, polyelectrolytes exhibiting various functional groups are available, which gives higher flexibility to optimize product properties. Although detergents such as SD, SP, and SO are expensive in a pure form (>99%), they are as cheap as US$32 per kilogram in the form of diluted soap. This low price of soaps makes them very attractive for exfoliation of GNPs on a mass scale.

4.3.2 Selection of Electrodes

Apart from HOPG, which is a standard reference material for graphite electrodes, cheaper graphite materials were tested. These materials are normally used in furnace structures, foundries, and aluminum processing. The main difference between these materials is in their processing and grain size. The MR190 and MR75 are isopressed graphite materials, whereas MREG-E is extruded graphite. The grain size increases from MR190 (25 μm) to MR75 (0.8 mm) and MREG-E (2 mm). Larger grains might be connected with possibly larger flake sizes of resultant GNPs. Preprocessing of the graphitic material may also have an impact on the product of electrolysis, as it introduces damage to grains and crystals. For an insight into the graphitic structure, see Figure 4.6 and Table 4.2.

The electrodes were characterized by Raman spectroscopy and SEM.

Raman spectrometry is a very powerful technique to investigate carbon nanostructures. The basic features of Raman spectra of graphitic materials and their interpretation are discussed in this section for easy reference. The position of the peaks and their width, shape, and height are correlated with the probed carbon structure. A graphitic peak at about 1590 cm^{-1} (the so-called G peak, G stands for graphitic) and another peak at about 2700 cm^{-1} (the so-called 2D peak) are typical for sp^2-hybridized carbon atoms, whereas the diamond peak at about 1335 cm^{-1} (D peak, D for diamond or defect) is typical for sp^3-hybridized carbon atoms. The defect mode at about 1620 cm^{-1} (D′ peak) can be observed as well. The more heterogeneous the investigated material the broader the peaks in a Raman spectrum. In a perfect graphitic structure, no D peak is present, in contrast to a large 2D peak.

Reference measurements of Raman spectra for carbon materials are presented in Figure 4.7, and Figure 4.8 relates a graphitic Raman spectrum to a graphene Raman spectrum. As the intensity of a spectrum is typically normalized by the height of the G peak, we speak of relative rather than absolute heights of individual peaks. For pure graphitic materials, the size of the crystallites is inversely related to the height relation between D and G peaks (I_D/I_G; Tuinstra and Koenig 1970). This relation is valid for a graphitic flake size of approximately 2 nm; below this limit, height of the D peak typically increases until fully amorphous carbon is reached, revealing a high D peak (Ferrari and Robertson 2000). Apart from pure graphitic materials, defects and functional groups indicate their presence in a higher D peak. For graphene, a shift of the position of the G peak occurs relative to the position of the G peak in graphite. (see Figure 4.8).

Regarding the selection of electrode materials, a low D peak is indicative of large crystallites and thus favorable for exfoliation of high-quality graphene materials. The HOPG (as seen in Figure 4.7f) shows the smallest, almost

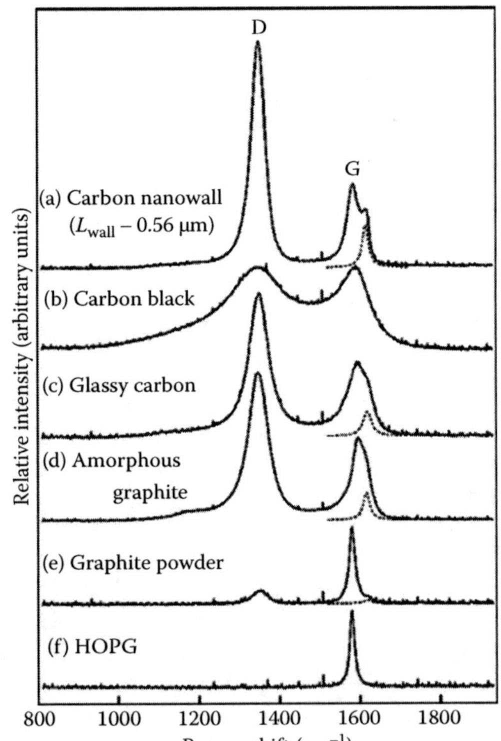

FIGURE 4.7
Raman spectra: Spectra of (a) carbon nanowalls, (b) amorphous carbon, and (c–f) graphite-like carbons. (From Kurita, S. et al. Raman spectra of carbon nanowalls grown by plasma-enhanced chemical vapor deposition, *J. Appl. Phys.* 97: 104320–104325, 2005. With permission.)

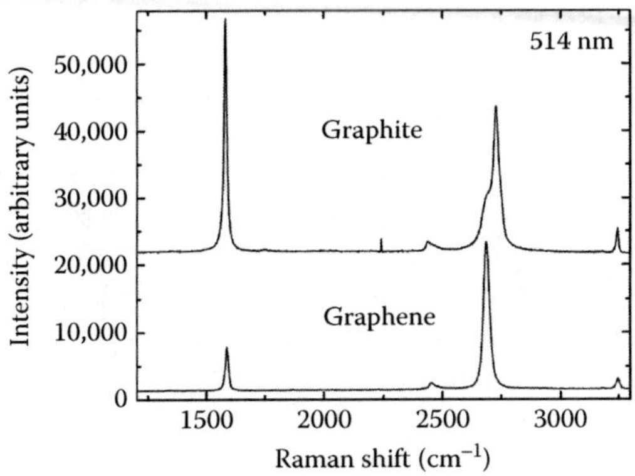

FIGURE 4.8
Raman spectra at 514 nm of graphite and graphene: Notice the absence of the D peak, a smaller G peak (left) resulting from scaling to 2D intensity, and the left-shifted 2D peak (right, at 2700 cm^{-1}) for graphene, indicative of fewer defects and fewer graphitic layers in relation to graphite. (From Ferrari, A. C. et al. Raman spectrum of graphene and graphene layers, *Phys. Rev. Lett.* 97: 187401–187404, 2006. With permission.)

absent, D peak, and graphite powder shows a larger D peak (cf. Figure 4.7e) than HOPG but smaller than amorphous graphite (cf. Figure 4.7d).

The properties and cost, and images of the graphitic materials used in the study are given in Table 4.2 and Figure 4.6, respectively. Whereas natural graphite shows high crystallinity and anisotropy, technical graphite grades, which are processed by milling and pressing (isopressing), exhibit more isotropic and homogeneous forms. Less-processed graphites are typically cheaper and display larger grain size and higher crystallinity, which is favorable for the exfoliation of large flakes by electrolysis. The grain size from the data sheets in Table 4.2 is correlated with the crystal size from SEM images; accordingly, MREG-E is the tested product with the best performance/cost relation. We use HOPG as a well-controlled standard reference in graphite research.

4.3.3 Electrolytic Exfoliation Process

The electrolytic exfoliation process and its up-scaling are affected by the details of the experimental setup such as the choice and arrangement of electrodes and the conditions for electrolysis. This subsection (4.3.3) presents the effords towards the optimisation of the process and successful preparation of GNPs.

4.3.3.1 Design of Experimental Setup

For the first trial experiments, blocks of graphite materials were cut into pieces of approximately 5 mm × 5 mm × 50 mm. For up-scaling the process, larger electrodes were used (Figure 4.9).

To isolate the contribution of the anode to the process, two different setups were tested: (1) One with separation of the electrodes by a salt bridge (cf. Figure 4.9a) and (2) one with an inert Pt cathode (cf. Figure 4.9b). In the case of separated electrodes in a salt bridge setup, exfoliation occurred only on the anode. Additionally, it was shown that no exfoliation took place when replacing the graphite anode with the Pt electrode. On the other hand, exfoliation was still possible with a replaced Pt cathode, proving that the type of the cathode has little effect on exfoliation. Because metals react at the voltages used with electrolytes to produce unwanted by-products, metallic cathodes are not preferred for this process. Both electrodes may be graphitic without affecting the exfoliation process. Comparison of anode and cathode with respect to electrolytic erosion is shown in Figure 4.10, and it proves that only the anode is affected and loses mass. These findings resulted in the decision to use the setup with two graphitic electrodes for all follow-up experiments (see Figure 4.9c and d).

4.3.3.2 Conditions for the Electrolysis

Electrolysis is a process in which ionic chemical species react with electrons or holes on the surface of an electrode. In this case, anions change on the surface of the anode to neutral radicals that react with the electrode. As a result of a physical follow-up process, small flakes of graphite are detached from the surface of the electrode.

Throughout the experimental investigations, each electrolysis process was carried out at a voltage of 2–4 V. The DC was flowing for 4–12 hours through the

FIGURE 4.9
Different electrolytic cell setups. (a) Small setup with a salt bridge. (b) Small setup with Pt cathode. (c) Small setup with graphitic electrodes. (d) Middle-sized setup with a magnetic stirrer.

(a) (b)

FIGURE 4.10
Photographs of (a) cathode and (b) anode after electrolysis, showing decomposition on the bottom part of the anode but not on the cathode: Exfoliation takes place at the anode and no mass loss was detected on the cathode. The electrode material is MREG-E graphite.

electrolyte. In every case, gas was produced at the cathode and, in most cases, gas was produced at the anode as well, as a result of electrochemical decomposition of water and/or inert electrolytes. With increasing voltage, current was increasing as well. However, beyond some limits by-products were created in a larger amount, which resulted in decrease in efficiency of the process.

4.3.3.3 Up-Scaling

One of the biggest advantages of the electrochemical exfoliation process is the promise of mass production. However, the up-scaling of the process results in new challenges, which are not considered at laboratory-scale testing. In large systems, formerly negligible effects start to play an important role. Among such effects one can list higher current at the same voltage, overheating of the electrolytic cell, necessity of mixing and longer time of electrolysis.

From a technical point of view, up-scaling and performance improvement of the system could include, for example, parallel stacking of electrodes (see Figure 4.11), increasing electrodes mass, or ensuring constant flow of the electrolyte, to mention a few methods.

4.3.4 Postprocessing

The process of electrolysis results in a suspension, which except from GNPs contains electrolyte and sedimented graphite powder. For the reason of obtaining GNPs only, post-exfoliation processes have to be applied.

4.3.4.1 From Electrolyte Solution to Thin Films

After electrolysis, the suspension was transferred to the sonication processing vessel. The anode and the bottom of the electrolytic cell were thoroughly washed with deionized water and transferred to the sonication vessel as well. The vessel was filled with deionized water and the suspension was sonicated for 10 minutes at 60 W. After sonication, the suspension was transferred to

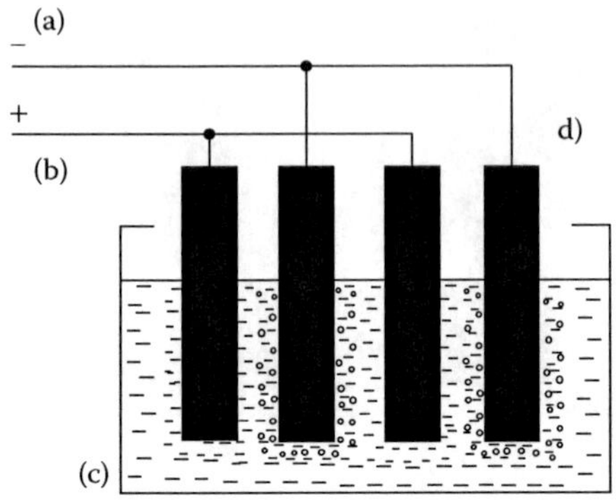

FIGURE 4.11
Scaled-up electrolytic cell with stacked electrodes. (a) Wires connected to cathodes. (b) Wires connected to anodes. (c) Electrochemical cell with an electrolyte. (d) Graphitic electrodes.

centrifugation tubes and centrifuged at 3000 rpm for 20 minutes. The centrifuged suspension was divided into three parts: (1) Top 10 mL, (2) the rest of the suspension, and (3) the precipitate. The top 10 mL of the suspension was diluted to 100 mL with deionized water. This suspension was divided into two parts, volumes of 25 and 75 mL in most cases, and subjected to filtration, which resulted in films of various thicknesses. By varying the amounts of GNP suspension used, films with various thicknesses (from tens of nanometers to several micrometers) could be achieved. However, in the case of film preparation with the use of polycarbonate membrane (pore size of 0.45 μm in diameter), only very thin films (of typically 100 nm) could be prepared. This is the result of pores getting covered by GNPs, which in turn disable the flow of the liquid and thus stop the filtration process. It is possible to evade problems of membrane fouling by changing the order of treatment steps. For instance, if the filtration is performed directly after electrolysis, large agglomerates can be separated first and the solution can be further subjected to the sonication and centrifugation steps. In this way, at least a decrease in the amount of electrolyte residues can be achieved at an early stage of the process.

It was observed that in most cases the GNP films were too thin and unstable to be detached from the membrane in the same way as with carbon nanotube buckypapers. Therefore, films together with membranes were deposited on the substrate; afterward the membranes were dissolved with chloroform and washed away. After the films were dry, they could be detached from the glass substrate to become freestanding films, much like carbon nanotube buckypapers. The GNP films having a thickness 10–100 times lower than that of the average carbon nanotube buckypaper (for buckypapers made of single-walled carbon nanotubes of 30–100 μm [Kaiser et al. 2008] and buckypapers made of multiwalled carbon nanotubes of up to 650 μm [Whitby et al. 2008]) were prepared. Due to the low thickness they exhibited poor mechanical stability.

The GNP suspensions contain residual electrolyte, which preferably should be removed from the sample. In order to do so, exfoliated GNPs were used to prepare thin films, which were washed several times with deionized water.

Some of these films were dissolved again in deionized water to obtain suspensions containing no residues of electrolyte. These suspensions could be used as a source of GNPs for composites.

As an example, it is possible to directly prepare composites with polyelectrolyte serving as a matrix. In such a case, excess polyelectrolyte is introduced to the GNP suspension, which is followed by the evaporation of water. We have prepared such a composite by mixing 5 mL of GNP suspension with 2 mL of PSS water solution. The water was evaporated. For comparison, 2 mL of pure PSS solution was evaporated under the same conditions, resulting in a reference sample. Resultant materials were investigated with SEM. Results of sample comparison are presented in Figure 4.12.

4.3.4.2 Annealing of GNP Films

The GNP films deposited on glass or quartz substrate prepared by the described method were of black color with a gray-brown tone, depending on film thickness and the electrolyte used. Their conductivity dropped dramatically compared to the original graphite. Some of the properties of the graphite were regained by annealing the films at 800°C for 1 hour under argon or nitrogen gas atmosphere.

4.3.5 Characterization of Products

The first product of the electrolysis process is a dispersion of graphite nanoplatelets in electrolyte solution. After filtration a filter cake results, consisting of a functionalized GNP, which can be transferred to glass substrates or washed with water to remove residual electrolyte and dissolved again at will. In the case when the electrolyte and its residues have no influence on the composite properties, the resultant product can be used directly for composite preparation. Composites of PSS and PSS-MREG-E GNPs at low concentrations (approximately 0.01%–0.1%) were prepared by solution processing and subjected to

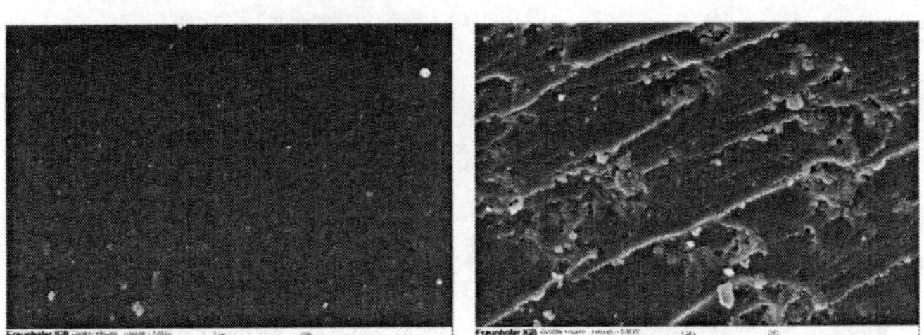

(a) (b)

FIGURE 4.12
The SEM images of cross sections of (a) a pure PSS sample and (b) a composite of GNPs with PSS (GNPs were obtained from MREG-E material in PSS solution electrolyte).

comparative characterization. The SEM images (cf. Figure 4.12) of cross sections show the different fracture mechanics of both materials. The fracture surface of pure PSS polymer is plane and nearly defect-free, whereas contrarily the fracture surface of the composite is rough with flaws around agglomerates of particles.

4.3.5.1 Characterization of Filtrated Products by Raman Spectroscopy and SEM

To find the best combination of electrode and electrolyte, on the one hand, electrode materials were varied with a well-operating electrolyte (PSS) and, on the other hand, electrolytes were varied for a fixed electrode (MREG-E). Raman spectra and SEM images were obtained to characterize the products in relation to their starting materials.

Raman spectra of the products resulting from electrolysis on HOPG, MREG-E, MR75, and MR190 electrodes are presented in Figure 4.13 in relation

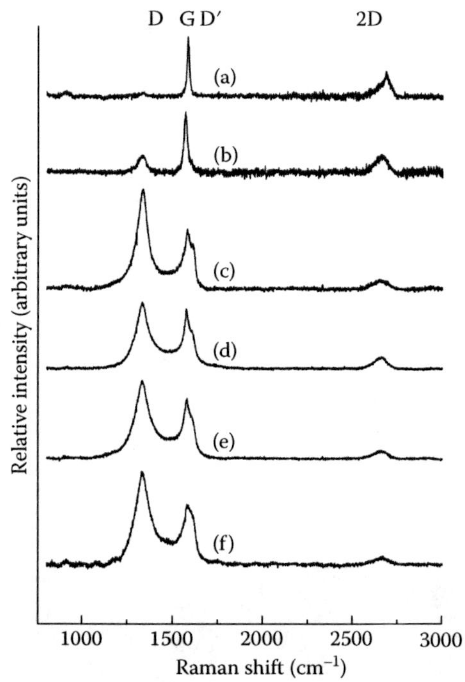

FIGURE 4.13
Raman spectra of electrode materials: Spectra of (a) HOPG and (b) MREG-E and products of exfoliation generated with the PSS polyelectrolyte and four different electrode materials, that is, (c) PSS-HOPG, (d) PSS-MREG-E, (e) PSS-MR75, and (f) PSS-MR190. After electrolysis, the D peak occurs (for HOPG electrode) or increases (all other electrodes) in all products of electrolysis. Larger D peaks are observed for (c) and (d) in relation to the base materials (a) and (b), respectively, indicating functionalization in the exfoliated product. All spectra of products (c–f) show high right shoulder (D′ peak) on the G peak. All spectra show a 2D peak, which indicates rigid graphitic structure in the products.

to the references of the electrodes HOPG and MREG-E. All products show an increase in the D peak intensity and an increase in the shoulder peak D′ on the G peak in relation to their reference. The D peak of electrolysis products increases from MREG-E to MR190, where the shape of the three products is constant. This effect results from the size of exfoliated flakes. The higher the D peak the smaller the flakes (Tuinstra and Koenig 1970). Size of flakes in electrode materials decreases from MREG-E to MR190, which is in agreement with the Raman spectra shown in Figure 4.13. The measured numerical data on peak frequencies, bandwidth, and relative intensities of different peaks is shown in Table 4.4. The increase in D peak intensity is interpreted as an increase in the amount of defects and, thus, a sign of functionalization (cf. Figure 4.14 for a plot of intensities of each product), involving covalent bonds with the anions or a by-product of electrolysis. All spectra show a 2D peak, which indicates rigid graphitic structure. Interestingly, the electrolysis of HOPG resulted in the product with the highest D peak but the lowest bandwidth. This feature indicates that the quality of the electrode influences the electrochemical process. A higher quality of electrode material leads to a higher degree of uniformity of functionalization. The position of the G peak is not altered significantly; any changes might be related to the occurrence of the D′ peak as a shoulder of the G peak. The width of the D peak (cf. Table 4.4) is increasing in relation to the reference. Occurrence of wider G and D peaks together with higher relative intensity of the D peak is normally related to higher disorder in graphite (Kudin et al. 2008).

Using the electrode material MREG-E, three different polyelectrolytes (PAMPS, PSS, and PAA) and a detergent (SD) resulted in products, which were characterized by Raman spectra, as shown in Figure 4.15 and Table 4.5. All used electrolyte materials cause a similar change in the spectra as already seen for the PSS-MREG-E sample. The main difference is the intensity relation of D peak to G peak, increasing in the order PAA, PAMPS, PSS, and SD,

TABLE 4.4

Peak Frequencies υ_D, υ_G, $\upsilon_{D'}$, and υ_{2D}; Bandwidth W_D; and Relative Intensities I_D/I_G, $I_{D'}/I_G$, and I_{2D}/I_G in the Raman Spectra from Figure 4.13 Estimated from Raw Data

Name of Product	υ_D (cm^{-1})	υ_G (cm^{-1})	$\upsilon_{D'}$ (cm^{-1})	υ_{2D} (cm^{-1})	I_D/I_G (arbitrary unit)	$I_{D'}/I_G$ (arbitrary unit)	I_{2D}/I_G (arbitrary unit)	W_D (cm^{-1})
HOPG (reference)	1336	1583	—	2684	0.07	—	0.42	27
MREG-E (reference)	1329	1573	1607	2660	0.28	0.16	0.30	52
PSS-HOPG	1336	1581	1609	2656	1.66	0.75	0.15	70
PSS-MREG-E	1333	1578	1604	2662	1.11	0.65	0.19	73
PSS-MR75	1333	1581	1604	2655	1.30	0.73	0.13	96
PSS-MR190	1333	1581	1607	2659	1.55	0.90	0.12	97

Note: W_D is the width measured at half peak intensity.

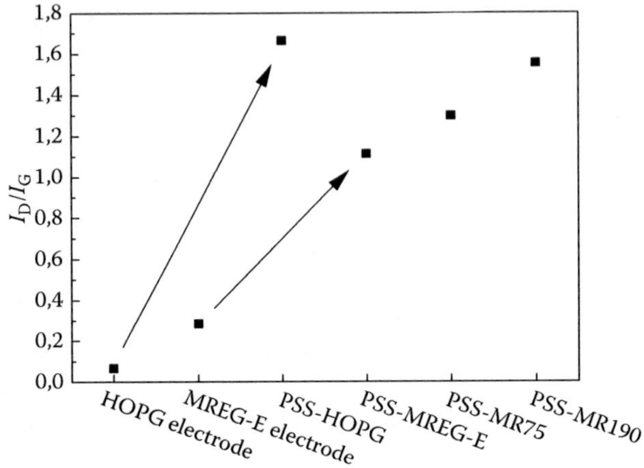

FIGURE 4.14
Intensity ratio I_D/I_G of original electrodes HOPG and MREG-E and the products with PSS, shown with two additional materials PSS-MR75 and PSS-MR190 without reference measurements: The large increase from HOPG to PSS-HOPG hints at a well-functionalized product, that is, PSS-HOPG. In relation to this, the increase from MREG-E to PSS-MREG-E is smaller. The PSS-MR75 and PSS-MR190 products display comparable functionalization levels.

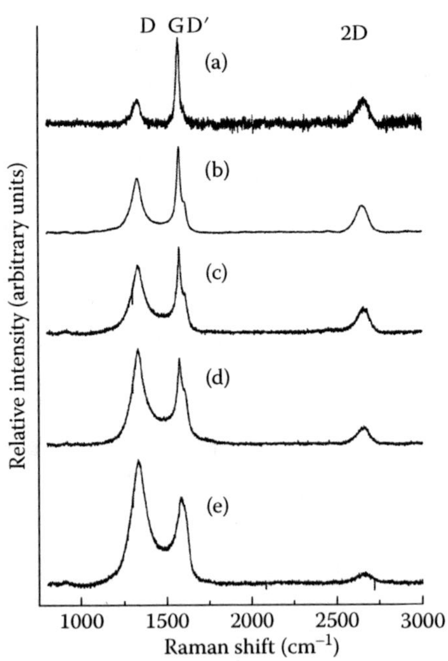

FIGURE 4.15
Raman spectra. Spectra of (a) MREG-E electrode as reference and its products with the following electrolytes: (b) PAA, (c) PAMPS, (d) PSS, and (e) SD. In relation to the original electrode material (a), the increasing height of D peaks signals functionalization and its increasing extend (cf. Figure 4.16).

TABLE 4.5

Peak Frequencies υ_D, υ_G, υ_D', and υ_{2D}; Bandwidth W_D; and Relative Intensities I_D/I_G, I_D'/I_G, and I_{2D}/I_G in the Raman Spectra from Figure 4.15 Estimated from Raw Data

Name of Product	υ_D (cm^{-1})	υ_G (cm^{-1})	υ_D' (cm^{-1})	υ_{2D} (cm^{-1})	I_D/I_G (arbitrary unit)	I_D'/I_G (arbitrary unit)	I_{2D}/I_G (arbitrary unit)	W_D (cm^{-1})
MREG-E (reference)	1329	1573	1607	2660	0.28	0.16	0.30	52
PSS-MREG-E	1333	1578	1608	2662	1.11	0.64	0.19	73
PAA-MREG-E	1329	1575	1609	2650	0.64	0.36	0.30	71
PAMPS-MREG-E	1331	1575	1610	2659	0.79	0.47	0.27	94
SD-MREG-E	1333	1582	1605	2667	1.42	0.90	0.11	91

Note: W_D is the width measured at half peak intensity.

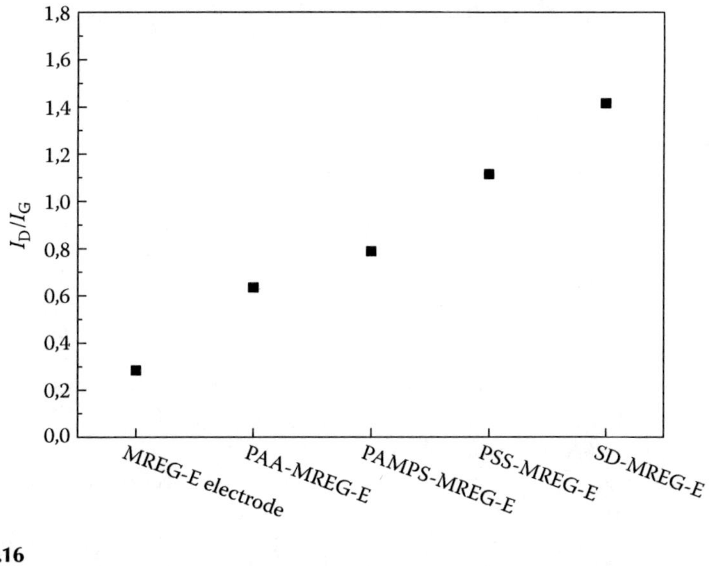

FIGURE 4.16
Intensity ratio I_D/I_G of original electrode and products of exfoliation for MREG-E electrode and the electrolytes PAA, PAMPS, PSS, and SD. Figure shows a clear increase in the depicted order.

as plotted in Figure 4.16. Higher relative intensities might be seen as a sign of there being a higher amount of functionalized defect sites.

It is noted that SEM imaging was used to visualize the shape and thickness of the resulting products. The SEM was performed on the graphene films deposited on the filter membrane. In most cases, the films were frozen in liquid nitrogen and fractured. The resulting cross section was examined. A top view was imaged in some cases to investigate flake sizes and possible inhomogeneities.

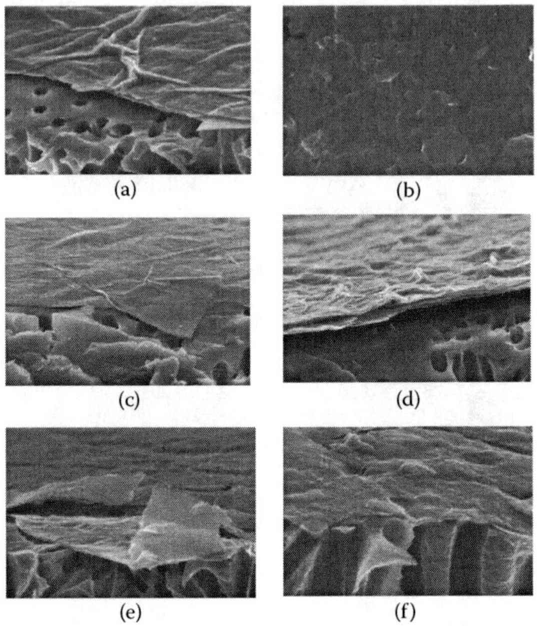

FIGURE 4.17
Comparison of different products filtrated on a polycarbonate filter. Horizontally different electrodes and vertically different electrolytes were used for exfoliation. The cross section displays separate layers of the filtration cake. On the bottom, some pores of the filter can be distinguished, and stacked on the top of the sample are numerous layers of large, folded graphene flakes. The folding indicates thin and, therefore, flexible layers of exfoliated graphene. Estimated thickness of GNP films is 100–400 nm. Note the dependence of the size of folds on the electrode (horizontally) and on the electrolyte (vertically). The flattest GNP film is in PAA-HOPG (c). Generally, HOPG films are flatter than MREG-E films. This is due to graphite crystalline size in materials (see Table 4.2).

Typical thin (100–200 nm) and crumpled films can be seen in Figure 4.17; the figure systematically compares the effect of three different polyelectrolytes, PSS, PAA, and PAMPS, and two electrode materials, HOPG and MREG-E, on the structure of films. These films consist of approximately 2–10 nm thick GNPs. In the case of PAA electrolyte product, a very flat, thin film, almost free of folds is created. For the products resulting from PSS and PAMPS, folds are more common. For all electrolytes, HOPG products show larger flakes than do MREG-E products. This is because there are larger monocrystals of graphite in HOPG than in MREG-E.

4.3.5.2 Characterization of Annealed Products by SEM and Raman Spectroscopy

To remove the functionalization, a sample of PSS-MREG-E was heated in argon atmosphere while measuring Raman spectra, until changes in spectra occurred. The spectra are shown in Figure 4.18 and numerically tabulated in Table 4.6. Until

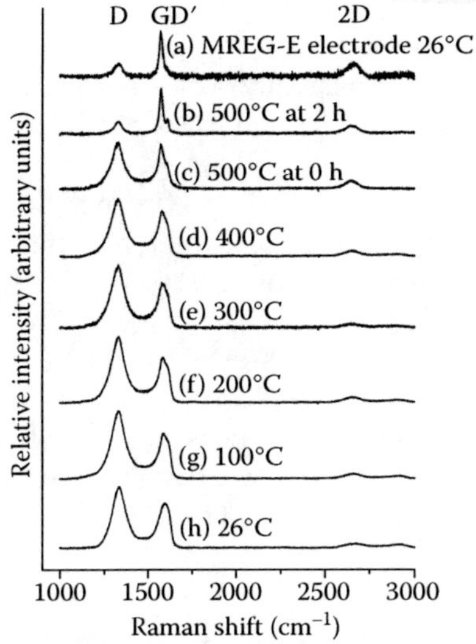

FIGURE 4.18
Raman spectra of annealed PSS-MREG-E sample and of MREG-E electrode as reference, done quasi static from 26°C to 500°C: Until heating to 400°C, only minor changes could be discerned; at 500°C, a fast change of D peak height occurred, related to the removal of functionalization. Finally, in the figure the spectrum shown in (b) is very similar to reference spectrum shown in (a) apart from the clearly remaining D′ peak.

TABLE 4.6

Peak Frequencies υ_D, υ_G, υ_D', and υ_{2D}; Bandwidth W_D; and Relative Intensities I_D/I_G, I_D'/I_G, and I_{2D}/I_G in the Raman Spectra from Figure 4.18 Estimated from Raw Data

Temperature of PSS-MREG-E	υ_D (cm^{-1})	υ_G (cm^{-1})	υ_D' (cm^{-1})	υ_{2D} (cm^{-1})	I_D/I_G (arbitrary unit)	I_D'/I_G (arbitrary unit)	I_{2D}/I_G (arbitrary unit)	W_D (cm^{-1})
26°C	1334	1595	1609	2670	1.36	0.93	0.09	92
100°C	1332	1587	1609	2658	1.50	0.86	0.10	94
200°C	1331	1584	1609	2651	1.45	0.81	0.12	92
300°C	1328	1580	1609	2650	1.39	0.75	0.07	90
400°C	1327	1579	1607	2650	1.27	0.74	0.11	95
500°C at 0 h	1328	1573	1608	2644	1.05	0.56	0.17	78
500°C at 2 h	1329	1573	1612	2649	0.25	0.32	0.16	50
MREG-E electrode reference at 26°C	1329	1573	1607	2660	0.28	0.16	0.30	52

Note: W_D is the width measured at half peak intensity. The steady decrease in υ_G and W_D with ongoing heating until almost the reference data is reached is noted.

a temperature of 400°C is reached, only a small decrease in D peak intensity can be seen. At 500°C, the peak intensity drops in the timescale of 2 hours to a value comparable to the reference. The position of G peak, which revealed high values at the beginning, drops steadily to the reference value. This might be seen as better separation of G and D′ peaks or decreasing D′ peak intensity combined with a thinning of both bands. The D′ peak, visible in the spectra below 400°C only as shoulder, might affect the position estimation of G peak. The spectrum of the product at high temperature (500°C at 2 hours) is distinguishable from the reference spectrum only by the well-separated D′ peak. The D′ band appears in graphitic carbons with relatively low disorder (Vidano and Fischbach 1978; Kurita et al. 2005) and is absent in disordered carbons such as carbon black (Cuesta et al. 1994). In essence, the heat-treated product shows graphite-like properties with a hint of restructuring or reordering, which might be further elucidated in future studies. In Section 4.3.5.3, it is shown that the electrical and optical properties are reverted to graphite-like properties.

4.3.5.3 Characterization of Electrical and Optical Properties of Annealed Products

Sample films resulting from electrolysis, centrifugation, and filtration appear translucent to the eye. After heat treatment, the sample films turned opaque. Correlated with this change in optical properties, there occurs change in conductivity of several orders of magnitude. In Table 4.7, changes in resistance of one to five orders of magnitude can be followed numerically. Thickness of investigated films ranged from 50 to 200 nm. Additional results were measured on films of thickness less than 20 nm, in which annealing in nitrogen forming gas resulted in a decrease in resistance by two orders of magnitude as well as a decrease in transparency by approximately 5%–10%.

Optical and electrical properties change as expected from an initially high-functionalized, low-conductive, and translucent sample to a low-functionalized, high-conductive, and opaque sample, resembling graphitic

TABLE 4.7

Approximate Values of Resistance of Preannealed and Annealed GNP Thin Films under Argon Atmosphere

Material	Resistance Before Heating (kΩ)	Resistance After Heating (kΩ)
PSS-MREG-E	12,600	0.9
PAMPS-MREG-E	>500,000	10
SD-MREG-E	3.33	0.1
PAA-HOPG	12.3	1.4
PAA-MREG-E	22,000	0.2
PAA-MR190	830	4.2

Note: Thickness of measured films is approximately 1 μm. The electrical resistance of the films decreased typically by several orders of magnitude after heating.

behavior again. The GNPs functionalized by polyelectrolytes show a very low conductivity, whereas the relatively high conductivity of SD-MREG-E-GNP might stem from ionic conduction or doping. Remarkably, there is a high conductivity for PAA-HOPG-GNPs, but not for PAA-MREG-E, exhibiting additional effects of the used electrode material (possibly flake size).

4.4 Summary and Future Outlook

By exfoliation of inexpensive electrode materials (such as MREG-E) and comparing the results with costly HOPG, we demonstrate that highly functionalized GNPs can be produced at potentially low cost. The choice of electrode mostly determines nanoplatelet size and graphitic quality, whereas the choice of electrolyte has an effect on the type of functionalization that occurs and process efficiency. Covalent bonds are formed between the electrolyte and the graphene sheets. The product is transparent and typically no longer conductive or loses orders of magnitude in conductivity. Annealing in inert atmosphere as posttreatment partially removes the functionalization and leads to the reconstruction of electrical properties. Thus, annealing can be used to tune the conductivity or functionalization properties to intended values.

Scaling up the process in order to produce bulk quantities of GNPs is currently limited by the duration of electrolysis and the cost of electrolyte that is used up in the reaction. Therefore, further optimization of the processing parameters as well as development of cheap and chemically effective electrolytes will facilitate scaling up of the process in the future. Further, up until now there is no efficient process for the removal of impurities and process residues; this limits the production of high-quality bulk GNPs.

After treatment, further processing of GNP suspensions involves applying deposition techniques such as filtration, spray coating, and dip coating. Films prepared by these techniques could be annealed to retain electrical conductivity. Future work on deposition techniques apart from filtration, which is covered in this chapter, might result in significant time and cost improvements.

Defects in graphite might play an important role in functionalization as reaction centers. For future efforts, we aim at tuning processing parameters so as to control defect concentration and achieve intended effects more systematically. Potential correlations between thickness of the nanoplatelets and size (or molar mass) of the polyelectrolyte may pave the road to optimized, designed electrolytes for exfoliation and appropriate functionalization. Further, other types of electrolytes will be tested to find the most effective ones for introducing various defects.

It is also of interest to analyze the by-products, especially gaseous ones, as they could provide additional value to the electrochemical exfoliation process. Possible gaseous by-products are hydrogen and oxygen.

By optimizing process parameters such as voltage, concentration, and current density, many unwanted by-products might be suppressed, because at high voltages water and electrolytes are decomposed into many concurrent electrochemical reactions. By suppressing concurrent reactions, exfoliation and charge efficiency will improve.

Functionalized GNP might prove valuable in polymer composites as filler, especially if transparency is desirable. It is expected that functional groups greatly improve the incorporation of GNPs into the polymer matrix by increasing the compatibility of both materials and thus providing improved interaction with a positive effect especially on the mechanical properties of the composite material.

In this chapter, it is shown that recent advances in the production of bulk quantities of exfoliated graphene and GNPs further facilitate and add value to the fabrication of graphene–polymer composites.

The electrolytic exfoliation of graphite is yet another method to contribute to this progress. This method allows the simplified production of functionalized GNPs. The oxygen-containing functionalities on graphene sheets facilitate their dispersion in polar solvents, which makes this process convenient for composite applications. The oxygen and hydroxyl functional groups are well suited to form composites with polar polymers such as poly(methyl methacrylate) or PMMA, polyacrylonitrile (PAN), and PAA, giving rise to intimate nanosheet–polymer interactions and a percolated interphase essential to mechanical enhancement (Ramanathan et al. 2008).

Distortions caused by oxygen functionalization and the resultant defects of exfoliation of graphite, as well as the small thickness of GNPs, lead to a wrinkled topology. This nanoscale surface roughness and wrinkled sheet morphology of GNPs is likely to result in enhanced mechanical interlocking with polymer chains and, consequently, afford better interaction and adhesion with the host matrix, thus imparting superior mechanical properties to GNPs.

Further, given the exceptional in-plane properties of graphene, it is anticipated that a more complete exfoliation of graphite toward single graphene sheets and their controlled functionalization will lead to proper dispersion of graphene in a polymer and result in economically viable nanocomposites with excellent properties.

Acknowledgments

We greatly appreciate and express our gratitude for the helpful discussions and support provided by Dr. Viera Skákalová and the whole team of Danubia NanoTech, Bratislava, Slovakia, as well as Prof. Siegmar Roth from Korea University, Seoul, South Korea. Our sincere thanks go to Armin Schulz from the Max Planck Institute for Metals Research for Raman spectroscopy, Stuttgart, Germany; Monika Riedl from Fraunhofer Institute for Interfacial Engineering and Biotechnology IGB, Stuttgart, Germany, for SEM images; and the company Müller & Rössner GmbH for providing us with their graphite materials.

References

Balandin, A. A., Ghosh, S., Bao, W., et al. 2008. Superior thermal conductivity of single-layer graphene. *Nano Lett.* 8: 902–907.

Cuesta, A., Dhamelincourt, P., Laureyns, J., et al. 1994. Raman microprobe studies on carbon materials. *Carbon* 32: 1523–1532.

Ferrari, A. C., Meyer, J. C., Scardaci, V., et al. 2006. Raman spectrum of graphene and graphene layers. *Phys. Rev. Lett.* 97: 187401–187404.

Ferrari, A. C., Robertson, J. 2000. Interpretation of raman spectra of disordered and amorphous carbon. *PRB* 61: 14095–14107.

Kaiser, A. B., Skakalova, V., Roth, S. 2008. Modelling conduction in carbon nanotube networks with different thickness, chemical treatment and irradiation. *Physica* E 40: 2311–2318.

Kudin, K. N., Ozbas, B., Schniepp, H. C., et al. 2008. Raman spectra of graphite oxide and functionalized graphene sheets. *Nano Lett.* 8: 36–41.

Kurita, S., Yoshimura, A., Kawamoto, T., et al. 2005. Raman spectra of carbon nanowalls grown by plasma-enhanced chemical vapor deposition. *J. Appl. Phys.* 97: 104320–104325.

Lee, C., Wei, X., Kysar, J. W., et al. 2008. Measurement of the elastic properties and intrinsic strength of monolayer graphene. *Science* 321: 385–388.

Li, X. L., Wang, X. R., Zhang, L., et al. 2008. Chemically derived, ultrasmooth graphene nanoribbon semiconductors. *Science* 319: 1229–1232.

Liu, N., Luo, F., Wu, H., et al. 2008. One-step ionic-liquid-assisted electrochemical synthesis of ionic-liquid-functionalized graphene sheets directly from graphite. *Adv. Funct. Mater.* 18: 1518–1525.

Meyer, J. C., Geim, A. K., Katsnelson, M. I., et al. 2007. The structure of suspended graphene nanosheets. *Nature* 446: 60–63.

Ramanathan, T., Abdala, A. A., Stankovich, S., et al. 2008. Functionalized graphene sheets for polymer nanocomposites, *Nat. Nanotechnol.* 3: 327–331.

Schedin, F., Geim, A. K., Morozov, S. V., et al. 2007. Detection of individual gas molecules adsorbed on graphene. *Nat. Mater.* 6: 652–655.

Schwandt, C., Dimitrov, A. T., Fray, D. J. 2010. The preparation of nano-structured carbon materials by electrolysis of molten lithium chloride at graphite electrodes. *J. Electroanal. Chem.* 647: 150–158.

Stoller, M. D., Park, S., Zhu, Y., et al. 2008. Graphene-based ultracapacitors. *Nano Lett.* 8: 3498–3502.

Tuinstra, F., Koenig, J. L. 1970. Raman spectrum of graphite. *J. Chem. Phys.* 53: 1126–1130.

Vidano, R., Fischbach, D. B. 1978. New lines in the raman spectra of carbons and graphite. *J. Am. Ceram. Soc.* 61: 13–17.

Wang, G., Wang, B., Park, J., et al. 2009. Highly efficient and large-scale synthesis of graphene by electrolytic exfoliation. *Carbon* 47: 3242–3246.

Weng, W.-G., Chen, G.-H., Wu, D.-J., et al. 2003. Preparation and characterizations of nanoparticles from graphite via an electrochemically oxidizing method. *Syn. Met.* 139: 221–225.

Whitby, R. L. D., Fukuda, T., Maekawa, T., et al. 2008. Geometric control and tuneable pore size distribution of buckypaper and buckydiscs. *Carbon* 46: 949–956.

Zhamu, A., Jang, J., Jang, B. Z. 2009. Electrochemical method of producing nano-scaled graphene platelets. Patent Application Publication, Pub. No. US2009/0026086 A1.

Zhang, Y., Tan, Y.-W., Stormer, H. L., et al. 2005. Experimental observation of the quantum hall effect and Berry's phase in graphene. *Nature* 438: 201–204.

5

Exfoliation of Graphite toward Graphene from Lab to Industry

Weifeng Zhao and Guohua Chen*

CONTENTS

5.1 Introduction

Graphite, the most thermodynamically stable form of carbon, has been extensively investigated in the past decades due to its unique layered structure that makes it highly anisotropic. This naturally abundant material is well known for its stiffness and good electrical and thermal properties, thus offering people a great opportunity for reinforcing polymers in a cost-effective way. In addition, the good powder conductivity of 10^4 S/cm at ambient temperature makes graphite an attractive additive for fabricating conductive

* Correspondence to: G. Chen (hdcgh@hqu.edu.cn)

polymer composites. By changing the content of graphite, the conductivity of polymers can be effectively modified in a wide range between the conductivities of neat polymers and that of the graphite fillers. In addition to excellent electrical conductivity, graphite is also a good solid lubricant and pencil material owing to its anisotropic layered structure. Due to the weak van der Waals-like coupling between graphite layers, the carbon sheets in graphite can easily slide past one another, giving a remarkable lubricant behavior. Graphite flakes have been typically employed in the production of rubber composites to provide conductivity and lubricant properties to the matrix.

It should be noted that conventional graphite powders are usually micrometer-thick flakes that have low reinforcement efficiency. Although other properties, such as electrical and thermal conductivities of polymer composites, increase with increasing graphite concentration, the maximum content of graphite is seriously limited by the poor mechanical properties of the final composites at high graphite filler loading. To yield higher reinforcement efficiency, the layered graphite flakes are expected to be exfoliated into thinner platelets that possess a higher aspect ratio and a larger specific surface area. In fact, the ratio of the surface area to volume for layered materials is inversely proportional to their breadth and thickness, which can be expressed as $2/t + 4/l$, where t is the thickness and l is the side length of a four-sided filler plane (Thostenson et al. 2005). Clearly, the transition of the thickness of layered materials from micrometer to nanometer will yield a dramatic increase in the surface area/volume ratio by three orders of magnitude. To minimize the needed filler concentration and to explore novel properties of graphite composites without compromising mechanical properties, intense research on polymer/graphite nanocomposites has been promoted during the past years (Du et al. 2004; Kim et al. 2007; Lu et al. 2006b; Mo et al. 2005; Wang and Chen 2010; Wang et al. 2005; Weng et al. 2003b; Xiao et al. 2001; Yang et al. 2007; Zheng et al. 2002; Zheng and Wong 2003). This rising surge in activity rapidly in composites research has been being greatly facilitated by the recent discovery of free-standing graphene by A. K. Geim of the University of Manchester, United Kingdom, together with his Russian colleagues, in 2004 (Novoselov et al. 2004).

Graphene is the name given to an atomically thick, two-dimensional (2D) graphite layer composed of sp^2 carbon atoms arranged in a six-numbered-ring plane. With a strictly 2D structure, it has been viewed as the building block of other graphitic carbon forms of different dimensionality. For example, a 3D graphite flake is composed of graphene sheets stacked on top of each other in a …ABAB… sequence, while 1D carbon nanotube is a scrolled graphene sheet with joined edges. Also, both diamond and 0D fullerenes, including five-mumbered rings, can be derived from a section of graphene sheet.

From the viewpoint of polymer composites, among the various carbon materials, graphene is the most efficient additive for polymer reinforcement. This is because of the fact that the graphene sheets not only have an extremely high surface area for a given volume (2630 m^2/g) but they also possess various fascinating properties. The Young's modulus of single-layer graphene reaches as high as ~1 TPa, and its ultimate strength is ~130 GP. Moreover, single-layer graphene has a very high electrical conductivity, up to 6000 S/cm, along with a thermal conductivity of 5000 W/m-K. Compared to carbon nanotubes, the disk-shaped graphene materials can be dispersed in polymers more easily because they are less entangled. These extraordinary properties, in addition to super specific surface area, establish graphene as an excellent candidate to modify polymers. Thus, to maximize the reinforcing efficiency to utilize graphite fillers fully, graphite flakes should be exfoliated giving atom-thick graphenes as sufficiently as possible and dispersed throughout polymer matrix properly.

In recent years, extensive efforts have been done by researchers to develop production methods for graphene sheets. Several routes are available now, such as micromechanical cleavage of graphite using Scotch tape (Novoselov et al. 2004), chemical vapor deposition (CVD) and epitaxial growth (Kim et al. 2009; Zhang and Feng 2010), chemical reduction of graphite oxide (GO) (Stankovich et al. 2007), solvothermal synthesis (Choucair et al. 2009), thermal exfoliation and reduction of GO (McAllister et al. 2007), and liquid-phase exfoliation of graphite by sonication (Gu et al. 2009; Hernandez et al. 2008; Li et al. 2008) or quenching (Tang et al. 2009). Of these approaches, only the top-down methods from GO are currently suitable for polymer composites that require a large amount of graphene.

In early 1997, our group conducted research on the ways of separating/exfoliating graphite by using top-down processing (Chen et al. 1999). Our work involved exfoliation of graphite by using techniques of electrical oxidation, ultrasonication, and wet ball milling. Extensive research on the resulting polymer/graphite nanocomposites was also conducted. Through the electrochemical oxidation method, GO nanoparticles (GNPs) containing single-layered and muitilayered graphene were obtained. Pulverization of expanded graphite (EG) using ultrasonication produced a thinner form of graphite known as graphite nanosheets (GNs) or graphite nanoplatelets (GNPs), which possess nanometer thickness. By using the wet ball milling method, single- and few-layer graphene sheets were attained. In addition, similar to international colleagues interested in single- and few-layer graphene materials, we are now committing ourselves to manufacturing graphene commercially. This chapter focuses on the studies carried out by our group and the developments achieved in our lab on the subjects of graphite exfoliation and polymer nanocomposites reinforced by the graphite nanomaterials produced.

5.2 Exfoliation of Graphite

Graphite is a 3D form of carbon with a sheetlike structure. Within each layer, the carbon atoms are arranged in a hexagonal pattern through σ bonding involving sp^2 hybridization. The layers, on the other hand, are held together in a staggered sequence ...ABAB... by weak van der Waals forces related to the 2pz electron of carbon, which forms a delocalized orbital of π symmetry. Corresponding to the anisotropic layered structure, graphite has different physical properties for in-plane and *c*-axis crystallographic directions. The loosely bound π electrons of high mobility play a dominant role in the conductivity of graphite.

Although having a layered structure similar to the geometry of silicate clay minerals, graphite is different from silicates, as it does not bear any net charge. Therefore, unlike the lamellar silicates, whose exfoliation can be achieved by ion exchange and redox reactions, graphite cannot be exfoliated in the same way. However, due to the weak coupling–layered structure, graphite can be intercalated with certain atoms, molecules, and ions to form graphite intercalation compounds (GIC) by allowing these guest species (called intercalate) to reside between the interlayer spaces of the graphite lattice (Robert et al. 1973; Shioyama 2000; Ubbelohde 1972). Depending on the particular intercalating substance, the bonding between the carbon atoms and the intercalate can be covalent or ionic. According to the bonding type, GICs can be classified into two groups: covalent intercalation compounds and ionic intercalation compounds.

In the first group, the bonding between carbon atoms and the intercalate is covalent, which is favored by the presence of conjugated double bonds within the carbon layers. The typical substance is GO, first prepared Brodie (1859). By treating graphite with strong oxidants such as $KMnO_4$, $KClO_3$, and $NaNO_2$ in concentrated sulfuric acid or its mixture with nitric acid, natural graphite can be heavily oxygenated, giving hydrophilic GO materials. Although possessing a layered structure, analogous to graphite, GO is an insulator, with an electrical conductivity of 10^{-7}–10^{-3} S/cm, depending on the oxygen content (Chung 2002). To restore the electrical conductivity, chemical reduction of the graphene oxide in GO is required.

In the second group, the bonding between carbon atoms and intercalate is partially ionic, which has a certain degree of ionic or polar character. The typical examples are graphite nitrate, graphite bisulfate, and graphite intercalated by alkali metal, halogen, metal chloride, and so on. In the ionic GIC, only a fraction of the interlayer spacing of graphite is filled with intercalates. The carbon and intercalate layers always stack on top of one another in a periodic sequence called staging phenomenon as shown in Figure 5.1. The definite number of graphene layers between two intercalated layers is defined as the stage of GIC, which can be varied from 1 to *n*, depending

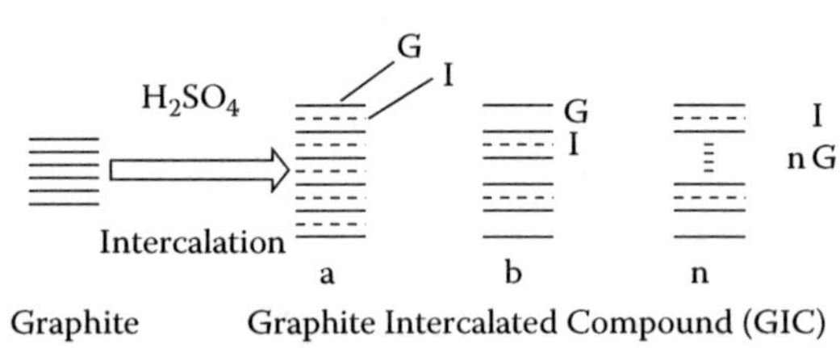

FIGURE 5.1
Schematic illustration of the stages 1 (a), 2 (b), and n (n) of graphite intercalation compounds (GIC).

on the particular intercalate and intercalation situations. The stage usually decreases with an increasing concentration of the intercalate (Sasa et al. 1971).

The intercalates in GICs occupy and expand the gallery spaces of the graphite crystal without disrupting the carbon layers, offering a reliable way to exfoliate graphite.

5.2.1 Exfoliation of Graphite via Electrochemical Oxidation

As mentioned earlier, graphite can be expanded by intercalation. Based on this reasoning, our group was the first to delaminate graphite layers (Weng et al. 2003a) by proposing an electrochemical oxidizing method. Briefly, two identical graphite electrodes were immersed in an aqueous electrolyzer solution as cathode and anode respectively. After a constant potential (DC, 15V) had been applied to the electrodes, the anodic graphite rod became gradually oxidized and exfoliated, giving a homogeneous colloidal dispersion, the color of which varied from brown to black during the oxidization process. Meanwhile, some coarse particles, originating from the graphite anode, precipitated. It was found that the colloidal dispersion was highly polar since the addition of external electrolytes into the dispersion led to flocculation, and the flocs could not be redispersed. Figure 5.2 shows the transmission electron microscopy (TEM) image of the dispersed substance. The micrograph clearly reveals that the GO materials in the supernatant are spherelike nanoparticles with a diameter of around 40 nm. A high-resolution TEM image in Figure 5.3 demonstrates that the GNPs obtained actually consisted of a considerable amount of single- and multi-layered graphene sheets of different sizes, which are assembled in a disorderly manner. Besides exfoliating graphite layers, the electrochemical oxidation also creates oxygen-containing polar groups such

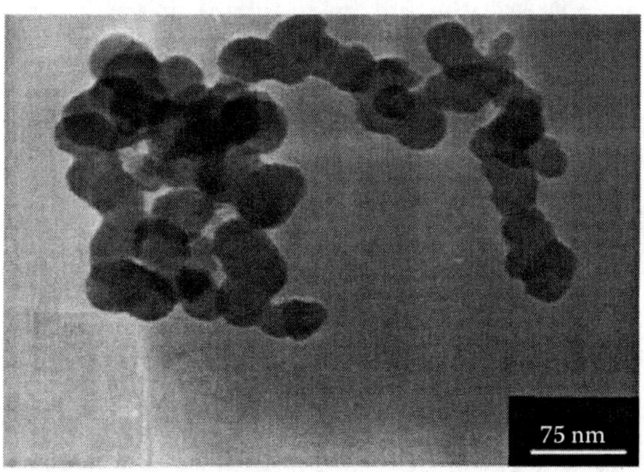

FIGURE 5.2
TEM image of nanoparticles from the supernatant. (Reprinted from Weng, W. G., Chen, G. H., Wu, D. J., Lin, Z. Y., and Yan, W. L. Preparation and characterizations of nanoparticles from graphite via an electrochemically oxidizing method, *Synthetic Met.* 139:22, 221–225, 2003. With permission from Elsevier.)

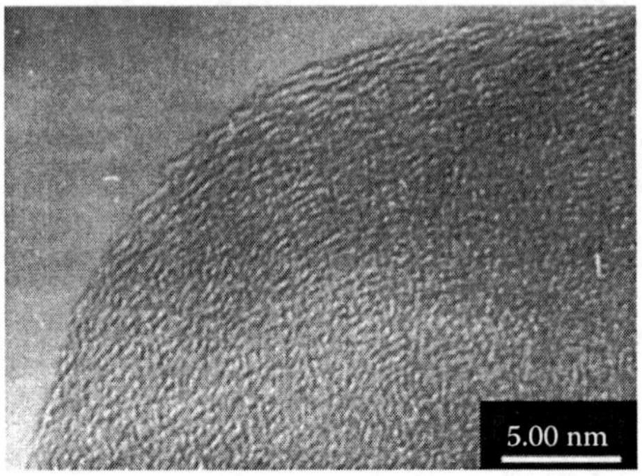

FIGURE 5.3
High-resolution TEM image of graphite oxide nanoparticle. (Reprinted from Weng, W. G., Chen, G. H., Wu, D. J., Lin, Z. Y., and Yan, W. L. Preparation and characterizations of nanoparticles from graphite via an electrochemically oxidizing method, *Synthetic Met.* 139:22, 221–225, 2003. With permission from Elsevier.)

as hydroxyl, epoxide, carbonyl, and carboxylic groups on the carbon sheets, making the GNPs readily dispersible in water.

Similar to the traditional GO, the GNP is an insulator, with an electrical conductivity of only 10^{-6} S/cm. After chemical reduction with hydrazine hydrate in water medium, the electrical conductivity can be improved to 10^{-3} S/cm owing to a certain restoration of graphitic sp^2-bonded network during the reaction (Hu et al. 2009).

For the precipitated coarse particles (Hu and Chen 2010), scanning electron microscope (SEM) images show that they display a loose structure due to the introduction of oxygen to the graphite layers and disruption of van der Waals forces between the graphite layers caused by the electrochemical oxidation (Figure 5.4a). Under higher magnification (Figure 5.4b), it is found that the coarse particles still contain thin graphite sheets with a lateral size of micrometers, which have not been exfoliated completely from the graphite matrix. After being treated by ultrasonication in water, these coarse particles were degraded, resulting in hydrophilic GNPs shown in Figure 5.5, which have good dispersibility in water medium.

5.2.2 Exfoliation of Graphite via Thermal Expansion

GICs, whether covalent or ionic type, are expandable upon heating (Schniepp et al. 2006; Thiele 1938). The mechanism lies in the vaporization of intercalate at a high temperature, leading to an elongation of graphite flake along the *c*-axis. However, the surface area of the exfoliated products varies depending on the source of GIC. Thermal expansion of GOs under inert gas produces thermally reduced graphene oxides that are mainly single layers (McAllister et al. 2007; Schniepp et al. 2006), while exfoliation of ionic GICs, such as alkali- and acid-intercalated graphite, creates EG, which mainly consists of thin graphite platelets with a nanometer thickness.

FIGURE 5.4
(a, b) SEM images of coarse particles from graphite anode. (Reprinted from, Hu, H. W., and Chen, G. H. Electrochemically modified graphite nanosheets and their nanocomposite films with poly(vinyl alcohol), *Polym. Compos.* 31(10): 1770–1775, 2010. With permission of Elsevier.)

Of the various acid-intercalated graphite compounds, graphite bisulfate (graphite-H_2SO_4) is the most popular one for expansion to prepare EG. It can be produced through spontaneous chemical interaction of graphite with a mixture of concentrated sulfuric acid and an oxidizing agent. The reaction taking place between graphite and concentrated sulfuric acid can be expressed as follows (Shioyama and Fujii 1987):

$$\mathrm{n(graphite) + nH_2SO_4 + n/2[O] \rightarrow n(graphite \cdot HSO_4) + n/2H_2O}$$
$$\mathrm{(O): Oxidant, (graphite \cdot HSO_4): GIC}$$

After experiencing an extensive thermal shock at 1050°C or after being heated in a domestic microwave oven, where the heating is excessive, the graphite bisulfate expands irreversibly by up to hundreds of times along the *c*-axis. The expansion ratio, defined as the reciprocal of the apparent density, can be as high as 300. As shown in Figure 5.6, the resulting EG is not compact. Rather, it has a porous wormlike shape with irregular honeycomb

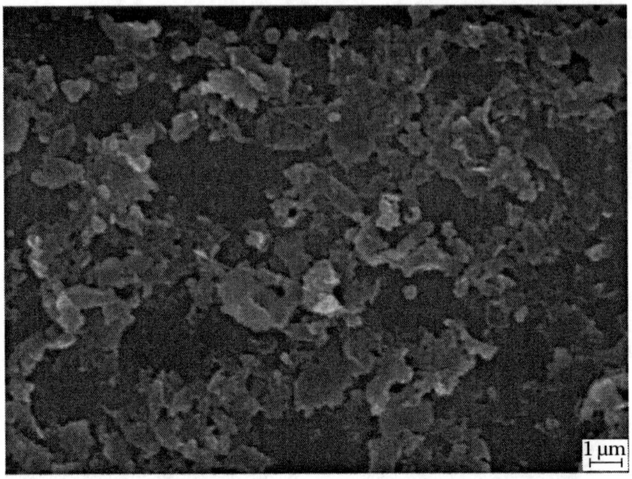

FIGURE 5.5
SEM image of graphite nanoplatelets prepared from the coarse graphite particles. (Reprinted from Hu, H. W., and Chen, G. H. Electrochemically modified graphite nanosheets and their nanocomposite films with poly(vinyl alcohol), *Polym. Compos.* 31(10): 1770–1775, 2010. With permission of Elsevier.)

FIGURE 5.6
(a, b) SEM images of expanded graphite. (Reprinted from Hu, H. W., and Chen, G. H. Electrochemically modified graphite nanosheets and their nanocomposite films with poly(vinyl alcohol), *Polym. Compos.* 31(10): 1770–1775, 2010. With permission of Elsevier.)

networks composed of thin graphite platelets. TEM imaging at a high magnification reveals that the EG worm contains a large number of delaminated graphite sheets that have an apparent thickness in the nanometer range, and some sheets thinner than 10 nm are also observed (Figure 5.7).

With remarkable nanostructures, EG has been proposed as a good candidate to develop polymer/graphite nanocomposite since the late 1990s (Celzard et al. 1996). Due to their puffed-up and porous structure, the enlarged galleries in EG can be easily occupied by monomers or polymer to achieve nanocomposites. Intense research has been conducted during the past years based on various polymers such as poly(methyl methacrylate) (PMMA) (Zheng and Wong 2003), poly(styrene-co-acrylonitrile) (Wang et al. 2005), and polystyrene (PS) (Xiao et al. 2001). All the EG-reinforced composites have a much lower percolation threshold than that of the polymers reinforced by graphite.

A critical issue for polymer/graphite nanocomposite fabrication is ensuring good exfoliation and dispersion of graphite in polymer matrices. Aggregates of fillers are proven to degrade the mechanical properties. Although some success has been achieved for polymer nanocomposites with EG, aggregated graphite may still exist in the composites. This is because GNPs in EG interlock with each other firmly, and the presence of closed pores in EG hinders the penetration of monomers or polymer solution into the EG galleries. As a result, the graphite sheets around such pores cannot be separated by polymer resin, and they will overlap with each other, forming aggregates during processing.

Thus, in an attempt to enhance the nanodispersion of graphite in a polymer matrix, a simple but effective procedure was developed by our group (Chen et al. 2003). By sonicating EG particles in an alcohol solution, isolated GNs with a thickness range from 30 to 80 nm and a diameter of 5 ~ 20μm were produced, as shown in Figure 5.8. The resulting GNs have a thickness that ranges from 30 to 80 nm, possessing a high aspect ratio (width-to-thickness) of around 100–500, which offers advantages in forming a conducting network in a polymer matrix.

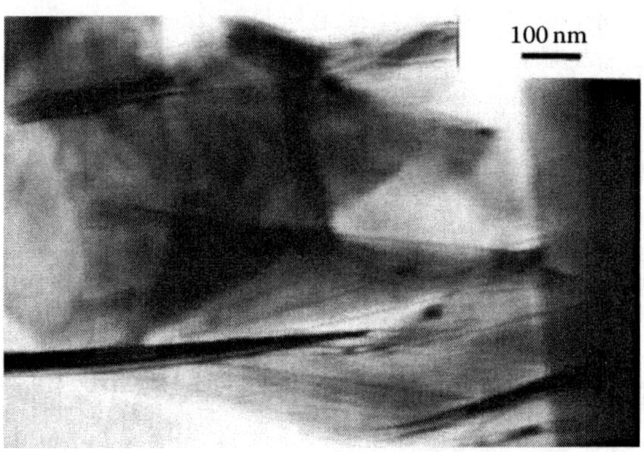

FIGURE 5.7
TEM of EG in slice showing their specific nanostructure. (From Hu, H. W., Chen, L. F., and Chen, G. H., *Mater. Manuf. Process.* 26(4), 2011. With permission.)

FIGURE 5.8
SEM of GNs. (Reprinted from Chen, L., Lu, L., Wu, D. J., and Chen, G. H. Silicone rubber/graphite nanosheet electrically conducting nanocomposite with a low percolation threshold, *Polym. Compos.* 28: 493–498, 2007. With permission of Elsevier.)

The powder had an apparent density of 0.015 g/cm^3, and this is much smaller than the density of the original flake graphite, 2.25 g/cm^3 (Chen et al. 2007).

TEM observation at a high magnification demonstrates that the GNs factually contain a large number of thinner sheets, even mutilayered graphenes, with thicknesses less than 5 nm, as shown in Figure 5.9. This is in agreement with the result from Figure 5.7. From this viewpoint, the GNs should be considered as less-exfoliated graphene stacks that are loosely coupled compared with the interaction between adjacent carbon layers in graphite rather than graphite crystal, even though an appearance of 3D platelet is still maintained. The preserved expansion taken over from EG is caused by the thermal shock expansion of GIC during the preparation of EG. This is understandable. According to the mechanism for the formation of GICs, when the GIC is of stage 2 to stage 5, the thickness of the sheets in their expanded forms should be 0.5–2.5 nm, assuming the thickness of a single carbon layer was 0.5 nm. In view of the detection limitation, it is reasonable to deduce that a number of isolated sheets with monolayers or a few layers must exist among the exfoliated GNs, establishing the GNs as a good parent material for graphene sheets.

Similar to the relationship between single-walled and multiwalled carbon nanotubes, the GNs can be considered as the multilayered form of single-layer graphene. Although the physical properties are somewhat different from the monolayer sheet, GNs possess a large diameter while still preserving the rigidity and conductivity of graphite flakes, offering people a good

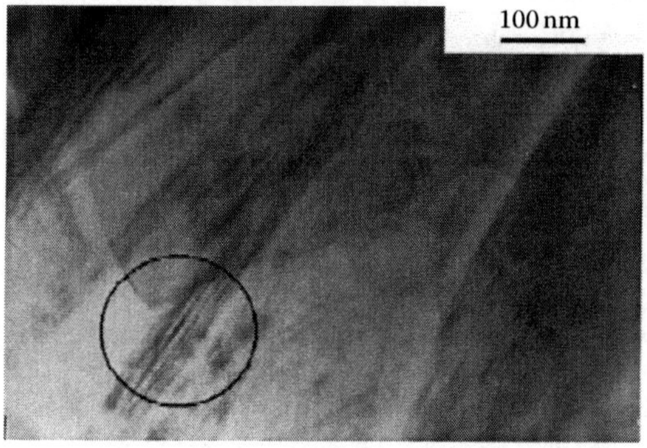

FIGURE 5.9
Specific microstructure of GNs showing partially exfoliated graphene stacks. (Reprinted from Chen, G. H., Wu, D. J., Weng, W. G., and Wu, C. L. Exfoliation of graphite flake and its nanocomposites, *Carbon* 41(3): 619–621, 2003. With permission of Elsevier.)

choice of conductive filler to reinforce polymers. GNs can improve both electrical and mechanical properties of polymers at a concentration substantially lower than conventional graphite, EG, or carbon blacks (CBs).

5.3 Modification of Graphite Nanoparticles and Their Polymer Composites

5.3.1 Modification of Graphite Oxide Nanoparticles with Surfactant

The GNPs produced by electrochemical oxidation discussed in Section 5.2.1 are highly hydrophilic (Figure 5.10a). Even the reduced samples are able to form good dispersions in water (Figure 5.10d). By mixing with certain surfactants, the surface properties of these GO materials can be modified. As shown in Figure 5.10b and e, addition of cetyltrimethylammonium bromide (CTAB) into the aqueous dispersions (a) and (d) makes either GNPs or their reduced samples change from hydrophilic to hydrophobic, leading to flocculation and precipitation of the particles in water medium (Figure 5.10b and e). The precipitate could later be extracted from water by adding toluene, forming a stable, black dispersion in the organic phase, as shown in Figure 5.10c and f.

5.3.2 Polymer/Graphite Oxide Nanoparticles Nanocomposites

Possessing a certain number of oxygen function groups, the electrochemically modified graphite materials were found to be effective in polymer reinforcement (Hu and Chen 2010; Hu et al. 2010). For example, the thermal stability of epoxy resin was enhanced significantly with additions of the GNPs (Figure 5.11a). The degradation temperature of the epoxy composite

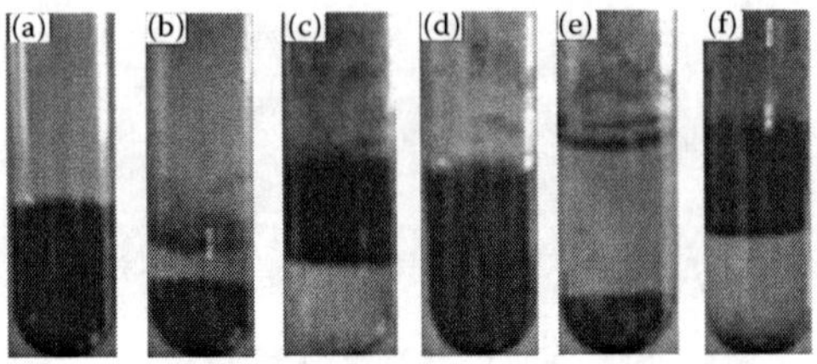

FIGURE 5.10
(a, b, c, d, e, f) Dispersion properties of electrochemically derived graphite oxides before and after organic modification with CTAB. (Reprinted from Hu, H. W., Chen, G. H., Fang, M., and Zhao, W. F. Modification of graphite oxide nanoparticles prepared via electrochemically oxidizing method, *Synthetic Met.*, 59: 1505–1507, 2009. With permission of Elsevier.)

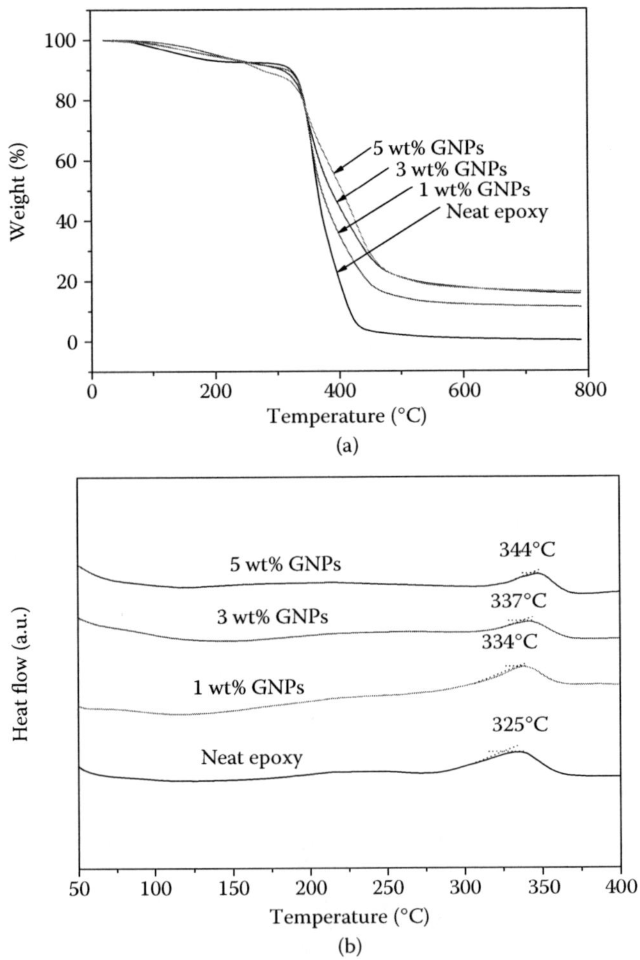

FIGURE 5.11
TGA (a) and DSC (b) patterns of epoxy/GNP nanocomposites. (From Hu, H. W., Chen, L. F., and Chen, G. H., *Mater. Manuf. Process* 26(4), 2011. With permission.)

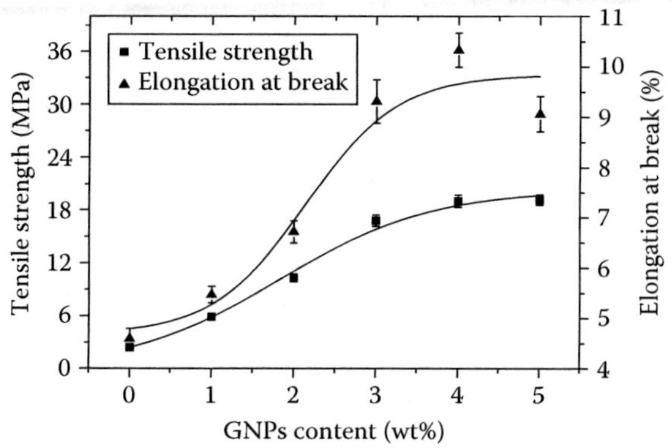

FIGURE 5.12
Tensile strength and elongation at break of epoxy nanocomposites as a function of GNPs content. (From Hu, H. W., Chen, L. F., and Chen, G. H., *Mater. Manuf. Process* 26(4), 2011. With permission.)

containing 5.0 wt% GNPs reaches 334°C, 19°C higher than that of the neat epoxy (Figure 5.11b). Moreover, an interesting synergistic effect of reinforcing and toughening of the GNPs on epoxy is observed, probably originating from the good filler/epoxy interactions related to the introduction of oxygen groups by the electrochemical oxidation (Figure 5.12).

The enhancement of thermal stability of polymers by the GO materials was also observed for poly(vinyl alcohol)-based composites. PVA loaded with 5 wt% electrochemically modified GNPs has an onset degradation temperature of 45°C, and this is higher than that of the neat PVA (Hu and Chen 2010).

5.4 Polymer/Graphite Sheets Nanocomposites

During the past years, polymer/graphite nanocomposites have been widely studied to develop polymer materials for applications where good electrical and mechanical properties are often required. Similar to the silicate clay minerals, exfoliation of graphite layers is needed to achieve nanocomposites. How to reach a good exfoliation and dispersion of graphite in polymer matrices is the most critical issue for the polymer/graphite nanocomposite production. The layered graphites are expected to be exfoliated giving atom-thick graphenes as sufficiently as possible and properly dispersed throughout polymer matrix. A combination of sufficiently exfoliated graphite layers, good filler dispersion, and a strong graphite/matrix interfacial interaction will allow graphite to provide the maximum reinforcement efficiency at a given loading level.

5.4.1 Polymers Reinforced with EG

Thanks to the remarkable expansion ability of graphite bisulfate, EG can be easily derived from it. The nanostructured EG particles provide people great opportunities to incorporate GNPs in polymer matrixes to achieve polymer/graphite nanocomposites. Since the early 1990s, when EG was proposed as a good GNP precursor to develop polymer/graphite nanocomposites, intense researches have been conducted for developing and optimizing the processing method to achieve good exfoliation and dispersion of graphite and strong graphite–matrix adhesion at the interface to improve the strength and conductivity of the final composites.

Based on thermally exfoliated graphite, polymer nanocomposites can be prepared by direct incorporation of these nanostructured fillers into polymers. The methods of filler incorporation can be classified into three classes: in situ polymerization, solution compounding, and melt blending, depending on the state of organics during blending process.

In situ polymerization is also known as intercalation polymerization. The compounding process involves inhabitancy or "intercalation" of monomer into the interplanar spacing of layered fillers and a subsequent in situ polymerization of the monomers in the presence of the intercalated fillers. Compared with the methods of solution compounding and melt blending, in situ polymerization gives more satisfactory dispersion of fillers and stronger interactions between the reinforcements and the polymeric phase. Therefore, composites fabricated by in situ processing tend to display better mechanical properties and a lower percolation threshold than those composites made by melt mixing and other ex situ methods. Through in situ polymerization, we have successfully incorporated EG particles into PMMA (Chen et al. 2001), poly(vinyl chloride) (PVC) (Chen et al. 2001), poly(styrene-methyl methacrylate) (Chen et al. 2001), and PS (Chen et al. 2001) respectively in the form of delaminated GNPs that possess a thickness of 10–30 nm. Because of the occurrence of delamination during monomer polymerization, the EG worms are much more efficient for polymer reinforcement compared to the conventional micrometer-size graphite particles. The graphite nanocomposites obtained always display a markedly low percolation threshold and high electrical conductivity due to the high aspect ratio (ratio of diameter to thickness) of the GNPs embedded in the polymer. However, this approach is not feasible in real industrial production. This is because in situ polymerization usually consumes electrical energy as heating due to the need to promote polymerization of monomers. Also, the active center responsible for the growth of macromolecular chains can be easily terminated by some reactive species such as radicals and ions derived from EG, making the resulting polymers have a low molecular weight and broad molecular weight distribution, which will seriously deteriorate the properties of the final composites.

Different from in situ polymerization, melt compounding starts from polymers and is a traditional technique most commonly used for polymer blending in the plastics industry. To meet industrial applications, we therefore managed to disperse EG into polymers by using melt extrusion (Li and Chen 2007; She et al. 2007). EG particles were first modified into an electrically conducting masterbatch with certain polymer carriers such as high density polyethylene (HDPE). The plastic concentrates were then introduced into the chosen matrix (e.g., HDPE) through melt extrusion. Experimental results suggested that melt mixing employing a masterbatch could give more intensive flake delamination of EG and better nanosheets dispersion than direct melt extrusion. As a result, the EG nanocomposites prepared by the masterbatch filling method exhibit a lower percolation threshold compared with the composites made by direct melt blending, and an enhancement in tensile strength was observed. However, the overall enhancement in mechanical properties was not found to be exciting while using EG as the reinforcement phase. The mechanical properties of the final composites strongly depend on the degree of dispersion of EG and the compatibility between the masterbatch and the polymer matrix. A lower interfacial interaction among composite components will result in a sharp decrease in mechanical behavior.

Melt blending is unquestionably a feasible technique to fabricate polymer nanocomposites on an industrial scale. Our studies confirmed the advantages of masterbatch processing in polymer/graphite nanocomposites production. A broad future in industrialization was then established. How to ensure a constant homogeneous dispersion of GNPs in polymers by using the existing melt blending devices is still a challenge. Working toward improvements either in processability or in compatibility of the polymer/graphite nanosystems is recommended for the future.

5.4.2 Polymers Reinforced with GNs

Being different from the long, porous, distorted cylinder EG in which constitutive nanosheets stick to each other, GNs are isolated graphite platelets with a nanometer-sized thickness. The worm features of EG are completely lost in GNs. Therefore, compared with EG particles, the shapely GNs can be dispersed into polymers much more favorably. The good homogeneity of the dispersion of GN in polymer matrix can be ensured with the help of ultrasonication. Due to their high aspect ratio between the diameter and the thickness, GNs possess remarkable advantages in forming conductive network in a polymer matrix, leading to nanocomposites with very low values of the percolation threshold. The polymer/GN nanocomposites also exhibit many attractive properties that are very important for both theoretical and practical reasons.

Through in situ polymerization, GNs have been successfully incorporated into PMMA (Chen et al. 2003), PS (Chen et al. 2003), unsaturated polyester (Lu et al. 2006), nylon 6 (Weng et al. 2004), and epoxy resin (Lu et al. 2006). As some enlarged spacing is still maintained in the thinner sheets, monomers/oligomers are able to intercalate into the nanospace in GNs, further facilitating the exfoliation and dispersion of graphite flakes, as shown schematically in Figure 5.13. The percolation values of the resulting nanocomposites are listed in Table 5.1.

The conduction behavior of the polymer/GN nanocomposites is nonlinear. Several theories can be applied to interpret the electrical conductivity, namely, percolation theory, mean field theory, and excluded volume theory. For example, we found that the low volume concentration for percolation can be interpreted by the mean field theory and the excluded volume theory, indicating that the morphologies of the GNs with a high aspect ratio plays an active role in minimizing the Pc value (Lu et al. 2006).

Apart from in situ polymerization, solution compounding is also an effective approach to disperse GNs into polymers. Using this method, we prepared silicon rubber (SR)/GN nanocomposite by blending liquid SR with GNs in hexane medium (Chen et al. 2007a). A uniform dispersion of GNs into SR was accomplished with the help of ultrasonication. Due to the high aspect ratio of GNs, the nanocomposite displays a very low percolation threshold, 0.9 vol%, which is far less than 5.3 and 7.0 vol% of the composites made

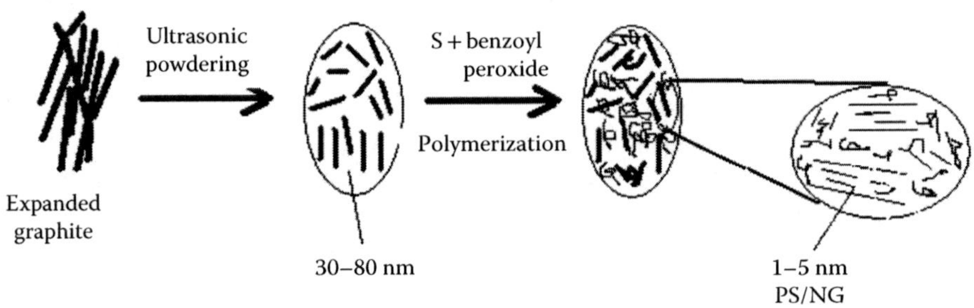

FIGURE 5.13
A schematic illustration of the process for the formation of polystyrene/GN nanocomposite. (Reprinted from Chen, G. H., Wu, C. L., Weng, W. G., Wu, D. J., and Yan, W. L. Preparation of polystyrene/graphite nanosheet composite, *Polymer* 44: 1781–1784, 2003. With permission of Elsevier.)

TABLE 5.1

Percolation Values of Polymer/GN Nanocomposites

Polymer/GN Nanocomposites	Compounding Method	Percolation Value (Pc)
Unsaturated polyester/GNs	In situ polymerization	Pc = 0.64 vol%
PMMA/GNs	In situ polymerization	Pc = 0.31 vol%
PS/GNs	In situ polymerization	Pc = 0.44 vol%
Nylon 6/GNs	In situ polymerization	Pc = 0.74 vol%
Epoxy/GNs	In situ polymerization	Pc = 1.5 vol%

Sources: Data from Chen, G. H., Wu, C. L., Weng, W. G., Wu, D. J., and Yan, W. L., *Eur. Polym. J.*, 39 (12), 2003; Chen, G. H., Wu, C. L., Weng, W. G., Wu, D. J., and Yan, W. L., *Polymer*, 44 (6), 2003; Lu, W., Lin, H. F., and Chen, G. H., *Polymer*, 47, 2006; Lu, W., Weng, J. X., Wu, D. J., and Chen, G. H., *Mater. Manuf. Process*, 21, 2006.

with 8000 and 2000 mesh graphite respectively. What is most interesting is that when the GNs loading reaches 1.36 vol%, which is near the percolation threshold value, the resistivity of the SR/GN nanocomposites increases markedly with increasing loading pressure. The positive pressure coefficient (PPC) effects caused by compressive stress–induced deformation of the conducting network has a good repeatability in the finger-pressure range from 0 to 0.8 MPa, which can be regarded as the main advantage of this new kind of nanocomposite.

Piezoresistivity was also observed in HDPE/GN nanocomposites prepared by melt blending GNs with HDPE resins on a two-roll mill (Lu et al. 2005). The resulting HDPE/GN nanocomposites exhibit both a negative pressure coefficient and a PPC of resistivity with the PPC effect much more pronounced. Under a fixed pressure, the resistivity of the composites varies as a function of time owing to the deformation of the polymer matrix caused by the creep of polymer chains under an external load (Chen et al. 2005a; Lu et al. 2006).

Similar to PTC/NPC effect, the arrangement of the conductive networks in the polymer matrix can also be altered by external heating. Due to the temperature-dependent expansion of the polymer matrix, the HDPE/GN nanocomposites consequently show a sharp increase in electrical resistivity at a transition temperature, as shown in Figure 5.14a (Chen et al. 2005b). This phenomenon is defined as positive temperature coefficient (PTC) effect. Following the PTC transition, the resistivity decreases with increasing temperature, which is known as negative temperature coefficient (NTC) effect. The ratio of the peak resistivity to the room temperature resistivity is defined as the PTC intensity. As shown in Figure 5.14a, the PTC intensity decreases with increasing GNs content. The HDPE/GN nanocomposites exhibit higher PTC intensity than the HDPE composites reinforced by traditional graphite powder of 2000 mesh in the vicinity of the respective percolation thresholds (Figure 5.14b).

The properties of polymer/GN composites depend not only on the nature of the components but also on the GNs alignment in the matrix, because the GNs are anisotropic. It is expected that the orientation of GNs in a certain direction in the polymer matrix could result in composite materials with unique properties. We have accomplished the alignment of GNs under shear forces as well as under electric and magnetic fields (Chen et al. 2007; Lu et al. 2005).

Owing to the particular high-aspect-ratio morphology and the flakelike shape, the GNs could be preferentially arranged along the flow direction under shear stress. For example, when blending GNs with HDPE resins on a two-roll mill, the GNs tend to get reoriented with their main plane parallel to the plane of the resulting composite plate, leading to polymer materials with significant conductive anisotropy. As shown in Figure 5.15, the composite resistivity in the plane of the plate is much lower than that along the thickness direction.

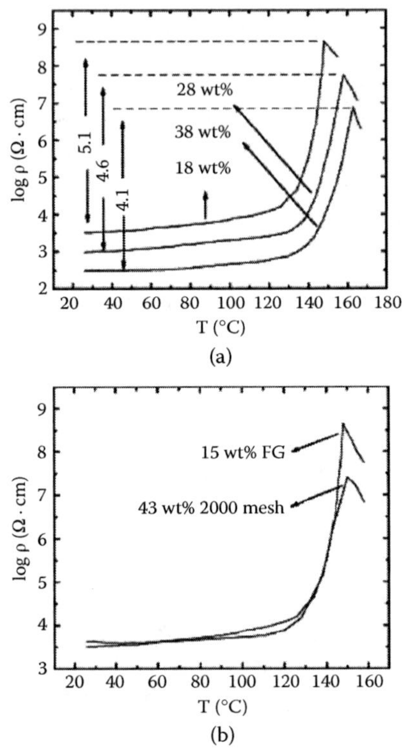

FIGURE 5.14
(a, b) Temperature dependence of electrical resistivity of HDPE/GN nanocomposites. (With kind permission from Springer Science+Business Media: Chen, G. H., Lu, J. R., and Wu, D. J. PTC effect of polyethylene/foliated graphite nanocomposites, *J. Mater. Sci.* 40(18): 5041–5043, 2005.)

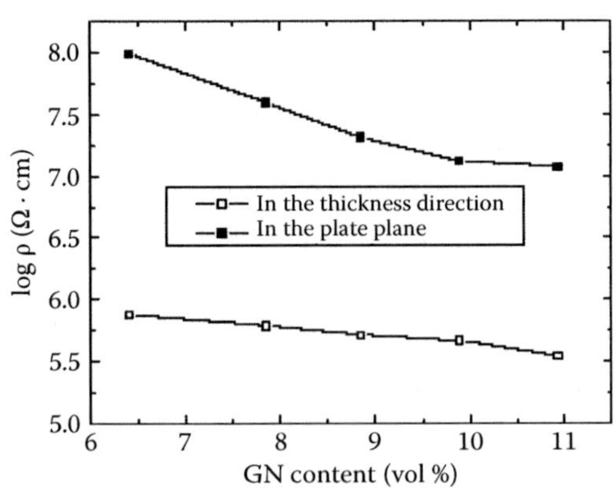

FIGURE 5.15
Anisotropic resistivity of HDPE nanocomposite with GNs oriented along shear forces. (From Lu, J. R., Weng, W. G., Chen, X. F., Wu, D. J., Wu, C. L., Chen, G. H. Novel piezoresistive material from directed shear-induced assembly of graphite nanosheets in polyethylene, *Adv. Funct. Mater.* 15: 358–1363, 2005. Copyright Wiley-VCH Verlag GmbH & Co. KGaA. Reproduced with permission.)

Applying an electric field to the dispersion of GNs in liquid unsaturated polyester resin, the randomly arranged GNs begin to undergo a polarization. Since the GNs are of flakelike shape having crystalline anisotropy, polarization parallel to the plane of the sheet is much larger than that perpendicular to the fake. This polarization moment leads to a field-induced torque applied on the dispersed GNs, orienting GNs against the viscous drag of the surrounding medium to align along the electric field direction. In addition, coulombic attractions are generated between the oppositely charged ends of different sheets, and dielectrophoresis then occurs. Under the external inducement, the randomly dispersed GNs in the liquid prepolymer begin to reorient and align with their 002 planes parallel to the electric field. By applying the electric inducement on the compounding suspension until the mixture becomes solidified, the alignment was frozen, giving birth to an ordered nanocomposite of GNs. Due to the GNs orientation, the optical properties of the electrical field–induced composite film showed a significant improvement in light transmittance in the range of visible wavelength from 400 to 750 nm, as shown in Figure 5.16 (Wang et al. 2008). Moreover, resulting from dielectrophoresis, the embedded GNs in the polymer matrix tend to form conductive connections in the electric field direction. Therefore, compared with the composite reinforced by randomly arranged GNs, the ordered composites possess a lower percolation threshold along the electric field direction, as shown in Figure 5.17 (Wang et al. 2007).

By depositing magnetite nanoparticles onto the GNs surface, we conveyed notable magnetic functions to GNs. Combining the magnetic function of

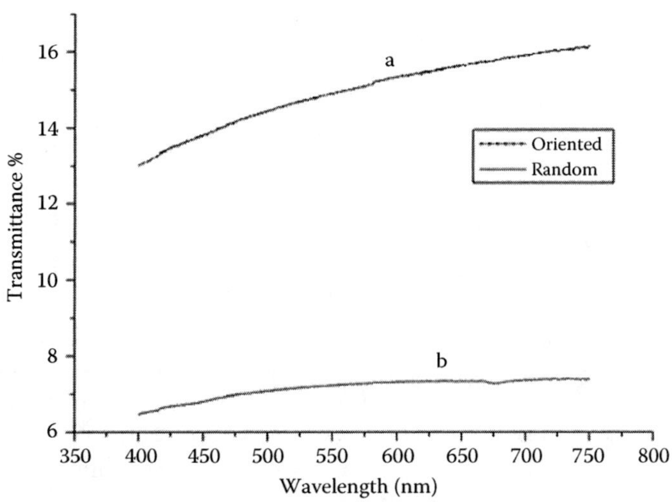

FIGURE 5.16
Transmittances of unsaturated polyester resin composite films with oriented (a) and randomly dispersed (b) GNs. (Reprinted from Chen, G. H., Zhao, W. F., Tang, H. T., and Wang, H. Q. Preparation and surface characterization of highly ordered polymer/graphite nanosheet composites, *Mater. Manuf. Process* 22, 2007. With permission from Elsevier.)

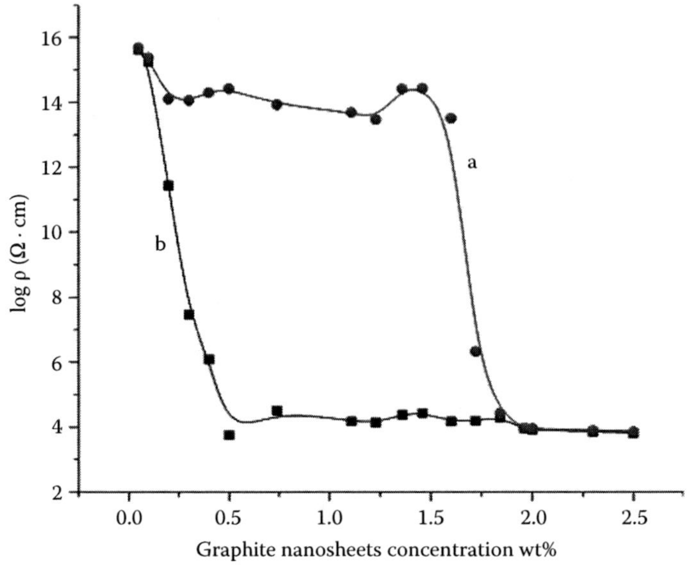

FIGURE 5.17
Percolation threshold of the unsaturated polyester resin/GN composites without (a) and with (b) electric field inducement. (Reprinted from Wang, H. Q., Zhang, H. Y., Chen, G. H. Preparation of unsaturated polyester/graphite nanosheet conducting composite under electric field, *Compos. Part A: Appl. Sci. Manuf.* 38: 2116–2120, 2007. With permission of Elsevier.)

nano-magnetite and an anisotropic structure of graphite, the functionalized GNs possess a remarkable magnetic susceptibility anisotropy, which makes it possible for them to be oriented under an external magnetic field. Epoxy composites with aligned GNs were then fabricated under a magnetic field (Zhao et al. 2006). The alignment of GNs results in interesting anisotropies in the final epoxy composite.

5.5 Modification of GNs

Compared with the conventional micrometer-size graphite particles, GNs have been proven much more efficient in reinforcement. When GNs are incorporated into polymers properly, they are capable of imparting interesting electrical properties to insulating polymeric materials. With the benefit of high aspect ratio of up to 250, GNs tend to more easily form conducting networks in polymers leading to electrically conductive composites with a low percolation threshold. However, the inherently limited conductivity of the GNs limits further increase in the composite conductivity. To overcome this obstacle, metallization of graphite surface is deemed to be an effective way to improve the conductivity of GNs and endow them with additional metallic characteristics.

Therefore, we coated GNs with silver nanoparticles by electroless deposition (Sun et al. 2008). A uniform silver film, about 50 nm thick, was obtained on GNs surfaces with help of 3-aminepropyltrimethoxysilane (APTMS), which acts as an encapsulator to limit nano-silver growth and aggregate formation. The silver-coated GNs exhibit excellent conductivity of up to 10^6 S/m, which is equivalent to the conductivity of silver powder.

Through the electroless plating method, we also prepared nickel-coated GNs by using polyvinylpyrrolidone (PVP) to control the size of nickel particles attached onto GNs surfaces (Li et al. 2010). Hysteresis loops demonstrate that the nickel-coated GNs are hard magnetic materials with saturation magnetization of about 0.6 emu/g and 105 Oe of coercive force.

In addition to the metallization modification, we have also treated GNs on a planetary mill through wet ball milling in ethanol medium (Li and Chen 2009). Thirty hours of the ball milling treatment gives GNs surface more defects and phenolic groups compared to that of EG and neat GNs, meaning that the modified GNs can be used as an effective adsorbent to remove 1,2-dichlorobenzene from water. The adsorption capacity reaches 2.83 mg/g, which is comparable to that of carbon nanotubes.

5.6 Exfoliation of Graphite through Ball Milling

Thermal expansion of acid-intercalated graphite is unquestionably a usable method to exfoliate graphite. The resulting products, either EG or GNs derived from GIC, are good fillers for polymer nanocomposites. Even without complete exfoliation, they can improve electrical conductivity of polymers at substantially smaller loading than conventional graphite flakes. However, although possessing a nanometer thickness, the GN materials are very different from single- and few-graphene platelets, because they are still composed of hundreds of single carbon layers held together by van der Waals-like forces. This makes GNs themselves similar to graphite flakes in electric or mechanical properties.

In recent years, extensive efforts have been made by researchers to develop production methods for graphene sheets (Choucair et al. 2009; Gu et al. 2009; Hernandez et al. 2008; Kim et al. 2009; Li and Chen 2009; McAllister et al. 2007; Novoselov et al. 2004; Stankovich et al. 2007; Tang et al. 2009; Zhang and Feng 2010). For the realization of practical applications, however, one must overcome a great obstacle: it is very difficult to produce graphene materials of good crystal quality in large quantities, especially in terms of industry scale and value for commercialization. To increase the supply of graphene sheets and to satisfy their industrial applications, new routes with an improved monolayer graphene yield that suit industrial production are strongly expected.

Starting from GNs, we managed to adopt ball milling to exfoliate thick graphite flakes mechanically into graphenes in a wide variety of organic solvents including ethanol, formamide, N′,N-dimethylformamide (DMF), tetramethyluren (TMU), tetrahydrofuran (THF) acetone, and *N*-methylpyrrolidone (NMP) (Zhao et al. 2010a, 2010b). To avoid any intense shock stress that will cause defects in graphene plane, our wet ball milling process was carried out on a planetary mill that mainly exerts shear forces on the materials, and the rotating tray was controlled at a low speed to further limit the shock forces applied by the ball impact. Due to the weak van der Waals-like coupling between graphite layers, the graphene sheets in GNs can slide and then separate easily with respect to one another under the applied shear forces, giving rise to isolated graphene sheets in the solvent medium in situ, as illustrated in Figure 5.18.

After removing large particles from suspensions prepared by using an intensive centrifugation procedure, single- and few-layer graphene sheets with desirable crystal structure were obtained. These were homogeneously and stably suspended in a good solvent medium such as DMF, which possesses a surface tension value close to 40 mJ/m^2. With the help of carefully selected solvents, the graphene yield was able to reach more than 32.0 wt% without any recycling; this is listed in Table 5.2. As a common industrial technique, the wet ball milling method can be easily scaled up into mass production. It, therefore, holds great promise to become a realizable way for industrial production of graphene. Furthermore, the ability to provide stable suspensions of single-layer graphene in organic solvents with such a high concentration should facilitate the processing of graphene materials remarkably. In particular, processing of the graphene materials, such as film and paper depositions, surface modifications and functionalization chemistry of the edges, and defects of graphene sheets, is best conducted in the liquid phase. We believe that both theoretical investigations and practical

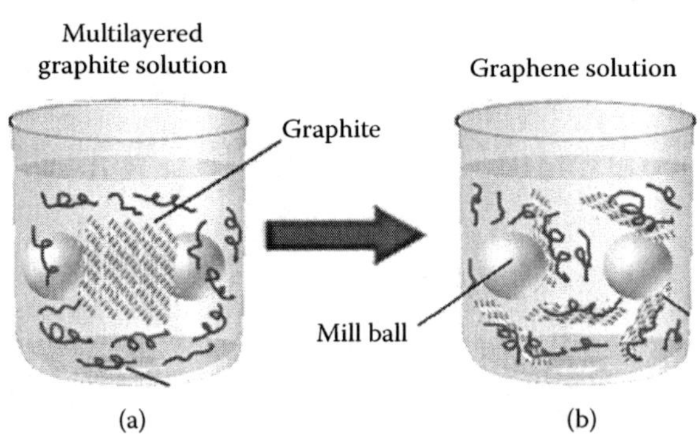

FIGURE 5.18
(a, b) Schematic illustration for the creation of graphene from multilayered graphite by using wet ball milling.

TABLE 5.2

Fraction of Graphene Remaining after Centrifugation in Different Solvents

	Remaining after 4000 rpm of Centrifugation		
Solvent	**Concentration (μg/mL)**	**Surface Tension (mJ/m²)**	**wt% Remaining**
DMF	96.67	35.2	38.67
TMU	88.28	34.7	35.31
NMP	88.9	41	35.56
Acetone	66.6	23.7	26.64
THF	76.03	26.4	30.41
Formamide	3.67	58.35	1.468
Ethanol	10.32	22.27	4.128

applications related to graphene, such as functionalization chemistry of the sheet edges and graphene surface modification, will benefit a lot from this bulk exfoliation technique.

5.7 Graphene Industries

As discussed previously, GNs, the multilayered form of single-layer graphene, possess remarkable advantages in forming a conductive network in a polymer. Therefore, the percolation transition of polymer composites can be easily attained with only a slight amount of GNs. GNs combine low cost with excellent mechanical, electrical, and thermal properties equal to pristine graphite. It should be an ideal alternative to CB most commonly used in recent polymer modifications. There is little doubt that the polymer materials filled with GNs will be superior to the traditional polymer composites filled with micron-sized natural graphite powders or CB, and they have tremendous potential to replace current conductive polymer composites that are commercially available. To confirm this, we prepared HDPE/GN and HDPE/CB nanocomposites by direct melting mixing and carried out a typical comparison study on their electrical, mechanical, and melt flow behaviors (Wang et al. 2010). As expected, the HDPE/GN nanocomposites displayed a much lower percolation threshold (6 wt%) than that of HDPE/CB nanocomposites (22 wt%), as shown in Figure 5.19. For the mechanical properties, a marked increase in Young's modulus with increasing filler content was observed for both GN and CB nanocomposites. With an increase in GNs concentration, the tensile and impact strength of HDPE/GN nanocomposites increased, especially at a low GNs loading. Then a sharp reduction took place when the GNs content was raised to a high value. In contrast, the tensile strength of HDPE/CB nanocomposites increased gradually and their impact strength decreased at the same filler content. The reduced strengths at high filler content for the GN composites are closely related to the inevitable aggregation of GNs at a high concentration.

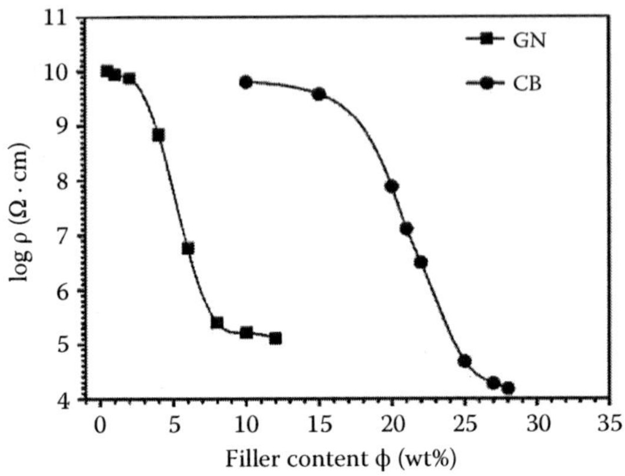

FIGURE 5.19
Electrical conductivities of HDPE reinforced by GNs and CB. (From Wang, L. W., Hong, J. B., and Chen, G. H., *Polym. Eng. Sci.* 50(11), 2010. With permission.)

TABLE 5.3
Melt Flow Rate of HDPE Composites with GNs and CB

GN Melt Flow Rate (MFR) of HDPE/GN		CB MFR of HDPE/CB	
Content (%)	**g/10 minutes**	**Content (%)**	**g/10 minutes**
0.5	0.912	10	0.466
1	0.874	15	0.178
2	0.801	20	0.121
4	0.738	21	0.104
6	0.557	22	0.093
8	0.483	25	0.052

Source: Wang, L. W. et al., *Polym. Eng. Sci.*, 50, 2176–2181, 2010. With permission.

Melt flow rates of the GN- and CB-reinforced HDPE composites with different filler concentration are listed in Table 5.3. With a filler concentration at the percolation threshold 6 and 22 wt% for HDPE/GNs and HDPE/CB respectively, the melt flow rate value of HDPE/GN composite is up to six times greater than that of HDPE/CB system. The much better melt flow property suggests that the HDPE/GN nanocomposite has better processability, which will be favored for future practical applications.

In fact, like EG and most nanomaterials, with a nanometer-scale thickness, the GNs hold great tendency to accumulate/restack when they are blended with polymer resins directly. Effective dispersions of the 2D nanoflakes throughout polymer matrix by using the normal direct melt blending process can be hardly achieved especially when the filler loading is high, thus seriously limiting the industrial applications of GNs.

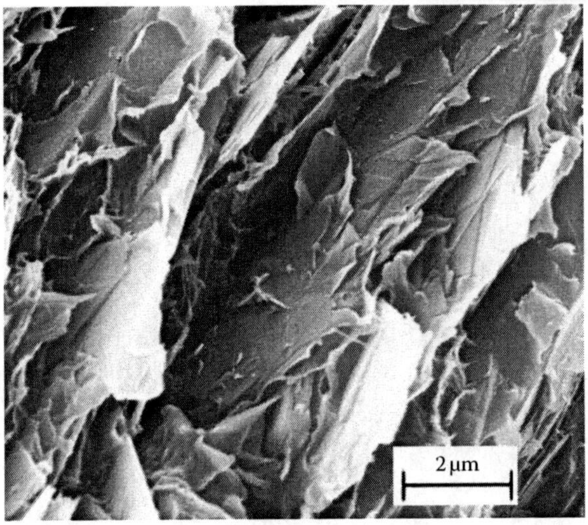

FIGURE 5.20
SEM image of the AS/GN masterbatch.

To improve the GNs dispersion degree, we managed to incorporate GNs into polymers through masterbatch technique by using poly(acrylonitrile-styrene) (AS) as the coating resin (Chen et al. 2007). As shown in Figure 5.20, the GNs are encapsulated very well by the AS resin, and the aggregates of GNs are greatly limited. The AS-coated masterbatch containing 70–80 wt% of GNs were then blended with AS and HDPE respectively employing a single-screw extruder. Experimental results suggested that the dispersion state of GNs in the final composites was strongly dependent on the miscibility of the masterbatch with the polymer matrix. Good compatibility between the two will lead to a much finer GNs dispersion. When the polymeric phases in the composites are immiscible, such as HDPE/EVA system in which the two polymer matrices have great difference in crystallinity, the GNs tend to locate in a certain polymer phase where the interfacial tension between them is low (Chen et al. 2007c).

Based on the current state of the art as achieved by our group, a start-up advanced technology enterprise, Xiamen Knano Graphite Technology Co., Ltd (http://www.knano.com.cn/) has been established. Here, mass production of GNs has been achieved and atomically thin graphene crystallites are also commercially available now.

As the first large batch supplier of graphene and GNP materials located in mainland China, Xiamen Knano Graphite Technology Co., Ltd currently provides graphene and GNP materials of KNG™ series with varying thickness to satisfy industry and lab requirements. The available KNG™ series products are listed in Table 5.4.

TABLE 5.4
KNG™ Series Products Available in KNANO

Product Number	Product Name	Thickness (nm)	Diameter	Package
KNG-150	Graphite nanoplatelet	5–50	0.5–20 μm	20 kg
KNG-105S	Graphite nanoplatelet	<100	0.5–20 μm	20 kg
KNG-110	Graphene	2–10	0.5–10 μm	10 kg
KNG-102	Graphene	<2	100–200 nm	200 mg
KNG-21	Graphene oxide	<1	1–100 μm	200 mg

FIGURE 5.21
KNG-150 products.

The current throughput of GNP has reached 10 tons a year, and the products KNG-150 have been successfully used in the plastics industry (Figure 5.21), demonstrating the feasibility of commercial applications. What the users have to do is to mix the powder GNPs with target polymers through double-roller, banbury-mixer, twin screw extruder, and other mixers that are being commonly used in the plastics industry. The blending processing is very simple and cost-effective. For better dispersions of the product powder in the target polymer matrix, some surface modifiers, such as the coupling agents of silane, titanate, and aluminate, are recommended to be used before mixing the powder with plastics.

Acknowledgments

The authors thank the National Natural Science Foundation of China (NO.29874016, 20174012, 5037301520574025) and the Natural Science Foundation of Fujian Province (E0820001) for support of this work.

References

Brodie, B. 1859. On the atomic weight of graphite. *Phil. Trans. Roy. Soc. Lond.* 149:249–259.

Celzard, A., Mcrae, E., Mareche, J. F., Furdin, G., Dufort, M., and Deleuze, C. 1996. Composites based on micron-sized exfoliated graphite particles: Electrical conduction, critical exponents and anisotropy. *J. Phys. Chem. Solids* 57:715–718.

Chen, L., Chen, G. H., and Lu, L. 2007a. Piezoresistivie behaviors study on finger-sensing silicone rubber/graphite nanosheet nanocomposite. *Adv. Funct. Mater.* 17(6):898–904.

Chen, G. H., Chen, X. F., Wang, H. Q., and Wu, D. J. 2007b. Dispersion of graphite nanosheets in polymer resins via masterbatch technique. *J. Appl. Polym. Sci.* 103:3470–3475.

Chen, G. H., Lu, J. R., Lu, W., Wu, D. J., and Wu, C. L. 2005a. Time-dependence of piezo-resistive behavior for polyethylene/foliated graphite nanocomposites. *Polym. Int.* 54:1689–1693.

Chen, G. H., Lu, J. R., and Wu, D. J. 2005b. PTC effect of polyethylene/foliated graphite nanocomposites. *J. Mater. Sci.* 40(18):5041–5043.

Chen, G. H., Lu, J. R., and Wu, D. J. 2007c. The electrical properties of graphite nanosheet filled immiscible polymer blends. *Mater. Chem. Phys.* 104:240–243.

Chen, L., Lu, L., Wu, D. J., and Chen, G. H. 2007d. Silicone rubber/graphite nanosheet electrically conducting nanocomposite with a low percolation threshold. *Polym. Compos.* 28:493–498.

Chen, G. H., Weng, W. G., Wu, D. J., and Wu, C. L. 2003a. PMMA/graphite nanosheets composite and its conducting properties. *Eur. Polym. J.* 39(12):2329–2335.

Chen, G. H., Wu, D. J., Weng, W. G., He, B., and Yan, W. L. 2001a. Preparation of polystyrene-graphite conducting nanocomposite via intercalation polymerization. *Polym. Int.* 50(9):980–985.

Chen, G. H., Wu, D. J., Weng, W. G., and Wu, C. L. 2003b. Exfoliation of graphite flake and its nanocomposites. *Carbon* 41(3):619–621.

Chen, G. H., Wu, C. L., Weng, W. G., Wu, D. J., and Yan, W. L. 2003c. Preparation of polystyrene/graphite nanosheet composite. *Polymer* 44:1781–1784.

Chen, G. H., Wu, D. J., Weng, W. G., and Yan, W. L. 2001b. Dispersion of graphite nanosheets in a polymer materix and the conducting property of nanocomposite. *Polym. Eng. Sci.* 41(12):2148–2154.

Chen, G. H., Wu, D. J., Weng, W. G., and Yan, W. L. 2001c. Preparation of polymer graphite conducting nanocomposite by intercalation polymerization. *J. Appl. Polym. Sci.* 82:2506–2513.

Chen, G. H., Wu, D. J., Ye, W., and Yan, W. L. 1999. Delamination of graphite by electrochemical intercalation. *New Carbon Mater.* (Chinese) 14:59–62.

Chen, G. H., Zhao, W. F., Tang, H. T., and Wang, H. Q. 2007e. Preparation and surface characterization of highly ordered polymer/graphite nanosheet composites. *Mater. Manuf. Process.* 22:733–736.

Choucair, M., Thordarson, P., and Stride, J. A. 2009. Gram-scale production of graphene based on solvothermal synthesis and sonication. *Nat. Nanotechnol.* 4:30–33.

Chung, D. D. L. 2002. Review graphite. *J. Mater. Sci.* 37:1475–1489.

Du, X. S., Xiao, M., Meng, Y. Z., and Hay, A. S. 2004. Synthesis and properties of poly(4,4′-oxybis(benzene)disulfide)/graphite nanocomposites via in situ ring-opening polymerization of macrocyclic oligomers. *Polymer* 45:6713–6718.

Gu, W. T., Zhang, W., Li, X. M. et al. 2009. Graphene sheets from worm-like exfoliated graphite. *J. Mater. Chem.* 19:3367–3369.

Hernandez, Y., Nicolosi, M., Blighe, V. et al. 2008. High yield production of graphene by liquid phase exfoliation of graphite. *Nat. Nanotechnol.* 3:563–568.

Hu, H. W., and Chen, G. H. 2010. Electrochemically modified graphite nanosheets and their nanocomposite films with poly(vinyl alcohol). *Polym. Compos.* 31(10):1770–1775.

Hu, H. W., Chen, L. F., and Chen, G. H. 2011. Reinforcement of epoxy with graphite oxide nanoparticles prepared via electrochemical route. *Mater. Manuf. Process.* 26(4):618–622.

Hu, H. W., Chen, G. H., Fang, M., and Zhao, W. F. 2009. Modification of graphite oxide nanoparticles prepared via electrochemically oxidizing method. *Synthetic Met.* 59:1505–1507.

Kim, H., Hahn, H. T., Viculis, L. M., Gilje, S., and Kaner, R. B. 2007. Electrical conductivity of graphite/polystyrene composites made from potassium intercalated graphite. *Carbon* 45:1578–1582.

Kim, K. S., Zhao, Y., Jang, H. et al. 2009. Large-scale pattern growth of graphene films for stretchable transparent electrodes. *Nature* 457:706–710.

Li, Y. C., and Chen, G. H. 2007. HDPE/EG nanocomposites prepared via a masterbatch process. *Polym. Eng. Sci.* 47:882–888.

Li, X., and Chen, G. H. 2009. Surface modified graphite nanosheets used as adsorbent to remove 1,2-dichlorobenzene from water. *Mater. Lett.* 63:930–932.

Li, X. L., Wang, X. R., Zhang, L., Lee, S. W., and Dai, H. J. 2008. Chemically derived, ultrasmooth graphene nanoribbon semiconductors. *Science* 319:1229–1232.

Li, Q., Zeng, G. Z., Zhao, W. F., and Chen, G. H. 2010. Preparation and characterization of nickel-coated graphite nanosheets. *Synthetic Met.* 160:200–202.

Lu, J. R., Chen, X. F., Lu, W., and Chen, G. H. 2006a. The piezoresistive behaviors of polyethylene/foliated graphite nanocomposites. *Eur. Polym. J.* 42(5):1015–1021.

Lu, W., Lin, H. F, and Chen, G. H. 2006b. Non-universal transport behavior in heterogeneous high-density polyethylene/graphite nanosheets composites. *J. Polym. Sci. Part B: Polym. Phys.* 44:1846–1852.

Lu, W., Lin, H. F., Wu, D. J., and Chen, G. H. 2006c. Unsaturated polyester resin/graphite nanosheet conducting composite and its electrical properties. *Polymer* 47:4440–4444.

Lu, J. R., Weng, W. G., Chen, X. F., Wu, D. J., Wu, C. L., and Chen, G. H. 2005. Novel piezoresistive material from directed shear-induced assembly of graphite nanosheets in polyethylene. *Adv. Funct. Mater.* 15:358–1363.

Lu, W., Weng, J. X., Wu, D. J., Wu, C. L., and Chen, G. H. 2006d. Epoxy resin/graphite nanosheets electrically conductive nanocomposite. *Mater. Manuf. Process.* 21:167–171.

McAllister, M. J., Li, J. L., Adamson, D. H. et al. 2007. Single sheet functionalized graphen by oxidation and thermal expansion of graphite. *Chem. Mater.* 19:4390–4404.

Mo, Z. L., Sun, Y. X., Chen, H. et al. 2005. Preparation and characterization of a PMMA/$Ce(OH)_3$, Pr_2O_3/graphite nanosheet composite. *Polymer* 46:12670–12676.

Novoselov, K. S., Geim, A. K., Morozov, S. V. et al. 2004. Electric field effect in atomically thin carbon films. *Science* 306:666–669.

Robert, M. C., Oberlin, M., and Mering, J. 1973. *Chemistry and Physics of Carbon*. Pergamon Press, Oxford.

Sasa, T., Takahashi, Y., and Mukaibo, T. 1971. Crystal structure of graphite bromine lamellar compounds. *Carbon* 9(4):407–416.

Schniepp, H. C., Li, J. L., McAllister, M. J. et al. 2006. Functionalized single graphene sheets derived from splitting graphite oxide. *J. Phys. Chem. B*. 110:8535–8539.

She, Y. H., Chen, G. H., and Wu, D. J. 2007. Fabrication of polyethylene/graphite nanocomposite from modified expanded graphite. *Polym. Int*. 56:679–685.

Shioyama, H. 2000. The interactions of two chemical species in the interlayer spacing of graphite. *Synthetic Met*. 114:1–15.

Shioyama, H., and Fujii, R. 1987. Electrochemical reactions of stage 1 sulfuric acid-graphite intercalation compound. *Carbon* 25(6):771–774.

Stankovich, S., Dikin, D. A., Piner, R. D. et al. 2007. Synthesis of graphene-based nanosheets via chemical reduction of exfoliated graphite oxide. *Carbon* 45:1558–1565.

Sun, W. F., Chen, G. H., and Zheng, L. L. 2008. Electroless deposition of silver particles on graphite nanosheets. *Scripta Mater*. 59:1031–1034.

Tang, Y. B., Lee, C. S., Chen, Z. H. et al. 2009. High-quality graphenes via a facile quenching method for field-effect transistors. *Nano Lett*. 9:1374–1377.

Thiele, H. 1938. Advances in electrochemical science and engineering. *Trans. Faraday Soc*. 34:1033.

Thostenson, E., Li, C., and Chou, T. 2005. Nanocomposites in context. *Comp. Sci. Technol*. 65:491–516.

Ubbelohde, A. R. 1972. Intercalation overpotentials for compound formation by graphite. *Carbon* 10:201–206.

Wang, L. W., and Chen, G. H. 2010. Dramatic improvement in mechanical properties of GNs-reinforced HDPE nanocomposites. *J. Appl. Polym. Sci*. 116:2029–2034.

Wang, L. W., Hong, J. B., and Chen, G. H. 2010. Comparison study of graphite nanosheets and carbon black as fillers for high density polyethylene. *Polym. Eng. Sci*. 50(11):2176–2181.

Wang, W. P., Pan, C. Y., and Wu, J. S. 2005. Electrical properties of expanded graphite/poly(styrene-*co*-acrylonitrile) composites. *J. Phys. Chem. Solids* 66:1695–1700.

Wang, H. Q., Zhang, H. Y., and Chen, G. H. 2007. Preparation of unsaturated polyester/graphite nanosheet conducting composite under electric field. *Compos. Part A: Appl. Sci. Manuf*. 38:2116–2120.

Wang, H. Q., Zhang, H. Y., Zhao, W. F., Zhang, W., and Chen, G. H. 2008. Preparation of polymer/oriented graphite nanosheet composite by electric field-inducement. *Compos. Sci. Technol*. 68:238–243.

Weng, W. G., Chen, G. H., Wu, D. J., Chen, X. F., Lu, J. R., and Wang, P. P. 2004. Fabrication and characterization of nylon 6/foliated graphite electrically conducting nanocomposite. *J. Polym. Sci. Part B: Polym. Phys*. 42:2844–2856.

Weng, W. G., Chen, G. H., Wu, D. J., Lin, Z. Y., and Yan, W. L. 2003a. Preparation and characterizations of nanoparticles from graphite via an electrochemically oxidizing method. *Synthetic Met*. 139:221–225.

Weng, W. G., Wu, D. J., and Chen, G. H. 2003b. Crystallization kinetics and melting behaviors of nylon 6/foliated graphite nanocomposites. *Polymer* 44:8119–8132.

Xiao, P., Xiao, M., and Gong, K. 2001. Preparation of exfoliated graphite/polystyrene composite by polymerization-filling technique. *Polymer* 42:4813–4816.

Yang, J., Tian, M., Jia, Q. X. et al. 2007. Improved mechanical and functional properties of elastomer/graphite nanocomposites prepared by latex compounding. *Acta Mater*. 55:6372–6382.

Zhang, H. X., and Feng, P. X. 2010. Fabrication and characterization of few-layer graphene. *Carbon* 48:359–364.

Zhao, W. F., Fang, M., Wu, F. R., Wu, H., Wang, L. W., and Chen, G. H. 2010a. Preparation of graphene by exfoliation of graphite using wet ball milling. *J. Mater. Chem*. 20:5817–5819.

Zhao, W. F., Wang, H. Q., Tang, H. T., and Chen, G. H. 2006. Facile preparation of epoxy-based composite with oriented graphite nanosheets. *Polymer* 47:8401–8405.

Zhao, W. F., Wu, F. R., Wu, H., and Chen, G. H. 2010b. Preparation of colloidal dispersions of graphene sheets in organic solvents by using ball milling. *J. Nanomater*. doi:10.1155/2010/528235.

Zheng, W. G., and Wong, S. C. 2003. Electrical conductivity and dielectric properties of PMMA/expanded graphite composites. *Compos. Sci. Technol*. 63:225–235.

Zheng, W. G., Wong, S. C., and and Sue, H. J. 2002. Transport behavior of PMMA/expanded graphite nanocomposites. *Polymer* 43:6767–6773.

6

Graphene-Based Materials for Clean Energy Applications

Bin Luo, Minghui Liang, Michael Giersig, and Linjie Zhi*

CONTENTS

* Correspondence to: Linjie Zhi (zhilj@nanoctr.cn)

6.1 Introduction

Graphene, consisting of a single atomic layer of graphite, has generated enormous excitement because of its extraordinary physical and chemical properties, such as high electrical conductivity, high specific surface area, and surface functionality, and therefore its great potential in various applications (Elias et al. 2009; Geim 2009; Geim and Novoselov 2007; Kim et al. 2009; Li et al. 2008b; Nair et al. 2008; Service 2009; Stankovich et al. 2006a; Tung et al. 2009a). In particular, graphene is a suitable candidate for energy-related applications including manufacture of photovoltaic (PV) devices, lithium-ion batteries (LIBs), supercapacitors, and catalysis.

Because of the low cost of production and unusual properties of graphene, current interest in it arises from its potential to be used in flexible and highly efficient heterojunction PV devices.

An outstanding example is the use of graphene as a thin film for transparent and conductive electrodes of organic (Wang et al. 2008d; Wu et al. 2008) and dye-sensitized (Wang et al. 2008c) solar cells. Further, graphenes have been incorporated into conjugated polymers to improve exciton dissociation and charge transport property of the materials (Hong et al. 2008; Liu et al. 2008b,c). In addition, graphene can be potentially used as a photoactive material (Yong and Tour 2010) since its bandgap and band position can be induced and tuned through chemical fictionalization (Cervantes-Sodi et al. 2008) or it can be applied in its modified form as nanoribbons (Li et al. 2008b; Owens 2008).

Rechargeable LIBs, which are used widely in portable electronics, such as cell phones, laptop computers, and digital cameras, are believed to be promising choices as energy-effective and environment-friendly devices (Tarascon and Armand 2001).

Concerning the carbon emission of fossil-based energy sources, especially in transportation systems, it is highly desirable that LIBs have higher energy density, faster charge/discharge rate, and more durable cycling performance, which may meet the requirements of high power facilities, such as vehicles. Graphite is a commonly selected anode material in LIBs due to its high coulombic efficiency (Kaskhedikar and Maier 2009; Ng et al. 2006). However, the specific capacity of graphite is relatively low (theoretical value: 372 mA·h/g) since every six carbon atoms can host only one lithium ion by intercalation (LiC6). In the past decade, many new anode materials, such as nanosize alloys, intermetallic alloys (Boukamp et al. 1981; Deng and Lee 2009; Kasavajjula et al. 2007; Wang and Lee 2006), and transition metal oxides (Nam et al. 2006; Poizot et al. 2000; Zhi et al. 2008), have been explored for LIBs with the aim to improve the capacity and energy density of the battery system. A major drawback of these substances as anode materials is their huge volume variation during the charge/discharge process, which causes pulverization of the electrode, resulting in poor reversibility. Reducing the size of these materials to nanometer scale can partially suppress the expansion and shrinkage of anode materials (Barsukov and North Atlantic Treaty Organization 2006; Chen et al. 2003), and designing these materials as nanowires; nanorods; nanotubes; or hollow spherical, core/shell, or porous particles can improve lithium storage capacity and their initial coulombic efficiency (Han et al. 2005; Kim and Cho 2008; Yu et al. 2007; Zhi et al. 2008). However, capacity decaying of the electrode along with charge/discharge cycling still cannot be completely avoided. Recently, metal oxides nanoparticles encapsulated by graphene layers or dispersed uniformly in a graphene-based carbonaceous framework have been reported to display high specific capacity and excellent cycling performance as anode materials in LIBs compared to naked particles (Paek et al. 2009; Zhi et al. 2008). These structures, in which graphene layers act as both a "buffer zone" of volume variation of nanoparticles and a good electron transfer media, were believed to be the key issue, suggesting that a buffer layer around the electrode materials having both excellent mechanical properties and good electron transportation behaviors is of great importance for effectively solving the aforementioned problems.

Supercapacitors, also called "electrochemical capacitors" and "ultracapacitors," are a commonly used name for a class of electrochemical energy storage devices that are ideally suited to the rapid storage and release of energy. Advantages of supercapacitors over most of the traditional energy storage devices include high power capability, long life, wide thermal operating range, low weight, flexible packaging, and low maintenance (Kotz and Carlen 2000).

This device has generated much interest for a wide and growing range of applications, including load cranes, forklifts, electric vehicles, electric utilities, aerospace systems, factory power backup, and so on (Miller and Simon 2008; Simon and Gogotsi 2008; Stoller et al. 2008). So far, different materials such as various carbon materials, (Kaempgen et al. 2009; Pandolfo and Hollenkamp 2006; Shaijumon et al. 2008) mixed metal oxides (Miller et al. 1997; Toupin et al. 2004), and conducting polymers (CPs) (Rudge et al. 1994) have been used as supercapacitor electrode materials. Particularly, carbon in its various forms (Pandolfo and Hollenkamp 2006) has been used as an electrode material in supercapacitors, aiming at high specific capacitance together with high power density. However, a major shortcoming of current supercapacitors is their low energy density (typically 5–10 W·h/kg). The CPs such as polyanilines (PANIs) (Arbizzani et al. 2001) have been shown to exhibit high pseudocapacitance, but they exhibit poor stability during charge/discharge cycling. Graphene has recently been found to exhibit exceptionally high thermal conductivity, electrical conductivity, strength, and specific surface area (up to 2675 m^2/g; Balandin et al. 2008; Becerril et al. 2008; Geim and Novoselov 2007; Park and Ruoff 2009; Stankovich et al. 2006, 2007). Most significantly, the intrinsic capacitance of graphene was recently found to be 21 $\mu F/cm^2$, which sets the upper limit of electrical double layer (EDL) capacitance for all carbon-based materials (Xia et al. 2009). These results assert that graphene is the ideal carbon electrode material for EDL supercapacitors because it is capable of storing an EDL capacitance value of up to 550 F/g. Compared with other carbonaceous materials, another advantage of graphene in a supercapacitor electrode is that both major surfaces of a graphene sheet (GS) are exterior surfaces readily accessible by electrolytes (Liu et al. 2010a).

In the case of catalysis applications, various carbonaceous materials such as activated carbon, carbon black, graphite, carbon nanotubes (CNTs), nanofibers, and graphitic materials with much higher specific surface areas than other conventional catalyst supports (e.g., alumina, silica) have been used for decades in catalysis (Brandon et al. 2003; Cui et al. 2007c; Zhong et al. 2010). Carbon-supported electrocatalysts are effective in maximizing the performance of proton exchange membrane (PEM)-based fuel cells and minimizing the use of precious metals (Costamagna and Srinivasan 2001). Such support materials not only maximize the availability of nanosized electrocatalyst surface area for electron transfer but also provide better mass transport of reactants to the electrocatalyst. In addition, the conductive support facilitates efficient collection and transfer of electrons to the collecting electrode surface. Carbon black, because of its high surface area, high conductivity, and low cost, has been extensively used as a support for PEM-based fuel cells. One issue that calls for new and improved carbon materials is the long-term stability of carbon supports under fuel cell operation (Meyers and Darling 2006). Durability tests have indicated that conventional carbon black supports undergo morphological changes and induce catalyst aggregation during extended use (Kongkanand et al. 2006). The recent emergence of graphene has opened a new avenue for utilizing two-dimensional (2D) carbon material as a support in fuel cells (Li et al. 2009g; Seger and Kamat 2009) or in traditional catalytic reactions (Li et al. 2010b) due to its unique structure with high surface area as well as its special electronic and ballistic transport properties.

6.2 Graphene-Based Materials Applied in Solar Cells

Efforts have been made recently on developing graphene-based solar cells, in which graphene is used in different parts of the cells, such as transparent and conductive electrode and charge collection or transportation materials. In Sections 6.2.1 and 6.2.2, the potential roles that graphene might serve in a solar cell and the methods developed for preparation of graphene and graphene-based materials are discussed briefly.

6.2.1 Graphene-Based Films as Transparent Conductive Electrodes

Materials with a remarkable combination of high electrical conductivity and optical transparency are of great importance in various optoelectronic devices such as organic light-emitting diodes (OLEDs) and solar cells (Granqvist 2007; Haacke 1977). Presently, doped metal oxide films such as indium–tin oxide (ITO) and fluorine-doped tin oxide (FTO) single-handedly dominate the field (Edwards et al. 2004). In the case of solar cells, these components work as anodes to extract separated charge carriers from the absorbing region. The ability to deposit these materials with controlled thickness and controlled doping concentration has significantly contributed to their widespread application. However, the next generation of optoelectronic devices requires transparent conductive electrodes (TCEs) to be lightweight, flexible, and cheap devices that are compatible with large-scale manufacturing methods, in addition to being conductive and transparent. These requirements severely limit the use of ITO as transparent conductors because of the brittleness of ITO films, restricting their use in flexible optoelectronic devices (Chen et al. 2001). In addition, the instability of ITO at acid or base conditions, its susceptibility to ion diffusion into polymer layers, and the limited availability of indium sources create an urgent need to find other materials that can work as transparent electrodes for future optoelectronic devices.

Many research communities have made significant advances in finding potential candidate materials. The most important materials among these are CNT films (Nirmalraj et al. 2009; Wu et al. 2004; Zhang et al. 2005), metal gratings (Kang and Guo 2007), networks of metallic nanowires (De et al. 2009a; Lee et al. 2008), and graphene films (Eda et al. 2008a; Tung et al. 2009b; Wang et al. 2008c). To date, nanotube films have been prepared with direct current conductivities close to $6 \times 10^5\ S \cdot m^{-1}$ (Geng et al. 2007; Wu et al. 2004). In addition, polymer–nanotube composite films (De et al. 2009) have been shown to be electrically stable under flexing. However, the high surface roughness of the CNT film may hinder its further application in commercial products. On the other hand, the process of CNT preparation is highly energy consuming irrespective of the preparation method of chemical vapor deposition (CVD) or arc discharge used. Use of surfactant-free silver nanowires eliminates the need for high-temperature annealing, thus enabling their deposition on flexible substrates for subsequent application in flexible optoelectronic devices (De et al. 2009). Similarly, a nanowire grating structure with well-defined dimensions represents an ideal configuration for device applications (Kang and Guo 2007). Although this is a good start for metal nanowire networks, research in this direction is still in its infancy and requires scalable approaches to fabricate such films.

Graphene, a 2D monolayer of sp^2-bonded carbon atoms, as a rising star in material science has been attracting much interest owing to its high optical transmittance and conductivity (Geim 2009; Service 2009). As a zero-bandgap semiconductor, its electronic structure is unique in the sense that charge carriers are delocalized over large areas, making it a scattering-free platform for carrier transport (Geim and Novoselov 2007). A single sheet of graphene is a zero-bandgap semiconductor having extremely high carrier mobility; it only absorbs 2.3% of visible light (Nair et al. 2008). Hence, graphene may be a viable candidate for applications as a transparent conductor.

Compared to CNTs, graphene is an ideal 2D material with excellent electron transport properties (Novoselov et al. 2004), which can be assembled into a film electrode with low roughness (Wang et al. 2008c). Due to its excellent electrical conductivity, graphene has been incorporated with polymers (Stankovich et al. 2006) or silica (Watcharotone et al. 2007) to obtain conductive composites. In practice, however, large-area graphene films produced through solution processing of functionalized graphene contain multiple grain boundaries and incorporate lattice defects and oxidative traps that increase the electrical resistance of the material. As a result, the films must be made thicker than one atomic layer to obtain practical sheet resistances. Recently, many efforts have been made to improve the conductivity of graphene thin films used as electrodes in optoelectronics. Experimental investigations disclosed that a graphene film as prepared has excellent conductivity, good transparency in both visible and near-infrared regions, ultrasmooth surface with tunable wettability, high chemical and thermal stabilities, and flexibility for transfer between alternative substrates (Wang et al. 2008c), which render the graphene-based window electrodes versatile for applications in various optoelectronic devices. Solar cells, including organic solar cells and dye-sensitized solar cells (DSSCs), based on these graphene films as anode materials have been fabricated in our group. Comparable cell performance to the ITO-based cell under monochromatic light of 510 nm was achieved (Table 6.1) (Wang et al. 2008d).

Obviously, mass production of graphene is crucial for practical applications. Since peeling of pyrolytic graphite does not appear to be a readily scalable process, several alternatives have been investigated to prepare graphenes, including thermal decomposition of silicon carbide and chemical manipulation of graphite (Hummers and Offeman 1958; Kedzierski et al. 2008; Li et al. 2008a; Nam et al. 2006; Stankovich et al. 2007). Considering the facile solution-processing procedure, which enables film fabrication through relatively low-cost and large-scale production methods such as spin coating, roll-to-roll processing, and printing, graphene oxide (GO) prepared by oxidative exfoliation of graphite is preferred, since oxygen-containing functional groups render the GO hydrophilic and dispersible in water and other polar solvents.

Several solar cells have been fabricated using solution-processable graphene-based TCEs (De Arco et al. 2010; Eda et al. 2008b; Wang et al. 2008c,d; Wu et al. 2008). Recently, we used exfoliated graphite oxide as the starting

TABLE 6.1

Performance of a Solar Cell Based on Transparent Graphene Film and ITO as Anodes

Light	Anode	I_{sc} (mA·cm^{-2})	V_{oc} (V)	FF	η
Monochromic light of 510 nm	ITO	0.04	0.14	0.3	1.50
	TGF	0.05	0.13	0.23	1.53
Simulated solar light	ITO	1.00	0.41	0.48	1.17
	TCCF	0.36	0.38	0.25	0.29

Source: Wang, X., L. J. Zhi et al. *Angewandte Chemie-International Edition*. 2008. 47(16). 2990–2992. Copyright Wiley-VCH Verlag GmbH & Co. KGaA. Reproduced with permission.

Note: I_{sc} is short-circuit photocurrent density; V_{oc} is open-circuit voltage; FF is the calculated filling factor; and η is the overall PCE.

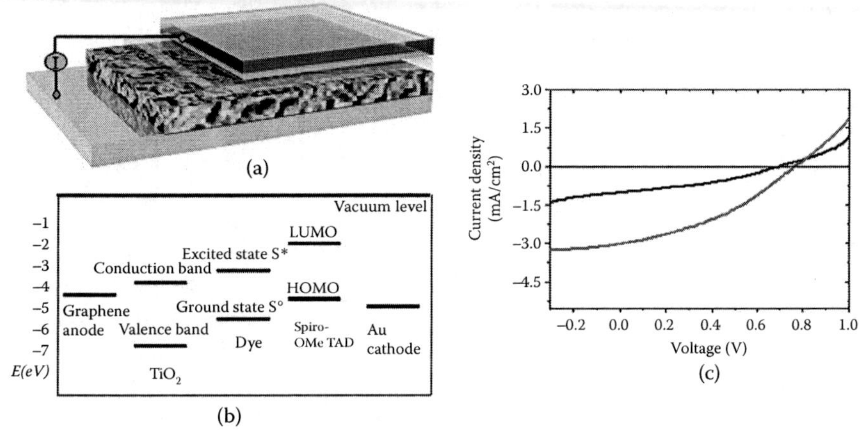

FIGURE 6.1
Illustration and performance of a solar cell based on graphene electrodes. (a) A DSSC using graphene film as electrode; the four layers from bottom to top are Au, dye-sensitized heterojunction, compact TiO_2, and graphene film. (b) The energy-level diagram of graphene/TiO_2/dye/spiro-OMeTAD (2,2′,7,7′-tetrakis-(N,N-di-p-methoxyphenylamine)-9,9′-spirobifluorene)/Au device. (c) The *I-V* curve of the graphene-based cell (black) and the FTO-based cell (gray), illuminated under amplitude-modulated solar light (1 sun). (Reprinted with permission from Wang X., Zhi L. J. et al. *Nano Letters*. 2008, 8, 323. Copyright 2008 American Chemical Society.)

material and a dip coating process and thermal reduction to obtain a graphene film having a thickness of 10 nm (atomic force microscopy [AFM] characterization), transparency higher than 70%, and conductivity of 550 S/cm (Wang et al. 2008c). For the first time, DSSCs were fabricated successfully using graphene film as the window electrode (Figure 6.1). The current–voltage (*I-V*) characteristics of the device (the efficient area of the device is 0.11 cm^2) under illumination by simulated solar light (light intensity: 98.3 mW/cm^2, calibrated against silicon PV solar cells) showed a short-circuit photocurrent density (I_{sc}) of 1.01 mA/cm^2 with an open-circuit voltage (V_{oc}) of 0.7 V, calculated filling factor (FF) of 0.36, and overall power conversion efficiency (PCE) of 0.26%. The somewhat lower I_{sc} and efficiency of a graphene film–based cell might be due to the series resistance of the device, the relatively lower transmittance of the electrode, as well as the electronic interfacial change. Obviously, the graphene electrode–based cell fabrication and study is in its early stages and there is still lots of room for further improvement, such as increasing the conductivity of graphene films by the use of high-quality GSs with lateral dimensions on the micrometer scale and exploitation of novel cell structure based on graphene electrodes. The experimental results demonstrate a new direction for searching an alternative window electrode of ITO, although the PCE of this graphene-based solar cell is not as good as that of the FTO-based reference device. Similarly, Zhu et al. (2009) produced graphene-based transparent conducting films via dip-coating process as well; but in this instance, they reduced GO (RGO) in a solution of water and hydrazine at 80°C before casting a film. They observed that RGO assembles at the surface of a water–air interface and utilized this behavior by dipping quartz substrates into the solution to collect these films (Zhu et al. 2009). From the RGO films prepared, a low sheet resistance of 11.3 kΩ/sq was obtained with an optical transmittance of 87% at 550 nm.

Another standard method to fabricate thin films from a solution-processable material is using spin coating, which utilizes centrifugal force to cast away excess material from the surface of a flat substrate. Graphene film prepared by spin coating exfoliated graphite oxide on a quartz slice and subsequent reduction has been used as window electrode in the fabrication of organic solar cell by Chen et al. (Becerril et al. 2008; Wu et al. 2008). The graphene films were reduced either by vacuum annealing at 1100°C or by a combination of hydrazine treatment and Ar annealing at 400°C. The AFM characterization revealed that film thickness could be less than 10 nm and surface roughness of the film was approximately 3 nm. Further investigation demonstrated that transparency of the film was better than 80% when the film thickness was less than 20 nm, but the resistance of the film ranged from 5 kΩ/sq to 1 MΩ/sq. The organic solar cell fabricated using this graphene film as electrode had a PCE of 0.4%; that is, a value still lower than that of a cell with ITO as electrode (0.84%), which was probably due to the larger resistance of the graphene electrode.

Large-area graphene films produced via solution processing of functionalized graphene contain multiple grain boundaries and lattice defects that increase the electrical resistance of the material as discussed earlier in this section. One approach to improve the conductivity of GO films without severely diminishing optical transparency is to incorporate a conductive filler material into them. For example, Watcharotone et al. (2007) used a conductive silica matrix, which they blended with GO and spin-casted into films that were then thermally annealed. A conductivity of 0.45 S/cm at 94% transparency was obtained, which is comparable to the values reported using multiwalled CNTs (MWCNTs) instead of reduced graphite oxide. Recently, several attempts were made to fabricate hybrid films of graphene/CNTs mostly from homogeneously mixed solutions of the two carbon-based materials (Cai et al. 2008; Tung et al. 2009b). Tung et al. (2009) reported the formation of a nanocomposite comprising chemically converted graphene (CCG) and CNTs, delivering a film resistance of 240 Ω/sq at 86% transmittance. This low-temperature process is completely compatible with flexible substrates and does not require a sophisticated transfer process. A proof-of-concept application in a polymer solar cell with a PCE of 0.85% is demonstrated. Kim et al. (Kim and Min 2009) reported a graphene/CNT double-wall film was developed with a resistivity of 151 kΩ/sq at 93% transparency via self-assembly process. The adsorption of MWCNTs onto RGO films considerably decreased sheet resistance of the films without compromising much on transparency.

For the top-down solution-processed graphene, the structural defects and lateral disorder of chemically exfoliated graphite affects the carrier mobility of the film negatively (Gomez-Navarro et al. 2007). Besides incorporating conductive fillers into graphene nanosheets (GNSs), another approach to decrease the resistance of GO films is to repair these defects using carbon-rich materials as building units through a chemical or thermal process (Liang et al. 2009b; Lopez et al. 2009; Su et al. 2009). For example, acetylene was used as a carbon source to repair substantial defects within GO sheets using a thermal reduction process (Liang et al. 2009b). Conductivities of 1425 S/cm were reported with an optical transmittance of more than 70% with the claim that this method is 5 times more effective in increasing the conductivity of reduced graphite oxide when compared to thermal reduction performed alone. Further, large aromatic molecules are also chosen as building units since they can attach chemical functions through van der Waals forces or π–π interactions without destroying the electronic π-system. Su et al. (2009) presented an approach to noncovalently functionalize graphene with large aromatic donor and acceptor molecules, such as pyrene-1-sulfonic acid sodium salt (PyS) and the disodium salt of 3,4,9,10-perylenetetracarboxylic diimide bisbenzenesulfonic acid (PDI). These planar aromatic structures can be anchored onto the hydrophobic surface of reduced graphene (RG) sheets using π–π interactions. Considering the fact that large aromatic molecules serve as nanographene building blocks, further thermal annealing may enable a bottom-up approach to construct graphene films. Obvious enhancement in conductivity was observed after thermal treatment on films of RG-PyS and RG-PDI, suggesting that large aromatic molecules may heal the defective vacancies in GSs under high temperatures. This process gives rise to better π-conjugation of graphene basal plane and, therefore, a dramatic increase in film conductivity. Using RG-based or RG-PyS-based graphene films bulk-heterojunction (BHJ) solar cells were fabricated, and an overall power efficiency of 1.12% was achieved using RG-PyS as electrode, whereas RG showed a lower efficiency of 0.78%.

The graphene-based films fabricated using the aforementioned approaches are mainly derived from top-down exfoliation of graphite, which results in mostly graphenes containing defects due to the chemical exfoliation process. Recently, a bottom-up approach using aromatic carbon-rich molecules as building units to construct 2D graphene structures opened a new window toward the synthesis of transparent graphene-constructed films (Wang et al. 2008d). This was achieved by the thermal reaction of synthetic nanographene molecules of giant polycyclic aromatic hydrocarbons (PAHs), which were cross-linked with each other and further fused into larger GSs. The prepared graphene film showed stronger interactions with the substrate than those shown by a graphite oxide–induced film. The film of 4 nm thickness holds a transparency of 90% at a wavelength of 500 nm. Organic solar cells have been successfully fabricated using such a window electrode of graphene film with 85% transparency. The highest external quantum efficiency (EQE) of the cell with a graphene film window electrode was observed as 43% at the illumination of 520 nm monochromatic light and, at the same condition, the EQE of the organic solar cell with ITO as electrode was found to be 47% (Figure 6.2). Under monochromatic light of wavelength 510 nm, the

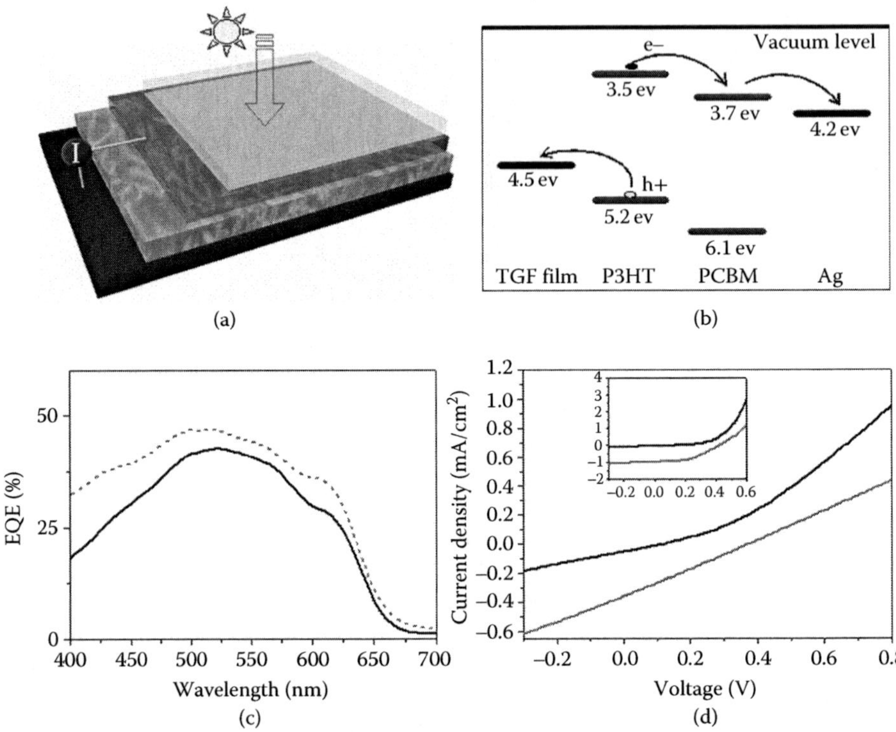

FIGURE 6.2
(a) Illustration of a solar cell: The four layers from bottom to top in the figure are Ag, a blend of P3HT and PCBM, TGF, and quartz. (b) A schematic representation of charge transfer and transport as an energy-level diagram. (c) The EQE of a TGF-based cell (solid black curve) and an ITO-based cell (dashed curve). (d) The *I-V* curve of a TGF-based cell (center) and an ITO-based cell (inset) illuminated by simulated solar light (light gray curve) and monochromatic light of wavelength 510 nm (black curve). The calibrated light intensity of the simulated solar light is 167 $W{\cdot}m^{-2}$ for ITO and 119 $W{\cdot}m^{-2}$ for TGF as anode. (From Wang, X., L. J. Zhi et al. Transparent carbon films as electrodes in organic solar cells. *Angewandte Chemie-International Edition*. 2008. 47(16). 2990–2992. Copyright Wiley-VCH Verlag GmbH & Co. KGaA. Reproduced with permission.)

PCE of this graphene-based cell was 1.53%, which was similar to that of the ITO-based cell (1.5%). However, under the illumination of simulated solar light, the cell with graphene film and that with ITO as electrode materials had overall efficiencies of 0.29% and 1.17%, respectively. The output voltage of graphene-based cells is comparable to that of ITO-based cells, suggesting that graphene has an appropriate work function for this type of organic solar cells. Further, the relatively low I_{sc}, FF, and EQE of the cell may also result from the high resistance of the as-prepared graphene films.

Compared to solution processes, another promising method to synthesize high-quality graphene is CVD (De Arco et al. 2009; Kim et al. 2009; Li et al. 2009d,e,f; Reina et al. 2009; Yu et al. 2008). Researchers have found optimal catalysts for growing large-area graphene in a fashion similar to that used to synthesize CNTs. Particularly, films with sheet resistance of 280 Ω/sq (80% transparent; Kim et al. 2009) and 770 Ω/sq (90% transparent; Reina et al. 2009) have been reported for graphene synthesized on Ni films, whereas sheet resistance of 350 Ω/sq (90% transparent) has been reported for CVD graphene grown on Cu films (Li et al. 2009f), which represents significant advance in the use of graphene as TCEs.

The PV devices using CVD graphene as TCEs have also been constituted (De Arco et al. 2010; Shi et al. 2010; Wang et al. 2009h). Wang et al. (2009) synthesized a graphene film on Ni film–coated SiO_2/Si wafer using the CVD process and then transferred it to another substrate directly using the dry-transfer technology. For a graphene film of thickness 6–30 nm, average sheet resistance varies from 1350 to 210 Ω/sq and optical transparency varies from 91% to 72% in the visible light wavelength range. These values are lower than the sheet resistance of chemically processed GO films by at least 2–3 orders of magnitude. The fabricated device with the structure graphene/PEDOT:PSS/P3HT:PCBM/LiF/Al showed an improved PCE of 1.71% when the hydrophilicity of graphene films was enhanced by noncovalent modification with pyrene butanoic acid succinimidyl ester.

Similarly, graphene films were synthesized by CVD, transferred to transparent substrates, and evaluated in organic solar cell heterojunctions with a structure of CVD graphene/PEDOT/CuPc/C_{60}/BCP/Al for CVD graphene organic PVs (OPVs; Figure 6.3a) (De Arco et al. 2010). A schematic illustration of the typical transfer process is shown in Figure 6.3b. The continuous nature of the

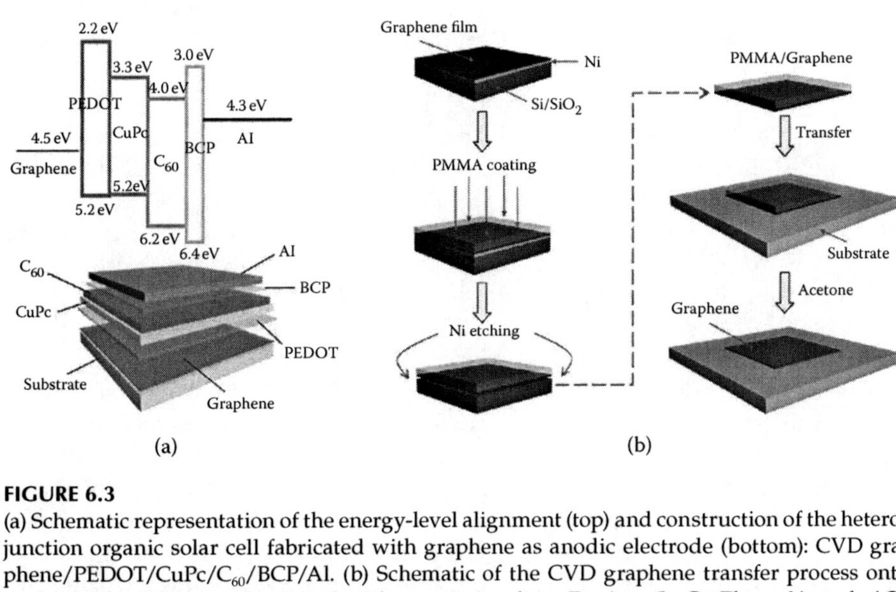

FIGURE 6.3
(a) Schematic representation of the energy-level alignment (top) and construction of the heterojunction organic solar cell fabricated with graphene as anodic electrode (bottom): CVD graphene/PEDOT/CuPc/C_{60}/BCP/Al. (b) Schematic of the CVD graphene transfer process onto transparent substrates. (Reprinted with permission from De Arco L. G., Zhang Y. et al. *ACS Nano.* 2010. 4(5): 2865–2873. Copyright 2010 American Chemical Society.)

CVD graphene films led to minimal surface roughness (~0.9 nm) and offered sheet resistance down to 230 Ω/sq (at 72% transparency), which is much lower than that of stacked graphene flakes at a similar transparency. Additionally, solar cells made with CVD graphene exhibited performance that is comparable to ITO devices and surpasses the performance of ITO devices under bending conditions, exhibiting PCEs of 1.18%; such solar cells are operational under bending conditions up to 138°, indicating the great potential of CVD graphene films in flexible PV applications. Although CVD graphene appears to yield high-quality graphenes, for applications related to TCEs, scaled-up production of films as well as feasible solution-processing techniques need to be explored.

6.2.2 Graphene-Based Materials as Other Components in Solar Cells

For applications related to solar cells, graphene-based materials are still in the initial stage. Besides their potential to be TCEs, graphenes also have other attractive properties for PVs. For example, graphenes have been shown to have very large charge mobilities, which could be useful for charge collection and transportation in solar cells. Additionally, their tunable bandgap and large optical absorptivity are desirable in efficient light harvesting. Their electronic levels and interactions with other materials for charge transfer processes can both be tuned with well-developed carbon chemistry. Several types of solar cells using graphene-based materials as the charge collection or transportation part have been reported (Chang et al. 2010; Guo et al. 2010a; Liu et al. 2009a; Liu et al. 2008c; Sun et al. 2010; Yan et al. 2010e,f).

The OPV devices have attracted considerable attention due to their potential as a renewable, alternative source of electricity and their advantages in terms of low cost, low weight, high mechanical flexibility, and ease of fabrication as compared to conventional inorganic solar cells (Gunes et al. 2007). With recent advances in both theoretical and experimental research, organic solar cells have achieved significant progress in terms of cell performance. Their PCEs have reached over 6% with the use of BHJ structures (Forrest 2005; Thompson and Frechet 2008). The commonly accepted optoelectric conversion process is as follows: The active material absorbs photons and generates excitons; the excitons diffuse into the donor/acceptor (D/A) interface and dissociate into electrons and holes, which travel along donor and acceptor phases toward the anode and cathode electrodes, respectively, where they are collected to generate power. (Brabec et al. 2001).

Based on the solar energy conversion mechanism understood so far, the most important intrinsic requirements for active materials used in BHJ PVs include (Dennler et al. 2008; Scharber et al. 2006; Thompson and Frechet 2008) strong light absorption over the whole solar spectrum, appropriate energetic distance between the highest occupied molecular orbital (HOMO)/the lowest unoccupied molecular orbital (LUMO) of the donor/acceptor for a high open circuit voltage, and high and balanced electron and hole mobility in the active layer. In addition to these intrinsic electronic characteristics of the individual components, morphological issues and device structure are also critically important.

Recently, (Yong and Tour 2010) Yong et al. reported the efficiency prediction of nanostructured graphene-based PVs, which showed the potential for single-cell efficiencies exceeding 12% (and 24% in a tandem structure). The use of graphene as a photoactive material in OPV devices requires an optimum bandgap of approximately 1.4–1.9 eV. Since graphene can be solution-processed and made to be either metallic (in the pristine state) or semiconducting (through chemical functionalization [Cervantes-Sodi et al. 2008] or quantum confinement [Li et al. 2008; Owens 2008]), this could lead to all-graphene OPV devices in which semiconducting graphene is used as the photoactive material and metallic graphene is used as the conductive electrodes. The tunable bandgap and band-position features of graphene, as well as its superior thermal stability and mechanical integrity, render graphene a credible material for the construction of next-generation low-cost, high-efficiency, thermally stable, environmentally friendly, lightweight, and flexible solar energy conversion devices.

The functionalization of graphene is believed to be important for improving its solubility and self-assembly properties so that it is a competitive candidate as acceptor material in OPV applications (Si and Samulski 2008b). So far, chemical functionalization of graphene has mainly focused on improving the solubility or processability of graphene in both water and organic solvents by attaching various soluble groups to graphene (Eda et al. 2008; Hao et al. 2008; Niyogi et al. 2006; Stankovich et al. 2006; Tung et al. 2009a; Xu et al. 2008). Organic solution-processed functionalized graphene material has been used as an acceptor material in BHJ OPV devices with P3HT or P3OT as electron donor (Liu et al. 2008b,c, 2009a). Controlled annealing of the material improves considerably the performance of as-fabricated OPV devices, and PCEs of 1.1% and 1.4% were obtained under a standard AM 1.5G illumination condition (100 mW cm^{-2}). The functionalized graphene material can be dispersed into an organic solvent with P3OT or P3HT to form a homogeneous solution.

Yin, Liu et al. (2010) modified PEDOT: PSS with GO and used it as the hole-collecting buffer layer in OPV cells. By introducing graphene into the PEDOT: PSS buffer layer of the PV device having a structure of ITO/PEDOT: PSS (40 nm)/P3HT: PCBM (1:0.6, 100 nm)/LiF (1 nm)/Al (70 nm), the energy conversion efficiency increased from 2.1% to 3.8%.

Starting from the research work done by Grätzel's group (Oregan and Gratzel 1991; they introduced nanostructured TiO_2 film into the anode), DSSCs have attracted considerable interest because of their relatively low cost and high efficiency of photoelectrical conversion. Although a conversion efficiency of approximately 11% has been obtained (Chiba et al. 2006; Nazeeruddin et al. 2005), further improvements are necessary. The major bottleneck of DSSCs is the transport of photogenerated electrons across the TiO_2 nanoparticle network, which competes with charge recombination.

To suppress the recombination and improve the electron transport, charge carriers are introduced into the cells (Brown et al. 2008; Kongkanand et al.

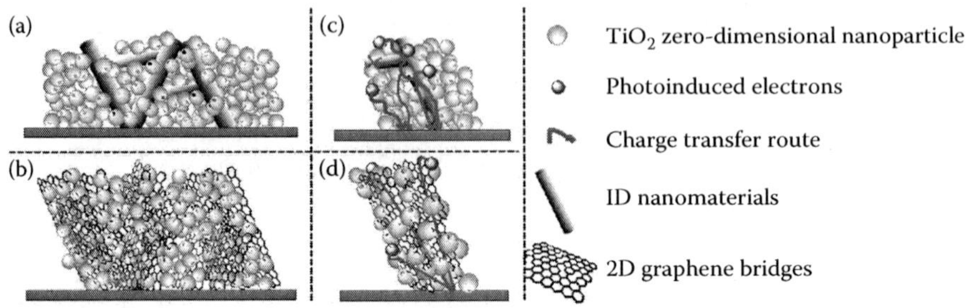

FIGURE 6.4
Differences between 1D (a and c) and 2D (b and d) nanomaterial composite electrodes. In 2D nanomaterial composite electrodes (graphene bridges), the TiO_2 particles can be anchored better in the graphene, and the photoinduced electrons can be captured and transferred by the graphene. However, with 1D nanomaterial compositing, there are less intermolecular forces and connection between TiO_2 and the 1D nanomaterial; therefore, the transfer barrier is larger and the recombination occurs more easily. (Reprinted with permission from Yang N. L., Zhai J. et al. *ACS Nano*. 2010. 4(2): 887–894. Copyright 2010 American Chemical Society.)

2007). One-dimensional (1D) nanomaterials, such as CNTs, are beneficial for this purpose, as shown in Figure 6.4a. However, there is only limited improvement in transport efficiency due to the point contact between TiO_2 nanosphere and columniform 1D nanomaterials. On the contrary, 2D graphene affords a possibility of surface contact, instead of point contact, to TiO_2 nanoparticles (Figure 6.4b), which enhances significantly electron transport from the conduction band (CB) of TiO_2 at the anchor position and suppresses the recombination of electrons and holes.

A DSSC fabricated using graphene/TiO_2 as electrode showed improved performance (Yang et al. 2010a). The short-circuit current density was increased by 45% without sacrificing the open-circuit voltage; the total conversion efficiency was 6.97%, which was increased by 39% when compared with the cell based on pure nanocrystalline titanium dioxide photoanode. The DSSC was also much better than the cell fabricated with 1D nanomaterial/TiO_2 composite electrode.

6.2.3 Summary

In summary, graphene has emerged as one of the interesting nanomaterials for application in solar cells, especially as an alternative to conventional electrode materials. Although solution methods have advantages in large-scale production of graphene, CVD methods produce graphene with better qualities. Whether or not graphene leads to a paradigm shift in transparent conductor technology, there is clearly a requirement for ITO replacement and carbon-based systems that utilize graphene are likely to fit the bill in the near future since they are both cost-effective and flexible. Substantial improvements in preparation and fabrication methods must be made before graphene can have serious practical applications. Further research work will therefore be focused mainly on the improvement of electron transfer behavior of graphene films.

Besides their potential to be TCEs, graphenes also have other attractive properties such as high charge carrier mobilities, tunable bandgap, and broad optical absorptivity for PVs. These properties make graphenes promising candidates for charge collection and transportation or efficient light harvesting in solar cells.

6.3 Graphene-Based Materials for Libs

Generally, an LIB is a lithium-ion device comprising an anode, an electrolyte, and a cathode (Figure 6.5). Since recharging of LIBs is completed with the lithium-ion insertion/extraction process in the electrodes, the nature of the two electrode materials is crucial to the performance of the battery (Winter et al. 1998). Although a freestanding graphene was observed only some years ago (Novoselov et al. 2004), the influence of graphene structure on the improvement of lithium storage capacity and cycling performance of the anode was realized more than 10 years ago (Xing et al. 1996). Very recently, experimental investigations demonstrated the great potential of using graphene-based materials in LIBs (Ding et al. 2010; Geim and Novoselov 2007; Kaskhedikar and Maier 2009; Lee et al. 2010; Li et al. 2009b; Liang and Zhi 2009; Paek et al. 2009; Wang et al. 2009a,e; Yang et al. 2010b; Yoo et al. 2008). In Sections 6.3.1 through 6.3.3, we try to summarize briefly the recent advances of using graphene-containing anode materials in LIBs. Two aspects on graphene-containing carbonaceous materials (Section 6.3.1) and graphene-containing metal/oxides composites (Section 6.3.2) are discussed.

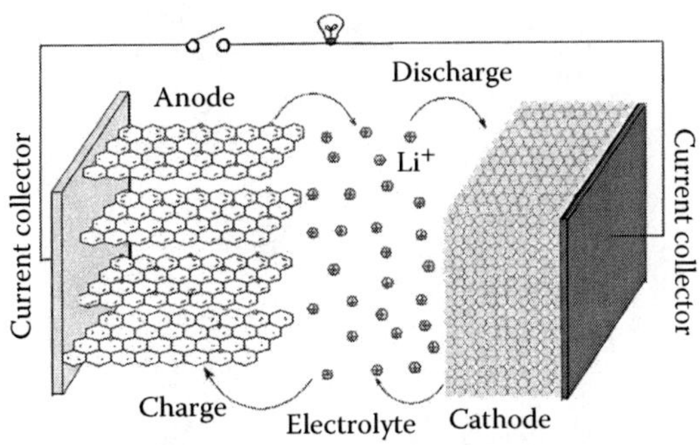

FIGURE 6.5
Schematic illustration of a rechargeable lithium battery comprising a cathode, an anode, and an electrolyte. (From Liang M. H. and Zhi L. J., *Journal of Materials Chemistry*, 19(33), 5871–5878, 2009. Reproduced by permission of The Royal Society of Chemistry.)

6.3.1 Graphene-Containing Carbonaceous Materials

To simplify their description, graphene-containing carbonaceous materials can be classified into two types: (1) One type is ordered graphene material in which graphenes are densely stacked, forming a highly ordered crystalline structure (graphitic carbon, as shown in Figure 6.6a through c); and (2) the second type is disordered graphene material in which graphenes are aggregated disorderly into a "house of cards" architecture (Zheng et al. 1996; as shown in Figure 6.6d through f). Since the electrochemical performance of these two carbon materials as anode materials is quite different, they are introduced separately in the following discussion.

Graphite is one of the most commonly used anode materials in commercialized LIBs. However, to further improve battery performance, at least two problems need to be considered: (1) One is the relatively low theoretical capacity (372 mA·h/g) of graphite resulting from the dense packing of the graphene layers. (2) Another problem is the long diffusion distances of lithium ion into its host position between the graphene layers, attributed normally to the large graphene size, and the resulting long-range crystalline order (Figure 6.6a). These two problems have a close relationship with the results of low specific capacity and slow charge/discharge rate of graphite-based LIBs.

One way to improve the charge/discharge rate is to shorten the diffusion distance of lithium ion in the graphitic materials. A more efficient approach is to arrange the graphene layers into a nanoscale 1D fiber with graphene layers oriented perpendicular to the fiber axis (Figure 6.6b), in which the diffusion distance of lithium ion is only half the diameter of the nanofibers. These

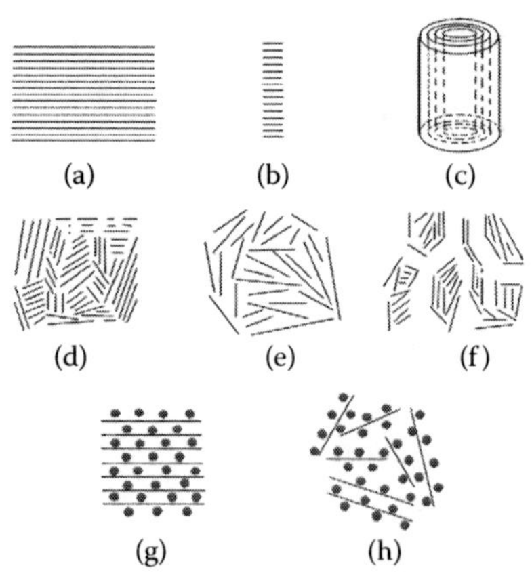

FIGURE 6.6
Schematic illustration of graphene-based carbonaceous materials with ordered graphene structures (a through c) and disordered graphene structures (d through f), and intercalation of lithium ions in the ordered (g) and disordered (h) graphene structures. (From Liang M. H. and Zhi L. J., *Journal of Materials Chemistry*, 19(33), 5871–5878, 2009. Reproduced by permission of The Royal Society of Chemistry.)

carbon nanofibers (CNFs) have been prepared by the CVD method (Yoon et al. 2004), as well as methods using porous alumina membrane as template (Jian et al. 2003). An improvement in rate capability was observed with these CNFs as anode materials. The reversible capacity was 461 mA·h/g at the rate of 0.1°C and it was maintained at 170 mA·h/g even at the rate of 10°C; the coulombic efficiency reached 95% (Subramanian et al. 2006). Graphitic CNTs with graphene layers in the tube wall oriented perpendicularly to the tube axis may supply even better structures for this purpose (Zhi et al. 2005). However, the large-scale production of these graphitic 1D nanocarbons still poses a challenge.

Very recently, a novel type of nanographene-constructed hollow carbon spheres (NGHCs) was designed and fabricated through the precursor-controlled pyrolysis approach (as shown in Figure 6.7) (Yang et al. 2010c). In this work, discotic nanographenes and silica/space/mesoporous-silica spheres were used as building blocks and template, respectively. These NGHCs exhibit uniform size and consist of dual walls; that is, thick exterior mesoporous walls and thin interior solid walls. Nanochannels constructed by nanographene in the exterior walls arrange perpendicular to the surface of the curved spheres, which is favorable for lithium-ion diffusion from different orientations, whereas the interior graphitic solid walls facilitate the collection and transport of electrons during the cycling process. Such a unique microstructure of NGHCs provides favorable transport kinetics for

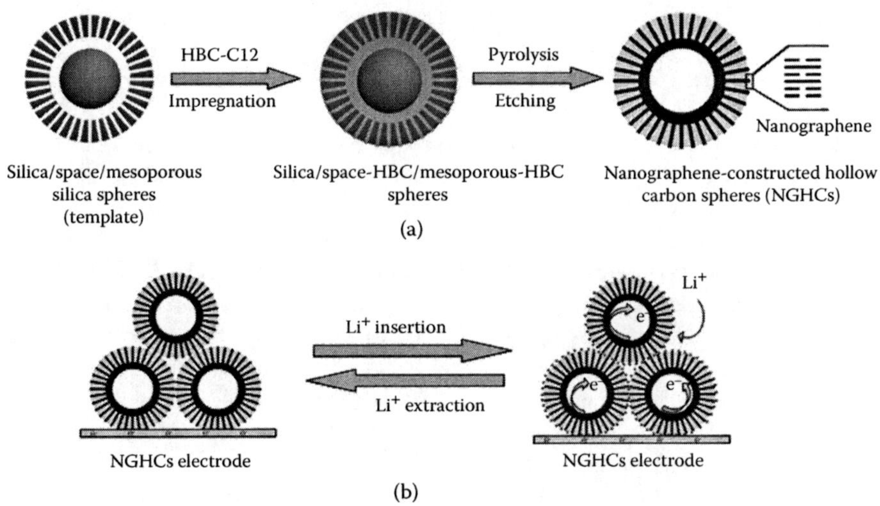

FIGURE 6.7
(a) Schematic illustration of the fabrication of NGHCs and (b) diffusion of lithium ions and electrons during the discharge (insertion) and charge (extraction) processes of the NGHC electrode. (From Yang, S. B., Feng, X. L., Zhi, L. J. et al. Nanographene-constructed hollow carbon spheres and their favorable electroactivity with respect to lithium storage. *Advanced Materials*. 2010. 22. 838–842. Copyright Wiley-VCH Verlag GmbH & Co. KGaA. Reproduced with permission.)

both lithium ions and electrons. As a result, high reversible capacity (about 600 mA·h/g) and excellent high-rate capability (about 200 mA·h/g at the rate of 10°C) were achieved when NGHCs were used as anode material for LIBs.

Besides rate performance, increasing the specific capacity of anode material is always a key issue in LIB improvement. To overcome the low capacity of graphitic structure, various strategies for reversible lithium-ion insertion and extraction on graphite have been considered. One interesting example is the use of graphitic CNT. Theoretical study suggests that the C/Li ratio for lithium storage capability of CNTs could be two or three times higher than that of normal graphite, since both the exterior and interior parts of the tube can host lithium ions (Zhao et al. 2000). However, primary experimental investigation disclosed only a poor increase of approximately 25% in the reversible capacity of lithium uptake relative to graphite (Gao et al. 1999; Maurin et al. 1999). This was probably due to the difficulty of direct lithium-ion diffusion through the sidewall of pristine CNT into its interior. Instead, the ions prefer to enter the tube through defects or open ends of the tube (Meunier et al. 2002). Hence, modification of CNT structure is essential to improve its electrochemical performance. For instance, ball milling of CNT bindles introduced not only defects but also fragmentation on the tubes, resulting in CNTs with dramatically reduced irreversible capacity (Gao et al. 2000). In addition, short CNTs were observed to hold a capacity almost two times higher than that of long CNTs (Wang et al. 2009g; Yang et al. 2008). All these suggest that producing convenient ion transfer channels for lithium-ion diffusion is effective in improving the performance of graphitic anode materials.

Other strategies for structure modification of graphitic carbon as anode material include oxidizing graphite (Wu and Holze 2003) and adjusting the synthetic process of graphite formation so that various defects can be introduced into the graphene structure (Takamura et al. 2007). One typical example was the oxidation of graphite into exfoliated graphite oxide nanoplatelets; the latter showed very high reversible specific capacity of more than 1000 mA·h/g when they were incorporated with polyelectrolytes to form a layer-by-layer structure (Cassagneau and Fendler 1998). The exfoliation and chemical modification of graphene were believed to be the main reasons for the performance enhancement. However, in this case, the modification of graphite actually led to disordered carbon, for which the electrochemical mechanism was believed to be different from that of graphitic carbon.

For a long time, researchers have realized that disordered carbons show higher specific capacity compared to ordered graphitic carbons (Sato et al. 1994). The single-layer graphene in disordered carbon was proposed to form Li_2C_6 since both the graphene surfaces can host lithium ions (Dahn et al. 1995). These suggestions have been proved reasonable by further experimental investigations (Gautier et al. 2001; Liu et al. 1996). Efforts have therefore been devoted to obtain disordered carbon comprising as much single-layer graphenes as possible (Figure 6.6e). Ideal graphene is a flat monolayer of

sp^2-bonded carbon atoms tightly packed into a 2D honeycomb lattice. Nearly perfect graphene has been prepared by mechanical exfoliation, which has opened promising avenues in the field of graphene-based electronics (Geim and Novoselov 2007; Novoselov et al. 2004). Theoretical studies suggested that the dispersion of graphenes in anode materials should be appropriate, with, for example, a distance in the range of 0.77–0.83 nm between two graphene layers, so that a double layer of lithium ions can be intercalated between neighboring graphenes but the intercalation of electrolyte molecules is prevented, which might be helpful in decreasing the irreversible capacity (Suzuki et al. 2003). However, a controlled arrangement of graphene layers in the material was a huge challenge, especially with the conventional pyrolysis approaches.

Concerning the possibility of small-sized graphene layer formation and distribution in a carbon material, well-defined molecular precursors such as hexaphenylbenzene (HPB) and its bromide, which can be easily converted into hexabenzocorene by dehydrogenation or debromination reactions, have been pyrolyzed at 600°C for 108 hours to obtain a carbonaceous product having a carbon content of 98.93% and hydrogen content of 1.03% (Renouard et al. 2005). The reversible specific capacity of this anode material could be maintained at 500 mA·h/g even after 60 cycles. Further improvement of the material resulted in a reversible capacity of 550 mA·h/g (Bonino et al. 2005).

Another class of disordered graphene materials can be prepared through solution chemistry involving oxidation of natural graphite into layered GO and subsequent reduction (Lomeda et al. 2008; Paek et al. 2009; Stankovich et al. 2006, 2007; Wang et al. 2008c; Yoo et al. 2008). Such a chemical route has attracted much attention because of the significant defects, large specific surface area, and facile functionalization through oxygen-containing functional groups in the GO precursor (Radovic and Bockrath 2005; Szabo et al. 2006; Wang et al. 2008b). All these factors have ever been proposed to be related to additional Li storage capacity in non-graphite carbonaceous materials (Winter et al. 1998). Disordered GNSs have been obtained from GO using different reduction methods including hydrazine reduction, low-temperature pyrolysis, rapid expansion, and electron beam irradiation (Guo et al. 2009; Pan et al. 2009; Wang et al. 2009d; Yoo et al. 2008). As the anode material in LIBs, these GSs exhibited significantly higher lithium storage capacitance than graphite. As discussed earlier in this section, a high specific capacity is expected in GNSs having large interlayer spacing. Indeed, when CNTs or C_{60} molecules were used as spacers to prevent restacking of GSs, the capacity of the battery was increased even further, to 730 or 784 mA·h/g, respectively (Yoo et al. 2008). Besides CNTs or C_{60} molecules, several kinds of metallic or metal oxides nanoparticles, such as Sn, Si, SnO_2, and TiO_2, have been combined with GNSs to be anode electrode materials for LIBs, in which GNSs act as not only lithium storage electrodes but also electronic conductive channels to improve the electrochemical performances. These are discussed in Section 6.3.2.

6.3.2 Graphene-Containing Metal/Oxide Composites

Owing to their high theoretical capacity, some metals are of great interest as anode materials in LIBs, since they can alloy with lithium to form Li_xM_y, which enables the storage of large amounts of lithium. For example, theoretical capacities of $Li_{4.4}Sn$ and $Li_{4.4}Si$ can reach 993 and 4200 mA·h/g, respectively, which are significantly higher than the theoretical capacity of conventional graphite. A major drawback of these alloy anodes is the large volume expansion accompanying the alloy of lithium into the metal matrix, resulting in cracks and eventually pulverization of the electrode, which leads to serious limitation of the electrode's cycling life. Much effort has been devoted to solve this problem. One of the effective approaches involves encapsulation of nanoscale metal particles with carbonaceous materials, as illustrated in Figure 6.8. The graphene-containing carbon capsule serves not only as a mechanically stable buffer but also as a conductive layer for electron and ion transportation (Barsukov and North Atlantic Treaty Organization 2006; Derrien et al. 2007).

Tin is one of the most commonly studied metals for alloying with lithium due to its low cost and high theoretical capacity. Various fabrication routes have been used to encapsulate tin with graphene-containing carbons. A relatively easy way is to prepare porous carbon followed by the introduction of tin nanoparticles into the pores of carbon to form a carbon/Sn hybrid in which the tin particles are surrounded by carbon frameworks (with a similar architecture shown in Figure 6.8d) (Fan et al. 2004; Grigoriants et al. 2005; Zhao et al. 2008). However, in these cases, the amount of tin loading in the pores is limited, resulting in relatively low capacity. A two-step approach involving preparation of tin particles first followed by coating of carbon layers has been reported to produce a carbon/Sn composite (Figure 6.8a) (Noh et al. 2005). A reversible capacity of 664 mA·h/g was obtained after 50 cycles, although the tin particles

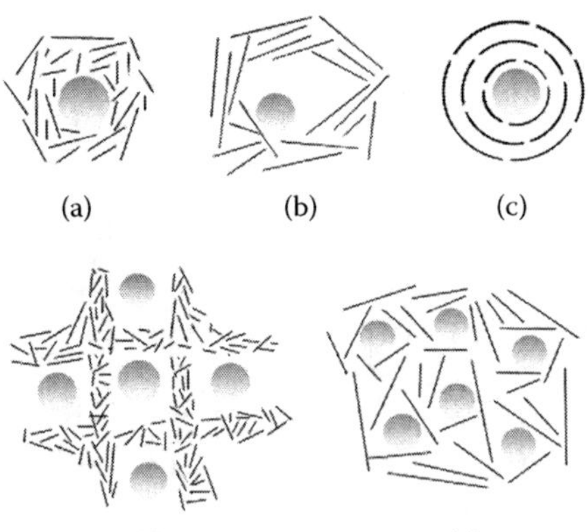

FIGURE 6.8
Schematic illustration of metal (oxides) nanoparticles encapsulated by graphenes with various structures and alignments. (From Liang M. H. and Zhi L. J., *Journal of Materials Chemistry*, 19(33), 5871–5878, 2009. Reproduced by permission of The Royal Society of Chemistry.)

obtained using this method were large. Alternatively, smaller tin nanoparticles were encapsulated successfully into a hollow carbon sphere (Figure 6.8b; Lee et al. 2003; Zhang et al. 2008b). The disadvantage of these materials is the complicated fabrication process, which usually needs several steps to encapsulate tin particles into the hollow carbon spheres, and the adjustment of graphene arrangement in the carbon is restricted. A simpler method has been developed for synthesizing hollow tin nanoparticles encapsulated with carbon through the one-step pyrolysis of organic-tin precursor (Figure 6.8e; Cui et al. 2007). Hollow tin nanoparticles having an average diameter of less than 20 nm were covered using a disordered carbon layer of 2–3 nm thickness (Figure 6.9). The material with a tin content of approximately 43 wt% exhibited high stability and a reversible capacity of 550 mA·h/g after 20 cycles, which was suggested to be closely related to the graphene-containing carbon capsule that can effectively relieve the expansion and shrinkage of tin particles during cycling.

Owing to its high theoretical capacity (4200 mA·h/g) when alloying with lithium, silicon is another interesting anode material for LIBs, although it is not actually a metal. Similar to tin, alloying of silicon with lithium is associated with a volume dilation of more than 300%, leading to pulverization and poor cyclability of the electrode (Besenhard et al. 1990). To solve this problem, the most common attempt is to encapsulate the Si structure with a conducting carbonaceous layer in the hope that this would better retain the Si fragments from being disconnected from the conducting electrode. Various carbon-rich precursors, such as resorcinol-formaldehyde gel (Lee et al. 2009),

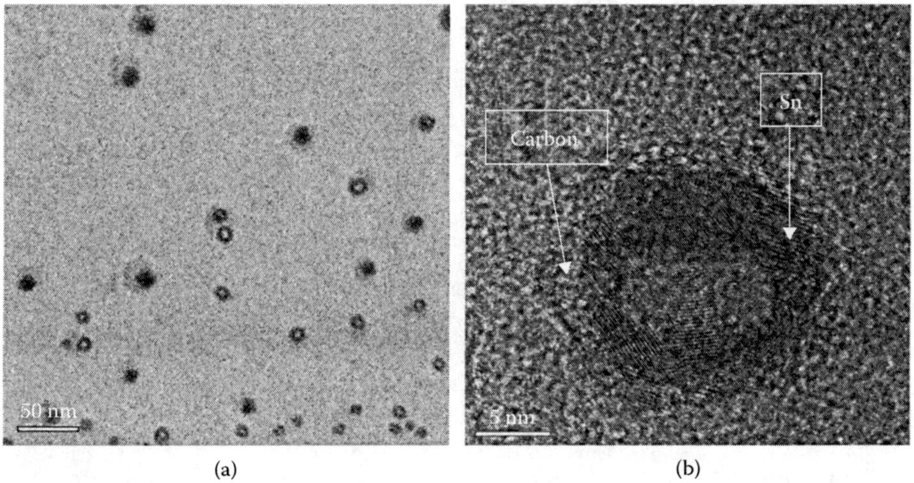

FIGURE 6.9
Typical TEM images of (a) a carbon/Sn composite and (b) its enlarged images, showing crystallized hollow tin nanoparticles coated with disordered carbon. (From Cui G. L. et al. A one-step approach towards carbon-encapsulated hollow tin nanoparticles and their application in lithium batteries. *Small*. 2007. 3(12). 2066–2069. Copyright Wiley-VCH Verlag GmbH & Co. KGaA. Reproduced with permission.)

poly(vinyl chloride)-co-vinyl acetate (Liu et al. 2009b) or polyvinyl chloride and chlorinated polyethylene (Si et al. 2009), and glucose (Hu et al. 2008), have been used for encapsulation. Although noticeable improvements were achieved, capacity degradation was not eliminated.

Since GNSs are flexible and naturally corrugated, deposition of metallic nanoparticles on graphene forms a three-dimensional (3D) nanocomposite architecture, which can not only effectively prevent the restacking of GNSs but also circumvent the aggregation of metallic atoms on cycling. Tin/graphene nanocomposite with such a 3D nanoarchitecture has been prepared; it has delivered enhanced electrochemical performance as an anode material in LIBs (795 mA·h/g in the second cycle and 508 mA·h/g in the hundredth cycle; Wang et al. 2009e), which represents a much improved performance compared to bare graphene electrodes and bare Sn electrodes. Silicon/graphene composites were fabricated as well by the high dispersion of Si nanoparticles between GNSs (Lee et al. 2010). This electrode material exhibits a high Li-ion storage capacity of >2200 mA·h/g after 50 cycles and >1500 mA·h/g after 200 cycles. However, the composition and structures of these nanocomposites still remain to be tuned to further improve their electrochemical performance.

With graphene-containing materials as the buffer matrix, other metals such as Ge, Sb, and alloys of two different metals (Kwon et al. 2007; Rock and Kumta 2007) are attractive candidates for use in LIBs as well. Germanium nanoparticles encapsulated with carbon have been synthesized by solid-state pyrolysis of organic-germanium precursors (Cui et al. 2008). The prepared carbon/Ge composite offered an initial discharge capacity of 1190 mA·h/g as anode material for LIBs, and its reversible capacity reached 920 mA·h/g. Modification of alloys with graphene-containing materials is expected to afford similar property improvement as that for the aforementioned metals. However, from the viewpoint of functions of graphene structure in the materials, further experimental studies are necessary.

Tin oxide (SnO_2) could be a good substitute for the carbon anode in LIBs because its theoretical capacity (782 mA·h/g) is much larger than that of graphite (372 mA·h/g). On the first discharge, the lithium bonds with the oxygen in SnO_2, forming tin metal and Li_2O. Then, the tin alloys/dealloys up to the theoretical limit of $Li_{4.4}Sn$ (capacity of 783 mA·h/g) (Winter and Besenhard 1999). However, some critical problems, such as large irreversible capacity (~700 mA·h/g) and severe fading capacity, exist, which limit the application of SnO_2 as anode material. The large irreversible capacity of SnO_2 electrode is associated with both the bulk formation of Li_2O and the intense surface reactions taking place between the Li-Sn compounds and the electrolyte solution.

On the other hand, the reaction of lithium with transition metal oxides is a conversion reaction. For instance, Co_3O_4 and Fe_2O_3 can be reversibly reduced and oxidized, coupled with the formation and destruction of lithium oxide, respectively (Poizot et al. 2001). The excellent reversibility of this reaction provides an outstanding lithium storage capability, which is usually three or four times more than that of graphite. A major drawback of these metal oxides as anodes in LIBs is their poor kinetics, which originates mainly from the cracking of metal oxide by conversion reaction into nanoparticles through which subsequent lithium-ion and/or electron transportation becomes worse and worse with cycling.

Similar to the metallic anode materials, one effective way to improve the electrochemical performance of metal oxides is to encapsulate them with carbonaceous materials that serve as not only a buffer to relieve the cracking of metal oxides but also a good ion and/or electron transport media in electrodes (see the illustration in Figure 6.8). Various methods have been developed to incorporate metal oxides with graphene-containing materials. A simple approach is to mix mechanically carbonaceous materials with metal oxides (Lee et al. 2000). However, the degree of hybridization is low with this method. A more efficient way to encapsulate metal oxides with graphene-containing carbon materials is to first prepare metal oxides colloids or dispersion with as small a particle size as possible, followed by solvothermal treatment under the presence of carbon resources to encapsulate the particles with carbon-rich materials. Subsequent carbonization affords metal oxides nanoparticles covered with graphene-containing carbon materials. Carbon/SnO_2 and carbon/Fe_3O_4 nanocomposite are obtained through this approach using sucrose and glucose as carbon sources, respectively (Zhang et al. 2008b,c). It should be noted that these carbon capsules are disordered carbons; hence, the composites display usually low initial coulombic efficiency compared to composites with graphitic materials (Read et al. 2001).

An alternative strategy for fabricating carbon-encapsulated metal oxides nanoparticles is an in situ method in which the formation of metal oxide particles and encapsulation are occured simultaneously. For example, electrospinning of a mixture of polyacrylonitrile (PAN) and $Fe(acac)_3$ followed by preoxidation and carbonization produced 1D carbon/Fe_3O_4 nanocomposite with small Fe_3O_4 particles encapsulated homogeneously into the carbon (Wang et al. 2008a). Recently, we developed a one-step approach toward Co_3O_4 nanoparticles encapsulated with thin graphene layers (Zhi et al. 2008). The carbon/Co_3O_4 composite was produced by solid-state pyrolysis of structurally defined organic-cobalt precursors. The prepared graphene-coated Co_3O_4 (~10 wt% carbon) exhibited an initial coulombic efficiency of 73%, and a subsequent coulombic efficiency of over 99%. Its initial reversible capacity reached 920 mA·h/g, and the capacity was still maintained at 940 mA·h/g even after 20 cycles (Figure 6.10). This dramatic improvement in cycling performance compared with that of naked Co_3O_4 nanoparticles was probably due to the efficient encapsulation of graphene on the particle as well as the existence of defects on the graphene structure, with perhaps a structure similar to that shown in Figure 6.8c.

It is noted that GNSs hold promise as matrices for metals and metal oxides to improve their electrochemical performance. Several reports demonstrate that the incorporation of GNSs indeed enhance the electrochemical performance of metal oxides or hydroxides, including SnO_2 (Paek et al. 2009; Wang et al. 2010b; Yao et al. 2009), TiO_2 (Wang et al. 2009a), Co_3O_4

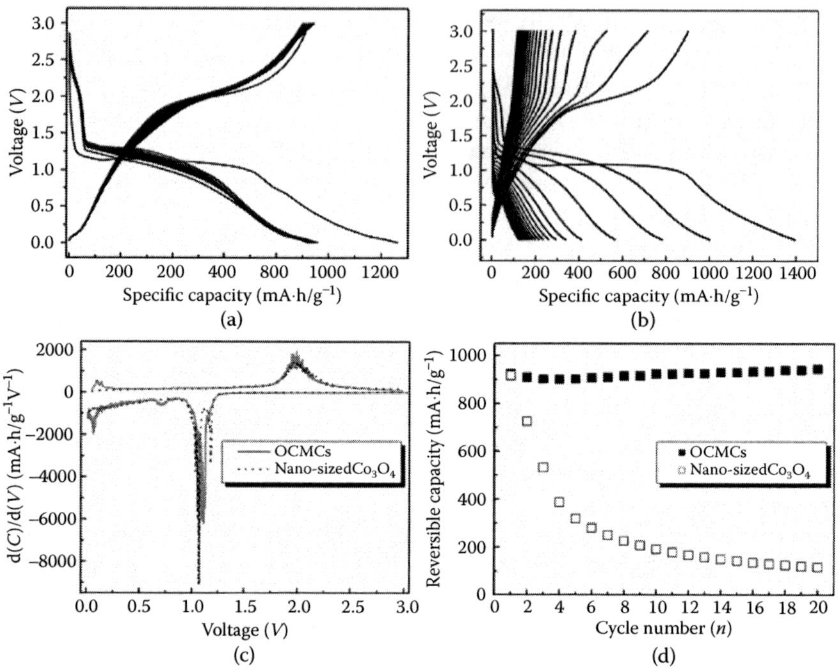

FIGURE 6.10
Galvanostatic discharge/charge curves of (a) graphene-covered Co_3O_4 nanoparticles (*O*-carboxymethyl chitosans [OCMCs]) and (b) naked nanosized Co_3O_4 electrodes cycled at the rate of *C*/5; (c) first-cycle differential capacity plots of OCMC electrode (solid line) and nanosized Co_3O_4 electrode (dotted line; note: *C* is capacity, *V* is cell voltage, and d(*C*)/d(*V*) is differential capacity); and (d) comparison of cycling performance between OCMCs and nanosized Co_3O_4 electrodes. (From Zhi L. J. et al. Precursor-controlled formation of novel carbon/metal and carbon/metal oxide nanocomposites. *Advanced Materials*. 2008. 20, 1727–1731. Copyright Wiley-VCH Verlag GmbH & Co. KGaA. Reproduced with permission.)

(Yang et al. 2010b), NiO (Wang et al. 2010b), and $Co(OH)_2$ (He et al. 2010b). For instance, Seung-Min Paek et al. (2009) reassembled the GNSs in ethylene glycol solution in the presence of rutile SnO_2 nanoparticles, as illustrated in Figure 6.11. The cycling performance of the SnO_2/GNS thus obtained is drastically enhanced in comparison with that of the bare SnO_2 nanoparticle. The charge capacity of SnO_2/GNS remained at 570 mA·h/g after 30 cycles; that is, about 70% retention of the reversible capacity (810 mA·h/g), whereas the specific capacity of the bare SnO_2 nanoparticle on the first charge was 550 mA·h/g, which dropped rapidly to 60 mA·h/g only after 15 cycles. The enhanced electrochemical performance is mainly attributed to the three-dimensionally delaminated flexible structure, in which the dimensional confinement of SnO_2 nanoparticles by the surrounding GNSs limits the volume expansion on lithium insertion, and the pores developed between SnO_2 and GNS can be used as buffered spaces during charge/discharge, resulting in superior cyclic performance. Further, GNSs in hybrid materials can act as not

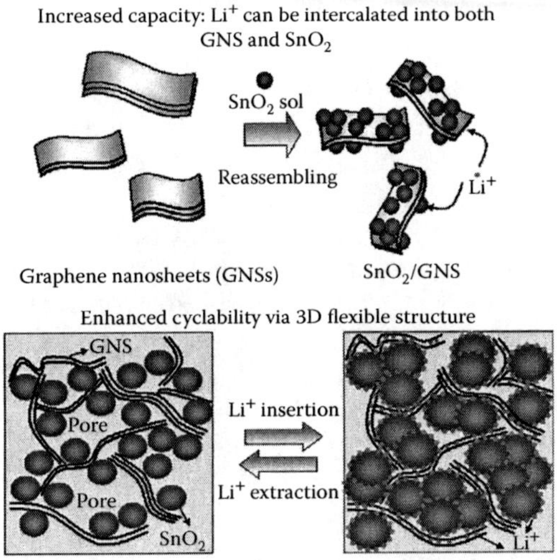

FIGURE 6.11
Schematic illustration of the synthesis and structure of SnO_2/GNS. (Reprinted with permission from S. M. Paek, E. Yoo et al. *Nano Letters*. 2009. 9(1): 72–75. Copyright 2009 American Chemical Society.)

only lithium storage electrodes but also electronic conductive channels to improve the electrochemical performance. For the same reasons, nanostructured TiO_2–graphene hybrid materials show significantly enhanced Li-ion insertion/extraction in TiO_2, especially at high charge rates, when compared with the pure TiO_2 phase (Wang et al. 2009a). However, those two composites were prepared by mechanically mixing nanoparticles with graphene dispersions, which limits the homogeneous dispersion of nanoparticles and the separation of GSs. An in situ chemical synthesis approach has been developed to prepare an SnO_2/graphene nanocomposite having a homogeneous distribution of SnO_2 nanoparticles (4–6 nm in size) on a graphene matrix. The as-synthesized SnO_2/graphene nanocomposite exhibits a reversible lithium storage capacity of 765 mA·h/g in the first cycle and enhanced cyclability.

Most of the aforementioned composites were fabricated by the assembly of GSs in the presence of inorganic precursors. Highly reversible capacities and good cycle performances were achieved when using these composites as anode materials in LIBs. Nevertheless, these inorganic precursors are still prone to strong aggregation and thus preclude homogenous dispersion in graphene-based composites. Thus, a new strategy was put forward for the purpose of fabricating organic metal/graphene composites using planar organometallic molecules, such as cobalt phthalocyanine (CoPc), as cobalt precursors (Yang et al. 2010b). Due to their pronounced π-interactions with GSs, they enable a homogenous dispersion of cobalt and cobalt oxide into/onto GSs by a pyrolysis and oxidation process. The Co_3O_4/graphene composites exhibit improved lithium storage performance than commercially available Co_3O_4 nanoparticles.

6.3.3 Summary

Carbon-based materials have been the dominant anode material in LIBs for a long time and may continue to play an important role in LIB manufacture in the future owing to their structural variability. Such carbon materials offer a great opportunity for optimizing micro- and nanostructuring in terms of not only capacity but also rate performance. From both technological and scientific points of view, disordered partially graphitic carbon looks most promising, as defects provide large excess capacities. Graphene holds great potential for developing LIBs due to its 2D nanostructure. A series of graphene-based materials, including GNSs and the composites of graphene with other nanophase materials, have been investigated and proved to be promising alternatives as anode materials in LIBs. However, research efforts for understanding the relationship between graphene and the LIB electrode are still in their early stages, for example, the relationship between the composition, structure, and electrochemical performance of graphene-based electrodes are still remaining for further investigations. For instance, the sheer increase in surface area of graphene-based materials has to compete with increased irreversible capacity resulting from electrolytic interaction. Therefore, control of structure and the alignment of graphene layers are essential to the improvement of electrochemical performance of the materials, whether pure graphene or graphene/metal (oxide) composites. In addition, it is reasonable to believe that graphene-based electrode materials with high energy density, good safety, and excellent performance will become one of the most competitive candidates as LIB electrodes commercialized in the near future.

6.4 Graphene-Based Materials for Supercapacitors

Supercapacitors can be divided into two types according to the mechanism of charge storage: (1) Electrochemical double-layer capacitors (EDLCs) and (2) pseudocapacitors. Figure 6.12 shows the schematic diagram of an EDLC consisting of a single cell with a high-surface-area electrode material, which is loaded with electrolyte. Recently, it was proposed that graphene should be a competitive material for supercapacitor application; several studies on graphene-based supercapacitors have been reported (Stoller et al. 2008; Vivekchand et al. 2008; Wang et al. 2009i; Zhao et al. 2009b). A number of reviews discuss the science and technology of supercapacitors for various configurations and electrode materials (Burke 2000; Conway 1999; Kotz and Carlen 2000; Sarangapani et al. 1996; Simon and Gogotsi 2008). In Sections 6.4.1 through 6.4.3, we focus on the recently developed applications of graphene-based materials in the two aforementioned types of supercapacitors.

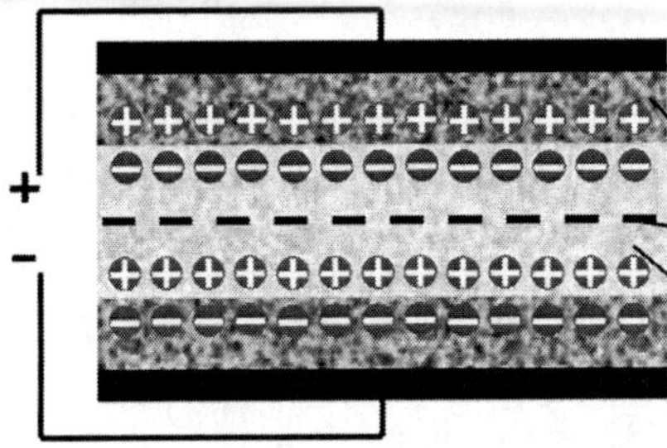

FIGURE 6.12
Principle of a single-cell double-layer capacitor.

6.4.1 Graphene-Based EDLCs

It is noted that EDLCs are electrochemical capacitors that store charge electrostatically using reversible adsorption of ions of the electrolyte onto active materials that are electrochemically stable and have high accessible specific surface area. Charge separation occurs on polarization taking place at the electrode–electrolyte interface, producing the double-layer capacitance C:

$$C = \frac{\varepsilon_r \varepsilon_0 A}{d}$$

where
- ε_r is the electrolyte dielectric constant,
- ε_0 is the dielectric constant of the vacuum,
- d is the effective thickness of the double layer (charge separation distance) and
- A is the electrode surface area

According to the reviews of Pandolfo and Hollenkamp (2006) and Obreja (2008), conventional carbon supercapacitors have a specific energy of 1–10 W·h/kg and specific power of 0.5–10 kW/kg. Tens or even hundreds of kilograms of supercapacitors are thus required for a single electric vehicle. A conventional lead-acid battery is typically 30–40 W·h/kg and modern LIBs are approximately 120 W·h/kg; therefore, it is important to improve the physical limits of EDLC and reduce its weight. Two factors limit the capacitance of supercapacitors: (1) Pore distribution and (2) resistance of the electrode material. To improve electrode performance, the pore distribution of the material should be optimized. To increase specific power, both the bulk resistance and interface resistance of electrode materials should be minimized.

Various forms of carbon are currently the most extensively examined and widely utilized electrode materials in EDLCs with development focusing on achieving high surface areas with low matrix resistivities. Graphitic carbon satisfies all the requirements for this application, including high conductivity, high electrochemical stability, and open porosity (Gogotsi 2006). Activated carbon and carbon fabrics, fibers, nanotubes, onions, and nanohorns have been tested for EDLC applications (Futaba et al. 2006; Portet et al. 2008; Yang et al. 2007). Activated carbons are the most widely used materials today because of their high specific surface area and moderate cost. A porous network comprising micropores (<2 nm in size), mesopores (2–50 nm), and macropores (>50 nm) can be created in the bulk of the carbon particles after activation. Longer activation time or higher temperature leads to larger mean pore size. The double-layer capacitance of activated carbon reaches 100–120 F/g in organic electrolytes; this value can exceed 150–300 F/g in aqueous electrolytes but at a lower cell voltage because the electrolyte voltage window is limited by water decomposition. Due to the relatively high electrical resistance of activated carbon materials, commercial electrodes are limited in thickness and usually contain conductive additives such as carbon black to enable rapid electrical charge transfer.

As previously mentioned in this section, many carbons have been tested for EDLC applications and a recent study provides an overview of what has been achieved to this end (Pandolfo and Hollenkamp 2006). One conventional 1D carbon material, CNTs, which have a tubular geometry comprising one or more graphene layers, has been fabricated for supercapacitors since 1997 (An et al. 2001; Diederich et al. 1999; Liu et al. 2008a; Niu et al. 1997). Both theoretical and experimental measurements have demonstrated that CNTs and graphene have ultralow resistivity for electron transport (Areshkin et al. 2007). However, CNT-based supercapacitors do not meet the expected performance; one possible reason is probably the observed contact resistance between the electrode and current collector (Shaijumon et al. 2008). Hence, many studies have focused on the morphology of carbon materials to boost the performance of the capacitor, such as growing CNTs directly on bulk metals to eliminate the contact resistance (Lota et al. 2008; Talapatra et al. 2006; Yang et al. 2007).

Besides conventional CNTs, several new styles of 1D carbon nanostructure have been designed and investigated as electrodes of supercapacitors. For instance, one type of graphene-based materials can be obtained by a bottom-up approach in a more controlled fashion. Highly porous carbonaceous materials with 1D nanostructures have been prepared using carbon-rich polyphenylene as precursor with subsequent pyrolysis at high temperatures (Liang et al. 2009a). Materials obtained at 800°C having a specific surface area as high as 1140 m^2/g have been selected as electrodes in EDLCs. A superior specific capacitance of up to 304 F/g was obtained at a scan rate of 5 mV/s, and the capacitance was 243 F/g at a scan rate of 100 mV/s (Figure 6.13). The enhanced capacitive behavior of this material might be attributed to the proper alignment of GSs as well as the interconnected nanoscale channels distributed in a 1D architecture.

Further, 1D fibers and tubes are constructed through the oriented carbon–carbon cross-linking reactions to form rigid conjugated polymer networks (Feng et al. 2009). As a result, template-free, one-step synthesis of CNTs and CNFs is achieved through simple carbonization of the as-formed carbon-rich tubular and fiber-like polyphenylene precursors under argon. Microporous

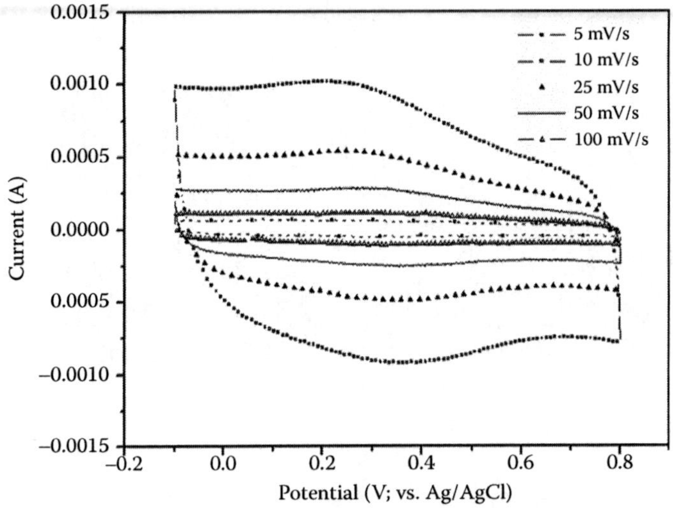

FIGURE 6.13
Cyclic voltammograms of 1D mesoporous nanocarbon at sweep rates of 5, 10, 25, 50, and 100 mV/s in 1 M H_2SO_4 electrolyte. (From Y. Y. Liang, X. L. Feng et al. A simple approach towards one-dimensional mesoporous carbon with superior electrochemical capacitive activity. *Chemical Communications*. 2009. 7. 809–811. Copyright Wiley-VCH Verlag GmbH & Co. KGaA. Reproduced with permission.)

CNTs and CNFs with surface areas up to 900 m^2/g are obtained, together with high-resolution transmission electron microscopy (HR-TEM) characterizations indicating the formation of intrinsic microporous structure in these rigid carbon-rich networks. The primary electrochemical experiments reveal their promising applications as advanced electrodes in EDLCs.

Recently, it was proposed that graphene should be a competitive material for supercapacitor application (Si and Samulski 2008a) and several studies on supercapacitor application have been reported based on a multilayered graphene material (Stoller et al. 2008; Vivekchand et al. 2008; Wang et al. 2009i). In contrast to the conventional high-surface-area materials, the effective surface area of graphene materials as capacitor electrodes does not depend on the distribution of pores in the solid state (Stoller et al. 2008), which is different from the current supercapacitors fabricated with activated carbons and CNTs (Diederich et al. 1999; Pandolfo and Hollenkamp 2006; Yang et al. 2007). Obviously, the effective surface area of graphene materials should depend highly on the layers. Single- or few-layered graphene with less agglomeration should be expected to exhibit higher effective surface area and thus better supercapacitor performance (Wang et al. 2009b).

Graphene-based supercapacitors reported by Ruoff et al. (Stoller et al. 2008) were based on chemically modified graphenes (CMGs) that have good electrical conductivities ($\sim 2 \times 10^2$ S/m) and large surface areas (705 m^2/g). Specific capacitances of 135 and 99 F/g in aqueous and organic electrolytes, respectively,

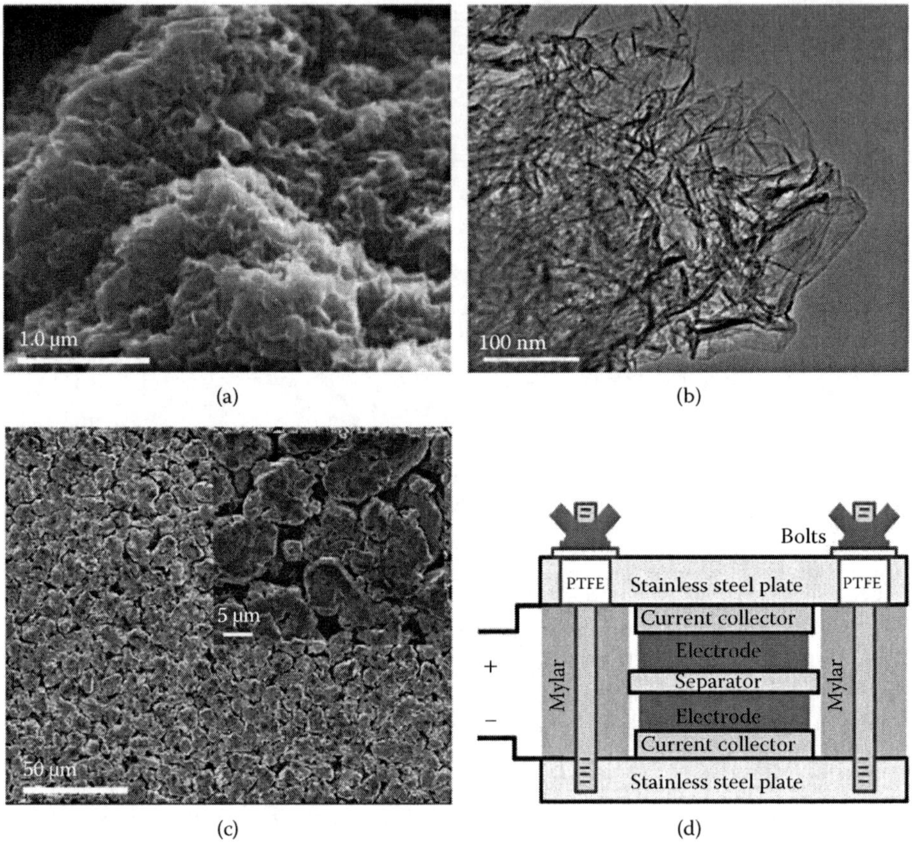

FIGURE 6.14
(a) Scanning electron microscopy (SEM) image of CMG particle surface, (b) TEM image showing individual GSs extending from the CMG particle surface, (c) low-magnification and high-magnification (inset) SEM images of the CMG particle electrode surface, and (d) a schematic of the test cell assembly. (Reprinted with permission from M. D. Stoller, S. J. Park et al. *Nano Letters*. 2008. 8(10): 3498–3502. Copyright 2008 American Chemical Society.)

have been measured (for schematics of graphene-based electrodes for ultracapacitors and for an ultracapacitor design; see Figure 6.14). The main innovative idea behind CMG electrodes is that the supercapacitor does not depend on the rigid porous structure of activated carbon to deliver its large surface area but on the flexible GSs, which adjust their position depending on the electrolyte used. Yan Wang et al. reduced the single-layered GO sheets using gas-based hydrazine reduction at room temperature (Wang et al. 2009i) as a result of which the reduced graphene materials have a lower degree of agglomeration than the CMGs prepared in aqueous solution at 100°C (Stoller et al. 2008). In comparison, graphene materials prepared by thermal treatment of graphite oxide have been reported as electrodes for EDLCs (Vivekchand et al. 2008). The capacitance of this material was 117 F/g at a scan rate of 100 mV/s and

100 F/g at a scan rate of 1000 mV/s, with H_2SO_4 aqueous solution as electrolyte. The highest power density reached by this material was 32 kW/kg. Using ionic liquid as electrolyte, a capacitance of 75 F/g was obtained.

The functional groups on the surface of functionalized GSs have significant influence on supercapacitive behaviors. Du et al. (2010) produced two kinds of functionalized GSs by thermal exfoliation of graphite oxide: (1) One was produced at a low temperature in air and (2) the other was produced at a higher temperature in N_2. The electrochemical results indicate that the specific capacitance values of the first kind of functionalized GSs derived from low temperature (230 F/g) are almost twice as much as those of the second kind (100 F/g). Lv et al. (2009) fabricated graphene by vacuum-promoted low-temperature (200°C) exfoliation of graphite oxide, and a high capacitance of up to 264 F/g in aqueous system was obtained. This is due to the fact that a lot of functional groups, such as hydroxyl, carboxyl, and epoxy groups, exist on the surface of the first kind of sheets. The groups result in fast redox processes occurring at or near the electrode surface, which can provide pseudocapacitance and enhance charge storage capability.

Recent efforts have led to solution-processable GNSs by chemical exfoliation of graphite into GOs, followed with solution reduction and functionalization (Li et al. 2008a). The availability of solution-processable GNSs allowed the formation of GN-based films through various solution-processing methods, including solution casting (Becerril et al. 2008; Cai et al. 2008), filtration (Xu et al. 2008), electrophoretic deposition (Wu et al. 2009), and Langmuir-Blodgett deposition (Li et al. 2008c). However, most of the aforementioned techniques cannot prevent the aggregation of graphene effectively, leading to the loss of surface area for energy storage.

In order to solve this problem, one feasible approach is to use other objects such as CNTs (Yan et al. 2010d; Yu and Dai 2010), carbon black (Yan et al. 2010c), and metal oxide nanoparticles (Li et al. 2009b; Lu et al. 2010b; Zhang et al. 2009b) to physically separate 2D GNSs to preserve graphene's high surface area. The GNSs/carbon black composites were prepared by ultrasonication and in situ reduction methods (Yan et al. 2010c). The electrochemical performances of these hybrid materials are superior to those of pure graphene material, demonstrating that carbon black particles as spacers ensure high electrochemical utilization of graphene layers as well as the open nanochannels provided by the 3D hybrid material. Dingshan Yu and Liming Dai (2010) prepared water-soluble polyethyleneimine (PEI)-modified GSs and used them for sequential self-assembly with acid-oxidized MWCNTs, forming hybrid carbon films. These hybrid films exhibited a nearly rectangular cyclic voltammogram even at an exceedingly high scan rate of 1 V/s with an average specific capacitance of 120 F/g due to its interconnected network of carbon structures with well-defined nanopores.

Very recently, a novel strategy to prepare 3D CNT/graphene sandwich (CGS) with CNT pillars grown in the graphene layers by CVD was reported (Figure 6.15) (Fan et al. 2010). In this structure, CNTs are grown between the GSs

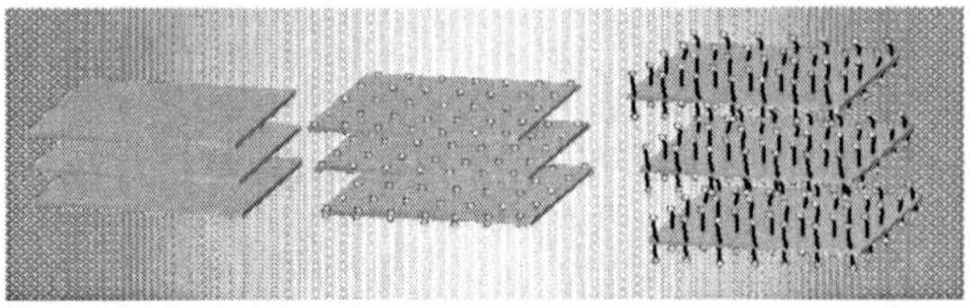

FIGURE 6.15
Illustration of the formation of hybrid materials with CNT grown onto layered GNSs, showing stacked layers of GNSs (left panel), catalyst particles adhered onto the layer surface after deposition (middle panel), and aligned CNTs between layers after growth (right panel). (From Z. J. Fan, J. Yan et al. A three-dimensional carbon nanotube/graphene sandwich and its application as electrode in supercapacitors. *Advanced Materials*. 2010. 22(33). 3723–3728. Copyright Wiley-VCH Verlag GmbH & Co. KGaA. Reproduced with permission.)

and distributed uniformly but sparsely on the whole sheet surface. The CGS, with an improved BET surface area of 612 m^2/g compared with that of graphene (202 m^2/g), has been used successfully as electrode in supercapacitors. Due to the rapid transport of electrolyte ions or electrons throughout the sandwich structure and comprehensive utilization of pseudocapacitance from catalyst and double-layer capacitance from graphene, a maximum specific capacitance of 385 F/g has been obtained at a scan rate of 10 mV/s in 6 M KOH solution.

6.4.2 Graphene-Based Pseudocapacitors

In addition to EDLCs (described in Section 6.4.1), another class of supercapacitors that is based on pseudocapacitance can be used. Whereas the charge storage mechanism of EDLCs is non-faradic, pseudocapacitance is based on faradic, redox reactions using electrode materials such as CPs and metal oxides. The energy densities of pseudocapacitance-based devices can be greater than those of EDLCs; however, phase changes within the electrode resulting from the faradic reaction limit the lifetime and power density of the devices.

The most commonly investigated classes of pseudocapacitive materials are transition metal oxides (notably, ruthenium oxide [Cui et al. 2007]) and CPs (Simon and Gogotsi 2008) such as PANI, polypyrrole, or derivatives of polythiophene. Particularly, PANI is a typical CP with good environmental stability, interesting electroactivity, and unusual doping/dedoping chemistry. Further, nanostructured PANIs such as PANI–nanofibers (PANI-NFs) can be facilely prepared by interfacial or rapid-mixing polymerization; they are applied in fabricating chemical sensors, actuators, memory devices, batteries, and supercapacitors (Huang et al. 2003; Li et al. 2009a). However, chemically prepared nanostructured CPs including PANI-NF are usually powdery and insulating in their dedoped states (Huang and Kaner 2004, 2006). Thus, various porous carbon materials (e.g., activated carbon, mesoporous carbon, and CNTs) and a polymer binder (e.g., Nafion) are usually used as additives for fabricating CP-based electrodes (Beguin et al. 2005; Chen et al. 2008; Ci et al. 2007; Futaba et al. 2006; Jang et al. 2005; Kaempgen et al. 2007;

Li et al. 2009c; Peng et al. 2008; Pushparaj et al. 2007; Zhang et al. 2010c). The incorporation of an electrochemically active second phase in a carbon-based freestanding electrode can dramatically enhance electrode capacitance (Chou et al. 2008). Nevertheless, the less active surface of carbon materials always prevents them from exhibiting high capacitance performance. Graphene, as a new member of the carbon nanomaterials family, has been applied as well in fabricating composites with polymers (Stankovich et al. 2006). Composites of graphene and polymers are of scientific and industrial interest because of their enhanced properties arising from the high conductivity and reinforcement of graphene nanofillers (Rao et al. 2009; Stankovich et al. 2006). For instance, composites of PANI and graphene, graphite oxide, and graphene-based nanosheets or graphene paper were successfully prepared by in situ chemical or electrochemical polymerization, covalent or noncovalent functionalization, and self-assembly (Bai et al. 2009; Bissessur et al. 2006; Higashika et al. 1999; Murugan et al. 2009; Wang et al. 2009c,f; Wu et al. 2010a; Xiao et al. 2000; Yan et al. 2010d; Zhang et al. 2010b; Zhao et al. 2009).

Various CPs have been widely studied as electrode materials for supercapacitors because of their high capacitance, ease of production, and low cost (Simon and Gogotsi 2008). However, poor conductivity and weak flexibility of CPs limit them from being used in high-performance flexible supercapacitors. Graphene paper shows excellent tensile modulus up to 35 GPa and room-temperature electrical conductivity of 7200 S/m (Li et al. 2008a). These intriguing characteristics enable graphene paper to be a freestanding electrode.

To exploit the potential of graphene-based materials for supercapacitor applications, graphene/PANI composite paper was prepared by in situ anodic electropolymerization of aniline on graphene paper; as a result, electrochemical capacitance as high as 233 F/g was achieved (Wang et al. 2009c). However, the capacitance was mainly dominated by the pseudocapacitance from the PANI films coated on the graphene paper surface, and the electric double-layer capacitance from the GS was limited due to the agglomerated layer-like structure in the graphene paper. Meanwhile, GO and GNs have also been doped into the PANI matrix (Murugan et al. 2009; Wang et al. 2009f; Yan et al. 2010d). They synthesized serious composites under different mass ratios, using graphite of two different sizes (12,500 and 500 mesh) as starting materials (Wang et al. 2010d). The result shows that the electrochemical performance of the prepared composites was influenced dramatically by the size of the raw materials and the mass ratio of GO over PANI. The enhanced specific capacitance and cycling life implies a synergistic effect between the two components. In the former case, only a small amount of insulating GO (e.g., 1%) was used because excessive GO reduces the conductivity of the composite (Wang et al. 2009f). Its high specific capacitance of 531 F/g was obtained in the potential range from 0 to 0.45 V at 200 mA/g by charge/discharge analysis compared to 216 F/g of individual PANI. It is noted that doping and the ratio of GO have a pronounced effect on the electrochemical capacitance of the composites. In the latter cases, GNSs were first prepared by reduction of GO followed by the polymerization of aniline, and the GSs were in a partially agglomerated form (Murugan et al. 2009; Yan et al. 2010d). Kai Zhang et al. (2010) prepared graphene/PANI composites using an in situ polymerization method, aiming at achieving a homogeneous dispersion of individual GSs within the polymer matrix. The aniline monomer was first mixed with GO sheets in aqueous solution by sonicating to form a homogeneous suspension before adding oxidant to this mixture. Finally, the composites were reduced by hydrazine, followed by reoxidation and reprotonation to obtain GSs/PANI-NFs composites. The mass ratio of graphene over PANI was tuned such that the composites had either PANI as the main component (i.e., graphene-doped PANIs) or GSs as the main component (i.e., PANI-doped graphenes). A specific capacitance as high as 480 F/g at a current density of 0.1 A/g was achieved with a PANI-doped graphene composite.

In most of the previous studies, graphite oxide or aggregated graphene rather than stable dispersions of GSs is used as the starting material. However, graphite oxide is an insulator and aggregated graphene sacrifices both large specific surface area and outstanding single-layer electric property of graphene (Li et al. 2008c; Stankovich et al. 2007; Wang et al. 2009b). Very recently, Qiong Wu et al. (2010) reported a novel method for preparing stable aqueous dispersions of CCG/PANI-NFs composites. By filtrating the mixed dispersions, paper-like composite films of CCG and PANI-NFs (G-PNF) were produced. Supercapacitor devices based on this conductive, flexible composite film exhibited large electrochemical capacitance (210 F/g at a discharge rate of 0.3 A/g) and improved electrochemical stability and rate performances.

Besides CPs, it is well known that metal oxides such as RuO_2, IrO_2, MnO_2, and NiO_x can improve the capacitance of carbon-based supercapacitors as they can contribute pseudocapacitance to the total capacitance in addition to the double-layer capacitance from carbon materials. At the same time, loading of metal oxides can avoid or decrease the possibility of serious agglomeration and restacking of GNSs and consequently provide a higher available electrochemically active surface area (ECSA) for increasing the energy storage capability of EDLCs. There have been several reports related to graphene–metal oxide or hydroxide (SnO_2 [Li et al. 2009b; Lu et al. 2010], ZnO [Lu et al. 2010; Zhang et al. 2009b], RuO_2 [Wu et al. 2010b], Mn_3O_4 [Wang et al. 2010a], Co_3O_4 [Yan et al. 2010b], $Ni(OH)_2$ [Wang et al. 2010c]) composites as electrode materials used in supercapacitors.

For instance, Li et al. (2009b) reduced GO with $SnCl_2$ to produce graphene–SnO_2 composite, which showed better capacitive behavior with a specific capacitance of 34.6 F/g in 1 M H_2SO_4 solution. A graphene–ZnO composite film was fabricated by ultrasonic spray pyrolysis (USP); it had a higher specific capacitance (about 11.3 F/g) in comparison with pure graphene or ZnO electrode (Zhang et al. 2009b). Further, Zhang et al. (2009) synthesized graphene, graphene–ZnO, and graphene–SnO_2 films and used them as electrode materials for electrochemical supercapacitors. The screen-printing approach was used to fabricate graphene film on a graphite substrate, whereas ZnO and SnO_2 were deposited on graphene films by USP. The results showed that the incorporation of ZnO or SnO_2 improved the capacitive performance of graphene electrodes. The graphene–ZnO composite electrode exhibited a higher capacitance value (61.7 F/g) and the maximum power density (4.8 kW/kg) compared to graphene–SnO_2 and pure graphene electrodes. Hydrous RuO_2-anchored GSs were prepared by combining sol-gel synthesis and low-temperature annealing and used as a composite electrode in supercapacitors (Wu et al. 2010b). The specific capacitance is greatly increased from 148 $F{\cdot}g^{-1}$ for the as-prepared GSs to 570 $F{\cdot}g^{-1}$ for the RuO_2-GSs composite with 38.3 wt% Ru. The RuO_2-GSs-based supercapacitors exhibit enhanced rate capability, excellent electrochemical stability (~97.9% after 1000 cycles), and high energy density (20.1 $W{\cdot}h{\cdot}kg^{-1}$) and power density (10,000 $W{\cdot}kg^{-1}$).

Recently, a two-step method to grow $Ni(OH)_2$ nanocrystals on graphene with various degrees of oxidation was reported by Dai et al.(Wang et al. 2010e). Further, the composites were investigated as electrochemical pseudocapacitor materials for potential energy storage applications; they exhibited a high specific capacitance of approximately 1335 F/g at a charge/discharge current density of 2.8 A/g and approximately 953 F/g at 45.7 A/g with excellent cycling ability (Wang et al. 2010c). However, a simple physical mixture of presynthesized $Ni(OH)_2$ nanoplates and GSs shows lower specific capacitance, which highlights the importance of direct growth of nanomaterials on graphene to enable intimate interactions and efficient charge transport between the active nanomaterials and the conducting graphene network.

6.4.3 Summary

Graphene is emerging as a unique morphological carbon material with potential applications in electrochemical energy storage devices. Regarding supercapacitor applications, current researches demonstrate that the effective surface area of graphene materials as capacitor electrode materials does not depend on the distribution of pores at solid state but on the layers. Electrode materials comprising single- or few-layered graphenes having less agglomeration are expected to exhibit higher effective surface area and, thus, better supercapacitor performance. Efforts are needed to further improve the performance characteristics by modifying or improving the quality of the graphene samples as well as the fabrication of graphene into well-defined structures so that its surface and electrical properties can be fully utilized. Since incorporation of an electrochemically active second phase, such as CPs and metal oxides, in a carbon-based freestanding electrode can dramatically enhance electrode capacitance, the composites of graphene and polymers or metal oxides are of scientific and industrial interest and may become competitive materials for supercapacitor application. However, the investigation on graphene-based hybrid electrode materials for application in supercapacitors is not nearly enough so far.

6.5 Graphene-Based Materials for Catalysis

Various graphene-based materials, such as functionalized GSs, doped graphene, and graphene/metal or metal oxide composites, are emerging and being investigated as catalysts for electrocatalytic reactions in fuel cells or for other traditional catalytic reactions. In Sections 6.5.1 through 6.5.3, we briefly summarize the recent progress made on the investigation of graphene-based materials that serve as either catalysts or catalyst supports in catalytic reactions.

6.5.1 Graphene-Based Electrocatalysts for Fuel Cells

Fuel cells are devices that generate electricity from chemical energy produced on the electrodes where electrochemical reactions of fuels and oxygen occur. Among all kinds of fuel cells, the PEM fuel cell (PEMFC) is attractive due to its high energy conversion efficiency, low pollution, low weight, high power density, and low operating temperature and it is used for a wide range of applications in power supplies from automobiles and automatic bicycles to electronic devices such as notebooks and cellular phones (Acres et al. 1997; Brandon et al. 2003; Zhong et al. 2010). Generally, the PEMFC comprises anode, cathode, electrolyte, and a membrane to separate the fuels and oxygen. Catalysts are loaded on the electrode to catalyze electrochemical reactions between fuel molecules and oxygen. On the anode, fuels such as H_2 and methanol are converted into protons and electrons. The protons produced pass through the membrane to reach the cathode, whereas electrons flow from the anode to the cathode through the outside circuit. At the same time, the oxygen molecules on the surface of catalysts loaded on the cathode are reduced to form water by combining with the protons. As a result of this process, the fuel burns out gradually and external current is produced.

Factors influencing the output potential of PEMFCs involve the operating temperature and concentrations of fuels and oxygen, the activation overvoltage from both anode and cathode, and the ohmic overvoltage. The oxygen reduction reaction (ORR) on the cathode plays a predominant role in determining the activation overvoltage because the anode overvoltage is very small when pure hydrogen is used. Thus, the catalysts for ORR are critical in the performance of PEMFCs. Diffusion and cell resistance play important roles as well in determining overall cell performance. Direct methanol fuel cells (DMFCs), having a theoretical energy density of 6094 W·h/kg and a theoretical potential of 1.2 V, are a kind of PEMFCs having methanol as the fuel. However, the practical output cell voltage of DMFC is 0.4 V and the energy density is only 1500–3100 W·h/kg as a result of the overpotential on cathode and the irreversible methanol oxidation and ORRs taking place. During the catalytic oxidation of methanol on the usually used platinum-based catalysts, some intermediates, such as carbon monoxide and formic acid, may poison the catalysts and reduce the catalytic activity. Modification of Pt with other metals, such as ruthenium, can resist the poison somewhat. In addition, for practical applications it is necessary to overcome the drawbacks of catalysts such as low activity, poor stability, and high cost.

Graphene is naturally a good support for electrocatalysts due to its excellent electrical conductivity and high specific surface area (Li et al. 2009g; Seger et al. 2009; Wang et al. 2010). The metal-support interaction can be modified by conveniently doping graphene with other elements such as nitrogen or boron to ensure better performance of fuel cells (Huang et al. 2009; Qu et al. 2010.) Chemically exfoliated graphene can be dispersed in solvents, such as water or ethanol, which provides the possibility of assembling the desired composites with well-defined structures. There are some exciting experimental results regarding graphene-based electrocatalysts for fuel cells. These research works can be introduced as two sorts: (1) One is the oxidative reaction of fuels on anode and (2) the other is the ORR on cathode. Because performances of the catalysts depend highly on their preparation methods and process, an introduction on the research work involves the preparation of catalysts to know the difference between them and their performances.

6.5.1.1 Graphene-Based Catalysts on Anode

The preparation and electrochemical performance of methanol oxidation of Pt/graphene nanocomposites were reported in 2009 (Li et al. 2009g). The Pt/graphene catalyst with 45 wt% Pt loading was prepared by reducing a mixture of GO and H_2PtCl_6 with $NaBH_4$ under basic conditions (Figure 6.16).

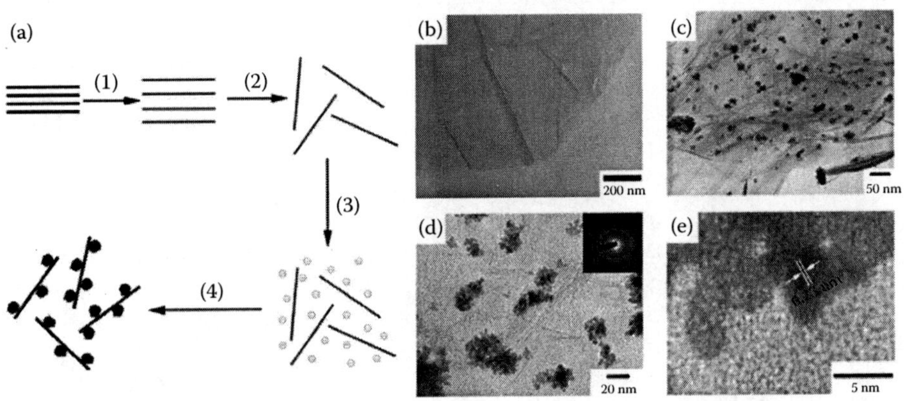

FIGURE 6.16
(a) Scheme of the route of synthesis of Pt/graphene nanocomposites: (1) Oxidation of graphite to graphite oxide; (2) exfoliation of graphite oxide in water by sonication; (3) addition of Pt ions to graphite oxide solution; and (4) chemical reduction of graphite oxide and Pt ions leading to the formation of Pt/graphene composites. The rest of the figure shows TEM images of graphite oxide (b), Pt/graphene composites at different magnifications (c and d), selected area electronic diffraction (SAED) of Pt/graphene (inset of d), and HR-TEM images of Pt/graphene (e). (Reprinted from *Electrochemistry Communications*, 11(4), Y. M. Li, L. H. Tang et al., Preparation and electrochemical performance for methanol oxidation of Pt/graphene nanocomposites, 846–849, 2009, with permission from Elsevier.)

The prepared catalyst has an ECSA of 44.6 m^2/g, which is higher than that of the reference sample Pt/Vulcan (30.1 m^2/g) according to electrochemical measurements. In the electrochemical test, Pt/graphene showed a lower current-peak potential than Pt/Vulcan by 12 mV, which indicated that Pt/graphene has a higher catalytically oxidative capacity than Pt/Vulcan. The catalytic stabilities of Pt/graphene and Pt/Vulcan in a solution of 0.5 M H_2SO_4 and 0.5 M CH_3OH were also tested. For both catalysts, a decrease in potentiostatic current occurred, which may be caused by the effect of formation of intermediate species, such as CO and CHO, on catalytic performance. However, the current density of Pt/graphene was higher than that of Pt/Vulcan, which means that the methanol oxidation stability of Pt/graphene was higher than that of Pt/Vulcan. All these results reveal that the methanol electrochemical oxidation performance of Pt/graphene is higher than that of Pt/Vulcan. The authors attributed the higher performance of Pt/graphene to the high specific surface area (SSA) of graphene, the good charge transport property of graphene, and the easy loading of catalyst nanoparticles on graphene.

A method was proposed to prepare graphene from polyacrylic weak acid cation–exchanged resin (AC resin) with exchanged Fe^{2+} as the precursor of the catalyst (Wang et al. 2010f). Under high temperature, AC-Fe^{2+} was converted to graphene material. The products resulting from this method consist of single-layer, few-layer, and multilayer graphene materials. The loading of Pt on this graphene material was realized by reducing H_2PtCl_6 with $NaBH_4$ in the base solution. The first methanol electroxidation peak on Pt/graphene in the solution of 1 M H_2SO_4 and 1 M CH_3OH was located at 0.748 V with a scan rate of 50 mV/s, which is lower than the peak of the reference catalysts of Pt/Vulcan (0.785 V). On the other hand, the higher current density (317.6 A/g Pt at 0.2 V) on Pt/graphene compared to that on Pt/Vulcan (193.2 A/g Pt at 0.28 V) can be observed from the electrochemical measurements. In the test on the durability of catalysts, the Pt/graphene catalyst displayed better methanol electroxidation properties than Pt/Vulcan. This conclusion is similar to that derived from the aforementioned experimental results.

Apart from the methods of preparing Pt/graphene nanocomposites by reducing noble metal salts with reducing reagents, a method to prepare Pt/graphene by electrodepositing Pt nanoparticles from a Pt salt on reduced expandable GO was reported (Liu et al. 2010b). The electrochemical tests on Pt/graphene and the reference catalyst Pt/glass carbon (Pt/GC) were executed in a solution of 0.5 M H_2SO_4 and 0.5 M CH_3OH. A forward peak current density of 7.41 mA/cm^2 at 0.92 V was obtained on the Pt/graphene catalyst, which can be attributed to the residual oxygen-containing functional groups. In all the tests, the Pt/graphene catalyst exhibited better performance than the Pt/GC catalyst, which may be caused by the synergistic effect between Pt nanoparticles and GSs.

Since Pt catalsts have weak ability to resist the poison caused by the intermediates, bimetal catalysts have been prepared in order to obtain better fuel oxidation performance. The comparison between Pt/graphene and Pt-Ru/graphene has been investigated (Dong et al. 2010). The Pt/graphene and Pt-Ru/graphene catalysts were prepared using the modified alkaline-ethylene glycol method; that is, reducing the metal oxide in a base solution in the presence of a support. The Pt/graphene showed a higher activity for methanol oxidation and a better tolerance to CO than the reference catalysts Pt/carbon black and Pt/graphite. Addition of Ru further enhanced the methanol oxidation performance of the Pt/graphene catalyst by decreasing the oxidation potential of methanol with 0.15 V in the solution of 0.5 M H_2SO_4 and 1 M CH_3OH at a scan rate of 50 mV/s. Further, the addition of Ru can release CO-like species adsorbed on the surface of Pt. For the ethanol oxidation reaction, the addition of Ru also lowers the potentials for ethanol oxidation with 0.06 V. These results reveal that modifying the Pt/graphene catalyst may be viable in improving the performance of fuel cells.

Another kind of metal, palladium, was also investigated to improve the catalytic performance of the Pt/graphene catalyst (Guo et al. 2010b). Before preparing the Pt-on-Pd/graphene catalyst, polyvinylpyrrolidone (PVP)-functionalized graphene was prepared. Then, deposition of Pd nanoparticles on graphene was achieved by reducing Pd salt with formic acid and, further, deposition of Pt was executed by reducing Pt salt with ascorbic acid. The obtained Pt-on-Pd nanoparticles have dendrite-like shapes which are of particular interest in catalysis due to their porous structure. This catalyst has a high ECSA of 81.6 m^2/g, which is higher than that of E-TEK catalyst (54.7 m^2/g). Methanol oxidation was performed on this catalyst in a solution of 0.5 M H_2SO_4 and 1 M CH_3OH. The mass current density for this catalyst is three times higher than that for the E-TEK catalyst. This result is disproportional with the ratio of the ECSAs of Pt-on-Pd/graphene catalyst and E-TEK catalyst, which may be caused by the synergic interaction between Pt and Pd. In addition, the Pt-on-Pd/graphene catalyst has better current density and stability than the E-TEK catalyst or other catalysts.

In addition to the experimental results, a theoretical prediction on the nanometer-sized hydrographene species as support material for fuel cells was made recently (Yumura et al. 2009). The small HOMO-LUMO gap for the nanometer-sized hydrographene species is preserved even after the Pt cluster binds to an edge of sp^2-hybridized carbon. The interaction energy between the Pt cluster and hydrographene is higher than that between the Pt cluster and infinitely sized graphene; that is, hydrographene is an appropriate support to load the Pt cluster. Moreover, the density functional theory (DFT) calculation revealed that spontaneous H-H bond cleavage mediated by bare Pt cluster is also expected, when a Pt cluster sits on nanometer-sized hydrographenes. These demonstrated that nanometer-sized hydrographene species are a promising candidate for an effective support for Pt catalysts of fuel cells.

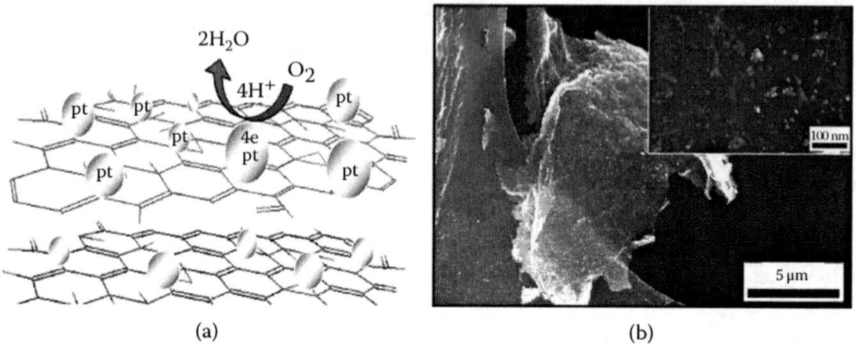

FIGURE 6.17
(a) Dispersion of Pt nanoparticles on a 2D carbon sheet (graphene) to facilitate an electrocatalytic reaction. (b) The SEM image of GO-Pt deposited on carbon Toray paper. This image shows a few stacked GSs lying across the carbon macrofiber of the Toray paper. The inset shows the magnified view of Pt nanoparticles dispersed on the GS. (Reprinted with permission from B. Seger and P. V. Kamat. *Journal of Physical Chemistry C*. 2009. 113(19): 7990–7995. Copyright 2009 American Chemical Society.)

6.5.1.2 Graphene-Based Catalysts on Cathode

The RGO-supported Pt catalysts applied in hydrogen/oxygen fuel cells were reported in 2009 (Seger et al. 2009). A GO-Pt catalyst (Figure 6.17) was prepared by reducing the mixed dispersion of GO and H_2PtCl_6 with $NaBH_4$ (GO: H_2PtCl_6 mass ratio of 1:1). The prepared catalyst has an ECSA of 16.9 m^2/g, which is comparable to that of carbon black–supported Pt catalysts (19 m^2/g). This reduced GO-Pt catalyst was fabricated into the PEM hydrogen/oxygen fuel cell as the cathode, and the output power had a maximum value of 161 mW/cm^2, which is higher than that of the unsupported Pt catalyst. However, the authors found that the hydrazine-reduced GO-Pt catalyst following heat treatment performed worse than the $NaBH_4$-reduced GO-Pt catalyst, which may be attributed to the loss of the proton conductivity property of the Nafion ionomer undergoing changes at high temperatures.

The electrocatalytic oxygen reduction performance of Pt catalysts supported on functionalized GSs (FGS) was reported in 2009 (Kou et al. 2009). An FGS with single-layer but wrinkled GSs was prepared by the chemical oxidation of graphite flakes and subsequent rapid thermal expansion. The Pt/FGS catalyst prepared using the impregnation method has an initial electrochemical surface area (ESA) of 108 m^2/g and a decreased ESA of 67.6 m^2/g after 5000 cycles of cyclic voltammetry. Further, the Pt sizes increased from 2 nm to 5.5 nm after the Pt/FGS catalyst experienced 5000 cycles. In comparison, the Pt sizes in E-TEK increased from 2.8 nm to 6.9 nm after 5000 cycles. The authors attributed the slower sintering of Pt nanoparticles in Pt/FGS to the interaction between Pt nanoparticles and the π-sites and functional groups in FGS.

The Pt/graphene nanoplatelet (Pt/GNP) catalyst was reported a durable catalyst for ORR (Shao et al. 2010). A GNP is produced by exfoliating intercalated graphite with a highly graphitic structure; thus, the properties of GNP are different from those of graphite and graphene. The Pt/GNP catalyst was prepared with the modified alkaline-ethylene glycol method, and the average size of Pt nanoparticles in this catalyst was 3.5 nm according to statistical results. The Pt/GNP has an ESA of about 40 m^2/g, which is similar to that of other catalysts of Pt/CNT and E-TEK Pt/C. However, the Pt/GNP catalyst showed enhanced durability (the loss of ESA: 40%, the loss of ORR: 20% after 44 hours) to other catalysts, such as Pt/CNT and E-TEK Pt/C (the loss of ESA and ORR: 75% after 44 hours). The authors also attributed the high stability of the catalyst to the interaction between Pt and GSs.

A composite of graphene and MWCNT (G-MWCNT) supporting Pt nanoparticles was prepared as an ORR catalyst with high performance (Jafri et al. 2010). The 3D structure of G-MWCNT is very different from the 2D structure of graphene and the 1D structure of CNT, which may be an efficient catalyst support for PEMFC and other catalytic reactions. Different compositions of graphene and MWCNT were investigated to prepare different catalysts. Dispersion of Pt nanoparticles on the surface of the supports was performed by reducing Pt salt with $NaBH_4$. The authors found that the ECSAs of Pt/MWCNT, Pt/(25% G+75% MWCNT), Pt/(50% G+50% MWCNT), Pt/(75% G+25% MWCNT), and Pt/G were 77, 100, 108, 98, and 95 m^2/g, respectively. The authors considered that the addition of MWCNT may change the morphology of GSs so that there is less possibility of the aggregation of Pt nanoparticles. The cathode electrocatalyst Pt/(50% G+50% MWCNT) performed well with a potential of 540 mV at 1 A/cm^2, and the highest output power density was 540 mW/cm^2. In addition, the low direct current resistance of the cathode catalyst of 0.21 Ω may be an important reason for the high output power density.

Because of the importance of the mechanism of OH radicals generation in ORR, the ORR activity of a Pt-Co/graphene catalyst was investigated (Yue et al. 2010). The Pt-Co/graphene catalyst was prepared by reducing a mixture of GO, Pt salt, and cobalt salt with $NaBH_4$, and the average size of the Pt-Co alloy was 15.4 nm. The ORR test of Pt-Co/graphene catalyst was performed in an alkaline solution of 0.1 M NaOH. The influence of Pt:Co ratios on ORR activity was studied, and the best Pt:Co atomic ratio was found to be 1:1. The OH radicals produced in ORR fluorometry were detected with scanning electrodes in an alkaline solution of terephthalic acid. The OH radicals are generated by the dissociation of HO_2^- according to the authors' deduction, and the content of Co is crucial for the generation of OH radicals.

A strategy was developed to prepare porous CNT from the carbon-rich molecules in the alumina template (Figure 6.18, Cui et al. 2007). The obtained CNT was used as the support to load the Pt nanoparticles with the electrodeposition method. (Cui et al. 2007). Mesoporous CNT (MPCNT; Figure 6.19) has a specific surface area of 1100 m^2/g. The performance of ORR on Pt/MPCNT

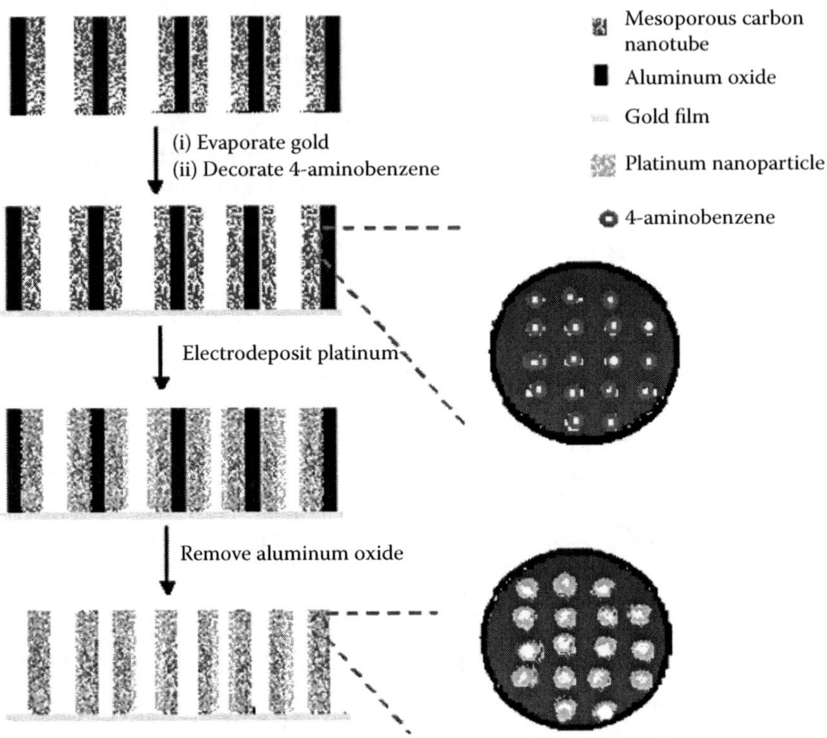

FIGURE 6.18
Preparation of MPCNT from carbon-rich molecules. (From G. L. Cui, L. J. Zhi et al. One-dimensional porous carbon/platinum composites for nanoscale electrodes. *Angewandte Chemie-International Edition*. 2007. 46(19). 3464–3467. Copyright Wiley-VCH Verlag GmbH & Co. KGaA. Reproduced with permission.)

FIGURE 6.19
(a) The TEM image of Pt/MPCNT; (b) the HR-TEM image of Pt/MPCNT; and the inset shows the SAED pattern. (From G. L. Cui, L. J. Zhi et al. One-dimensional porous carbon/platinum composites for nanoscale electrodes. *Angewandte Chemie-International Edition*. 2007. 46(19). 3464–3467. Copyright Wiley-VCH Verlag GmbH & Co. KGaA. Reproduced with permission.)

was tested with the Pt/C catalyst as the reference catalyst. It is found that the current density on Pt/MPCNT is about 1000 times as high as that on the Pt/C catalyst.

Developing noble metal–free catalysts will reduce the cost of catalysts, which may promote the commercialization of fuel cells. Some graphene-based noble metal–free catalysts have been prepared by treating the mixture of glucose, adenine, Fe (II) gluconate dehydrate, and Cu (II) gluconate (Maruyama et al. 2007). The prepared catalysts comprise carbon (a very large percentage [at least 85 wt%]), hydrogen, oxygen, nitrogen, Fe, and Cu. The x-ray diffraction (XRD) measurement of the prepared catalysts reveals that the formed carbon materials are in the form of graphene layers. With these catalysts, power density can reach up to 0.24 W/cm^2. Although the stability of these non–noble metal catalysts is lower than that reported for the Co–polypyrrole–carbon black composite catalyst, the output power density of the assembled PEMFC with the noble metal–free catalyst after 100 hours of operation was comparable to that of the initial performance. The PEMFC with noble metal–free catalyst still has a lower power output than that obtained with Pt-based catalysts; therefore, it is necessary to enhance the performance of the noble metal–free catalyst for practical applications.

Another example of using a graphene-based catalyst as a metal-free catalyst is nitrogen-doped graphene for ORRs in fuel cells. Nitrogen-doped graphene (N-graphene) was prepared using a modified CVD method, which involves treating the gas mixture of NH_3, CH_4, H_2, and Ar under high temperature (1000°C) so that the samples are deposited on a nickel film supported on a SiO_2/Si substrate (Qu et al. 2010). The graphene electrode having no dopants (C-graphene) prepared with the CVD method and the Pt/C electrode are the reference catalysts. The electric current density at the steady state for the ORR on N-graphene electrode is about three times higher than that of Pt/C or C-graphene over a large potential range, which indicates that N-graphene is a promising metal-free catalyst for ORR in alkaline solutions. The properties of resistance to methanol and CO of N-graphene are better than those of the Pt/C catalyst, which is determined by electrochemical tests in which methanol or CO is introduced into the systems. The stability of N-graphene for ORR is excellent so that no obvious decrease in current can be observed after 200,000 continuous cycles. These results reveal that doping on metal-free graphene is very helpful for improving ORR performance and that this doped graphene is a potential electrocatalyst that can replace noble metal catalysts if its catalytic activity is made comparable to that of noble metal catalysts.

In summary, graphene-based materials exhibit great potential for application in both anode and cathode in fuel cells. The high specific surface area of the active metal on graphene can be achieved, which is a critical factor influencing the performance of fuel cells. The strong interaction between metal nanoparticles and graphene containing some functional groups contributes to the high stability of the graphene-supported catalysts. The noble metal–free graphene-based catalyst is a material that can potentially replace noble metal catalysts for ORR. Future work on catalysts for fuel cells will be concentrated on increasing the catalytic performance of existing catalysts and searching for alternative catalysts having low cost and high activity.

6.5.2 Graphene-Based Catalysts for Use in Traditional Catalytic Reactions

In this section, graphene-based catalysts applied in traditional catalytic reactions are introduced. The graphene-based catalysts can be used not only as electrocatalysts for fuel cells but also as traditional catalysts with good performance. The GO was often used as the precursor of the catalyst support for its easy preparation.

The catalysts Pd/GO and Pd/graphene were prepared to test their catalytic performance on the Suzuki–Miyaura coupling reaction between aryl bromide molecules (Scheuermann et al. 2009). The Pd/GO was prepared by reducing the exchanged Pd^{2+} on GO (Pd^{2+}/GO) with hydrogen, and Pd/graphene was prepared by reducing Pd/GO with hydrazine hydrate. The Pd^{2+}/GO had the highest activity and Pd/graphene had the lowest activity for the Sizuki–Miyaura coupling reaction among the Pd/GO, Pd^{2+}/GO, Pd/graphene, and Pd/C catalysts. A turnover number (TON) of 6,700 and a turnover frequency (TOF) of 39,000 h^{-1} on the Pd^{2+}/GO catalyst can be obtained, and the leaching amount of Pd^{2+} was 1.1 ppm after the catalytic reaction was completed. The TEM of the used Pd^{2+}/GO catalyst shows the formed nanoparticles of small sizes, which may be a mixture of PdO and Pd^0 nanoparticles.

The graphene-supported copper–nickel bimetal catalysts were reported to be good catalysts for the direct synthesis of dimethyl carbonate (DMC) from methanol and carbon dioxide (Bian et al. 2009). The catalysts were synthesized by impregnating the salts of Cu and Ni onto the surface of GO and subsequently calcining at 400°C. The conversions of methanol on these bimetal catalysts were higher than 3.6%, and the highest conversion of 10.13% over the catalyst with a Cu/Ni molar ratio of 2:1 can be obtained. More than 85% of the converted methanol was converted to DMC over all the Cu-Ni/graphene catalysts. The stability test results revealed that the conversion of methanol and the selectivity and yield of DMC had a slight decrease after 10 hours of continuous reaction. The Cu-Ni alloy is the active compartment of the catalyst according to the reference experimental results on pure graphene, Ni/graphene, and Cu/graphene. The TEM and XRD characterization results of the Cu-Ni/graphene catalyst confirmed the formation of the Ni-Cu alloy in the catalysts.

The electrocatalytic oxidation of hydrazine in alkaline media into nitrogen was carried out on the reduced GSs-modified glassy carbon electrode (Wang et al. 2010g). The hydrazine RGO was fabricated on a glassy carbon electrode to be used as the active material for hydrazine electrocatalytic oxidation. For the reference electrode of glassy carbon, a large overpotential of 1.1 eV and a low anodic current from the oxidation of hydrazine can be observed. In comparison, a lower overpotential of 0.9 eV and a prominent anodic current of the hydrazine oxidation on the graphene-modified glassy carbon electrode can be obtained. These results indicate that graphene may be used as not only a catalyst for electrocatalytic oxidation of graphene but also an anode material for direct hydrazine fuel cells.

The applied field of graphene-supported catalysts has been extended to photocatalysis, such as the photodegradation of methylene blue on TiO_2 (P-25)/graphene (Zhang et al. 2010a). The TiO_2/graphene catalyst was prepared from GO and P-25 using the hydrothermal method. Compared with the reference catalysts P-25 and P-25/CNT, the TiO_2/graphene catalyst had higher efficiency on the photodegradation of methylene blue under irradiation of both ultraviolet and visible light. The better performance of TiO_2/graphene compared to P-25 may be due to its stronger interaction with methylene blue and broader light adsorption band.

Graphene can not only act as supports for metal or metal oxide catalysts but also play the catalytic role with its functionalized forms. Since the report of oxidized CNT as the active catalyst for the oxidative dehydrogenation of n-butane, the catalytic role of C = O binding with the conjugated system attracted great interest for developing the new types of catalysts (Zhang et al. 2008a). A molecule catalyst with six ketonic carbonyl groups per molecule (macrocyclic trimer of phenanthroquinone, MCT) was synthesized using a coupling reaction (Figure 6.20), which exhibited excellent performance for the ODH of ethyl benzene (Figure 6.21). Under the same reaction conditions, the reaction rate on MCT is 5–9 times as high as that on other nanocarbons, such as multiwalled nanotube, active carbon, and popcorn carbon. The catalyst experienced only a tiny change after catalyzing the ODH reaction of ethyl benzene according to the thermal analysis of fresh MCT and used MCT, which indicated that this MCT catalyst is very stable under the reaction conditions (Zhang et al. 2009a).

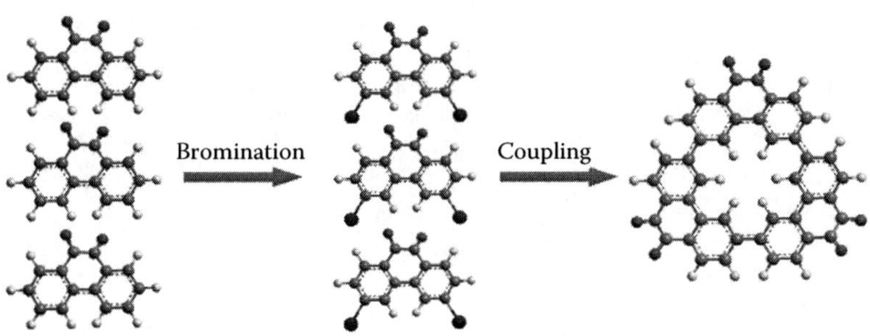

(1) (2) (3)

FIGURE 6.20
Schematic synthesis of the MCT oligomer. Compounds: (1) phenanthrenequinone; (2) 3,6-dibromo-phenanthrenequinone; and (3) macrocyclic trimer of phenanthroquinone, MCT. (Reprinted with permission from J. Zhang, X. Wang et al. *Journal of the American Chemical Society*. 2009. 131(32): 11296–11297. Copyright 2009 American Chemical Society.)

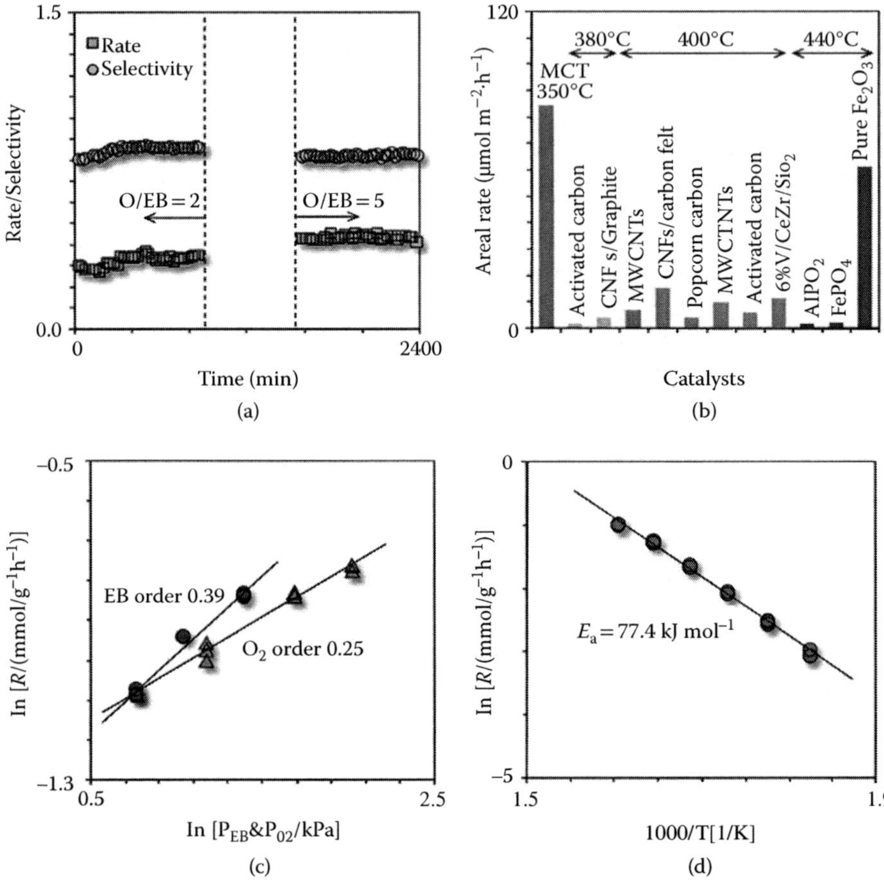

FIGURE 6.21
(a) Reaction rate (in millimole per gram hour) and styrene selectivity on MCT oligomer. The conditions are 20 mg, 2.1% EB (EB: ethylbenzene), O/EB = 2–5 (O/EB: O2/ethylbenzene), 12.5 mL·min^{-1}, and 350°C. The figure also shows a comparison of activities with typical catalysts in the literature (b), and dependencies of reaction rate on partial pressures of each reactant (c) and temperature (d). (Reprinted with permission from J. Zhang, X. Wang et al. *Journal of the American Chemical Society*. 2009. 131(32): 11296–11297. Copyright 2009 American Chemical Society.)

6.5.3 Summary

Obviously, the graphene-based catalysts, including graphene-supported catalysts and decorated graphene catalysts, have intensive application perspectives in various catalytic fields. Different from other carbon materials, graphene can be easily modified using various functional groups, which provides the possibility of occurrence of versatile and tunable interactions between graphene and other active species. In addition, the functional groups in decorated graphene can act as the catalytically active sites; hence, the huge surface area of graphene loads plenty of groups and gives a basis for high performance, which makes graphene a robust catalyst for multi-functional purposes.

6.6 Summary and Future Outlook

In summary, the examples discussed in this chapter illustrate the multiple application possibilities of graphene. As a unique morphological carbon material, graphene has triggered a gold rush in nanomaterial society. Nevertheless, methods of large-scale production of graphene and the quality of graphene materials still need to be improved. This is the challenge we face before graphene can be viewed as a serious candidate for various applications, particularly in solar cells.

Not only the classical form of graphene but also its morphologically modified form as a CNT in combination with CPs or metal and metal oxides can dramatically enhance the performance of graphene's electronic properties due to synergy effects. The industrial interest in such graphene-based composites as competitive materials in energy-related applications is growing rapidly. Considering the fact that use of graphene leads to a paradigm shift in transparent conductor technology, there is clearly a requirement for ITO replacement by carbon-based systems that utilize graphene, not only because of their cost-effective and flexible properties but also because of the great application potential of graphene as a base material for developing versatile optoelectronic devices.

In the case of LIBs and capacitors, graphene's most promising form is partially disordered graphitic carbon, as defects provide large excess capacities. Nonetheless, research directed at understanding the relationship between graphene and bulk material needs to be improved, and special emphasis must be given to structure control and the alignment of graphene layers. As has been previously shown in section 6.4.1, the application of graphene in supercapacitors requires further improvement of the performance characteristics by modifying or enhancing the quality of graphene samples as well as the fabrication of graphene into well-defined structures. It has also been shown that graphene-based materials exhibit great potential when used as both anodes and cathodes in LIBs. The porous structure and excellent conductivity of graphene-based electrodes supply ideal both ion and electron transport channels, which are crucial for their application in high-energy-density and high-power-density devices.

References

An, K. H., Kim, W. S., Park, Y. S. et al. 2001. Electrochemical properties of high-power supercapacitors using single-walled carbon nanotube electrodes. *Advanced Functional Materials* 11: 387–392.

Arbizzani, C., Mastragostino, M., and Soavi, F. 2001. New trends in electrochemical supercapacitors. *Journal of Power Sources* 100: 164–170.

Areshkin, D. A., Gunlycke, D., and White, C. T. 2007. Ballistic transport in graphene nanostrips in the presence of disorder: Importance of edge effects. *Nano Letters* 7: 204–210.

Bai, H., Xu, Y. X., Zhao, L., Li, C., and Shi, G. Q. 2009. Non-covalent functionalization of graphene sheets by sulfonated polyaniline. *Chemical Communications* 1667–1669.

Balandin, A. A., Ghosh, S., Bao, W. Z. et al. 2008. Superior thermal conductivity of single-layer graphene. *Nano Letters* 8: 902–907.

Barsukov, I. V., and North Atlantic Treaty Organization. 2006. *New carbon based materials for electrochemical energy storage systems: Batteries, supercapacitors and fuel cells;* Springer: Dordrecht.

Becerril, H. A., Mao, J., Liu, Z. et al. 2008. Evaluation of solution-processed reduced graphene oxide films as transparent conductors. *ACS Nano* 2: 463–470.

Beguin, F., Szostak, K., Lota, G., and Frackowiak, E. 2005. A self-supporting electrode for supercapacitors prepared by one-step pyrolysis of carbon nanotube/polyacrylonitrile blends. *Advanced Materials* 17: 2380–2384.

Besenhard, J. O., Hess, M., and Komenda, P. 1990. Dimensionally stable Li-alloy electrodes for secondary batteries. *Solid State Ionics* 40–1: 525–529.

Bian, J., Xiao, M., Wang, S. J., et al. 2009. Highly effective synthesis of dimethyl carbonate from methanol and carbon dioxide using a novel copper-nickel/graphite bimetallic nanocomposite catalyst. *Chemical Engineering Journal* 147: 287–296.

Bissessur, R., Liu, P. K. Y., White, W., and Scully, S. F. 2006. Encapsulation of polyanilines into graphite oxide. *Langmuir* 22: 1729–1734.

Bonino, F., Brutti, S., Reale, P. et al. 2005. A disordered carbon as a novel anode material in lithium-ion cells. *Advanced Materials* 17: 743–746.

Boukamp, B. A., Lesh, G. C., and Huggins, R. A. 1981. All-Solid Lithium Electrodes with Mixed-Conductor Matrix. *Journal of the Electrochemical Society* 128: 725–729.

Brabec, C. J., Sariciftci, N. S., and Hummelen, J. C. 2001. Plastic solar cells. *Advanced Functional Materials* 11: 15–26.

Brandon, N. P., Skinner, S., and Steele, B. C. H. 2003. Recent advances in materials for fuel cells. *Annual Review of Materials Research* 33: 183–213.

Brown, P., Takechi, K., and Kamat, P. V. 2008. Single-walled carbon nanotube scaffolds for dye-sensitized solar cells. *Journal of Physical Chemistry C* 112: 4776–4782.

Burke, A. 2000. Ultracapacitors: Why, how, and where is the technology. *Journal of Power Sources* 91: 37–50.

Cai, D. Y., Song, M., and Xu, C. X. 2008. Highly conductive carbon-nanotube/graphite-oxide hybrid films. *Advanced Materials* 20: 1706–1709.

Cassagneau, T., and Fendler, J. H. 1998. High density rechargeable lithium-ion batteries self-assembled from graphite oxide nanoplatelets and polyelectrolytes. *Advanced Materials* 10: 877–881.

Cervantes-Sodi, F., Csányi, G., Piscanec, S., and Ferrari, A. C. 2008. Edge-functionalized and substitutionally doped graphene nanoribbons: Electronic and spin properties. *Physical Review B* 77: 165427.

Chang, H. X., Lv, X. J., Zhang, H., and Li, J. H. 2010. Quantum dots sensitized graphene: in situ growth and application in photoelectrochemical cells. *Electrochemistry Communications* 12: 483–487.

Chen, D., Tang, L. H., and Li, J. H. 2010. Graphene-based materials in electrochemistry. *Chemical Society Reviews* 39: 3157–3180.

Chen, J., Minett, A. I., Liu, Y. et al. 2008. Direct growth of flexible carbon nanotube electrodes. *Advanced Materials* 20: 566–570.

Chen, W. X., Lee, J. Y., and Liu, Z. L. 2003. The nanocomposites of carbon nanotube with Sb and SnSb0.5 as Li-ion battery anodes. *Carbon* 41: 959–966.

Chen, Z., Cotterell, B., Wang, W., Guenther, E., and Chua, S. J. 2001. A mechanical assessment of flexible optoelectronic devices. *Thin Solid Films* 394: 202–206.

Chiba, Y., Islam, A., Watanabe, Y. et al. 2006. Dye-sensitized solar cells with conversion efficiency of 11.1%. *Japanese Journal of Applied Physics Part 2-Letters & Express Letters* 45: L638–L640.

Choi, E. K., Jeon, I. Y., Bae, S. Y. et al. 2010. High-yield exfoliation of three-dimensional graphite into two-dimensional graphene-like sheets. *Chemical Communications* 46: 6320–6322.

Chou, S. L., Wang, J. Z., Chew, S. Y., Liu, H. K., and Dou, S. X. 2008. Electrodeposition of MnO2 nanowires on carbon nanotube paper as free-standing, flexible electrode for supercapacitors. *Electrochemistry Communications* 10: 1724–1727.

Ci, L. J., Manikoth, S. M., Li, X. S., Vajtai, R., and Ajayan, P. M. 2007. Ultrathick freestanding aligned carbon nanotube films. *Advanced Materials* 19: 3300–3303.

Conway, B. E. 1999. *Electrochemical supercapacitors: Scientific fundamentals and technological applications*; Plenum Press: New York.

Costamagna, P., and Srinivasan, S. 2001. Quantum jumps in the PEMFC science and technology from the 1960s to the year 2000 Part I. Fundamental scientific aspects. *Journal of Power Sources* 102: 242–252.

Cui, G. L., Gu, L., Zhi, L. J. et al. 2008. A germanium-carbon nanocomposite material for lithium batteries. *Advanced Materials* 20: 3079–3083.

Cui, G. L., Hu, Y. S., Zhi, L. J. et al. 2007a. A one-step approach towards carbon-encapsulated hollow tin nanoparticles and their application in lithium batteries. *Small* 3: 2066–2069.

Cui, G. L., Zhi, L. J., Thomas, A. et al. 2007b. A novel approach towards carbon-Ru electrodes with mesoporosity for supercapacitors. *Chemphyschem* 8: 1013–1015.

Cui, G. L., Zhi, L. J., Thomas, A. et al. 2007c. One-dimensional porous carbon/platinum composites for nanoscale electrodes. *Angewandte Chemie-International Edition* 46: 3464–3467.

Dahn, J. R., Zheng, T., Liu, Y. H., and Xue, J. S. 1995. Mechanisms for lithium insertion in carbonaceous materials. *Science* 270: 590–593.

De Arco, L. G., Zhang, Y., Kumar, A., and Zhou, C. W. 2009. Synthesis, transfer, and devices of single- and few-layer graphene by chemical vapor deposition. *IEEE Transactions on Nanotechnology* 8: 135–138.

De Arco, L. G., Zhang, Y., Schlenker, C. W. et al. 2010. Continuous, highly flexible, and transparent graphene films by chemical vapor deposition for organic photovoltaics. *ACS Nano* 4: 2865–2873.

De, S., Higgins, T. M., Lyons, P. E. et al. 2009a. Silver nanowire networks as flexible, transparent, conducting films: Extremely high DC to optical conductivity ratios. *ACS Nano* 3: 1767–1774.

De, S., Lyons, P. E., Sorel, S. et al. 2009b. Transparent, flexible, and highly conductive thin films based on polymer—Nanotube composites. *ACS Nano* 3: 714–720.

Deng, D., and Lee, J. Y. 2009. Reversible storage of lithium in a rambutan-Like Tin-Carbon Electrode. *Angewandte Chemie-International Edition* 48: 1660–1663.

Dennler, G., Scharber, M. C., Ameri, T. et al. 2008. Design rules for donors in bulk-heterojunction tandem solar cells-towards 15% energy-conversion efficiency. *Advanced Materials* 20: 579–583.

Derrien, G., Hassoun, J., Panero, S., and Scrosati, B. 2007. Nanostructured Sn-C composite as an advanced anode material in high-performance lithium-ion batteries. *Advanced Materials* 19: 2336–2340.

Diederich, L., Barborini, E., Piseri, P. et al. 1999. Supercapacitors based on nanostructured carbon electrodes grown by cluster-beam deposition. *Applied Physics Letters* 75: 2662–2664.

Ding, Y., Jiang, Y., Xu, F. et al. 2010. Preparation of nano-structured LiFePO4/graphene composites by co-precipitation method. *Electrochemistry Communications* 12: 10–13.

Dong, L. F., Gari, R. R. S., Li, Z., Craig, M. M., and Hou, S. F. 2010. Graphene-supported platinum and platinum-ruthenium nanoparticles with high electrocatalytic activity for methanol and ethanol oxidation. *Carbon* 48: 781–787.

Du, X. A., Guo, P., Song, H. H., and Chen, X. H. 2010. Graphene nanosheets as electrode material for electric double-layer capacitors. *Electrochimica Acta* 55: 4812–4819.

Eda, G., and Chhowalla, M. 2010. Chemically derived graphene oxide: Towards large-area thin-film electronics and optoelectronics. *Advanced Materials* 22: 2392–2415.

Eda, G., Fanchini, G., and Chhowalla, M. 2008a. Large-area ultrathin films of reduced graphene oxide as a transparent and flexible electronic material. *Nature Nanotechnology* 3: 270–274.

Eda, G., Lin, Y. Y., Miller, S. et al. 2008b. Transparent and conducting electrodes for organic electronics from reduced graphene oxide. *Applied Physics Letters* 92: 2333051–2333053.

Edwards, P. P., Porch, A., Jones, M. O., Morgan, D. V., and Perks, R. M. 2004. Basic materials physics of transparent conducting oxides. *Dalton Transactions* 2995–3002.

Elias, D. C., Nair, R. R., Mohiuddin, T. M. G. et al. 2009. Control of graphene's properties by reversible hydrogenation: Evidence for graphane. *Science* 323: 610–613.

Fan, J., Wang, T., Yu, C. Z. et al. 2004. Ordered, nanostructured tin-based oxides/carbon composite as the negative-electrode material for lithium-ion batteries. *Advanced Materials* 16: 1432–1436.

Fan, Z. J., Yan, J., Zhi, L. J. et al. 2010. A three-dimensional carbon nanotube/graphene sandwich and its application as electrode in supercapacitors. *Advanced Materials* 22: 3723–3728.

Feng, X. L., Liang, Y. Y., Zhi, L. J. et al. 2009. Synthesis of microporous carbon nanofibers and nanotubes from conjugated polymer network and evaluation in electrochemical capacitor. *Advanced Functional Materials* 19: 2125–2129.

Forrest, S. R. 2005. The limits to organic photovoltaic cell efficiency. *Mrs Bulletin* 30: 28–32.

Futaba, D. N., Hata, K., Yamada, T. et al. 2006. Shape-engineerable and highly densely packed single-walled carbon nanotubes and their application as super-capacitor electrodes. *Nature Materials* 5: 987–994.

Gao, B., Bower, C., Lorentzen, J. D. et al. 2000. Enhanced saturation lithium composition in ball-milled single-walled carbon nanotubes. *Chemical Physics Letters* 327: 69–75.

Gao, B., Kleinhammes, A., Tang, X. P. et al. 1999. Electrochemical intercalation of single-walled carbon nanotubes with lithium. *Chemical Physics Letters* 307: 153–157.

Gautier, S., Leroux, F., Frackowiak, E. et al. 2001. Influence of the pyrolysis conditions on the nature of lithium inserted in hard carbons. *Journal of Physical Chemistry A* 105: 5794–5800.

Geim, A. K. 2009. Graphene: Status and prospects. *Science* 324: 1530–1534.

Geim, A. K., and Novoselov, K. S. 2007. The rise of graphene. *Nature Materials* 6: 183–191.

Geng, H. Z., Kim, K. K., So, K. P. et al. 2007. Effect of acid treatment on carbon nanotube-based flexible transparent conducting films. *Journal of the American Chemical Society* 129: 7758–7759.

Geng, J. X., Liu, L. J., Yang, S. B. et al. 2010. A simple approach for preparing transparent conductive graphene films using the controlled chemical reduction of exfoliated graphene oxide in an aqueous suspension. *Journal of Physical Chemistry C* 114: 14433–14440.

Gogotsi, Y. 2006. *Carbon nanomaterials;* CRC Press: Philadelphia.

Gomez-Navarro, C., Weitz, R. T., Bittner, A. M. et al. 2007. Electronic transport properties of individual chemically reduced graphene oxide sheets. *Nano Letters* 7: 3499–3503.

Granqvist, C. G. 2007. Transparent conductors as solar energy materials: A panoramic review. *Solar Energy Materials and Solar Cells* 91: 1529–1598.

Grigoriants, I., Sominski, L., Li, H. L. et al. 2005. The use of tin-decorated mesoporous carbon as an anode material for rechargeable lithium batteries. *Chemical Communications*: 921–923.

Gunes, S., Neugebauer, H., and Sariciftci, N. S. 2007. Conjugated polymer-based organic solar cells. *Chemical Reviews* 107: 1324–1338.

Guo, C. X., Yang, H. B., Sheng, Z. M. et al. 2010a. Layered graphene/quantum dots for photovoltaic devices. *Angewandte Chemie-International Edition* 49: 3014–3017.

Guo, P., Song, H. H., and Chen, X. H. 2009. Electrochemical performance of graphene nanosheets as anode material for lithium-ion batteries. *Electrochemistry Communications* 11: 1320–1324.

Guo, S. J., Dong, S. J., and Wang, E. W. 2010b. Three-Dimensional Pt-on-Pd Bimetallic Nanodendrites Supported on Graphene Nanosheet: Facile Synthesis and Used as an Advanced Nanoelectrocatalyst for Methanol Oxidation. *ACS Nano* 4: 547–555.

Haacke, G. 1977. Transparent conducting coatings. *Annual Review of Materials Science* 7: 73–93.

Han, S. J., Jang, B. C., Kim, T., Oh, S. M., and Hyeon, T. 2005. Simple synthesis of hollow tin dioxide microspheres and their application to lithium-ion battery anodes. *Advanced Functional Materials* 15: 1845–1850.

Hao, R., Qian, W., Zhang, L. H., and Hou, Y. L. 2008. Aqueous dispersions of TCNQ-anion-stabilized graphene sheets. *Chemical Communications* 6576–6578.

He, F. A., Fan, J. T., Ma, D. et al. 2010a. The attachment of Fe3O4 nanoparticles to graphene oxide by covalent bonding. *Carbon* 48: 3139–3144.

He, Y. S., Bai, D. W., Yang, X. W. et al. 2010b. A Co(OH)(2)-graphene nanosheets composite as a high performance anode material for rechargeable lithium batteries. *Electrochemistry Communications* 12: 570–573.

Higashika, S., Kimura, K., Matsuo, Y., and Sugie, Y. 1999. Synthesis of polyaniline-intercalated graphite oxide. *Carbon* 37: 354–356.

Hong, W. J., Xu, Y. X., Lu, G. W., Li, C., and Shi, G. Q. 2008. Transparent graphene/PEDOT-PSS composite films as counter electrodes of dye-sensitized solar cells. *Electrochemistry Communications* 10: 1555–1558.

Hu, Y. J., Jin, J. A., Zhang, H., Wu, P., and Cai, C. X. 2010. Graphene: Synthesis, functionalization and applications in chemistry. *Acta Physico-Chimica Sinica* 26: 2073–2086.

Hu, Y. S., Demir-Cakan, R., Titirici, M. M. et al. 2008. Superior storage performance of a Si@SiOx/C nanocomposite as anode material for lithium-ion batteries. *Angewandte Chemie-International Edition* 47: 1645–1649.

Huang, J. X., and Kaner, R. B. 2004. A general chemical route to polyaniline nanofibers. *Journal of the American Chemical Society* 126: 851–855.

Huang, J. X., and Kaner, R. B. 2006. The intrinsic nanofibrillar morphology of polyaniline. *Chemical Communications*: 367–376.

Huang, J. X., Virji, S., Weiller, B. H., and Kaner, R. B. 2003. Polyaniline nanofibers: Facile synthesis and chemical sensors. *Journal of the American Chemical Society* 125: 314–315.

Huang, S. F., Terakura, K., Ozaki, T., et al. 2009. First-principles calculation of the electronic properties of graphene clusters doped with nitrogen and boron: Analysis of catalytic activity for the oxygen reduction reaction. *Physical Review B* 80: 235410.

Hummers, W. S., and Offeman, R. E. 1958. Preparation of graphitic oxide. *Journal of the American Chemical Society* 80: 1339–1339.

Ishikawa, R., Bando, M., Morimoto, Y., Park, S. Y., and Sandhu, A. 2010. Patterning of two-dimensional graphene oxide on silicon Substrates. *Japanese Journal of Applied Physics* 49.

Jafri, R. I., Arockiados, T., Rajalakshmi, N., and Ramaprabhu, S. 2010. Nanostructured Pt Dispersed on Graphene-Multiwalled Carbon Nanotube Hybrid Nanomaterials as Electrocatalyst for PEMFC. *Journal of the Electrochemical Society* 157:B874–B879.

Jang, J., Bae, J., Choi, M., and Yoon, S. H. 2005. Fabrication and characterization of polyaniline coated carbon nanofiber for supercapacitor. *Carbon* 43: 2730–2736.

Jian, K. Q., Shim, H. S., Schwartzman, A., Crawford, G. P., and Hurt, R. H. 2003. Orthogonal carbon nanofibers by template-mediated assembly of discotic mesophase pitch. *Advanced Materials* 15: 164–167.

Kaempgen, M., Chan, C. K., Ma, J., Cui, Y., and Gruner, G. 2009. Printable thin film supercapacitors using single-walled carbon nanotubes. *Nano Letters* 9: 1872–1876.

Kaempgen, M., Ma, J., Gruner, G., Wee, G., and Mhaisalkar, S. G. 2007. Bifunctional carbon nanotube networks for supercapacitors. *Applied Physics Letters* 90: 641041–3.

Kang, M. G., and Guo, L. J. 2007. Nanoimprinted semitransparent metal electrodes and their application in organic light-emitting diodes. *Advanced Materials* 19: 1391–1396.

Kasavajjula, U., Wang, C. S., and Appleby, A. J. 2007. Nano- and bulk-silicon-based insertion anodes for lithium-ion secondary cells. *Journal of Power Sources* 163: 1003–1039.

Kaskhedikar, N. A., and Maier, J. 2009. Lithium storage ion carbon nanostructures. *Advanced Materials* 21: 2664–2680.

Kasry, A., Kuroda, M. A., Martyna, G. J., Tulevski, G. S., and Bol, A. A. 2010. Chemical doping of large-area stacked graphene films for use as transparent, conducting electrodes. *ACS Nano* 4: 3839–3844.

Kedzierski, J., Hsu, P. L., Healey, P. et al. 2008. Epitaxial graphene transistors on SIC substrates. *IEEE Transactions on Electron Devices* 55: 2078–2085.

Kim, H., Abdala, A. A., and Macosko, C. W. 2010a. Graphene/Polymer nanocomposites. *Macromolecules* 43: 6515–6530.

Kim, H., and Cho, J. 2008. Template synthesis of hollow Sb nanoparticles as a high-performance lithium battery anode material. *Chemistry of Materials* 20: 1679–1681.

Kim, J., Cote, L. J., Kim, F. et al. 2010b. Graphene oxide sheets at interfaces. *Journal of the American Chemical Society* 132: 8180–8186.

Kim, K. S., Zhao, Y., Jang, H. et al. 2009. Large-scale pattern growth of graphene films for stretchable transparent electrodes. *Nature (London)* 457: 706–710.

Kim, Y. K., and Min, D. H. 2009. Durable large-area thin films of graphene/carbon nanotube double layers as a transparent electrode. *Langmuir* 25: 11302–11306.

Knieke, C., Berger, A., Voigt, M. et al. 2010. Scalable production of graphene sheets by mechanical delamination. *Carbon* 48: 3196–3204.

Kongkanand, A., Dominguez, R. M., and Kamat, P. V. 2007. Single wall carbon nanotube scaffolds for photoelectrochemical solar cells. Capture and transport of photogenerated electrons. *Nano Letters* 7: 676–680.

Kongkanand, A., Kuwabata, S., Girishkumar, G., and Kamat, P. 2006. Single-wall carbon nanotubes supported platinum nanoparticles with improved electrocatalytic activity for oxygen reduction reaction. *Langmuir* 22: 2392–2396.

Kou, R., Shao, Y. Y., Wang, D. H., et al. 2009. Enhanced activity and stability of Pt catalysts on functionalized graphene sheets for electrocatalytic oxygen reduction. *Electrochemistry Communications* 11: 954–957.

Kotz, R., and Carlen, M. 2000. Principles and applications of electrochemical capacitors. *Electrochimica Acta* 45: 2483–2498.

Kwon, Y., Kim, H., Doo, S. G., and Cho, J. H. 2007. Sn0.9Si0.1/carbon core-shell nanoparticles for high-density lithium storage materials. *Chemistry of Materials* 19: 982–986.

Lee, J. K., Kung, M. C., Trahey, L., Missaghi, M. N., and Kung, H. H. 2009. Nanocomposites derived from phenol-functionalized Si nanoparticles for high performance lithium ion battery anodes. *Chemistry of Materials* 21: 6–8.

Lee, J. K., Smith, K. B., Hayner, C. M., and Kung, H. H. 2010. Silicon nanoparticles-graphene paper composites for Li ion battery anodes. *Chemical Communications* 46: 2025–2027.

Lee, J. Y., Connor, S. T., Cui, Y., and Peumans, P. 2008. Solution-processed metal nanowire mesh transparent electrodes. *Nano Letters* 8: 689–692.

Lee, J. Y., Zhang, R. F., and Liu, Z. L. 2000. Lithium intercalation and deintercalation reactions in synthetic graphite containing a high dispersion of SnO. *Electrochemical and Solid State Letters* 3: 167–170.

Lee, K. T., Jung, Y. S., and Oh, S. M. 2003. Synthesis of tin-encapsulated spherical hollow carbon for anode material in lithium secondary batteries. *Journal of the American Chemical Society* 125: 5652–5653.

Li, D., Huang, J. X., and Kaner, R. B. 2009a. Polyaniline nanofibers: A unique polymer nanostructure for versatile applications. *Accounts of Chemical Research* 42: 135–145.

Li, D., Muller, M. B., Gilje, S., Kaner, R. B., and Wallace, G. G. 2008a. Processable aqueous dispersions of graphene nanosheets. *Nature Nanotechnology* 3: 101–105.

Li, F. H., Song, J. F., Yang, H. F. et al. 2009b. One-step synthesis of graphene/SnO2 nanocomposites and its application in electrochemical supercapacitors. *Nanotechnology* 20: 455602.

Li, H., Pang, S., Feng, X., Mullen, K., and Bubeck, C. 2010a. Polyoxometalate assisted photoreduction of graphene oxide and its nanocomposite formation. *Chemical Communications* 46: 6243–6245.

Li, L. X., Song, H. H., Zhang, Q. C., Yao, J. Y., and Chen, X. H. 2009c. Effect of compounding process on the structure and electrochemical properties of ordered mesoporous carbon/polyaniline composites as electrodes for supercapacitors. *Journal of Power Sources* 187: 268–274.

Li, N., Wang, Z. Y., Zhao, K. K. et al. 2010b. Graphene-Pd composite as highly active catalyst for the suzuki-miyaura coupling reaction. *Journal of Nanoscience and Nanotechnology* 10: 6748–6751.

Li, X. L., Wang, X. R., Zhang, L., Lee, S. W., and Dai, H. J. 2008b. Chemically derived, ultrasmooth graphene nanoribbon semiconductors. *Science* 319: 1229–1232.

Li, X. L., Zhang, G. Y., Bai, X. D. et al. 2008c. Highly conducting graphene sheets and Langmuir-Blodgett films. *Nature Nanotechnology* 3: 538–542.

Li, X. S., Cai, W. W., An, J. H. et al. 2009d. Large-area synthesis of high-quality and uniform graphene films on copper foils. *Science* 324: 1312–1314.

Li, X. S., Cai, W. W., Colombo, L., and Ruoff, R. S. 2009e. Evolution of graphene growth on Ni and Cu by carbon isotope labeling. *Nano Letters* 9: 4268–4272.

Li, X. S., Zhu, Y. W., Cai, W. W. et al. 2009f. Transfer of large-area graphene films for high-performance transparent conductive electrodes. *Nano Letters* 9: 4359–4363.

Li, Y. M., Tang, L. H., and Li, J. H. 2009g. Preparation and electrochemical performance for methanol oxidation of pt/graphene nanocomposites. *Electrochemistry Communications* 11: 846–849.

Liang, M. H., and Zhi, L. J. 2009. Graphene-based electrode materials for rechargeable lithium batteries. *Journal of Materials Chemistry* 19: 5871–5878.

Liang, Y. Y., Feng, X. L., Zhi, L. J., Kolb, U., and Mullen, K. 2009a. A simple approach towards one-dimensional mesoporous carbon with superior electrochemical capacitive activity. *Chemical Communications* 809–811.

Liang, Y. Y., Frisch, J., Zhi, L. J. et al. 2009b. Transparent, highly conductive graphene electrodes from acetylene-assisted thermolysis of graphite oxide sheets and nanographene molecules. *Nanotechnology* 20: 434007.

Liu, C. G., Liu, M., Li, F. and Cheng, H. M. 2008a. Frequency response characteristic of single-walled carbon nanotubes as supercapacitor electrode material. *Applied Physics Letters* 92: 1431081–3.

Liu, C. G., Yu, Z. N., Neff, D., Zhamu, A., and Jang, B. Z. 2010a. Graphene-based supercapacitor with an ultrahigh energy density. *Nano Letters* 10: 4863–4868.

Liu, Q., Liu, Z. F., Zhang, X. Y. et al. 2008b. Organic photovoltaic cells based on an acceptor of soluble graphene. *Applied Physics Letters* 92: 2233031–3.

Liu, Q., Liu, Z. F., Zhong, X. Y. et al. 2009a. Polymer photovoltaic cells based on solution-processable graphene and P3HT. *Advanced Functional Materials* 19: 894–904.

Liu, S., Wang, J. Q., Zeng, J., et al. 2010b. "Green" electrochemical synthesis of Pt/graphene sheet nanocomposite film and its electrocatalytic property. *Journal of Power Sources* 195: 4628–4633.

Liu, Y., Wen, Z. Y., Wang, X. Y. et al. 2009b. Improvement of cycling stability of Si anode by mechanochemcial reduction and carbon coating. *Journal of Power Sources* 189: 480–484.

Liu, Y. H., Xue, J. S., Zheng, T., and Dahn, J. R. 1996. Mechanism of lithium insertion in hard carbons prepared by pyrolysis of epoxy resins. *Carbon* 34: 193–200.

Liu, Z. F., Liu, Q., Huang, Y. et al. 2008c. Organic photovoltaic devices based on a novel acceptor material: Graphene. *Advanced Materials* 20: 3924–3930.

Lomeda, J. R., Doyle, C. D., Kosynkin, D. V., Hwang, W. F., and Tour, J. M. 2008. Diazonium functionalization of surfactant-wrapped chemically converted graphene sheets. *Journal of the American Chemical Society* 130: 16201–16206.

Lopez, V., Sundaram, R. S., Gomez-Navarro, C. et al. 2009. Chemical vapor deposition repair of graphene oxide: A route to highly conductive graphene monolayers. *Advanced Materials* 21: 4683–4686.

Lota, G., Centeno, T. A., Frackowiak, E., and Stoeckli, F. 2008. Improvement of the structural and chemical properties of a commercial activated carbon for its application in electrochemical capacitors. *Electrochimica Acta* 53: 2210–2216.

Lu, P., Feng, Y. Y., Zhang, X. Q., Li, Y., and Feng, W. 2010a. Recent progresses in application of functionalized graphene sheets. *Science China-Technological Sciences* 53: 2311–2319.

Lu, T., Zhang, Y. P., Li, H. B. et al. 2010b. Electrochemical behaviors of graphene-ZnO and graphene-SnO2 composite films for supercapacitors. *Electrochimica Acta* 55: 4170–4173.

Lv, W., Tang, D. M., He, Y. B. et al. 2009. Low-temperature exfoliated graphenes: Vacuum-promoted exfoliation and electrochemical energy storage. *ACS Nano* 3: 3730–3736.

Maruyama, J., and Abe, I. 2007. Structure control of a carbon-based noble-metal-free fuel cell cathode catalyst leading to high power output. *Chemical Communications* 2879–2881.

Maurin, G., Bousquet, C., Henn, F. et al. 1999. Electrochemical intercalation of lithium into multiwall carbon nanotubes. *Chemical Physics Letters* 312: 14–18.

Meunier, V., Kephart, J., Roland, C., and Bernholc, J. 2002. Ab initio investigations of lithium diffusion in carbon nanotube systems. *Physical Review Letters* 88: 075506.

Meyers, J. P. and Darling, R. M. 2006. Model of carbon corrosion in PEM fuel cells. *Journal of the Electrochemical Society* 153: A1432–A1442.

Miller, J. M., Dunn, B., Tran, T. D., and Pekala, R. W. 1997. Deposition of ruthenium nanoparticles on carbon aerogels for high energy density supercapacitor electrodes. *Journal of the Electrochemical Society* 144: L309–L311.

Miller, J. R., and Simon, P. 2008. Materials science—Electrochemical capacitors for energy management. *Science* 321: 651–652.

Murugan, A. V., Muraliganth, T., and Manthiram, A. 2009. Rapid, facile microwave-solvothermal synthesis of graphene nanosheets and their polyaniline nanocomposites for energy strorage. *Chemistry of Materials* 21: 5004–5006.

Nair, R. R., Blake, P., Grigorenko, A. N. et al. 2008. Fine structure constant defines visual transparency of graphene. *Science* 320: 1308–1308.

Nam, K. T., Kim, D. W., Yoo, P. J. et al. 2006. Virus-enabled synthesis and assembly of nanowires for lithium ion battery electrodes. *Science* 312: 885–888.

Nazeeruddin, M. K., De Angelis, F., Fantacci, S. et al. 2005. Combined experimental and DFT-TDDFT computational study of photoelectrochemical cell ruthenium sensitizers. *Journal of the American Chemical Society* 127: 16835–16847.

Ng, S. H., Wang, J. Z., Wexler, D. et al. 2006. Highly reversible lithium storage in spheroidal carbon-coated silicon nanocomposites as anodes for lithium-ion batteries. *Angewandte Chemie-International Edition* 45: 6896–6899.

Nirmalraj, P. N., Lyons, P. E., De, S., Coleman, J. N., and Boland, J. J. 2009. Electrical connectivity in single-walled carbon nanotube networks. *Nano Letters* 9: 3890–3895.

Niu, C. M., Sichel, E. K., Hoch, R., Moy, D., and Tennent, H. 1997. High power electrochemical capacitors based on carbon nanotube electrodes. *Applied Physics Letters* 70: 1480–1482.

Niyogi, S., Bekyarova, E., Itkis, M. E. et al. 2006. Solution properties of graphite and graphene. *Journal of the American Chemical Society* 128: 7720–7721.

Noh, M., Kwon, Y., Lee, H. et al. 2005. Amorphous carbon-coated tin anode material for lithium secondary battery. *Chemistry of Materials* 17: 1926–1929.

Novoselov, K. S., Geim, A. K., Morozov, S. V. et al. 2004. Electric field effect in atomically thin carbon films. *Science* 306: 666–669.

Obreja, V. V. N. 2008. On the performance of supercapacitors with electrodes based on carbon nanotubes and carbon activated material—A review. *Physica E-Low-Dimensional Systems & Nanostructures* 40: 2596–2605.

Oregan, B., and Gratzel, M. 1991. A low-cost, high-efficiency solar-cell based on dye-sensitized colloidal Tio2 films. *Nature* 353: 737–740.

Owens, F. J. 2008. Electronic and magnetic properties of armchair and zigzag graphene nanoribbons. *Journal of Chemical Physics* 128.

Paek, S. M., Yoo, E., and Honma, I. 2009. Enhanced cyclic performance and lithium storage capacity of SnO2/graphene nanoporous electrodes with three-dimensionally delaminated flexible structure. *Nano Letters* 9: 72–75.

Pan, D. Y., Wang, S., Zhao, B. et al. 2009. Li storage properties of disordered graphene nanosheets. *Chemistry of Materials* 21: 3136–3142.

Pandolfo, A. G., and Hollenkamp, A. F. 2006. Carbon properties and their role in supercapacitors. *Journal of Power Sources* 157: 11–27.

Park, S., and Ruoff, R. S. 2009. Chemical methods for the production of graphenes. *Nature Nanotechnology* 4: 217–224.

Peng, C., Zhang, S. W., Jewell, D., and Chen, G. Z. 2008. Carbon nanotube and conducting polymer composites for supercapacitors. *Progress in Natural Science* 18: 777–788.

Poizot, P., Laruelle, S., Grugeon, S., Dupont, L., and Tarascon, J. M. 2000. Nano-sized transition-metaloxides as negative-electrode materials for lithium-ion batteries. *Nature* 407: 496–499.

Poizot, P., Laruelle, S., Grugeon, S., Dupont, L., and Tarascon, J. M. 2001. Searching for new anode materials for the Li-ion technology: Time to deviate from the usual path. *Journal of Power Sources* 97–8: 235–239.

Portet, C., Chmiola, J., Gogotsi, Y., Park, S., and Lian, K. 2008. Electrochemical characterizations of carbon nanomaterials by the cavity microelectrode technique. *Electrochimica Acta* 53: 7675–7680.

Pushparaj, V. L., Shaijumon, M. M., Kumar, A. et al. 2007. Flexible energy storage devices based on nanocomposite paper. *Proceedings of the National Academy of Sciences of the United States of America* 104: 13574–13577.

Qu, L. T., Liu, Y., Baek, J. B., and Dai, L. M. 2010. Nitrogen-Doped Graphene as Efficient Metal-Free Electrocatalyst for Oxygen Reduction in Fuel Cells. *ACS Nano* 4: 1321–1326.

Radovic, L. R., and Bockrath, B. 2005. On the chemical nature of graphene edges: Origin of stability and potential for magnetism in carbon materials. *Journal of the American Chemical Society* 127: 5917–5927.

Rao, C. N. R., Sood, A. K., Subrahmanyam, K. S., and Govindaraj, A. 2009. Graphene: The new two-dimensional nanomaterial. *Angewandte Chemie-International Edition* 48: 7752–7777.

Read, J., Foster, D., Wolfenstine, J., and Behl, W. 2001. SnO2-carbon composites for lithium-ion battery anodes. *Journal of Power Sources* 96: 277–281.

Reina, A., Jia, X. T., Ho, J. et al. 2009. Large area, few-layer graphene films on arbitrary substrates by chemical vapor deposition. *Nano Letters* 9: 30–35.

Renouard, T., Gherghel, L., Wachtler, M. et al. 2005. Pyrolysis of hexa(phenyl)benzene derivatives: A molecular approach toward carbonaceous materials for Li-ion storage. *Journal of Power Sources* 139: 242.

Rock, N. L., and Kumta, P. N. 2007. Synthesis and characterization of electrochemically active graphite-silicon-tin composite anodes for Li-ion applications. *Journal of Power Sources* 164: 829–838.

Rudge, A., Davey, J., Raistrick, I., Gottesfeld, S., and Ferraris, J. P. 1994. Conducting polymers as active materials in electrochemical capacitors. *Journal of Power Sources* 47: 89–107.

Sarangapani, S., Tilak, B. V., and Chen, C. P. 1996. Materials for electrochemical capacitors—Theoretical and experimental constraints. *Journal of the Electrochemical Society* 143: 3791–3799.

Sato, K., Noguchi, M., Demachi, A., Oki, N., and Endo, M. 1994. A mechanism of lithium storage in disordered carbons. *Science* 264: 556–558.

Scharber, M. C., Wuhlbacher, D., Koppe, M. et al. 2006. Design rules for donors in bulk-heterojunction solar cells—Towards 10% energy-conversion efficiency. *Advanced Materials* 18: 789–794.

Scheuermann, G. M., Rumi, L., Steurer, P., Bannwarth, W., and Mulhaupt, R. 2009. Palladium Nanoparticles on Graphite Oxide and Its Functionalized Graphene Derivatives as Highly Active Catalysts for the Suzuki-Miyaura Coupling Reaction. *Journal of the American Chemical Society* 131: 8262–8270.

Seger, B., and Kamat, P. V. 2009. Electrocatalytically active graphene-platinum nanocomposites. Role of 2-D carbon support in PEM fuel cells. *Journal of Physical Chemistry C* 113: 7990–7995.

Service, R. F. 2009. Materials science carbon sheets an atom thick give rise to graphene dreams. *Science* 324: 875–877.

Shaijumon, M. M., Ou, F. S., Ci, L. J., and Ajayan, P. M. 2008. Synthesis of hybrid nanowire arrays and their application as high power supercapacitor electrodes. *Chemical Communications* 2373–2375.

Shao, Y. Y., Zhang, S., Engelhard, M. H. et al. 2010. Nitrogen-doped graphene and its electrochemical applications. *Journal of Materials Chemistry* 20: 7491–7496.

Shao, Y. Y., Zhang, S., Wang, C. M., Nie, Z. M., Liu, J., Wang, Y., and Lin, Y. H. 2010. Highly durable graphene nanoplatelets supported Pt nanocatalysts for oxygen reduction. *Journal of Power Sources* 195: 4600–4605.

Shi, Y. M., Kim, K. K., Reina, A. et al. 2010. Work function engineering of graphene electrode via chemical doping. *ACS Nano* 4: 2689–2694.

Si, Q., Hanai, K., Imanishi, N. et al. 2009. Highly reversible carbon-nano-silicon composite anodes for lithium rechargeable batteries. *Journal of Power Sources* 189: 761–765.

Si, Y. and Samulski, E. T. 2008b. Synthesis of water soluble graphene. *Nano Letters* 8: 1679–1682.

Si, Y. C., and Samulski, E. T. 2008a. Exfoliated graphene separated by platinum nanoparticles. *Chemistry of Materials* 20: 6792–6797.

Simon, P., and Gogotsi, Y. 2008. Materials for electrochemical capacitors. *Nature Materials* 7: 845–854.

Stankovich, S., Dikin, D. A., Dommett, G. H. B. et al. 2006a. Graphene-based composite materials. *Nature* 442: 282–286.

Stankovich, S., Dikin, D. A., Piner, R. D. et al. 2007. Synthesis of graphene-based nanosheets via chemical reduction of exfoliated graphite oxide. *Carbon* 45: 1558–1565.

Stankovich, S., Piner, R. D., Nguyen, S. T., and Ruoff, R. S. 2006b. Synthesis and exfoliation of isocyanate-treated graphene oxide nanoplatelets. *Carbon* 44: 3342–3347.

Stoller, M. D., Park, S. J., Zhu, Y. W., An, J. H., and Ruoff, R. S. 2008. Graphene-based ultracapacitors. *Nano Letters* 8: 3498–3502.

Su, Q., Pang, S. P., Alijani, V. et al. 2009. Composites of graphene with large aromatic molecules. *Advanced Materials* 21: 3191–3195.

Subramanian, V., Zhu, H. W., and Wei, B. Q. 2006. High rate reversibility anode materials of lithium batteries from vapor-grown carbon nanofibers. *Journal of Physical Chemistry B* 110: 7178–7183.

Sun, S. R., Gao, L., and Liu, Y. Q. 2010. Enhanced dye-sensitized solar cell using graphene-TiO2 photoanode prepared by heterogeneous coagulation. *Applied Physics Letters* 96: 0831131–3.

Suzuki, T., Hasegawa, T., Mukai, S. R., and Tamon, H. 2003. A theoretical study on storage states of Li ions in carbon anodes of Li ion batteries using molecular orbital calculations. *Carbon* 41: 1933–1939.

Szabo, T., Berkesi, O., Forgo, P. et al. 2006. Evolution of surface functional groups in a series of progressively oxidized graphite oxides. *Chemistry of Materials* 18: 2740–2749.

Takamura, T., Endo, K., Fu, L. et al. 2007. Identification of nano-sized holes by TEM in the graphene layer of graphite and the high rate discharge capability of Li-ion battery anodes. *Electrochimica Acta* 53: 1055.

Talapatra, S., Kar, S., Pal, S. K. et al. 2006. Direct growth of aligned carbon nanotubes on bulk metals. *Nature Nanotechnology* 1: 112–116.

Tarascon, J. M., and Armand, M. 2001. Issues and challenges facing rechargeable lithium batteries. *Nature* 414: 359–367.

Thompson, B. C., and Frechet, J. M. J. 2008. Organic photovoltaics—Polymer-fullerene composite solar cells. *Angewandte Chemie-International Edition* 47: 58–77.

Toupin, M., Brousse, T., and Belanger, D. 2004. Charge storage mechanism of MnO2 electrode used in aqueous electrochemical capacitor. *Chemistry of Materials* 16: 3184–3190.

Tung, V. C., Allen, M. J., Yang, Y., and Kaner, R. B. 2009a. High-throughput solution processing of large-scale graphene. *Nature Nanotechnology* 4: 25–29.

Tung, V. C., Chen, L. M., Allen, M. J. et al. 2009b. Low-temperature solution processing of graphene-carbon nanotube hybrid materials for high-performance transparent conductors. *Nano Letters* 9: 1949–1955.

Vivekchand, S. R. C., Rout, C. S., Subrahmanyam, K. S., Govindaraj, A., and Rao, C. N. R. 2008. Graphene-based electrochemical supercapacitors. *Journal of Chemical Sciences* 120: 9–13.

Wang, B., Park, J., Wang, C. Y., Ahn, H., and Wang, G. X. 2010a. Mn3O4 nanoparticles embedded into graphene nanosheets: Preparation, characterization, and electrochemical properties for supercapacitors. *Electrochimica Acta* 55: 6812–6817.

Wang, D. H., Choi, D. W., Li, J. et al. 2009a. Self-assembled TiO2-graphene hybrid nanostructures for enhanced Li-ion insertion. *ACS Nano* 3: 907–914.

Wang, D. H., Kou, R., Choi, D. et al. 2010b. Ternary self-assembly of ordered metal oxide-graphene nanocomposites for electrochemical energy storage. *ACS Nano* 4: 1587–1595.

Wang, D. W., Li, F., Wu, Z. S., Ren, W. C., and Cheng, H. M. 2009b. Electrochemical interfacial capacitance in multilayer graphene sheets: Dependence on number of stacking layers. *Electrochemistry Communications* 11: 1729–1732.

Wang, D. W., Li, F., Zhao, J. P. et al. 2009c. Fabrication of graphene/polyaniline composite paper via in situ anodic electropolymerization for high-performance flexible electrode. *ACS Nano* 3: 1745–1752.

Wang, G. X., Shen, X. P., Yao, J., and Park, J. 2009d. Graphene nanosheets for enhanced lithium storage in lithium ion batteries. *Carbon* 47: 2049–2053.

Wang, G. X., Wang, B., Wang, X. L. et al. 2009e. Sn/graphene nanocomposite with 3D architecture for enhanced reversible lithium storage in lithium ion batteries. *Journal of Materials Chemistry* 19: 8378–8384.

Wang, H. L., Casalongue, H. S., Liang, Y. Y., and Dai, H. J. 2010c. Ni(OH)(2) nanoplates grown on graphene as advanced electrochemical pseudocapacitor materials. *Journal of the American Chemical Society* 132: 7472–7477.

Wang, H. L., Hao, Q. L., Yang, X. J., Lu, L. D., and Wang, X. 2009f. Graphene oxide doped polyaniline for supercapacitors. *Electrochemistry Communications* 11: 1158–1161.

Wang, H. L., Hao, Q. L., Yang, X. J., Lu, L. D., and Wang, X. 2010d. Effect of graphene oxide on the properties of its composite with polyaniline. *Acs Applied Materials & Interfaces* 2: 821–828.

Wang, H. L., Robinson, J. T., Diankov, G., and Dai, H. J. 2010e. Nanocrystal growth on graphene with various degrees of oxidation. *Journal of the American Chemical Society* 132: 3270–3271.

Wang, L., Yu, Y., Chen, P. C., Zhang, D. W., and Chen, C. H. 2008a. Electrospinning synthesis of C/Fe3O4 composite nanofibers and their application for high performance lithium-ion batteries. *Journal of Power Sources* 183: 717–723.

Wang, L., Tian, C. G., Wang, H., Ma, Y. G., Wang, B. L., and Fu, H. G. 2010f. Mass Production of Graphene via an in Situ Self-Generating Template Route and Its Promoted Activity as Electrocatalytic Support for Methanol Electroxidization. *Journal of Physical Chemistry C* 114: 8727–8733.

Wang, S., Chia, P. J., Chua, L. L. et al. 2008b. Band-like transport in surface-functionalized highly solution-processable graphene nanosheets. *Advanced Materials* 20: 3440–3446.

Wang, X., Zhi, L. J., and Mullen, K. 2008c. Transparent, conductive graphene electrodes for dye-sensitized solar cells. *Nano Letters* 8: 323–327.

Wang, X., Zhi, L. J., Tsao, N. et al. 2008d. Transparent carbon films as electrodes in organic solar cells. *Angewandte Chemie-International Edition* 47: 2990–2992.

Wang, X. X., Wang, J. N., and Su, L. F. 2009g. Preparation and electrochemical performance of ultra-short carbon nanotubes. *Journal of Power Sources* 186: 194–200.

Wang, Y., Chen, X. H., Zhong, Y. L., Zhu, F. R., and Loh, K. P. 2009h. Large area, continuous, few-layered graphene as anodes in organic photovoltaic devices. *Applied Physics Letters* 95.

Wang, Y., and Lee, J. Y. 2006. One-step, confined growth of bimetallic tin-antimony nanorods in carbon nanotubes grown in situ for reversible Li+ ion storage. *Angewandte Chemie-International Edition* 45: 7039–7042.

Wang, Y., Shi, Z. Q., Huang, Y. et al. 2009i. Supercapacitor devices based on graphene materials. *Journal of Physical Chemistry C* 113: 13103–13107.

Wang, Y., Wan, Y., and Zhang, D. 2010g. Reduced graphene sheets modified glassy carbon electrode for electrocatalytic oxidation of hydrazine in alkaline media. *Electrochemistry Communications* 12:187–190.

Watcharotone, S., Dikin, D. A., Stankovich, S. et al. 2007. Graphene-silica composite thin films as transparent conductors. *Nano Letters* 7: 1888–1892.

Wei, Z. Q., Wang, D. B., Kim, S. et al. 2010. Nanoscale tunable reduction of graphene oxide for graphene electronics. *Science* 328: 1373–1376.

Winter, M., and Besenhard, J. O. 1999. Electrochemical lithiation of tin and tin-based intermetallics and composites. *Electrochimica Acta* 45: 31–50.

Winter, M., Besenhard, J. O., Spahr, M. E., and Novak, P. 1998. Insertion electrode materials for rechargeable lithium batteries. *Advanced Materials* 10: 725–763.

Wu, J. B., Becerril, H. A., Bao, Z. N. et al. 2008. Organic solar cells with solution-processed graphene transparent electrodes. *Applied Physics Letters* 92: 2633021–3.

Wu, Q., Xu, Y. X., Yao, Z. Y., Liu, A. R., and Shi, G. Q. 2010a. Supercapacitors based on flexible graphene/polyaniline nanofiber composite films. *ACS Nano* 4: 1963–1970.

Wu, Y. P., and Holze, R. 2003. Anode materials for lithium ion batteries obtained by mild and uniformly controlled oxidation of natural graphite. *Journal of Solid State Electrochemistry* 8: 73–78.

Wu, Z. C., Chen, Z. H., Du, X. et al. 2004. Transparent, conductive carbon nanotube films. *Science* 305: 1273–1276.

Wu, Z. S., Pei, S. F., Ren, W. C. et al. 2009. Field emission of single-layer graphene films prepared by electrophoretic deposition. *Advanced Materials* 21: 1756–1760.

Wu, Z. S., Wang, D. W., Ren, W. et al. 2010b. Anchoring hydrous RuO2 on graphene sheets for high-performance electrochemical capacitors. *Advanced Functional Materials* 20: 3595–3602.

Xia, J. L., Chen, F., Li, J. H., and Tao, N. J. 2009. Measurement of the quantum capacitance of graphene. *Nature Nanotechnology* 4: 505–509.

Xiao, P., Xiao, M., Liu, P. G., and Gong, K. C. 2000. Direct synthesis of a polyaniline-intercalated graphite oxide nanocomposite. *Carbon* 38: 626–628.

Xing, W., Xue, J. S., and Dahn, J. R. 1996. Optimizing pyrolysis of sugar carbons for use as anode materials in lithium-ion batteries. *Journal of the Electrochemical Society* 143: 3046–3052.

Xu, Y. X., Bai, H., Lu, G. W., Li, C., and Shi, G. Q. 2008. Flexible graphene films via the filtration of water-soluble noncovalent functionalized graphene sheets. *Journal of the American Chemical Society* 130: 5856–5857.

Xu, Y. X., Sheng, K. X., Li, C., and Shi, G. Q. 2010. Self-assembled graphene hydrogel via a one-step hydrothermal process. *ACS Nano* 4: 4324–4330.

Yan, J., Wei, T., Fan, Z. J. et al. 2010a. Preparation of graphene nanosheet/carbon nanotube/polyaniline composite as electrode material for supercapacitors. *Journal of Power Sources* 195: 3041–3045.

Yan, J., Wei, T., Qiao, W. M. et al. 2010b. Rapid microwave-assisted synthesis of graphene nanosheet/Co3O4 composite for supercapacitors. *Electrochimica Acta* 55: 6973–6978.

Yan, J., Wei, T., Shao, B. et al. 2010c. Electrochemical properties of graphene nanosheet/carbon black composites as electrodes for supercapacitors. *Carbon* 48: 1731–1737.

Yan, J., Wei, T., Shao, B. et al. 2010d. Preparation of a graphene nanosheet/polyaniline composite with high specific capacitance. *Carbon* 48: 487–493.

Yan, X., Cui, X., Li, B. S., and Li, L. S. 2010e. Large, solution-processable graphene quantum dots as light absorbers for photovoltaics. *Nano Letters* 10: 1869–1873.

Yan, X., Cui, X., and Li, L. S. 2010f. Synthesis of large, stable colloidal graphene quantum dots with tunable size. *Journal of the American Chemical Society* 132: 5944–5945.

Yang, C. M., Kim, Y. J., Endo, M. et al. 2007. Nanowindow-regulated specific capacitance of supercapacitor electrodes of single-wall carbon nanohorns. *Journal of the American Chemical Society* 129: 20–21.

Yang, N. L., Zhai, J., Wang, D., Chen, Y. S., and Jiang, L. 2010a. Two-dimensional graphene bridges enhanced photoinduced charge transport in dye-sensitized solar cells. *ACS Nano* 4: 887–894.

Yang, S. B., Cui, G. L., Pang, S. P. et al. 2010b. Fabrication of cobalt and cobalt oxide/graphene composites: Towards high-performance anode materials for lithium ion batteries. *Chemsuschem* 3: 236–239.

Yang, S. B., Feng, X. L., Zhi, L. J. et al. 2010c. Nanographene-constructed hollow carbon spheres and their favorable electroactivity with respect to lithium storage. *Advanced Materials* 22: 838–842.

Yang, S. B., Huo, J. P., Song, H. H., and Chen, X. H. 2008. A comparative study of electrochemical properties of two kinds of carbon nanotubes as anode materials for lithium ion batteries. *Electrochimica Acta* 53: 2238–2244.

Yao, J., Shen, X. P., Wang, B., Liu, H. K., and Wang, G. X. 2009. In situ chemical synthesis of SnO2-graphene nanocomposite as anode materials for lithium-ion batteries. *Electrochemistry Communications* 11: 1849–1852.

Yeh, T. F., Syu, J. M., Cheng, C., Chang, T. H., and Teng, H. S. 2010. Graphite oxide as a photocatalyst for hydrogen production from water. *Advanced Functional Materials* 20: 2255–2262.

Yin, B., Liu, Q., Yang, L. Y. et al. 2010. Buffer layer of pedot:pss/graphene composite for polymer solar cells. *Journal of Nanoscience and Nanotechnology* 10: 1934–1938.

Yin, Z. Y., Wu, S. X., Zhou, X. Z. et al. 2010. Electrochemical deposition of ZnO nanorods on transparent reduced graphene oxide electrodes for hybrid solar cells. *Small* 6: 307–312.

Yong, V., and Tour, J. M. 2010. Theoretical efficiency of nanostructured graphene-based photovoltaics. *Small* 6: 313–318.

Yoo, E., Kim, J., Hosono, E. et al. 2008. Large reversible Li storage of graphene nanosheet families for use in rechargeable lithium ion batteries. *Nano Letters* 8: 2277–2282.

Yoon, S. H., Park, C. W., Yang, H. J. et al. 2004. Novel carbon nanofibers of high graphitization as anodic materials for lithium ion secondary batteries. *Carbon* 42: 21–32.

Yu, D. S., and Dai, L. M. 2010. Self-assembled graphene/carbon nanotube hybrid films for supercapacitors. *Journal of Physical Chemistry Letters* 1: 467–470.

Yu, Q. K., Lian, J., Siriponglert, S. et al. 2008. Graphene segregated on Ni surfaces and transferred to insulators. *Applied Physics Letters* 93: 1131031–3.

Yu, Y., Chen, C. H., and Shi, Y. 2007. A tin-based amorphous oxide composite with a porous, spherical, multideck-cage morphology as a highly reversible anode material for lithium-ion batteries. *Advanced Materials* 19: 993–997.

Yue, Q. L., Zhang, K., Chen, X. M., et al. 2010. Generation of OH radicals in oxygen reduction reaction at Pt-Co nanoparticles supported on graphene in alkaline solutions. *Chemical Communications* 46:3369–3371.

Yumura, T., Kimura, K., Kobayashi, H. et al. 2009. The use of nanometer-sized hydrographene species for support material for fuel cell electrode catalysts: A theoretical proposal. *Phys Chem Chem Phys* 11: 8275–8284.

Zhang, H., Lv, X. J., Li, Y. M., Wang, Y. and Li, J. H. 2010a. P25-Graphene Composite as a High Performance Photocatalyst. ACS Nano 4: 380–386.

Zhang, J., Wang, X., Su, Q. et al. 2009a. Metal-free phenanthrenequinone cyclotrimer as an effective heterogeneous catalyst. *Journal of the American Chemical Society* 131: 11296–11297.

Zhang, J., Liu, X., Blume, R., Zhang, A. H., Schlogl, R., and Su, D. S. 2008a. Surface-modified carbon nanotubes catalyze oxidative dehydrogenation of n-butane. *Science* 322: 73–77.

Zhang, K., Zhang, L. L., Zhao, X. S., and Wu, J. S. 2010b. Graphene/polyaniline nanofiber composites as supercapacitor electrodes. *Chemistry of Materials* 22: 1392–1401.

Zhang, L. L., Li, S., Zhang, J. T. et al. 2010c. Enhancement of electrochemical performance of macroporous carbon by surface coating of polyaniline. *Chemistry of Materials* 22: 1195–1202.

Zhang, M., Fang, S. L., Zakhidov, A. A. et al. 2005. Strong, transparent, multifunctional, carbon nanotube sheets. *Science* 309: 1215–1219.

Zhang, M., Parajuli, R. R., Mastrogiovanni, D. et al. 2010d. Production of graphene sheets by direct dispersion with aromatic healing agents. *Small* 6: 1100–1107.

Zhang, W. M., Hu, J. S., Guo, Y. G. et al. 2008b. Tin-nanoparticles encapsulated in elastic hollow carbon spheres for high-performance anode material in lithium-ion batteries. *Advanced Materials* 20: 1160–1165.

Zhang, W. M., Wu, X. L., Hu, J. S., Guo, Y. G., and Wan, L. J. 2008c. Carbon coated Fe3O4 nanospindles as a superior anode material for lithium-ion batteries. *Advanced Functional Materials* 18: 3941–3946.

Zhang, Y. P., Li, H. B., Pan, L. K., Lu, T., and Sun, Z. 2009b. Capacitive behavior of graphene-ZnO composite film for supercapacitors. *Journal of Electroanalytical Chemistry* 634: 68–71.

Zhao, J., Buldum, A., Han, J., and Lu, J. P. 2000. First-principles study of Li-intercalated carbon nanotube ropes. *Physical Review Letters* 85: 1706–1709.

Zhao, L., Xu, Y. X., Qiu, T. F., Zhi, L. J., and Shi, G. Q. 2009a. Polyaniline electrochromic devices with transparent graphene electrodes. *Electrochimica Acta* 55: 491–497.

Zhao, X., Tian, H., Zhu, M. Y. et al. 2009b. Carbon nanosheets as the electrode material in supercapacitors. *Journal of Power Sources* 194: 1208–1212.

Zhao, Z. W., Guo, Z. P., Yao, P., and Liu, H. K. 2008. Mesoporous carbon-tin nanocomposites as anode materials for Li-ion battery. *Journal of Materials Science & Technology* 24: 657–660.

Zheng, T., Xue, J. S., and Dahn, J. R. 1996. Lithium insertion in hydrogen-containing carbonaceous materials. *Chemistry of Materials* 8: 389–393.

Zhi L. J., Hu, Y. S., El Hamaoui, B., et al. 2007. From well-defined carbon-rich precursors to monodisperse carbon particles with hierarchic structures. *Advanced Materials*. 19: 1849–1853.

Zhi, L. J., Hu, Y. S., El Hamaoui, B. et al. 2008. Precursor-controlled formation of novel carbon/metal and carbon/metal oxide nanocomposites. *Advanced Materials* 20: 1727–1731.

Zhi, L. J., Wu, J. S., Li, J. X., Kolb, U., and Mullen, K. 2005. Carbonization of disclike molecules in porous alumina membranes: Toward carbon nanotubes with controlled graphene-layer orientation. *Angewandte Chemie-International Edition* 44: 2120–2123.

Zhong, C. J., Luo, J., Fang, B. et al. 2010. Nanostructured catalysts in fuel cells. *Nanotechnology* 21: 062001.

Zhu, Y. W., Cai, W. W., Piner, R. D., Velamakanni, A., and Ruoff, R. S. 2009. Transparent self-assembled films of reduced graphene oxide platelets. *Applied Physics Letters* 95: 1031041–3.

7

Electrochemistry of Graphene-Based Nanomaterials

Martin Pumera*

CONTENTS

7.1 Introduction

Electrochemistry is a very important discipline. In electrochemistry, the chemical state of a compound can be modulated by applying voltage and by injecting or withdrawing electrons from the molecule. Alternatively, by changing the chemical state of a particular compound, we can generate a flow of electrons/holes through a circuit (Bard and Faulkner 2001). The implications are clear: Electrochemistry is behind most energy storage and conversion processes, such as those taking place in batteries or fuel cells. It is also behind important applications in sensing and biosensing, such as the detection of glucose levels for people suffering from diabetes.

Graphene is a novel two-dimensional (2D) nanomaterial, a one-atom-thick sheet of sp^2-hybridized carbon (Novoselov et al. 2004). It can be viewed as molecular graphite or as an extralarge polycyclic aromatic molecule. It possesses several unique properties that are unseen in other materials, such as room-temperature quantum Hall effect, high mechanical strength, high elasticity and thermal conductivity, and tunable bandgap (Geim and Novoselov

* Correspondence to: Martin Pumera (pumera@ntu.edu.sg)

2007), as described in Chapters 2 and 3. As in other fields of research on graphene, we can rely to a large extent on older literature regarding electrochemistry of graphite and carbon nanotubes (CNTs). However, there are many cases in which such dependence can be misleading. In this chapter, we focus on the electrochemical properties of graphene.

7.2 Fundamental Electrochemistry of Graphene

Graphene, shown in Figure 7.1, has two main features: At first sight, the main feature is a basal plane, the surface of graphene. However, on second glance, because the dimensions of graphene are not unlimited, the material also features edges, where the basal plane of graphene ends. On a third, and even closer look, the ends of the basal plane can be seen to terminate in different ways, having carbon atoms arranged in a "zigzag" structure or in an "armchair" structure. This is directly related to the underlying geometry of graphene, where the hexagonal lattice cannot be equally terminated in a square format. Termination in a zigzag or armchair structure has a profound

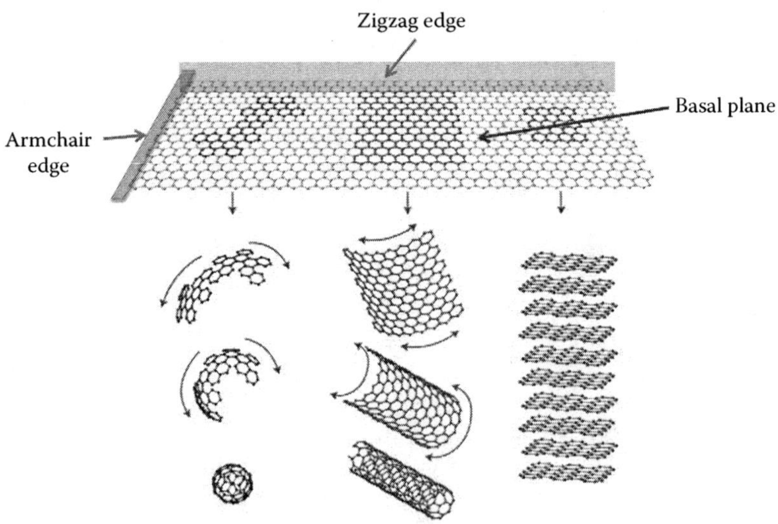

FIGURE 7.1
Graphene as the 2D building material for carbon materials of all other dimensionalities: It can be wrapped into zero-dimensional buckyballs, rolled into one-dimensional nanotubes, or stacked into three-dimensional graphite. We show that a graphene sheet exhibits two different types of edges, zigzag and armchair edges, and a basal plane, which is almost electrochemically inert. (From Geim, A. K., and K. S. Novoselov, *Nat. Mater.*, 6, 2007. With permission.)

influence on the electronic properties of graphene. However, its influence on electrochemical responses has not yet been proven (Pumera 2009a, 2010).

Depending on the dimensions of the basal plane, graphene with basal plane dimension is often called a "nanoribbon"; it possesses specific electronic properties. On the other hand, a graphene sheet can have dimensions on the order of millimeters or centimeters.

The closest and, in the past, well-studied analog of graphene is highly oriented pyrolytic graphite (HOPG). It is noted that HOPG has two features similar to graphene: (1) A basal plane and (2) edges. The edges possess a very high heterogeneous electron transfer (HET) rate, on the order of 10^{-2} $cm^2 \cdot s^{-1}$, which means the electrochemical reaction of HOPG material, or electron transfer from the HOPG material to another molecule, in solution is fast. However, the basal plane of HOPG possesses a much slower HET rate, on the order of about 10^{-9} $cm^2 \cdot s^{-1}$ (Davis et al. 2005). Although the magnitude of this rate is still being discussed in the literature, it is generally considered that the edges of HOPG provide a much easier site for the electrochemical reaction to occur, whereas the basal plane is practically electrochemically inert. The basal plane is electrochemically active only in cases where so-called edge-like defects are present, with some openings of the defects. It is very important to note that "chemically reduced" graphene has many defects (Gomez-Navarro et al. 2010), as shown in Figure 7.2. Such a conclusion about the electrochemistry of HOPG can easily be transferred to the electrochemistry of graphene, as we (Goh and Pumera 2010a) demonstrated that there is no significant difference among the electrochemical responses of single-, few-, and multilayer

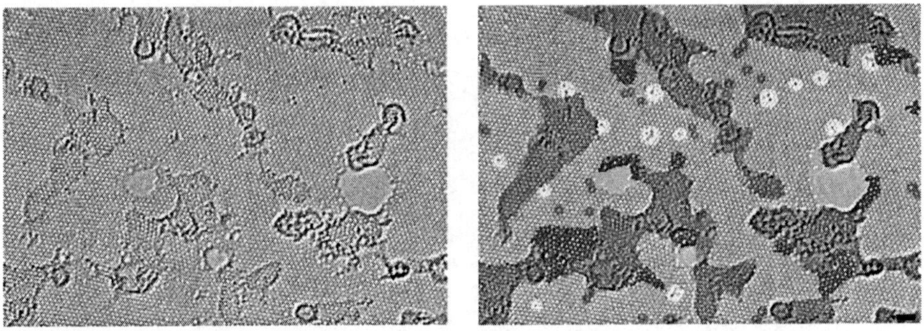

(a) (b)

FIGURE 7.2
Structure of chemically reduced graphene: Chemically reduced graphene exhibits a structure different from that of pristine graphene. The figure shows an atomic resolution, aberration-corrected transmission electron microscopy (TEM) image of a single-layer reduced–graphene oxide membrane. (a) Original image and (b) image with color added to highlight the different features. Scale bar is 1 nm. (Reprinted from Gomez-Navarro, C. et al., *Nano Lett.*, 10, 1144, 2010. With permission from American Chemical Society.)

graphene sheets (see Figure 7.3). This conclusion is also supported by findings about the electrochemistry of CNTs according to which pristine CNTs exhibit a sluggish electron transfer rate, whereas CNTs for which defects are introduced to their walls (which are the equivalent of a basal plane in HOPG) exhibit a much faster electron transfer rate (Pumera et al. 2008).

Using electrochemical impedance spectroscopy (EIS), we also demonstrated that multilayer graphene nanoribbons with cross sections of 100 × 100 nm provide larger capacitance (15.6 $F \cdot g^{-1}$) than do few-layer graphene nanoribbons (14.9 $F \cdot g^{-1}$) and far larger capacitance than single-layer graphene nanoribbons (10.9 $F \cdot g^{-1}$) with the same cross sections (Goh and Pumera 2010b). This is shown in Figure 7.4.

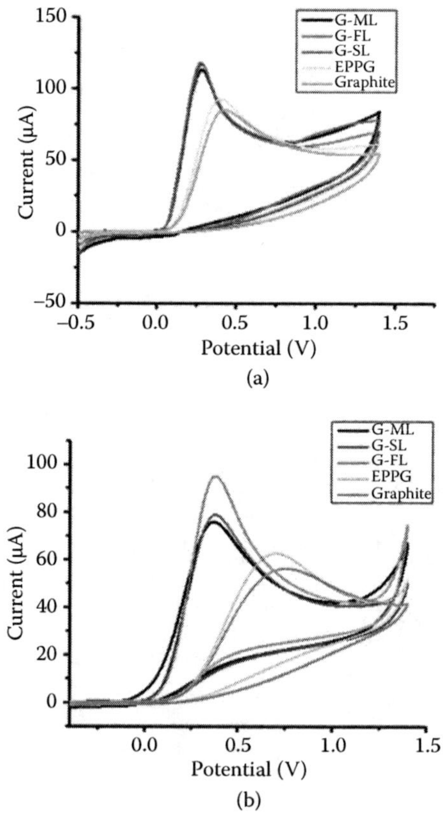

FIGURE 7.3
Cyclic voltammograms resulting from electrochemical oxidation of 5 mM dopamine (a) and ascorbic acid (b) on single-layer graphene nanoribbons (G-SL), few-layer graphene nanoribbons (G-FL), multilayer graphene nanoribbons (G-ML), graphite microparticles (GMPs), and edge plane pyrolytic graphite (EPPG) surfaces. Conditions are scan rate, 100 $mV \cdot s^{-1}$; background electrolyte, 50 mM phosphate buffer, with pH 7.4; and reference electrode, Ag/AgCl. (Reprinted from Chng, E. L. K., and M. Pumera, *Chem. Asian J.*, 6, 2304–2307, 2011. With permission from John Wiley and Sons.)

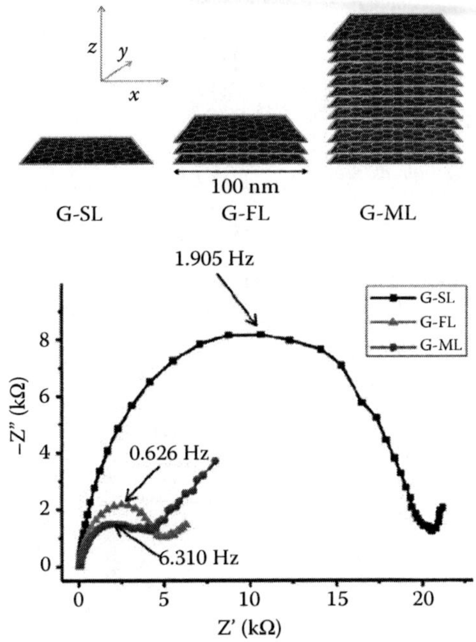

FIGURE 7.4
Figure showing schematics of used graphene-based nanomaterials (top) and Nyquist impedance plots (bottom). Single-layer graphene nanoribbons are denoted by G-SL, few-layer graphene nanoribbons by G-FL, and multilayer graphene nanoribbons by G-ML. The size of the base of all nanomaterials used is about 100 × 100 nm. Scheme is not shown to scale. The Nyquist impedance plots of G-SL-, G-FL-, and G-ML-modified electrodes for 10 mM ferricyanide solution are shown in the bottom part of the figure. The supporting electrolyte is 1 M KCl. Arrows indicate frequencies of the Z′–Z″ semicircle maxima. Z refers to "impedance." (Reprinted from Goh, M. S., and M. Pumera, *Electrochem. Commun.*, 12, 1375, 2010. With permission from Elsevier.)

7.3 Electrochemical Oxidation and Reduction of Graphene

Graphene itself can undergo electrochemical reactions. Pristine graphene sheets can be oxidized and functional groups, such as hydroxyl and carboxyl, can be introduced, creating graphene oxide. However, more often, graphene oxide is the final product of thermal and sono exfoliation of graphite oxide to graphene oxide. Graphene oxide can be reduced chemically or electrochemically to graphene at potentials ranging from −0.6 to −0.9 V (vs. Ag/AgCl; see Figure 7.5) (Zhou et al. 2009). Such an electrochemical reduction is very important as electrochemical reduction can be controlled very well.

By electrochemical reduction, it is possible to determine the number of oxygen-containing reducible groups, such as aldehyde, peroxide, and epoxide, in graphene oxides. The quantification of various oxygen-containing

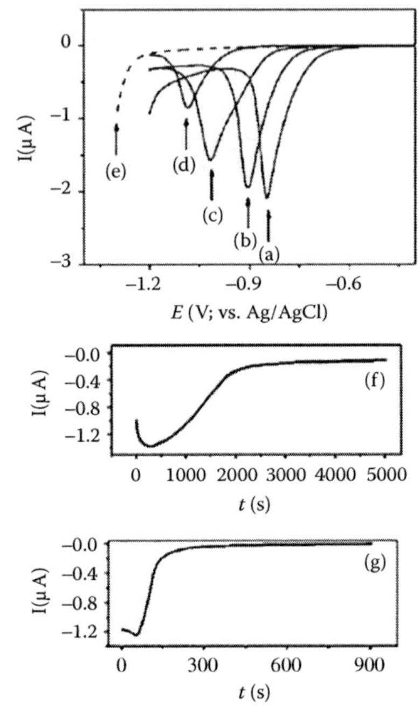

FIGURE 7.5
Electrochemical reduction of graphene: Linear sweep voltammograms of GC electrode in contact with approximately 7-μm-thick GO films on quartz (5 × 4 cm^2) at pH values of (a) 4.12, (b) 7.22, (c) 10.26, and (d) 12.11; (e) linear sweep voltammograms of GC electrode suspended in Na–phosphate buffered saline (PBS; 1 M, pH 4.12); and typical *i–t* curves for the electrolysis of approximately 7-μm-thick GO films on (f) quartz (5 × 4 cm^2) and (g) GC electrode (3 mm in diameter) with an applied potential of −0.90 V in Na–PBS (1 M, pH 4.12). (Reprinted from Zhou, M. et al., *Chem. Eur. J.*, 15, 6116, 2009. With permission from John Wiley and Sons.)

groups in graphene oxides is of very high importance as these groups have strong influences on the physical, chemical, and material properties of graphene oxides. All current methods to this end provide only relative quantification. Electrochemistry is a powerful electrochemical technique for the absolute quantification of selected oxygen-containing groups in graphene oxide (Chng and Pumera 2011).

7.4 Oxygen-Containing Groups

As noted in Section 7.3, graphene can contain oxygen-containing groups, such as hydroxyl or carboxyl groups. This is extremely important in the study of electrochemistry, as electrochemistry involves sensitivity of molecules to the presence of surface modifications; that is, oxygen-containing groups (Yuge et al. 2008) showed a method for the visualization of oxygen-containing groups on graphene by staining them with a Pt complex (see Figure 7.6). These oxygen-containing groups are located close to the edges of graphene sheets.

From the point of view of electrochemistry, such groups can enhance electrochemical performance by lowering the energy barrier for electron transfer and can contribute electrocatalytic properties (Crevillen et al. 2009) to

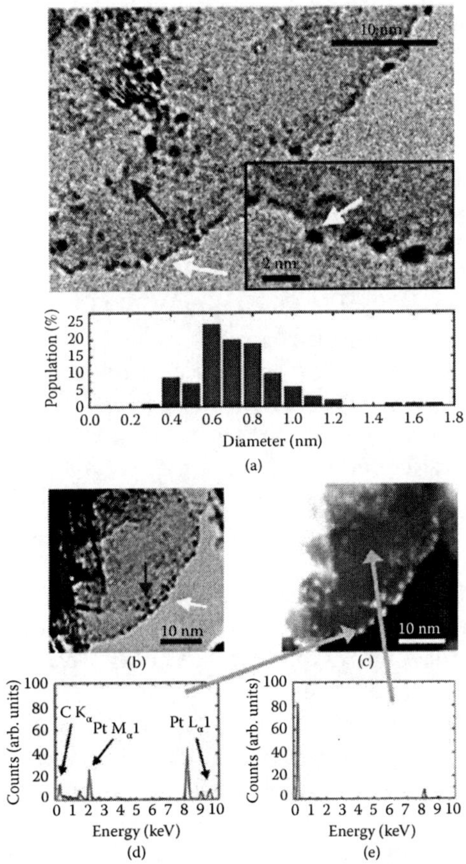

FIGURE 7.6
Location of oxygen-containing groups on graphene oxide sheets: Preferential adsorption of Pt–ammine complexes occurs at the carboxylic groups at edges and steps of oxidized graphene sheets. (a) The TEM images of Pt complex/oxidized graphene and their size distributions obtained from about 300 black spots observed in the TEM images. The Pt complex clusters attached to the edges are denoted by dark arrow and to the steps by light arrow. (b) Scanning TEM and (c) Z-contrast images of Pt complex/oxidized graphene are shown. The energy-dispersive x-ray (EDX) spectrum of Pt complex/oxidized graphene at a bright spot (arrow; d) and at an area (arrow) on the featureless surface (e) in the Z-contrast image is given. In the EDX spectrum (d), peaks originating from C, K, Pt M1, and Pt L1 are shown. Other peaks are associated with Cu and Al of the Cu grid disk (TEM sample holder) or detector bodies of the EDX analyzer. (Reprinted from Yuge, R. et al., *ACS Nano*, 2, 1747, 2008. With permission from American Chemical Society.)

graphene. However, on the other hand, many molecules that interact with adsorbed products often slow down electrochemical reactions. Such adsorption is often highly sensitive to the presence of oxygen-containing groups on the surface. This issue must be addressed on an individual basis. Here, we show an example of an important molecule for enzyme biosensors and biofuel cells employing dehydrogenase enzymes (Pumera et al. 2009). This molecule is β-nicotinamide adenine dinucleotide (NAD^+). There are major problems associated with the adsorption of NAD^+ on carbon electrodes made from sp^2-hybridized materials as this adsorption directly results in the passivation of such electrodes because no additional electroactive material can be oxidized at that particular site. This situation has been known for more than 20 years, but a clear explanation has never been given as to why such passivation occurs. We used high-resolution x-ray photoelectron spectroscopic measurements, cyclic voltammetry, amperometry, and molecular dynamics to tackle this problem. We investigated the basal and edge planes of graphene before and after electrochemical oxidation of NADH and discovered that this is due to the interaction of NAD^+, the product of the reaction with the oxygen-containing groups, with COO^- groups. Interestingly, the hydroxyl groups did not play any role in the adsorption process. Using Car-Parrinello simulation, we found the large interaction that agrees with what was found experimentally only occurs if a positive NAD^+ ion is placed close to a graphene sheet edge containing a $-COO^-$ group. If a NAD^+ ion is positioned close to the basal plane of graphene or near the edges of graphene sheets substituted with only hydrogen, no significantly strong interactions occur (see Figure 7.7).

There are various preparation methods of graphene. The structures of graphenes prepared by various methods are different. If chemical reagents are used for exfoliations, such graphenes are called "chemically modified graphenes" (CMGs; Figure 7.8).

We compared the electrochemical performance and properties of CMGs (Ambrosi et al. 2011), as shown in Figure 7.9. We showed that thermally reduced graphene oxide exhibits the lowest number of oxygen-containing groups and the fastest HET rate.

With increasing amounts of oxygen-containing groups on the graphene surface, in the row of electrochemically reduced graphene oxide, chemically reduced graphene oxide, graphene oxide, and graphite oxide, the HET rate determined by cyclic voltammetry decreases and resistance to charge transfer measured using impedance spectrometry increases. Therefore, we can conclude that with increasing amounts of oxygen-containing groups, HET rate decreases.

Oxygen-containing groups might pose additional challenge to the usability of graphene oxides for sensing. Even though the oxidation of molecules on graphene oxide surfaces does not pose problems, the reduction section of the electrochemical window does. As the oxygen-containing groups are reducible, the electrochemical window of graphene oxides is actually limited

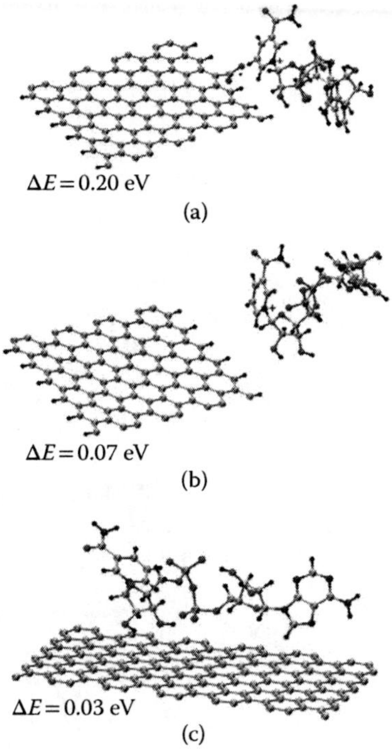

FIGURE 7.7
Adsorption of nicotinamide adenine on graphene sheets: Geometries of adsorption of NAD^+ on (a) graphene edge terminated by hydrogen atoms and containing one $-COO^-$ group, (b) graphene edge fully terminated by hydrogen atoms, and (c) basal plane of graphene sheet via Car-Parrinello molecular dynamics. (From Pumera, M., R. Scipioni, H. Iwai, et al. A mechanism of adsorption of β-nicotinamide adenine dinucleotide on graphene sheets: Experiment and theory. *Chem. Eur. J.*, 15, 10851, 2009. Copyright Wiley-VCH Verlag GmbH & Co. KGaA. Reproduced with permission.)

to −0.75 V (vs. Ag/AgCl) (Pumera, Scipiani et al. 2009). At any lower potential, the reduction peak of graphene oxide is the dominant signal and detection of analytes is not possible (Toh et al. 2011).

7.5 Graphene for Energy Storage

Graphene exhibits a very large surface area, about 2630 $m^2 \cdot g^{-1}$ (Geim and Novoselov 2007). This is double the surface area of single-walled CNTs. However, it should be understood that exfoliated single-sheet graphene tends to restack and, thus, the real surface area is significantly reduced if graphene is not stabilized by addition of surfactant or separated by nanoparticles. The very large surface area is highly beneficial for energy storage devices such as supercapacitors and batteries.

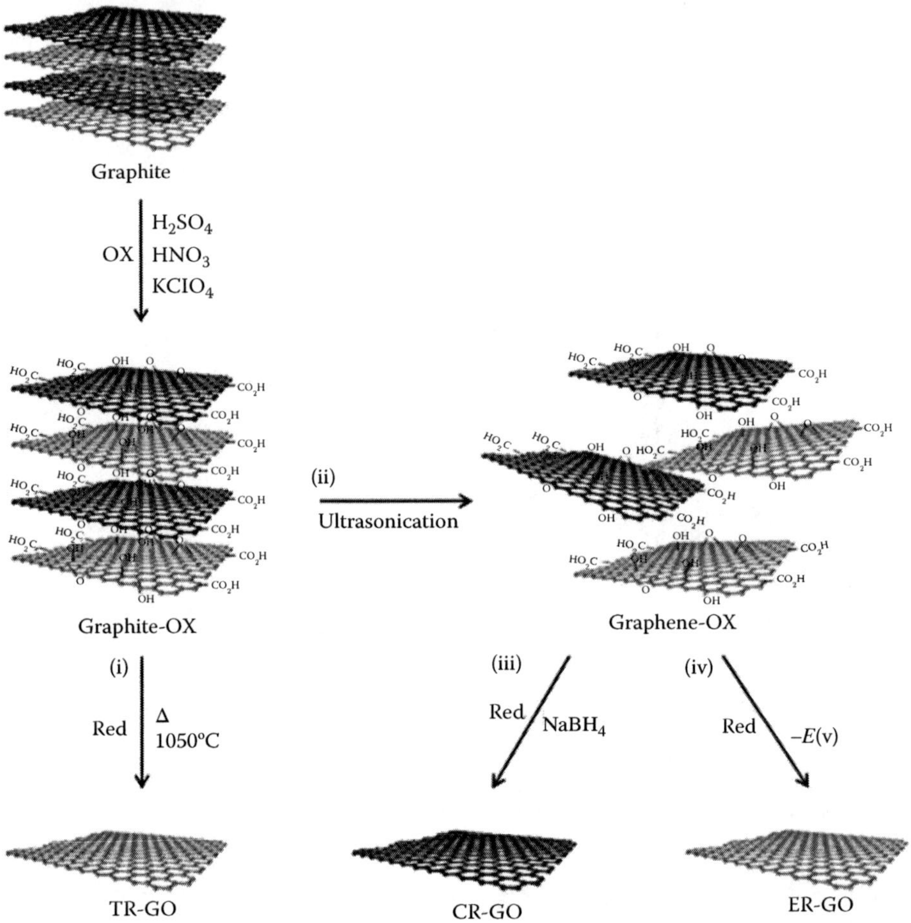

FIGURE 7.8
Preparation of various graphene oxide–related structures, the so-called CMGs: Schematic of the procedure followed for the production of CMGs using graphite as the starting material is shown in the figure. An oxidative treatment was preliminarily performed to generate graphite oxide. This is followed by (i) thermal reduction/exfoliation of graphite oxide to produce thermally reduced graphene oxide (TR-GO); (ii) exfoliation by ultrasonication to generate graphene oxide; (iii) chemical reduction of graphene oxide by means of $NaBH_4$ to produce chemically reduced graphene oxide (CR-GO); and (iv) electrochemical reduction of graphene oxide to produce electrochemically reduced graphene oxide (ER-GO). The labels "OX" and "RED" stand for oxidation and reduction, respectively. (Reprinted from Ambrosi, A. et al., *Chem. Eur. J.*, 17, 10763–10770, 2011. With permission from John Wiley and Sons.)

7.5.1 Supercapacitors

Supercapacitors are devices that store energy as electrical charges and are able to provide large amounts of energy on demand for short periods. Similar to typical "normal" capacitors, which store charges in an electrode/electrolyte double layer, supercapacitors also store energy in a double layer. However, in

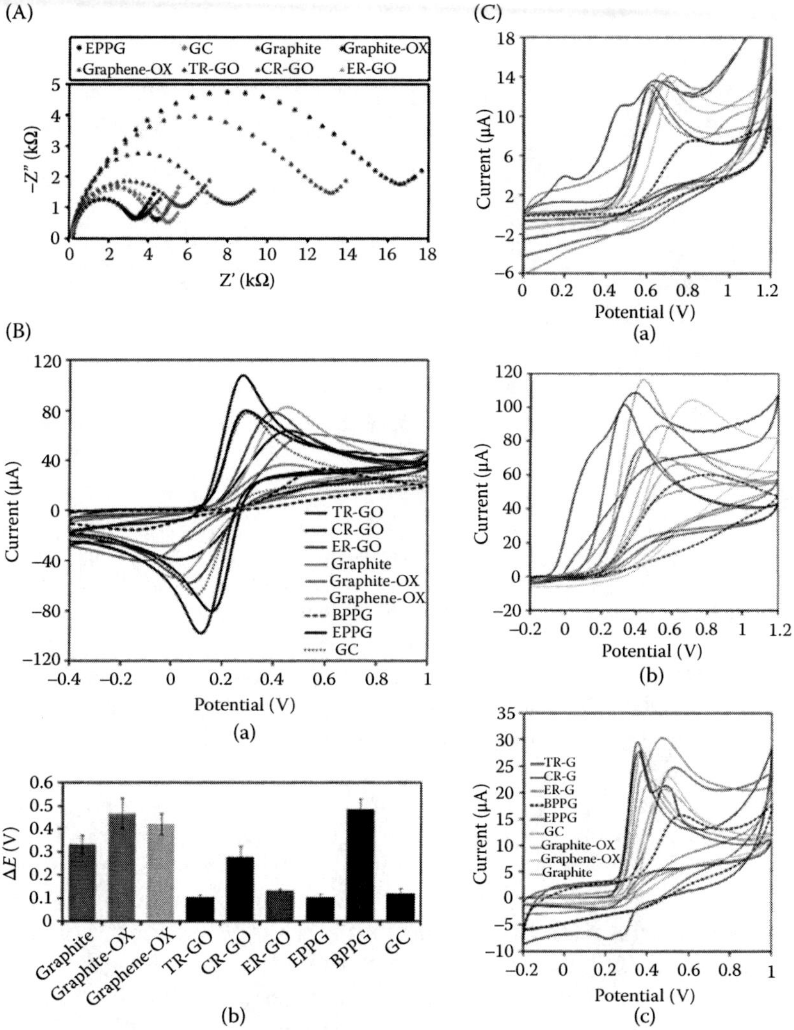

FIGURE 7.9
Electrochemical behavior of CMGs. (A) Nyquist diagrams for EIS measurements of graphite-modified (▲), graphite oxide–modified (▲), graphene oxide–modified (▲), TR-GO-modified (▲), ER-GO-modified (▲), and CR-GO-modified (▲) electrodes. For comparison, bare GC (◇) and EPPG (■) electrodes are also shown. Conditions are as follows: Background electrolyte, 50 mM PBS; and 10 mM $K_3[Fe(CN)_6]/K_4[Fe(CN)_6]$. Randles equivalent circuit used for data fitting is shown as inset. (B) (a) Cyclic voltammograms of 10 mM $[Fe(CN)_6]^{4-/3-}$ on graphite, graphite oxide, graphene oxide, TR-GO, ER-GO, CR-GO, EPPG, BPPG, and GC electrodes. Conditions are as follows: Background electrolyte, 0.1 M KCl; and scan rate, 100 mV·s^{-1}. (b) Peak-to-peak separation average values for all materials (n = 5) and corresponding standard deviations. (C) Cyclic voltammograms of (a) NADH, (b) ascorbic acid, and (c) uric acid on graphite, graphite oxide, graphene oxide, TR-GO, ER-GO, CR-GO, EPPG, BPPG, and GC electrodes. Conditions are as follows: Scan rate, 100 mV·s^{-1}; background electrolyte, 50 mM PBS; and pH, 7.2. (Reprinted from Ambrosi, A. et al., *Chem. Eur. J.*, 17, 10763–10770, 2011. With permission from John Wiley and Sons.)

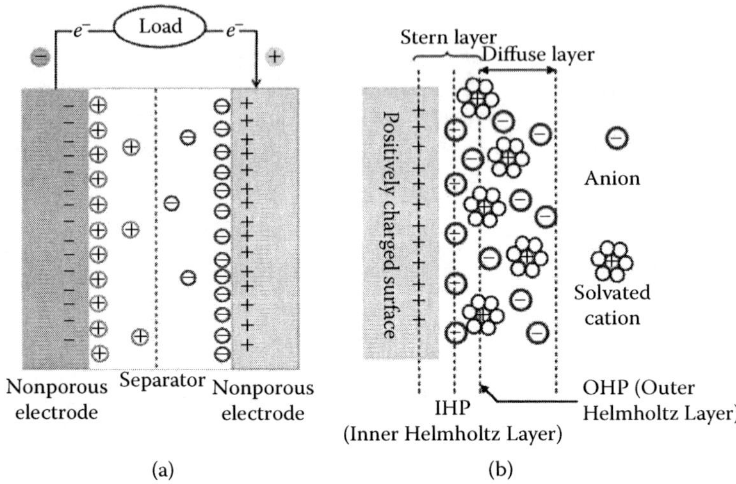

FIGURE 7.10
Schematic diagrams of (a) a two-cell supercapacitor device made of a nonporous electrode and (b) an electrical double-layer (EDL) structure based at a positively charged electrode surface. (Reprinted from Zhang, L. L., R. Zhou, and X. S. Zhao, *J. Mater. Chem.*, 20, 5983–59922, 2010. Reproduced by permission of The Royal Society of Chemistry.)

addition, supercapacitors can transfer a charge to a layer of redox molecules close to the interface (Zhang et al. 2010) (see Figure 7.10). Therefore, the amount of stored charge increases by several orders of magnitude.

The redox molecules on the surface of graphene can be polymers, such as polyaniline or polypyrrole, or oxides/hydroxides, such as MnO_x or $Ni(OH)_2$, having redox functionalities, such as oxygen-containing groups. The advantage of using graphene in supercapacitors is that it has a large surface area and the stored charge is proportional to the area. Ruoff and coworkers (Stoller et al. 2008) first used the high surface area of graphene for supercapacitors, showing that CMG has large capacitances of 135 and 99 $F{\cdot}g^{-1}$ for aqueous and organic electrolytes, respectively. Their setup is shown in Figure 7.11. This concept was further expanded by other groups, showing capacitances of about 200 $F{\cdot}g^{-1}$ (Wang, Shi, et al. 2009).

As mentioned in the paragraph above, in the paragraph above, graphene itself can be decorated with inorganic oxides to increase capacitance. It has been shown (Chen et al. 2010) that decorating graphene sheets with MnO_x nanocrystals can improve capacitance to the level of about 200 $F{\cdot}g^{-1}$, whereas a well-thought-out nanoarchitectonics design of graphene–nickel hydroxide composites can improve capacitance to the level of 1300 $F{\cdot}g^{-1}$ (Wang et al. 2010).

7.5.2 Batteries

Batteries are devices that store chemical energy and transform it to electricity by an electrochemical reaction. Batteries have two electrodes, a cathode

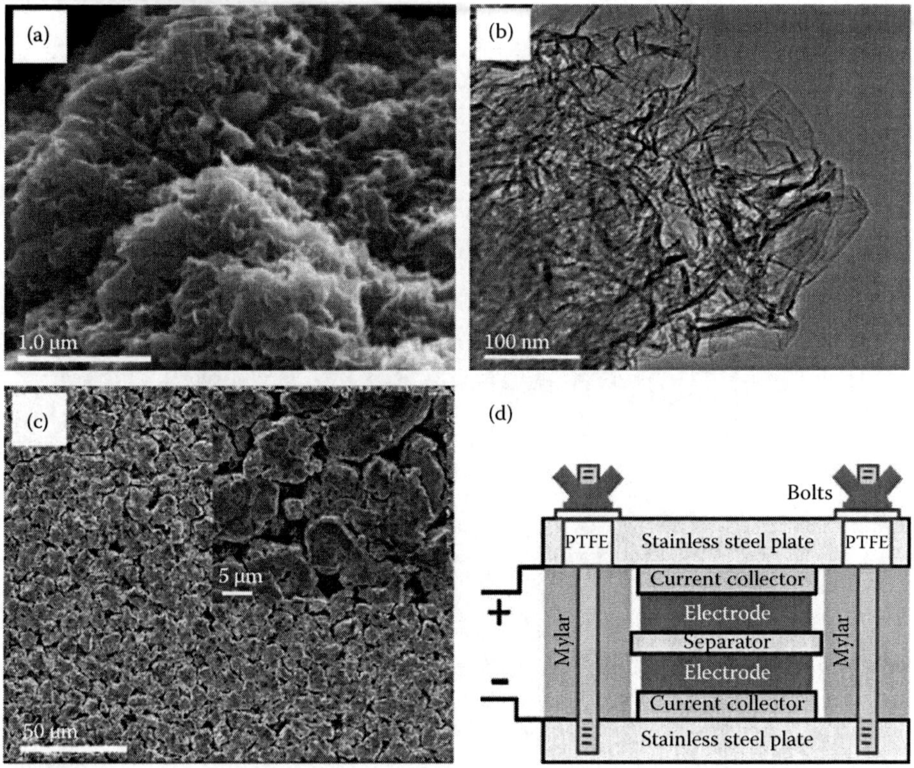

FIGURE 7.11
Supercapacitors. (a) A scanning electron microscopy (SEM) image of CMG particle surface. (b) A TEM image showing individual graphene sheets extending from the CMG particle surface. (c) Low-magnification and high-magnification (inset) SEM images of the CMG particle electrode surface. (d) Schematic of the test cell assembly. (From Stoller, M. D., Park, S., Zhu, Y., An, J., and Ruoff, R. S. *Nano Letters*, 8(10), 2008. With permission.)

and an anode, each of which is made of a different material. Graphite-based electrodes for batteries have been known for a long time. Graphite can store Li^+ ions, which can intercalate between layers of graphite, between graphene sheets, effectively creating the compound LiC_6. This leads to a relatively low capacity of 372 mA·h·g^{-1} (Bruce et al. 2008). It has been suggested that graphene can accommodate Li^+ atoms on both sides of graphene sheets, doubling the amount of Li^+ stored and also doubling the specific capacity of graphene-based lithium batteries to 744 mA·h·g^{-1} (Wang, Wang, et al. 2009). In order to create practical and stable graphene-based batteries, the graphene sheets must be separated. It was found that graphene sheets are prone to restacking when CNTs and fullerenes are employed as spacers of individual graphene sheets. The capacity of a Li^+ battery was increased even more to 730 and 784 mA·h·g^{-1} when CNTs and C_{60}, respectively, were used as separators of the graphene sheets (Yoo et al. 2008) (for an example, see Figure 7.12).

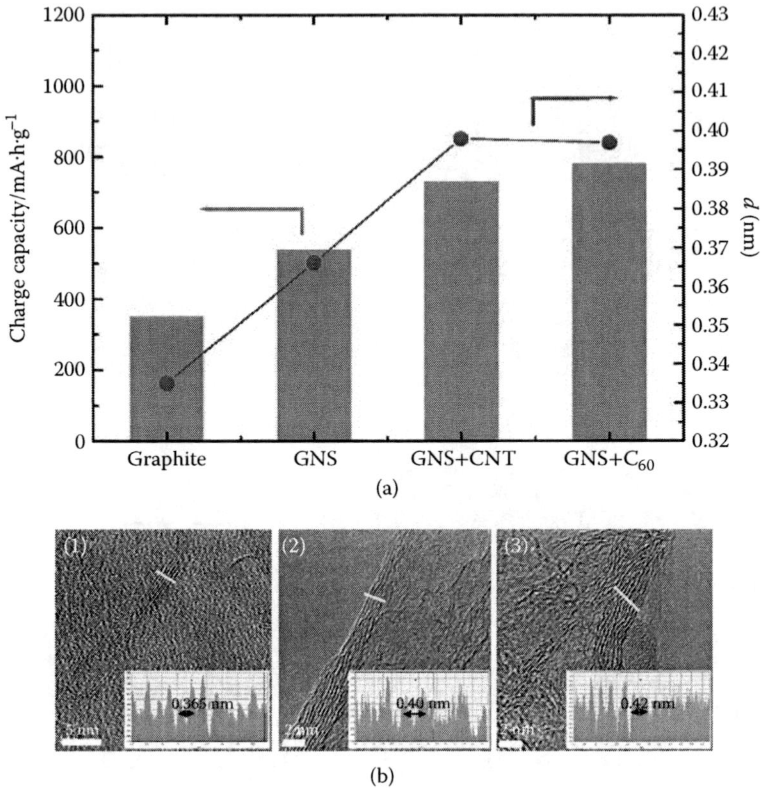

FIGURE 7.12
Graphene for batteries. (a) Relationship between *d* spacing and charge capacity of GNS families and graphite. (b) Cross-sectional TEM images of GNS families with almost the same number (5–6) of graphene stacking layers for (1) GNS, (2) GNS+CNT, and (3) GNS+C_{60}. (Reprinted from Yoo, E. J. et al., *Nano Lett.*, 8, 2277, 2008. With permission.)

7.6 Graphene for Sensing and Biosensing

Graphene nanomaterials are widely used for the construction of biosensors (Pumera et al. 2010). The main reasoning behind their use in this field is that carbon is often more prone to passivation than are metallic electrodes. In principle, carbon is cheaper than most metals and the small dimensions of graphene compared with graphite (graphite particles have dimensions of 10–20 μm) enable the construction of electrodes having a more uniform distribution of electroactive elements, leading to lower current densities at these sites and lower overpotentials. It is noted that previously CNTs were used for this task. However, because they are highly expensive and contain uncontrollable amounts of metallic impurities (Pumera 2009b), which are toxic and participate in electrochemical redox reactions even at amounts as low as 0.01 wt% (Pumera and

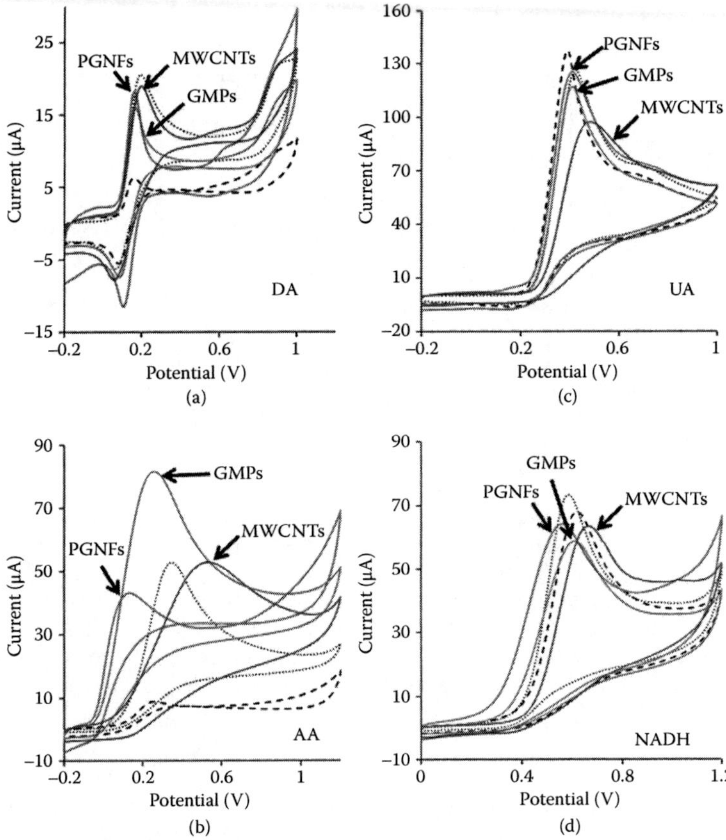

FIGURE 7.13
Stacked graphene nanofibers for biosensing. Cyclic voltammograms of (a) dopamine, (b) ascorbic acid, (c) uric acid, and (d) NADH on stacked platelet graphene nanofibers (PGNFs), graphite microparticles (GMPs), multiwalled CNTs (MWCNTs), EPPG (dotted black line), and bare GC electrodes (dashed black line). Conditions are as follows: Scan rate, 100 mV s^{-1}; background electrolyte, 50 mM phosphate buffer; and pH, 7.4. (From Ambrosi, A., T. Sasaki, and M. Pumera. Platelet graphite nanofibers for electrochemical sensing and bio-sensing: The influence of graphene sheet orientation. *Chem. Asian J.*, 5, 266, 2010. Copyright Wiley-VCH Verlag GmbH & Co. KGaA. Reproduced with permission.)

Miyahara 2009), graphene is preferred for use in biosensors. We have recently shown, however, that it is possible to use stacked graphene nanofibers (formerly called graphite nanofibers) with exactly the same effect as single- or few-layer nanoplatelets (Chng and Pumera 2011). We also demonstrate that stacked graphene platelets have better electrochemical performance than graphite microparticles (GMPs) and CNTs (Ambrosi et al. 2010) (see Figure 7.13).

Thus, we suggest that stacked graphene nanofibers can be used for sensing instead of graphene. We showed that stacked graphene nanofibers are excellent electrodes for DNA detection (Ambrosi and Pumera 2010) (see Figure 7.14).

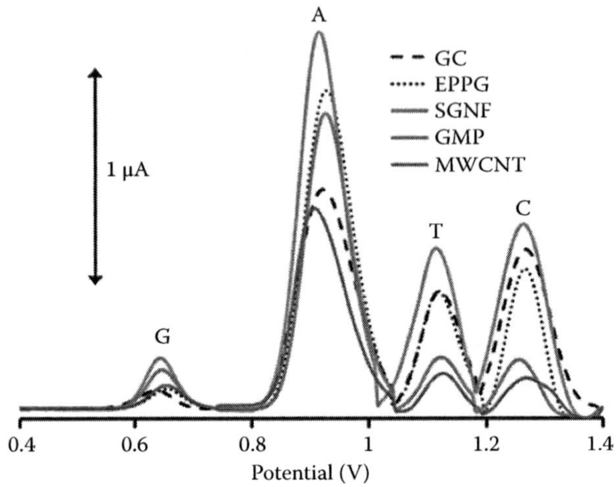

FIGURE 7.14
Graphene for biosensing. Differential pulse voltammetry of a mixture of guanine, adenine, thymine, and cytosine at SGNF (curve), GMP (curve), and MWCNT (curve) electrodes. For comparison, GC (dashed black line) and EPPG (dotted black line) electrode signals are also shown. Concentrations for the DNA bases are as follows: Guanine, 4 μM; adenine, 50 μM; thymine, 100 μM; and cytosine, 100 μM. Conditions are as follows: Background electrolyte, 50 mM phosphate buffer; and pH, 7.4. (Reprinted from Ambrosi, A., and M. Pumera, *Phys. Chem. Chem. Phys.*, 12, 8943–8947, 2010. Reproduced by permission of The Royal Society of Chemistry.)

We have also investigated the usability of single-, few-, and multilayer graphene nanosheets (GNSs) in DNA detection (Bonanni and Pumera 2011) based on hairpin oligonucleotides (Figure 7.15). The DNA hybridization signal was followed by EIS. We have found that even though single-layer graphenes provide a higher signal, it also exhibits a larger relative standard deviation (RSD); thus, few-layer graphenes with lower RSD of the signal is preferred for DNA detection on graphene platforms.

7.7 Concluding Remarks and Future Directions

If nothing else, it should be clear from this chapter that even though many fundamental properties of graphene can be derived from those of graphite, graphene offers significant advantages over graphite in terms of higher surface area and, thus, higher performance in energy storage, which is dependent on surface area. Graphene also possesses a smaller "particle" size for the construction of biosensors, leading to lower current densities and lower overpotentials. It becomes clear that with well-thought-out nanoarchitectonics

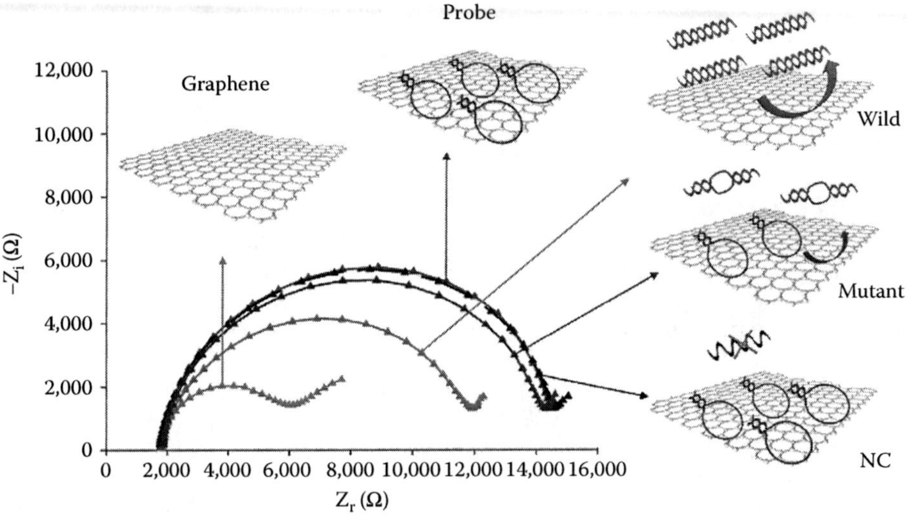

FIGURE 7.15
Impedimetric sensing of DNA on graphene. Schematic of the protocol and Nyquist plots, $-Z_i$ versus Z_r, of the graphene surface (gray); hpDNA; complementary target; one mismatch target; and negative control with a noncomplementary sequence (black; concentration of the DNA probes, 1×10^{-5} M; concentration of the DNA target, 3×10^{-8} M). All measurements were performed in 0.1 M PBS buffer solution containing 10 mM $K_3[Fe(CN)_6]/K_4[Fe(CN)_6]$. Z stands for impedance and Nc for noncomplementary. (Reprinted from Bonanni, A., and M. Pumera, *ACS Nano*, 5, 2356, 2011. With permission from American Chemical Society.)

design, we can achieve nearly the theoretical limits of graphene's electrochemical performance. Much work remains to be done in understanding the relation between graphene's structure and electrochemical function.

References

Ambrosi, A., A. Bonanni, Z. Sofer et al. 2011. Electrochemistry at chemically modified graphenes. *Chem. Eur. J.* 17: 10763–10770.

Ambrosi, A., and M. Pumera. 2010. Stacked graphene nanofibers for electrochemical oxidation of DNA bases. *Phys. Chem. Chem. Phys.* 12: 8943–8947.

Ambrosi, A., T. Sasaki, and M. Pumera. 2010. Platelet graphite nanofibers for electrochemical sensing and bio-sensing: The influence of graphene sheet orientation. *Chem. Asian J.* 5: 266.

Bard, A. J., and L. R. Faulkner. 2001. *Electrochemical methods: Fundamentals and applications*, 2nd ed. New York: Wiley.

Bonanni, A., and M. Pumera. 2011. Graphene platform for hairpin-DNA-based impedimetric genosensing. *ACS Nano* 5: 2356.

Bruce, P. G., B. Scrosati, and J.-M. Tarascon. 2008. Nanomaterials for rechargeable lithium batteries. *Angew. Chem. Int. Ed.* 47: 2930–2946.

Chen, S., J. Zhu, X. Wu, Q. Han, and X. Wang. 2010. Graphene oxide-MnO_2 nanocomposites for supercapacitors. *ACS Nano* 4: 2822–2830.

Chng, E. L. K., and M. Pumera. 2011. Metallic impurities are responsible for electrocatalytic behavior of carbon nanotubes towards sulfides. *Chem. Asian J.* 6: 2304–2307.

Crevillen, A. G., M. Pumera, M. C. Gonzalez, et al. 2009. The preferential electrocatalytic behavior of graphite and multiwalled carbon nanotubes on enediol groups and their analytical implications in real domains. *Analyst* 134: 657–662.

Davis, T. J., M. E. Hyde, and R. G. Compton. 2005. Fabrication of random assemblies of metal nanobonds: A general method. *Angew. Chem. Int. Ed.* 44: 6491–6496.

Geim, A. K., and K. S. Novoselov. 2007. The Rise of Graphene. *Nat. Mater.* 6: 183.

Goh, M. S., and M. Pumera. 2010a. The electrochemical response of graphene sheets is independent of the number of layers from a single graphene sheet to multilayer stacked graphene platelets. *Chem. Asian J.* 5: 2355.

Goh, M. S., and M. Pumera. 2010b. Multilayer graphene nanoribbons exhibit larger capacitance than their few layer and single layer graphene counterparts. *Electrochem. Commun.* 12: 1375.

Gomez-Navarro, C., J. C. Meyer, R. S. Sundaram, A. Chuvilin, S. Kurasch, M. Burghard, K. Kern, and U. Kaiser. 2010. Atomic structure of reduced graphene oxide. *Nano Lett.* 10: 1144.

Novoselov, K. S., A. K. Geim, S. V. Morozov, D. Jiang, Y. Zhang, S. V. Dubonos, I. V. Grigorieva, and A. A. Firsov. 2004. Electric field effect in atomically thin carbon films. *Science* 306: 666.

Pumera, M. 2009a. Electrochemistry of graphene: New horizons for sensing and energy storage. *Chem. Rec.* 9: 211.

Pumera, M. 2009b. The electrochemistry of carbon nanotubes: Fundamentals and applications. *Chem. Eur. J.* 15: 4970–4978.

Pumera, M. 2010. Graphene-based nanomaterials and their electrochemistry. *Chem. Soc. Rev.* 39: 4146–4157.

Pumera, M., A. Ambrosi, A. Bonanni, et al. 2010. Graphene for electrochemical sensing and biosensing. *Trends Anal. Chem.* 29: 954.

Pumera, M., and Y. Miyahara. 2009. What amount of metallic impurities in carbon nanotubes is small enough not to dominate their redox properties. *Nanoscale* 1: 260–265.

Pumera, M., T. Sasaki, and H. Iwai. 2008. Relationship between carbon nanotube structure and electrochemical behavior: Heterogeneous electron transfer at electrochemically activated carbon nanotubes. *Chem. Asian J.* 3: 2046–2055.

Pumera, M., R. Scipioni, H. Iwai, et al. 2009. A mechanism of adsorption of β-nicotinamide adenine dinucleotide on graphene sheets: Experiment and theory. *Chem. Eur. J.* 15: 10851.

Stoller, M. D., S. Park, Y. Zhu, J. An, and R. S. Ruoff. 2008. Graphene based ultracapacitors. *Nano Lett.* 8: 3498.

Toh, H. S., A. Ambrosi, C. K. Chua, and M. Pumera. 2011. Graphene oxides exhibit limited cathodic potential window due to their inherent electroactivity. *J. Phys. Chem. C.* 115: 17647–17650.

Wang, H., H. S. Casalongue, Y. Liang, and H. Dai. 2010. $Ni(OH)_2$ nanoplates grown on graphene as advanced electrochemical pseudocapacitor materials. *J. Am. Chem. Soc.* 132: 7472–7477.

Wang, Y., Z. Shi, Y. Huang, Y. Ma, C. Wang, M. Chen, and Y. Chen. 2009. Supercapacitor devices based on graphene materials. *J. Phys. Chem. C*. 113: 13103–13107.

Wang, G., B. Wang, X. Wang, J. Park, S. Dou, H. Ahn, and K. Kim, 2009. Sn/graphene nanocomposites with 3D architecture for enhanced reversible lithium storage in lithium ion batteries. *J. Mater. Chem*. 19: 8378–8384.

Yoo, E. J., J. Kim, E. Hosono, H.-S. Zhou, T. Kudo, and I. Honma. 2008. Large reversible Li storage of graphene nanosheet families for use in rechargeable lithium ion batteries. *Nano Lett*. 8: 2277.

Yuge, R., M. Zhang, M. Tomonari, T. Yoshitake, S. Iijima, and M. Yudasaka. 2008. Site identification of carboxyl groups on graphene edges with Pt derivatives. *ACS Nano* 2: 1747.

Zhang, L. L., R. Zhou, and X. S. Zhao. 2010. Graphene-based materials as supercapacitor electrodes. *J. Mater. Chem*. 20: 5983–59922.

Zhou, M., Y. Wang, Y. Zhai, J. Zhai, W. Ren, F. Wang, and S. Dong. 2009. Controlled synthesis of large area and patterned electrochemically reduced graphene oxide films. *Chem. Eur. J*. 15: 6116.

8

Graphene–Polymer Composites for Pulse Lasers

Qiaoliang Bao, Dingyuan Tang, Han Zhang, and Kian Ping Loh*

CONTENTS

8.1 Introduction

Polymer nanocomposites combine two concepts in material design; that is, composites and nanometer-size materials. Tailor-made nanocomposites that exploit the superlative properties of both components can show enhanced performance in wide-ranging applications from flexible packaging, printable electronics, and interlayer dielectrics to thermoplastics. Polymer nanocomposites that are reinforced with nanofillers such as one-dimensional nanowires (Chang et al. 2006; Vivekchand et al. 2006) or two-dimensional layered materials (Emmanuel 1996; Ray and Okamoto 2003; Sheng et al. 2004) exhibit enhanced mechanical, electrical, and thermal properties. Carbon-based

* Correspondence to: Kian Ping Loh (chmlohkp@nus.edu.sg)

nanofillers that include graphite nanoplatelets (Ramanathan et al. 2007; Wakabayashi et al. 2008; Zheng and Wong 2003), carbon nanotubes (Ajayan et al. 2000; An et al. 2004; Coleman et al. 2006; Grunlan et al. 2004; Munoz et al. 2005; Ramanathan et al. 2005), graphite oxide (Cassagneau et al. 2000; Kotov et al. 1996; Kovtyukhova et al. 1999), and functionalized graphene (Ramanathan et al. 2008; Stankovich et al. 2006; Verdejo et al. 2008; Vickery et al. 2009) have attracted considerable attention due to their intrinsic mechanical strength. In particular, graphene, a single layer of aromatic carbon, has one of the strongest in-plane bonds among all materials as well as superior thermal and electrical conductivity. The conjugated graphene sheet can be readily functionalized through noncovalent π–π stacking or covalent C–C coupling reactions. By derivatizing graphene with different organic moieties, the solubility of graphene can be tuned to suit different solvents needed for the processing of composite films. Certain functional groups can broaden the properties of the graphene through the formation of donor–acceptor complex with graphene, which affords tunability in electrical conductivity and optical and photovoltaic properties (Potts et al. 2010). It can be expected that based on the superlative properties of graphene and its derivatives, significant improvement in viscosity, electrical, mechanical, and thermal properties of the graphene–polymer composite can be achieved. For example, it has recently been demonstrated that the properties of functionalized graphene–polymer (i.e., polystyrene and poly(methyl methacrylate) [PMMA]) rival that of single-walled carbon nanotube (SWNT)–polymer composites (Ramanathan et al. 2008; Stankovich et al. 2006). Graphene should provide good integration with the polymer matrix due to its sheetlike properties, which provide maximum surface area for π–π stacking with polymer host. The carboxylic and hydroxyl functional groups on the edges of graphene oxide (GO) can also act as linkers for joining organic segments together and impart mechanical strength.

The incorporation of graphene into a polymer matrix can produce a new type of optical material by marrying the special optical properties of graphene with the structural properties of the polymer. A single atomic sheet of graphene can absorb $\pi\alpha = 2.3\%$ of incident white light, where α is the fine-structure constant (Nair et al. 2008). The universal wide band absorption of graphene renders the material attractive as a wavelength-independent saturable absorber; that is, an optical element used in mode-locking of lasers, based on its property of exhibiting optical transparency at high optical intensity (Bao et al. 2009). Its nonlinear optical behavior makes it potentially useful in ultrafast photonics (Dawlaty et al. 2008). The use of graphene is inspired by the fact that SWNTs have been used as effective saturable absorbers for the mode-locking of lasers to generate ultrashort pulses in the telecommunication bandwidth (Set et al. 2004; Yamashita et al. 2004). SWNTs dispersed in a polymer matrix have proven to be highly effective saturable absorbers due to its ultrafast recovery time and large modulation depth (Rozhin et al. 2006; Scardaci et al. 2008; Tausenev et al. 2008; Wang et al. 2008b).

Although the initial focus of the graphene community is on the wonderful electronic properties of graphene, the optical applications of graphene–polymer composites may emerge in the market before the high-end electronic application due to the relatively less stringent processing conditions of films for optical applications. Polymeric materials are the ideal choice because of their ease of fabrication by a low-cost room-temperature fabrication process as well as the customer-defined optical characteristics. Many photonic polymers, such as acrylates, silicones, and polyimides, have excellent thermal properties and good environmental and radiation stability (Ma, Jen et al. 2002; Ma, Liu et al. 2002). Moreover, the advantage of polymer encapsulation passivates graphene from ambient oxygen and prevents photooxidation and optical bleaching by high-powered lasers. This is very important for developing organic photonics components that do not photodegrade with time.

The use of graphene as a saturable absorber to generate sub-picosecond laser pulses in the telecommunication wavelength band was first demonstrated by Bao et al. (2009). Following them, Zhang et al. (Bao et al. 2010; Zhang, Bao et al. 2009) and another two groups (Song et al. 2010; Sun, Hasan et al. 2010) also confirmed the mode-locking of lasers using graphene–polymer composites and chemically processed graphene film as saturable absorbers. In this chapter, we will start with the fundamental linear and nonlinear optics of graphene, then we will introduce the fabrication of graphene nanocomposites as well as the materials characterizations; finally, detailed descriptions on graphene–polymer nanocomposite for photonics applications in enabling pulsed lasers will be provided. Specifically, the two main techniques to generate laser pulses, Q-switching and mode-locking, will be discussed.

8.2 Optical Properties of Graphene

8.2.1 Linear Optical Absorption

Light transmittance (T) through free-standing graphene can be derived using the Fresnel equation for a thin film with a universal optical conductance of $G_0 = e^2/4\hbar \approx 6.08 \times 10^{-5}\,\Omega^{-1}$. It gives the following:

$$T = (1 + \pi\alpha/2)^{-2} \approx 1 - \pi\alpha \approx 0.977 \tag{8.1}$$

where $\alpha = e^2/(4\pi\varepsilon_0\hbar c) = G_0/(\pi\varepsilon_0 c) \approx 1/137$ is the fine-structure constant. Since graphene reflects only a small portion (<0.1%) of the incident light in the visible region (Nair et al. 2008), single monolayer graphene should transmit ~97.7% of the incident light and absorb $a \approx 1 - T \approx \pi\alpha \approx 2.3\%$, independent of its wavelength. For graphene sample with few layers (<10), its absorption generally agrees with the Beer–Lambert law, which gives a nearly linear relationship between the absorption and number of layers. This is valid because the perturbation from the adjacent layers is negligibly small so that it is optically equivalent to a superposition of almost noninteracting monolayer graphene. The absorption spectrum of monolayer graphene is quite flat over a wide wavelength range from visible to infrared (IR) (from 300 to 2500 nm). Deviations were observed only in the far-IR and ultraviolet (UV). The absorption peak in the UV region (~270 nm) is due to the exciton-shifted van Hove singularity in the graphene density of states. Overall, the strong graphene–light interaction is remarkable considering its infinitesimal thickness. For comparison, a 20-nm-thick film of InGaAs would be required to absorb the same amount of 1.55-μm light as a single-layer graphene.

8.2.2 Saturable Absorption

Figure 8.1a shows the excitation processes responsible for absorption of light in monolayer graphene, in which electrons from the valence band (dark gray) are excited into the conduction band (light gray) (Bao et al. 2009). Shortly after photoexcitation (within 10–150 fs) (George et al. 2008), these hot electrons thermalize and cool down to form a hot Fermi–Dirac distribution (Figure 8.1b) with electronic temperature T_e (Moos et al. 2001; Tobias et al. 2005). These newly created electron–hole pairs could block some of the originally possible interband optical transitions in a range of $k_B T_e$ (k_B is the Boltzmann constant) around the Fermi energy E_F and decrease the

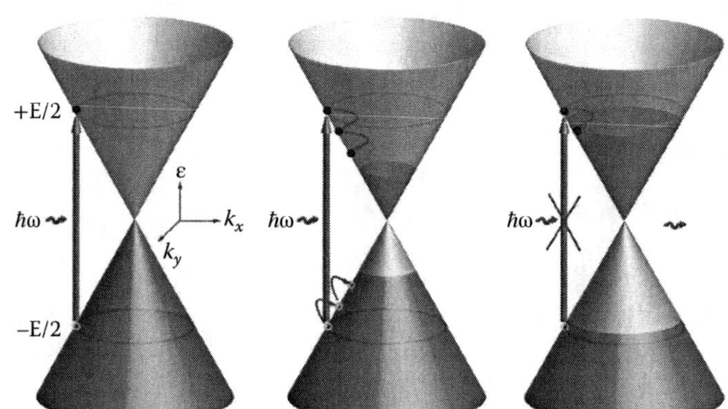

FIGURE 8.1
Absorption of light in graphene. (a) Schematic excitation process responsible for absorption of light in graphene. The arrow indicates optical interband transition. (b) The photogenerated carriers thermalize and cool down within subpicosecond to form a hot Fermi–Dirac distribution, an equilibrium electron and hole distribution could be finally approached through intraband phonon scattering and electron–hole recombination. (c) At enough high excitation intensity, the photogenerated carriers cause the states near the edge of the conduction and valence bands to fill, blocking further absorption, thus saturable absorption is achieved. (From Bao, Q. et al., *Adv. Funct. Mater.*, 19, 3077–3083, 2009. With permission.)

absorption of photons $\hbar\omega \sim k_B T_e$. In the following, ~1 ps, intraband phonon scattering further cools the thermalized carriers. After that, electron–hole recombination will dominate the process until the equilibrium electron and hole distribution is restored (Figure 8.1b) (Dawlaty et al. 2008; George et al. 2008). This describes the dynamics of linear optical absorption process. Nonlinear optical phenomenon could occur under high excitation intensity. As the excitation is increased to higher intensity, the photogenerated carriers increase in concentration (much larger than the intrinsic electron and hole carrier densities of ~$8 \times 10^{10}/\text{cm}^2$ in graphene at room temperature, Dawlaty et al. 2008) and cause the states near the edge of the conduction and valence bands to be filled, which blocks further absorption and thus imparts transparency to light at photon energies just above the band-edge (Figure 8.1c). Band-blocking occurs because no two electrons can fill the same state. Thus, saturable absorption or absorption bleaching is achieved due to this *Pauli blocking* process (Robert 1969).

Saturable absorption in direct bandgap semiconductors is generally considered to be associated with third-order optical nonlinearities (Boyd 2003; Shen 1984). Graphene's third-order susceptibility $|\chi^3|$ was measured to be ~10^{-7} esu, which is up to 1 order of magnitude larger than that reported so far for similar measurements on SWNTs (Hendry et al. 2010). Mishchenko (2009) suggested a simple quantitative model to describe the nonlinear response of suspended graphene in vacuum (or air) and determine its optical transparency. The incident radiation is assumed to be propagating along the negative direction of the z axis and normal to the graphene plane (xy plane). The electric field consists of the incident E_0 and reflected E_R waves above the graphene flake and of the transmitted wave E_T below it. By solving Maxwell's equations with the two-dimensional current provided by graphene sheet, one obtains the following algebraic equation for the transmission coefficient $T = E_T^2/E_0^2$,

$$1+\frac{\pi\alpha}{1+\sqrt{1+TI}}=\frac{1}{\sqrt{T}} \tag{8.2}$$

where α is the fine-structure constant and $I=\left(evE_0/\hbar\varpi\Gamma\right)^2$ is the dimensionless intensity of the incident wave. The dependence of the transmission coefficient on the magnitude of the incident wave is shown in Figure 8.2a. At low intensities, the transmission coefficient starts from the value of 97.7%, which corresponds to linear light transmittance given by the first-order perturbation theory. As the intensity increases, the transmission initially increases steeply, but eventually tends much more slowly toward $T=1$ according to $T \approx 1-2\pi\alpha/\sqrt{T}$ (Mishchenko 2009). This points to the saturation of the absorption in graphene. Most recently, Vasko (2010) gave a complete theoretical treatment of the saturation mechanism by considering the transient response of an intrinsic graphene. We will not involve the details here.

The experimental observation of saturable absorption in few-layer graphene was firstly demonstrated by Bao et al. (2009), as shown in Figure 8.2b.

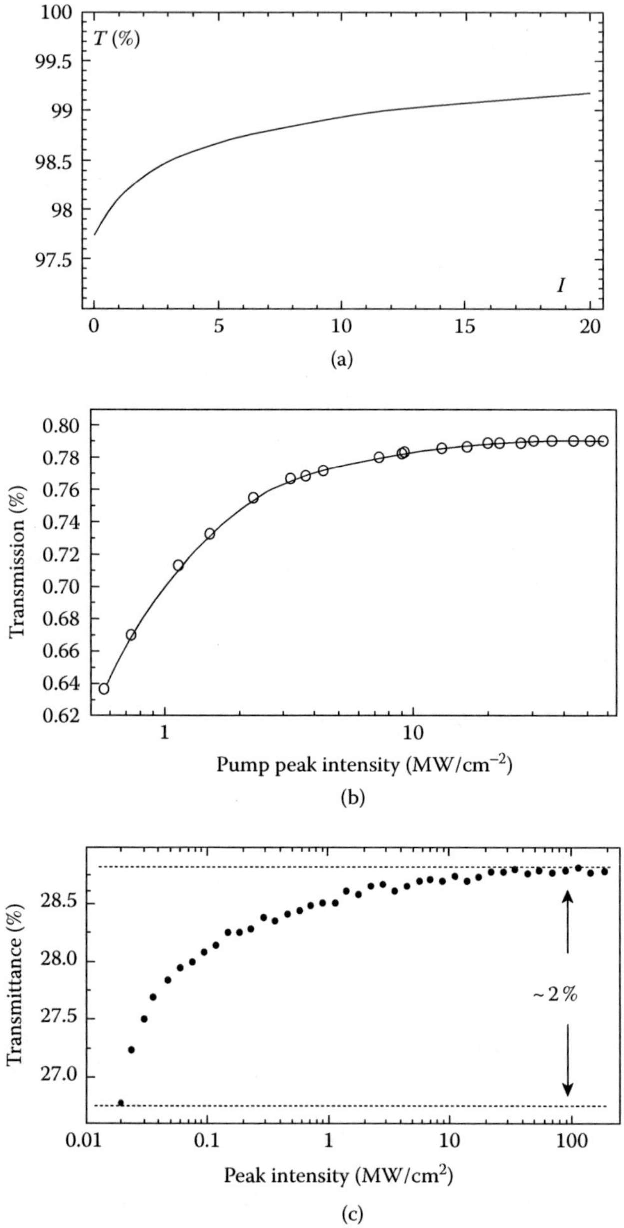

FIGURE 8.2
(a) Theoretically predicted dependence of the transmission coefficient of suspended graphene on the effective intensity of the incident radiation. (b) Experimentally observed saturable absorption in a few-layer graphene film covered on fiber pigtail and excited by ~1-ps pulse laser at 1550 nm. (c) Power-dependent nonlinear saturable absorption of graphene-based PVA nanocomposites. (From Mishchenko, E. G., *Phys. Rev. Lett.*, 103, 246802, 2009; Popa, D. et al., *Appl. Phys. Lett.*, 97, 203106, 2010. With permission.)

The graphene film grown by chemical vapor deposition (CVD) on Ni was transferred onto the end facet of a fiber pigtail. The graphene film consists of a few numbers of layers with a misoriented arrangement of each layer, leading to a similar electronic structure to monolayer graphene. For the power-dependent nonlinear absorption measurements, a stable and standard soliton mode-locked fiber laser working at 1550 nm with pulse width of 1 ps and repetition rate of ~5 MHz was used as input seed pulse source. Obviously, the experimental result agrees well with the theoretical prediction; that is, the transmission loss induced by absorption in graphene film tends to decrease as the input laser power grows. It suggests that the absorption in graphene is gradually saturated under higher laser pump. Similar saturable absorption phenomenon was also repeated on monlayer graphene (Bao et al. 2011) as well as confirmed in graphene–polymer composites (Figure 8.2c) by another group (Popa et al. 2010).

8.3 Graphene Nanocomposites and Characterizations

8.3.1 Synthesis of Graphene and Its Derivatives

There are mainly four approaches to synthesize graphene and its derivatives. The first approach is the micromechanical exfoliation of graphite, which is also known as the "Scotch tape" or peel-off method (Novoselov et al. 2004, 2005), followed on earlier work on micromechanical exfoliation from patterned graphite (Lu et al. 1999). This approach gives the best samples in terms of electronic properties, but the yield is quite low, only suitable for fundamental studies. The second method is epitaxial growth on electrically insulating surfaces such as SiC (Berger et al. 2006). The third was CVD on metals such as Ni (Kim et al. 2009) and Cu (Li et al. 2009) and epitaxial growth on certain facet of metal such as ruthenium (0001) (Sutter et al. 2008). These early efforts (which started in 1970) were followed by a large body of work by the surface-science community on "monolayer graphite" (Aizawa et al. 1990). The CVD method can produce large area uniform graphene film, which can be used for making optoelectronic devices (Bae et al. 2010).

The fourth approach was the creation of colloidal suspensions or liquid-phase exfoliation, which includes the production of graphene and chemically modified graphene from colloidal suspensions made from graphite, derivatives of graphite (such as graphite oxide), and graphite intercalation compounds. This approach is both scalable, affording the possibility of high-volume production, and versatile in terms of being well suited to chemical functionalization. These advantages mean that graphene and its derivatives produced by the colloidal suspension method are good nanofillers to incorporate into polymer to form functional composites. In the following, we will introduce more details regarding chemical approach to produce graphene and its derivatives.

Although GO sheets exfoliated from graphite can be dispersed as single sheets in aqueous solution, graphene generated from these GO precursors are generally defective and its electrical conductivity is several orders below that of graphite even after thermal and chemical reduction. A method that can bypass the oxidation of graphene while allowing individual graphene sheets to be dispersed well in solution is ideal, since this will allow the extended conjugation to be preserved and avoid the tedious postreduction treatments. Another fundamental reason is the issue of stoichiometric functionalization of graphene versus nonstochiometric functionalization, as discussed earlier. Chemical derivatization of GO, which relies on the linking of molecules to epoxide, carboxylic, or hydroxide functionalities, is inherently nonstoichiometric due to the random distributions of these groups. In principle, functionalization chemistry of pure graphene that involves binding to its π-conjugated network can be regioselective, thus stochiometric functionalization on the basal or edge planes may be achievable.

To develop alternative approaches to the oxidation–exfoliation method, chemists have revisited old chemistry developed for "expanded graphite" used in batteries and fuel cells, with the aim of uncovering recipes that allow graphene sheets to be directly obtained from expanded graphite using intercalation and exfoliation methods. There are mainly three types of methods, as schematically illustrated in Figure 8.3 (Ang et al. 2009; Loh et al. 2010). The first is graphene from graphite intercalation compounds and expandable graphite. In the past two decades, researchers had shown that although graphite intercalation compounds are excellent precursors for exfoliated graphite (Dresselhaus and Dresselhaus 2002), intercalation–exfoliation of graphite through thermal shock, acid treatment, and intercalation yielded only thin graphite nanoplatelets and not single sheets (Viculis et al. 2005). Following the generation of expanded graphite by acid treatments and thermal shocks, intercalants must be added to further expand the interlayer distance. Sonication of the resultant graphite intercalation compounds in stabilizers (or surfactants) results in stable colloidal suspension of graphene monolayer sheets in organic or aqueous solutions, as demonstrated in Li et al. 2008.

The second method is liquid-phase exfoliation of graphite. Colloidal suspension of graphene sheets in organic solvents such as DMF (Dimethylformamide) (Blake et al. 2008), NMP (N-Methyl-2-Pyrrolidone) (Hernandez et al. 2008), and chloroform (Biswas and Drzal 2008) have been reported. Novoselov and coworkers (Blake et al. 2008) reported that sonicating graphite in DMF (Dimethylformamide) returned a high yield of monolayer flakes up to 50%. DMF (Dimethylformamide) sufficiently prevents exfoliated graphene sheets from agglomerating due to its good wetting of individual graphene sheets. Direct dispersion of the graphene sheets in aqueous alkaline solution has also been achieved by Hao et al. (2008) through the use of 7,7,8,8-Tetracyanoquinodimethane (TCNQ) for the noncovalent functionalization of graphene.

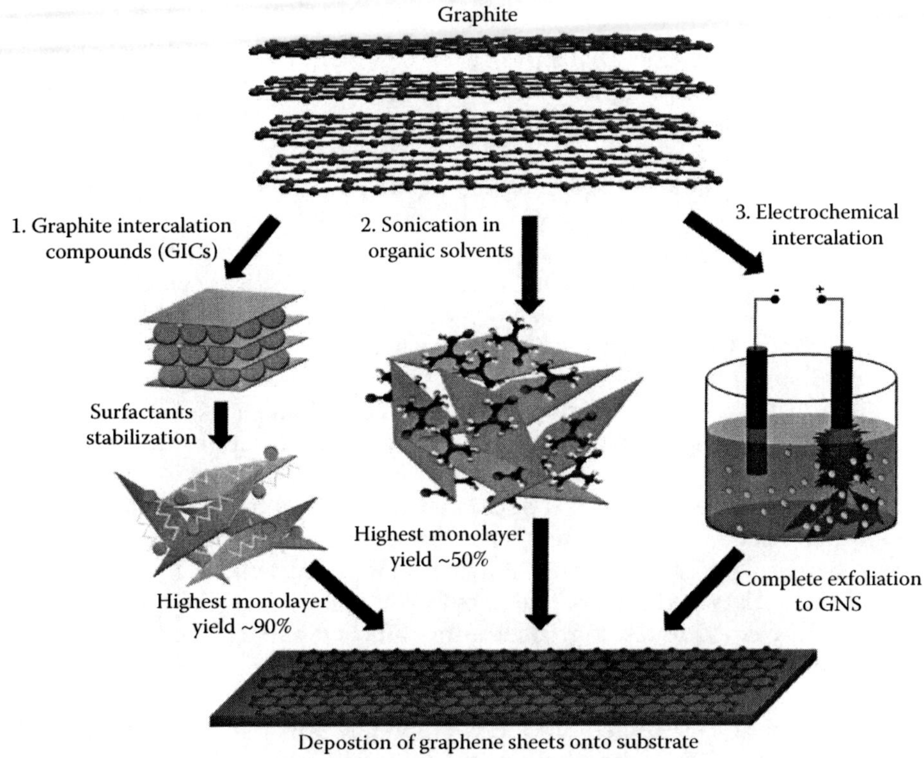

FIGURE 8.3
Schematic shows the various exfoliation routes to graphene. Route 1 shows graphene sheets obtained from graphite intercalation compounds in which the graphite interlayer distance is increased by intercalants. A high monolayer yield (~90%) of graphene sheets stabilized by surfactants can be achieved. Route 2 illustrates graphene sheets produced through direct sonication of graphite in organic solvents that yield a monolayer yield of 50%. Route 3 displays a recent method that involves ionic liquid–assisted electrochemical exfoliation of graphite anode to obtain graphene sheets with almost complete exfoliation to graphene nanosheets. (From Ang, P. K. et al., *ACS Nano.*, 3, 3587–3594, 2009; Loh, K. P. et al., *J. Mater. Chem.*, 20, 2277–2289, 2010. With permission.)

The third method is to produce graphene from electrochemical exfoliation of graphite. Liu et al. (2008) devised an electrochemical approach to obtain imidazolium-based ionic liquid–functionalized graphene sheets. Exfoliation of functionalized graphene (FG) sheets readily formed a stable colloidal suspension in DMF, dimethyl sulfoxide (DMSO), and NMP. Loh and coworkers (Longhua et al. 2009) demonstrated a facile one-pot ionic liquid–assisted electrochemical exfoliation approach to obtain fluorescent carbon nanoribbons, nanoparticles, and graphene from graphite. The mechanism of the exfoliation is due to a complex interplay of anodic oxidation of water and anionic intercalation from the ionic liquid.

8.3.2 Fabrication of Graphene–Polymer Composites

Similar to other carbon allotropes such as carbon nanotube and nanofiber, the commonly used methods including solution blending, melt blending, and in situ polymerization can also be applied to produce graphene–polymer composites (Moniruzzaman and Winey 2006). Solution-based processing methods (Potts et al. 2010; Ramanathan et al. 2005, 2007) are widely used to fabricate graphene–polymer composites due to its easy fabrication. The first step is to disperse graphene in aqueous or nonaqueous media through ultrasonication, following the dispersion mixed with the host polymer. Then, the mixtures are drop-cast, spin-coated, and vacuum filtrated to form a uniform film, depending on the application requirements. We will not provide details on the dispersion of graphene as it has been discussed in Chapters 4, 5, 11, 13, 14, and 16. A detailed review can be found in Potts et al. (2010), and this chapter will mainly focus on the graphene–polymer composites that are specifically designed from photonic aplications.

For the purpose of optical applications, the aggregates of graphene flakes and micro-size bubbles in composite film should be avoided to reduce possible scattering loss. Moreover, the thickness requirement for composites sandwiched between fiber connectors is ~10–30 μm to reduce insertion loss. Generally, even thinner composite films are favored for the same reason. However, the thin film might be very fragile to be freestanding. Our strategy is to incorporate functionalized graphene nanoflakes into polymer nanofiber by electrospinning. The as-produced graphene–polymer mat not only provides a freestanding membrane with appreciable mechanical strength for transferring onto any optical components for variant applications but also affords protections from photooxidation and optical bleaching.

Electrospinning provides a simple and versatile technique for producing networked membranes from various materials. Such network polymer membranes allow many engineering applications to be enabled, for example, mechanical reinforcement, water filtration, textile, tissue engineering, and drug release (Greiner and Wendorff 2007; Huang et al. 2003; Li and Xia 2004; Teo and Ramakrishna 2006). Electrospun polymer fibers can be ultrafine with diameters ranging from nanometers to sub-micrometers and possess high surface area to volume or mass ratios. The electrospinning process that can incorporate graphene sheets into a continuous nanoscale composite fiber is illustrated schematically in Figure 8.4 (Bao et al. 2010). A freestanding membrane that is composed of graphene–polymer nanofiber network could be fabricated and applied as an optical element in fiber lasers.

The selection of host polymers is very important for both processability and optical properties of composites. The most desirable characteristics of the host polymer matrices are stability against laser irradiation and humidity or other environmental factors. The polymers must also have low absorption losses at the operation wavelength. To directly generate high-energy single pulse with the saturable absorber, polymer host with high damage

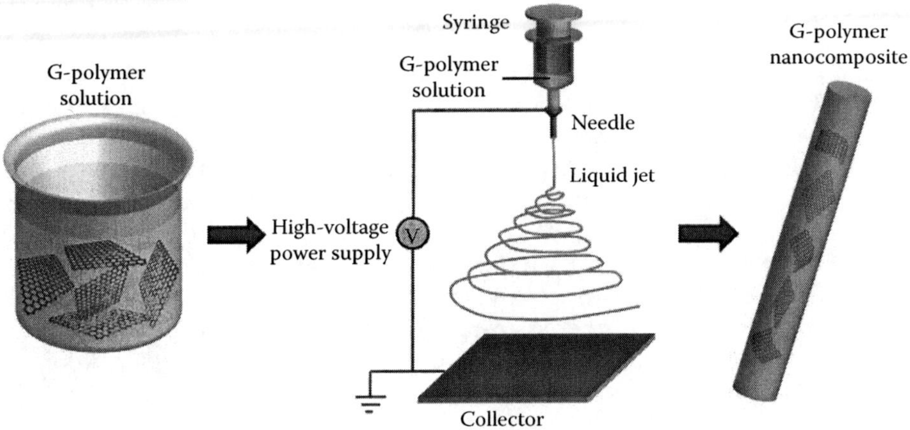

FIGURE 8.4
Schematic illustration of the fabrication of graphene–polymer nanofiber composite by electrospinning. (From Bao, Q. et al., *Adv. Funct. Mater.*, 20, 782–791, 2010. With permission.)

threshold should be selected. Here, we select polyvinylidene fluoride (PVDF) as polymer host, which has a high melting point at 170°C and decomposition temperature above 316°C. PVDF owns high resistance to radiation and chemicals. In a typical fabrication process, graphene nanosheets were first produced by chemically reducing the oxidized graphene exfoliated from graphite flakes. The graphene nanosheets were then noncovalently functionalized with 1-pyrenebutanoic acid, succinimidyl ester (PBASE) to improve their solubility in organic solvents (i.e., ethanol and acetone) and compatibility with polymers (i.e., PVDF). The as-produced graphene (2 mg) was further dispersed in ethanol (3 mL) and mixed with PVDF solution comprising 1.5 g PVDF dissolved in 10 mL dimethylacetamide/acetone (2:3). The graphene–PVDF solution was then stirred for 24 hours at 60°C in sealed bottle to form wet paste for electrospinning. The electrospinning was carried out in a MECC nanon electrospinning setup at a bias voltage of 30 kV and feeding rate of 0.5 mL/h.

In another parallel experiment to develop broadband saturable absorbers, we chose poly(vinyl acetate) (PVAc) as the polymer matrix because its refractive index of 1.47 is close to that of optics components (n = 1.45 for silica). Most importantly, PVAc has very good compatibility with graphene derivatives and excellent processability to form uniform nanofiber mats with less "beads" by electrospinning. To enhance the solubility, improve dispersion in polymer matrix, and enrich the optical properties, we chemically functionalized GO or graphene. Four types of graphene derivatives were used as nanofillers, namely, GO, GO noncovalently bonded with poly[(m-phenylenevinylene)-co-(2, 5-dioctoxy-p-phenylenevinylene)] (GO-PPV), graphene modified by 1-pyrenebutanoic acid, succinimidyl ester (G-PBASE), and graphene functionalized with 4-(2-(pyridin-4-yl)vinlyl) phenyl group

(G-dye). Figure 8.5a shows the chemical structure of these graphene derivatives. PPV has been widely investigated because of its nonlinear optical response and electroluminescent activity (Davey et al. 1999; Holzer et al. 1996). The functionalization of GO with PPV is expected to impart nonlinear optical properties on the composite. The PPV backbone can undergo π–π stacking on aromatic structures and enhance interfacial bonding in polymeric composites (Dalton et al. 2000). Similarly, PBASE has been used extensively for the noncovalent functionalization and solubilization of carbon nanotubes due to its hydrophilic character (Kim et al. 2006). The attachment of PBASE through π–π stacking imparts hydrophilic character on the otherwise hydrophobic graphene without disrupting its conjugated network. In the case of G-dye, the dye that is covalently coupled to G, has

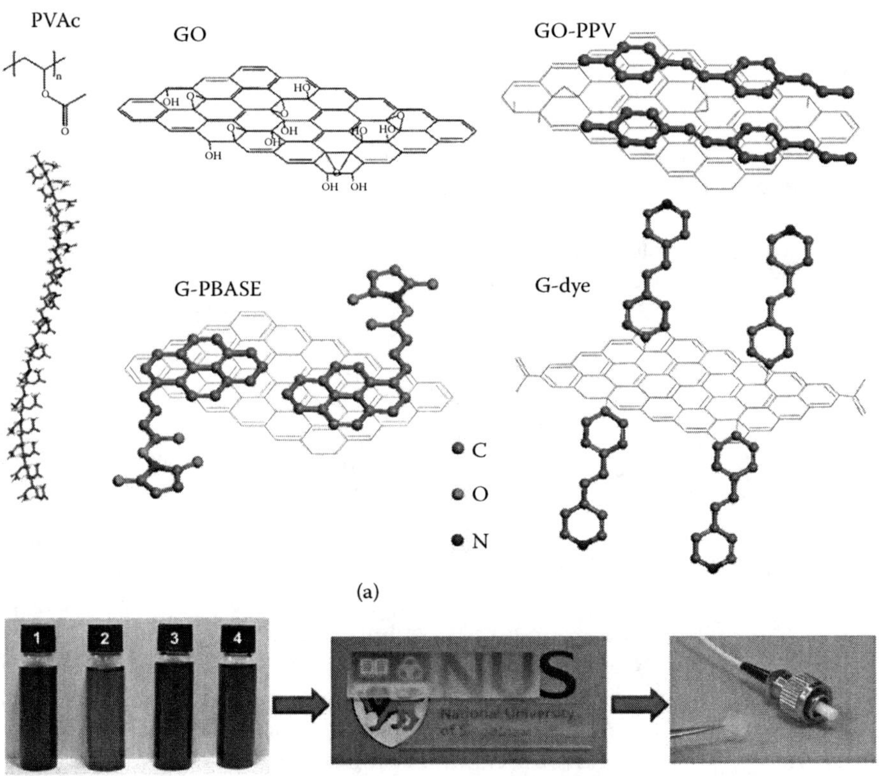

FIGURE 8.5
(a) Chemical structure of PVAc, graphene-based nanofillers (GO, GO-PPV, G-PBASE, G-dye). (b) The dispersions of graphene derivatives 5 hours after ultrasonication (1: GO, 2: GO-PPV, 3: G-PBASE, 4: G-dye; all in DMF, 0.15 mg/mL). (c) Typical picture of membranes electrospun on quartz substrates (loading of G-dye: up, 0.07 wt%; bottom, 0.035 wt%). (d) Transferring the freestanding membrane onto the end face of optical fiber for mode-locking. (From Bao, Q. et al., *Adv. Funct. Mater.*, 20, 782–791, 2010. With permission.)

a donor-π-acceptor structure. It is a highly π-conjugated chromophore containing a pyridium moiety (acceptor). The pyridium salt helps to improve the solubility of the systems by introducing electron-withdrawing group that stabilizes the electron-rich phenylethyl core and prevents aggregation. This would further broaden the scope of applications such as graphene-based organic and polymer composite materials. Figure 8.5b–d shows the procedures from dispersion of graphene derivatives to form semitransparent thin film and transfer the film onto the end face of optical fiber for mode-locking.

Sun et al. (Popa et al. 2011; Sun, Hasan et al. 2010; Sun, Popa et al. 2010) chose polyvinyl alcohol (PVA) as host materials to produce graphene–polymer composites for mode-locked and Q-switched lasers. In the typical processes, natural graphite is first exfoliated in water without functionalization under mild sonication to generate single-layer or few-layer graphene flakes. To obtain an aqueous dispersion, 1.2 wt% of micronized graphite flakes are sonicated for ~3 h in a bath sonicator with 0.5 wt% sodium deoxycholate (SDC), a bile salt surfactant. To prepare the composite, aqueous solutions, containing 120 mg PVA, are sonicated with 4 mL of centrifuged graphene dispersion. The water is then slowly evaporated in a desiccator at room temperature, resulting in a ~50-μm-thick graphene–PVA composite, which is a few times thicker than those made by electrospinning (~10 μm). Recently, Kim et al. (2011) demonstrated a simple deformation-suppressed optical deposition process, that is, the deposition of the graphene–PVAc composite on an optical fiber by laser radiation, in which there are three intercorrelated mechanisms of optical trapping effect, thermally induced convection flow, and thermodiffusion. The authors claimed that PVAc can provide immunity to graphene against deleterious degradation of morphology and optical nonlinearity in their composites.

8.3.3 Materials Characterizations

It is well known that parameters of the polymer solution such as molecular weight, solution viscosity, surface tension, solution conductivity, and dielectric constant are critical factors that affect the morphology of electrospun mats (Greiner and Wendorff 2007; Li and Xia 2004; Shui and Li 2009). Generally, the formation of "beads" in the fiber is due to the instability of a liquid jet, which has a tendency of reducing its surface energy (Shui and Li 2009). The solution properties (i.e., viscosity, surface chemistry, and electrical properties) of PVAc are changed remarkably by adding a small amount (0.07 wt%) of graphene derivatives (SWNT was used as reference). For pristine PVAc without any nanofillers, the electrospun fibers were generally smooth and uniform with very less beads, as shown in Figure 8.6a and e. A small amount of GO, GO-PPV, and SWNT could cause increased "beads" formation, as illustrated in Figure 8.6b–d. The formation of beads when plain GO (shown in

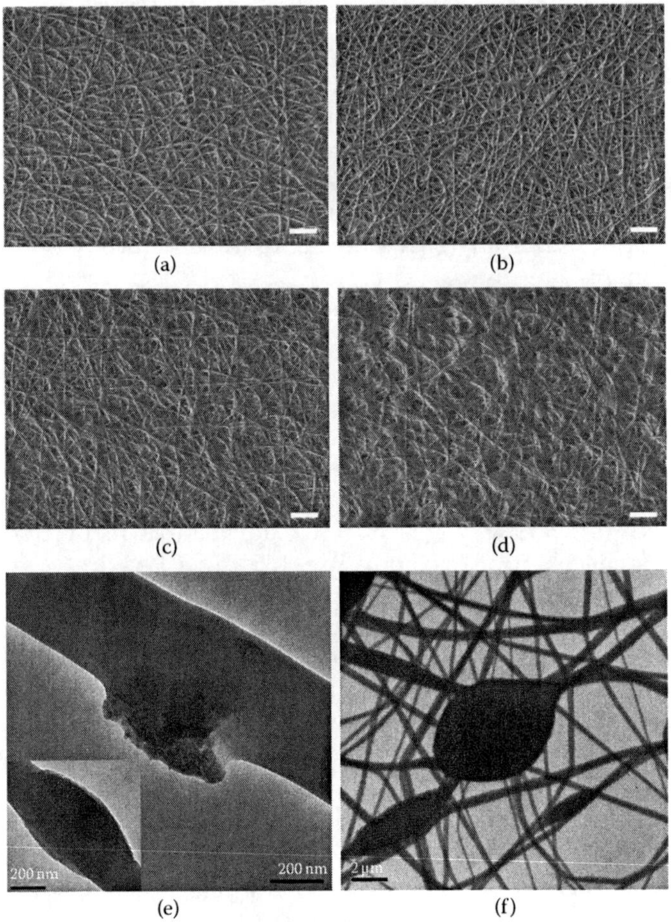

FIGURE 8.6
SEM images of electrospinning graphene–polymer nanocomposite membrane. (a) Pristine PVAc, (b) GO–PVAc, (c) GO-PPV–PVAc, (d) SWNT–PVAc. Scale bars in all images: 10 μm. (e) TEM image of GO-PPV–PVAc shows aggregation of GO inside nanofiber. The inset shows one bead. (f) TEM image of SWNT–PVAc shows micron-size beads incorporated in nanofibers. (From Bao, Q. et al., *Adv. Funct. Mater.*, 20, 782–791, 2010. With permission.)

Figure 8.6b) is used as the nanofiller is due to the poor dispersion (Stankovich et al. 2006) of GO in DMF, the solvent used in electrospinning. The bundling and aggregation of SWNTs also result in the formation of numerous big size beads (Ko et al. 2003), as shown in Figure 8.6d and f respectively.

In contrast, uniform and smooth nanofibers were readily obtained when G-PBASE and G-dye were used as nanofillers under the same process parameters, as shown in Figure 8.7a and b. Figure 8.7c and d shows representative transmission electron microscopy (TEM) images of G-PBASE–PVAc nanofibers with diameters in the range of 300~500 nm. The high magnification TEM image in Figure 8.7d shows that individual graphene sheet is well

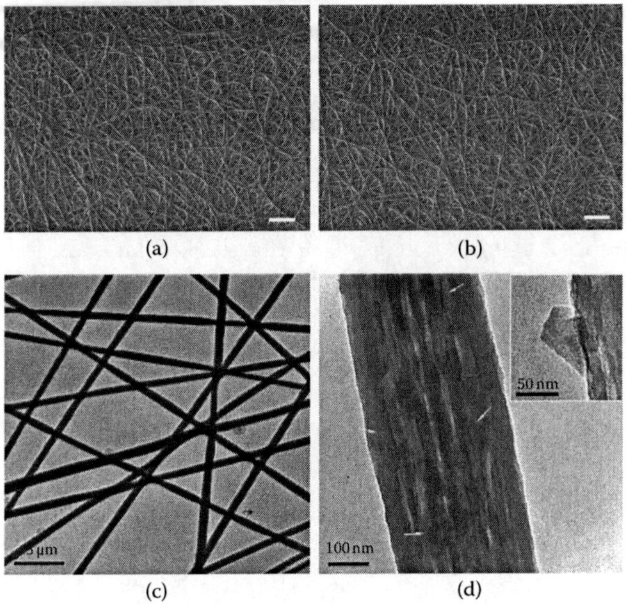

FIGURE 8.7
SEM images of the graphene–polymer nanocomposite membrane fabricated by electrospinning. (a) G-PBASE–PVAc, (b) G-dye–PVAc (scale bars: 10 μm). (c) TEM image of G-PBASE–PVAc nanofibers. (d) High-magnification TEM image of G-PBASE–PVAc nanofiber. The arrows indicate the graphene flakes inside nanofiber. The inset shows enlarged image of G-PBASE embedded in the sidewall of PVAc nanofiber. (From Bao, Q. et al., *Adv. Funct. Mater.*, 20, 782–791, 2010. With permission.)

dispersed in the polymer nanofiber matrix. Compared with electrospinning graphite nanoplatelets in the polymer matrix (Mack et al. 2005), the good affinity between PVAc and functionalized graphene (G-PBASE and G-dye) results in uniform fibers with no beading. The less content of beads and good uniformity of the membrane suggest smaller scattering loss while incorporating into laser cavity for mode-locking.

The mechanical properties of the electrospun nanocomposite membranes were investigated with the tensile stress–strain testing method. G-PBASE–PVAc and G-dye–PVAc were found to be reinforced a lot compared with pristine PVAc (Bao et al. 2010). The improved dispersion of the functionalized graphene in the polymer matrix coupled with an increase in the interfacial bond strength is likely the mechanism for the increased load transfer effectiveness. Improved thermal properties of these graphene-based nanocomposites are also observed by thermogravimetric analysis (TGA) and differential thermogravimetric (DTG) analysis owing to the presence of interfacial bonding. A qualitative analysis of the decomposition behavior reveals that G-PBASE–PVAc and G-dye–PVAc exhibit higher decomposition temperature of carbon skeleton and larger amount of residue (Bao et al. 2010).

8.3.4 Linear Optical Properties and Saturable Absorption

The linear and nonlinear optical absorption of electrospun membrane was investigated. Figure 8.8 compares the absorption spectra of graphene-based polymer composite thin films. The absorption spectra of pure PVDF or PVAc and SWNT–PVAc are also shown for comparison. Figure 8.8a shows that pure PVDF has a relatively low absorbance of ~35% in the L-band of the optical communication windows, while graphene-based PVDF nanocomposites have an enhanced absorbance of ~52%. Figure 8.8b shows the linear absorption spectra of graphene-based PVAc nanocomposites. Due to the transparency of GO, adding 0.07 wt% GO as nanofiller only slightly enhanced the total absorption. The functionalization of GO by PPV and PBASE enhanced the absorption of the composites relatively, which is even higher than

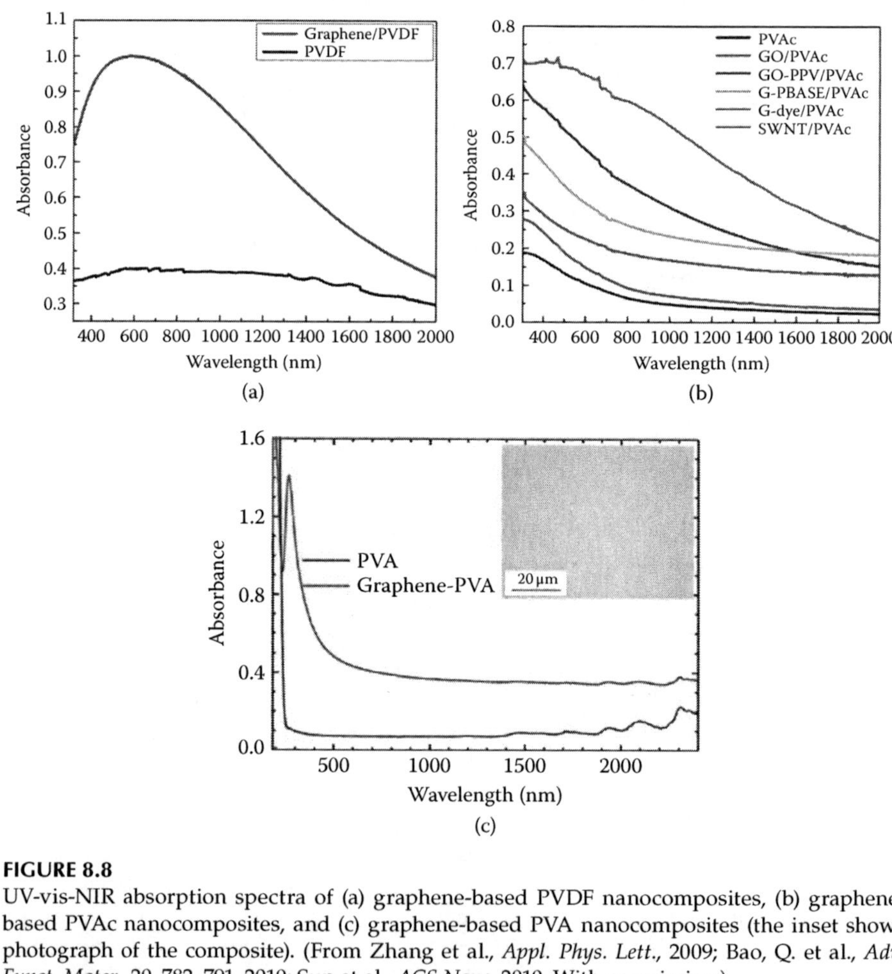

FIGURE 8.8
UV-vis-NIR absorption spectra of (a) graphene-based PVDF nanocomposites, (b) graphene-based PVAc nanocomposites, and (c) graphene-based PVA nanocomposites (the inset shows photograph of the composite). (From Zhang et al., *Appl. Phys. Lett.*, 2009; Bao, Q. et al., *Adv. Funct. Mater.*, 20, 782–791, 2010; Sun et al., *ACS Nano*, 2010. With permission.)

the control sample of SWNT–PVAc, but it seems that the absorption maximum does not shift too much. In contrast, adding 0.07 wt% G-dye results in strong wideband absorption from the visible to near IR region because of the strong absorption of G-dye. Charge transfer interaction causes the absorption maxima of polymer composite to redshift from that of the G-dye by as much as 200–470 nm. Pure PVAc is highly transparent in the telecommunication window (around 1550 nm) with a relatively low absorbance of 0.03 (equal to transmission loss of 0.3 dB). Remarkably, the graphene–polymer composites exhibit enhanced absorbance at this region compared with pure polymer system. For example, at 1550 nm, the absorbance is 1.5 times for GO–PVAc, 6.4 times for GO-PPV–PVAc, 6.2 times for G-PBASE–PVAc, and 10.5 times for G-dye–PVAc (4.4 times for SWNT–PVAc) compared to pure PVAc. The enhanced broadband absorption is also observed in graphene–PVA composites fabricated by solvent evaporation method, as shown in Figure 8.8c (Bonaccorso et al. 2010; Sun, Hasan et al. 2010). The wideband strong IR absorbance makes these composites attractive as saturable absorbers for fiber lasers.

The nonlinear absorption properties of our composites are studied by power-dependent measurements, as shown in Figure 8.9. The saturation absorption could be seen quite clearly with the increase of incident light intensity. The absorption as a function of increasing light intensity can be fitted by Garmire (2000):

$$\alpha(I) = \frac{\alpha_S}{1 + I/I_S} + \alpha_{NS} \tag{8.3}$$

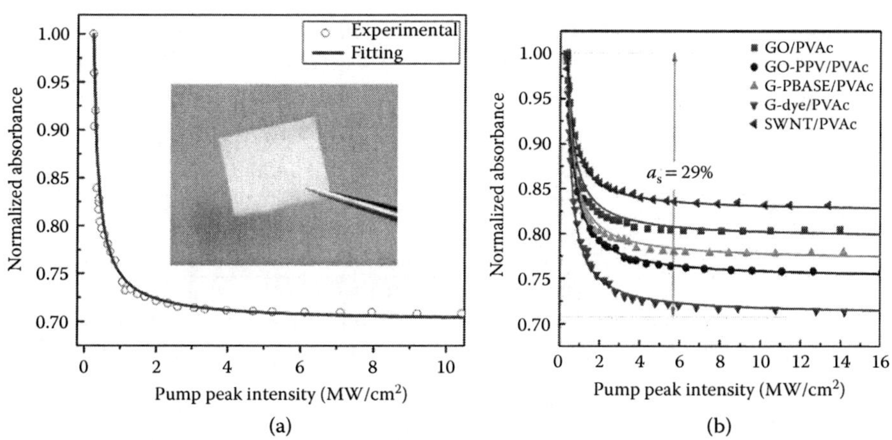

FIGURE 8.9
(a) Power-dependent nonlinear saturable absorption of graphene-based PVDF nanocomposites. Inset shows the photograph of a freestanding graphene–PVDF composite thin film. (b) Power-dependent nonlinear saturable absorption of graphene-based PVAc nanocomposites. (From Zhang et al., *Appl. Phys. Lett.*, 2009; Bao, Q. et al., *Adv. Funct. Mater.*, 20, 782–791, 2010. With permission.)

where α_S and α_{NS} are the saturable and the nonsaturable absorptions respectively and I_s is the saturation intensity, defined as the optical intensity required in a steady state to reduce the absorption to half of its unbleached value.

The nonlinear absorption curve of Figure 8.9a (measured at the wavelength of $\lambda = 1590$ nm with 1-ps pulse laser) shows that the graphene-based PVDF nanocomposite has a normalized modulation depth of ~28.3% and a saturable fluence of 0.75 MW/cm^2, which is about an order of magnitude smaller than that of the SWCNT-based saturable absorber. Moreover, the insertion loss of the graphene-based PVDF nanocomposite was as low as ~1.5 dB. For graphene-based PVAc nanocomposites (shown in Figure 8.9b), the saturation intensity ranges from 0.5 to 0.7 MW/cm^2, which is also much smaller than that reported for SWNT (Scardaci et al. 2008; Set et al. 2004; Wang et al. 2008b) and semiconductor saturable absorber mirrors (SESAMs) (Keller et al. 1996). Most importantly, the graphene-based polymer composites show larger modulation depth (saturable absorption component) than the SWNT–PVAc composite, as summarized in Table 8.1. For example, among the several types of graphene–polymer composites, G-dye–PVAc composite shows the largest saturable absorption component of 29%, which is two times that of SWNT–PVAc composite. The large modulation depth and the high ratio of saturable to nonsaturable absorption indicate stronger noise suppression and pulse shaping ability (Sakakibara et al. 2005), making these composites promising as noise suppressor and saturable absorber.

The large modulation depth of graphene–polymer composites is related to their relatively low nonsaturable loss. Nonsaturable loss generally arises from scattering losses by defects or aggregates, nonsaturable absorption from the polymer host, reflection at the membrane surface, and coupling loss in fibers. The selection of PVAc as the polymer host is guided by its matching refractive index with the optical fiber, which reduces reflection at the membrane surface. The low scattering loss of graphene–polymer composites is attributed to the good dispersion of the two-dimensional graphene nanosheets in the polymer nanofibers. The fabrication of composite nanofibers with diameters

TABLE 8.1

Optical Properties of Graphene-Based PVAc Nanocomposites

	Absorbance (at 1550 nm)	Saturable Absorption (%)	Saturation Intensity (MW/cm^2)
G-PBASE–PVDF	0.52	0.283	0.75
GO–PVAc	0.046	20.5	0.62
GO-PPV–PVAc	0.198	25	0.68
G-PBASE–PVAc	0.194	23	0.59
G-dye–PVAc	0.326	29	0.60
SWNT–PVAc	0.138	17.5	0.71

Source: Bao, Q. et al., *Adv. Funct. Mater.*, 20, 782–791, 2010. With permission.

much smaller than the working wavelength (1550 nm) also helps to reduce scattering loss. In contrast, the micron-size beads occurring in SWNT–PVAc are responsible for inducing large nonlinear scattering and contribute to significant nonsaturable loss.

8.4 Q-Switched Laser

In addition to mode-locking, which will be discussed in detail in Section 8.5, Q-switching is known for generating a pulse with a small width but a high peak power and a much larger pulse energy. Q-switching is a modulation of the qualify factor Q of a laser cavity, where Q is defined as the ratio between the energy stored in the gain medium and that lost per oscillation cycle. In Q-switching, initially, the active medium is pumped while the Q-switch is set to prevent feedback of light into the gain medium by a low Q factor. This produces a population inversion, the amount of energy stored in the gain medium increases as the active medium is pumped. While the stored energy reaches some maximum level (the medium is gain saturated), the Q-switch device is quickly changed from low to high Q, allowing feedback and the process of optical amplification by stimulated emission to begin. The energy is then released in terms of giant pulse with a short pulse width (from microsecond to nanosecond) and a very high peak intensity. Compared with mode-locking, Q-switching leads to much lower pulse repetition rates and much longer pulse durations, because they are related to the recovery time of the gain medium, which is typically about in milliseconds for erbium-doped fibers (EDFs). Q-switching can be active or passive, the Q-switch is an externally controlled variable attenuator for the former type, and the Q-switch is a saturable absorber in the latter type. In this chapter, we will discuss the use of graphene or graphene–polymer composite as a saturable absorber for passive Q-switching.

In the earliest efforts to make graphene-based fiber laser, chemically exfoliated GO, reduced GO (RG), and dye-molecule functionalized graphene were used as the working materials (Bao et al. 2008). Laser Q-switching operation was observed using GO and RG as saturable absorbers. Figure 8.10 shows a typical Q-switched pulse train and the enlarged pulse feature. The pulse separation is about 10.4 μs with a full width at half maximum (FWHM) of 1.90 μs. Although such pulse separation can be varied by rotating the paddles of polarization controllers (PCs) and varying the pump strength, the pulse separation is always in the same order of magnitude (10 μs), due to the fact that it is inherently determined by the nonlinear interaction of the light with the gain medium and the absorber (Keller 2003; Keller et al. 1996). The Q-switched lasers oscillated around 1530 nm. Due to the fact that EDF is a quasi-two-level gain medium, it has strong reabsorption at 1530 nm, therefore, no mode-locking could be obtained in this case (Bao et al. 2008).

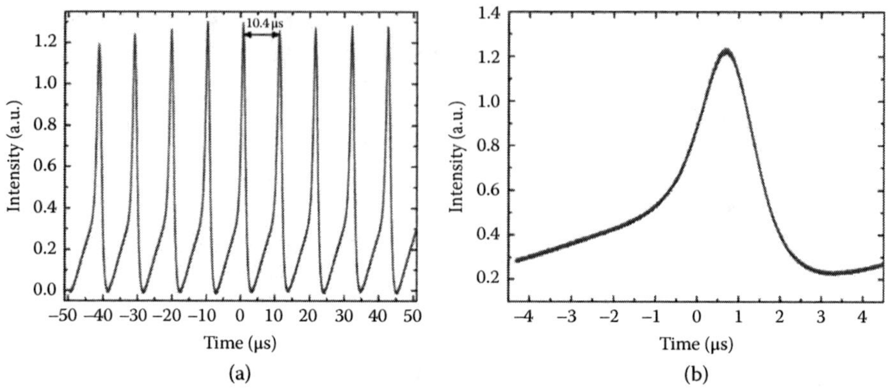

FIGURE 8.10
Typical Q-switching state from GO and RG. (a) Output pulse train. (b) enlarged pulse feature. (From Bao, Q. et al., Graphene related materials for pulsed laser generation, unpublished. With permission.)

Recently, passively Q-switched fiber lasers with different gain medium (Nd:YAG, Yu et al. 2010; Nd:$LuVO_4$, Yu et al. 2011; and Nd:$GdVO_4$, Li et al. 2011) were fabricated to generate large energy pulses at ~1. 06 μm. It is worth noting that a maximum pulse energy up to 3.2 μJ is observed in a passively Q-switched Ng:$GdVO_4$ laser with an energy conversion slope efficiency of ~40% (Li et al. 2011). In this work, Q-switching operation turned to a stable regime under an incident pump power level higher than 2.9 W, corresponding to an intracavity intensity of ~0.926 MW/cm^2 on the graphene sheets, which agrees with the saturation intensity of 0.87 MW/cm^2 reported in Bao et al. (2009).

Apart from the usage of epitaxial graphene grown on SiC (Yu et al. 2011) and chemically processed graphene (Li et al. 2011) in fiber lasers, Popa et al. (2011) demonstrated graphene–PVA composite as an effective fiber-compatible saturable absorber in fiber laser working at telecommunication C band. Figure 8.11a shows the schematic setup of Q-switched fiber laser with a ring cavity (Popa et al. 2011), with 1.25-m EDF as gain medium (from 1520 to 1560 nm), pumped with a 980-nm laser diode (LD), coupled through a wavelength division multiplexer (WDM). An in-line tunable optical band-pass filter is inserted after the optical isolator (ISO). The rest of the cavity consists of a combination of single-mode fiber (SMF) and SMF-28 with a total cavity length of ~10.4 m. The pulses generated from such laser have a width of ~2 μs with up to ~40 nJ energy for ~60 kHz repetition rate, as shown in Figure 8.11b. The pulse energy is approximately six times larger than graphene mode-locked fiber lasers (Song et al. 2010; Sun, Hasan et al. 2010; Zhang, Tang, Zhao et al. 2009) and much larger than thus far achieved in SWNT Q-switched lasers (Dong et al. 2010; Zhou et al. 2010). Most importantly, the laser is tunable in a wideband from ~1522 to 1555 nm (Figure 8.11c), which is comparable to the 31-nm range reported for doped crystal Q-switched tunable lasers (Laroche et al. 2002) and much larger than the 5 nm thus far achieved for

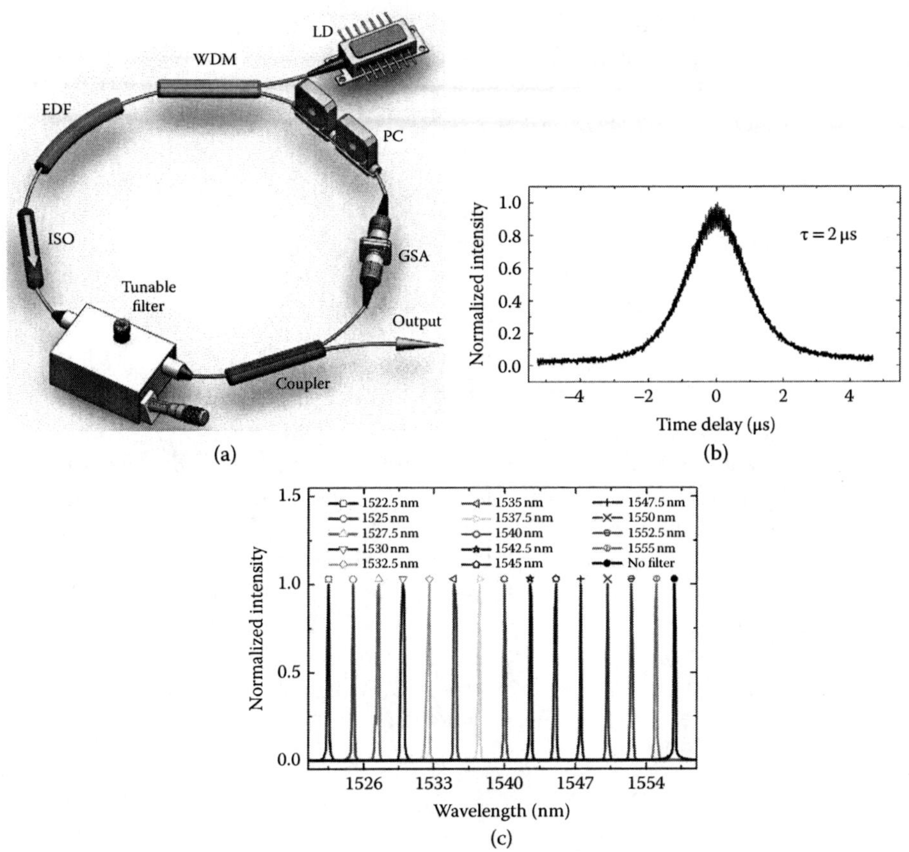

FIGURE 8.11
Wideband tunable Q-switched laser using graphene–PVA composite as saturable absorber. (a) Schematic setup of graphene composite–based Q-switched laser:laser diode (LD), wavelength division multiplexer (WDM), erbium-doped fiber (EDF), isolator (ISO), graphene-based saturable absorber (GSA), and polarization controller (PC). (b) Single-pulse envelope. (c) Output spectra for 14 tuning wavelengths. (From Sun, Z. et al., *ACS Nano*, 4, 803–810, 2010. With permission.)

SWNT Q-switched lasers (Dong et al. 2010; Zhou et al. 2010). Such wideband Q-switched lasers are ideal candidates for applications where ultrafast pulses are not necessary, such as material processing, environmental sensing, and biomedical diagnostics.

8.5 Mode-Locked Laser

Fiber lasers operate simultaneously in a large number of longitudinal modes falling within the gain bandwidth. Mode-locking occurs when phases of various longitudinal modes are synchronized such that the phase difference between any two neighboring modes is locked to a constant value (Haring et al. 2002). In mode-locking, a single pulse with typical duration ranging from tens of picoseconds to sub-tens of femtoseconds is preferred because the random phase relation originating from the interference of cavity modes is fixed. The repetition rate of mode-locked laser corresponds to the inverse of the cavity round-trip time. To obtain stable mode-locking state, the balance between the self-phase modulation (SPM) and the negative group velocity dispersion (GVD) is generally required. Due to the ability to generate extremely short pulses, mode-locked lasers can produce pulses that achieve an extremely high peak power.

8.5.1 Large Energy Mode-Locked Laser

A graphene–PVDF nanocomposite membrane with a thickness of ~10 μm inserted between two ferules was used as the saturable absorber for the laser mode-locking. The configuration of the fiber laser is schematically shown in Figure 8.12. A piece of 5.0-m, 2880-ppm EDF with GVD of −32 (ps/nm)/km was used as the gain medium, and 23.5-m SMF with GVD of 18 (ps/nm)/km was employed in the cavity to compensate the normal dispersion of the EDF and ensure that the cavity has net anomalous GVD. The net cavity dispersion is estimated as $-0.3419\ ps^2$. A 30% fiber coupler was used to output the signal, and the laser was pumped by a high-power Raman fiber laser source (KPS-BT2-RFL-1480-60-FA) of wavelength 1480 nm. The maximum pump power can be as high as 5 W. A polarization independent isolator was spliced in the

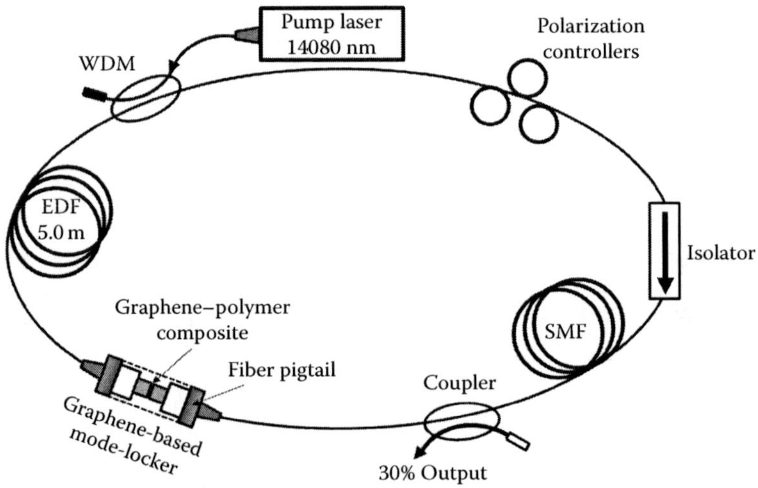

FIGURE 8.12
Schematic setup of the ring laser cavity configuration. This schematic shows the standard fiber-optic components such as wavelength division multiplexer (WDM), polarization controller, coupler, optical isolator, erbium-doped fiber (EDF), and single-mode fiber (SMF).

cavity to force the unidirectional operation of the ring cavity, and an intracavity PC was used to fine-tune the linear cavity birefringence. An optical spectrum analyzer (Ando AQ-6315B) and a 350-MHz oscilloscope (Agilent 54641A) combined with a 2-GHz photodetector was used to simultaneously monitor the spectra and the pulse train respectively.

Mode-locking of the laser self-started at a pump power of ~400 mW. Figure 8.13a shows the typical optical spectra of the mode-locked laser emissions (Zhang, Bao et al. 2009). The spectra have a broad spectral bandwidth with obvious Kelly spectral sidebands, characterizing that the mode-locked pulses are optical solitons. Soliton refers to an optical field that does not change during propagation because of a delicate balance between nonlinear and linear effects in the medium. It is well known that soliton pulse in fiber optics generated by the balance between the SPM and the negative GVD is short and stable (Kartner et al. 1996). The solitonic sidebands are unambiguous evidences of stable mode-locking of fiber lasers in the tuning range. The central wavelength of the spectra is at 1589.68 nm, which is in the L-band of the optical communication windows. The 3-dB bandwidth of the spectra is

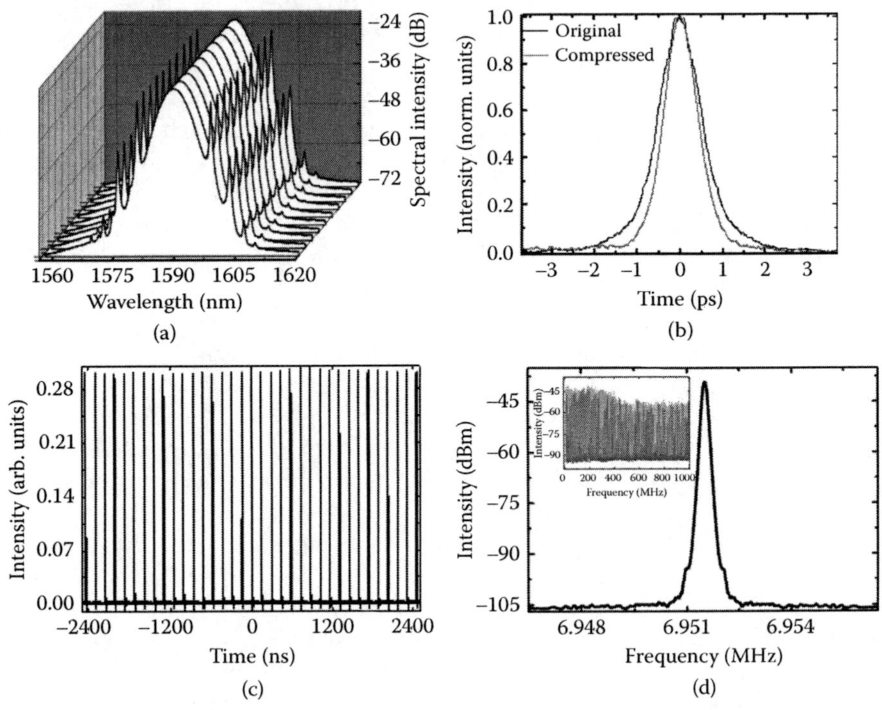

FIGURE 8.13
Soliton operation of the fiber laser. (a) Multiple soliton spectra measured at a 4-hour interval over 2 days. (b) Autocorrelation traces of the solitons. (c) An oscilloscope trace of the laser emission. (d) The fundamental radio frequency (RF) spectrum of the laser output. Inset: RF spectrum up to 1 GHz. (From Zhang, H. et al., *Optics Express*, 17, 12692–12697, 2009. With permission.)

about 5.0 nm. Figure 8.13b shows the measured autocorrelation (AC) trace of the solitons. It has a sech-profile with a FWHM width of about 1.07 ps, which, divided by the decorrelation factor of 1.54, corresponds to a pulse width of 694 fs. The time–bandwidth product (TBP) of the pulses is 0.412, showing that the solitons are slightly chirped. We also measured the radio frequency (RF) spectrum of the mode-locking state. The typical oscilloscope trace of the laser emission is shown in Figure 8.13c. Figure 8.13d shows a measurement made at a span of 10 kHz and a resolution bandwidth of 10 Hz. The fundamental peak located at the cavity repetition rate of 6.95 MHz has a signal-to-noise ratio of 65 dB. The insert of Figure 8.13d shows the wideband RF spectrum up to 1 GHz. The absence of spectral modulation in RF spectrum indicates that the laser operates well in the continuous wave (CW) mode-locking regime.

Different from the soliton operation of the EDF lasers mode-locked with the nonlinear polarization rotation (NPR) technique or a SESAM, no multiple solitons were formed in the cavity immediately after the mode-locking. Instead, only one soliton was always initially formed. This experimental result shows that the laser has a much lower effective mode-locking threshold than those mode-locked with SESAMs, which is traced back to the much smaller saturation fluence of the graphene–polymer nanocomposite than the SESAMs. The single soliton operation could be maintained in the laser as the pump power was gradually increased to 2 W. Pulse breaking occurred on further increasing the pump power. Eventually, multiple solitons were formed in the laser. Under multiple soliton operations, occasionally, harmonic mode-locking was also observed. We have focused on the single soliton operation of the laser. The energy of the soliton increased with the pump power. A maximum output power of ~13.1 dBm had been obtained under the pump power of ~2 W, which indicates the single soliton energy as high as ~3 nJ. Experimentally, we found that the pump power could be increased to as high as 3.2 W and the laser output power could be as large as 17 dBm. Under the pump and laser operation condition, the mode-locking of the laser could still maintain for hours, which indicates that the graphene–polymer composite could endure at least an optical fluency as high as 21.4 mJ/cm^2. After the operation, we had also checked the graphene–polymer composite film using the optical microscopy and found that its morphology was kept intact, which verified its strong thermal stability.

To investigate the long-term stability of the mode-locking, we have recorded the soliton spectra of the laser in a 4-hour interval over 2 days, as shown in Figure 8.13a. It shows that the central wavelength, 3-dB spectral bandwidth, Kelly sideband positions and the spectral peak powers, are kept reasonably unchanged. Experimentally, it was also found that the soliton pulse width could be compressed to ~524 fs after passing through a 10-m dispersion compensation fiber (DCF) of GVD approximately −4 (ps/nm)/km. The result shows that the output solitons were negatively chirped. We note that the current experimental results were obtained by the first try of the mode-locking technique. It is expected that through further careful design of the laser cavity and optimization on the saturable absorption parameters, mode-locked pulses with even larger pulse energy and narrower pulse width could be generated.

8.5.2 Broadband Tuneable Mode-Locked Laser

The enhanced broadband absorption of the graphene–polymer composites motivated us to investigate its applications in ultrafast photonics with broadband features. The intensity-dependent attenuation discussed earlier has important application in ultrafast photonics because of its pulse shaping; that is, a saturable adsorber is transparent to the high-intensity components of an optical pulse, but opaque to the lower-intensity components of the pulse (Keller 2003). Therefore, a saturable absorber suppresses CW radiation and enhances short-pulse generation when it is incorporated into a laser cavity.

The experimental configuration of the ring fiber laser for broadband mode-locking is schematically shown in Figure 8.12. A series of experiments show that mode-locking can be achieved with each of the graphene-based PVAc–polymer composites as saturable absorber. For composites with larger modulation depth, the mode-locking status is more stable. Here, we focus our discussion on the mode-locking performance of G-dye–PVAc since it has the highest modulation depth among the graphene-based polymer composites (Bao et al. 2010). The typical pulse train of the laser output depicted in Figure 8.14a verifies the pulse formation with a repetition rate of 6.95 MHz, which matches with the cavity roundtrip time. The AC trace of the mode-locked pulses is shown in Figure 8.14b, which is well fitted by a sech-squared function with FWHM = 1.67 ps. This gives a pulse width of 1.08 ps by multiplying the FWHM of the AC trace with the decorrelation factor (0.648). The TBP of the mode-locked pulses is about 0.4, indicating that the pulses are slightly chirped. The RF spectrum of the mode-locked pulses was also measured after optical-to-electrical conversion using a fast photodiode, as shown in Figure 8.14c. A 58-dB peak-to-background ratio (~10^6 contrast) is observed for the fundamental peak centered at the cavity repetition rate (6.95 MHz).

Most importantly, a tuning range of 30 nm (1570–1600 nm) of the laser output was achieved from G-PBASE–PVAc and G-dye–PVAc, as shown in Figure 8.14d. All the output spectra as a function of wavelength show two discrete spectral sidebands, indicating the soliton operation of the laser (Keller et al. 1996). As mentioned earlier, the solitonic sidebands are unambiguous evidences of stable mode-locking of the fiber lasers. In addition, the mode-locking can be operated for 3 days continuously and the output laser pulses maintain high stability; that is, the fluctuation of peak position is within 0.3 nm and the peak intensity varies within 0.15 dB (Figure 8.14e).

At present, there are a lot of interests in using carbon nanotube–based nanocomposites as saturable absorbers. The optical properties of the graphene-based nanocomposites in this work compete favorably with the

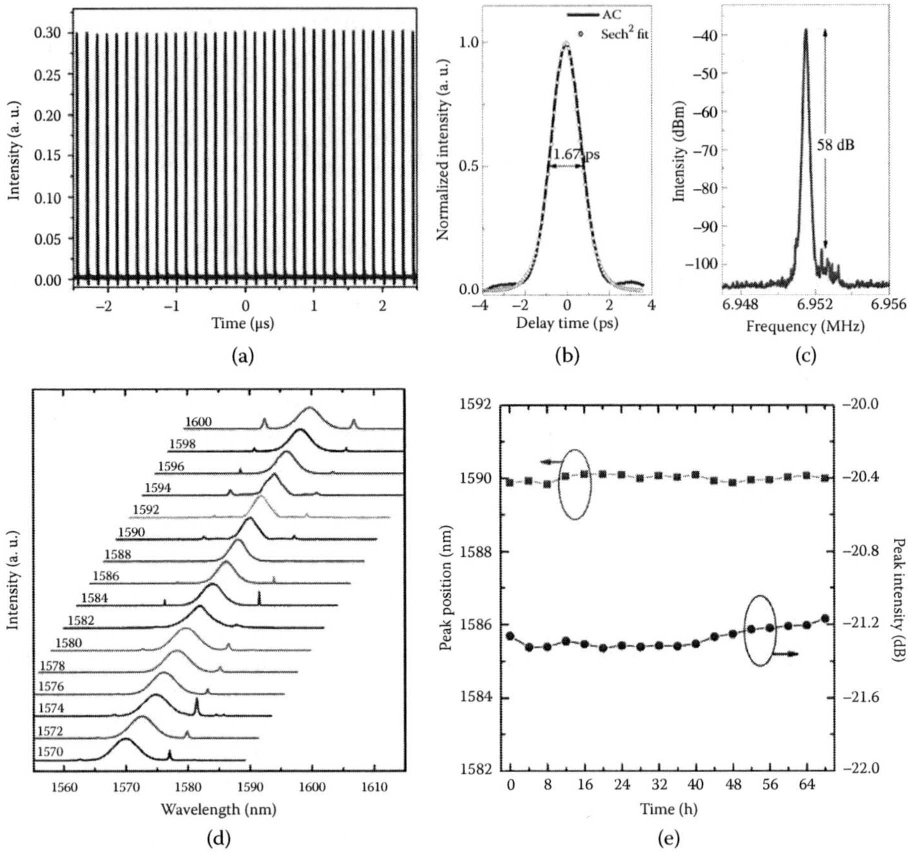

FIGURE 8.14
Mode-locking characteristics at 1590 nm. (a) Output pulse train. (b) AC trace and sech-squared fitting curve. (c) Radio frequency spectrum of mode-locking pulses. (d) Wideband output spectra from 1570 to 1600 nm. (e) The long-term stable operation of mode-locking at 1590 nm. The data were collected every 4 hours for 68 hours. (From Bao, Q. et al., *Adv. Funct. Mater.*, 20, 782–791, 2010. With permission.)

best performance reported in the literature for SWNT–polymer composites (Scardaci et al. 2008; Wang et al. 2008b). The saturable absorption modulation depth of our composites with 0.07 wt% loadings range from 20% to 29%, which is generally higher than SWNT–polymer composites (4~17%) (Scardaci et al. 2007, 2008; Wang et al. 2008b, 2008). The modulation depth can be tuned further by varying the concentration of graphene derivatives in the composite. The broadband operation was further confirmed by graphene mode-locked fiber laser at ~1060 nm (Zhao et al. 2010). Similar broadband tunability is also reported in (Wang et al. 2008a), where the mixture of SWNTs with a wide diameter distribution is used. However, it pales in comparison with graphene because SWNTs not in resonance are not used and give insertion losses. In principle, the graphene–polymer composites can function better than SWNT as full-waveband saturable absorber due to its strong broadband absorption, which is independent of wavelength.

Finally, the mode-locking performance of the graphene–polymer composites was compared with that of the conventional saturable absorber SESAMs. It was found that with graphene–polymer composites as the mode-locker, mode-locking could be easily achieved in a wide wavelength range; that is, continuous soliton central wavelength tuning as large as 30 nm could be obtained, and the mode-locked pulses have very large single pulse energy (as large as several nanoJoules). With a SESAM as the mode-locker (Zhang, Tang, Wu et al. 2009), mode-locking could only be achieved in a much narrower wavelength range (only ~10 nm), and the mode-locked pulses could only have much smaller single pulse energy due to breakup into multiple pulses in cavity.

8.6 Conclusions and Perspective

In summary, graphene-based polymer nanocomposites are remarkable materials that combine the chemically tunable optoelectronic properties of graphene derivatives with the mechanical, thermal properties and processing advantages of plastics. Chemically functionalized graphene derivatives are incorporated into polymer matrix to reinforce the mechanical, thermal, and optical properties. The results show that (1) graphene-based polymer composites have desired saturable absorption for Q-switched and mode-locked lasers; (2) the incorporation of graphene affords broadband operation of these pulse lasers; (3) the polymer matrix enhances the performance in terms of good processability and large damage threshold, making graphene–polymer composite a practical and efficient photonic material for the generation of ultrashort pulses or high-energy pulses in fiber lasers as well as solid-state lasers. Coupled with intelligent laser cavity design and fiber gain engineering, we believe that even larger energy and full-waveband ultrafast laser can be realized using graphene–polymer composites.

References

Aizawa, T., Souda, R., Otani, S., Ishizawa, Y., and Oshima, C. 1990. Anomalous bond of monolayer graphite on transition-metal carbide surfaces. *Physical Review Letters*, 64, 768–771.

Ajayan, P. M., Schadler, L. S., Giannaris, C., and Rubio, A. 2000. Single-walled carbon nanotube-polymer composites: Strength and weakness. *Advanced Materials*, 12, 750–753.

An, L. N., Xu, W. X., Rajagopalan, S. et al. 2004. Carbon-nanotube-reinforeed polymer-derived ceramic composites. *Advanced Materials*, 16, 2036–2040.

Ang, P. K., Wang, S., Bao, Q., Thong, J. T. L., and Loh, K. P. 2009. High-throughput synthesis of graphene by intercalation-exfoliation of graphite oxide and study of ionic screening in graphene transistor. *ACS Nano*, 3, 3587–3594.

Bae, S., Kim, H., Lee, Y. et al. 2010. Roll-to-roll production of 30-inch graphene films for transparent electrodes. *Nature Nanotechnology*, 5, 574–578.

Bao, Q., Zhang, H., Ni, Z. et al. 2011. Monolayer graphene as a saturable absorber in a mode-locked laser. *Nano Research*, 4, 297–307.

Bao, Q., Zhang, H., Tang, A. L. L. et al. unpublished. Graphene related materials for pulsed laser generation.

Bao, Q., Zhang, H., Wang, Y. et al. 2009. Atomic-layer graphene as a saturable absorber for ultrafast pulsed lasers. *Advanced Functional Materials*, 19, 3077–3083.

Bao, Q., Zhang, H., Yang, J. X. et al. 2010. Graphene-polymer nanofiber membrane for ultrafast photonics. *Advanced Functional Materials*, 20, 782–791.

Berger, C., Song, Z., Li, X. et al. 2006. Electronic confinement and coherence in patterned epitaxial graphene. *Science*, 312, 1191.

Biswas, S., and Drzal, L. T. 2008. A Novel approach to create a highly ordered monolayer film of graphene nanosheets at the liquid-liquid interface. *Nano Letters*, 9, 167–172.

Blake, P., Brimicombe, P. D., Nair, R. R. et al. 2008. Graphene-based liquid crystal device. *Nano Letters*, 8, 1704–1708.

Bonaccorso, F., Sun, Z., Hasan, T., and Ferrari, A. 2010. Graphene photonics and optoelectronics. *Nature Photonics*, 4, 611–622.

Boyd, R. 2003. *Nonlinear Optics*. London: Academic Press.

Cassagneau, T., Guerin, F., and Fendler, J. H. 2000. Preparation and characterization of ultrathin films layer-by-layer self-assembled from graphite oxide nanoplatelets and polymers. *Langmuir*, 16, 7318–7324.

Chang, C. Y., Tsao, F. C., Pan, C. J. et al. 2006. Electroluminescence from ZnO nanowire/polymer composite p-n junction. *Applied Physics Letters*, 88, 173503.

Coleman, J. N., Khan, U., and Gun'ko, Y. K. 2006. Mechanical reinforcement of polymers using carbon nanotubes. *Advanced Materials*, 18, 689–706.

Dalton, A., Stephan, C., Coleman, J. et al. 2000. Selective interaction of a semiconjugated organic polymer with single-wall nanotubes. *Journal of Physical Chemistry B*, 104, 10012–10016.

Davey, A. P., Drury, A., Maier, S., Byrne, H. J., and Blau, W. J. 1999. Synthesis and optical properties of phenylene-vinylene copolymers. *Synthetic Metals*, 103, 2478–2479.

Dawlaty, J. M., Shivaraman, S., Chandrashekhar, M., Rana, F., and Spencer, M. G. 2008. Measurement of ultrafast carrier dynamics in epitaxial graphene. *Applied Physics Letters*, 92, 042116.

Dong, B., Liaw, C. Y., Hao, J., and Hu, J. 2010. Nanotube Q-switched low-threshold linear cavity tunable erbium-doped fiber laser. *Applied Optics*, 49, 5989–5992.

Dresselhaus, M. S., and Dresselhaus, G. 2002. Intercalation compounds of graphite. *Advances in Physics*, 51, 1–186.

Emmanuel, P. G. 1996. Polymer layered silicate nanocomposites. *Advanced Materials*, 8, 29–35.

Garmire, E. 2000. Resonant optical nonlinearities in semiconductors. *IEEE Journal of Selected Topics in Quantum Electronics*, 6, 1094–1110.

George, P. A., Strait, J., Dawlaty, J. et al. 2008. Ultrafast optical-pump terahertz-probe spectroscopy of the carrier relaxation and recombination dynamics in epitaxial graphene. *Nano Letters*, 8, 4248–4251.

Greiner, A., and Wendorff, J. H. 2007. Electrospinning: A fascinating method for the preparation of ultrathin fibres. *Angewandte Chemie-International Edition*, 46, 5670–5703.

Grunlan, J. C., Mehrabi, A. R., Bannon, M. V., and Bahr, J. L. 2004. Water-based single-walled-nanotube-filled polymer composite with an exceptionally low percolation threshold. *Advanced Materials*, 16, 150–153.

Hao, R., Qian, W., Zhang, L., and Hou, Y. 2008. Aqueous dispersions of TCNQ-anion-stabilized graphene sheets. *Chemical Communications*, 2008, 6576–6578.

Haring, R., Paschotta, R., Aschwanden, A. et al. 2002. High-power passively mode-locked semiconductor lasers. *IEEE Journal of Quantum Electronics*, 38, 1268–1275.

Hendry, E., Hale, P., Moger, J., Savchenko, A., and Mikhailov, S. 2010. Coherent nonlinear optical response of graphene. *Physical Review Letters*, 105, 97401.

Hernandez, Y., Nicolosi, V., Lotya, M. et al. 2008. High-yield production of graphene by liquid-phase exfoliation of graphite. *Nature Nanotechnology*, 3, 563–568.

Holzer, W., Penzkofer, A., Gong, S. H., Bleyer, A., and Bradley, D. D. C. 1996. Laser action in poly (m-phenylenevinylene-co-2, 5-dioctoxy-p-phenylenevinylene). *Advanced Materials*, 8, 974–978.

Huang, Z. M., Zhang, Y. Z., Kotaki, M., and Ramakrishna, S. 2003. A review on polymer nanofibers by electrospinning and their applications in nanocomposites. *Composites Science and Technology*, 63, 2223–2253.

Kartner, F. X., Jung, I. D., and Keller, U. 1996. Soliton mode-locking with saturable absorbers. *IEEE Journal of Selected Topics in Quantum Electronics*, 2, 540–556.

Keller, U. 2003. Recent developments in compact ultrafast lasers. *Nature*, 424, 831–838.

Keller, U., Weingarten, K. J., Kartner, F. X. et al. 1996. Semiconductor saturable absorber mirrors (SESAM's) for femtosecond to nanosecond pulse generation in solid-state lasers. *IEEE Journal of Selected Topics in Quantum Electronics*, 2, 435–453.

Kim, H., Cho, J., Jang, S.-Y., and Song, Y.-W. 2011. Deformation-immunized optical deposition of graphene for ultrafast pulsed lasers. *Applied Physics Letters*, 98, 021104.

Kim, J. W., Kotagiri, N., Kim, J. H., and Deaton, R. 2006. In situ fluorescence microscopy visualization and characterization of nanometer-scale carbon nanotubes labeled with 1-pyrenebutanoic acid, succinimidyl ester. *Applied Physics Letters*, 88, 213110.

Kim, K. S., Zhao, Y., Jang, H. et al. 2009. Large-scale pattern growth of graphene films for stretchable transparent electrodes. *Nature*, 457, 706–710.

Ko, F., Gogotsi, Y., Ali, A. et al. 2003. Electrospinning of continuons carbon nanotube-filled nanofiber yarns. *Advanced Materials*, 15, 1161–1165.

Kotov, N. A., Dekany, I., and Fendler, J. H. 1996. Ultrathin graphite oxide-polyelectrolyte composites prepared by self-assembly: Transition between conductive and non-conductive states. *Advanced Materials*, 8, 637–641.

Kovtyukhova, N. I., Ollivier, P. J., Martin, B. R. et al. 1999. Layer-by-layer assembly of ultrathin composite films from micron-sized graphite oxide sheets and polycations. *Chemistry of Materials*, 11, 771–778.

Laroche, M., Chardon, A. M., Nilsson, J. et al. 2002. Compact diode-pumped passively Q-switched tunable Er-Yb double-clad fiber laser. *Optics Letters*, 27, 1980–1982.

Li, D., and Xia, Y. N. 2004. Electrospinning of nanofibers: Reinventing the wheel? *Advanced Materials*, 16, 1151–1170.

Li, X., Zhang, G., Bai, X. et al. 2008. Highly conducting graphene sheets and Langmuir–Blodgett films. *Nature Nanotechnology*, 3, 538–542.

Li, X. L., Xu, J. L., Wu, Y. Z., He, J. L., and Hao, X. P. 2011. Large energy laser pulses with high repetition rate by graphene Q-switched solid-state laser. *Optics Express*, 19, 9950–9955.

Li, X. S., Cai, W. W., An, J. H. et al. 2009. Large-area synthesis of high-quality and uniform graphene films on copper foils. *Science*, 324, 1312–1314.

Liu, N., Luo, F., Wu, H. et al. 2008. One-step ionic-liquid-assisted electrochemical synthesis of ionic-liquid-functionalized graphene sheets directly from graphite. *Advanced Functional Materials*, 18, 1518–1525.

Loh, K. P., Bao, Q., Ang, P. K., and Yang, J. 2010. The chemistry of graphene. *Journal of Materials Chemistry*, 20, 2277–2289.

Longhua, T., Ying, W., Yueming, L. et al. 2009. Preparation, structure, and electrochemical properties of reduced graphene sheet films. *Advanced Functional Materials*, 19, 2782–2789.

Lu, X., Yu, M., Huang, H., and Ruoff, R. 1999. Tailoring graphite with the goal of achieving single sheets. *Nanotechnology*, 10, 269–272.

Ma, H., Jen, A. K. Y., and Dalton, L. R. 2002. Polymer-based optical waveguides: Materials, processing, and devices. *Advanced Materials*, 14, 1339–1365.

Ma, H., Liu, S., Luo, J. et al. 2002. Highly efficient and thermally stable electro©\ optical dendrimers for photonics. *Advanced Functional Materials*, 12, 565–574.

Mack, J. J., Viculis, L. M., Ali, A. et al. 2005. Graphite nanoplatelet reinforcement of electrospun polyacrylonitrile nanofibers. *Advanced Materials*, 17, 77–80.

Mishchenko, E. G. 2009. Dynamic conductivity in graphene beyond linear response. *Physical Review Letters*, 103, 246802.

Moniruzzaman, M., and Winey, K. I. 2006. Polymer nanocomposites containing carbon nanotubes. *Macromolecules*, 39, 5194–5205.

Moos, G., Gahl, C., Fasel, R., Wolf, M., and Hertel, T. 2001. Anisotropy of quasiparticle lifetimes and the role of disorder in graphite from ultrafast time-resolved photoemission spectroscopy. *Physical Review Letters*, 87, 267402.

Munoz, E., Suh, D. S., Collins, S. et al. 2005. Highly conducting carbon nanotube/polyethyleneimine composite fibers. *Advanced Materials*, 17, 1064–1067.

Nair, R. R., Blake, P., Grigorenko, A. N. et al. 2008. Fine structure constant defines visual transparency of graphene. *Science*, 320, 1308–1308.

Novoselov, K. S., Geim, A. K., Morozov, S. V. et al. 2004. Electric field effect in atomically thin carbon films. *Science*, 306, 666–669.

Novoselov, K. S., Jiang, D., Schedin, F. et al. 2005. Two-dimensional atomic crystals. *Proceedings of the National Academy of Sciences of the USA*, 102, 10451–10453.

Popa, D., Sun, Z., Hasan, T. et al. 2011. Graphene Q-switched, tunable fiber laser. *Applied Physics Letters*, 98, 073106.

Popa, D., Sun, Z., Torrisi, F. et al. 2010. Sub 200 fs pulse generation from a graphene mode-locked fiber laser. *Applied Physics Letters*, 97, 203106.

Potts, J. R., Dreyer, D. R., Bielawski, C. W., and Ruoff, R. S. 2010. Graphene-based polymer nanocomposites. *Polymer*, 52, 5–25.

Ramanathan, T., Abdala, A. A., Stankovich, S. et al. 2008. Functionalized graphene sheets for polymer nanocomposites. *Nature Nanotechnology*, 3, 327–331.

Ramanathan, T., Liu, H., and Brinson, L. C. 2005. Functionalized SWNT/polymer nanocomposites for dramatic property improvement. *Journal of Polymer Science Part B: Polymer Physics*, 43, 2269–2279.

Ramanathan, T., Stankovich, S., Dikin, D. A. et al. 2007. Graphitic nanofillers in PMMA nanocomposites: An investigation of particle size influence on nanocomposite and dispersion and their properties. *Journal of Polymer Science Part B: Polymer Physics*, 45, 2097–2112.

Ray, S. S., and Okamoto, M. 2003. Polymer/layered silicate nanocomposites: A review from preparation to processing. *Progress in Polymer Science*, 28, 1539–1641.

Robert, N. Z. 1969. Saturated optical absorption through band filling in semiconductors. *Applied Physics Letters*, 14, 73–74.

Rozhin, A. G., Sakakibara, Y., Namiki, S. et al. 2006. Sub-200-fs pulsed erbium-doped fiber laser using a carbon nanotube-polyvinylalcohol mode locker. *Applied Physics Letters*, 88, 051118.

Sakakibara, Y., Rozhin, A. G., Kataura, H., Achiba, Y., and Tokumoto, M. 2005. Carbon nanotube-poly(vinylalcohol) nanocomposite film devices: Applications for femtosecond fiber laser mode lockers and optical amplifier noise suppressors. *Japanese Journal of Applied Physics Part 1*, 44, 1621–1625.

Scardaci, V., Rozhin, A. G., Tan, P. H. et al. 2007. Carbon nanotubes for ultrafast photonics. *Physica Status Solidi B*, 244, 4303–4307.

Scardaci, V., Sun, Z. P., Wang, F. et al. 2008. Carbon nanotube polycarbonate composites for ultrafast lasers. *Advanced Materials*, 20, 4040–4043.

Set, S. Y., Yaguchi, H., Tanaka, Y., and Jablonski, M. 2004. Ultrafast fiber pulsed lasers incorporating carbon nanotubes. *IEEE Journal of Selected Topics in Quantum Electronics*, 10, 137–146.

Shen, Y. 1984. *The Principles of Nonlinear Optics*. New York: Wiley.

Sheng, N., Boyce, M. C., Parks, D. M. et al. 2004. Multiscale micromechanical modeling of polymer/clay nanocomposites and the effective clay particle. *Polymer*, 45, 487–506.

Shui, J. L., and Li, J. C. M. 2009. Platinum nanowires produced by electrospinning. *Nano Letters*, 9, 1307–1314.

Song, Y.-W., Jang, S.-Y., Han, W.-S., and Bae, M.-K. 2010. Graphene mode-lockers for fiber lasers functioned with evanescent field interaction. *Applied Physics Letters*, 96, 051122.

Stankovich, S., Dikin, D. A., Dommett, G. H. B. et al. 2006. Graphene-based composite materials. *Nature*, 442, 282–286.

Sun, Z., Hasan, T., Torrisi, F. et al. 2010. Graphene mode-locked ultrafast laser. *ACS Nano*, 4, 803–810.

Sun, Z. P., Popa, D., Hasan, T. et al. 2010. A stable, wideband tunable, near transform-limited, graphene-mode-locked, ultrafast laser. *Nano Research*, 3, 653–660.

Sutter, P., Flege, J., and Sutter, E. 2008. Epitaxial graphene on ruthenium. *Nature Materials*, 7, 406–411.

Tausenev, A. V., Obraztsova, E. D., Lobach, A. S. et al. 2008. 177 fs erbium-doped fiber laser mode locked with a cellulose polymer film containing single-wall carbon nanotubes. *Applied Physics Letters*, 92, 171113.

Teo, W. E., and Ramakrishna, S. 2006. A review on electrospinning design and nanofibre assemblies. *Nanotechnology*, 17, R89–R106.

Tobias, K., Luca, P., Florian, S., Christian, F., and Martin, W. 2005. Strongly coupled optical phonons in the ultrafast dynamics of the electronic energy and current relaxation in graphite. *Physical Review Letters*, 95, 187403.

Vasko, F. 2010. Saturation of interband absorption in graphene. *Physical Review B*, 82, 245422.

Verdejo, R., Barroso-Bujans, F., Rodriguez-Perez, M. A., De Saja, J. A., and Lopez-Manchado, M. A. 2008. Functionalized graphene sheet filled silicone foam nanocomposites. *Journal of Materials Chemistry*, 18, 2221–2226.

Vickery, J. L., Patil, A. J., and Mann, S. 2009. Fabrication of graphene-polymer nanocomposites with higher-order three-dimensional architectures. *Advanced Materials*, 21, 2180–2184.

Viculis, L. M., Mack, J. J., Mayer, O. M., Hahn, H. T., and Kaner, R. B. 2005. Intercalation and exfoliation routes to graphite nanoplatelets. *Journal of Materials Chemistry*, 15, 974–978.

Vivekchand, S. R. C., Ramamurty, U., and Rao, C. N. R. 2006. Mechanical properties of inorganic nanowire reinforced polymer-matrix composites. *Nanotechnology*, 17, S344–S350.

Wakabayashi, K., Pierre, C., Dikin, D. A. et al. 2008. Polymer-graphite nanocomposites: Effective dispersion and major property enhancement via solid-state shear pulverization. *Macromolecules*, 41, 1905–1908.

Wang, F., Rozhin, A., Scardaci, V. et al. 2008a. Wideband-tuneable, nanotube mode-locked, fibre laser. *Nature Nanotechnology*, 3, 738–742.

Wang, F., Rozhin, A. G., Scardaci, V. et al. 2008b. Wideband-tuneable, nanotube mode-locked, fibre laser. *Nature Nanotechnology*, 3, 738–742.

Wang, F., Rozhin, A. G., Sun, Z. et al. 2008. Soliton fiber laser mode-locked by a single-wall carbon nanotube-polymer composite. *Physica Status Solidi B*, 245, 2319–2322.

Yamashita, S., Inoue, Y., Maruyama, S. et al. 2004. Saturable absorbers incorporating carbon nanotubes directly synthesized onto substrates and fibers and their application to mode-locked fiber lasers. *Optics Letters*, 29, 1581–1583.

Yu, H., Chen, X., Hu, X. et al. 2011. Graphene as a Q-switcher for neodymium-doped lutetium vanadate laser. *Applied Physics Express*, 4, 022704.

Yu, H. H., Chen, X. F., Zhang, H. J. et al. 2010. Large energy pulse generation modulated by graphene epitaxially grown on silicon carbide. *Acs Nano*, 4, 7582–7586.

Zhang, H., Bao, Q., Tang, D., Zhao, L., and Loh, K. 2009. Large energy soliton erbium-doped fiber laser with a graphene-polymer composite mode locker. *Applied Physics Letters*, 95, 141103.

Zhang, H., Tang, D. Y., Wu, X., and Zhao, L. M. 2009. Multi-wavelength dissipative soliton operation of an erbium-doped fiber laser. *Optics Express*, 17, 12692–12697.

Zhang, H., Tang, D. Y., Zhao, L. M., Bao, Q. L., and Loh, K. P. 2009. Large energy mode locking of an erbium-doped fiber laser with atomic layer graphene. *Optics Express*, 17, 17630–17635.

Zhao, L., Tang, D., Zhang, H. et al. 2010. Dissipative soliton operation of an ytterbium-doped fiber laser mode locked with atomic multilayer graphene. *Optics Letters*, 35, 3622–3624.

Zheng, W., and Wong, S. C. 2003. Electrical conductivity and dielectric properties of PMMA/expanded graphite composites. *Composites Science and Technology*, 63, 225–235.

Zhou, D. P., Wei, L., Dong, B., and Liu, W. K. 2010. Tunable passively Q-switched erbium-doped fiber laser with carbon nanotubes as a saturable absorber. *IEEE Photonics Technology Letters*, 22, 9–11.

9

Epoxy Nanocomposites: Graphene a Promising Filler

Iti Srivastava, Mohammad A. Rafiee, Fazel Yavari, Javad Rafiee, and Nikhil Koratkar*

CONTENTS

9.1 Introduction

The purpose of this chapter is to discuss the potential of graphene as a nanofiller in epoxy-based composite materials. Our focus will be on mechanical properties enhancement. First, we provide a general introduction to epoxies and their composites and then we describe the various mechanical properties that can be significantly enhanced by the use of graphene additives. Throughout we make an attempt, wherever possible, to compare graphene

* Correspondence to: Nikhil Koratkar (koratn@rpi.edu)

with other forms of nanofillers in order to clarify the potential advantages/disadvantages of graphene over other competing nanofillers.

Epoxy and its composites are used extensively for applications ranging from floor coatings to aircraft fuselages. Since their commercial inception in 1947 by the *Devoe-Raynolds* Company, epoxy has become the most extensively used, studied, and researched thermoset polymer (Delmonte et al. 1988). The superior mechanical properties, excellent adherence, chemical and corrosion resistance, high specific strength, ease of processing, and multiple curing options have made it a widely discussed technical material (Pilato and Michno 1994; Schab-balcerzak 2006). Due to low shrinkage and a multitude of available curing options, epoxy is one of the most commonly used adhesives. Its major application is in the field of structural metal bonding in the aerospace industry in addition to small-part assemblies (May 1988). They also make excellent coating materials due to their high corrosion resistance, strong adhesion, and good physical properties. Most commonly used epoxy coatings are bisphenol A products (compounds with two phenol functional groups). Epoxies are also a commonly used encapsulant for circuit components, as the application requires high electrical insulation, isolation from adverse environmental conditions of temperature, moisture, and chemicals (Lee 1988). Epoxy polymers are also extensively used as a stabilizer for plastics (Fisch and Hegranes 1988).

Many structural applications of epoxy require a reinforcing fiber network, for example, glass, carbon, and polyaromatic nylon (Kevlar) fibers, increasing the polymer strength. Fiber-reinforced epoxy composites are extensively used for structural applications especially in the aerospace, marine, and building construction industries, due to their high strength-to-weight and stiffness-to-weight ratios (Delmonte et al. 1988). Currently, more than 500 kg per aircraft of fiber-reinforced epoxy composites are used by the United States Air Force and Navy (Delmonte et al. 1988). The aerospace giant Boeing is currently using carbon-fiber-reinforced epoxies extensively in the fuselage and other structural component of their next-generation jetliner, the Boeing 787 *Dreamliner* (Miller 2007).

9.2 Epoxy Structure/Properties

The thermoset epoxy polymer consists of two basic components: epoxy resin and curing agent or hardener.

9.2.1 Epoxy Resin

Epoxy resin is a molecule containing more than one 1,2-epoxide group. The 1,2-epoxide group, as shown in Figure 9.1a, is a cyclic ether group. It is also called glycidyl group. This group can undergo various reactions to give different kinds of epoxy resin structures.

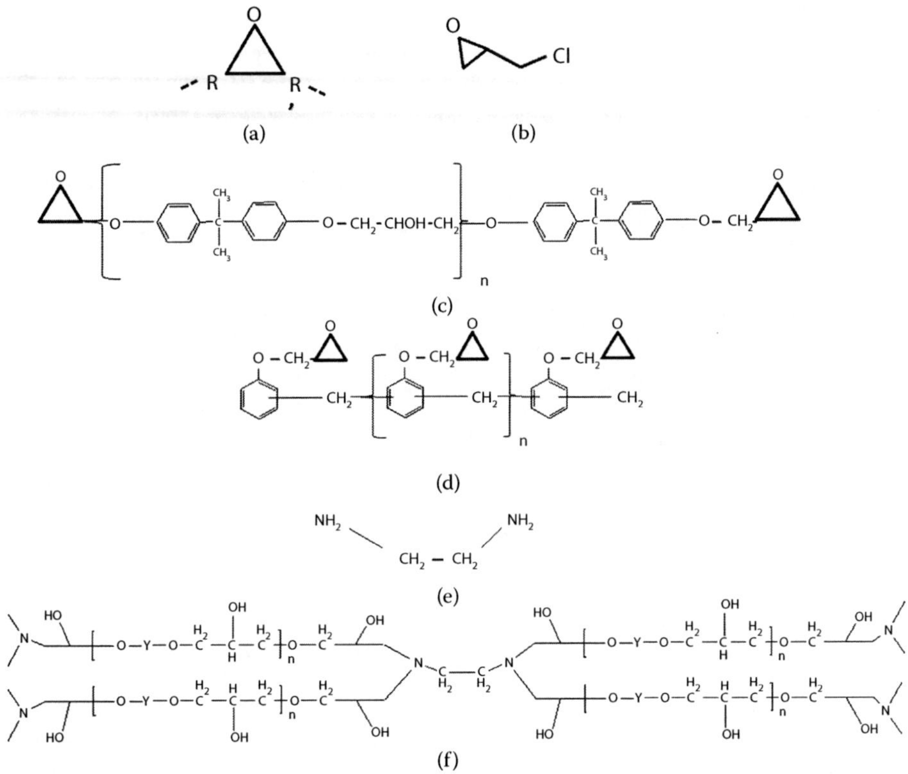

FIGURE 9.1
Schematic of (a) epoxide group, (b) epicholorohydrin group, (c) DGEBA (Adapted from Delmonte, J. et al., *Epoxy Resins: Chemistry and Technology*, Marcel Dekker, New York, 1988), (d) novolac epoxy resin (Adapted from Hadad, D. K., *Epoxy Resins: Chemistry and Technology*, CRC Press, Boca Raton, FL, 1988), (e) curing agent, ethylene diamine, and (f) network structure of cured DGEBA epoxy.

There are two basic classes of epoxy resin depending on their preparation techniques: glycidyl epoxy and nonglycidyl epoxy resins. Glycidyl epoxy could be further classified according to the attached groups as ether (R–O–R′), ester (R–CO–OR′), and amine group (–NH_2). Glycidyl epoxies are prepared by the condensation reaction of dihydroxy (two –OH groups), dibasic acid (acid with two replaceable –H atoms) or diamine (two –NH_2 groups) and epichlorohydrin (both organochlorine and epoxide groups attached, as shown in Figure 9.1b). Nonglycidyl epoxies are formed by a peroxidation reaction of olefinic (unsaturated open-chain hydrocarbons) double bonds (Dunn 2003). The most frequently used forms of glycidyl epoxides are bisphenol-A and novolac epoxy resins. Diglycidyl ether of bisphenol A (DGEBA) solely accounts for 85% of the world's epoxy resin production (Schab-balcerzak 2006). As shown in Figure 9.1c and d, DGEBA and novolac epoxy resin have aromatic rings and oxygen moieties in its chain, which increase the polymer stiffness.

n in the polymer structure shown in Figure 9.1c and d determines whether the resin exists in liquid or solid state. For n between 0 and 3, the resin is generally liquid and for n between 3 and 30, it is in solid state (Dunn 2003).

9.2.2 Curing Options

There are various kinds of hardeners used for the condensation curing of epoxy resin. The chemical nature of the hardener controls the cross-link nature, and hence the cure kinetics and glass transition temperature (T_g) of the epoxy (Hale et al. 1991). Some examples of hardeners are aliphatic amine, amido amine, aromatic amine, polyamine, and anhydrides (Dunn 2003).

The resin and hardener are mixed in predetermined ratios and used; frequent heating is necessary to obtain the best polymer properties. The versatility of epoxy lies in the fact that there are multiple combinations of resin and hardener available, yielding wholly different polymer structures with robust but flexible mechanical properties. In Figure 9.1e, a curing agent, ethylene diamine is shown. Figure 9.1f is a schematic of DGEBA epoxy network polymer formed when the resin and curing agent react.

9.2.3 Mechanical Properties of Interest

This section gives a brief introduction on the diverse mechanical properties relevant to the performance of epoxy composites before delving into the effect of graphene on mechanical properties.

Young's modulus is the ability of the material to resist elastic deformation under applied stress, given by the slope of the stress–strain curve. The yield point on the stress–strain curve indicates the endpoint of the material's elastic regime. The elastic/Young's modulus is calculated in the linear portion of the plot below the yield point, after which it varies. For bending tests, the modulus obtained is called flexural modulus.

Another important mechanical property that is of great practical relevance is buckling resistance. Buckling is a structural instability failure mode under compressive stresses. The higher the compressive load that the structure can withstand before it buckles (or bows in an unstable fashion), the more is its buckling resistance. The resistance to buckling is a function of the sample geometry and its elastic modulus.

Generally, increasing the elastic modulus of a material increases its buckling resistance. Euler's equation is used to predict the commencement of instability for a column under an axial compressive load; the least load for the instability to initiate is given by

$$P = C\pi^2 IE/L^2$$

where P is the ultimate buckling load, E is the elastic modulus, C is the function of geometry, L is the effective length, and I is the moment of inertia of the cross section.

The toughness of a material, given by the total area under the stress–strain curve, is equivalent to the energy absorbed by the material before failure. The greater the energy the material absorbs before failure, the tougher it is. The strain energy absorbed is not completely recoverable when plastic deformation occurs or in the presence of hysteresis in the elastic regime as for elastomers.

Ductility is another important parameter describing the total amount of plastic strain in the material before failure. Elongation to failure (e_f) is the maximum strain the material can withstand before failure and is used to judge a material's ductility. Yield strength (σ_y) is the maximum stress that a material can undertake at a specified value of permanent strain (usually 0.2%). The tensile strength is the maximum strength a material can withstand (σ_{UTS}). For tests in bending mode, the strength of the material is called the flexural strength. These parameters have been shown in Figure 9.2a. Another essential property of a material not shown in the figure is its impact strength. It is the ability of a material to withstand high strain rate and shock loading.

Generally, material characteristics, which enhance ductility, reduce the tensile strength and vice versa. For example, although rigid plastics such as epoxy and polystyrene can withstand a large amount of stress, they do not show much elongation before breaking. Hence, though they are strong and stiff, they are not tough. In contrast, elastomers undergo large elongation before failure and hence are very ductile but have small stiffness. The most optimum condition for high toughness is to have high strength and good ductility. High toughness also leads to high damage tolerance of the material. It is defined as the material's ability to safely sustain unforeseen defects. Figure 9.2b is a comparison of stress–strain curves for brittle, ductile, and tough materials.

Two other material properties, which cannot be obtained from a typical stress–strain curve, are the *fracture toughness* and *fatigue resistance* of a material. Both these properties are experimentally characterized in the presence

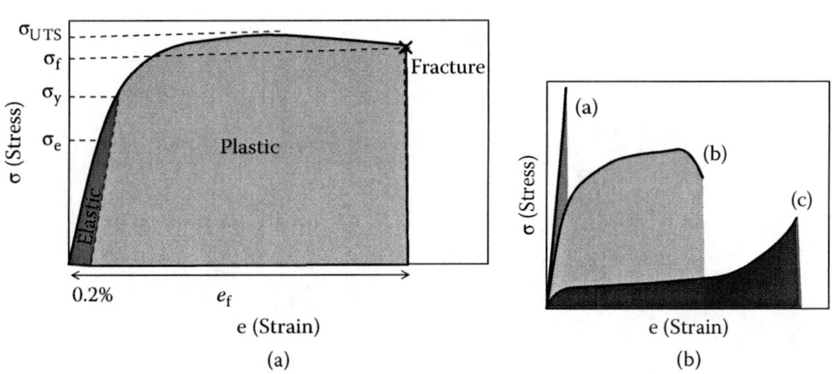

FIGURE 9.2
(a) Typical stress–strain curve showing elastic and plastic regime. (b) Stress–strain curve for (a) brittle, (b) tough, and (c) ductile material.

of a well-defined crack in the sample; these are critical for structural applications as virtually all materials have some preexisting flaws and under practical loading conditions inevitably suffer frequent stress fluctuations.

Under a given applied stress, there are three basic crack opening modes: mode I in normal tension, mode II in-plane shear, and mode III out-of-plane shear, as shown in Figure 9.3.

Presence of a notch or crack acts as a stress raiser; to account for the localized increase in the stress, a parameter, the stress concentration factor k_t, is defined. Using Griffiths crack model (Perez 2004) it is also given by

$$k_t = \frac{\sigma_t}{\sigma_o} = 2\sqrt{a/\rho}$$

where σ_t is the stress at the crack tip, σ_o is the applied stress, a is the crack length (half of an internal crack length), and ρ is the radius of curvature at the crack tip. The value of k_t is a function of the crack opening mode. The sharper the crack tip or smaller the ρ, the higher is the k_t. Ductile materials yield under the high local stress at the crack tip, leading to an increase in ρ, decreasing the stress concentration and hence the debilitating effect of the crack. In contrast, a brittle material does not yield significantly and hence displays high k_t.

Linear elastic fracture mechanics has been very helpful in understanding material failure and fracture at the microscopic level. The stress intensity factor k describes the material condition, as a function of the applied stress under a preexisting crack by the following equation:

$$k = Y\sigma_o\sqrt{\pi a}$$

where σ_o is the applied stress, a is the crack length (half of an internal crack length), and Y is the function of loading condition, crack geometry, and crack length. The critical value of k, which causes material failure under critical stress (σ_o), is called the fracture toughness (k_c) of the material, given by

$$k_c = Y\sigma_{oc}\sqrt{\pi a_c}$$

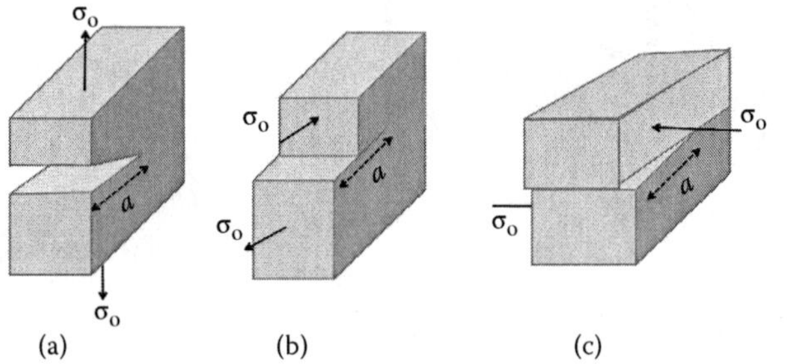

FIGURE 9.3
Three modes of failure: (a) mode I, (b) mode II, and (c) mode III.

Its value is reported under plain strain condition (condition for the sample with thickness $\geq 2.5\left(\frac{k_{IC}}{\sigma_y}\right)^2$), for mode I loading, indicated as k_{IC}. k_{IC} is defined as the ability of a material containing a preexisting crack to resist failure. The k_{IC} is related to the critical energy release rate G_{IC}, defined as the amount of energy released per unit area as the crack propagates, which is given by

$$G_{IC} = k_{IC}^2 (1-\nu^2)/E$$

Fatigue resistance is the resistance offered by the material to a subcritical crack that propagates under alternating stresses lower than the material's ultimate tensile stress (Suresh 1998). Fatigue life measurement and fatigue crack propagation rate (FCPR) measurement based on fracture mechanics models are two popular methods to characterize the fatigue resistance or the fatigue life of material. The FCP life of a material is divided into three stages, as shown in Figure 9.4a. In stage I, the crack growth is below the threshold value of the stress intensity factor. The threshold stress intensity factor (ΔK_{th}) is defined at the stress range below which cracks will not propagate in the material. In the fatigue life diagram, it is indicated by the fatigue limit. In stage II, the size of a crack grows steadily, and its growth is governed by the Paris law given by (Paris et al. 1961)

$$\frac{da}{dN} = C(\Delta K)^n$$

The Paris law states that the crack propagation rate per cycle (da/dN) is directly related to the stress intensity factor range

$$\Delta K = (K_{max} - K_{min}) = Y(\sigma_{max} - \sigma_{min})\,(\pi a)^{1/2}$$

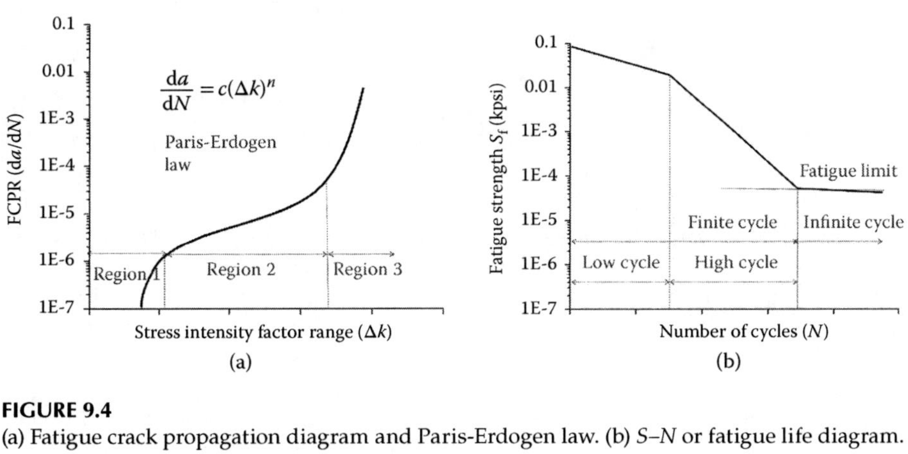

FIGURE 9.4
(a) Fatigue crack propagation diagram and Paris-Erdogen law. (b) S–N or fatigue life diagram.

and on external testing conditions described by two constants *C* and *n* (Mai and Williams 1979). The higher the crack growth rate (d*a*/d*N*), the shorter would be the material's life. Stage III is the terminal regime, where the crack growth is unstable and the material finally fails.

In the fatigue life diagrams or *S*–*N* (stress–number of cycles) curves, the measurements are made to estimate the expected number of life cycles (N_f) a material can withstand under a given stress (*S*). At high *S* values, N_f is small and the region is called low cycle fatigue, while at low *S* values, N_f is large corresponding to the high cycle fatigue region. The FCPR diagram and *S*–*N* curve of a typical material are shown in Figure 9.4b.

The fatigue life of a material is especially difficult to predict as nearly all cracks propagate under mixed loading throughout the life. Ductile materials generally show both modes I and II of crack propagation. The initial propagation of fatigue cracks in materials has been observed to be under shear mode while in the later stages of crack propagation, it propagates under tensile mode. Various approaches like critical plane model (Socie 1993) and plastic strain energy model have been used to predict the life and fatigue crack growth under mixed mode conditions (Stephens and Fuchs 2001). Compact tension-based geometries are most widely used for fracture toughness and fatigue resistance studies. ASTM D5045 is the commonly used standard for k_{IC} testing of plastic materials (D 54045 Standard Test Methods for Plain Strain Fracture Toughness and Strain Energy Release Rate of Plastic Materials Philadelphia, 1999). ASTM E647 (E 647 Standard Test Method for Measurement of Fatigue Crack Growth Rates 1997) and taper double cantilever beam (DCB) geometry (Beres et al. 1997) are typically used for fatigue testing.

Epoxy, in general, lacks the crucial attribute of damage tolerance, which is important for structural applications. In contrast to other polymers, the high cross-linking density of epoxy chains results in brittle failure (Levita et al. 1991). Epoxy has a fracture energy of less than 0.3 kJ/m^2 making it a brittle material (Garg and Mai 1988) for aircraft and other basic engineering applications (Schab-balcerzak 2006). Increasing the cross-link density of epoxy increases its strength but at the cost of its ductility (Argon and Cohen 2003). Although the use of plasticizers increases the ductility of epoxy, it also leads to poor high temperature performance (Delmonte et al. 1988).

The mechanical properties of epoxy polymers are a strong function of temperature. In addition to the melting point (T_m), polymers possess another important transition temperature called the glass transition temperature (T_g), or the "alpha peak," T_α. Below T_g, the polymer behaves like a rigid solid, such as metals, although with a lower elastic modulus. This is called the glassy region. The value of T_g differs depending on the polymer's molecular structure; for example, T_g of most common epoxies varies between room temperature and 260°C, whereas for natural rubbers it is around −85°C (Sircar and Chartoff 1994). Other temperature peaks at temperature lower than T_α indicate small changes in molecular conformation and are termed "beta peaks," T_β (Moalli 2001).

9.3 Epoxy Composites

Various sizes of filler materials ranging from micron to nanoscale fillers can be introduced into epoxies forming composite materials with enhanced mechanical properties as compared to the neat epoxy (Srivastava and Koratkar 2010). Filler particle size has a great influence on the composite properties (Nielsen 1974). To enhance the mechanical properties of the polymer matrix without attenuating its ductility, the use of nanometer-sized fillers has proved far more effective than micron-sized fillers. Nanoparticles possess higher specific surface area than micron-sized particles, resulting in greater interfacial contact area with the polymer matrix and hence nanoparticle-reinforced composites show relatively superior mechanical properties (Abramoff and Covino 1992; Ajayan et al. 2003; Bartczak et al. 1999; Landry et al. 1993; Sumita et al. 1983; Wetzel et al. 2003). For example, the specific surface area of single-walled carbon nanotubes (SWNTs) is ~1315 m^2/g (Peigney et al. 2001; Schlapbach and Zuttel 2001), of Al_2O_3 nanoparticles with 13 nm diameter is ~100 m^2/g, and for $CaSiO_3$ rods of diameter 3–5 μm and length 12–20 μm it is ~5 m^2/g (Wetzel et al. 2003). Micron-sized particles are also prone to cracking due to high volume of stress concentration regions (Nielsen 1974). They can display a greater tendency to debond on application of load compared to nanoparticles (Móczó and Pukánszky 2008). On debonding of the particle, the void created in the matrix is larger for micron-sized fillers and hence more likely to be of the critical size required for crack propagation and brittle failure. Another advantage of having nanofillers is that for the same volume fraction of fillers added, the number of particles will be higher than for microfillers (Wetzel et al. 2002), and therefore the interparticle distance will be smaller for nanofillers, which is an important parameter in determining various mechanical properties. Here we review some of the common filler materials that have been explored for epoxies before we discuss the role of graphene as potential nanofiller for epoxy composites.

Rubber is one of the most extensively studied and well-understood fillers for toughening of epoxies. Some commonly used rubbers are liquid functional rubbers such as carboxyl terminated butadiene acrylonitrile (CTBN) and copolymers of amino-terminated butadiene and acrylonitrile (ATBN) (Kinloch et al. 1983; Kunz and Beaumont 1981). Rubber–polymer composites on stressing undergo rubber cavitations, leading to localized yielding. This mechanism results in increased polymer toughness.

A similar approach for increasing the toughness of epoxy is to add aliphatic epoxy resins or thermoplastics to aromatic epoxy resins or reduce the cross-link density in epoxy using different curing agents or chemistry. Inclusion of hyperbranched polymers (HBP) (highly branched dendritic polymers) in epoxy has also improved epoxy toughness. HBP have been reported to increase the fracture toughness of DGEBA epoxy by 100% at around 20 parts per hundred of resin (phr). However, they also increase the viscosity of epoxy reducing the workable region and deteriorate the Young's modulus and the thermomechanical properties of the resin including T_g. Mixtures of DGEBA and Novalac epoxy have also been studied and found to improve epoxy properties (Schab-balcerzak 2006).

Silica (SiO_2) (Kinloch et al. 2005; Nakamura et al. 1991; Young and Beaumont 1975), alumina (Al_2O_3) (Lange and Radford 1971; Wetzel et al. 2003; Zhao et al. 2008a), glass beads (Spanoudakis and Young 1984a, 1984b), and other ceramic nanoparticles have proven very useful in reinforcing epoxy without sacrificing its thermomechanical properties. Improving the filler interaction with the matrix by surface functionalization has improved fracture toughness as well as fatigue resistance (Nielsen 1975; Zhao et al. 2008a). For 20.2 wt.% SiO_2–epoxy nanocomposites, the mode I fracture toughness (k_{IC}) and modulus (E) increase to ~0.88 $MPa^{1/2}$ from ~0.5 $MPa^{1/2}$ and ~3.85 GPa from ~2.96 GPa, respectively (Blackman et al. 2007). Note that the fracture toughness increases achieved with ceramic nanoparticles have not been as high as rubber-reinforced epoxy. A combination of inorganic and organic fillers, called hybrid composites, has caused synergistic effects in improving epoxy properties (Kinloch, Maxwell, and Young 1985).

Nanofibers are whiskers having diameter <100 nm. Different combinations of nanofibers and polymers have been investigated for nanocomposite applications because of immense advantages of high surface area and low density. Polymer nanofibers of diameter >10 nm commonly obtained through the electrospinning process (Huang et al. 2003), graphitic carbon nanofibers (GCNFs) (Wei Hong et al. 2005), vapor grown carbon nanofibers (VGCNFs) (Mordkovich 2003), and cellulose nanofibers (Shimazaki et al. 2007), obtained from chemical vapor deposition (CVD) of hydrocarbons have been used in epoxy matrices for advanced applications in the aviation and marine industries. VGCNFs have been observed to increase T_g of an epoxy matrix from ~80°C to ~100°C at ~10 wt.%, in addition to increasing its storage modulus (E'), tensile strength, and Young's modulus (Choi et al. 2005). The morphological and topological characteristics of VGCNFs favor their twisting and intertwining, and hence they tend to agglomerate. Sonication (Choi et al. 2005), functionalization (Teyssier et al. 2006), and use of surfactants (Ying et al. 2002) have proven useful for obtaining homogeneous dispersions, followed by an increase in fracture toughness and fatigue resistance (Miyagawa and Drzal 2005; Zhou et al. 2007).

The interest in fibrous carbon was renewed by the discovery of carbon nanotubes (CNTs) by Ijiama (Iijima 1991). CNTs have higher aspect ratios and hence higher specific surface area (>1000 m^2/g) than carbon fibers, higher E (~1 TPa) than ceramics with ultimate tensile strength as high as 200 GPa and ductility of ~6% at room temperature with the capacity to exhibit superplasticity at higher temperatures (Huang et al. 2006). Single-walled nanotubes (SWNTs) and multi-walled nanotubes (MWNTs) have become among the most actively studied fillers for nanocomposites. The biggest challenge encountered in fabricating CNT composites is obtaining their uniform dispersion in polymer matrices. Large van der Waals forces of attraction in SWNTs and MWNTs obstruct the formation of homogeneous dispersions. Functionalization and use of surfactants and organic solvents are several means that improve the dispersion in polymer matrices (Gojny et al. 2003; Pötschke et al. 2004; Shaffer et al. 1998). Functionalization of CNTs with amine groups have proven particularly useful in improving CNT adhesion to epoxy (Chen et al. 2008; Zhang et al. 2009).

Although the use of fibrous nanofillers has led to significant enhancements in the properties of epoxy, planar fillers have proven superior to them. Two important examples of planar nanofillers are inorganic silicate sheets, that is, nanoclays and graphene platelets, which have proven to be more efficient than either fibers or particulates as fillers for mechanical property enhancement of epoxy. In a properly exfoliated clay–polymer composite, the clay sheets have thickness as small as ~10 Å (Lan and Pinnavaia 1994). The resulting high aspect ratio provides large interfacial contact area with the polymer matrix (LeBaron and Pinnavaia 2001); the stiffness, toughness, and fatigue resistance of nanoclay/epoxy increase with increasing filler weight fraction (Becker et al. 2002). Further achievements have been reported through chemical treatment of nanoclays (Battistella et al. 2008; Stankovich et al. 2007). We now direct our attention to *Graphene,* which is another highly promising planar (or sheet-like) nanofiller.

9.4 Graphene as a Nanofiller

Graphene, a potentially very promising nanofiller for epoxy systems, is a single-atom thick sheet of sp^2-hybridized carbon atoms packed in a honeycomb crystal lattice. Graphene has attracted much attention for use in composites. This interest in graphene stems from their high aspect ratio (few-atom thickness) and specific surface area, strong sp^2 carbon–carbon bonds, full organic compatibility, and potentially low cost of production due to top-down approach for production using bulk graphite. Moreover, graphene sheets produced by exfoliation and reduction of graphite oxide show enhanced interfacial adhesion with the epoxy matrix due to the presence of hydroxyl, carboxyl, and epoxide functional groups on the graphene surface.

The various methods of graphene synthesis can be broadly classified into top-down (mechanical exfoliation) and bottom-up (chemical synthesis, vapor deposition) approaches. The top-down approach offers abundance of starting material, whereas the bottom-up approach allows extensive tuning of the properties. Graphene can be fabricated by CVD, micromechanical exfoliation of graphite, epitaxial growth on electrically insulating surfaces, and by creation of colloidal suspensions (Park and Ruoff 2009). Micromechanical exfoliation of graphite, also known as the "Scotch tape" method was the very first method used to obtain graphene. Being a tedious and manual process, the graphene yield is less extensive, but obtained graphene tends to be of the highest quality. In the process highly oriented pyrolytic graphite is used and graphene sheets are peeled using razor blade or tape (Novoselov et al. 2004; Sutter et al. 2008). Vacuum graphitization has also been used to produce an ultrathin epitaxial graphite layer on atomically smooth single crystal surfaces (Berger et al. 2006).

Graphene has been produced on a large scale with and without substrates using CVD and microwave-plasma deposition techniques, respectively (Dato et al. 2008; Eizenberg and Blakely 1979). In the substrate method, nickel is used to dissolve carbon and release it when cooled and the graphene layer is extracted by chemical etching (Obraztsov 2009). The use of single crystal substrate minimizes the number of defects in the layer due to lattice mismatch between graphene and substrate (Obraztsov 2009). In the substrate-free growth method, hydrocarbons like ethanol are evaporated in an argon plasma at atmospheric-pressure using a microwave plasma reactor and the solid reaction products are collected on membrane filters to yield graphene sheets (Dato et al. 2008).

Chemical methods are popular bottom-up techniques where graphene is obtained from colloidal suspensions made from graphite and/or graphite oxide (Park and Ruoff 2009). Wet chemical techniques for obtaining graphene from graphite oxide have been recently studied in detail due to its ease of scalablility and versatile nature. In the presence of strong oxidants such as mixtures of concentrated nitric and sulfuric acids, graphite is oxidized to graphite oxide, with carboxylic and carbonyl groups and other carbon- and oxygen-based groups attached in-between sheets and at sheet edges. The graphite to graphite oxide conversion increases the interlayer sheet distance to 6–12 Å from 3–4 Å. Graphite oxide can be directly dispersed in polar solvents leading to homogeneous suspensions and are ultimately reduced by chemical and thermal means to yield graphene. The use of reducing agents such as hydrazine and dimethylhydrazine has proven successful. An alternative to colloidal suspensions is the direct thermal reduction of graphite oxide to yield graphene sheets. In this method, the exfoliation and reduction steps are combined into a single step that involves giving a thermal shock (rapid rate of heating ~2000°C/min) to graphite oxide (McAllister et al. 2007; Rafiee et al. 2010b; Schniepp et al. 2006). The exfoliation is caused by pressure build up due to the rapid release of gases that have insufficient time to diffuse out of the intercalated graphite oxide structure. Graphene sheets produced by this method are called functionalized graphene sheets (FGS) due to residual oxygen in the form of trace amounts of hydroxyl, epoxide, carbonyl, and carboxylic groups that remains on the graphene in spite of the thermal shock. FGS have shown exceptional mechanical behavior in polymer matrices. Addition of FGS in quantities as small as 0.05 wt.% leads to a large increase in the T_g of around 30°C in poly-methyl-methacrylate (PMMA) (Ramanathan et al. 2008); 1 wt.% of FGS improved the Young's modulus and ultimate tensile strength of PMMA by factors of 1.8 and 1.2, respectively. The exceptional mechanical and thermomechanical properties of FGS–PMMA composites is attributed to the highly wrinkled surfaces of graphene sheets concurrent with the presence of defects and hydroxyl groups, which are believed to increase matrix-nanofiller adherence without the need for additional functionalization.

9.5 Graphene/Epoxy Composites

To our knowledge, the only studies that report mechanical properties of graphene/epoxy composite materials have been performed by our group (Rafiee et al. 2009, 2010a; Yavari et al. 2010). In this section, we will summarize our main findings.

9.5.1 Graphene Synthesis in Bulk Quantities

For composite applications, it is critical to produce bulk quantities of graphene. Here we describe our approach for bulk production of graphene sheets from graphite. Natural graphite flakes with an average diameter of 48 μm were supplied from Huadong Graphite Factory (Pingdu, China). Concentrated sulfuric acid (95–98%), concentrated nitric acid (68%), and hydrochloric acid (36–38%) were purchased from Beijing Chemical Factory, China. Potassium chlorate (99.5%) was provided from Fuchen Chemical Reagents (Tianjin, China). Graphite oxide was prepared by oxidizing graphite in a solution of sulfuric acid, nitric acid, and potassium chlorate for 96 hours. The graphite oxide powder (200 mg) was placed in a quartz tube of 200 mm inner diameter and 1 m long, which was sealed at one end. The other end of the glass tube was closed by using a rubber stopper and an argon inlet was passed through the rubber stopper. After flushing the sample with argon for 10 minutes, the quartz tube was rapidly inserted into a tube furnace (Thermolyne 79300, Thermo Fisher Scientific, Inc., USA), which was preheated to 1050°C, and held in the furnace for about 30 seconds. This generates a thermal shock (heating rates of up to 2000°C/min) in the sample. Because of the temperature rise, the oxygen present in the graphite oxide is converted primarily to CO_2 gas; however, because of the rapid rate of temperature increase, the trapped CO_2 gas is unable to diffuse out of the graphite structure. This generates large pressure in the graphite structure, which overcomes the van der Waals forces that are responsible for binding the graphene sheets. The graphite oxide structure can, therefore, be exfoliated into individual graphene sheets. This method is broadly classified in the literature as the "thermal reduction" of graphite oxide as described in Section 9.4. This technique was pioneered by the Aksay group (McAllister et al. 2007; Schniepp et al. 2006) at Princeton. The process was found to give a high yield of graphene, around 50%–60% as confirmed by thermal gravimetric analysis (TGA).

Figure 9.5a shows a concept schematic of this process, while Figure 9.5b and c show transmission electron microscopy (TEM) images of graphene platelets manufactured in our laboratory by using this process. Note that although we are able to exfoliate the graphite, we do not get individual graphene sheets. We typically get platelets comprised of three to four individual graphene sheets (Figure 9.5c). The electron diffraction pattern in the inset of Figure 9.5c confirms the hexagonal-packed lattice structure of graphene. A very interesting feature of our graphene sheets is the highly wrinkled (rough) surface texture of the graphene platelets, as indicated in Figure 9.5b. Such wrinkles are indicative of defects in graphene, suggesting a high density of structural and topological defects such as vacancies, dangling bonds, and five- to seven-membered ring defects in our graphene samples. The high defect density appears to be an artifact of the oxidation

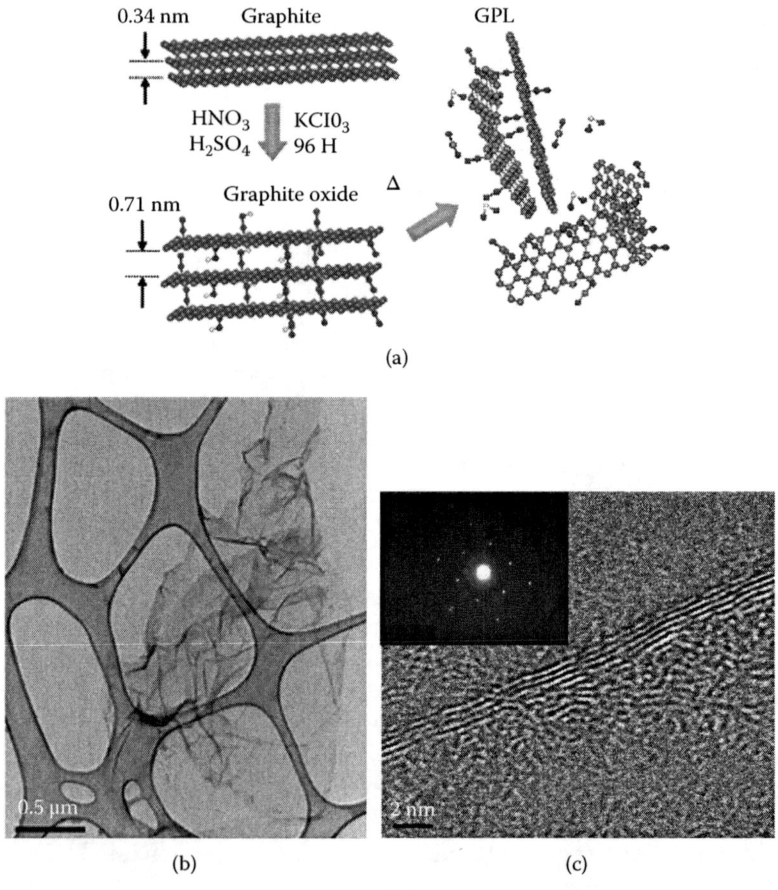

FIGURE 9.5
(a) Graphene synthesis process from bulk graphite. (b) Transmission electron microscopy (TEM) image of individual graphene platelet. (c) High-resolution TEM indicating that each platelet is comprised of ~3–4 individual graphene sheets. Inset shows electron diffraction pattern confirming the crystallographic signature of few-layered graphene. (Reprinted from Rafiee, M. A. et al., *Appl. Phys. Lett.*, 95, 223103, 2009b. With permission.)

of graphite and the thermal shock that was employed to exfoliate graphite oxide to graphene. Although these defects are undesirable for enhancing electrical/thermal properties, they play a critical role in improving the interfacial adhesion (interlocking) at the graphene–matrix interface, as will be discussed later. Raman spectroscopy of our graphene platelets confirmed the highly defective nature of the graphene as evidenced by a higher D band peak relative to the G band and significant broadening of both the G and D band peaks. We have also performed nitrogen cryosorption experiments and the corresponding specific surface area of our graphene samples was calculated using Brunauer–Emmett–Teller (BET) theory as ~800 m^2/g. This is ~3.5 times smaller than the specific surface area (~2630 m^2/g) of an idealized (single) graphene sheet, which confirms the high-resolution transmission electron microscopy (HRTEM) observation (Figure 9.5c) that the graphene platelets are composed of, on average, ~3–4 individual graphene layers. This is because while an ideal (single) graphene sheet provides a theoretical specific surface area of ~2630 m^2/g, this will reduce as the number of graphene layers in the platelet is increased, since nitrogen cannot penetrate into the ~0.34 nm spacing between the graphene planes.

From here onwards, the graphene platelets shown in Figure 9.5b and c will be referred to as GPL for the sake of brevity. For mixing GPL in thermosetting epoxy resins, we first dispersed GPL in acetone (ratio of 100 mL of acetone to 0.1 g of GPL) using an ultrasonic probe sonicator at high amplitude (Sonics Vibracell VC 750, Sonics and Materials, Inc., USA) for 1.5 hours in an ice bath. The epoxy (System 2000 Bisphenol-A based Epoxy, Fibreglast, Inc., USA) was added to the mixture and sonicated following the same procedure for another 1.5 hours. Next, the acetone is evaporated off by heating the mixture on a magnetic stir plate using teflon-coated magnetic bar for 3 hours at 70°C. The mixture is then placed in a vacuum chamber for 12 hours at 70°C to ensure that all of the acetone have been removed. After allowing the GPL/epoxy slurry to cool down to room temperature to prevent any premature curing, a low-viscosity curing agent (2120 Epoxy Hardener, Fibreglast, Inc., USA) is added and mixed using a high-speed shear mixer (ARE-250, Thinky, Japan) for 4 minutes at 2000 rpm. The mixture is again placed in a vacuum chamber to degas the epoxy for approximately 30 minutes. Finally, the mixture is poured into silicone molds and the nanocomposite is cured at room temperature and 90 psi pressure for 24 hours, followed by 4 hours of postcure at 90°C.

Figure 9.6a shows a schematic of graphene dispersion process, while Figure 9.6b shows a scanning electron micrograph (SEM) of the fracture surface of a typical graphene/epoxy sample. The SEM image in Figure 9.6b clearly indicates epoxy-coated GPL flakes on the fracture surface of the sample. The inset in Figure 9.6b depicts a typical low-magnification (zoom-out) optical image in transmission mode indicating a fairly uniform dispersion of graphene in the matrix. Figure 9.6c shows a typical high-magnification view of the graphene–epoxy interface after mechanically fracturing a sample under uniaxial tensile loading. Note that the integrity of the interface is maintained

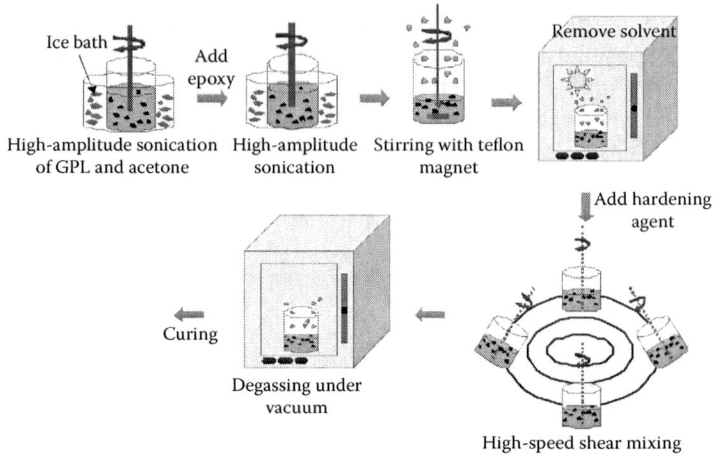

(a)

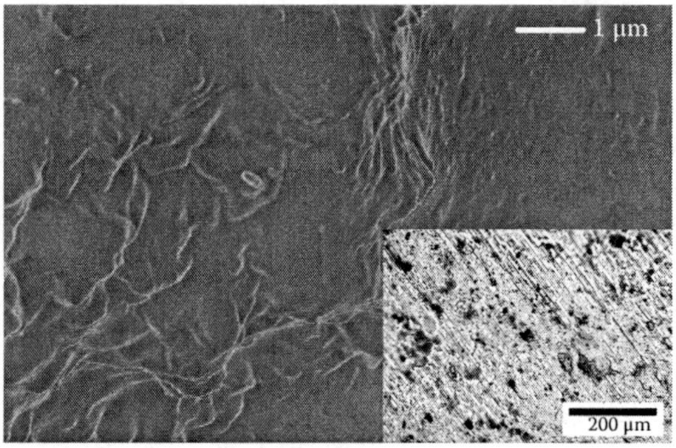

(b)

FIGURE 9.6
(a) Process for graphene dispersion in epoxy. (Reprinted from Rafiee, M. A. et al., *ACS Nano*, 3, 3884–3890, 2009b. With permission.) (b) Scanning electron microscopy (SEM) image showing graphene platelets dispersed in the epoxy matrix. Inset shows low-magnification optical image. (Reprinted from Rafiee, M. A. et al., *Small* 6, 179–183, 2010a. With permission.) (c) High-magnification SEM image of typical grapheme–epoxy interface. (Reprinted from Yavari, F. et al., *ACS Appl. Mater. Interfaces*, 2010. With permission.) (d) Raman study of characteristic G band shift of graphene filler in polystyrene (PS) and polydimethylsiloxane (PDMS) polymer. Results for single-walled carbon nanotubes at the same loading fraction in PDMS are also shown for comparison. (From Srivastava, I. et al., *Appl. Phys. Lett.*, 98, 2011. With permission.)

with no sign of debonding of the graphene suggesting a strong interface. This is further supported by differential scanning calorimetry measurements, which show over 10°C rise in the glass transition temperature of the neat epoxy resin with only ~0.1% weight of graphene additives. Raman analysis of nanocomposites with graphene additives also indicate large frequency shifts in the Raman modes of vibration of graphene (induced by tensile/

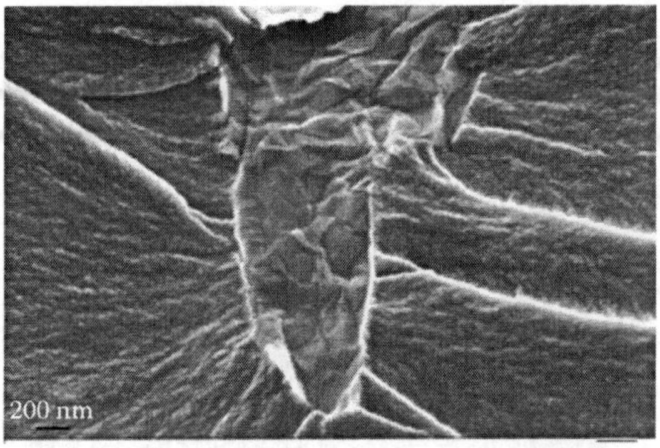

(c)

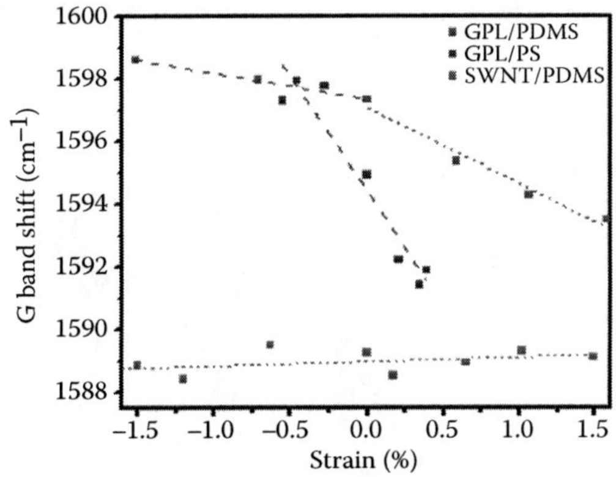

(d)

FIGURE 9.6
(*Continued*)

compressive loading) further confirming strong interfacial load transfer at the graphene–matrix interfaces. This is indicated in Figure 9.6d for a thermoplastic as well as an elastomeric polymer matrix (Srivastava et al. 2011).

9.5.2 Tensile Testing

We first prepared dog-bone-shaped coupons for tensile testing of the GPL/epoxy composites along with baseline epoxy samples for comparison. The results are summarized in Figure 9.7a through c. Figure 9.7a depicts the measured modulus for the pristine epoxy and GPL/epoxy nanocomposites. At 0.125% GPL weight fraction, the modulus of the nanocomposite is ~50% greater than the baseline epoxy. Above 0.125% weight fraction, the effectiveness of GPL diminishes, possibly due to the degradation in dispersion quality of GPL in the epoxy matrix. This highlights one of the disadvantages of graphene in that it is significantly more difficult to disperse at the higher weight fractions

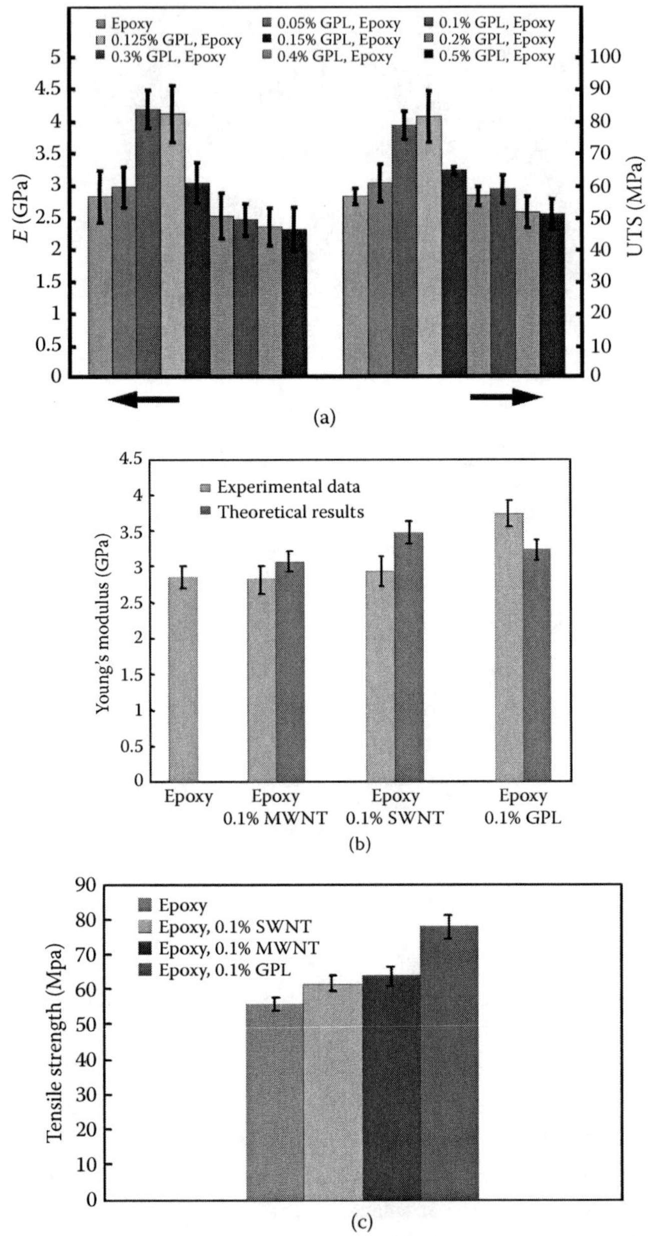

FIGURE 9.7
(a) Young's modulus (*E*) and ultimate tensile strength (UTS) for baseline epoxy and graphene/epoxy composites with various weight fractions of graphene. (Reprinted from Rafiee, M. A. et al., *Small*, 6, 179–183, 2010a. With permission.) (b) Comparison of modulus of graphene, SWNT, and MWNT at 0.1% weight fraction of nanofillers. Predictions of the well-known Halpin-Tsai theory are also plotted. (c) Comparison of ultimate tensile strengths at 0.1% weight fraction of nanofillers. (Reprinted from Rafiee, M. A. et al., *ACS Nano*, 3, 3884–3890, 2009b. With permission.)

compared to other nanofillers. This is probably an artifact of their enhanced specific surface area and two-dimensional geometry. Figure 9.7a also shows the ultimate tensile strength measurements; we observe ~45% enhancement in the ultimate tensile strength at ~0.125% weight fraction of GPL. These results are far superior to other competing nanofillers such as CNTs (SWNT and MWNT). We compared the strength and modulus of GPL, SWNT, and MWNT epoxy composites at the same filler loading fraction of ~0.1%. Clearly, the GPL far outperform (Figure 9.7b and c) both the SWNT and the MWNT in terms of their reinforcement capability. We believe this occurs due to the following reasons:

1. Enhanced interfacial contact area since both surfaces of GPL are in contact with the matrix, while in the case of CNTs only the outside nanotube surface contacts the matrix.
2. Rough/wrinkled surface topology of GPL (Figure 9.5b), which enhances mechanical interlocking at the nanofiller–matrix interface resulting in improved interfacial load transfer. This is in contrast to SWNT and MWNT fillers, which are atomically smooth tubes that can easily slip and slide within the matrix unless covalently bonded to the epoxy resin system. For our system, we do not see any evidence of graphene pullout and the interface appears to be very strong, as shown in Figure 9.6c.
3. Finally, for one-dimensional fiber additives such as CNTs, orientation of the fibers relative to the loading direction is critical. Misaligned fibers are ineffective at carrying load. By contrast, for two-dimensional additives such as graphene, lack of alignment (in-plane) with the loading axis does not substantially limit the reinforcement capability of the graphene platelets.

The experimental Young's modulus of the graphene/epoxy and nanotube/epoxy composites showed reasonable correlation with the predictions of the well-established Halpin-Tsai model for fiber-reinforced composite materials (Figure 9.7b). For details regarding how the model was adapted to CNTs, the reader is referred to Rafiee et al. (2009a). For graphene, we assumed that the GPL acts as an effective rectangular solid fiber with width (W), length (L), and thickness (t). To predict elastic properties, the well-known Halpin–Tsai equations were modified for the GPL/Epoxy nanocomposite as follows:

$$E_C = \frac{3}{8}\frac{1+\xi\eta_L V_{eff,fib}}{1-\eta_L V_{eff,fib}} \times E_M + \frac{5}{8}\frac{1+2\eta_W V_{eff,fib}}{1-\eta_W V_{eff,fib}} \times E_M \quad (9.1)$$

$$\eta_L = \frac{(E_{eff,fib}/E_M)-1}{(E_{eff,fib}/E_M)+\xi} \quad (9.2)$$

$$\eta_W = \frac{\left(E_{\text{eff,fib}}/E_M\right) - 1}{\left(E_{\text{eff,fib}}/E_M\right) + 2} \tag{9.3}$$

where E_C is the composite elastic modulus, $V_{\text{eff,fib}}$ is the volume fraction of the effective fiber, and $E_{\text{eff,fib}}$ and E_M are the effective fiber and matrix moduli, respectively. $E_{\text{eff,fib}}$ is assumed as the GPL modulus (~1.01 TPa). The parameter ξ depends on the geometry and boundary conditions of the effective fiber. For rectangular filaments, the parameter ξ can be expressed as

$$\xi = 2\left(\frac{(W + L)/2}{t}\right) \tag{9.4}$$

where L, W, and t indicate the length, width, and thickness of the GPL sheet. Hence, the nanocomposite elastic modulus can be defined in terms of the epoxy matrix properties and the GPL reinforcement. Assuming that $V_{\text{eff,fib}} = V_{\text{GPL}}$ and by substituting Equations 9.2, 9.3, 9.4 into 9.1 we have

$$E_C = \frac{3}{8}\frac{1+((W+L)/t)\left(\dfrac{(E_{\text{GPL}}/E_M)-1}{(E_{\text{GPL}}/E_M)+(W+L)/t}\right)V_{\text{GPL}}}{1-\left(\dfrac{(E_{\text{GPL}}/E_M)-1}{(E_{\text{GPL}}/E_M)+(W+L)/t}\right)V_{\text{GPL}}}\times E_M + \frac{5}{8}\frac{1+2\left(\dfrac{(E_{\text{eff,fib}}/E_M)-1}{(E_{\text{eff,fib}}/E_M)+2}\right)V_{\text{GPL}}}{1-\left(\dfrac{(E_{\text{eff,fib}}/E_M)-1}{(E_{\text{eff,fib}}/E_M)+2}\right)V_{\text{GPL}}}\times E_M \tag{9.5}$$

where E_C is the elastic modulus of the nanocomposite. The densities of the GPL and epoxy were estimated to convert weight fraction to volume fraction, required to predict the elastic properties. Additional details regarding the model implementation and validation with test data are available in Rafiee et al. (2009a).

Having measured the elastic modulus of nanocomposites, we prepared rectangular bars, which were buckled by the application of a compressive displacement to the specimen (at the rate of ~0.1 mm/min). The resulting typical load–displacement response was used to determine the critical buckling load. For the baseline epoxy specimen, the average buckling load (Figure 9.8) is ~1285 N, which is in reasonable agreement with the theoretical prediction (1316 N) based on classical Euler buckling mode. Figure 9.8 indicates that with the addition of ~0.1% weight MWNT to the epoxy matrix, the average buckling load is increased by ~6.2% to 1363 N. For 0.1% weight of SWNT, the critical buckling load increases to 1477 N (a 15% increase). At the same nanofiller loading fraction of 0.1%, the GPL far out-performs the SWNT and MWNT additives and shows ~52% increase in the critical buckling load to 1947 N. The superiority of GPL over SWNT/MWNT additives from a buckling standpoint can again be attributed to enhanced interfacial contact area and adhesion at the GPL–matrix interface. This improved adhesion arises from the wrinkled surface texture of GPL as well as the presence of epoxide and hydroxyl functional groups that can interact chemically with epoxy

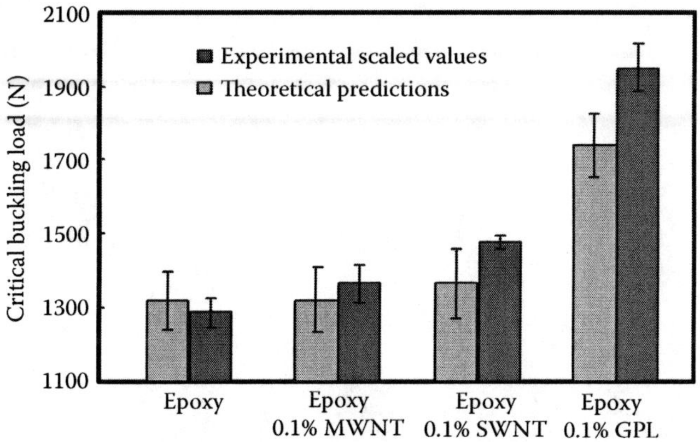

FIGURE 9.8
Critical buckling load of rectangular bars in mode I (clamped–clamped boundary conditions). Data are shown for baseline epoxy bar as well as for epoxy with 0.1% weight of MWNT, SWNT, and graphene (GPL) additives. Predictions of the classical Euler buckling theory are also shown in the figure for comparison. The GPL fillers significantly enhance the buckling stability of the composite and far out-perform carbon nanotube additives. (Reprinted from Rafiee, M. A. et al., *Appl. Phys. Lett.*, 95, 223103, 2009b. With permission.)

chains. This is in contrast to the smooth and defect-free (pristine) surface of CNTs, however, more in-depth modeling is needed to establish this.

9.5.3 Fracture Toughness

Crack opening tests on compact tension samples were performed to measure the mode I fracture toughness (k_{IC}) of the pure epoxy matrix and the GPL–epoxy nanocomposites at various weight fraction of GPL. The tests were conducted using a MTS-858 material testing system following ASTM standard D5045. An initial precrack was created in the compact tension samples by gently tapping a fresh razor blade over a molded starter notch. The radius at the tip of the precrack was similar for all the samples tested, which was confirmed by optical microscopy prior to testing. At each weight fraction of GPL additives, we tested between 4 and 6 different samples to check for reproducibility of the results. Figure 9.9a shows that the mode I fracture toughness (k_{IC}) of the baseline epoxy (without GPL) is ~1.03 MPa $m^{1/2}$, which correlates well with published literature for epoxy materials. Addition of GPL into the epoxy matrix causes a sharp increase in the nanocomposite's k_{IC} to ~1.75 MPa $m^{1/2}$ at 0.125% weight fraction of GPL, which corresponds to a ~65% increase in fracture toughness. For higher loading fractions, the enhancement in k_{IC} diminishes and finally begins to approach the pure epoxy value at a GPL weight fraction of ~0.5%. This might be a result of degradation in the dispersion of the GPL additives above a weight fraction of 0.125%. In spite of this, the maximum enhancement in k_{IC} (~65% at 0.125% weight fraction of GPL) is impressive. To achieve comparable increase (~62%) in k_{IC}, the required

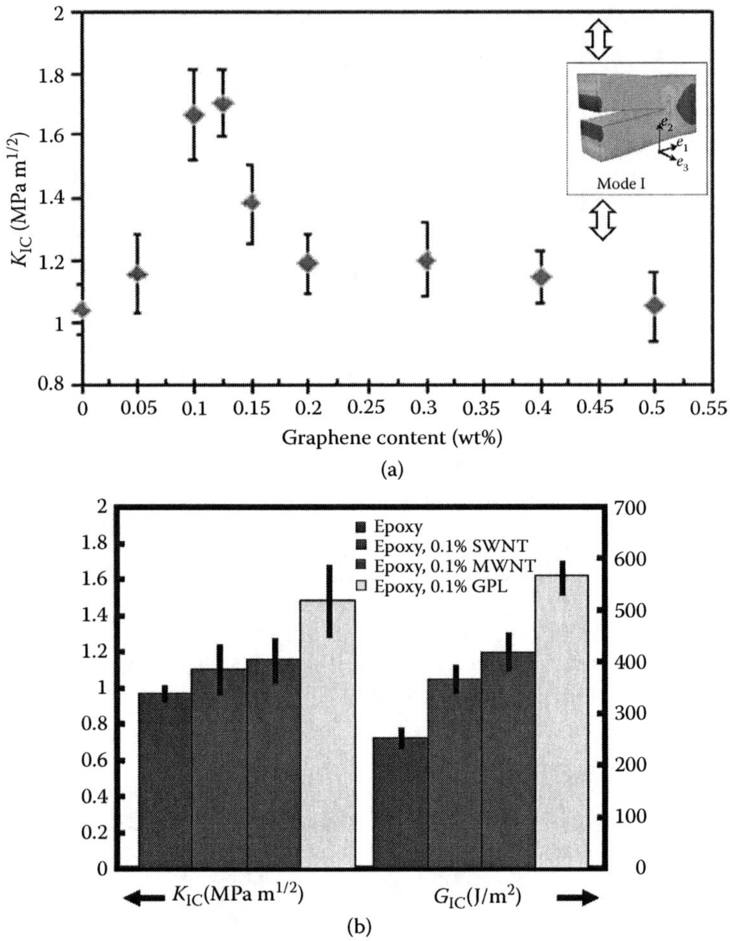

FIGURE 9.9
(a) Mode I fracture toughness (k_C) for baseline epoxy and epoxy/graphene composites at various weight fractions of graphene (GPL). Inset shows the compact tension sample geometry. (Reprinted from Rafiee, M. A. et al., *Small*, 6, 179–183, 2010a. With permission.) (b) Comparison of GPL with SWNT and MWNT at a filler weight fraction of 0.1%. The fracture toughness (k_{IC}) is shown on the left, while the fracture energy (G_{IC}) is shown on the right. GPL additives are found to significantly out-perform carbon nanotubes. (Reprinted from Rafiee, M. A. et al., *ACS Nano*, 3, 3884–3890, 2009b. With permission.)

weight fraction (~14.8%) of SiO_2 nanoparticles in epoxy is ~120-fold larger than GPL (Blackman et al. 2007). Similarly, to obtain a 65% increase in k_{IC}, the volume fractions of Al_2O_3 (~5%) and TiO_2 (~10%) nanoparticles in epoxy are ~30- to 60-fold larger than GPL (Wetzel et al. 2002). We also compared the performance of GPL with SWNT and MWNT at the same nanofiller loading fraction of 0.1% by weight. windicate that GPLs are highly effective in suppressing crack propagation in epoxies. However, there is a clear need to develop improved techniques for GPL dispersion in polymer matrices in order to realize their full potential. For example, premodifying GPL by the use of specially designed surfactants may enable enhanced dispersion at higher GPL loading fractions (>0.125%).

9.5.4 Fatigue Crack Propagation

Fatigue tests were conducted using an MTS-858 material testing system following ASTM standard E647. All tests were performed under load control at a constant load ratio R of 0.1 ($R = K_{min}/K_{max}$) and at a test frequency of 5 Hz. The crack length was measured using the compliance method and was confirmed by a high-resolution video monitoring system. Figure 9.10a shows the measured crack propagation rate (da/dN) versus the applied stress intensity factor amplitude (ΔK). A substantial lowering in the crack growth rate over the full range of stress intensity factor amplitudes can be observed for the GPL nanocomposite compared to the baseline epoxy. For example, at $\Delta K = 0.58$ MPa m$^{1/2}$, the da/dN for the nanocomposite (~1×10^{-4} mm/cycle) is ~100-fold lower than the baseline epoxy (~1×10^{-2} mm/cycle). Figure 9.10a also compares the fatigue suppression performance of GPL with SWNT and MWNT additives at the same nanofiller weight fraction of ~0.1%. The performance of the GPL is clearly superior

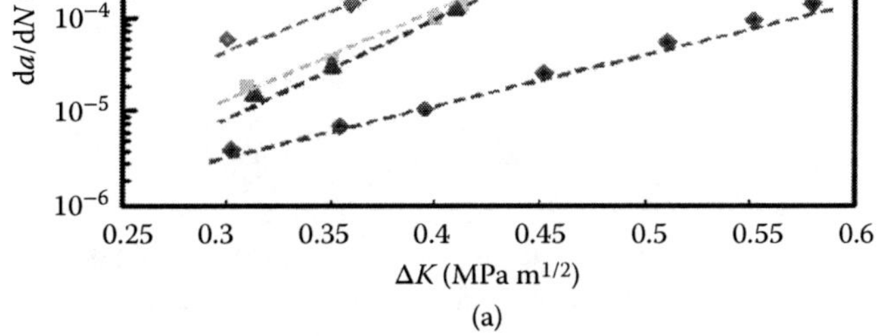

FIGURE 9.10
(a) Fatigue crack propagation testing; y-axis shows crack growth rate, while x-axis shows stress intensity factor amplitude. Data are shown for baseline epoxy as well as nanocomposites with 0.1% weight of GPL, SWNT, and MWNT. (Reprinted from Rafiee, M. A. et al., *ACS Nano*, 3, 3884–3890, 2009. With permission.) (b) Surface roughness changes to the fracture surface of compact tension samples (tested as part of the fracture toughness study in Figure 9.9) as a function of the graphene weight fraction. The surface roughness was measured on the whole fracture surface and not just the initial crack propagation region. (c) SEM image of the fracture surface showing direct evidence of crack deflection by the graphene additives. (Reprinted from Rafiee, M. A. et al., *Small*, 6, 179–183, 2010a. With permission.)

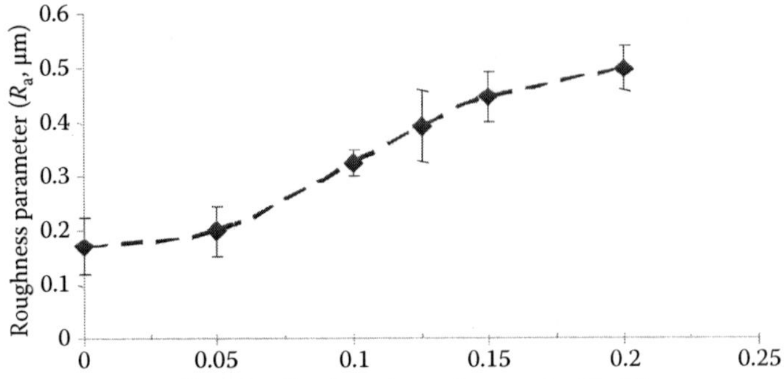

(b)

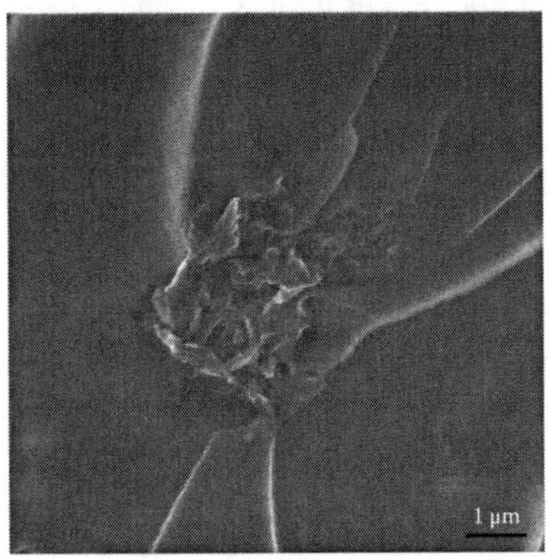

(c)

FIGURE 9.10
(Continued)

to the nanotubes, particularly, as the stress intensity factor amplitude (ΔK) is increased. For the case of the nanotubes, we observe a substantial degradation in the fatigue suppression with increasing ΔK. This is because in the case of nanotubes, the dominant toughening and fatigue suppression mechanism is crack bridging. Previous work by our group has shown that the fatigue crack is bridged by high aspect ratio nanotubes generating a fiber-bridging zone in the wake of the crack tip. As the crack advances, energy is dissipated by the frictional pullout of the bridging nanotubes from the epoxy matrix, which slows the crack propagation speed. However, this crack bridging effect loses effectiveness at high ΔK due to progressive shrinkage in the size of the fiber-bridging zone as ΔK is increased. The fact that such behavior is not observed in GPL/epoxy nanocomposites indicates that the toughening mechanism for GPL is different from nanotubes. Unlike nanotubes, frictional pullout of GPL from the matrix is less likely given the strong interfacial adhesion of GPL with epoxy matrices.

A possible toughening mechanism for GPL is crack deflection, which is the process by which an initial crack tilts and twists when it encounters a rigid inclusion. This generates an increase in the total fracture surface area resulting in greater energy absorption as compared to the unfilled polymer material. The tilting and twisting of the crack front as it is forced to move out of the initial propagation plane also forces the crack to grow locally under mixed-mode (tensile/in-plane shear and tensile/anti-plane shear) conditions. Crack propagation under mixed-mode conditions requires a higher driving force than in mode I (tension), which also results in higher fracture toughness of the material. Crack deflection processes may be highly effective for GPL because of its planar (two-dimensional) geometry and large aspect ratio. The fatigue suppression response for the GPL/epoxy nanocomposite shown in Figure 9.10a is consistent with crack deflection. To study the contribution of crack deflection to the observed toughening, we measured the average surface roughness of the fracture surface as a function of the GPL weight fraction. We observed a significant increase in the average surface roughness of the fracture surface with GPL content as indicated in Figure 9.10b. The average surface roughness (R_a) of the fracture surfaces was measured using a Dektak Surface Profiler (from VEECO); at least six separate measurements were performed for each sample for statistics. The data indicate a doubling in the average surface roughness with increase in GPL content from 0 to 0.125% weight. This roughening effect begins to saturate with further increase in the GPL content probably due to poor dispersion of GPL at higher weight fractions. The increasing roughness of the fracture surface with GPL content suggests that the crack deflection appears to play a significant role in the observed toughening. Further fractographic analysis of the fracture surface was performed using an environmental SEM (in collaboration with researchers at SUNY Albany) indicating significant out-of-plane deflection of the crack when it encounters a graphene inclusion (see Figure 9.10c). Scanning electron microscopy (SEM) analysis of the fracture surface did not reveal any direct evidence of crack pinning, crack bridging, or GPL pullout from the epoxy matrix. These results suggest that the crack deflection appears to be the dominant toughening mechanism in epoxy/GPL composites.

9.5.5 Damping Properties

The damping properties of the composite are also a function of the adhesion between the filler and the matrix. Strong filler–matrix interactions reduce energy dissipation via damping. An ~17% increase in the loss modulus has been reported for GPL–epoxy nanocomposites at a GPL content as low as 0.0125 vol.%, highlighting the significant, although not enormous, ability of GPL to dissipate energy under dynamic (vibratory) loading. The energy dissipation effects are attributed to the tendency of the weak interlayer forces in GPL that allow the layers to slide easily (Xiao et al. 2010). For CNTs, an increase in the damping ratio by 700% was observed for both 5 wt.% of SWNT and MWNT in epoxy. The damping occurs due to weak adherence of the CNT to the polymer matrix, resulting in debonding at high stress values, by a "stick-slip" mechanism (Rajoria and Jalili 2005). For SiO_2–epoxy nanocomposites, varying results have been reported, with an increase in the loss factor from 0.67 to 0.70 at 10 phr, followed by a decrease to 0.51 at higher filler content of 30 phr. The increase is attributed to a reduction in the epoxy cross-link density, while the increase is likely due to increased filler–matrix interaction at higher filler contents (Preghenella et al. 2005). The damping properties have been reported to increase for both organoclay and unmodified clay–epoxy nanocomposites (Mohan et al. 2006).

9.5.6 Thermomechanical Properties

Nanofillers have the most effect on the thermomechanical properties of the matrix at a temperature $T < T_g$ of the polymer, when sufficient interaction between the polymer chains and the filler occurs. At $T > T_g$, due to the enhanced mobility of epoxy chains, nanofiller interactions are diminished (Yao et al. 2006). An increase in T_g is expected when the filler–matrix adhesion is strong, whereas a decrease in T_g is expected for weakly adhering fillers due to decreased and enhanced mobility of polymer chains in the interfacial region, respectively (Kotsilkova and Pissis 2007). A T_g increase of ~8–10°C in GPL–epoxy nanocomposites has been measured by our group (unpublished work) at ~0.1% weight fraction of GPL in the epoxy matrix. This is an expected outcome of the strong filler–matrix adhesion and the conformational changes of the epoxy matrix at the GPL–epoxy interface (Rafiee et al. 2010a). A number of investigations have reported contradicting results of the effect on epoxy T_g with nanoclay addition (Kotsilkova and Pissis 2007; Lu and Nutt 2003). A diminution in T_g by 10°C has been reported for 6 phr loading of organoclay (Weiping et al. 2004), whereas an increment of 6°C for 10 phr loading has also been reported (Lu and Nutt 2003). Al_2O_3–epoxy nanocomposites do not exhibit any significant change in the T_g compared to the epoxy matrix (Zhao et al. 2008a). Fluorinated SWNTs are observed to decrease epoxy T_g by 30°C at 0.2 wt.%. For SiO_2 nanoparticles in epoxy, the T_g of the nanocomposite decreases with respect to neat epoxy. Lower cross-link density, plasticization, and modified stoichiometry of the epoxy are possible causes for the reduction of the T_g (Kotsilkova and Pissis 2007).

9.6 Hierarchical Graphene/Epoxy/Microfiber Composites

The results shown in Section 9.5 indicate that the addition of graphene fillers into a thermosetting epoxy resin generates very substantial improvement in mechanical properties including tensile strength, Young's modulus, fracture toughness, as well as resistance to fatigue. However, in a realistic structural application, this modified (improved) nanocomposite epoxy matrix material must be combined with conventional glass or carbon fibers to construct a three-phase composite structure. The three phases here comprise the nano-filler (e.g., graphene), the microfiber (e.g., glass or carbon fibers), and the epoxy resin system. It should be noted that for high-performance structural applications (e.g., in the aerospace or automotive industry), a conventional two-phase nanocomposite without continuous microfibers to carry the load appears unlikely to provide sufficient mechanical stiffness or strength to be competitive. Therefore, it is important to investigate three-phase nanocomposites involving the combination of a graphene-modified matrix together with conventional (continuous) microfiber reinforcement and compare its performance to the traditional microfiber-reinforced epoxy composites that are presently being used by industry.

The hierarchical composites developed by our group consist of traditional E-glass fibers paired with the epoxy matrix infused with a graphene network. Fabricating such nanocomposites is a detailed process. The first step is to disperse the GPL in the epoxy resin system as described previously. Then the epoxy is applied to the reinforcing glass fibers by being painted on layer by layer, rolling it with a deep slotted roller between each layer to help push out any entrapped air bubbles and to ensure proper wetting of the fibers. After all of the plies are laid up with the GPL-infused epoxy, a vacuum bag is placed over the system, and the sample is allowed to cure under vacuum for 24 hours at room temperature. During this process, excess epoxy is extracted out of the composite into an absorbent cloth material inside the vacuum bag. Lastly, the composite is taken out of the vacuum bag and placed in an oven at 90°C for high-temperature cure for 4 hours. The composite is then ready to be cut into sizes by either a wet tile saw with a diamond blade or an abrasive water jet cutter. Figure 9.11 shows a representation of the hierarchical composite with the three main phases: the E-glass fibers, the epoxy resin, and the graphene platelets interlinking the E-glass fibers.

The hierarchical glass-fiber/epoxy/GPL laminates were fatigue tested in a three-point bend test configuration as shown schematically in Figure 9.12a. In Figure 9.12b, we plot the bending stress (S) versus the number of cycles to failure (N) for the hierarchical composite for various GPL loading fractions ranging from 0.05% to 0.2%. The results indicate a dramatic increase in the number of cycles to failure across the entire range of applied stresses. Increasing the GPL weight fraction increased the fatigue life of the hierarchical system. The best results were observed at a GPL weight fraction of ~0.2% in the epoxy resin. For this case at a stress level of ~350 MPa, the fatigue life of the composite is enhanced by ~1000-fold as compared to the baseline composite without GPL reinforcement. Figure 9.12c compares the performance of GPL with CNT reinforcement at the same weight fraction of ~0.1%; as before, GPL greatly out-performs the nanotube additives. We also performed fatigue tests on the hierarchical composites in the uniaxial tensile mode (no compressive loading). For these cases, we do observe an

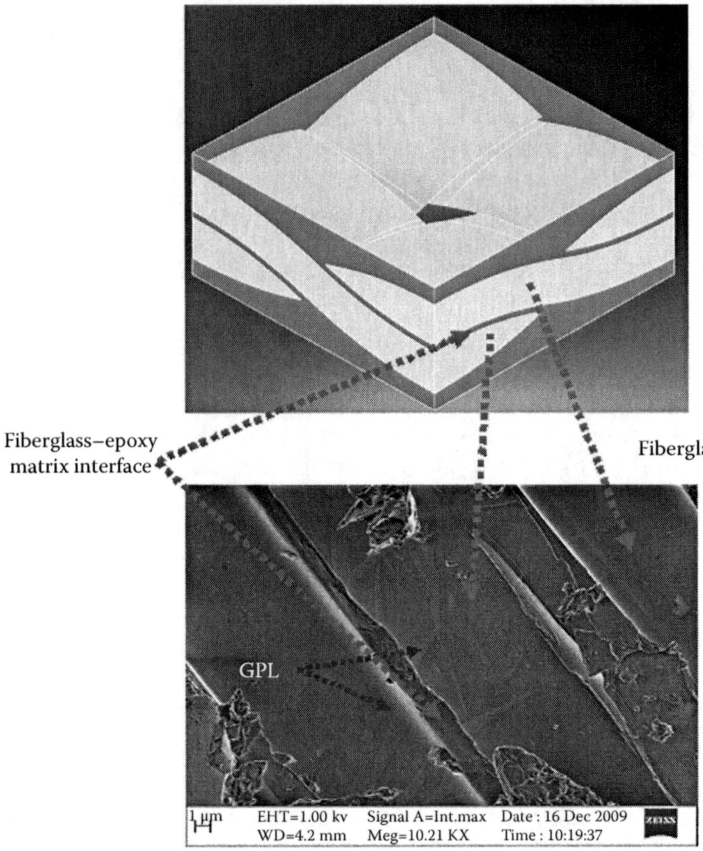

FIGURE 9.11
Schematic of the hierarchical, three-phase composite with conventional E-glass fibers, epoxy matrix binder, and graphene (GPL) platelets dispersed into the epoxy matrix. Also provided are SEM images of the composite showing glass fibers interlaced (or interconnected) by the GPL platelets that are mixed in the epoxy matrix. (Reprinted from Yavari, F. et al., *ACS Appl. Mater. Interfaces*, 2010. With permission.)

enhancement in fatigue life, but the improvements were relatively modest (~3–5 times increase) compared to the bending fatigue results shown in Figure 9.12.

We also measured static properties (such as flexural modulus and flexural strength) of the hierarchical and baseline composites. There was a ~20%–30% increase in flexural strength, while the flexural modulus remained unchanged for the hierarchical and baseline composites. The static tensile modulus and tensile strength of the hierarchical and baseline composites also showed no significant differences. Many studies (Rafiee et al. 2009a, 2010a) have shown that there is a huge difference between static and dynamic response in nanocomposites especially with respect to fatigue. For example, in Rafiee et al. (2009a), epoxy/graphene composites with ~0.1% weight of multilayer graphene

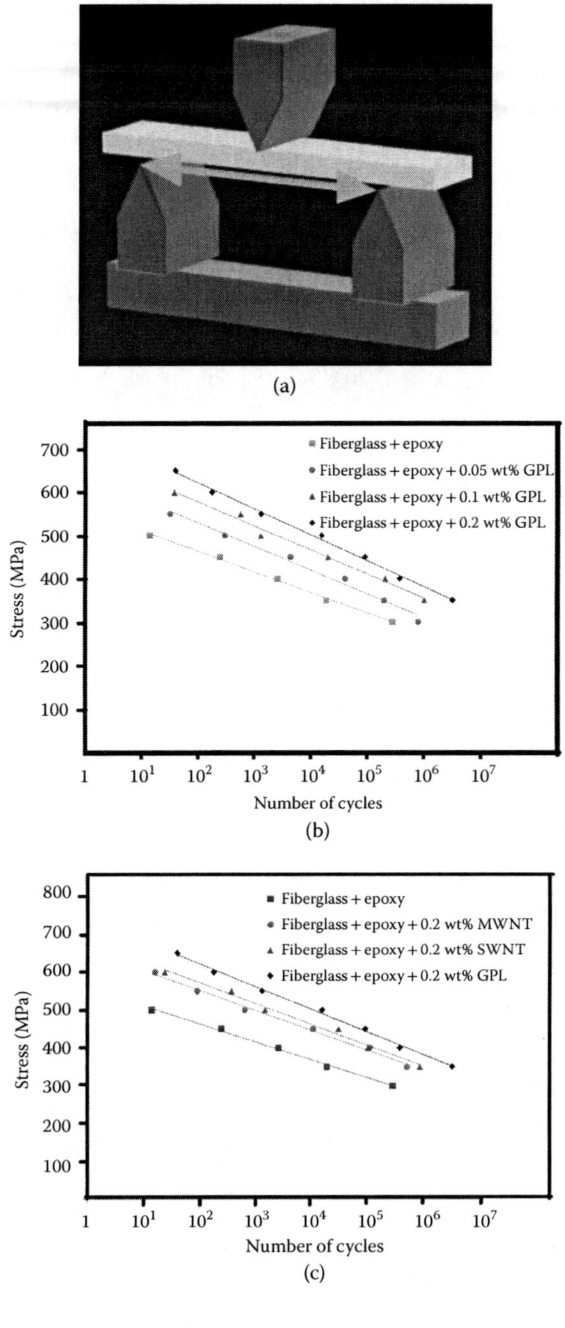

FIGURE 9.12
(a) Schematic showing the three-point bend test arrangement used for fatigue testing of glass-fiber/epoxy/GPL composites. (b) Stress (S) versus number of cycles to failure (N) for various weight fractions of graphene (GPL). (c) S–N curve comparing GPL with SWNT and MWNT at ~0.2% wt fraction. (Reprinted from Yavari, F. et al., *ACS Appl. Mater. Interfaces*, 2010. With permission.)

platelets enhanced the fracture toughness of the baseline epoxy by only ~53%, while the dynamic FCPR was suppressed by up to two orders of magnitude (~100-fold). In the hierarchical system, if the GPLs are suppressing the propagation rate of interlaminar fatigue cracks by ~100-fold, then it is conceivable that the total lifetime of the component (which involves the dynamic growth of such fatigue cracks and delaminations to critical dimensions) can also be enhanced by two to three orders of magnitude. Therefore, the dominant fatigue suppression mechanism for the hierarchical system appears to lie in the ability of the graphene network to slow the rate of interlaminar fracture in the composite, particularly, under compressive and shear loading. Under pure tensile fatigue, the microfiber–matrix interface is less important because unlike in compression, the microfibers are more effective than the matrix in carrying tensile loads. Consequently, maintaining the integrity of the microfiber–matrix interface while critical to boosting the fatigue life of the composite in bending/shear or direct compression is relatively less important under pure tension. Note that these are our preliminary findings and in-depth theoretical modeling in conjunction with fractography and ultrasound analysis is needed to establish this.

The nearly three orders of magnitude enhancement in the flexural bending fatigue life of conventional fiber-reinforced composites by the use of GPL reinforcement can translate into very significant weight/cost saving, enhanced reliability, and safety for a variety of structural components used in aeronautical, space, automotive, and marine applications.

9.7 Summary and Concluding Remarks

Graphene–epoxy nanocomposites have shown great potential for both strengthening and toughening epoxy systems, a combination rarely achieved. Graphene has proved superior to other nanofillers, including CNTs, in improving the mechanical properties of epoxy. Graphene's rough and wrinkled surface interlocks well with the surrounding matrix improving load transfer. Its large, planar surface area increases the contact with the matrix. The two-dimensional plate geometry of graphene provides greater tolerance to in-plane misalignment between the loading direction and the nanofiller orientation. These two-dimensional platelets have very high aspect ratio, which renders them highly effective at deflecting propagating cracks and absorbing greater amounts of energy. This enables very significant improvements in fracture and fatigue properties of graphene/epoxy composite materials. However, there are a number of fundamental scientific issues that need to be addressed in order to realize the full impact of this enabling materials technology. These include the following:

- Improved techniques for the dispersion of graphene fillers in epoxy resins at high weight fractions are necessary. Moreover, the stability of the dispersion over long time durations must be investigated. The use of surfactants, acid treatment, amine functionalization, or other novel routes (e.g., using attached ligands to improve their dispersion) must be investigated. Better exfoliation, if easily achieved, would lead to the full realization of graphene's potential as a reinforcement fiber for epoxy composites.
- Micromechanics and high-fidelity multiscale models are necessary to predict the reinforcement mechanisms of graphene fillers.
- Experiments and modeling studies are necessary to understand the effect of graphene geometry (sheet size, thickness, etc.) and the graphene sheet orientation (alignment) on mechanical performance.
- Characterization and modeling of other relevant mechanical properties such as impact strength, damage tolerance, damping response, high strain rate response, torsional/flexural response, and so on are needed to complete the picture.
- The use of graphene in epoxy resins can significantly increase their viscosity, which can be problematic for composite processing. Therefore, techniques that overcome this problem such as direct spray coating of microfiber plies with graphene need to be investigated.
- It is important to determine the effect of graphene on the thermal and electrical properties of epoxy composites. In particular, enhanced electrical conduction may afford opportunities for real-time strain sensing and in situ health monitoring in composites.
- Finally, it is important to study flammability properties of epoxy/graphene composites since flammability is one of the main obstacles to replacement of metal components by epoxy-type polymeric materials.

To conclude, the graphene content required to significantly boost the mechanical properties of epoxy systems has been as small as ~0.1 wt%, which is one to two orders of magnitude lower than other competing nanofillers such as CNTs, nanoclays, as well as silica/alumina/titania nanoparticles. This ultralow weight fraction requirement of graphene coupled with the inherently lower production costs associated with exfoliating graphite (a top-down process) as compared to the bottom-up processes that are used to produce other competing nanofillers suggests that graphene shows great promise as nanofiller for the next generation of advanced composite materials.

References

Abramoff, Bennet, and Josephine Covino. 1992. Transmittance and Mechanical Properties of Pmma-Fumed Silica Composites. *Journal of Applied Polymer Science* 46:1785–1791.

Ajayan, P. M., L. S. Schadler, and P. V. Braun. 2003. *Nanocomposite Science and Technology*. 1 ed. Weinheim, Germany: Wiley-VCH.

Argon, A. S., and R. E. Cohen. 2003. Toughenability of Polymers. *Polymer* 44:6013–6032.

Bartczak, Z., A. S. Argon, R. E. Cohen, and M. Weinberg. 1999. Toughness Mechanism in Semi-Crystalline Polymer Blends: II. High-Density Polyethylene Toughened with Calcium Carbonate Filler Particles. *Polymer* 40:2347–2365.

Battistella, M., M. Cascione, B. Fiedler, M. H. G. Wichmann, M. Quaresimin, and K. Schulte. 2008. Fracture Behaviour of Fumed Silica/Epoxy Nanocomposites. *Composites Part A: Applied Science and Manufacturing* 39:1851–1858.

Becker, O., R. Varley, and G. Simon. 2002. Morphology, Thermal Relaxations and Mechanical Properties of Layered Silicate Nanocomposites Based Upon High-Functionality Epoxy Resins. *Polymer* 43:4365–4373.

Beres, W., A. K. Koul, and R. Thamburaj. 1997. A Tapered Double-Cantilever-Beam Specimen Designed for Constant-K Testing at Elevated Temperatures. *Journal of Testing and Evaluation* 25:536–542.

Berger, Claire, Zhimin Song, Xuebin Li, Xiaosong Wu, Nate Brown, Cecile Naud, Didier Mayou, et al. 2006. Electronic Confinement and Coherence in Patterned Epitaxial Graphene. *Science* 312:1191–1196.

Blackman, B., A. Kinloch, J. Sohn Lee, A. Taylor, R. Agarwal, G. Schueneman, and S. Sprenger. 2007. The Fracture and Fatigue Behaviour of Nano-Modified Epoxy Polymers. *Journal of Materials Science* 42:7049–7051.

Chen, Wei, Hongbin Lu, and Steven R. Nutt. 2008. The Influence of Functionalized MWCNT Reinforcement on the Thermomechanical Properties and Morphology of Epoxy Nanocomposites. *Composites Science and Technology* 68:2535–2542.

Choi, Young-Kuk, Koh-ichi Sugimoto, Sung-Moo Song, Yasuo Gotoh, Yutaka Ohkoshi, and Morinobu Endo. 2005. Mechanical and Physical Properties of Epoxy Composites Reinforced by Vapor Grown Carbon Nanofibers. *Carbon* 43:2199–2208.

D 54045 Standard Test Methods for Plain Strain Fracture Toughness and Strain Energy Release Rate of Plastic Materials. 1999. *American Society for Testing Materials*. Philadelphia.

Dato, Albert, Velimir Radmilovic, Zonghoon Lee, Jonathan Phillips, and Michael Frenklach. 2008. Substrate-Free Gas-Phase Synthesis of Graphene Sheets. *Nano Letters* 8:2012–2016.

Delmonte, J., J. T. Hoggatt, and C. A. May. 1988. Fiber-Reinforced Epoxy Composites. In *Epoxy Resins: Chemistry and Technology*, edited by C. A. May. New York: Marcel Dekker.

Dunn, D. J. 2003. *Adhesives and Sealants—Technology, Applications and Markets*. Shrewsbury: Rapra Technology Ltd.

E 647 Standard Test Method for Measurement of Fatigue Crack Growth Rates. 1997. *American Society for Testing and Materials*, www.astm.org/Standards/E647.htm.

Eizenberg, M., and J. M. Blakely. 1979. Carbon Monolayer Phase Condensation on Ni(111). *Surface Science* 82:228–236.

Fisch, M. H., and B. A. Hegranes. 1988. Epoxides as Polymer Stabilizers and Plasticizers. In *Epoxy Resins: Chemistry and Technology*, edited by C. A. May. New York: Marcel Dekker.

Garg, Amar C., and Yiu-Wing Mai. 1988. Failure Mechanisms in Toughened Epoxy Resins—A Review. *Composites Science and Technology* 31:179–223.

Gojny, F. H., J. Nastalczyk, Z. Roslaniec, and K. Schulte. 2003. Surface Modified Multi-Walled Carbon Nanotubes in Cnt/Epoxy-Composites. *Chemical Physics Letters* 370:820–824.

Hadad, D. K. 1988. Physical and Chemical Characterization of Epoxy Resin. In *Epoxy Resins: Chemistry and Technology*, edited by C. A. May. New York: CRC Press.

Hale, Arturo, Christopher W. Macosko, and Harvey E. Bair. 1991. Glass Transition Temperature as a Function of Conversion in Thermosetting Polymers. *Macromolecules* 24:2610–2621.

Huang, J. Y., S. Chen, Z. Q. Wang, K. Kempa, Y. M. Wang, S. H. Jo, G. Chen, M. S. Dresselhaus, and Z. F. Ren. 2006. Superplastic Carbon Nanotubes. *Nature* 439:281–281.

Huang, Zheng-Ming, Y. Z. Zhang, M. Kotaki, and S. Ramakrishna. 2003. A Review on Polymer Nanofibers by Electrospinning and Their Applications in Nanocomposites. *Composites Science and Technology* 63:2223–2253.

Iijima, S. 1991. Helical Microtubules of Graphitic Carbon. *Nature* 354:56–58.

Kinloch, A. J., D. L. Maxwell, and R. J. Young. 1985. The Fracture of Hybrid-Particulate Composites. *Journal of Materials Science* 20:4169–4184.

Kinloch, A. J., R. D. Mohammed, A. C. Taylor, C. Eger, S. Sprenger, and D. Egan. 2005. The Effect of Silica Nano Particles and Rubber Particles on the Toughness of Multiphase Thermosetting Epoxy Polymers. *Journal of Materials Science* 40:5083–5086.

Kinloch, A. J., S. J. Shaw, and D. L. Hunston. 1983. Deformation and Fracture Behaviour of a Rubber-Toughened Epoxy: 2. Failure Criteria. *Polymer* 24:1355–1363.

Kotsilkova, R., and P. Pissis. 2007. *Thermoset Nanocomposites for Engineering Applications*. Shrewsbury: Smithers Rapra Technology.

Kunz, S. C., and P. W. R. Beaumont. 1981. Low-Temperature Behaviour of Epoxy-Rubber Particulate Composites. *Journal of Materials Science* 16:3141–3152.

Lan, Tie, and Thomas J. Pinnavaia. 1994. Clay-Reinforced Epoxy Nanocomposites. *Chemistry of Materials* 6:2216–2219.

Landry, C. J. T., B. K. Coltrain, M. R. Landry, J. J. Fitzgerald, and V. K. Long. 1993. Poly(Vinyl Acetate)/Silica Filled Materials: Material Properties of in situ Vs Fumed Silica Particles. *Macromolecules* 26:3702–3712.

Lange, F. F., and K. C. Radford. 1971. Fracture Energy of an Epoxy Composite System. *Journal of Materials Science* 6:1197–1203.

LeBaron, P. C., and T. J. Pinnavaia. 2001. Clay Nanolayer Reinforcement of a Silicone Elastomer. *Chemistry of Materials* 13:3760–3765.

Lee, S. M. 1988. Electrical and Electronic Applications. In *Epoxy Resins: Chemistry and Technology*, edited by C. A. May. New York: Marcel Dekker.

Levita, G., S. Depetris, A. Marchetti, and A. Lazzeri. 1991. Cross-Link Density and Fracture-Toughness of Epoxy-Resins. *Journal of Materials Science* 26:2348–2352.

Lu, Hongbin, and Steven Nutt. 2003. Restricted Relaxation in Polymer Nanocomposites Near the Glass Transition. *Macromolecules* 36:4010–4016.

Mai, Y. W., and J. G. Williams. 1979. Temperature and Environmental Effects on the Fatigue Fracture in Polystyrene. *Journal of Materials Science* 14:1933–1940.

May, C. A. 1988. *Epoxy Resins: Chemistry and Technology*. 2 ed. New York: Marcel Dekker.

McAllister, M. J., Li, J.-L, Adamson, D. H., Schniepp, H. C., Abdala, A. A., Liu, J., Herrera-Alonso, M., Milius, D. L., Car, R., Prud'homme, R. K., Askay, I. A. 2007. Single sheet functionalized grapheme by oxidation and thermal expansion of graphite. *Chem Mater.*, 19, 4396.

Miller, A. G. 2007. Keynote Address, the Boeing 787 Dreamliner. Paper read at 48th AIAA/ASME/ASCE/AHS/ASC Structures Structural Dynamics, and Materials Conference, Waikiki, HI.

Miyagawa, H., and L. T. Drzal. 2005. Effect of Oxygen Plasma Treatment on Mechanical Properties of Vapor Grown Carbon Fiber Nanocomposites. *Composites Part A: Applied Science and Manufacturing* 36:1440–1448.

Moalli, J. 2001. *Plastics Failure: Analysis and Prevention*. Norwich, NY : William Andrew Publishing.

Móczó, János, and Béla Pukánszky. 2008. Polymer Micro and Nanocomposites: Structure, Interactions, Properties. *Journal of Industrial and Engineering Chemistry* 14:535–563.

Mohan, T., M. Ramesh Kumar, and R. Velmurugan. 2006. Thermal, Mechanical and Vibration Characteristics of Epoxy-Clay Nanocomposites. *Journal of Materials Science* 41:5915–5925.

Mordkovich, V. Z. 2003. Carbon Nanofibers: A New Ultrahigh-Strength Material for Chemical Technology. *Theoretical Foundations of Chemical Engineering* 37:429–438.

Nakamura, Yoshinobu, Miho Yamaguchi, Akiko Kitayama, Masayoshi Okubo, and Tsunetaka Matsumoto. 1991. Effect of Particle Size on Fracture Toughness of Epoxy Resin Filled with Angular-Shaped Silica. *Polymer* 32:2221–2229.

Nielsen, L. E. 1974. *Mechanical Properties of Polymers and Composites*. New York: Marcel Dekker.

Nielsen, L. E. 1975. Fatigue Behavior of Some Filled Polymers. *Journal of Composite Materials* 9:149–156.

Novoselov, K. S., A. K. Geim, S. V. Morozov, D. Jiang, Y. Zhang, S. V. Dubonos, I. V. Grigorieva, and A. A. Firsov. 2004. Electric Field Effect in Atomically Thin Carbon Films. *Science* 306:666–669.

Obraztsov, Alexander N. 2009. Chemical Vapour Deposition: Making Graphene on a Large Scale. *Nat Nano* 4:212–213.

Paris, P. C., M. P. Gomez, and W. P Anderson. 1961. A Rational Analytic Theory of Fatigue. *The Trend in Engineering* 13:6.

Park, Sungjin, and Rodney S. Ruoff. 2009. Chemical Methods for the Production of Graphenes. *Nature Nanotechnology* 4:217–224.

Peigney, A., C. Laurent, E. Flahaut, R. R. Bacsa, and A. Rousset. 2001. Specific Surface Area of Carbon Nanotubes and Bundles of Carbon Nanotubes. *Carbon* 39:507–514.

Perez, N. 2004. *Fracture Mechanics*. New York: Springer.

Pilato, Louis A., and Michael J. Michno. 1994. *Advanced Composite Materials*. 1st ed. Berlin: Springer.

Pötschke, P., A. R. Bhattacharyya, and A. Janke. 2004. Melt Mixing of Polycarbonate with Multiwalled Carbon Nanotubes: Microscopic Studies on the State of Dispersion. *European Polymer Journal* 40:137–148.

Preghenella, Michele, Alessandro Pegoretti, and Claudio Migliaresi. 2005. Thermo-Mechanical Characterization of Fumed Silica-Epoxy Nanocomposites. *Polymer* 46:12065–12072.

Rafiee, M. A., J. Rafiee, I. Srivastava, Z. Wang, H. H. Song, Z. Z. Yu, and N. Koratkar. 2010a. Fracture and Fatigue in Graphene Nanocomposites. *Small* 6:179–183.

Rafiee, M. A., J. Rafiee, Z. Wang, H. H. Song, Z. Z. Yu, and N. Koratkar. 2009a. Enhanced Mechanical Properties of Nanocomposites at Low Graphene Content. *ACS Nano* 3:3884–3890

Rafiee, M. A., J. Rafiee, Z. Z. Yu, and N. Koratkar. 2009b. Buckling Resistant Graphene Nanocomposites. *Applied Physics Letters* 95:223103.

Rafiee, M. A., J. Rafiee, Z. Z. Yu, and N. Koratkar. 2010b. Super-Hydrophobic to Superhydrophilic Wetting Control in Graphene Films. *Advanced Materials* 22:2151–2154.

Rajoria, Himanshu, and Nader Jalili. 2005. Passive Vibration Damping Enhancement Using Carbon Nanotube-Epoxy Reinforced Composites. *Composites Science and Technology* 65:2079–2093.

Ramanathan, T., A. A. Abdala, S. Stankovich, D. A. Dikin, M. Herrera Alonso, R. D. Piner, D. H. Adamson, et al. 2008. Functionalized Graphene Sheets for Polymer Nanocomposites. *Nature Nanotechnology* 3:327–331.

Schab-balcerzak, E. 2006. Modifications of Epoxy Resins' Properties. In *Frontal Polymer Research*, edited by R. K. Bregg. New York: Nova Science Publishers.

Schlapbach, Louis, and Andreas Zuttel. 2001. Hydrogen-Storage Materials for Mobile Applications. *Nature* 414:353–358.

Schniepp, H. C., Li, J.-L., McAllister, M. J., Sai, H., Herrera-Alonso, M., Adamson, D. H., Prud'homme, R. K., Car, R., Saville, D. A., Askay, I. A. 2006. Functionalized single grapheme sheets derived from splitting graphite oxide. *J Phys Chem B*, 110, 8535.

Shaffer, M. S. P., X. Fan, and A. H. Windle. 1998. Dispersion and Packing of Carbon Nanotubes. *Carbon* 36:1603–1612.

Shimazaki, Yuzuru, Yasuo Miyazaki, Yoshitaka Takezawa, Masaya Nogi, Kentaro Abe, Shinsuke Ifuku, and Hiroyuki Yano. 2007. Excellent Thermal Conductivity of Transparent Cellulose Nanofiber/Epoxy Resin Nanocomposites. *Biomacromolecules* 8:2976–2978.

Sircar, A. K., and R. P. Chartoff. 1994. Measurement of the Glass Transition Temperature of Elastomer System. In *Assignment of the Glass Transition*, edited by R. J. Seyler. Philadelphia: ASTM Special Technical Publication 1249.

Socie, D. 1993. Critical Plane Approaches for Multiaxial Fatigue Damage Assessment. In *Advances in Multiaxial Fatigue*, edited by D. L. MacDowell. Ann Arbor: ASTM.

Spanoudakis, J., and R. J. Young. 1984a. Crack Propagation in a Glass Particle-Filled Epoxy Resin—Part 1 Effect of Particle Volume Fraction and Size. *Journal of Materials Science* 19:473–486.

Spanoudakis, J., and R. J. Young. 1984b. Crack Propagation in a Glass Particle-Filled Epoxy Resin—Part 2 Effect of Particle-Matrix Adhesion. *Journal of Materials Science* 19:487–496.

Srivastava, I., and N. Koratkar. 2010. Fatigue and Fracture Toughness of Epoxy Nanocomposites. *JOM* 62:50–57.

Srivastava, I., R. Mehta, L. Schadler, and N. Koratkar. 2011. Raman Study of Interfacial Load Transfer in Graphene Nanocomposites. *Applied Physics Letters* 98, 063102.

Stankovich, Sasha, Dmitriy A. Dikin, Richard D. Piner, Kevin A. Kohlhaas, Alfred Kleinhammes, Yuanyuan Jia, Yue Wu, et al. 2007. Synthesis of Graphene-Based Nanosheets Via Chemical Reduction of Exfoliated Graphite Oxide. *Carbon* 45:1558–1565.

Stephens, R. I., and H. O. Fuchs. 2001. *Metal Fatigue in Engineering*. 2 ed. New York: Wiley.

Sumita, Masao, Tadao Shizuma, Keizo Miyasaka, and Kinzo Ishikawa. 1983. Effect of Reducible Properties of Temperature, Rate of Strain, and Filler Content on the Tensile Yield Stress of Nylon 6 Composites Filled with Ultrafine Particles. *Journal of Macromolecular Science, Part B: Physics* 22:601–618.

Suresh, S. 1998. *Fatigue of Materials*. 2 ed. Cambridge: Cambridge University Press.

Sutter, Peter W., Jan-Ingo Flege, and Eli A. Sutter. 2008. Epitaxial Graphene on Ruthenium. *Nature Materials* 7:406–411.

Teyssier, O. M., S. S. Valdes, and L. F. Ramos-de Valle. 2006. Effect of Carbon Nanofiber Functionalization on the Dispersion and Physical and Mechanical Properties of Polystyrene Nanocomposites. *Macromolecular Materials and Engineering* 291:1547–1555.

Wei Hong, Zhong, Li Jiang, M. Lukehart Charles, and R. Xu Luoyu. 2005. Graphitic Carbon Nanofiber (GCNF)/Polymer Materials. Ii. Gcnf/Epoxy Monoliths Using Reactive Oxydianiline Linker Molecules and the Effect of Nanofiber Reinforcement on Curing Conditions. *Polymer Composites* 26:128–135.

Weiping, Liu, V. Hoa Suong, and Pugh Martin. 2004. Morphology and Performance of Epoxy Nanocomposites Modified with Organoclay and Rubber. *Polymer Engineering and Science* 44:1178–1186.

Wetzel, B., F. Haupert, K. Friedrich, M. Q. Zhang, and M. Z. Rong. 2002. Impact and Wear Resistance of Polymer Nanocomposites at Low Filler Content. *Polymer Engineering and Science* 42:1919–1927.

Wetzel, Bernd, Frank Haupert, and Ming Qiu Zhang. 2003. Epoxy Nanocomposites with High Mechanical and Tribological Performance. *Composites Science and Technology* 63:2055–2067.

Xiao, X. C., T. Xie, and Y. T. Cheng. 2010. Self-Healable Graphene Polymer Composites. *Journal of Materials Chemistry* 20:3508–3514.

Yao, X. F., H. Y. Yeh, D. Zhou, and Y. H. Zhang. 2006. The Structural Characterization and Properties of Sio2-Epoxy Nanocomposites. *Journal of Composite Materials* 40:371–381.

Yavari, F., M. A. Rafiee, J. Rafiee, Z.-Z. Yu, and N. Koratkar. 2010. Dramatic Increase in Fatigue Life in Hierarchical Graphene Composites. *ACS Applied Materials and Interfaces* 2: 2738–2743.

Ying, Z., J. H. Du, S. Bai, F. Li, C. Liu, and H. M. Cheng. 2002. Mechanical Properties of Surfactant-Coating Carbon Nanofiber/Epoxy Composite. *International Journal of Nanoscience* 1(5–6): 425–430.

Young, R. J., and P. W. R. Beaumont. 1975. Failure of Brittle Polymers by Slow Crack Growth - Part 2 Failure Processes in a Silica Particle-Filled Epoxy Resin Composite. *Journal of Materials Science* 10:1343–1350.

Zhang, W., I. Srivastava, Y. F. Zhu, C. R. Picu, and N. A. Koratkar. 2009. Heterogeneity in Epoxy Nanocomposites Initiates Crazing: Significant Improvements in Fatigue Resistance and Toughening. *Small* 5:1403–1407.

Zhao, S., L. S. Schadler, R Duncan, H Hillborg, and T Auletta. 2008a. Mechanisms Leading to Improved Mechanical Performance in Nanoscale Alumina Filled Epoxy. *Composites Science and Technology* 68:2965–2975.

Zhou, Y., F. Pervin, and S. Jeelani. 2007. Effect Vapor Grown Carbon Nanofiber on Thermal and Mechanical Properties of Epoxy. *Journal of Materials Science* 42:7544–7553.

10

Nanoparticles and Polymer Nanocomposites

Musa R. Kamal* and Jorge Uribe-Calderon

CONTENTS

10.1 Introduction

Polymers are used in innumerable applications and fields. However, it is common to modify the properties or upgrade the performance of polymer systems to satisfy specific processing or application requirements. Thus, additives are commonly used to modify polymer properties or upgrade the performance of polymer systems. The term "additive" is commonly used to identify a minor component of the polymer formulation, which modifies the properties of the polymer (e.g., antioxidants, plasticizers, flame retardants, processing aids, other polymers, colorants, UV absorbers, and extenders). On the other hand, a "filler" is a solid material or additive added to a polymer system to modify its mechanical, optical, electrical, thermal, barrier, or

* Correspondence to: Musa R. Kamal (musa.kamal@mcgill.ca)

flammability properties of the bulk material (Work et al. 2004). Usually, fillers are used in larger proportions than additives. In some cases, the low cost of some additives is a major consideration for incorporating them in the formulation. Also, fillers' contribution to the properties of the system is mainly due to physical factors, while chemical factors play an important part in the role of additives. Fillers intended to improve the mechanical properties are known as "reinforcement" or "reinforcing fillers."

The class of polymers that contain a significant amount of solid additives is usually referred to as "filled polymers." The additives could range from pigments, such as titanium dioxide, to glass fibers for reinforcement of the mechanical behavior. In this sense, composites are a subgroup of filled polymers, where the filler imparts significant improvement to some of the mechanical characteristics of filled polymer system. While mechanical improvement is the main objective of fiber-filled composites, this does not preclude the possibility that other properties of the system are not modified or upgraded. Moreover, although fiber-reinforced systems are the most commonly used polymer composites, other forms of particles, such as platelets or spherical particles, could be used in making composites.

Depending on the size of the solid additives, the composites can be divided into microcomposites (filler size in microns) and nanocomposites (filler size less than 100 nm). In fact, the term "molecular composite" is used to refer to systems where molecular entities, such as oriented polymer chains or liquid crystalline polymers, serve as the reinforcing agent. In the case of the microcomposites, the common fillers are granules (talc, calcium carbonate, etc.) and fibers (glass fibers). Conventional composites or microcomposites generally require a large amount of additives to affect important changes in the properties of the system, which presents both cost implications and technical challenges. Also, while such composites contribute to enhancement of the modulus and strength of the system, they could contribute to causing embrittlement and loss of toughness in the system, in addition to introducing processing complexities (Tjong 2006). On the other hand, nanocomposites are formed by the dispersion of a nanofiller into the polymer matrix.

Nanofillers are a class of new generation of fillers, exhibiting at least one characteristic length of the order of nanometer(s). The shape of nanofillers varies from isotropic to highly anisotropic. The large interfacial area between a polymer and the nanofiller, when well dispersed, differentiates polymer nanocomposites from traditional composites (microcomposites). The nanoscale-confined matrix polymer, nanoscale inorganic and organic filler, and nanoscale arrangements of these constituents contribute to the improvement of properties. Nanocomposite materials could be moldable to near-final shape. Nanocomposites exhibit four major advantages over conventional composites: (1) lighter weight (i.e., low filler loading), (2) low cost (i.e., lesser amount of filler required), (3) improved properties compared to conventional filler-based composites, and (4) a combination of specific properties (Bhowmick 2008).

The high specific interfacial area of the nanofillers offers the possibility for polymer–filler interactions at the molecular level, yielding substantial modification of system properties. The potential property improvements could be reflected in mechanical, thermal, barrier, electrical, and flammability behavior, with only small changes in density and optical properties.

The nanofillers used in the preparation of polymer nanocomposites can be classified following several criterions. One criterion is the shape of the nanofiller. So, nanofillers can be distinguished as (1) nanospheres, in which the three dimensions of the filler are in the nanoscale regime (as well as known as zero dimension, equiaxed [isodimensional] nanoparticles or nanogranules or nanocrystals); (2) nanotubes, nanowires, or nanofibers, having two dimensions in the nanometer range; and (3) nano-platelets, where at least the thickness of the filler is in the nanometer scale (Kumar et al. 2009).

One advantage of using nanofillers is the possibility to achieve significant improvements in properties by incorporating small amounts of nanofiller (≤1%–5% mass fraction). However, the performance of nanocomposites depends on the degree of dispersion and the quality of adhesion among the constituents, which depend in part on chemical factors of components and the processing factors.

There are four approaches for the preparation of polymer nanocomposites: (1) in situ polymerization of monomers in the presence of nanoparticles; (2) in situ formation of the nanoparticles and in situ monomer polymerization in the same medium; this method is suitable for preparing metal nanoparticles from metal precursors; (3) dissolution of the polymer in a solvent followed by mixing with nanoparticles in dispersion; and (4) melt intercalation, in which the particles are incorporated in polymer in the melt or rubbery state with or without shear (Bhattacharya et al. 2008; Matras-Postolek and Bogdal 2010). Each preparation method is associated with different challenges, depending on the nature of the components along with the technical limitations of each method.

Regardless of the preparation method, there are two important aims for the compounding process: (1) the deagglomeration of nanoparticles and (2) the uniform distribution of the nanoparticles in the polymer matrix. The former process is directly related to the interparticles forces. For this purpose, the applied processing forces should overcome the cohesive forces between filler particles.

Particle–particle and particle–polymer surface interaction forces are very important factors in determining both the ease of formation of nanocomposites and the performance of the final nanocomposite material. The interfacial area between the nanoparticles and the matrix should be maximized to maximize the benefits of the particle–matrix interactions. This means that agglomerates should be broken down to the smallest unit, which is the equivalent of exfoliation. Alternatively, intercalation provides an opportunity for a significant interfacial area. In either case, it is important that good adhesion exists at the interface between the phases. As it is not always possible to have sufficient compatibility between polymers and nanoparticles, it is common to use surfactants and compatibilizers that enhance these interactions.

Thus, surfactants and compatibilizers may be considered as processing aids that help the matrix and nanoparticles to generate nanocomposites with the desired enhanced performance characteristics (Bhattacharya et al. 2008; Kamal et al. 2009). Compatibilizer, which is a polymer or copolymer, is added to the nanocomposite formulation to modify the interfacial characteristics of the mixture and stabilizes its morphology. Compatibilizers can stabilize morphologies over distances of the order of micrometers (Work et al. 2004).

Extensive experimental work has been reported regarding the effect of processing conditions on the dispersion and exfoliation of clay nanoparticles. It has been reported that a combination of shear and elongation flow is desirable to achieve good dispersion. Additionally, longer residence time or mixing time could contribute to enhancing exfoliation and dispersion of particles in melt processing. Higher polymer molecular weight, which raises the melt viscosity and thus processing stresses, contributes to the breakup and delamination of clay agglomerates (Fornes et al. 2001; Nassar et al. 2005; Tanoue et al. 2006).

Three different phases can be distinguished in nanocomposites: matrix, nanoparticles, and the interfacial region. The interfacial region is responsible for the transfer of some of the filler properties to the matrix. The material adjacent to the interface exhibits properties different from those of the matrix, because of the proximity to the filler surface. When nanoparticles are well dispersed in the polymer matrix, several physical properties of the matrix can be modified (e.g., dielectric constant; electrical, ionic, and thermal conductivity; refractive index; UV resistance; rate of photooxidation; wear resistance; mechanical and barrier properties; and biodegradability) (Kumar et al. 2009; Supova et al. 2011).

10.2 Polymer Nanocomposites Based on Different Nanofillers

Another way to classify the nanoparticles is based on the chemical composition of the materials. In this sense, nanoparticles can be classified as element-based nanoparticles, nanoparticles from metal oxides and other derivatives, carbon nanotubes (CNTs), and natural source nanoparticles. Some examples of commonly used nanoparticles will be discussed, with special emphasis on CNTs and graphene nanocomposites.

10.2.1 Metallic Nanoparticles

Due to the development of methods to control the size, morphology, and aggregation of inorganic nanoparticles, the use of metallic nanoparticles is of particular interest for the production of nanocomposites for optical applications (Ciardelli et al. 2008). There is a wide variety of elemental nanoparticles available (i.e., Ag, Al, Au, C [activated carbon], C [diamond], Co, Cr, Cu, Fe, Mo, Ni, Si, Ti, W, Zn; nanoamor.com, Kumar et al. 2009). Metal nanoparticles have been used to prepare nanocomposites using different polymeric matrices. The electrical properties of the metal nanoparticles make their nanocomposites good candidates to produce materials for electrical applications, as in the design of materials for electromagnetic interference (EMI) shielding, energy storage, and charge dissipation. Colloids of polymer–metal nanoparticles exhibit rheological, optical, magnetic, or electrical properties useful in therapeutic and diagnostic applications in the biomedical field (Thomas et al. 2007). The use of compatibilizers (surfactants or copolymers) or surface modification of the dispersed phase is usually required to obtain functional nanodispersions (Ciardelli et al. 2008).

Silver nanomaterials have been of interest in a wide range of applications, including disinfecting of medical devices and home appliances and water treatment (Tolaymat et al. 2010). A recent example of Ag and Pt nanoparticles incorporation in thermoplastic matrices has been reported. Noble metal precursors ($AgNO_3$ or H_2PtCl_6) have been incorporated in poly(ethylene terephthalate) (PET), poly(vinyl chloride) (PVC), and polypropylene (PP), after in situ reduction of the precursors yielded metal–polymer nanocomposites (Volynskii et al. 2010). Silver precursors were reduced during the polymerization of several monomers to produce polymer nanocomposite films (polymethyl methacrylate [PMMA], polystyrene [PS], and poly-butyl methacrylate [PBMA]) using specific amphiphilic block copolymers as stabilizers. Silver nanoparticles enhance the mechanical properties and thermal stability of the nanocomposite significantly (Chatterjee et al. 2009). The use of silver nanowires as nanofiller is also reported in the literature (White et al. 2010). Figure 10.1 shows transmission electron microscopy (TEM) images of Ag–PMMA nanocomposites.

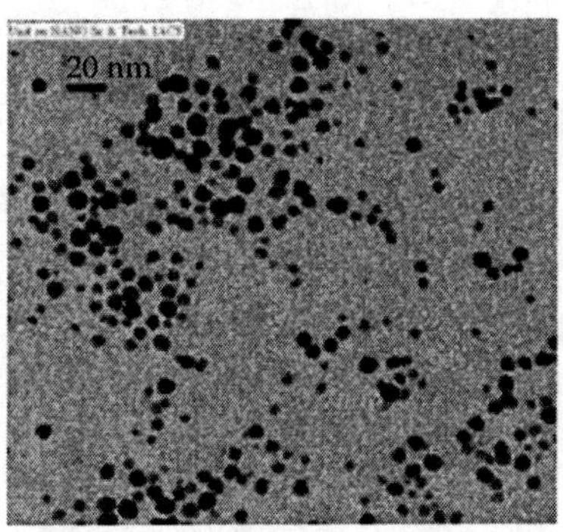

FIGURE 10.1
TEM images of section (70 nm diameter) of PBMA-b-PMMA-b-PBMA-Ag–PMMA nanocomposite film containing 20 wt% PBMA-b-PMMA-b-PBMA-Ag. (From Chatterjee, U. et al., *Polym. Compos.*, 30, 827–834, 2009. With permission.)

Gold (Au) nanoparticles have been used in the preparation of polymer nanocomposites. Thus, PP nanocomposites containing uniformly dispersed nanogold particles have been synthesized through metallocene catalysis with a selective chain transfer (Yeh et al. 2010). An alternative approach is the use of copolymers that are used as compatibilizers to enhance the dispersion of gold nanoparticles. Figure 10.2 shows a dispersion of Au particles in PS, using polyethylene glycol methacrylate (PEGMA) as a compatibilizer. The chemical functionalization of gold surface yielded has been used to improve dispersion of gold nanorods in the polymer matrix. For example, colloidal Au nanorods, functionalized with amino groups, were incorporated in a PMMA copolymer for use in plasmatic devices (Panniello et al. 2010). Gold (Au) nanorods were functionalized with poly(ethylene glycol) (PEG) to produce thin films based on PMMA and poly(ethylene oxide) for use in coatings and sensing devices (Hore and Composto 2010).

The high surface energy of inorganic nanoparticles makes them unstable, reactive, and self-aggregate. The aggregation of nanoparticles into regular aggregates (at certain compositions, controlled size, and shape) provides opportunities to design nanoparticle assemblies for advanced applications. Figure 10.3 shows the self-assembly of Au nanorods (Rozenberg and Tenne 2008).

Other surface modification techniques applicable to metal nanoparticles (Au or Ag) involve the use of organic acids or thiols. Another approach is to functionalize the surface by attaching reactive groups onto the metal surface, which would serve as anchors for polymers obtained by in situ radical polymerization techniques (free radical polymerization, or atom transfer radical polymerization [ATRP]) (Rozenberg and Tenne 2008).

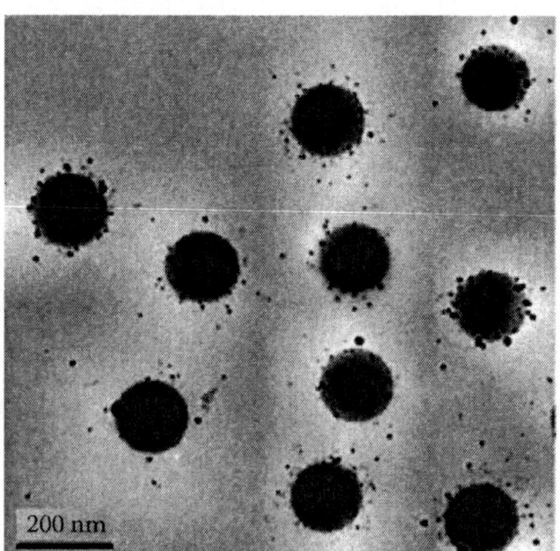

FIGURE 10.2
Cryo-TEM images for PS-PEGMA-Ag composite particles. (From Thomas, V. et al., *J. Macromol. Sci., Part A*, 45, 107–119, 2007. With permission.)

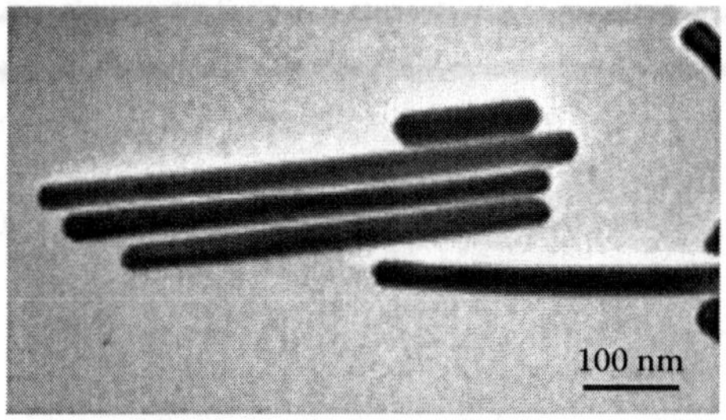

FIGURE 10.3
TEM micrographs of streptavidin-mediated 3D assembly of biotinylated gold nanorods. (From Rozenberg, B. A. and R. Tenne, *Prog. Polym. Sci.*, 33, 40–112, 2008. With permission.)

Cu nanoparticles have been used extensively to prepare conductive nanocomposites. Thus, Cu-low density polyethylene (LDPE) nanocomposites were prepared by melt processing to incorporate 50-nm Cu nanoparticles into the matrix. The incorporation of Cu nanoparticles improved electrical properties and lowered the melting temperature and the crystallinity while raising the crystallization temperature. Only marginal improvement in thermal stability of the polymer was observed in the nanocomposites (Xia et al. 2006). Improvement in thermal stability could be higher, when the nanoparticles are functionalized with a coupling agent. However, thermal stability usually suffers, probably due to the potential role of Cu as a catalyst for LDPE oxidation (Rozenberg and Tenne 2008).

A different approach to prepare Cu nanocomposites involves the production of nanoparticles during the synthesis of the nanocomposites. Thus, Cu nanoparticles were produced by reducing proper salts during the free radical polymerization of polyacrylamide grafted poly(vinyl alcohol) (PAM-g-PVA). Such nanocomposites could be used in development of chemical sensor applications. The electrical resistivity of the nanocomposites is sensitive to the presence of solvent vapors that result in matrix swelling and specific interactions between solvent vapor and nanocomposites (Wei et al. 2008).

Highly conductive copper nanowire (CuNW)–PS composites were produced by solution mixing for use in applications requiring shielding from EMI. The formation of a conductive network yields a percolation threshold at low concentration (0.24 vol% CuNW). Figure 10.4 shows the percolation behavior and the distribution of CuNW along with a TEM image of the composite (Al-Saleh et al. 2011; Gelves et al. 2008).

A combination of nickel nanoparticles/nickel nanowires (both modified with allyltrimethoxysilane [ATS]) was used to prepare polydimethylsiloxane (PDMS) nanocomposites. Nanowires were aligned by exposing the uncured system to a magnetic field. The synergism of using both nanofillers led to substantial improvement in the mechanical properties. The potential applications of the materials range from active structural components to sensing elements in macrosystem and microsystem applications (Denver et al. 2009).

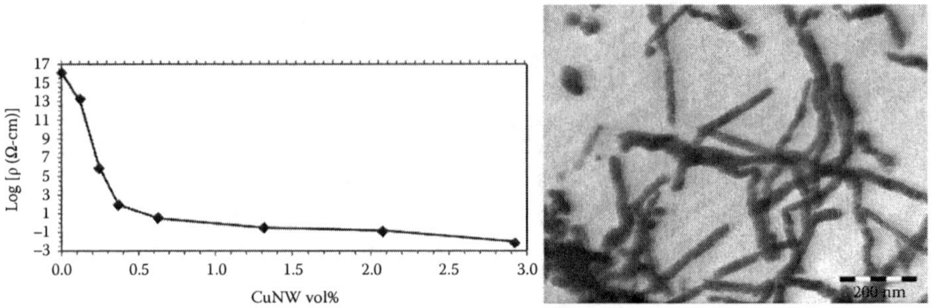

FIGURE 10.4
Percolation curve of conductive composite powder (CCP)–PS composite and TEM micrographs of the hot molded CCP film. (From Al-Saleh, M. H. et al., *Compos. Part A: Appl. Sci. Manuf.*, 42, 92–97, 2011. With permission.)

10.2.2 Nanoparticles Based on Metal Oxides and Derivatives

Many nanoparticles based on metal oxides are available for incorporation into polymer systems for various purposes (i.e., Al_2O_3, B_2O_3, CeO_2, CoO, Co_3O_4, Cr_2O_3, CuO, Dy_2O_3, Er_2O_3, Eu_2O_3, Fe_2O_3, Fe_3O_4, FeOOH, Gd_2O_3, HfO_2, In_2O_3, $In(OH)_3$, La_2O_3, MgO, $Mg(OH)_2$, MoO_3, Mn_3O_4, Nd_2O_3, NiO, Pr_6O_{11}, SiO_2, Sm_2O_3, SnO_2, Tb_4O_7, TiO_2, WO_3, Y_2O_3, ZnO, ZrO_2; nanoamor.com) (Kumar et al. 2009). Some effects of the oxide nanoparticles include surface conductivity and electronic surface, enhanced photocatalytic reactivity, supermagnetism, superhard coatings of metal alloys, and size-dependent electrical properties (Rozenberg and Tenne 2008). In some cases, there are concerns regarding the release of some hazardous nanoparticles to the environments after their use or the lifetime. To reduce the dissipation to the environment after use or lifetime, it is necessary to develop improved fixation methods to keep the nanoparticles in the nanocomposites (Reijnders 2009).

Nanoparticles of metal oxides and hydroxides (Fe_2O_3, ZnO, and $Mg(OH)_2$) or a mixture of the abovementioned nanoparticles have been used as anti-flame agents for PVC (Sertsova et al. 2010). Similarly, Ba-TiO_3, TiO_2 and ZrO_2 nanoparticles, and TiO_2 nanorods have been used to prepare PP nanocomposites with desirable dielectric properties (Guo et al. 2010). Nanocrystalline TiO_2 can be used for numerous applications, including photocatalysis, solar cells, membranes, sensors, nanoceramics, and for degrading environmentally hazardous chemicals (Rozenberg and Tenne 2008). To improve dispersion, nanoparticles are usually treated with brominated aryl diazonium salt, (1-phenylethyl bromide), or sodium lauryl sulfate surfactants and (3-(trimethoxysilyl) propylmethacrylate) coupling agent (Rozenberg and Tenne 2008). Moreover, nanocomposites with high optical transparency have been prepared based on onoleic acid–capped TiO_2 nanorods/PMMA-co-methacrylic acid (PMMA-co-MA) (Sciancalepore et al. 2008). The nanocomposites are useful in optical devices operating in the visible region.

Nanocomposites based on radial ZnO (R-ZnO) nanowires/polyvinylidene fluoride (PVDF) (Figure 10.5) exhibited high dielectric constants and good thermal stability at low percolation threshold (Wang et al. 2008). Nanowire functionalization improves the dispensability of nanofillers. Alkylthiol-modified ZnO nanorods were used to prepare shape memory thermoplastic polyurethane (PU) nanocomposites by solution blending. Tensile modulus and shape recovery characteristics were improved (Koerner et al. 2009).

Inorganic rodlike nanofillers are interesting materials for the preparation of nanocomposites. Hematite nanorods (α-Fe_2O_3, 8 nm in diameter and an average aspect ratio of 25) were dispersed in epoxy resin to produce nanocomposites with enhanced electrical properties. Figure 10.6 shows a TEM image of α-Fe_2O_3 nanorods (Dudic et al. 2008). Cadmium sulfide (CdS)

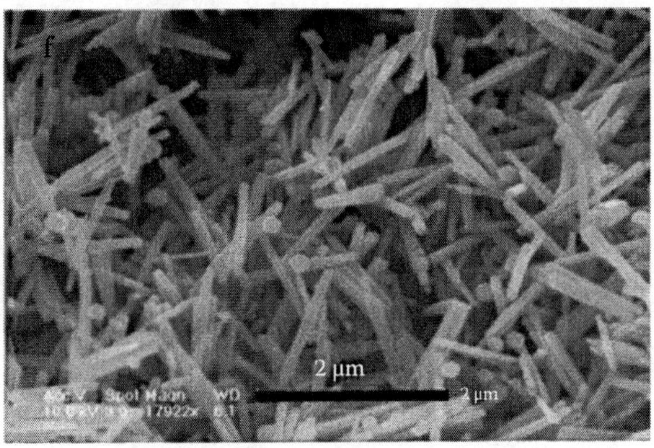

FIGURE 10.5
SEM image of radial ZnO (R-ZnO) nanowires prepared at molar ratio of dodecyl benzene sulfonic acid sodium salt (DBS)/$nZn(CH_3COO)_{22}H_2O$ = 0.4, as nanowire precursors. (From Wang, G. et al. *Adv. Funct. Mater.*, 18, 2584–2592, 2008. With permission.)

FIGURE 10.6
TEM micrograph of as-prepared α-Fe_2O_3 nanorods. (From Dudic, D. et al., *Polymer*, 49, 4000–4008, 2008. With permission.)

nanorods were dispersed in oligo(p-phenylenevinylene) to produce nanocomposites for applications on electronic and photonic devices (Guo et al. 2008). Nanotubular structures have been reported for NbO_xS_x, MgO, SiO_x TiO_2, ZrO_2, Co, Ni, WS_2 NbS_2, ReS_2, Cu_2S, CdS, and AgS (Whitby et al. 2004). Other systems include nitride nanowires, silicon nitride nanotubes, and boron nitride nanotubes (Kumar et al. 2009), in addition to layered nanofillers (MoS_2 and WS_2) (Rozenberg and Tenne 2008).

10.2.3 Polyhedral Oligomeric Silsesquioxane

Synthetic polyhedral oligomeric silsesquioxane (POSS) is another important type of nanofiller with the formula $(RSiO_{3/2})_n$, where R is hydrogen, alkyl, alkylene, aryl, arylene, cycloalkyl, or their organofunctional derivative groups, and $n = 8$, 10, 12, and so on. These particles are in the range 1–3 nm in diameter (Figure 10.7) (Zhao et al. 2008).

POSS nanoparticles have been successfully dispersed or incorporated in various polymers such as polyolefins, PMMA, cyanate and phenolic resins, epoxies, polysiloxanes, vinyl esters and PUs, polycarbonate (PC), PET, and polyimide by physical blending, chemical grafting, copolymerization, surface bonding, or other transformations (Zhao et al. 2008).

Property enhancements include increases in the glass transitional temperature (T_g), decomposition temperature (T_d), toughness, and oxidation resistance and reductions in hydrophobicity, dielectric constant, flammability, and viscosity during processing. Nanoparticles (organoclay, CNTs, or POSS) reduce polymer flammability by several mechanisms (limiting fuel transfer to the flame, formation of a protective char layer, etc.). The dramatic increases in mechanical properties are mainly attributed to the formation of star and

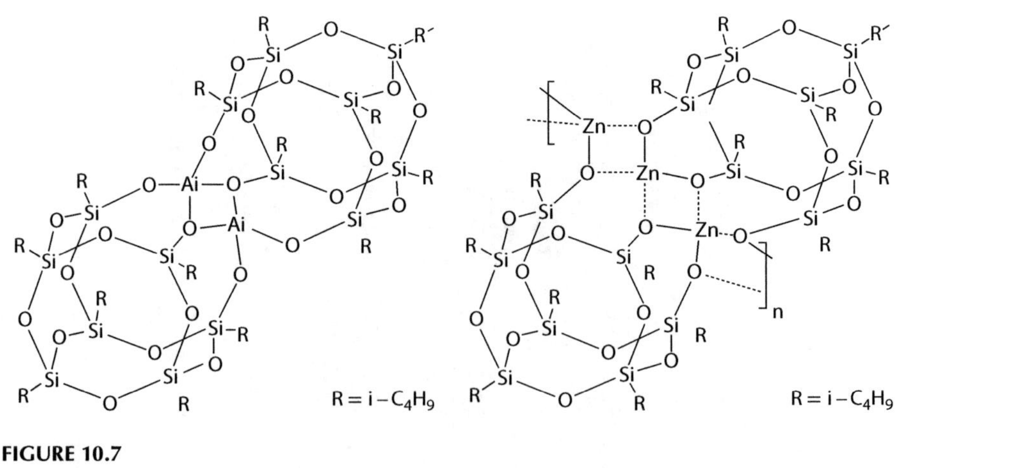

FIGURE 10.7
Examples of silsesquioxanes. (From Zhao, J. et al., *Polym. Polym. Compos.*, 16, 18, 2008. With permission.)

low cross-linking structures in the nanocomposites. POSS has been proved to be an effective nucleating agent, a chain extender, and a plasticizer (Laoutid et al. 2009; Zhao et al. 2008). Only the addition of a suitable type of POSS to a specific polymer offers opportunities for creating high-performance nanocomposites such as dental filling and environmentally friendly packaging materials or for biomedical applications (Zhao et al. 2008).

10.2.4 Silicates

Clay minerals, in particular smectites, are preferred as nanoparticles for reinforcing polymers, because of their availability, low cost, relatively simple processability, swelling properties, high cation exchange capacities, high aspect ratio, large surface area, chemical and thermal stability, and high mechanical strength and modulus. The weak bonds between the silicate layers in the stack (weak dipolar or van der Waals forces) allow the intercalation of polar organic molecules (surfactants) for surface modification. Clay minerals are naturally formed from volcanic glasses and rock-forming minerals under different formation mechanisms (inheritance, neoformation, and transformation) and different geological environments (weathering, sedimentary, and diagenetic-hydrothermal) (Bhattacharya et al. 2008). In nature, clay minerals are associated with crystalline materials, such as quartz, calcite, dolomite, feldspars, oxides, hydroxides, and organic phases, or with noncrystalline materials, such as colloidal silica, iron hydroxide gels, and organic gels (Guggenheim and Martin 1995).

Clays are formed by tetrahedron silicate units (SiO_4). Tetrahedrons can form single units (nesosilicates), chains (sorosilicates/inosilicates), sheets (phyllosilicates), rings (cyclosilicates), and framework structures (tectosilicates). Clay contains aluminosilicate phyllosilicates thus, the crystal lattice is composed of silica tetrahedral sheets and alumina or magnesia octahedral sheet (Sposito 2008).

Montmorillonite (MMT), a phyllosilicate mineral, is commonly used as nanofiller in the preparation of polymer nanocomposites. Figure 10.8 shows the crystal structure of MMT. The unbalanced substitution of ions of lesser charge for ions of higher charge results in a net negative electrostatic charge (called cation exchange capacity [CEC], expressed in meq/100 g). The exchangeable cations are randomly located in the interlayer positions and offer the possibility to be exchanged by organic cations (surfactants) to modify the surface properties of clay. Synthetic clays (e.g., fluoro mica, fluorohectorite, saponite, laponite, or hydrocalcite) show some advantages over natural clay (i.e., high purity, transparency, and high aspect ratio). Additionally, there are clay minerals with fiberlike structure (attapulgite–palygorskite, sepiolite) (Bhattacharya et al. 2008; Pan et al. 2008).

MMT platelets (~0.96 nm) exhibit large specific area (~657 m^2/g) and high aspect ratios (>200), and they are arranged in stacks with a regular gap (interlayer distance) between the individual layers. After organic modification, the

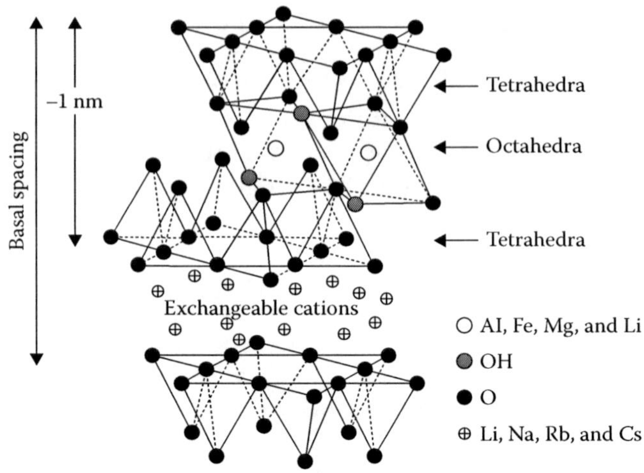

FIGURE 10.8
Crystal structure of montmorillonite. (From Sinha Ray et al., *Polymer*, 44, 6633–6646, 2003. With permission.)

original basal spacing (~1.1–1.33 nm) increases depending on the size and packing arrangement of the surfactant molecules. On the other hand, layered double hydroxides (LDHs), a class of anionic clays, are based on ($Mg(OH)_2$)-like layers, in which some of the divalent cations have been substituted by trivalent ions to form positively charged sheets. Morphologically, LDHs consist of stacks of positively charged mixed metal hydroxide layers with hydrated anions in-between. They can occur naturally or can be synthesized. Organic modification is possible for both silicate layers and LDHs (Ciardelli et al. 2008).

There is a vast array of both natural and synthetic-layered crystalline fillers that can, under specific conditions, undergo polymer intercalation. Table 10.1 shows some examples of layered nanoparticles with potential for use in the preparation of polymer nanocomposites.

The main issues associated with the preparation of polymer–clay nanocomposites are to promote the complete clay delamination and a uniform distribution of nanoparticles (Golubeva and Gusarov 2007). Pristine nanoclay is hard to disperse in most polymers, due to its stack configuration and large surface area that promotes agglomeration, in addition to the intrinsic incompatibility of hydrophilic layered silicates and mainly hydrophobic polymers. To improve the compatibility among the phases, layered silicates need to be organically modified using the clay mineral capacity for replacing inorganic exchangeable cations in the clay spacing or galleries of the pristine clay by surfactant molecules. Compatibility can be tailored by choosing both the chemical structure and the concentration of the surfactant. This process could lead to improved compatibility between the organoclay and the polymer, thus leading to the formation of nanocomposites with enhanced properties (Murray 2000).

TABLE 10.1
Different Groups of Layered Nanoparticles

Group (Type)	Examples
1. Element	Graphite
2. Oxides	Graphite oxide, V_6O_{13}, $HTiNbO_5$, $W_{0.2}V_{2.8}O_7$
3. Phyllosilicates (pyrophyllite–talc, smectite, vermiculite, illite, mica, palygorskite–sepiolite, chlorite)	Kaolinite, dickite, nacrite, halloysite, chrysotile, antigorite, lizardite, montmorillonite, bentonite, nontronite, beidellite, colknskonite, hectorite, pyrophyllite, talc, vermiculite, illite, glauconite, muscovite, celadonite, phlogopite, taenolite, margarite (brittle mica), palygorskite, sepiolite, clinochlore
4. Layered silicic acids	Kanemite, makamite, octosilicate, magadite, kenyaite, layered organosilicates
5. Mineral-layered hydroxides	Brucite ($Mg(OH)_2$), gibbsite ($Al(OH)_3$)
6. Layered double hydroxides (LDHs)	$M_6Al_2(OH)_{16}CO_3{\cdot}nH_2O$ (M: Al, Zn)
7. Layered aluminophosphates	Berlinite, vantasselite ($Al_4(PO_4)_3(OH)_3{\cdot}9H_2O$)
8. Metal (M4+) phosphates	M4+: Zr, Ti or Sn; $Zr(HPO_4)$
9. Chlorides	$FeCl_3$, FeOCl, CdI_2, $CdCl_2$
10. Metal chalcogenides	$(PbS)_{1.18}(TiS_2)_2$, TiS_2, MoS_2, MoS_3
11. Cyanides	$Ni(CN)_2$

Source: Kumar, A. P. et al., *Prog. Polym. Sci.* 34, 479–515, 2009. With permission.

There are several factors influencing the preparation of polymer–clay nanocomposites related to the organoclays and polymer matrix. Organoclay characteristics, including thermal stability, surfactant chemistry, surface energy, basal spacing, and aspect ratio, play an import role in the preparation and final performance of nanocomposite. Polymer characteristics, including molecular weight, polydispersity, formulation, chemical composition, viscosity, and so on, are important parameters in the preparation and performance of polymer–clay nanocomposites (Hussain et al. 2006; Uribe-Calderon and Kamal 2010).

The exchange of sodium or calcium cations with organic cations (commonly called surfactants, modifiers, or intercalants) causes a lowering of the surface energy of MMT and expansion of the interlayer spacing. The resulting material is called organoclay. Surface energy, basal spacing, and thermal stability of organoclays depend strongly on the chemical structure, grafting density, and the type of cation head included in the surfactant (Beyer et al. 2002). Organoclays have been used in other applications such as adsorbents, rheological control agents, paints, grease, cosmetics, personal care products, and oil well drilling fluids (Beyer et al. 2002; Lagaly 1986; Vaia et al. 1996; Xie et al. 2003).

Surfactants for clay modification usually include long aliphatic chains in their molecular structure. Aliphatic chains arrange themselves, depending on the size and concentration of surfactant molecules, into monolayer, bilayer, pseudo-trimolecular layer, or an inclined paraffin or interdigitated structure.

Clay has been treated with several organic compounds, including amines (primary, secondary or tertiary) and quaternary surfactants (ammonium, sulfonium, phosphonium, and imidazolium surfactants) to modify surface characteristics, expand the interlayer spacing, and improve compatibility with polymers. Processing conditions and the use of compatibilizers are important to improve miscibility and stability of nanostructures (Dennis et al. 2001; Nassar et al. 2005; Tanoue et al. 2006).

10.2.5 Biosource Nanoparticles

Cellulose fibers are low cost and widely available materials. Microfibrils and whiskers are two types of common reinforcements obtained from cellulose. Nanofibers from plants or animals consist of elongated cellulose chains stabilized by hydrogen bonding. Generally, they are composed of crystalline and amorphous parts. Isolated crystalline products are mainly obtained by acid hydrolysis to obtain whiskers, cellulose nanocrystals (CNCs), nanorods, or rodlike cellulose microcrystals. The typical length ranges from 500 nm to 2 μm and the diameter 8–20 nm. Theoretical calculations predict high values for elastic modulus (~150–200 GPa) and strength (~10 GPa) (Azeredo 2009; Kumar et al. 2009). Some sources of CNCs include bacterial cellulose, microcrystalline cellulose (MCC), sugar beet, cotton, tunicate cellulose, softwood pulp, hardwood pulp, and wheat straw. Geometric characteristics of the CNC depend on the source of the cellulose (Bai et al. 2009).

CNCs obtained by hydrochloric acid hydrolysis exhibit limited dispersibility in water. However, hydrolysis with sulfuric acid yields surface esterified nanocrystals that are stable in aqueous solution (Hasani et al. 2008). Hydrolysis with sulfuric acid can result in the introduction of sulfate esters at the surface of the cellulose crystallites, leading to added electrostatic stabilization of the CNC suspensions. At sufficiently high concentrations of such suspensions, birefringent and liquid ordered phases may be observed. The phase-forming ability of a cellulose crystallite suspension depends on the mineral acid chosen for the initial hydrolysis. The use of either sulfuric acid or phosphoric acid yields a chiral nematic phase, but hydrochloric acid will not yield because it cannot form the required ester at the crystallite surface (Hamad and Hu 2010).

Because of the hydrophilic surface, cellulose fibrils are well dispersed in hydrophilic matrices, but the filler–matrix interactions are weak in hydrophobic matrices. Surface modification is an alternative for improving the compatibility with nonpolar polymers. Chemical modification of the cellulose surface can include esterification or acylation. Chemical modification results in improved fiber–matrix interfacial adhesion; thus mechanical properties and thermal stability are enhanced (Azeredo 2009). Figure 10.9 shows CNC stabilized with epoxy-propyltrimethylammonium chloride (EPTMAC). A different approach is to use surfactant or polymer grafting to the fiber surface to improve the compatibility.

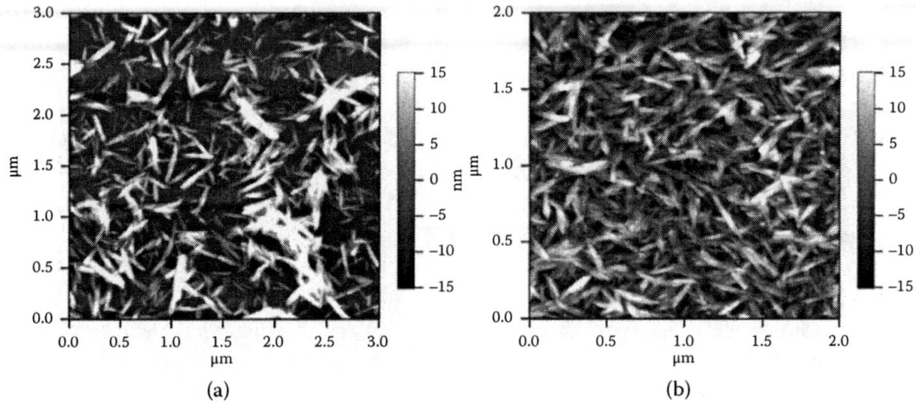

FIGURE 10.9
AFM images of CNC on mica (a) before and (b) after functionalization with EPTMAC. (From Hasani, M. et al., *Soft Matt.*, 4, 2238–2244, 2008. With permission.)

Factors affecting performance of cellulose as nano-reinforcement are the distribution, orientation, and aspect ratio of nanofiller. CNCs have been reported to improve thermal properties of polymers. The solid surface of cellulose whiskers could restrict mobility of polymer chains in the vicinity of the interfacial area, thus resulting in an increase in T_g (Azeredo 2009). Because of its useful characteristics, including zero toxicity, good hygroscopicity, chemical inactivity, and reversible absorbency, CNC has found wide use in pharmaceuticals as a tablet excipient and in food production (Hasani et al. 2008).

Other cellulose-based nanocomposites have been used as nanofillers, such as starch crystalline particles (exhibiting platelet morphology with thicknesses of 6–8 nm), or chitin whiskers (prepared by acid hydrolysis of chitin). The average dimensions of the whiskers are around 500 nm (length) and 50 nm (diameter) (Azeredo 2009).

10.2.6 Carbon Nanotubes

CNTs constitute another family of potential nanofillers. CNTs are tubular derivatives of fullerenes, but they exhibit properties that are quite different from those of the closed cage fullerenes such as C_{60}, C_{70}, and C_{76} (Kumar et al. 2009). CNTs are formed by graphene layers wrapped into a cylindrical shape. Hemispherical caps seal both ends of the tube. The carbon atom forming the graphene sheet is covalently bonded to three neighboring carbon atoms (Tjong 2006). This new allotropic form of carbon is built up of carbon atoms arranged in hexagons and pentagons, forming very long cylinders (Dubois and Alexandre 2006).

CNTs exhibit outstanding physical properties; for example, high elastic modulus (1 TPa), tensile strength (150 MPa), and electrical conductivity (ranging close to metallic conduction). The outstanding mechanical strength, electrical conductivity, and corrosion resistance make them attractive nanofillers for use in the production of nanocomposites (Tjong 2006). They are good candidates for applications requiring high electrical and thermal conductivity (Dubois and Alexandre 2006). However, since the surface of the graphene sheet exhibits low interaction with polymers, surface modification is required to promote dispersion and to enhance interfacial adhesion between the matrix and the CNT (Tjong 2006).

There are two types of CNTs: single-wall nanotubes (SWNTs) and multiwall nanotubes (MWNTs). SWNTs with a diameter of 1–2 nm consist of a single graphene layer. Depending on the rolling direction of the carbon hexagons in the graphene layer, SWNTs are classified as armchair, zigzag, and chiral (Figure 10.10). On the contrary, MWNTs are composed of a number of graphene layers coaxially rolled together to form a cylindrical tube. The outer diameter is usually in the range of 3–10 nm. Graphene layers interact with other layers by means of van der Waals forces. They could form intermeshing bundles that have a strong tendency to agglomerate. Effective dispersion of nanotubes in the polymer is required for optimum nanocomposite performance.

CNTs can be synthesized from hydrocarbons by means of the arc discharge, laser ablation, and chemical vapor deposition (CVD). In most cases, the CNTs thus produced contain impurities (amorphous carbon, nontubular fullerenes, graphite particles, or metal catalysts). Impurities are removed by an oxidation process, acid treatment, or by a combined oxidation and acid treatment. After the purification process, the CNTs are agglomerated into

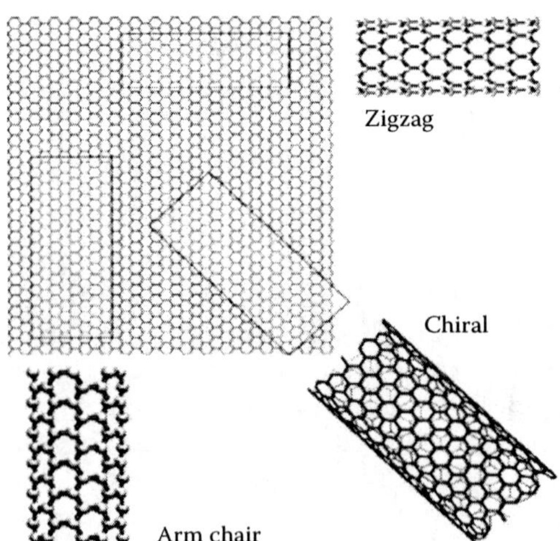

FIGURE 10.10
Different structures of carbon layer. (From Tjong, S. C., *Mater. Sci. Eng. R Rep.*, 53, 73–197, 2006. With permission.)

bundles, which need to be disaggregated by ultrasonication, ball milling, grinding and high-speed shearing, chemical oxidation, or reduction followed by chemical modification of the nanotube surface, to promote dispersion in polymer matrices (Dubois and Alexandre 2006; Tjong 2006).

CNTs exhibit extraordinary mechanical properties, with tensile modulus higher than 640 GPa and tensile strength up to 180 GPa, and thermal conductivity higher than 3000 W/m/K. Depending on the rolling direction of the graphene layer, different CNT structures may be generated showing either metallic or semiconducting characteristics. Among those, armchair SWNTs exhibit metallic conductions, while the others behave as semiconductors (Ciardelli et al. 2008). Electrical conductivity of CNT (higher than 10^{-2}–10^{-1} S/cm) and low percolation threshold (less than 0.1 wt%) have been reported in nanocomposites based on epoxy, PMMA, PC, and poly(vinyl acetate) (PVAc). Higher percolation threshold (2.6 wt%) has been observed in systems based on polyolefins. Mechanical properties and electrical conductivity of CNT–polymer nanocomposites depend on nanotube aspect ratio, degree of dispersion, and anisotropic distribution within the matrix. The strong van der Waals interactions between individual nanotubes make them difficult to separate since they tend to form microsized aggregates consisting of bundles of 100–500 CNTs packed together (Ciardelli et al. 2008). Research on CNT nanocomposites has focused on the development of robust and lightweight composites, conducting paints, fabrication of flat panel displays, supports for metals in the field of heterogeneous catalysis, Li+ batteries, gas (hydrogen) storage devices, toxic gas sensors, electronic nanodevices, and for immobilization of proteins and enzymes (Kumar et al. 2009).

Three main factors greatly influence the reinforcement behavior of the CNT in polymer composites: (1) good dispersion of the CNTs in the polymer matrix, (2) strong interfacial bonding between CNTs and polymer, and (3) good alignment of the CNTs in the polymer matrix (Rozenberg and Tenne 2008). To perform as reinforcement, CNTs should be wet by the surrounding matrix for load transfer, otherwise CNT agglomerate could act like a void, thus reducing the mechanical properties (Wu et al. 2010). Unsatisfactory performance of CNT nanocomposites is associated with poor dispersion of entangled CNT during processing, poor interfacial interaction between CNTs and polymer, low interfacial strength transfer, process-related deficiencies, undesirable alignment, filler agglomeration, and the fractal nature of filler clusters (Schaefer and Justice 2007). The difficulty for CNT dispersion is caused by the large aspect ratio (>1000), the extremely high surface area, and the highly entangled state of commercial CNT along with the low affinity with most polymer (Ma et al. 2010).

CNT–polymer nanotube nanocomposites have been produced by conventional techniques such as melt compounding, in situ polymerization, or solution methods. In the solution method, the CNTs are dispersed in a proper solvent (by mechanical mixing, magnetic agitation or sonication) and combined with a polymer solution at room or elevated temperatures. After solvent elimination, the nanocomposites are obtained. In melt compounding, high temperature and a high shear force are required to disperse CNTs in a thermoplastic polymer matrix. This technique is less effective to disperse CNTs in polymers than that of solution mixing. In in situ polymerization, CNTs are mixed with monomers, and after polymerization (addition or condensation reactions), the nanocomposites are formed. CNT–thermoset nanocomposites are obtained using this technique (Ma et al. 2010).

Various mixing techniques have been reported to disperse nanoparticles (multiwall CNT [MWCNT] or graphene) in the polymer matrix, in addition to conventional polymer processing techniques. The so-called solid-state shear pulverization (SSSP) is a continuous, scalable process and free of solvent use. It has been used to produce well-dispersed polymer–MWCNT nanocomposites in conjunction with melt mixing. By using this technique, heavily entangled and unmodified MWCNTs can be dispersed in polyolefins such as PP, resulting in nanocomposites with improved mechanical properties (Masuda and Torkelson 2008; Wakabayashi et al. 2008).

A new approach to disperse CNTs in a polymer is the so-called latex technology. In this technique, SWCNT or MWCNT are dispersed in polymers produced by emulsion polymerization. The first step is the dispersion/stabilization of nanoparticles (SWCNT or MWCNT) in an aqueous surfactant solution. Then, the stable dispersion is mixed with polymer latex. After processing (freeze-drying and melt-processing), nanocomposites are obtained. This technique is relatively easy, versatile, reproducible, and reliable (Ma et al. 2010).

There are two major issues related to the dispersion of CNT: lack of solubility and dispersion in polymer resins and poor interfacial adhesion in CNT–polymer systems. To overcome those difficulties, two different approaches have been employed to improve compatibility between the two components: chemical and physical modification. Chemical modifications of the CNT surface include the use of surfactants, functionalization of endcaps, and functionalization of sidewalls (Kumar et al. 2009). Nanotubes can exhibit surface functionalities (carbonyl or carboxyl moieties, and double or triple bonds generated during the production process), which could be used as precursors for surface modification (Tjong 2006). Additionally, reactive moieties can be produced on the surface of the CNT by an oxidation treatment using HCl, H_2SO_4/H_2O_2, H_2O_2/NH_4OH, HNO_3, $KMnO_4$, ozone, or reactive plasma. However, overoxidation could damage the CNT structure, which could affect the mechanical and electrical properties of CNT and in turn those of the CNT nanocomposites. For example, the conductivity of poly-1,8-diaminonaphthalene filled with SWCNT is enhanced by increasing the CNT polarity due to the presence of –OH and –COOH groups. However, excessive oxidation could lower conductivity of MWCNT–epoxy composite, either by modifying the structure of CNT or by removing the metal (conductive) impurities. On the contrary, the use of oxidant or other agents could reduce the length of CNT, which could favor the percolation network formation at lower CNT content (Spitalský et al. 2009).

Chemical functionalization occurs by covalent linking of functional entities onto CNT structure, either on the end of the tubes or at the sidewalls promoted by reactive sites on the CNT structure. Sidewall functionalization is due to the change of hybridization of the graphene layer from sp^2 to sp^3. High chemical reactivity molecules are capable to react with the CNT wall (e.g., fluorination, cycloaddition, such as Diels–Alder reaction, carbene and nitrene addition, chlorination, bromination, hydrogenation, and azomethineylids). Additionally, defect functionalization is carried out by the reaction of defects on CNTs (open ends and/or holes in the sidewalls, pentagon, or heptagon irregularities in the hexagon graphene framework) with chemical moieties. Defects on the CNT can be created by an oxidation process as mentioned earlier. Carboxylic acid (–COOH) or hydroxyl (–OH) groups result from the oxidation process, which can be used as precursors for further chemical reactions (e.g., silanation, polymer grafting, esterification, thiolation, alkylation, and arylation) (Ma et al. 2010).

The attachment of polar groups changes the hydrophobic nature of CNTs making functionalized CNTs soluble in many organic solvents. Additionally, functionalized CNTs can form strong interfacial bonds with many polymers, resulting in nanocomposites with improved mechanical properties. Besides the promising advantages from chemical modification of CNT, the functionalization process can damage the structure of CNT. This could create a larger number of defects, in addition to destroying the CNT and, thus, reducing the transport properties (electrical or thermal conductivity). Additionally, chemical treatments may employ environmentally unfriendly materials.

In the physical modification approach, polymers (e.g., poly(phenylene vinylene) or PS) and surfactants (i.e., nonionic surfactants, anionic surfactants, or cationic surfactants) are used to cover the outer surface of CNT to modify the interfacial properties of CNT. In the so-called endohedral method, molecules are introduced inside the CNT structure (capillary effect) (Ma et al. 2010).

Notwithstanding the considerable advances in the production of CNT nanocomposites, issues remain in need of further consideration. For example, techniques for quantitative analysis of the degree of dispersion or agglomeration are needed to relate the degree of dispersion to CNT and nanocomposite characteristics (particle size, aspect ratio, surface functionalities, or surface energies). Furthermore, modification methods for CNT functionalization should be optimized. Systematic studies are required to establish the correlations between surface treatment, degree of functionalization, processing and fabrication conditions and the mechanical, electrical, and thermal properties, failure mechanisms, and fracture performance of CNT nanocomposites. Fundamental studies regarding the stress transfer between matrix and CNTs are required to understand the mechanical behavior and transport properties of nanocomposites. Studies on interfacial bond strength and prediction and measurement of interfacial adhesion could provide useful information regarding the performance of nanocomposites.

Another interesting approach involves the production of hybrid nanocomposites, combining low-cost nanofillers along with CNT. However, dispersion of different nanofiller represents a significant challenge since each nanofiller has its own unique geometric physical, chemical, and dispersion characteristics.

10.2.7 Graphene

10.2.7.1 Physical and Chemical Characteristics

Carbon nanofillers can be divided according to the dimensionality of the material: 0D carbon molecules such as fullerenes, 1D CNTs, 2D graphene macromolecules, and 3D carbon materials such as diamond and graphite (Steurer et al. 2009).

Graphene is a one-atom-thick layer of graphite with sp^2-hybridized carbon atoms arranged in a six-numbered-ring plane, which can be produced by CVD, micromechanical exfoliation of natural graphite, growth on crystalline silicon carbide, highly oriented pyrolytic graphite (HOPG), or from graphite oxide (GO). Sheets can be stacked parallel to each other (Chen and Zhao 2010; Potts et al. 2011; Steurer et al. 2009).

It is possible that the use of graphene nanofiller would be preferred over CNT because of the low cost and availability of graphite. Additionally, as in the case of CNTs, the use of the graphene-based nanoparticles enhances mechanical and electrical properties of nanocomposites. Moreover, barrier properties and thermal conductivity are also enhanced. The multifunctional property enhancements realized in grapheme systems, coupled with its potential for low cost and large-scale production, make it an attractive candidate for research and development effort (Potts et al. 2011).

Graphene nanoribbons (GNRs) are derivative nanoparticles from graphene. They are produced using several techniques, including lithographic, chemical, sonochemical, and CVD methods, or by unzipping CNTs using the plasma etching route in the gas phase. The plasma etching technique involves the use of an oxidative mechanism to unzip MWNT, followed by a thermal reduction process. As in the case of nanotubes, nanoribbons face dispersion impediments, which make it difficult to disperse nanoribbons beyond 0.3% in weight fraction, due to the large contact surface area and geometrical factors of nanofiller. It is easier for polymers to adhere to flat surfaces than to hollowed structures (Rafiee et al. 2009).

As other nanofillers, graphene faces many challenges to yield its full potential benefits. Among the challenges to be faced are the poor dispersibility of neat graphene within almost polymeric matrix and the lack of the nanofiller–polymer adhesion, which compromise the final composite properties (Potts et al. 2011).

Various graphene-based nanoparticles are eligible for nanocomposite production. These nanoparticles include nanofillers derived from graphite nanoplatelets (GNPs), graphite intercalation compounds (GICs), and graphene

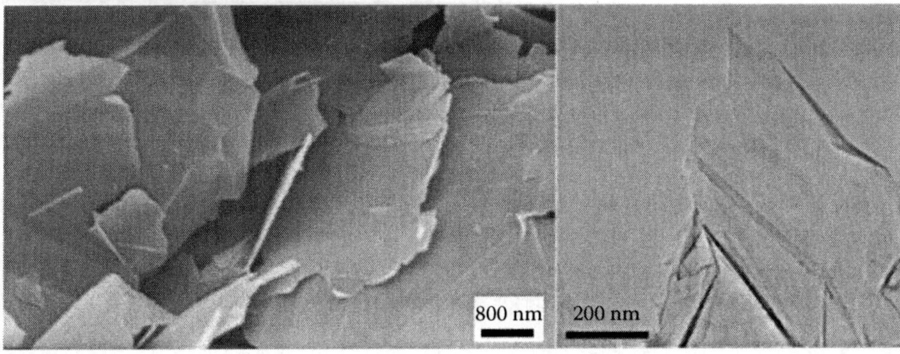

FIGURE 10.11
SEM micrograph of graphite nanosheets obtained by ultrasonic powdering technique and TEM image of exfoliated GO. (From Lu, J. R. et al., *Adv. Funct. Mater.*, 15, 1358–1363, 2005; Yang, Y. et al., *Langmuir*, 25, 11808–11814, 2009. With permission.)

oxide (GO). GICs are obtained by the intercalation of alkali metals or mineral acids between the carbon layers. This process increases the interlayer spacing and weakens the interlayer interactions between graphene layers. It makes it more likely to achieve the exfoliation platelets by mechanical or thermal methods (Potts et al. 2011). Figure 10.11 shows SEM and TEM images of graphite nanosheets and the exfoliated GO respectively.

Pristine graphite does not have reactive ion groups. Thus, intercalation by surfactants, as in the case of clay, is not possible. Since the graphite layers are held together in stacks by weak van der Waals forces, pristine graphite allows the intercalation by atoms, molecules, metal complexes, and salts between the graphene sheets, resulting in GICs (Chen and Evans 2009; Chen and Zhao 2010; Steurer et al. 2009).

Graphite forms intercalation compounds with several acids (e.g., nitric acid, sulfuric acid, perchloric acid, and selenic acid). Among the treated graphites, graphite bisulfate (graphite-H_2SO_4) is the most common. It is prepared by the intercalation of a combination of concentrated sulfuric acid and nitric acid. Upon heating, intercalated graphite can be exfoliated, due to the vaporization of intercalate. This produces the expanded or exfoliated graphite (EG) (Chen and Zhao 2010). EG, a highly porous, accordion-like, and low apparent density material, exhibits high temperature resistance and a specific surface area above 100 m^2/g (Steurer et al. 2009). Figure 10.12 shows several SEM images of EG. By intercalation of sulfuric acid or oxidation, the thick graphite sheet is reduced to graphite platelets. Usually, EG is dispersed in aqueous solution or polar solvents. Dispersion techniques include high-speed mixing, sonication, and high shear–strain rate (Potts et al. 2011; Wu et al. 2010). Figure 10.13 shows a schematic representation of the exfoliation process of graphite.

EG, showing a large number of delaminated graphite sheets, is a GNP precursor to develop polymer–graphite nanocomposites. This structure allows the intercalation of monomers or polymers to prepare nanocomposites.

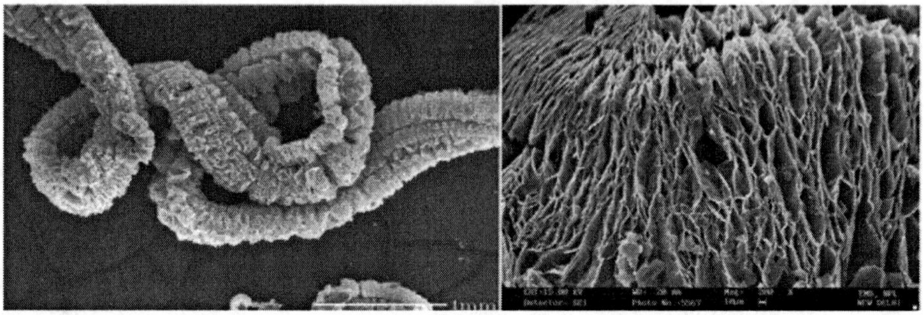

FIGURE 10.12
SEM images of expanded graphite. (From Dhakate, S. R. et al., *Energ. Fuels*, 22, 3329–3334, 2008; Wong, S.-C. et al., *Mater. Manuf. Process.*, 21, 159–166, 2006. With permission.)

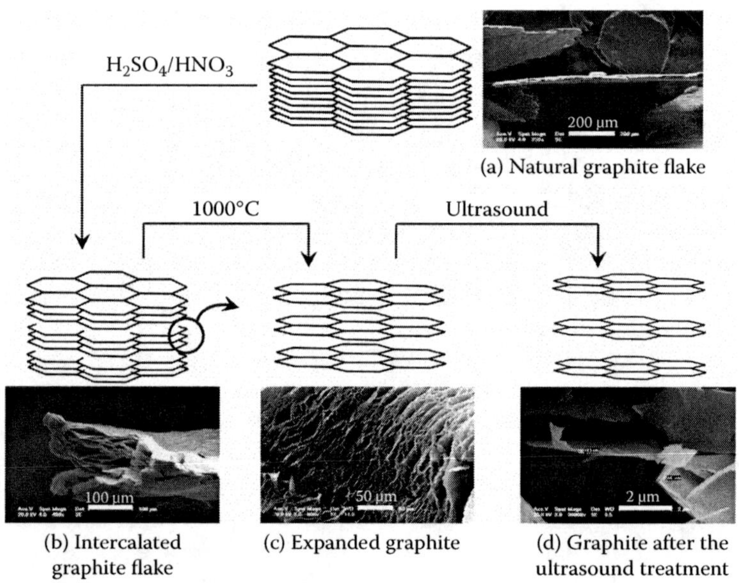

FIGURE 10.13
Schematic representations and SEM pictures of graphite structures after different treatments: (a) no treatment, (b) oxidation, (c) thermal expansion, and (d) ultrasonic dispersion. (From Fim, F. d. C. *J. Polym. Sci. Part A: Polym. Chem.*, 48, 692–698, 2010. With permission.)

Under proper mixing conditions, EG particles are dispersed in polymers (e.g., PMMA, poly(styrene-co-acrylonitrile), PS, or PS-PMMA) to obtain nanosheets with thicknesses varying from about 10 to 50 nm (Chen and Zhao 2010). Additionally, EG can be intercalated with organic surfactant (e.g., ammonium surfactants under sonification) to produce graphene-like sheets.

Other inorganic intercalants can be used to exfoliate graphite. For example, potassium-intercalated graphite was used as a source of GNP precursor for graphite nanocomposite preparation by in situ polymerization. Figure 10.14 shows a schematic representation of the process (Chen and Zhao 2010).

GO is produced by treating graphite with strong mineral acids and oxidizing agents (i.e., $KMnO_4$, H_2SO_4, $KClO_3$, and HNO_3). This oxidation treatment produces hydroxyl and epoxy groups on the surface and carboxylic acid groups on the edges of the sheets (Figure 10.15). Additional processing (chemical reduction or thermal shock) can produce exfoliated platelets similar to that of pristine grapheme (Potts et al. 2011).

Thermally reduced GO (TRGO), produced by thermal expansion, can exhibit large specific surface area (~600–950 m^2/g). Examples of nanocomposites based on TRGO include PC, isotactic PP (iPP), styrene-acrylonitrile polymer (SAN), and polyamide-6 (PA-6) (Steurer et al. 2009).

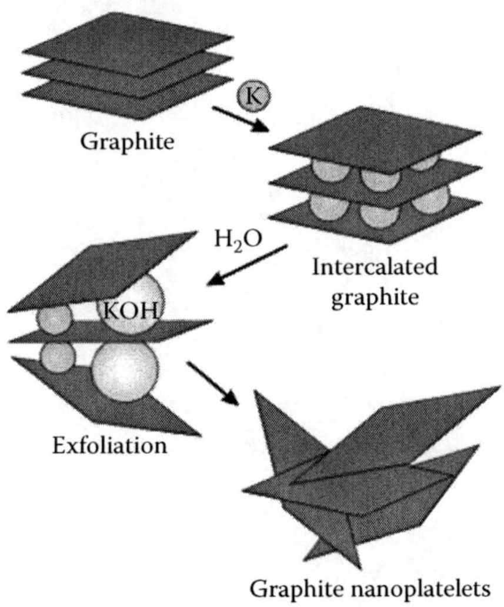

FIGURE 10.14
Simplified schematic of the intercalation/exfoliation process. Graphite is intercalated with potassium to form the first-stage intercalation compound and then exfoliated in aqueous solvent to produce GNPs that are 40 layers thick. (From Mack, J. J. et al., *Adv. Mater.* 17, 77–80, 2005. With permission.)

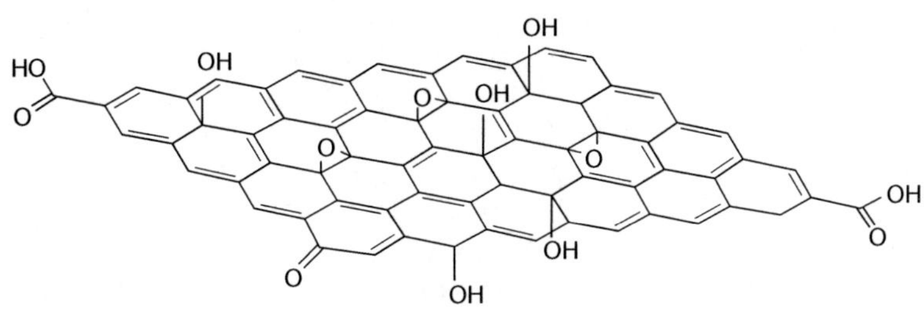

FIGURE 10.15
Schematic representation of graphite oxide. (From Sauter, C. et al., *Ultra. Sonochem.*, 15, 517–523, 2008. With permission.)

Chemical reduction agents have been used to react GO, such as hydrazine, sodium hydride, hydroquinone, or vitamin C (Wissert et al. 2010). Graphene nanocomposites have been prepared by chemical reduction of GO dispersion in the presence of SAN latex followed by melt compounding. Surfactants or polymeric dispersing agents can prevent agglomeration of graphene during the reduction process and, at the same time, to improve filler dispersion (Wissert et al. 2010).

Graphite exhibits different values of physical properties depending on the planar orientation (Chen and Zhao 2010). The high in-plane elastic modulus of pristine graphene (~1.1 TPa measured by micromechanical test) suggests its use in polymer nanocomposites to upgrade mechanical properties (Potts et al. 2011). Chemically modified graphene platelets exhibit an appreciably lower in-plane stiffness, which decreases with increasing level of oxidation and the presence of defects on the platelets (250 GPa for reduced GO against 650 GPa for GO) (Potts et al. 2011). Moreover, the overall mechanical properties of nanocomposites depend on the degree of nanofiller dispersion, exposed surface area, the adhesion at the interface, and the spatial configuration of the individual graphene platelet (Potts et al. 2011).

Graphene layers exhibit a high specific surface area (2600 m^2/g), which is higher than the surface area of exfoliated clay (Bhattacharya et al. 2008; Steurer et al. 2009). Thermal conductivity of the graphene nanosheet has been reported to be 1200 W/m-K, which is lower than that of CNTs (3000 W/m K) (Wu et al. 2010). It has been reported that the filler size and the aspect ratio affect heat propagation in the composite and the overall thermophysical behavior (Tlili et al. 2010). Graphene is optically transparent and electrically conducting. Thus, ultrathin films are useful as transparent electrodes (Steurer et al. 2009).

10.2.7.2 Nanocomposite Characterization

Similar characterization techniques to those used for clay–polymer nanocomposites have been used to determine the structure and morphology of graphene–polymer nanocomposites. Extensive effort has been made to characterize the degree of dispersion and structure of nanolayers in polymer–clay nanocomposites. TEM is useful to assess the dispersion and morphology of thick graphene tactoids, whereas single-layer platelets could be difficult, due to the lack of contrast (Figure 10.16). X-ray diffraction (XRD) is used as a complementary technique to assess the state of dispersion (Potts et al. 2011).

Structure characterization of nanocomposites has been studied by wide-angle XRD (WAXD), small-angle X-ray scattering (SAXS), X-ray photoelectron spectroscopy (XPS), small-angle neutron scattering, scanning (SANS), electron microscopy (SEM), and TEM. XRD (Figure 10.17) allows a quantitative understanding of the degree of intercalation (mean diffractive peak is shifted to lower angles) and a qualitative understanding of exfoliation (mean diffractive peak tends to disappear). Dispersion and exfoliation of the nanostructured graphite is observed through direct visualization by TEM, as

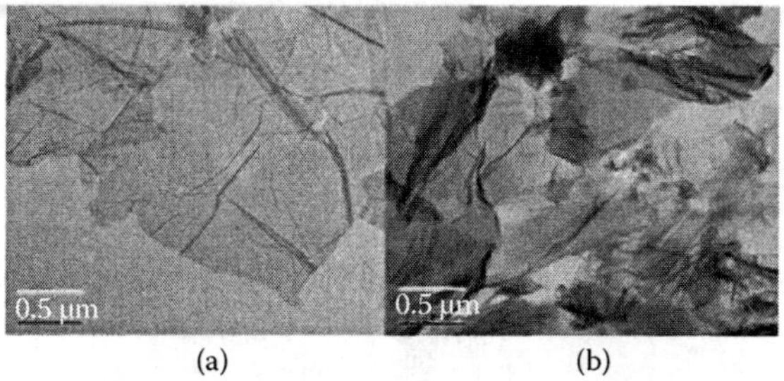

FIGURE 10.16
TEM images of (a) colloidal graphene oxide (CGO) and (b) CGO-Polyaniline (PANI) nanocomposite. Nanocomposite of CGO and PANI was fabricated through in situ oxidation polymerization in the presence of CGO prepared through a modified Hummers method without any other dopant. (From Zhang, W. L. et al., *Chem. Commun.* 46, 5596–5598, 2010a. With permission.)

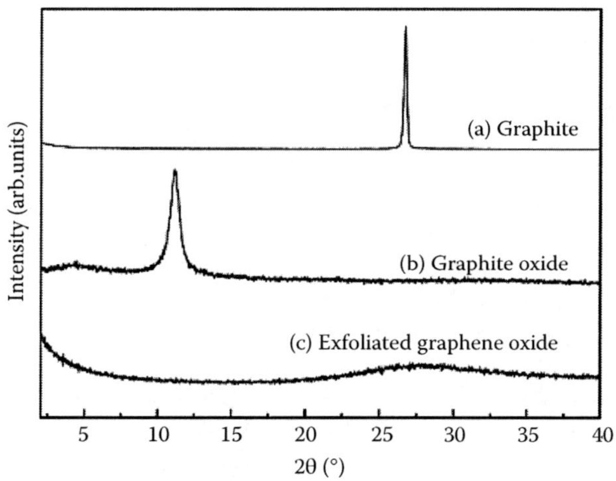

FIGURE 10.17
XRD patterns of (a) graphite, (b) graphite oxide (GO), and (c) the aggregate obtained by centrifugation from colloidal suspension of exfoliated GO nanosheets. (From Chen, D. et al., *ACS Appl. Mater. Interfaces* 2, 2005–2011, 2010. With permission.)

mentioned earlier (Chen and Zhao 2010). Other analytical techniques, such as atomic force microscopy (AFM), have been used to characterize the structure and morphology of graphene-based nanocomposites (Potts et al. 2011).

Three different morphologies have been observed in polymer–graphene nanocomposites: immiscible, intercalated, or exfoliated. The immiscibility of the components produces large agglomerates (stacked platelets) when observed by TEM, and XRD patterns reveal the characteristic peaks for the graphene nanoparticles. In intercalated systems, the interlayer spacing increases. The ideal nanocomposite is obtained when the graphene platelets are exfoliated and evenly dispersed in the polymer matrix. In that state, the XRD pattern corresponds to the neat polymer matrix (Potts et al. 2011).

Rheological and viscoelastic properties of graphene-based nanocomposites can be used to indicate their morphological characteristics. For example, the presence of plateau in the storage modulus for melt rheology measurements at low frequency can be used as an indication of filler dispersion due to the formation of a "solidlike" elastic network of filler (Potts et al. 2011).

Thermal, mechanical, and flow characteristics of graphene nanocomposites (crystallization behavior, glass transition and degradation temperatures, and rheological properties) have been assessed by differential scanning calorimetry (DSC), thermogravimetric analysis (TGA), thermomechanical analysis (TMA), dynamic mechanical analysis (DMA), and rheometry (Chen and Zhao 2010; Zhang et al. 2010a; Zhang et al. 2010b).

10.2.7.3 Factors Influencing Performance

There are several factors that could interfere with the effectiveness of graphene platelets as nanofiller. One of the factors is poor interfacial adhesion in graphene–polymer composites, caused by absence of strong interactions such as covalent or hydrogen bonding between the polymer and the filler. This reduces the efficiency for stress transfer, thus resulting in poor mechanical properties. Another important factor is the presence of defects on the structure of graphene sheets by the oxidation process. Changes in the structure could interfere with the electrical and mechanical properties of nanofillers. In consequence, less aggressive purification or exfoliation methods should be used with pristine graphite. Moreover, the flammability of graphene nanofiller is of concern in susceptible applications (Potts et al. 2011).

Adhesion was improved by using acid treatment of natural graphite to generate polar moieties on the platelets. This facilitates both physical and chemical interactions between graphite and the polymer (Soltani and Katbab 2010). Oxidation generates hydroxyl and epoxide groups on graphite layers and carbonyl and carboxyl groups on the edges. Hydrogen bonding tends to increase the adhesion with certain polymer matrices. As a result, the mechanical properties are improved. However, better stress transfer is observed in systems exhibiting covalent bonding between the polymer and the nanofiller surface (Chen and Zhao 2010; Potts et al. 2011).

Functionalized moieties on the graphene sheets have been used to attach chemical surface precursors such as isocyanate, porphyrin, or diazonium salts or amines for covalent bonding. Chemical modification/functionalization of graphene is essential to improve solubility/processability and to prevent reaggregation (Pramoda et al. 2010; Wissert et al. 2010).

Exfoliated graphene-based materials are observed as bent or crumpled platelets depending on the affinity among the components. Systems exhibiting high affinity among the components show extended platelets, whereas reduced affinity systems show crumpled platelets (Potts et al. 2011).

10.2.7.4 Nanocomposite Preparation Methods

There are three different approaches to prepare graphene-based nanocomposites: in situ polymerization, solution blending, and melt compounding. The in situ polymerization technique involves the intercalation of monomers into grapheme interplanar spacing and subsequent polymerization. Two types of nanocomposites are obtained using this approach: systems exhibiting covalent bonding and systems without bonding. The former is preferred for improved performance. Nanocomposites obtained with this method may be based on both thermoplastics and thermosets, including epoxy resins, α-caprolactam, nylon-6, PS, PMMA, PVC, poly(styrene-methyl methacrylate), aromatic polydisulfide, and poly(methylacrylic acid) (Chen and Zhao 2010).

In solution compounding, filler dispersion in the polymers is carried out in a solvent medium. Melt compounding usually follows after solvent elimination. PMMA, maleic anhydride–grafted PP, PC, ethylene vinyl acetate (EVA), silicon rubber (SR), PS, PAM, and polyimides are examples of matrices used in the preparation of graphene nanocomposites by this technique. The degree of dispersion of graphene platelets by solution mixing methods depends on the level of exfoliation of the initial platelets before mixing. In this sense, functionalized graphene exhibits stable highly exfoliated platelets. The main limitations of this approach are the large amounts of solvent and the high temperatures required to dissolve the polymer during the process, in addition to concerns regarding environmental issues (Chen and Zhao 2010; Potts et al. 2011).

Melt compounding is carried out using a twin-screw extruder, injection molding, twin-roller, or internal mixer using thermoplastics or thermoelastomers, as polymer matrices under high shear conditions. Some examples of graphene nanocomposites obtained with this method are based on acrylonitrile-styrene (AS), high-density polyethylene (HDPE), and PP. While this technique has economic and environmental advantages over solution mixing, it usually yields a lower level of filler dispersion (Chen and Zhao 2010; Potts et al. 2011).

An interesting approach to improve nanoparticle dispersion is to apply hydrostatic pressure on cavitated systems under the influence of ultrasound devices. Cavitation is generated by the action of ultrasound waves. Hydrostatic pressure causes collapse of cavitation bubbles and dispersion of the nanoparticles, especially in systems with low viscosity (Cai et al. 2007; Sauter et al. 2008).

10.2.7.5 Property Enhancements with Graphene

One of the interesting potential applications of graphene-based nanocomposites is the production of materials with high electrical conductivity. In the case of polymer nanocomposites, the concentration of graphene nanofiller should be sufficient to produce percolation, in which state, a conductive

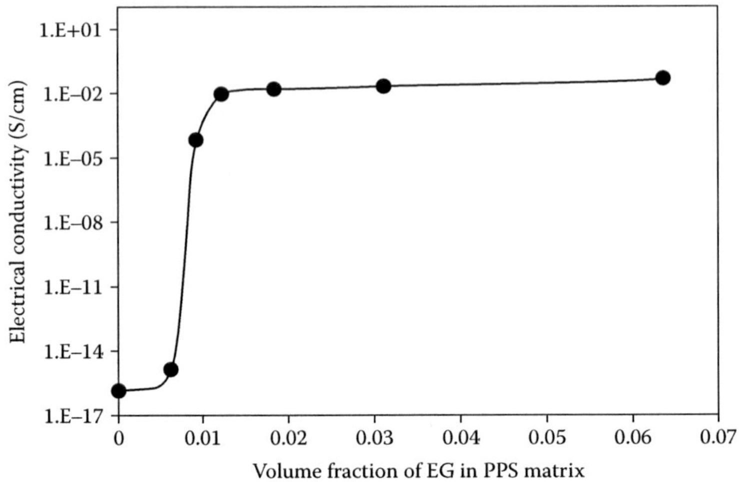

FIGURE 10.18
Electrical conductivity for poly(phenylene sulfide) (PPS)/EG composites. (From Goyal, R. K. et al., *J. Phys. D: Appl. Phys.* 43, 365404, 2010. With permission.)

continuous network of filler is created. The percolation threshold is the concentration at which the electrical conductivity of an insulating polymer matrix increases dramatically. Beyond this concentration, the conductivity of the composite increases marginally (Figure 10.18). Electrons can be transported by direct contact among nanofiller particles. Otherwise, conduction is through tunneling through the thin polymer layers surrounding the conductive filler particles. Percolation depends strongly on the intrinsic electrical properties of nanofiller, along with the geometry of the filler and the filler orientation. Thus, elongated and aligned nanofillers have higher probabilities to percolate at lower volumetric concentrations than spherical nanofillers. Lower filler concentration for the percolation threshold is preferred for economic and functional (i.e., low density of the composites, easy processing, etc.,) reasons. Lower threshold concentrations have been observed in systems with inhomogeneous dispersion of nanofillers. Percolation can be enhanced by the combination of conductive nanofillers to promote the electrical contact among the particles (Chen and Zhao 2010; Potts et al. 2011).

Since the electrical conduction in a composite is due to the formation of a continuous conductive network formed by the fillers, it is possible to interrupt the continuous network by the action of temperature, pressure, and solvent. This behavior can be used for a wide range of potential applications such as pressure measurement and sensors to measure gas or liquid concentration and current and temperature sensors (Chen and Zhao 2010; He et al. 2009).

Semiconductive nanofillers can be used in specific applications that require a certain degree of conductivity, such as air ducts in mines, pipes for solvent transport, EMI shielding, and some other areas (Móczó and Pukánszky 2008). Sensors based on nanomaterials (particles, wires/fibers, tubes, and belts) have potential for detection of chemical and biological species with high sensitivity and selectivity. Advances in these materials could lead to development of miniaturized devices for use in healthcare, pollution control, defense, and so on (Patra et al. 2008). Additional potential applications have been proposed for graphene-based nanocomposites, including field-effect transistors (FETs), batteries, solar cells, and other optoelectronic applications; other potential applications for graphene nanocomposites, supercapacitors, nanoelectronics devices, chemical sensors, and mechanically reinforced composite materials or temperature sensors include EMI shielding and antistatic coatings; packaging material; weight-sensitive aerospace and automotive applications; capacitive pressure sensors in microelectromechanical systems (MEMS); self-healing materials; and biomedical applications (biosensor applications) (Chen and Zhao 2010; Sauter et al. 2008).

The incorporation of graphene nanofiller tends to enhance mechanical properties of nanocomposites, such as the modulus, strength, fracture toughness, and fatigue strength. Improvements in mechanical properties are highly dependent on the degree of dispersion, intrinsic mechanical properties, filler orientation, and the aspect ratio of nanofillers (Potts et al. 2011; Steurer et al. 2009; Zhao et al. 2010). Figure 10.19 shows the effect of graphene content on the mechanical properties of nanocomposites. Various mechanisms have been proposed to explain the enhancement of toughness of nanocomposites. These include the contributions of nanoparticles mobility and orientation as an energy-dissipating mechanism. Other mechanisms suggest that nanoparticles can modify the cracks or delay their growth, resulting in enhanced toughness (Shah et al. 2005).

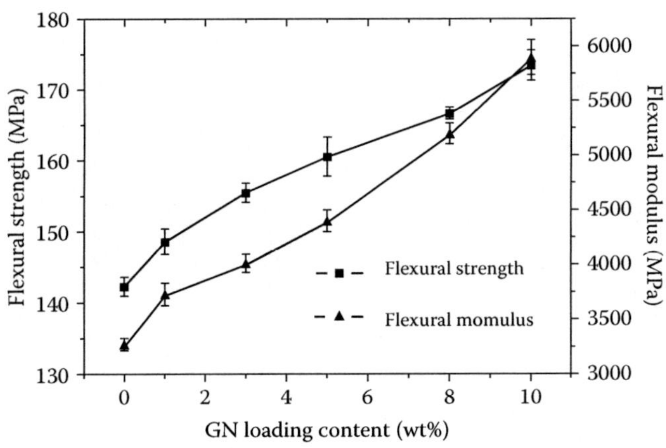

FIGURE 10.19
Effect of graphite nanosheets (GN) content on the flexural strength and flexural modulus of polyarylene ether nitriles (PEN)–GN nanocomposites prepared by masterbatch route. (From Zhan, Y. et al., *J. Mater. Sci.* 46, 824–831, 2011. With permission.)

Other physical properties of polymers can be improved by incorporating graphene-based nanofillers, such as thermal conductivity and thermal and dimensional stability. The outstanding thermal conductivity of pristine graphene and graphene-based nanofillers (3000 W/m K) can impart good thermal conduction to polymer nanocomposites. A key factor to improve the thermal conduction is the physical contact of the nanofiller (degree of dispersion) along with the presence of covalent bonding at the polymer–filler interface. In this sense, phonons are less scattered at the matrix–filler interface. The thermal conductivity of graphene-based nanocomposites can be highly enhanced. For example, the thermal conductivity of neat epoxy (0.2 W/m K) has been raised to 3–6 W/m K, depending on the content and the type of nanofiller (Figure 10.20) (Potts et al. 2011; Steurer et al. 2009; Sun et al. 2010).

Graphene nanoparticles contribute to the improvement of thermal stability. Improvements in thermal stability are reflected by shifting TGA curves to higher temperatures (Figure 10.21). Improved dimensional stability of graphene-based nanocomposites is because graphene nanoparticles exhibit negative coefficient of thermal expansion (CTE). This is supported by the high contact area and the high stiffness (Liu et al. 2011a; Potts et al. 2011; Steurer et al. 2009).

Polymer crystallinity is affected by the presence of graphene-based filler in different ways, causing increase, reduction, or no change in the crystallinity. Moreover, nanofillers modify the rate of crystallization by acting as a heterogeneous nucleation agents. It has been observed that the glass transition temperature (T_g) is raised by incorporation of graphene-like platelets.

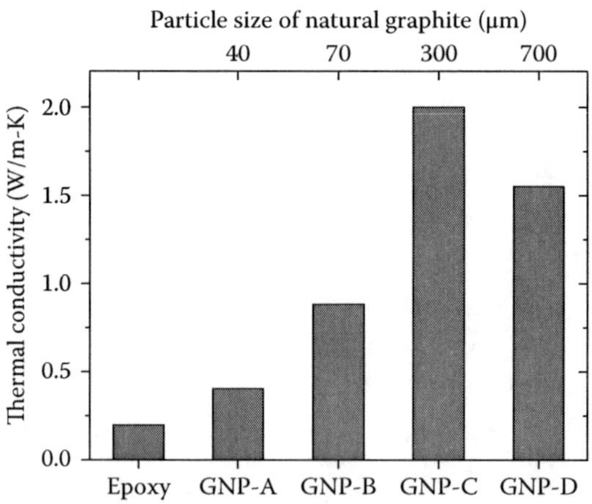

FIGURE 10.20
Thermal conductivities of neat epoxy and GNP–epoxy composites with different GNP fillers at 10 wt% filler loading. Dependence of the thermal conductivity of two-dimensional graphite nanoplatelet-based composites on the nanoparticle size distribution. (From Sun, X. et al., *J. Phys.: Cond. Matt.* 22, 334216, 2010. With permission.)

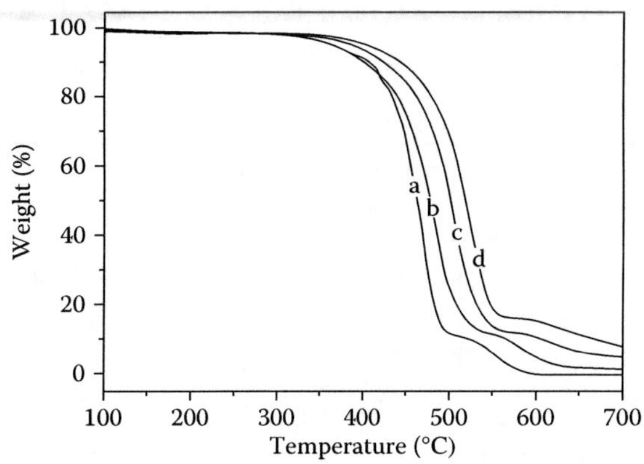

FIGURE 10.21
Effect of graphene content on the thermal stability of polyamide-6–graphite oxide nanocomposites. (From Liu, Y. et al., *J. Mater. Sci.* 46, 882–888, 2011a. With permission.)

The reduced mobility of the polymer chains on the filler surface, due to the attractive forces at the polymer–particle interface, tends to increase T_g. Contrarily, unstable interfaces facilitate the chain polymer mobility, thus lowering the T_g. This effect can be observed not only for polymer chains at the filler surface but also extends away from the interface and gradually diminishes with distance (Potts et al. 2011; Steurer et al. 2009).

Surface modification of graphene can result in improved dispersion of the nanosheets and strong interaction with the matrix. As a consequence, the mobility of the matrix polymer is altered in the interphase zone surrounding the individual sheets. This change in the polymer mobility could result in the improvement of physical properties of nanocomoposites, such as mechanical or barrier properties (Ramanathan et al. 2008).

The combination of premixing procedures (solution, melt, or solid) along with melt compounding can result in improved graphite dispersion. Surface modification of nanofillers strengthens the interaction between polymer matrix and nanofillers. As a result, the dispersion quality and performance are improved. Hydroxyl, epoxides, and carbonyl groups are attached to the graphite surface after the oxidation process. Such chemical moieties could form strong interaction with some polymers or can provide reactive sites for further surface treatments. As mentioned earlier, however, the oxidation process modifies the structure of graphite rings, thus reducing its electrical conductivity. It has been shown that additional thermal reduction of oxidized graphene can restore, in part, the electrical conductivity of treated graphene (Li and Zhong 2011). Grafting techniques for improving miscibility involve the use of polymers with reactive end groups to be attached at the surface of the nanofiller or the use of reactive groups covalently attached to the nanoparticle surfaces (Rozenberg and Tenne 2008).

10.3 Challenges in the Preparation of Polymer Nanocomposites

Significant property enhancement is achieved when nanoparticles are evenly distributed within the polymer matrix. On the contrary, phase-separated structures contribute only slightly to upgrading the properties of the matrix or, in some cases, they could lead to a loss of performance. To achieve the maximum improvement in properties, nanocomposites should satisfy the following requirements: (1) the system must exhibit intercalation/exfoliation characteristics for platelet-based nanocomposites or agglomerates should be reduced to individual particles or agglomerates consisting of very few particles that should be covered by the polymer matrix; (2) polymer/clay structure should be thermodynamically and chemically stable to prevent possible changes that might compromise material performance; (3) there should be good adhesion or other interaction forces at the interface between the nanofiller and the matrix to transfer the benefits of clay presence to the matrix; and (4) the nanofiller should be dispersed uniformly (distributive mixing) in the matrix. Other requirements may be necessary, depending on the expected application (Bhattacharya et al. 2008).

Since nanofillers could influence several properties at the same time, it is essential to separate the effects of (1) crystalline fraction, (2) crystallite morphology, (3) spherulite size, (4) extent of exfoliation, (5) preferred orientation, (6) interaction (adsorption) energy, (7) temperature of test in relation to T_g, (8) properties of surfactant added to the nanofiller and its interaction with the polymer, (9) thermal degradation of additives during processing, (10) processing variables that have allowed agglomeration, and (11) intrinsic properties of polymers and nanofiller and their volume fractions (Chen and Evans 2009).

The success in preparing a good nanocomposite depends on the degree and quality of the nanofiller dispersion. In this sense, nanocomposites exhibiting uniformly dispersed nanofiller would render optimum performance. Two different mixing mechanisms should be considered in the preparation of nanocomposites based on the combination of solid nanofiller and rubbery polymer matrix: dispersive mixing, defined as the breakup of agglomerates to the desired size of solid, and distributive mixing, which involves the uniform distribution of nanocomposites in the matrix. The original agglomerates of nanofiller (organoclays, CNT, metal particles, grapheme, etc.) are broken down to smaller agglomerates or individual particles during mixing by the action of the flow forces or ultrasonic waves. Individual particles of nanofillers, exhibiting large cohesive forces and huge surface area, require considerable energy or mechanical stress to achieve full deagglomeration or exfoliation.

In the pristine state, nanoparticles tend to form clusters, due to strong adhesive bonds acting on the corners, edges, and surfaces, which are not readily broken by hydrodynamic stresses. Aggregates can be broken only if the applied processing stresses exceed the adhesive forces binding the agglomerates. Agglomerate breakup may occur by a rupture and erosion mechanism. The former occurs when the original agglomerate is divided into several small agglomerates with similar or different sizes. The latter occurs as small fragments and is gradually peeled off from large agglomerates. Hydrodynamic stresses cannot break the resulting smaller fragments. Rupture requires more energy in comparison to erosion. Some useful guiding principles for enhancing nanoparticles dispersion include: (1) choosing the polymer–filler pairs with favorable polymer–filler interactions; (2) modifying nanofillers or the polymers (surface modification or polymer grafting); (3) using external fields for mixing (electric, magnetic, ultrasonic, or flow fields) (Özcan-Taskin et al. 2009).

Nanofiller dispersion is favored by the polymer–nanofiller interaction during mixing. Surface modification of nanofiller and/or the use of compatibilizer can contribute to better dispersion. In the case of melt compounding, nanoparticles separation occurs if the hydrodynamic dispersive force is greater than the attractive force between the nanoparticles. Dispersive forces depend on shear rate, viscosity of the matrix, basal spacing, value of the Hamaker constant of clay, and geometrical variables (Cho and Kamal 2004). Other mechanisms such as erosion or surface peeling (exfoliation) are involved in the dispersion of nanoparticles by the effect of hydrodynamic stresses (Bhattacharya et al. 2008; Borse and Kamal 2009). During nanocomposite processing, nanoparticles can also undergo a reversible and continuous transformation from dispersed to clustered states (Starr et al. 2003).

Concerning the surface properties of the nanofiller, various factors influence the occurrence of exfoliation: (1) the chemical structure of the surfactant, (2) the chemical structure and architecture of the compatibilizer (e.g., type block of copolymers), (3) the molecular weight of the polymer matrix, (4) chemical functionalization of the polymer matrix, and (5) the distribution of the functionalization groups on the polymer chain (Han 2010). These factors are important for practically all nanofillers.

Aggregation of particles during processing is a recurrent phenomenon. The probability for particle agglomeration increases with decreasing particle size of the filler. The extent of particle aggregation is a balance between the attractive forces inherent in the particles (mechanical interlocking, electrostatic forces, van der Waals forces, liquid bridges, and solid bridging) and the forces for separating them (shear and elongational forces) during processing. The presence of aggregates is generally detrimental to the performance of composites. In the case of mechanical properties, aggregates could act as fracture initiation sites (Móczó and Pukánszky 2008).

Efforts have been made to quantify the degree of clay dispersion in clay–polymer nanocomposites (Mollet and Kamal 2006). Another proposed method is based on measurement of the free-path spacing distance between clay particles. The dispersion value (D0.1) is estimated from the distribution of the free-path spacing. This dimensionless parameter is independent of filler size, shape, aggregation, concentration, or the existence of additional filler (Luo et al. 2008).

Thermodynamic and transport phenomena–based models provide insight into the effects of various factors (e.g., aspect ratio, interactions, and processing conditions, among others) on the dispersion and stability of the systems. The mean field approach has been helpful in establishing some of the key material characteristics that determine the stability of certain nanocomposites (Balazs et al. 1998; Kim et al. 2004; Kuznetsov and Balazs 2000; Lyatskaya and Balazs 1998). Other models indicate that thermodynamically stable dispersions of nanoparticles in a polymeric liquid can be achieved in systems where the radius of gyration of the linear polymer is greater than the radius of the nanoparticle. Miscibility of nanofiller is promoted when radius of gyration of the linear polymer is larger than the nanoparticle radius (Mackay et al. 2006).

A microscopic model of polymer nanocomposites considers de-mixing over a wide range of interfacial chemistry, chain length, and particle size conditions. A homogeneous mixture (nanofiller–polymer) is the result of a steric stabilization mechanism, due to the formation of thin, thermodynamically stable bound polymer layers on the surface of fillers. The second de-mixing transition occurs at relatively high monomer–particle adsorption energy and is inferred to involve the formation of an equilibrium physical network phase with local bridging of particles by polymers. The miscibility is reduced by increasing polymer chain length, van der Waals attractions among filler particles, and/or particle–monomer size asymmetry ratio (Hooper and Schweizer 2006).

Particle–polymer interactions are not sufficient to predict the dispersion quality. Computer simulation may provide deeper insight regarding dispersion and phase behavior of particles in homopolymer matrices, homopolymers with grafted nanoparticles, or self-assembly and organization of nanoparticles in block copolymer matrices. However, the time and length scales employed in the calculations may not necessarily be possible to assess some of the predictions by experimental observations (Ganesan et al. 2010). A variety of approaches have been proposed to apply multiscale modeling techniques to nanocomposite synthesis (Supova et al. 2011).

Filler dispersion and aggregation in the nanocomposites are directly related to the size, shape, concentration, and filler–filler and polymer–filler interactions. Molecular dynamics simulation results suggest that nanofiller–matrix interaction plays an important role in the dispersion of nanofiller. Macroscopic phase separation occurs at low nanofiller–matrix interaction. Moreover, polymer chains can locally bridge nanofiller particles through these interactions. Results indicate that rapid cooling below T_g of nanocomposites with good dispersion reduces the chances for nanofiller agglomeration, especially for nanocomposites incorporating glassy polymer (Liu et al. 2011b).

References

www.nanoamor.com.

Al-Saleh, M. H., G. A. Gelves et al. (2011). "Copper nanowire/polystyrene nanocomposites: Lower percolation threshold and higher EMI shielding." *Composites Part A: Applied Science and Manufacturing* **42**(1): 92–97.

Azeredo, H. M. C. d. (2009). "Nanocomposites for food packaging applications." *Food Research International* **42**(9): 1240–1253.

Bai, W., J. Holbery et al. (2009). "A technique for production of nanocrystalline cellulose with a narrow size distribution." *Cellulose* **16**(3): 455–465.

Balazs, A. C., C. Singh et al. (1998). "Modeling the interactions between polymers and clay surfaces through self-consistent field theory." *Macromolecules* **31**(23): 8370–8381.

Beyer, F. L., N. C. Beck Tan et al. (2002). "Polymer-layered silicate nanocomposites from model surfactants." *Chemistry of Materials* **14**(7): 2983–2988.

Bhattacharya, S. N., M. R. Kamal et al. (2008). *Polymeric Nanocomposites: Theory and Practice*. Munich, Hanser Gardner Publications.

Bhowmick, A. K., Ed. (2008). *Current Topics in Elastomers Research*. Boca Raton, CRC Press.

Borse, N. K., and M. R. Kamal (2009). "Estimation of stresses required for exfoliation of clay particles in polymer nanocomposites." *Polymer Engineering & Science* **49**(4): 641–650.

Cai, L. F., Y. L. Mai et al. (2007). "Interfacial effects in nano-silica/polypropylene composites fabricated by in-situ chemical blowing." *Express Polymer Letters* **1**(1): 6.

Chatterjee, U., S. K. Jewrajka et al. (2009). "Dispersion of functionalized silver nanoparticles in polymer matrices: Stability, characterization, and physical properties." *Polymer Composites* **30**(6): 827–834.

Chen, B., and J. R. G. Evans (2009). "Impact strength of polymer-clay nanocomposites." *Soft Matter* **5**(19): 3572–3584.

Chen, D., X. Wang et al. (2010). "Electrically conductive poly(vinyl alcohol) hybrid films containing graphene and layered double hydroxide fabricated via layer-by-layer self-assembly." *ACS Applied Materials & Interfaces* **2**(7): 2005–2011.

Chen, G., and W. Zhao (2010). "Polymer/graphite nanocomposites." In: A. K.-T. Lau, F. Hussain, K. Lafdi (Eds.), *Nano- and Biocomposites*, chap. 4. Boca Raton, CRC Press Taylor & Francis Group.

Cho, Y. G., and M. R. Kamal (2004). "Estimation of stress for separation of two platelets." *Polymer Engineering & Science* **44**(6): 1187–1195.

Ciardelli, F., S. Coiai et al. (2008). "Nanocomposites based on polyolefins and functional thermoplastic materials." *Polymer International* **57**(6): 805–836.

Dennis, H. R., D. L. Hunter et al. (2001). "Effect of melt processing conditions on the extent of exfoliation in organoclay-based nanocomposites." *Polymer* **42**(23): 9513–9522.

Denver, H., T. Heiman et al. (2009). "Fabrication of polydimethylsiloxane composites with nickel nanoparticle and nanowire fillers and study of their mechanical and magnetic properties." *Journal of Applied Physics* **106**(6): 5.

Dhakate, S. R., S. Sharma et al. (2008). "Development and characterization of expanded graphite-based nanocomposite as bipolar plate for polymer electrolyte membrane fuel cells (PEMFCs)." *Energy & Fuels* **22**(5): 3329–3334.

Dubois, P., and M. Alexandre (2006). "Performant clay/carbon nanotube polymer nanocomposites." *Advanced Engineering Materials* **8**(3): 147–154.

Dudic, D., M. Marinovic-Cincovic et al. (2008). "Electrical properties of a composite comprising epoxy resin and [alpha]-hematite nanorods." *Polymer* **49**(18): 4000–4008.

Fim, F. d. C., J. M. Guterres et al. (2010). "Polyethylene/graphite nanocomposites obtained by in situ polymerization." *Journal of Polymer Science Part A: Polymer Chemistry* **48**(3): 692–698.

Fornes, T. D., P. J. Yoon et al. (2001). "Nylon 6 nanocomposites: The effect of matrix molecular weight." *Polymer* **42**(25): 09929–09940.

Ganesan, V., C. J. Ellison et al. (2010). "Mean-field models of structure and dispersion of polymer-nanoparticle mixtures." *Soft Matter* **6**(17): 4010–4025.

Gelves, G. A., B. Lin et al. (2008). "Enhancing dispersion of copper nanowires in melt-mixed polystyrene composites." *Journal of Polymer Science Part B: Polymer Physics* **46**(19): 2064–2078.

Golubeva, O., and V. Gusarov (2007). "Layered silicates with a montmorillonite structure: Preparation and prospects for the use in polymer nanocomposites." *Glass Physics and Chemistry* **33**(3): 237–241.

Goyal, R. K., S. D. Samant et al. (2010). "Electrical properties of polymer/expanded graphite nanocomposites with low percolation." *Journal of Physics D: Applied Physics* **43**(36): 365404.

Guggenheim, S., and R. T. Martin (1995). "Definition of clay and clay minerals: Joint report of the AIPEA nomenclature and CMS nomenclature committiees." *Clay and Clay Minerals* **43**(2): 2.

Guo, N., S. A. DiBenedetto et al. (2010). "Nanoparticle, size, shape, and interfacial effects on leakage current density, permittivity, and breakdown strength of metal oxide–polyolefin nanocomposites: Experiment and theory." *Chemistry of Materials* **22**(4): 1567–1578.

Guo, Y., Y. Li et al. (2008). "Fabrication of homogeneous hybrid nanorod of organic/inorganic semiconductor materials." *The Journal of Physical Chemistry C* **112**(22): 8223–8228.

Hamad, W. Y., and T. Q. Hu (2010). "Structure–process–yield interrelations in nanocrystalline cellulose extraction." *The Canadian Journal of Chemical Engineering* **88**(3): 392–402.

Han, C. D. (2010). "On the mechanisms leading to exfoliated nanocomposites prepared by mixing." *Advances in Polymer Science* **231**: 75.

Hasani, M., E. D. Cranston et al. (2008). "Cationic surface functionalization of cellulose nanocrystals." *Soft Matter* **4**(11): 2238–2244.

He, F., S. Lau et al. (2009). "High dielectric permittivity and low percolation threshold in nanocomposites based on poly(vinylidene fluoride) and exfoliated graphite nanoplates." *Advanced Materials* **21**(6): 710–715.

Hooper, J. B., and K. S. Schweizer (2006). "Theory of phase separation in polymer nanocomposites." *Macromolecules* **39**(15): 5133–5142.

Hore, M. J. A., and R. J. Composto (2010). "Nanorod self-assembly for tuning optical absorption." *ACS Nano* **4**(11): 6941–6949.

Hussain, F., M. Hojjati et al. (2006). "Review article: Polymer-matrix nanocomposites, processing, manufacturing, and application: An overview." *Journal of Composite Materials* **40**(17): 1511–1575.

Kamal, M. R., J. U. Calderon et al. (2009). "Surface energy of modified nanoclays and its effect on polymer/clay nanocomposites." *Journal of Adhesion Science and Technology* **23**(5): 663–688.

Kim, K., L. A. Utracki et al. (2004). "Numerical simulation of polymer nanocomposites using self-consistent mean-field model." *Journal of Chemical Physics* **121**: 10766.

Koerner, H., J. Kelley et al. (2009). "ZnO nanorod–thermoplastic polyurethane nanocomposites: Morphology and shape memory performance." *Macromolecules* **42**(22): 8933–8942.

Kumar, A. P., D. Depan et al. (2009). "Nanoscale particles for polymer degradation and stabilization—Trends and future perspectives." *Progress in Polymer Science* **34**(6): 479–515.

Kuznetsov, D. V., and A. C. Balazs (2000). "Scaling theory for end-functionalized polymers confined between two surfaces: Predictions for fabricating polymer/clay nanocomposites." *Journal of Chemical Physics* **112**(9): 11.

Lagaly, G. (1986). "Interaction of alkylamines with different types of layered compounds." *Solid State Ionics* **22**(1): 43–51.

Laoutid, F., L. Bonnaud et al. (2009). "New prospects in flame retardant polymer materials: From fundamentals to nanocomposites." *Materials Science & Engineering: R: Reports* **63**(3): 100–125.

Li, B., and W.-H. Zhong (2011). "Review on polymer/graphite nanoplatelet nanocomposites." *Journal of Materials Science* **46**(17): 5595–5614.

Liu, Y., Z. Chen et al. (2011a). "Synthesis and characterization of polyamide-6/graphite oxide nanocomposites." *Journal of Materials Science* **46**(4): 882–888.

Liu, J., Y. Gao et al. (2011b). "Nanoparticle dispersion and aggregation in polymer nanocomposites: Insights from molecular dynamics simulation." *Langmuir* **27**(12): 7926–7933.

Lu, J. R., W. G. Weng et al. (2005). "Piezoresistive materials from directed shear-induced assembly of graphite nanosheets in polyethylene." *Advanced Functional Materials* **15**(8): 1358–1363.

Luo, Z., R. Littleton et al. (2008). "Quantifying the layer dispersion degree in polymer layered silicate nanocomposites by quantitative transmission electron microscopy." *Microscopy and Microanalysis* **14**(Supplement S2): 594–595.

Lyatskaya, Y., and A. C. Balazs (1998). "Modeling the phase behavior of polymer–clay composites." *Macromolecules* **31**(19): 6676–6680.

Ma, P.-C., N. A. Siddiqui et al. (2010). "Dispersion and functionalization of carbon nanotubes for polymer-based nanocomposites: A review." *Composites Part A: Applied Science and Manufacturing* **41**(10): 1345–1367.

Mack, J. J., L. M. Viculis et al. (2005). "Graphite nanoplatelet reinforcement of electrospun polyacrylonitrile nanofibers." *Advanced Materials* **17**(1): 77–80.

Mackay, M. E., A. Tuteja et al. (2006). "General strategies for nanoparticle dispersion." *Science* **311**(5768): 1740–1743.

Masuda, J., and J. M. Torkelson (2008). "Dispersion and major property enhancements in polymer/multiwall carbon nanotube nanocomposites via solid-state shear pulverization followed by melt mixing." *Macromolecules* **41**(16): 5974–5977.

Matras-Postolek, K., and D. Bogdal (2010). "Polymer nanocomposites for electro-optics: Perspectives on processing technologies, material characterization, and future application polymer characterization." *Advances in Polymer Science* **230**: 221–282.

Móczó, J., and B. Pukánszky (2008). "Polymer micro and nanocomposites: Structure, interactions, properties." *Journal of Industrial and Engineering Chemistry* **14**(5): 535–563.

Mollet, V. A., and M. R. Kamal (2006). "Quantitative characterization of exfoliation in polyamide-6/layered silicate nanocomposites." *Journal of Polymer Engineering* **26**(8–9): 757–782.

Murray, H. H. (2000). "Traditional and new applications for kaolin, smectite, and palygorskite: A general overview." *Applied Clay Science* **17**(5–6): 207–221.

Nassar, N., L. A. Utracki et al. (2005). "Melt intercalation in montmorillonite/polystyrene nanocomposites." *International Polymer Processing* **20**(4): 9.

Özcan-Taskin, N. G., G. Padron et al. (2009). "Effect of particle type on the mechanisms of break up of nanoscale particle clusters." *Chemical Engineering Research and Design* **87**(4): 468–473.

Pan, Y., Y. Xu et al. (2008). "Hybrid network structure and mechanical properties of rodlike silicate/cyanate ester nanocomposites." *Macromolecules* **41**(23): 9245–9258.

Panniello, A., M. L. Curri et al. (2010). "Au based nanocomposites towards plasmonic applications." *AIP Conference Proceedings* **1255**(1): 92–94.

Patra, M. K., K. Manzoor et al. (2008). "Nanotechnology applications for chemical and biological sensors." *Defence Science Journal* **58**(5): 14.

Potts, J. R., D. R. Dreyer et al. (2011). "Graphene-based polymer nanocomposites." *Polymer* **52**(1): 5–25.

Pramoda, K. P., H. Hussain et al. (2010). "Covalent bonded polymer–graphene nanocomposites." *Journal of Polymer Science Part A: Polymer Chemistry* **48**(19): 4262–4267.

Rafiee, M. A., J. Rafiee et al. (2009). "Enhanced mechanical properties of nanocomposites at low graphene content." *ACS Nano* **3**(12): 3884–3890.

Ramanathan,T., A. A. Abdala et al. (2008). "Functionalized graphene sheets for polymer nanocomposites." *Nature Nanotechnology* **3**(6): 327–331.

Reijnders, L. (2009). "The release of TiO_2 and SiO_2 nanoparticles from nanocomposites." *Polymer Degradation and Stability* **94**(5): 873–876.

Rozenberg, B. A., and R. Tenne (2008). "Polymer-assisted fabrication of nanoparticles and nanocomposites." *Progress in Polymer Science* **33**(1): 40–112.

Sauter, C., M. A. Emin et al. (2008). "Influence of hydrostatic pressure and sound amplitude on the ultrasound induced dispersion and de-agglomeration of nanoparticles." *Ultrasonics Sonochemistry* **15**(4): 517–523.

Sciancalepore, C., T. Cassano et al. (2008). "TiO_2 nanorods/PMMA copolymer-based nanocomposites: Highly homogeneous linear and nonlinear optical material." *Nanotechnology* **19**(20): 205705.

Schaefer, D. W., and R. S. Justice (2007). "How nano are nanocomposites?" *Macromolecules* **40**(24): 8501–8517.

Sertsova, A., M. Koroleva et al. (2010). "Fire-resistant polymer nanocomposites based on metal oxides and hydroxides." *Theoretical Foundations of Chemical Engineering* **44**(5): 772–777.

Shah, D., P. Maiti et al. (2005). "Effect of nanoparticle mobility on toughness of polymer nanocomposites." *Advanced Materials* **17**(5): 525–528.

Sinha Ray, S., K. Yamada et al. (2003). "New polylactide/layered silicate nanocomposites. 5. Designing of materials with desired properties." *Polymer* **44**(21): 6633–6646.

Soltani, R., and A. A. Katbab (2010). "The role of interfacial compatibilizer in controlling the electrical conductivity and piezoresistive behavior of the nanocomposites based on RTV silicone rubber/graphite nanosheets." *Sensors and Actuators A: Physical* **163**(1): 213–219.

Spitalský, Z., C. A. Krontiras et al. (2009). "Effect of oxidation treatment of multiwalled carbon nanotubes on the mechanical and electrical properties of their epoxy composites." *Composites Part A: Applied Science and Manufacturing* **40**(6–7): 778–783.

Sposito, G. (2008). *The Chemistry of Soils*. New York, Oxford University Press.

Starr, F. W., J. F. Douglas et al. (2003). "Origin of particle clustering in a simulated polymer nanocomposite and its impact on rheology." *Journal of Chemical Physics* **119** (3): 12.

Steurer, P., R. Wissert et al. (2009). "Functionalized graphenes and thermoplastic nanocomposites based upon expanded graphite oxide." *Macromolecular Rapid Communications* **30**(4–5): 316–327.

Sun, X., P. Ramesh et al. (2010). "Dependence of the thermal conductivity of two-dimensional graphite nanoplatelet-based composites on the nanoparticle size distribution." *Journal of Physics: Condensed Matter* **22**(33): 334216.

Supova, M., G. S. Martynkov et al. (2011). "Effect of nanofillers dispersion in polymer matrices: A review." *Science of Advanced Materials* **3**(1): 1–25.

Tanoue, S., A. Hasook et al. (2006). "Effect of screw rotation speed on the properties of polystyrene/organoclay nanocomposites prepared by a twin-screw extruder." *Journal of Applied Polymer Science* **101**(2): 1165–1173.

Thomas, V., M. Namdeo et al. (2007). "Review on polymer, hydrogel and microgel metal nanocomposites: A facile nanotechnological approach." *Journal of Macromolecular Science, Part A* **45**(1): 107–119.

Tjong, S. C. (2006). "Structural and mechanical properties of polymer nanocomposites." *Materials Science and Engineering: R: Reports* **53**(3–4): 73–197.

Tlili, R., A. Boudenne et al. (2010). "Thermophysical and electrical properties of nanocomposites based on ethylene–vinylacetate copolymer (EVA) filled with expanded and unexpanded graphite." *International Journal of Thermophysics* **31**(4): 936–948.

Tolaymat, T. M., A. M. El Badawy et al. (2010). "An evidence-based environmental perspective of manufactured silver nanoparticle in syntheses and applications: A systematic review and critical appraisal of peer-reviewed scientific papers." *Science of the Total Environment* **408**(5): 999–1006.

Uribe-Calderon, J., and M. R. Kamal (2010). "Evaluation of various surfactants and compatibilizers for preparation of PS/Clay nanocomposites by melt compounding." *Journal of Polymer Engineering* **30**(5–7): 377–412.

Vaia, R. A., K. D. Jandt et al. (1996). "Microstructural evolution of melt intercalated polymer–organically modified layered silicates nanocomposites." *Chemistry of Materials* **8**(11): 2628–2635.

Volynskii, A., N. Nikonorova et al. (2010). "Preparation method for noble metal-polymer matrix nanocomposites." *Colloid Journal* **72**(4): 464–470.

Wakabayashi, K., C. Pierre et al. (2008). "Polymer–graphite nanocomposites: Effective dispersion and major property enhancement via solid-state shear pulverization." *Macromolecules* **41**(6): 1905–1908.

Wang, G., Y. Deng et al. (2008). "Fabrication of radial ZnO nanowire clusters and radial ZnO/PVDF composites with enhanced dielectric properties." *Advanced Functional Materials* **18**(17): 2584–2592.

Wei, Q.-B., Y.-L. Luo et al. (2008). "Assembly of Cu nanoparticles in a polyacrylamide grafted poly(vinyl alcohol) copolymer matrix and vapor-induced response." *Sensors & Actuators B: Chemical* **134**(1): 49–56.

Whitby, R. L. D., W. K. Hsu et al. (2004). "Novel nanoscale architectures: Coated nanotubes and other nanowires." *Philosophical Transactions of the Royal Society of London. Series A: Mathematical, Physical and Engineering Sciences* **362**(1823): 2127–2142.

White, S. I., R. M. Mutiso et al. (2010). "Electrical percolation behavior in silver nanowire–polystyrene composites: Simulation and experiment." *Advanced Functional Materials* **20**(16): 2709–2716.

Wissert, R., P. Steurer et al. (2010). "Graphene nanocomposites prepared from blends of polymer latex with chemically reduced graphite oxide dispersions." *Macromolecular Materials and Engineering* **295**(12): 1107–1115.

Wong, S.-C., E. M. Sutherland et al. (2006). "Materials processes of graphite nanostructured composites using ball milling." *Materials and Manufacturing Processes* **21**(2): 159–166.

Work, W. J., K. Horie et al. (2004). "Definitions of therms related to polymer blends, composites, and multiphase polymeric materials." *Pure and Applied Chemistry* **76**(11): 23.

Wu, T.-L., T.-S. Lo et al. (2010). "Effect of dispersion on graphite nanosheet composites." *Polymer Composites* **31**(2): 292–298.

Xia, X., S. Cai et al. (2006). "Preparation, structure and thermal stability of Cu/LDPE nanocomposites." *Materials Chemistry and Physics* **95**(1): 122–129.

Xie, W., J. M. Hwu et al. (2003). "A study of the effect of surfactants on the properties of polystyrene-montmorillonite nanocomposites." *Polymer Engineering & Science* **43**(1): 214–222.

Yang, Y., J. Wang et al. (2009). "Exfoliated graphite oxide decorated by PDMAEMA chains and polymer particles." *Langmuir* **25**(19): 11808–11814.

Yeh, J.-R., F.-Y. Tseng et al. (2010). "Hard nanocomposite comprising stereoregular syndiotactic polypropylene with chemically bonded gold nanoparticles." *Journal of Polymer Science Part A: Polymer Chemistry* **48**(20): 4400–4407.

Zhan, Y., Y. Lei et al. (2011). "Electrical, thermal, and mechanical properties of polyarylene ether nitriles/graphite nanosheets nanocomposites prepared by masterbatch route." *Journal of Materials Science* **46**(3): 824–831.

Zhang, W. L., B. J. Park et al. (2010a). "Colloidal graphene oxide/polyaniline nanocomposite and its electrorheology." *Chemical Communications* **46**(30): 5596–5598.

Zhang, J., G. Shen et al. (2010b). "Individual nanocomposite sheets of chemically reduced graphene oxide and poly(N-vinyl pyrrolidone): Preparation and humidity sensing characteristics." *Journal of Materials Chemistry* **20**(48): 10824–10828.

Zhao, J., Y. Fu et al. (2008). "Polyhedral oligomeric silsesquioxane (poss)-modified thermoplastic and thermosetting nanocomposites: A review." *Polymers and Polymer Composites* **16**(8): 18.

Zhao, X., Q. Zhang et al. (2010). "Enhanced mechanical properties of graphene-based poly(vinyl alcohol) composites." *Macromolecules* **43**(5): 2357–2363.

11

Polymer-Functionalized Graphene via Atom Transfer Radical Polymerization Route for Graphene-Based Polymer Nanocomposites

Arun K. Nandi*, Rama K. Layek, Sanjoy Samanta, and Dhruba P. Chatterjee

CONTENTS

* Correspondence to: Arun K. Nandi (psuakn@mahendra.iacs.res.in; psuakn@yahoo.com)

11.1 Introduction

A great deal of research has recently been initiated on "graphene," a two-dimensional (2D) hexagonally arranged sp^2-hybridized carbon nanostructure, for its promising applications in the fields of sensors (Schedin et al. 2007), batteries (Geim and Novoselov 2007), supercapacitors (Wang et al. 2009), hydrogen storage (Du et al. 2010; Patchkovskii et al. 2005), nanoelectronics (Du et al. 2010; Geim and Novoselov 2007), and polymer nanocomposites (Geim and Novoselov 2007; Ramanathan et al. 2008; Stankovich et al. 2006a). Graphene sheets offer extraordinary electronic (Gómez-Navarro et al. 2007), thermal (Balandin et al. 2008), and mechanical properties (Gómez-Navarro et al. 2008), and to fully exploit these properties graphene should be incorporated into polymer matrices. High-performance polymer–graphene composites may then be fabricated by uniform filler dispersion and fine interfacial control, which are difficult due to the presence of a strong cohesive force among the layers of graphene. This difficulty in the preparation of processable graphene sheets hinders the application of graphene in various fields tempting an increase in research activity, which is documented in the recent literature. So far, two main methods (physical and chemical) have been developed for the dispersion of graphene sheets. In the physical method, thermal expansion of graphite oxide (produced by oxidation) followed by ultrasonication in polar organic solvents (McAllister et al. 2007) and ultrasonication of graphite oxide in the presence of a surfactant in water (Lotya et al. 2010) are the general procedures followed for the dispersion of graphene. In the chemical method, oxidation/reduction techniques (Balazs et al. 2006; Gao et al. 2009; Pradhan et al. 2009) are used to obtain dispersible graphene sheets.

11.2 Dispersion of Graphene

The dispersion of graphene in solvents with low Hildebrand solubility parameters and low boiling points is poor, and the best solvent for graphene dispersion is a solvent having a Hildebrand solubility parameter of approximately 23 $MPa^{1/2}$ and a high boiling point. It is noted that *N*-methyl-pyrrolidone (NMP) has a solubility parameter of about 23 $MPa^{1/2}$ and a boiling point close to 200°C, which makes it possible to form a good and stable dispersion of graphene (O'Neill et al. 2011). It is the probable reason for the difficulty of dispersion of graphene in a majority of solvents usually used to make polymer composites. Graphite oxide can be dispersed in almost all of the solvents for the just-sonicated samples, except dichloromethane and *n*-hexane and, to a lesser extent, methanol and *o*-xylene. However, many of these dispersions (in acetone, ethanol, 1-propanol, dimethyl sulfoxide

[DMSO], and pyridine) display only short-term stability and precipitate completely in a matter of hours or a few days (Paredes et al. 2008). But the graphite oxide dispersions in four organic solvents (ethylene glycol; *N,N'*-dimethyl formamide [DMF], NMP, and tetrahydrofuran [THF]) exhibit long-term stability comparable to that of the same material in water.

To produce high-performance polymer nanocomposites of graphene uniform dispersion, delicate morphological organization, fine interface control, and ease of processing are very essential (Putz et. al. 2006; Schadler et al. 2007). Surfactants and polyelectrolytes are often used to enhance exfoliation of graphene through electrostatic and physical interactions (Stankovich et al. 2006b,c). These techniques rely on establishing noncovalent intermolecular interactions, such as hydrogen bonding or electrostatic interactions, to adhere molecular systems with 2D graphene or a graphene oxide (GO) monolayer. The solubility of a noncovalently attached conjugated triblock copolymer (PEG-OPE-PEG)/reduced GO (RGO; by π–π interaction) is tested in a series of solvents, for example, toluene, THF, chloroform, acetone, DMF, ethanol, methanol, DMSO, and water. It is found to disperse well in these solvents, resulting in clear and homogeneous solutions ([PEG-OPE-rGO] = 1.22 mg/mL) that can last for months exhibiting excellent stability of dispersion (Qi et al. 2010). Although this strategy of exfoliation of graphene sheets are rather relatively easy, the inherent instability of these supramolecular systems makes it difficult to control and quantify the nanomaterials produced (Stankovich et al. 2006b). On the other hand, covalent attachment of small molecules (Kamaras et al. 2003; Qi et al. 2010; Stankovich et al. 2006c) or polymers to the extended aromatic surfaces of graphene and the resulting formation of composites generate materials that are much more stable under variable external conditions. Also, the ability to introduce a band gap in graphene in a simple manner by covalent functionalization facilitates the fabrication of graphene-based devices. The transformation of carbon centers from sp^2 to sp^3 hybridization introduces a barrier to electron flow by opening a band gap and allows the generation of insulating and semiconducting regions in graphene (Kamaras et al. 2003).

Organic molecules such as phenyl isocyanate and porphyrin are also attached to graphene surfaces to improve its dispersion in nonpolar polymers (Fang et al. 2009; Stankovich et al. 2006c; Xu et al. 2009). Sulfonation of GO is also an important route to produce water-soluble graphene (Si et al. 2008), and it can supramolecularly interact with many polymers to form a good dispersion in the composite. But to achieve a uniform stable dispersion of graphene and to induce different nanostructures of graphene, polymer functionalization is necessary (Chang et al. 2011; Cote et al. 2009; Goncalves et al. 2010; He and Gao 2010; Kan et al. 2011; Kim and Macosko 2008; Layek et al. 2010; Lee et al. 2010; Lego et al. 2009; Li et al. 2008, 2009, 2010; Liu et al. 2009, 2010; Prasad et al. 2009; Salavagione et al. 2009; Shen et al. 2009; Sun et al. 2010; Tkalya et al. 2010; Vickery et al. 2009; Yang et al. 2009). The action of polymer functionalization is threefold: (1) It increases the spacing between individual graphene sheets facilitating exfoliation, (2) the functional groups of the grafted polymers may assist with the miscibility of modified graphene in the matrix of other polymers through specific interactions yielding a high-performance composite, and (3) the formation of a band gap due to covalent functionalization of the polymer produces new materials suitable for electronic/optoelectronic devices.

In the literature, there are a few reports on the surface modification of graphene sheets through polymer grafting. Although both "grafting-to" (Liu et al. 2009; Salavagione et al. 2009; Shen et al. 2009; Sun et al. 2010) and "grafting-from" (Chang et al. 2011; Goncalves et al. 2010; He and Gao 2010; Layek et al. 2010; Lee et al. 2010; Lego et al. 2009; Li et al. 2009; Yang et al. 2009) approaches have been successfully applied, for proper modification the latter is somewhat more advantageous than the former due to lesser steric hindrance. During chain propagation in the grafting-from technique, monomers enter the gallery of graphene sheets; the growing chain increases the interlayer spacing and helps to exfoliate the graphene sheets in a thin layer, which is rather difficult using the grafting-to approach. The physicochemical and optoelectronic properties of graphene sheets depend on the nature of surface modification of graphene; hence, tuning of modification is essential in order to achieve the optimum properties for proper applications. Surface-initiated (SI) controlled polymerization is an ideal and advanced technique for the modification of graphene sheets. Graphene sheets covalently functionalized with poly-L-lysine (PLL) are soluble in water and biocompatible, which makes such graphene a novel material showing much promise in biological applications (Shan et al. 2009a). Polyacetylene-functionalized graphene is easily soluble in common organic solvents (Xu et al. 2011) and is useful in the electronics industry. The covalent functionalization of graphene by poly(methyl methacrylate) [PMMA], poly(butyl acrylate) [PBA], and polystyrene (PS; Fang et al. 2009; Lee et al. 2010) are very important to produce polymer composites with highly improved mechanical and conducting properties. The polymer chain length of the grafted polymer also has much influence on the solubility of the functionalized graphene, for example, at the same concentration of PS-grafted graphene in the sample the shortest polymer chains contained the largest amount of GO yielding the darkest colored solution (Lee et al. 2010).

11.3 Graphene–Polymer Composites

Presently a significant amount of research on polymer–graphene nanocomposites is initiated for dramatically improving their mechanical and electrical properties (Cote et al. 2009; Lu et al. 2006; Prasad et al. 2009; Ramanathan et al. 2008; Stankovich et al. 2006a; Tkalya et al. 2010; Vickery et al. 2009). The dispersion of graphene in these composites is primarily observed using scanning electron microscopy (SEM). The SEM images of expanded graphite–PMMA and

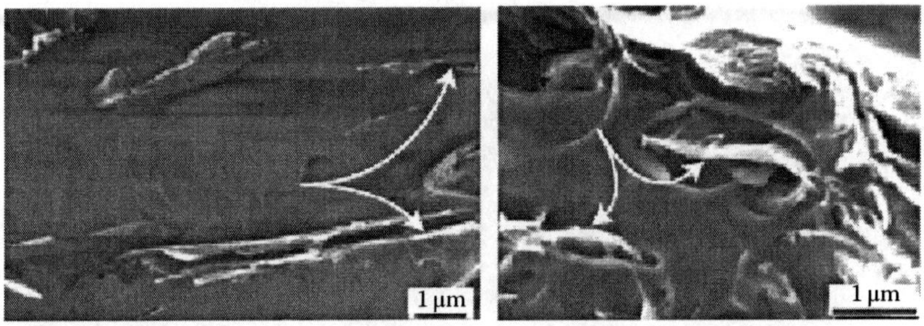

(a) (b)

FIGURE 11.1
(a) Expanded graphite exhibits thicker protruding plates with poor bonding to the polymer matrix. (b) The size scale (nanosheet thickness) and morphology (wrinkled texture) of functionalized graphene sheet (FGS) lead to strong interfacial interaction with the host polymer, as illustrated by polymer adhesion to the pulled-out FGS. (Reprinted by permission from *Nature Nanotechnology*, Ramanathan et al. 3, 327, 2008. Copyright 2008.)

functionalized graphene sheet (FGS)-PMMA obtained under an acceleration voltage of 3 kV, showing fracture surface topography, is given in Figure 11.1 (Ramanathan et al. 2008). The size, scale (nanosheet thickness), and morphology (wrinkled texture) of FGSs as well as their surface chemistry indicate strong interfacial interaction with the host polymer, as illustrated by polymer adhesion to the pulled-out FGS sheet. In contrast, expanded graphite–PMMA exhibits thicker protruding plates with poor bonding to the host polymer.

The major emphasis in polymer composite research is on increasing the mechanical property of the host polymer. The PMMA/graphene nanocomposites (Ramanathan et al. 2008) exhibit dramatic increase in modulus, ultimate strength, and thermal stability, which are comparable to those of single-walled carbon nanotube (CNT) composites. Graphene/poly(vinyl alcohol) [PVA] composite films exhibit significant improvement in mechanical properties, for example, a 150% increase in tensile strength for 1.8 vol% graphene nanosheets, which indicates efficient load transfer between the graphene nanosheets and the matrix. The Young's modulus of the composite is more than 10 times greater than that of PVA (Zhao et al. 2010). With the addition of only 0.9 wt% PS-modified graphene sheets to PS, the resulting PS composite film shows 57% and 70% increases in Young's modulus and tensile strength, respectively, over PS (Fang et al. 2009). In poly(vinylidene fluoride) [PVDF], sample storage modulus (G′) increases with increase in PMMA-functionalized graphene (MG) content and the highest increase of G′ is 124%, which is comparable to that of PVDF–multiwalled nanotube (MWNT) nanocomposites. The stress at break increases dramatically from 39 to 100 MPa for 5% (w/w) MG with a 157% increase and the Young's modulus of PVDF increases from 1.44 to 6.07 GPa for 5% MG showing a percentage increase of 321% (Layek et al. 2010). The Young's modulus increases by 31% for 0.1 wt% thermally exfoliated graphene/epoxy (Rafiee et al. 2009), 120% for 1 wt% chemically reduced graphene/thermoplastic polyurethane (Liang et al. 2009), and 200% for 0.25 wt% for thermally exfoliated graphene/silicone foam (Verdejo et al. 2008). Prasad et al. (2009) observed synergy in the mechanical property enhancement of PVA composite with a mixture of two different carbon nanomaterials. In a recent report, an oriented structure of GO is produced at high GO concentrations (40–60 wt%) in the PVA composite showing a significant enhancement in mechanical properties (1019% in G′, 1300% in Young's modulus, and 186% in tensile strength) over PVA in the plane of alignment (Putz et al. 2010).

The electrical conductivity of the graphene–polymer composite is also greatly increased compared to that of pristine polymer. Stankovich et al. (2006a) reported a PS–graphene nanocomposite showing 10^{-3} S/cm conductivity, which is sufficient for many electrical applications at only 1% (v/v) of graphene. The system exhibits the lowest percolation threshold of 0.1% (v/v) for room temperature conductivity among different carbon nanostructures. In the epoxy resin/graphite nanocomposite, the electrical conductivity is shown to increase by 12 orders (Lu et al. 2006). High conductivity is achieved using latex-based technology of composite formation of graphene (1.6 wt%) in PS (Tkalya et al. 2010). In polyester/exfoliated graphite nanocomposites, the electrical percolation data compared well with the rigidity percolation from rheological measurements (Kim and Macosko 2008). Pure PVDF is an insulator having a direct current (dc) conductivity of 5.3×10^{-13} S/cm, and on addition of MG the dc conductivity increases sharply showing a percolation threshold at 3.8% (w/w) filler containing neat 1.9% (w/w) graphene. The highest dc conductivity achieved is 10^{-7} S/cm for the addition of 20% filler at 30°C (Layek et al. 2010). The conductivity of the electrochemically modified graphene (ECG)/PVA composite is 10^{-7} S/cm for 6 wt% filler (Hu and Chen 2010). In the graphene/PS composite, the conductivity is 0.15 S/cm for 1.2 wt% graphene (Tkalya et al. 2010); for graphene/polycarbonate composites, it is 0.5 S/cm for 2.2 vol% graphene (Yoonessi and Gaier 2010); and for RGO/PVDF composites, it is 10^{-4} S/cm for 4 wt% RGO (Ansari and Giannelis 2009). Thus, research efforts on graphene–polymer nanocomposites have gained considerable momentum to increase their mechanical properties for household applications and to induce electronic property in the composites for application in the electronic/optoelectronic industry.

11.4 Atom Transfer Radical Polymerization

The last two decades have witnessed rapid research developments and initiatives in the field of controlled radical polymerizations (CRPs). Various CRPs, such as stable free radical polymerization (SFRP) (Georges et al. 1993), reversible addition-fragmentation chain transfer (RAFT) (Chiefari et al. 1998), and atom transfer radical polymerization (ATRP) (Kato et al. 1995; Wang and

Matyjaszewski 1995), have been explored during this period; among them, ATRP has emerged as the most robust and simple technique for controlled polymerization. This process is widely applied for the synthesis of various polymeric systems, keeping precise control on their structure, molecular weight, and polydispersity. Its tolerance to diverse functional groups and applicability in rather a broad range of reaction conditions (solvents, temperatures, or catalysts) has forced the scientific community to accept this technique from both academic and industrial perspectives. The ATRP allows the synthesis of various industrially attractive block copolymers of amphiphilic, thermoplastic, elastomeric, or ionomeric nature and makes possible the synthesis of various polymeric architectures as star, hyperbranched, or dendrimers, which are very attractive commercially. Particularly, the advent of "initiators for continuous activator regeneration" (ICAR) (Matyjaszewski et al. 2006), "activator generated by electron transfer" (AGET) (Matyjaszewski et al. 2007), "activator regenerated by electron transfer" (ARGET) (Jakubowski et al. 2006), single-electron-transfer living radical polymerization (SET-LRP) (Percec et al. 2006), and electrochemically mediated ATRP (eATRP) (Magenau et al. 2011) have made it possible to limit the use of copper catalyst in a few parts-per-million levels without losing control over the polymerization and have thus improved the cost–performance balance to make the process even more attractive commercially.

After tuning a standard protocol for SI-ATRP (Edmondson et al. 2004) from the graphene-based system and ensuring appreciable control over molecular weight, polydispersity index, and polymer end groups, one might think about performing block copolymerization. Block copolymers are always very interesting as they present diverse properties in a single polymer chain almost without affecting the properties of individual blocks. Synthesis of block copolymers by SI-ATRP from CNTs has already been reported (Xu et al. 2008). Considering its cost of production and widespread applicability, graphene shows brighter promise than CNTs. Therefore, future research on graphene may well be directed toward preparation of graphene materials that show compatibility in various solvents or polymeric matrices along with responsiveness toward external stimuli. To this end, block copolymer–grafted graphene sheets may be very useful. Therefore, selection of suitable "smart" polymeric blocks useful for better realization of applications of graphene-based systems, such as electronic, optical, mechanical, or biomedical applications, and their successful attachment from graphene surfaces may be the main thrust of future research.

As discussed earlier in this section, ATRP should be quite helpful in covalent polymer functionalization of graphene sheets through the application of SI-ATRP. The following section gives a brief introduction on the chemistry involved in ATRP. Scheme 11.1 is a general scheme for Cu-based ATRP as proposed by Matyjaszewski (Matyjaszewski and Xia 2001) and Sowamoto (Kamigaito et al. 2001).

An alkyl halide initiator reacts with Cu^{I} complex to produce alkyl radical and a Cu^{II}-based complex, which acts as a persistent radical. Therefore,

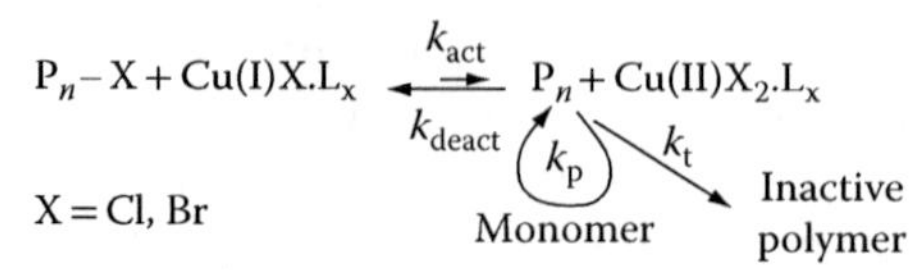

SCHEME 11.1
Reaction mechanism of ATRP.

a relatively low concentration of it remains in the medium, reacting with monomers in the propagation step with minimum terminations. Transition metal complexes other than copper-based systems such as Ru, Fe, and Mn may also be used. However, Cu-based ATRP has received maximum attention so far. In the literature, ATRP reactions have been used well in bulk monomer, solution, and dispersion or from surface (Kato et al. 1996).

11.5 Grafting Processes

Bulk quantities of graphene can be easily processed by chemical oxidation, yielding GO. The resulting oxygenated graphene sheets covered with hydroxyl, epoxy, and carboxyl groups offer tremendous opportunities for further functionalization by either grafting-to or grafting-from techniques, which are discussed in Sections 11.5.1 through 11.5.3.

11.5.1 Grafting-to Methods

In these methods, graphene is first oxidized to GO using Hummer's method (Hummer and Offeman 1958) or Staudenmain method (Schniepp et al. 2006; Staudenmaier 1898), followed by attachment with polymers. In the grafting-to technique, a simplest method is the direct covalent linking of the polymer chain on the GO surface (Salavagione et al. 2009). Direct esterification reaction between the carboxylic acid group of GO and PVA has been achieved in the presence of *N,N'*-dicyclohexylcarbodiimide (DCC) and 4-dimethylaminopyridine (DMAP) catalyst. In another modified procedure, the carboxylic group of GO is transformed into acid chloride followed by esterification with PVA. The resultant ester is then reduced with hydrazine to produce a molecularly linked graphene/PVA system. A recent approach of grafting-to techniques is the use of click chemistry using azidoterminated PS (Sun et al. 2010). Oligothiophenes with amino terminal groups are attached to the GO surface by grafting-to technique to produce a donor (thiophene)–acceptor (graphene) nanohybrid (Liu et al. 2009). A physical method of grafting-to approach involves the use of π–π stacking of pyrene-terminated poly(*N*-isopropyl acrylamide) produced by the RAFT polymerization technique; the pyrene and graphene interact with each other through π–π interaction (Liu et al. 2010).

11.5.2 Grafting-from Method

In a simple method of grafting-from technique, RGO and the monomers such as acrylic acid and acryl amide are polymerized in presence of $(NH_4)_2S_2O_8$ by free radical polymerization technique (Shen et al. 2009). Recently, 2D graphene polymer brush was reported from the polymerization of glycidyl methacrylate with azobis isobutyro nitrile (AIBN) on the GO surface. Other vinyl monomers can also be polymerized in a similar fashion (Kan et al. 2011). After polymerization, bundles of hairy polymers are grafted on the graphene sheets. It is believed that polymer chains are directly grafted onto the graphitic surface by such radical addition/coupling reaction between polymer radicals and the double bonds of graphene surfaces. But normal free radical polymerization has a few disadvantages of its own, such as formation of polymers having very high molar mass, gelation, and so on, which may have quite deleterious effects on the polymer-grafted graphene sheets. For example, if a nonconducting polymer chain having a very high molecular weight is grafted onto graphene sheets, then the conductivity of the polymer-grafted graphene sheets will be decreased to a very high extent, which is really very much undesirable. Again, the physicochemical, optical, and electrical properties of graphene are largely affected by this method of surface modification as normal free radical polymerization yields polymers of broad molecular weight distribution, affecting the properties of graphene sheets. Hence, application of controlled free radical polymerization initiated from graphene surfaces is more acceptable. Controlled free radical polymerization provides a method of generating polymer molecules with the desired degree of chain length, thereby avoiding undesired loss of properties. Thus, this method is very effective as it provides different approaches for tailoring graphene materials with required properties.

The most important method of grafting-from technique is ATRP, and it is noted that maximum number of reports in the literature dealing with polymer functionalization of graphene by ATRP has adopted grafting-from approach (Goncalves et al. 2010; Lee et al. 2010; He and Gao 2010; Lego et al. 2009; Li et al. 2009). The ATRP has been used in both strategies, in case of either graphene-based filler modification for composite formation or in situ composite formation where both the graphene-anchored polymer and the matrix polymer are the same and synthesized in one pot. Because of its robust nature, it is applicable to a variety of monomers and also at different reaction conditions. So far, methyl methacrylate (MMA) (Goncalves et al. 2010; He and Gao 2010; Layek et al. 2010; Lee et al. 2010; Lego et al. 2009; Li et al. 2009), *t*-butyl acrylate (tBA) (Li et al. 2010), 2-dimethyl amino ethyl methacrylate (DMAEMA) (Yang et al. 2009), methacrylamide (MAAM) (Chang et al. 2011), and styrene (Fang et al. 2009) are polymerized by ATRP techniques on the GO surface.

In an approach, ATRP grafting of poly(dimethyl aminoethyl methacrylate) [PDMAEMA] has been reported (Yang et al. 2009) and, accordingly, an amphiphilic graphene-based system is obtained following the amphiphilic nature of the PDMAEMA polymer. In another very interesting study, Layek et al. (2010) prepared a difficult PVDF/graphene nanocomposite. The major bottleneck in this case is the high degree of noncompatibility of these materials. In this study, graphene nanosheets are first modified with PMMA chains by SI-ATRP; then by exploiting the PMMA-PVDF compatibility, the desired nanocomposite is prepared. Lee et al. (2010) used ATRP by the grafting-from approach for polymer functionalization with MMA, butyl acrylate, and styrene, and after polymerization they cleaved the polymer chains from the graphene surface, characterized them thoroughly, and claimed a controlled polymerization. As expected, they reported appreciable development in the solubility of GO in nonpolar solvents.

11.5.3 Grafting-to versus Grafting-from Techniques

The grafted polymer is well controlled by virtue of its chain length, molecular weight, and so on, by grafting-to (Liu et al. 2009; Salavagione et al. 2009; Shen et al. 2009; Sun et al. 2010) or by grafting-from (Chang et al. 2011; Fang et al. 2009; Goncalves et al. 2010; He and Gao 2010; Layek et al. 2010; Lee et al. 2010; Lego et al. 2009; Li et al. 2009; Yang et al. 2009) techniques. The grafting-to approach involves covalent binding of presynthesized and end-functionalized polymer chains to the surface of graphene sheets of appropriate functionality under suitable conditions. Since the polymer is presynthesized, one can prepare it with the help of the so-called control living polymerization technique. For example, Salavagione et al. (2009) used PVA to graft with GO through esterification reaction. An alternative method has been used by Sun et al. (2010) in which azide-functionalized PS (PSt) was synthesized first using the ATRP technique followed by coupling with proparzyl alcohol–modified GO. In this case, a five-member ring acts as a bridge to covalently connect GO and PSt.

However, it is very difficult to control graft density in the grafting-to technique and it always yields low graft density mainly due to steric hindrance. The attachment of presynthesized polymer moiety with graphene is generally associated with long reaction times because of the low diffusion constant of the polymer. From these points of view, the grafting-from approach is a prospective technique and, contrary to the grafting-to approach, it involves polymerization of monomers starting from the graphene's surface previously modified with a suitable initiator moiety. As the chains are growing from the surface of graphene, the rate-controlling step for propagation depends on the diffusion of monomers to the chain ends, thereby yielding a well-defined brush-like structure with high grafting density and thus overcoming the problems associated with the grafting-to approach. Another promising benefit of the grafting-from approach is the creation of thin graphene sheets. In the grafting-from technique, the initiation starts from the initiator attached to the surface of graphene followed by propagation. During the chain propagation, the interlayer spacing of graphene gradually increases with increase in size of the growing polymer chain. When the distances between two graphene sheets are large, thin graphene sheets are generated, which are not possible to implement in the grafting-to approach. Thus in case of graphene modification, the grafting-from technique is more beneficial than the grafting-to approach.

11.6 Functionalization of Graphene by ATRP

At first, it is necessary to attach an ATRP initiator on the graphene sheet followed by polymerization from the anchored initiator by ATRP technique.

11.6.1 Attachment of ATRP Initiator on Graphene

Growth of polymers by ATRP may be done from both graphene and GO sheets. Majority of researchers have used a straightforward strategy by attachment of ATRP initiating sites either directly from the GO surface (basal plane or periphery) on the functional groups (–OH, –COOH, etc.) (Chang et al. 2011; Goncalves et al. 2010; He and Gao 2010; Layek et al. 2010; Lee et al. 2010; Lego et al. 2009; Li et al. 2009; Yang et al. 2009) or by coupling a small molecule first from the graphene surface (containing pendant functional groups) and then anchoring the ATRP initiator at the other end (Fang et al. 2009). The GO sheets by virtue of a good number of polar functionalities attached on the surface show appreciable degree of exfoliation in polar/polar protic solvents. Therefore, a much better dispersion of GO is produced, which helps in achieving better reactivation during initiator attachment for ATRP. Hence, a control over ATRP reactions from GO dispersion is likely to be better realized, compared to SI polymerizations (SIPs) on the graphene surface in which attachment of a spacer molecule between the ATRP initiator and graphene sheets is helpful to maintain a better degree of exfoliation.

The attachment of ATRP initiator on GO is usually made either from the periphery or from the whole surface, that is, basal plane and periphery. From solubility or dispersion points of view, attachment of ATRP initiators followed by polymerization from the whole surface of graphene is fruitful, although in this case the conductivity of the material is low enough in the reduced state due to the absence of conjugation. On the other hand, connection of initiator and subsequent polymerization from the periphery of GO makes the material moderately dispersible with high conductivity in the reduced state because of the unaltered conjugated structure of basal planes.

11.6.2 Polymerization

The general procedure of polymer grafting on graphene surfaces by ATRP is shown in Scheme 11.2. First, graphite is converted into GO (step 1), which is modified with an ATRP initiator (step 2). The ATRP initiator–modified GO is used to polymerize the monomer in the presence of CuCl/CuBr

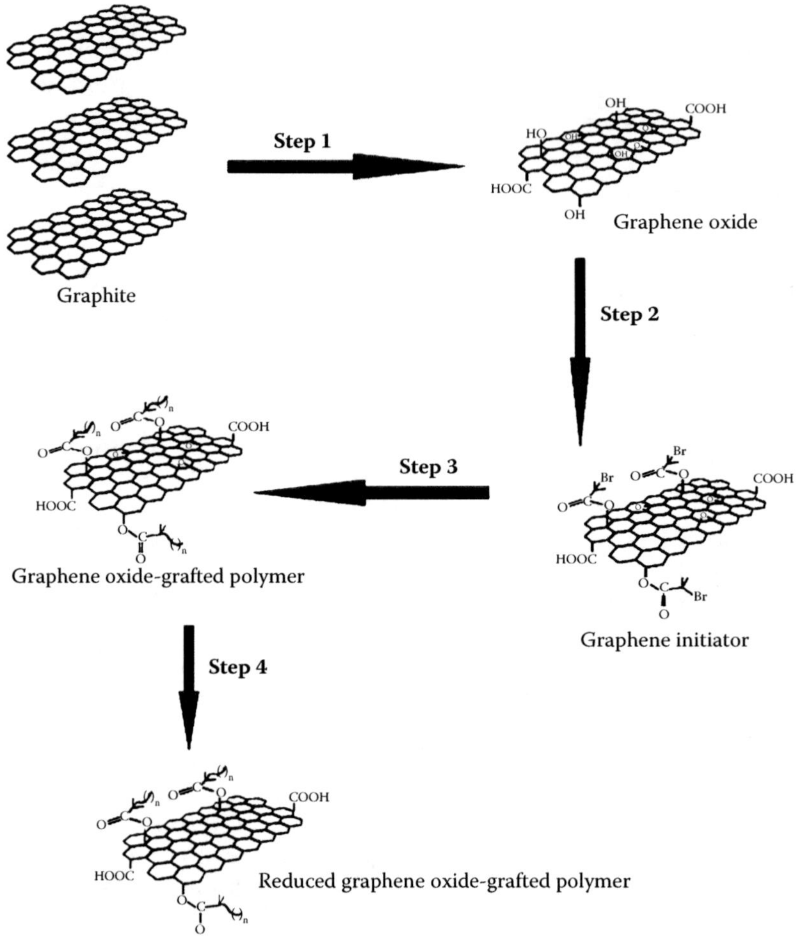

SCHEME 11.2
Reaction scheme for the preparation of polymer-modified graphene via ATRP route: Reagents used in step 1 are H_2SO_4. $NaNO_3$ and $KMnO_4$; reagent used in step 2 is initiator; reagents used in step 3 are catalyst, ligand, monomer, and solvent; and reagent used in step 4 is hydrazine.

catalyst and ligand to produce a GO–polymer (step 3), which is then reduced with hydrazine to produce polymer-modified graphene (step 4) (Chang et al. 2011; Goncalves et al. 2010; He and Gao 2010; Layek et al. 2010; Lee et al. 2010; Lego et al. 2009; Li et al. 2009; Yang et al. 2009). The choice of ATRP initiator used to modify graphene may be different for polymerization of different monomers; for example, GO-Br, GO-Cl, GO–bromoisobutyrate, GO–bromomethylpropionate, and so on. The ligand generally used in ATRP is pentamethyldiethylenetriamine (PMDETA), and the reaction medium may be any of tetrahydrofuran (THF), acetonitrile, or *N,N'*-dimethyl formamide (DMF).

The GO is prepared from graphite powder by oxidizing it with $KMnO_4$/ $NaNO_3$ mixture in concentrated H_2SO_4 medium using Hummer's method (Hummer and Offeman 1958). In an inert reaction vessel (nitrogen atmosphere), 250 mg of GO is mixed with 15 mL pyridine, and the GO is dispersed by sonication. Then, 2-bromobutyryl bromide (BIB) is added slowly at 0°C and the reaction is continued for 3 hours at 0°C and then for 24 hours at 30°C. The product is collected by repeated washing and centrifugation with water. Finally, BIB-attached graphene (BIBG) is obtained and is dried in vacuum at 40°C for 2 days. To prepare graphene oxide–polymethyl methacrylate or GO-PMMA, 100 mg of BIBG, 10 mg of CuCl, 20.4 μL of PMDETA, and 1 mL of DMF are taken in an inert reaction vessel, purged with nitrogen gas, and sealed. It is noted that 1 mL of nitrogen-purged MMA is injected into the vessel using a syringe and the system is sealed. The reaction vessel is kept at 60°C for 48 hours. At the end of the reaction, it is seen that viscosity of the medium has increased. The mixture is diluted with chloroform and precipitated with methanol. Finally, the product obtained is washed with $CHCl_3$ to remove non-grafted PMMA, if formed. The product obtained is dried in vacuum for 2 days. To prepare RGO-PMMA (MG), 100 mg of GO-PMMA is dispersed in 100 mL of water by sonication to obtain a yellowish brown–colored dispersion. To this solution, 1 mL of hydrazine hydrate solution is added and the mixture is refluxed at 100°C for 24 hours (Stankovich et al. 2007). The product is collected from centrifugation by washing with methanol for several times. The black solid is dried in vacuum overnight to obtain RGO-PMMA (MG), and gravimetric analysis suggests that MG samples contain 50% graphene by weight.

11.6.3 Advantages of ATRP

The use of ATRP for polymer functionalization of graphene has some advantages over other methods of graphene functionalization: As ATRP produces polymers with very low polydispersity, the chain lengths of grafted polymers are the same, separating graphene sheets in a uniform manner. The molecular weight of the grafted polymer can be easily changed, thereby controlling the degree of exfoliation; the grafted polymer can be made a block copolymer and structural hierarchy in the composite may be introduced by ATRP, which is really important in tailoring the property of the composite. Apart from the aforementioned general advantages of ATRP, in a recently published study (Layek et al. 2010) ATRP was used in a completely new manner, in which the graphene–PVDF composite was prepared with a predominance of the piezoelectric polymorph of PVDF. The problem of incompatibility of graphene and PVDF has been solved by suitable grafting of PMMA by ATRP on graphene and then making use of the interaction between >C=O groups of PMMA (grafted on graphene) and $>CF_2$ dipole of PVDF. In this study, under very simple and less stringent conditions, ATRP allows the synthesis of low-molecular-weight PMMA on the graphene surface in presence of various active functional groups, which is required for maintaining an appreciable property enhancement of the composite. Such reports establish the indispensability of ATRP in graphene chemistry.

11.6.4 Disadvantages of ATRP

Despite there being much room for future research and the huge potential of ATRP in graphene research, ATRP has some associated problems, which are required to be stated at this stage. It is noted that ATRP has been applied so far for the successful polymerization of vinyl monomers such as methacrylics, acrylamides, and vinyl pyridines. There are many other very interesting polymers that cannot be obtained by ATRP from their corresponding monomers such as olefins, halogenated alkenes, and vinyl acetate. However, in the ATRP of vinyl acetate some success is now claimed (Kamigaito et al. 2001; Kato et al. 1995; Wang and Matyjaszewski 1995). The GO sheets usually have an appreciable number of functional groups attached with it; therefore, after initiator attachment, some possibility of cross-chain termination via radical–radical coupling leading to the formation of cross-linked insoluble polymer material exists. Also, the stringent separation of Cu ions from the materials synthesized is required for their practical applications. In this regard, it is worth mentioning that introduction of relatively newly developed ATRP processes such as ARGET (Matyjaszewski et al. 2007), ICAR (Matyjaszewski et al. 2006), and SET-LRP (Percec et al. 2006), requiring parts-per-million level of copper concentration, in graphene-based ATRP synthesis should be helpful in solving the problem of residual Cu catalyst in the synthesized material. Particularly, techniques such as eATRP recently developed by Matyjaszewski et al (Magenau et al. 2011) reveal precise control on polymerization using very low concentrations of the Cu catalyst through in situ electrochemical regeneration of the active form and it would be very much effective to produce highly pure polymer-functionalized graphene.

11.7 Characterization of ATRP-Functionalized Graphenes

The ATRP-functionalized graphenes are usually characterized by microscopy (transmission electron microscopy [TEM], atomic force microscopy [AFM]), spectroscopic techniques (Fourier transform infrared [FT-IR] spectroscopy, nuclear magnetic resonance [NMR] spectroscopy, Raman, x-ray photoelectron spectroscopy [XPS]), and thermal (thermogravimetric analysis [TGA]) techniques. Short accounts on these techniques are presented in Sections 11.7.1 through 11.7.7.

11.7.1 Transmission Electron Microscopy

The TEM is an important tool to characterize graphene-grafted polymers. Fang et al. (2009) have reported a density difference between the central and peripheral areas of PS-grafted graphene sheets (Figure 11.2). The PS–graphene sheets

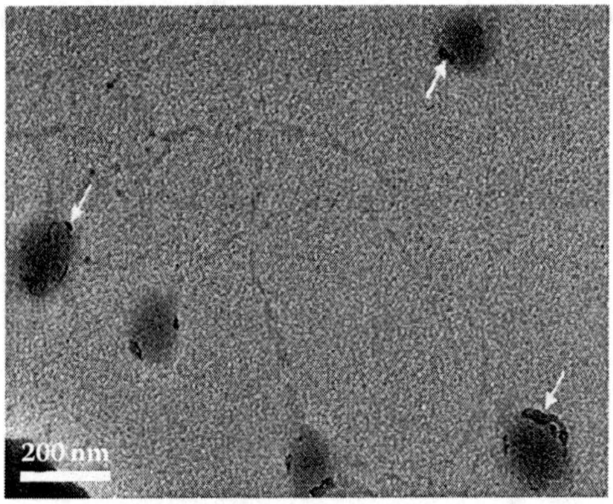

FIGURE 11.2
The TEM image of PS-grafted graphene sheets: Sporadic black points on the periphery of PS-graphene sheets (as denoted by arrows) should not be considered because they are supposed to be fragments of filter paper. (From Fang et al. *J. Mater. Chem.*, 19, 7098, 2009. Reproduced by permission of The Royal Society of Chemistry.)

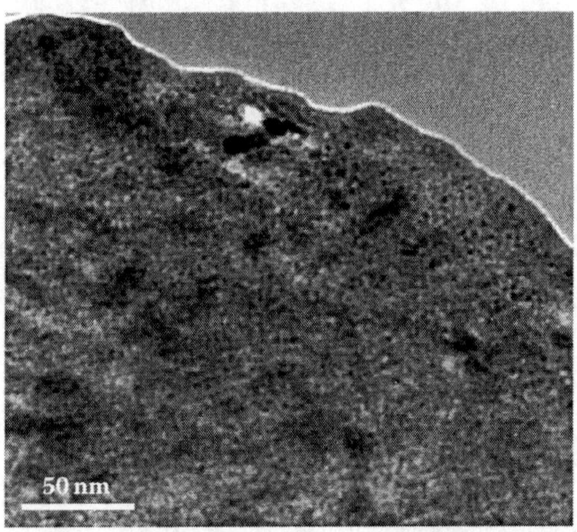

FIGURE 11.3
The TEM image of GO/PDMAEMA nanocomposite after being stained by osmium tetroxide: Black spots on graphene sheets are the result of decoration of PDMAEMA. (Reprinted with permission from Yang et al. *Langmuir*, 2009, 25, 11808. Copyright 2009 American Chemical Society.)

have an average lateral size of 100 nm and each sheet shows a color difference between the central and peripheral areas. Surface modification of graphene with PMMA also shows a density difference between the central and peripheral areas of graphene sheets. From the TEM image of GO-grafted PDMAEMA, Yang et al. (2009) have shown that polymer particles are decorated on GO sheets as dark nanodots produced during drying of the solution (Figure 11.3) (Yang et al. 2009). However, Li et al. (2010) have observed that the sheet morphology of GO is unchanged in graphene–grafted poly(*t*-butyl acrylate).

11.7.2 Atomic Force Microscopy

The AFM image and the corresponding height profile diagrams are the key tools used to characterize MG sheets. It is noted that AFM gives information about the thickness and length, that is, aspect ratio, of graphene sheets. For AFM measurements, GO is dispersed in water (0.01%) by sonication, and film is cast on a freshly cleaved mica surface and observed through the AFM instrument. The thickness of graphene sheets is measured from AFM topography. In a similar process, the initiator-attached sample and polymer-modified samples are made and their thicknesses measured (Fang et al. 2009; Layek et al. 2010). The thicknesses of initiator-attached GO and polymer-modified samples are more than that of pristine GO (Layek et al. 2010). But the use of RGO followed by diazonium coupling reaction of 2-(4-aminophenyl)ethanol provides hydroxyl group for initiator attachment showing the height of the graphene sheets similar to that of pure GO (Fang et al. 2009). The height of the graphene sheet after polymer grafting is higher than that of the initiator-attached graphene sheet (Figure 11.4). The average thickness of graphene-grafted polymer is –3 nm (Fang et al. 2009). The MG has a height profile showing an increase in height compared to the heights of initiator-attached samples (Figure 11.5). On polymerization with MMA, the height increases to 2 and 4.5 nm for two major populations of MG. Two different sites of polymerization (side and basal planes of graphene) are probably responsible for the two different height distributions. The height of

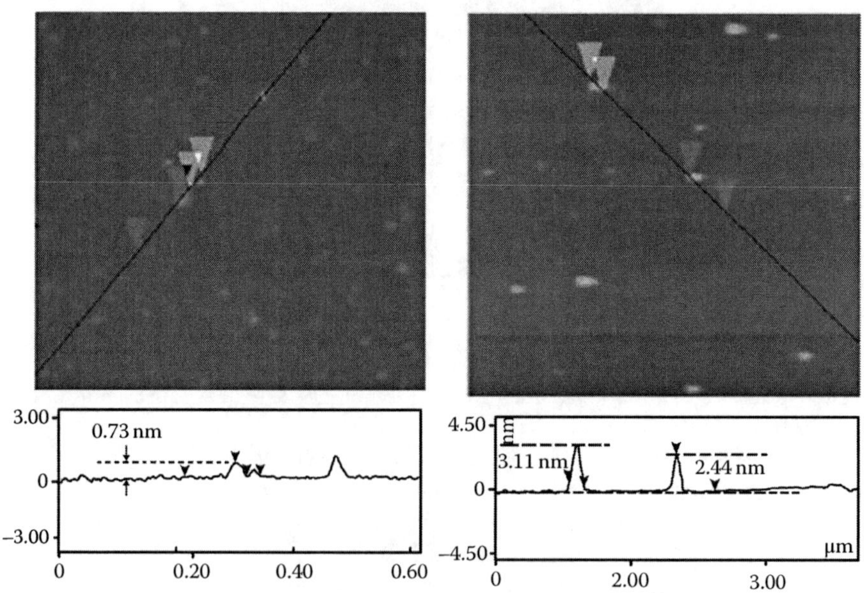

FIGURE 11.4
Tapping mode AFM images of initiator-grafted (left) and PS-grafted (right) graphene. (From Fang et al. *J. Mater. Chem.*, 19, 7098, 2009. Reproduced by permission of The Royal Society of Chemistry.)

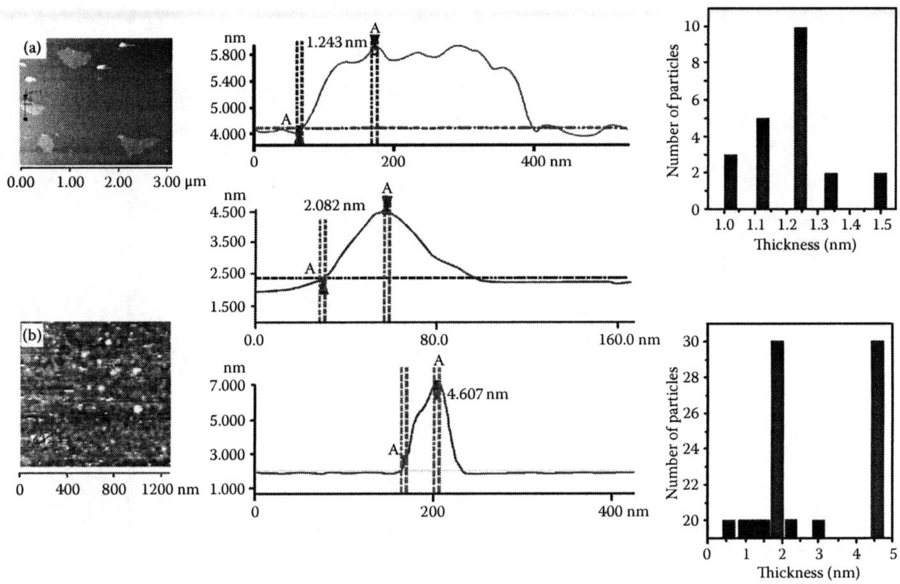

FIGURE 11.5
Tapping mode AFM images and height profiles of (a) BIBG and (b) MG, and corresponding histograms at the right side (obtained from nearly 30 objects). The distance between two "A" marked in each figure indicates the thickness of graphene sheet. (Reprinted from *Polymer*, 51, Layek et al., Physical and mechanical properties of poly(methyl methacrylate)–functionalized graphene/poly(vinylidine fluoride) nanocomposites: Piezoelectric b polymorph formation, 5846. Copyright 2010, with permission from Elsevier.)

2 nm may be the result of polymerization on the side of the graphene sheet (Scheme 11.2) producing PMMA that coats over the graphene sheet increasing the thickness. The increase of 4.5 nm thickness from that of BIBG is for polymerization from the initiator attached to the basal plane of graphene sheet increasing the gallery spacing.

11.7.3 Fourier Transform Infrared Spectroscopy

To elucidate the formation and modification of GO with initiator followed by polymer functionalization, FTIR spectra of samples are essential. Pristine graphite shows two strong peaks at 3430 cm^{-1} for O–H stretching and at 1610 cm^{-1} for skeletal vibrations from graphitic domains of adsorbed water and aromatic domain (C=C), respectively. In GO, there are some new peaks compared to pristine graphite due to oxygen-containing moieties, such as 1728 cm^{-1} for C=O stretching, 3406 cm^{-1} for O–H stretching, and 1052 cm^{-1} for C–O stretching (Ren et al. 2011). A shift of aromatic C=C peak from 1610 cm^{-1} in graphite to 1625 cm^{-1} in GO occurs, which indicates some electron-withdrawing groups (such as oxygen-containing functional groups) become bonded to the aromatic rings after oxidation. Besides, the peaks at

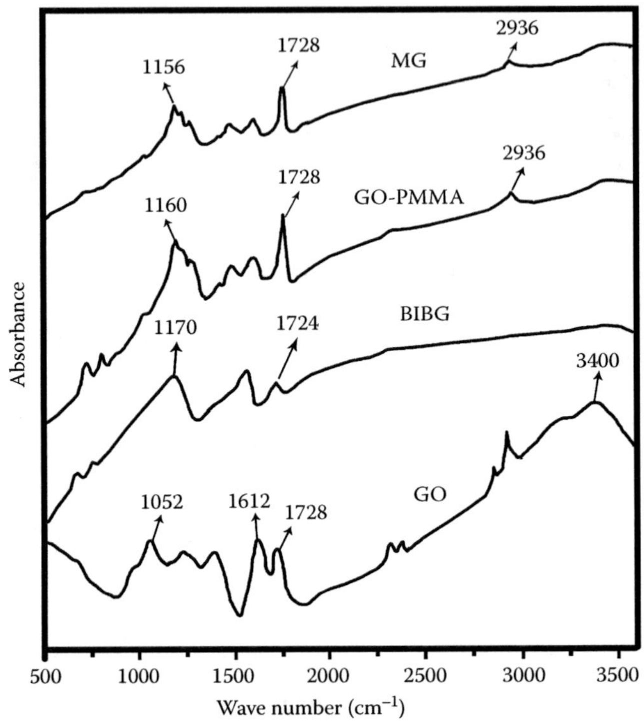

FIGURE 11.6
The FTIR spectra of GO, BIBG, GO-PMMA, and MG. (Reprinted from *Polymer*, 51, Layek et al., Physical and mechanical properties of poly(methyl methacrylate)–functionalized graphene/poly(vinylidine fluoride) nanocomposites: Piezoelectric b polymorph formation, 5846. Copyright 2010, with permission from Elsevier.)

2925/2855 cm^{-1} for sp^3 C–H stretching indicate the presence of some CH_2 or CH groups in the GO samples. In case of RGO samples, the peak of epoxide group (1052 cm^{-1}) disappears; further, the peak intensity of hydroxyl groups (3430 cm^{-1}) is reduced, but it still exists indicating that hydrazine treatment cannot restore all sp^2 carbons (Ren et al. 2011). Due to initiator attachment, the intensity ratio, $I_{OH}/I_{C=O}$, is reduced to half that of GO (Layek et al. 2010) (Figure 11.6). The FTIR spectra of RGO-PMMA shows characteristic peaks at 1728 cm^{-1} and 1160 cm^{-1} for the ester group of PMMA and 2936 cm^{-1} for sp^3 C–H stretching, which is more intense than that of GO and GO-Br (Layek et al. 2010).

11.7.4 Nuclear Magnetic Resonance Spectroscopy

It is noted that NMR spectra yield information about the formation of products obtained at different steps. For GO-PMMA (Figure 11.7), the peak at 3.59 ppm is for O–CO–CH_3 proton, the peak at 1.8 ppm is for C–CH_2–C proton,

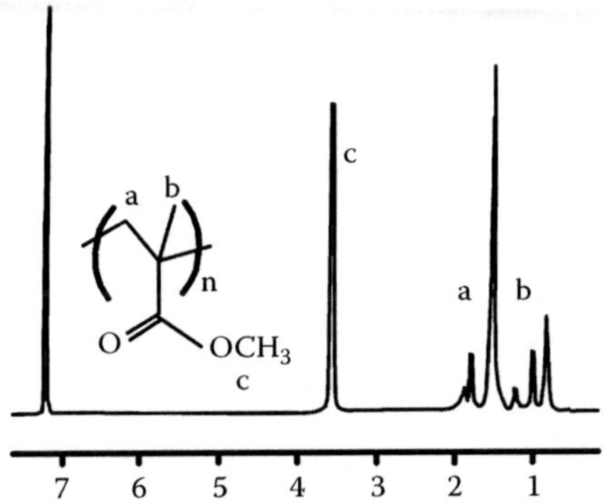

FIGURE 11.7
The HNMR spectra of GO-PMMA (a,b,c indicate the NMR peaks of the marked protons of PMMA attached to GO). (Reprinted from *Polymer,* 51, Layek et al., Physical and mechanical properties of poly(methyl methacrylate)–functionalized graphene/poly(vinylidine fluoride) nanocomposites: Piezoelectric b polymorph formation, 5846. Copyright 2010, with permission from Elsevier.)

and the peak at 0.73 ppm is for $-CH_3$ proton, which confirm that PMMA has been grafted from the graphene surface (Layek et al. 2010). Similar analysis of proton NMR (HNMR) spectra of graphene-grafted polymer chains in solution provides information about the polymerizations of styrene, MMA, and butyl acrylate indicating that ATRP operates successfully from initiator-attached GO surfaces (Lee et al. 2010).

11.7.5 Raman Spectroscopy

Carbonaceous materials such as nanodiamond, CNT, and graphene show high Raman intensities, which enable them to be used as a powerful tool to characterize MGs (Ferralis 2010). The D-band of GO occurs at 1366 cm^{-1} for the double-resonance excitation of phonons close to the *K* point in the Brillouin zone, and a G-band peak position occurs at 1582 cm^{-1} for the Brillouin zone–centered LO phonon (Kotov et al. 1996; Thomsen and Reich 2000). Normally, the D-band arises for the sp^3-hybridized carbons of graphene sheets, the intensity is relatively weak, and a strong G-band arises for the stretching of sp^2-hybridized carbon (Ferrari and Robertson 2000). From the spectrum, the intensity ratio (D/G) can be calculated and it is a measure of the extent of disorder present within graphene (Choi et al. 2006). Chang et.al. (2011) have reported that D/G ratios of GO, GO-Br, and GO–molecularly imprinted polymer (MIP) are 0.670, 0.884, and 1.169, respectively, indicating an increase in disorder due to graphene modification. The formation of covalent bonds between GO and initiator and/or MIP is also confirmed from the change in the D/G ratio (Figure 11.8) (Chang et al. 2011). Not only the D/G ratio but also the interaction between MIP and GO can

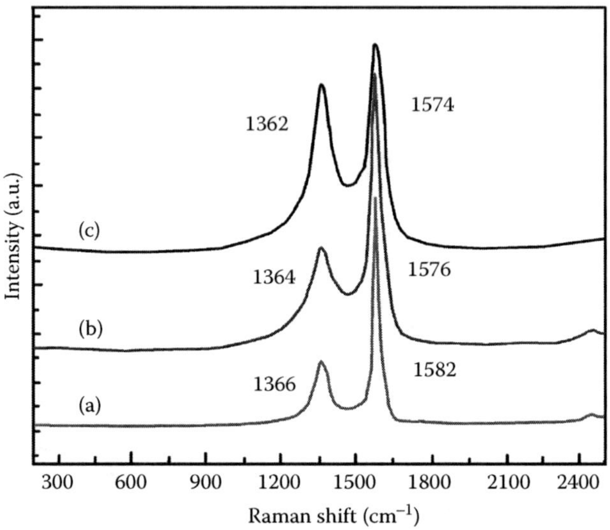

FIGURE 11.8
Raman spectra of (a) GO, (b) GO-Br, and (c) GO-MIP. (With kind permission from Springer Science+Business Media: *J. Mater. Sci.*, Preparation of graphene oxide–molecularly imprinted polymer composites via atom transfer radical polymerization, 46, 2011, 2024, Chang et al.)

be investigated from the Raman shift (Manna and Pati 2009). The G-band of GO is 1582 cm^{-1}; but in GO-MIP, it appears at 1574 cm^{-1} (Figure 11.8). This shift of G-band is due to the change in the electronic structure of graphene. Usually, the G-band is shown to shift to lower frequencies for electron-donor dopants and to higher frequencies for electron-acceptor dopants (Bendiab et al. 2001; Rao et al. 1997; Shin et al. 2008). Probably, the increased electron density in graphene ring caused by electron-donor dopants results in better delocalization of its π electrons facilitating the stretching vibration of sp^2 carbon and shifting the G-band to a lower frequency. In case of electron-acceptor dopants, the ring electrons are localized to a direction causing difficulty in the stretching vibration of sp^2 carbon of graphene. The D/G ratios are 1.73, 1.71, and 1.02 for GO, graphene-initiator, and graphene-grafted PS, respectively, indicating an increase in disorder (Fang et al. 2009). Sun et al. (2010) have reported that the D/G band ratio in PS-grafted graphene sheets does not change much probably due to the decrease in size of sp^2 domains.

11.7.6 X-Ray Photoelectron Spectroscopy

It is noted that XPS deals with elemental composition, empirical formula, chemical state, and electronic state of elements existing within a material. The GO/ATRP initiator composite is characterized by XPS (Figure 11.9).

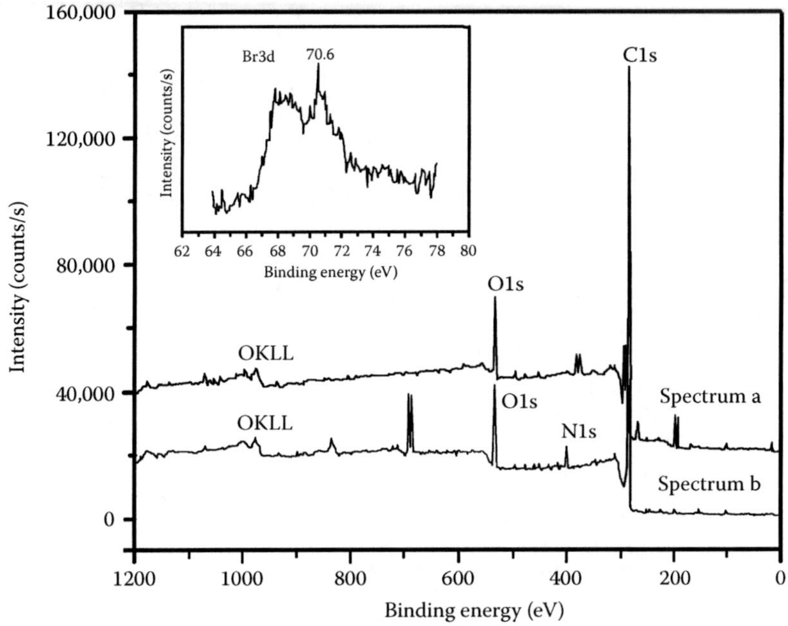

FIGURE 11.9
The XPS results of GO (spectrum a) and Go/PDMAEMA nanocomposite (spectrum b). In the inset, the spectrum of ATRP initiator attached to graphene is presented. (Reprinted with permission from Yang et al. *Langmuir*, 2009, 25, 11808. Copyright 2009 American Chemical Society.)

The peak at 70.5 eV corresponds to the binding energy of Br3d (bromine, 3d state), indicating attachment of an initiator to the GO surface. The percentage of grafting of PDMAEMA has been calculated by XPS by means of binding energies of N1s and C1s (Yang et al. 2009). In the spectrum of PDMAEMA-modified graphene obtained by ATRP, N1s has a binding energy of 399.8 eV; this peak is absent in GO (Figure 11.9). The percentage of nitrogen calculated from XPS is about 3.19 wt% in the grafted sample; therefore, the PDMAEMA percentage is about 35 wt%. Figure 11.10 shows C1s binding energy of GO and the GO/PDMAEMA nanocomposite. The C1s spectrum of GO have binding energies of 284.6 eV (C–C in GO), 285.7 eV (COH), 286.7 eV(C–O), 288 eV (C=O), and 289.1 eV (O–C = O) (Shan et al. 2009b; Park et al. 2008). The C1s XPS spectrum of the nanocomposite shows all the energy peaks of C1s, but the percentage of the area of the peak at 289.1 eV (O–C = O) increased from 0.97% to 3.85%, which indicates the grafting of PDMAEMA chains onto the surface of GO sheets. The PDMAEMA percentage calculated from the increased area of C1s peak is 32 wt%, which is very similar to the results obtained from the XPS calculation of N1s.

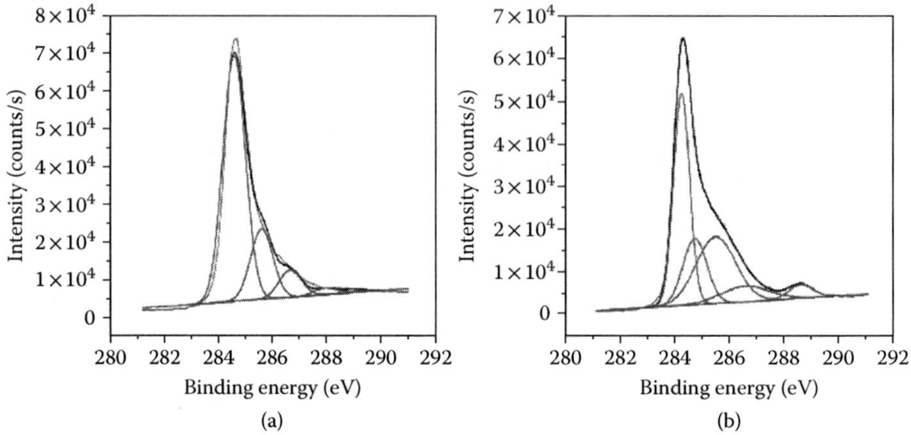

FIGURE 11.10
The C1s XPS spectra of (a) GO and (b) GO/PDMAEMA nanocomposites: The dark lines are experimental lines, dark gray lines are fitting lines, and light gray lines represent XPS spectra of C1s corresponding to different binding energies. (Reprinted with permission from Yang et al. *Langmuir*, 2009, 25, 11808. Copyright 2009 American Chemical Society.)

11.7.7 Thermogravimetric Analysis

It is noted that TGA of GO and polymer-modified graphene samples are usually performed under nitrogen atmosphere. The GO is unstable and loses its mass below 100°C (Shen et al. 2009). This is due to the degradation of easily labile oxygen-containing groups such as –OH, epoxide, and COOH. After reduction with hydrazine hydrate, a large fraction of oxygen-containing moieties is removed. Therefore, the thermal stability of the reduced graphene sheets increases compared to GO (Stankovich et al. 2007). In the case of polymer-grafted samples, there occurs more weight loss due to degradation of the polymer associated with the graphene in these samples. This difference in weight losses between these two materials is used to calculate the percentage of grafting of polymer in graphene surfaces. The initiator attached to the graphene surface followed by diazonium addition and ATRP of styrene monomer shows 82 wt% grafting to the graphene sheets determined from TGA (Fang et al. 2009).

Exfoliated GO, initiator-modified GO, and GO/PDMAEMA nanocomposites are analyzed by TGA (Figure 11.11) (Yang et al. 2009). The TGA of exfoliated GO revealed 13 wt% weight loss in the temperature range between 100°C and 800°C (curve a in Figure 11.11), which is attributed to the loss of functional groups such as COOH and OH groups on the sheets. On grafting of the ATRP initiator, the composite shows 32% weight loss (curve b in Figure 11.11). The TGA result of GO/PDMAEMA nanocomposites shows 56% weight loss (curve c in Figure 11.11), which indicates that the weight percentage of PDMAEMA is about 24% (Yang et al. 2009).

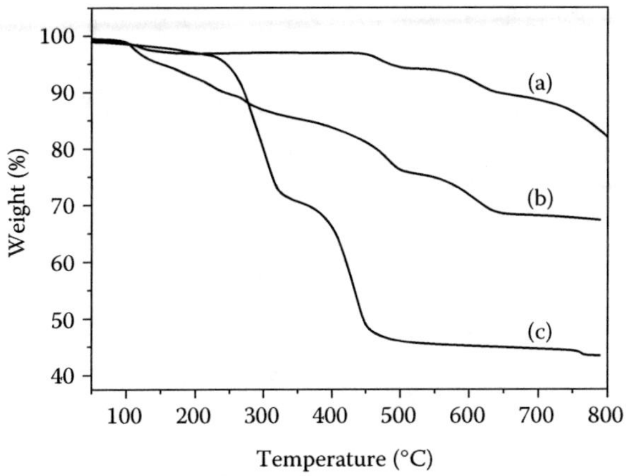

FIGURE 11.11
The TGA curves of (a) exfoliated GO, (b) GO/ATRP initiator, and (c) GO sheets with PDMAEMA chains on the surface. (Reprinted with permission from Yang et al. *Langmuir*, 2009, 25, 11808. Copyright 2009 American Chemical Society.)

11.8 Conclusions

Thus, ATRP technique of polymer functionalization of graphene is gaining considerable momentum in the field of graphene-based polymers and their composites. This is not only due to the easier exfoliation of graphene sheets but for the control of polymer molecular weight grafted to the graphene sheet and also for its ability to introduce different polymeric architecture. This is useful in producing tailor-made materials fulfilling the demands of optoelectronic, energy, and engineering materials. The grafting-from technique is superior to the grafting-to technique in this regard. This controlled polymer structure yields a good control over the physical and mechanical properties of graphene-based polymers and also their composites. The characterization technique of these materials is now well established. The new ATRP methods (e.g., ICAR, AGET, ARGET, SET-LRP, and eATRP) are expected to yield almost ATRP catalyst free product in the modified graphene suggesting better control of physical and electronic properties of graphene-based polymers and their composites.

11.9 Future Scope

Polymer-functionalized graphene generated by ATRP has wide scope to tune the properties of graphene and also its nanocomposites with commodity polymers. A few examples are as follows: Grafting of different molecular weight polymer with a narrow polydispersity on the graphene sheet would help to get desired mechanical and conductivity property, and Grafting of block copolymers can help to induce properties related to multiphase morphology of the block yielding different physical and mechanical property. Further grafting of the anchored polymer with another amphiphilic or hydrophobic monomer may also import new property in the material. All these graphene-based polymers may be used to prepare nanocomposites with commodity plastics not only to improve their mechanical and conducting properties but also to prepare different functional materials. Apart from the mixing with different carbon nanostructures, decoration of metal or semiconductor nanoparticles in graphene-based polymers yields new material properties and such composites may be suitable for energy harvesting and different sensor applications.

References

Ansari, S., and E. P. Giannelis. 2009. Functionalized Graphene Sheet—Poly(vinylidene fluoride) Conductive Nanocomposites. *J Polym Sci Part B Polym Phys*, 47:888–897.

Balandin, A., S. Ghosh, W. Bao, I. Calizo, D. Teweldebrhan, F. Miao et al. 2008. Superior Thermal Conductivity of Single-Layer Graphene. *Nano Lett*, 8:902–907.

Balazs, A. C., T. Emrick, and T. P. Russell. 2006. Nanoparticle Polymer Composites: Where Two Small Worlds Meet. *Science*, 314:1107–1110.

Bendiab, N. E., J. Anglaret, L. Bantignies, A. Zahab, and J. L. Sauvajol. 2001. Stoichiometry Dependence of the Raman Spectrum of Alkali-Doped Single-Wall Carbon Nanotubes. *Phys Rev B*, 64:245424.

Chang, L., S. Wu, S. Chen, and X. Li. 2011. Preparation of Graphene Oxide–Molecularly Imprinted Polymer Composites via Atom Transfer Radical Polymerization. *J Mater Sci*, 46:2024–2029.

Chiefari, J., Y. K. Chong, F. Ercole, J. Krstina, J. Jeffry, T. P. T. Le et al. 1998. Living Free Radical Polymerization by Reversible Addition-Fragmentation Chain Transfer: The RAFT Process. *Macromolecules*, 31:5559–5562.

Choi, W. S., S. H. Choi, B. Hong, D. G. Lim, K. J. Yang, and J. H. Lee. 2006. Effect of Hydrogen Plasma Pretreatment on Growth of Carbon Nanotubes by MPECVD. *Mater Sci Eng C*, 26:1211–1214.

Cote, L. J., R. Cruz-Silva, and J. Huang. 2009. Flash Reduction and Patterning of Graphite Oxide and Its Polymer Composite. *J Am Chem Soc*, 131:11027–11032.

Du, A., Z. Zhu, and S. C. Smith. 2010. Multifunctional Porous Graphene for Nanoelectronics and Hydrogen Storage:New Properties Revealed by First Principle Calculations. *J Am Chem Soc*, 132:2876–2877.

Edmondson, S., V. L. Osborne, and W. T. S. Huck. 2004. Polymer Brushes *via* Surface-Initiated Polymerizations. *Chem Soc Rev*, 33:14–22.

Fang, M., K. Wang, H. Lu, Y. Yang, and S. Nutt. 2009. Covalent Polymer Functionalization of Graphene Nanosheets and Mechanical Properties of Composites. *J Mater Chem*, 19:7098–7105.

Ferralis, N. 2010. Probing Mechanical Properties of Graphene with Raman Spectroscopy. *J Mater Sci*, 45:5135–5149.

Ferrari, A. C., and J. Robertson. 2000. Interpretation of Raman Spectra of Disordered and Amorphous Carbon. *Phys Rev B*, 61:14095–14107.

Gao, W., L. B. Alemany, L. Ci, and P. M. Ajayan. 2009. New Insights into the Structure and Reduction of Graphite Oxide. *Nat Chem*, 1:403–408.

Geim, A. K., and K. S. Novoselov. 2007. The Rise of Graphene. *Nat Mater*, 6:183–191.

Georges, M. K., R. P. N. Veregin, P. M. Kazmaier, and G. K. Hamer. 1993. Narrow Molecular Weight Resins by a Free-Radical Polymerization Process. *Macromolecules*, 26:2987–2988.

Gómez-Navarro, C., M. Burghard, and K. Kern. 2008. Elastic Properties of Chemically Derived Single Graphene Sheets. *Nano Lett*, 8:2045–2049.

Gómez-Navarro, C., R. T. Weitz, A. M. Bittner, M. Scolari, A. Mews, M. Burghard, et al. 2007. Electronic Transport Properties of Individual Chemically Reduced Graphene Oxide Sheets. *Nano Lett*, 7:3499–3503.

Goncalves, G., P. A. A. P. Marques, A. Barros-Timmons, I. Bdkin, M. K. Singh, N. Emami, et al. 2010. Graphene Oxide Modified with PMMA via ATRP as a Reinforcement Filler. *J Mater Chem*, 20:9927–9934.

He, H., and C. Gao. 2010. General Approach to Individually Dispersed, Highly Soluble, and Conductive Graphene Nanosheets Functionalized by Nitrene Chemistry. *Chem Mater*, 22:5054–5064.

Hu, H., and G. Chen. 2010. Electrochemically Modified Graphite Nanosheets and Their Nanocomposite Films with Poly(vinyl alcohol). *Polymer composite*, 31:1771–1775.

Hummer, W. S., and R. Offeman. 1958. Preparation of Graphitic Oxide. *J Am Chem Soc*, 80:1339.

Jakubowski, W., K. Min, and K. Matyjaszewski. 2006. Activators Regenerated by Electron Transfer for Atom Transfer Radical Polymerization of Styrene. *Macromolecules*, 39:39–45.

Kamaras, K., M. E. Itkis, H. Hu, B. Zhao, and R. C. Haddon. 2003. Covalent Bond Formation to a Carbon Nanotube Metal. *Science*, 301:1501.

Kamigaito, M., T. Ando, and M. Sawamoto. 2001. Metal-Catalyzed Living Radical Polymerization. *Chem Rev*, 101:3689–3745.

Kan, L., Z. Xu, and C. Gao. 2011. General Avenue to Individually Dispersed Graphene Oxide-Based Two-Dimensional Molecular Brushes by Free Radical Polymerization. *Macromolecules*, 44:444–452.

Kato, M., M. Kamigaito, M. Sawamoto, and T. Higashimura. 1995. Polymerization of Methyl Methacrylate with the Carbon TetrachloridelDichlorotris (triphenylphosphine)ruthedum(II)/Methylaluminum Bis(2,6-di-tert-butylphenoxide) initiating, System: Possibility of Living Radical Polymerization. *Macromolecules*, 28:1721–1728.

Kato, M., M. Kamigaito, M. Sawamoto, and T. Higashimura. 1996. Polymerization of Methyl Methacrylate with the Carbon TetrachloridelDichlorotris (triphenylphosphine)ruthedum(II)/Methylaluminum Bis(2,6-di-tert-butylphenoxidel) initiating System: Possibility of Living Radical Polymerization. *Macromolecules*, 28:1721–1723.

Kim, H., and C. W. Macosko. 2008. Morphology and Properties of Polyester/Exfoliated Graphite Nanocomposites. *Macromolecules*, 41:3317–3327.

Kotov, N. A., I. Dekany, and J. H. Fendler. 1996. Ultrathin Graphite Oxide-Polyelectrolyte Composites Prepared by Self-Assembly: Transition Between Conductive and Non-Conductive States. *Adv Mater*, 8:637–641.

Layek, R. K., S. Samanta, D. P. Chatterjee, and A. K. Nandi. 2010. Physical and Mechanical Properties of Poly(methyl methacrylate)–Functionalized Graphene/ poly(vinylidine fluoride) nanocomposites: Piezoelectric b Polymorph FORMATION. *Polymer*, 51:5846–5856.

Lee, S. H., D. R. Dreyer, J. An, A. Velamakanni, R. D. Piner, S. Park et al. 2010. Polymer Brushes via Controlled, Surface-Initiated Atom Transfer Radical Polymerization (ATRP) from Graphene Oxide. *Macromol Rapid Commun*, 31:281–288.

Lego, B., M. Francois, W. G. Skene, and S. Giasson. 2009. Polymer Brush Covalently Attached to OH-Functionalized Mica Surface via Surface-Initiated ATRP: Control of Grafting Density and Polymer Chain Length. *Langmuir*, 25:5313–5321.

Li, G. G. L., G. Liu, M. Li, D. Wan, K. G. Neoh, and E. T. Kang. 2010. Organo- and Water-Dispersible Graphene Oxide-Polymer Nanosheets for Organic Electronic Memory and Gold Nanocomposites. *J Phys Chem C*, 114:12742–12748.

Li, D., M. B. Muller, S. Gilje, R. B. Kaner, and G. G. Wallace. 2008. Processable Aqueous Dispersions of Graphene Nanosheets. *Nat Nanotechnol*, 3:101–105.

Li, L., G. Yan, and J. Wu. 2009. Modification of Polysulfone Membranes via Surface-Initiated Atom Transfer Radical Polymerization and Their Antifouling Properties. *J Appl Polym Sci*, 111:1942.

Liang, J. J., Y. F. Xu, Y. Huang, L. Zhang, Y. Wang, Y. F. Ma et al. 2009. Infrared-Triggered Actuators from Graphene-Based Nanocomposites. *J Phys Chem C*, 113:9921–9927.

Liu, J., W. Yang, L. Tao, D. Li, C. Boyer, and T. P Davis. 2010. Thermosensitive Graphene Nanocomposites Formed Using Pyrene-Terminal Polymers Made by RAFT Polymerization. *J Polym Sci, PART A: Polym Chem*, 48:425–433.

Liu, Y., J. Zhou, X. Zhang, Z. Liu, X. Wan, J. Tian, et al. 2009. Synthesis, Characterization and Optical Limiting Property of Covalently Oligothiophene-Functionalized Graphene Material. *Carbon*, 47:3113–3121.

Lotya, M., P. J. King, U. Khan, S. De, and J. N. Coleman. 2010. High-Concentration, Surfactant- Stabilized Graphene Dispersions. *ACS Nano*, 4:3155–3162.

Lu, W., J. Weng, D. Wu, C. Wu, and G. Chen. 2006. Epoxy Resin/Graphite Electrically Conductive Nanosheet Nanocomposite. *Mater Manuf Process*, 21:167–171.

Magenau, A. J., N. C. Strandwitz, A. Gennaro, and K. Matyjaszewski. 2011. Electrochemically Mediated Atom Transfer Radical Polymerization. *Science*, 332:81–84.

Manna, A. K., and S. K. Pati. 2009. Tuning the Electronic Structure of Graphene by Molecular Charge Transfer: A Computational Study. *Chem Asian J*, 4:855–860.

Matyjaszewski, K., W. Jakubowski, K. Min, W. Tang, J. Huang, W. A. Braunecker et al. 2006. Diminishing Catalyst Concentration in Atom Transfer Radical Polymerization with Reducing Agents. *Proc Natl Acad Sci U S A*, 103:15309–15314.

Matyjaszewski, K., N. V. Tsarevsky, W. A. Braunecker, H. Dong, J. Huang, W. Jakubowski et al. 2007. Role of Cu0 in Controlled/"Living" Radical Polymerization. *Macromolecules*, 40:7795–7806.

Matyjaszeski, K., and J. Xia. 2001. Atom Transfer Radical Polymerization. *Chem Rev*, 101:2921–2990.

McAllister, M. J., J. Li, D. H. Adamson H. C. Schniepp, A. A. Abdala, J. Liu, et al. 2007. Single Sheet Functionalized Graphene by Oxidation and Thermal Expansion of Graphite. *Chem Mater*, 19:4396–4404.

O'Neill, A., U. Khan, P. N. Nirmalraj, J. Boland, and J. N. Coleman. 2011. Graphene Dispersion and Exfoliation in Low Boiling Point Solvents. *J Phys Chem C*, 115:5422–5428.

Paredes, J. I., S. Villar-Rodil, A. Martínez-Alonso, and J. M. D. Tascónlvents. 2008. Graphene Oxide Dispersions in Organic Solvents. *Langmuir*, 24:10561–10564.

Park, S., J. An, R. Piner, I. Jung, D. Yang, A. Velamakanni, S. Nguyen, R. Ruoff. 2008. Aqueous Suspension and Characterization of Chemically Modified Graphene Sheets. *Chem Mater*, 20:6592–6594.

Patchkovskii, S., J. S. Tse, S. N. Yurchenko, L. Zhechkov, T. Heine, and G. Seifertes. 2005. Graphene Nanostructures as Tunable Storage Media for Molecular Hydrogen. *PNAS*, 102:10439–10444.

Percec, V., T. Guliashvili, J. S. Ladislaw, A. Wistrand, A. Stjerndahl, M. J. Sienkowska et al. 2006. Ultrafast Synthesis of Ultrahigh Molar Mass Polymers by Metal-Catalyzed Living Radical Polymerization of Acrylates, Methacrylates, and Vinyl Chloride Mediated by SET at 25°C. *J Am Chem Soc*, 128:14156.

Pradhan, S. K., B. B. Nayak, S. S. Sahay, and B. K. Mishra. 2009. Mechanical Properties of Graphite Flakes and Spherulites Measured by Nanoindentation. *Carbon*, 47:2290–2229.

Prasad, E., B. Das, U. Maitra, U. Ramamurty, and C. N. R. Rao. 2009. Extraordinary Synergy in the Mechanical Properties of Polymer Matrix Composites Reinforced with 2 Nanocarbons. *PNAS*, 106:13186–13189.

Putz K. W., O. C. Compton, M. J. Palmeri, S. T. Nguyen L. C. Brinson. 2010. High-Nanofi ller-Content Graphene Oxide–Polymer Nanocomposites via Vacuum-Assisted Self-Assembly. *Adv Funct Mater*, 20:3322–3329.

Qi, X., K. Pu, H. Li, X. Zhou, S. Wu, Q. Fan, et al. 2010. Amphiphilic Graphene Composites. *Angew. Chem Int Ed*, 49:9426–9429.

Rafiee, M. A., J. Rafiee, Z. Wang, H. Song, Z. Yu, and N. Koratkar. 2009. Enhanced Mechanical Properties of Nanocomposites at Low Graphene Content. *ACS Nano*, 3:3884–3890.

Ramanathan, T., A. A. Abdala, S. Stankovich, D. A. Dikin, M. Herrera-Alonso, R. D. Piner, et al. 2008. Functionalized Graphene Sheets for Polymer Nanocomposites. *Nat Nanotechnol*, 3:327–331.

Rao, A. M., P. C. Eklund, S. Bandow, A. Thess, and R. E. Smalley. 1997. Evidence for Charge Transfer in Doped Carbon Nanotube Bundles from Raman Scattering. *Nature*, 388:257–259.

Ren, P., D. Yan, X. Ji, T. Chen, and Z. Li. 2011. Temperature Dependence of Graphene Oxide Reduced by Hydrazine Hydrate. *Nanotechnology*, 22:055705.

Salavagione, H. J., M. A. Gomez and G. Martínez. 2009. Polymeric Modification of Graphene through Esterification of Graphite Oxide and Poly(vinyl alcohol). *Macromolecules*, 42:6331–6331.

Schadler, L. S., S. K. Kumar, B. C. Benicewicz, S. L. Lewis, and S. E. Hartonn. 2007. Designed Interfaces in Polymer Na ocomposites: A Fundamental Viewpoint. *MRS Bull*, 32:335–340.

Schedin, F., A. K. Geim, S. V. Morozov, E. W. Hill, P. Blake, M. I. Katsnelson, et al. 2007. Detection of Individual Gas Molecules Adsorbed on Graphene. *Nat Mater*, 6:652–655.

Schniepp, H. C., J. Li, M. J. McAllister, and S. Dong. 2006. In situ Synthesis and Characterization of Multiwalled Carbon Nanotube/Au Nanoparticle Composite Materials. *J Phys Chem B*, 110:853–857.

Shan, C., H. Yang, D. Han, Q. Zhang, A. Ivaska, and L. Niu. 2009a. Water-Soluble Graphene Covalently Functionalized by Biocompatible Poly-L-lysine. *Langmuir*, 25:12030–12033.

Shan, C., H. Yang, J. Song, D. Han, A. Ivaska, and L. Niu. 2009b. Direct Electrochemistry of Glucose Oxidase and Biosensing for Glucose Based on Graphene. *Anal Chem*, 81:2378–2382.

Shen, J., Y. Hu, C. Li, C. Qin, M. Shi, and M. Ye. 2009. Layer-by Layer Self-Assembly of Graphene Nanoplatelets. *Langmuir*, 25:6122–6128.

Shin, H., S. M. Kim, S. M. Yoon, A. Benayad, K. K. Kim, S. J. Kim, et al. 2008. Tailoring Electronic Structures of Carbon Nanotubes by Solvent with Electron-Donating and Withdrawing Groups. *J Am Chem Soc*, 130:2062–2066.

Si, Y., and E. T Samulski. 2008. Synthesis of Water Soluble Graphene. *Nano Lett*, 8:1679–1682.

Stankovich, S., A. D. Dikin, G. H. B. Dommett, K. M. Kohlhaas, E. J. Zimney, E. A. Stach et al. 2006a. Graphene-based Composite Materials. *Nature*, 442:282–286.

Stankovich, S., D. A. Dikin, R. D. Piner, K. A. Kohlhaas, A. Kleinhammes, Y. Jia, et al. 2007. Synthesis of Graphene-based Nanosheets via Chemical Reduction of Exfoliated Graphite Oxide. *Carbon*, 45:1558–1565.

Stankovich, S., R. D. Piner, X. Chen, N. Wu, S. T. Nguyen, and R. S. Ruoff. 2006b. Stable Aqueous Dispersions of Graphitic Nanoplatelets via the Reduction of Exfoliated Graphite Oxide in the presence of poly(sodium 4-styrenesulfonate). *J Mater Chem*, 16:155–158.

Stankovich, S., R. D. Piner, S. T. Nguyen, and R. S. Ruoff. 2006c. Synthesis and Exfoliation of Isocyanate-Treated Graphene Oxide Nanoplatelets. *Carbon*, 44:3342–3347.

Staudenmaier, L. 1898. Verfahren zur Darstellung der Graphitsaure. *Ber Dtsch Chem Ges*, 31:1481–1487.

Su, Q., S. Pang, V. Alijani, C. Li, X. Feng, and K. Mullen. 2009. Composites of Graphene with Large Aromatic Molecules. *Adv Mater*, 21:3191–3195.

Sun, S., Y. Cao, J. Feng, and P. Wu. 2010. Click Chemistry as a Route for the Immobilizationof Well-Defined Polystyrene onto Graphene Sheets. *J Mater Chem*, 20:5605–5607.

Thomsen, C., and S. Reich. 2000. Double Resonant Raman Scattering in Graphite. *Phys Rev Lett*, 85:5214.

Tkalya, E., M. Ghislandi, A. Alekseev, C. Koning, and J. Loos. 2010. Latex-based Concept for the Preparation of Graphene-based Polymer Nanocomposites. *J Mater Chem*, 20:3035–3039.

Verdejo, R., F. Barroso-Bujans, M. A. Rodriguez-Perez, J. A. de Saja, and M. A. Lopez. Manchado. 2008. Functionalized Graphene Sheet Filled Silicone Foam Nanocomposites. *J Mater Chem*, 18:2221–2226.

Vickery, L. J., A. J. Patil, and S. Mann. 2009. Fabrication of Graphene–Polymer Nanocomposites With Higher-Order Three-Dimensional Architectures. *Adv Mater*, 21:2180–2184.

Wang, J. S., and K. Matyjaszewski. 1995. Controlled "Living" Radical Polym-erization. Atom Transfer Radical Polymerization in the Presence of Transition-Metal Complexes. *J Am Chem Soc*, 117:5614–5615.

Wang, Y., Z. Shi, Y. Huang, Y. Ma, C. Wang, M. Chen et al. 2009. Supercapacitor Devices Based on Graphene Materials. *J Phys Chem C*, 113:13103–13107.

Xu, Y., H. Bai, G. Lu, C. Li, and G. Shi. 2008. Flexible Graphene Films via the Filtration of Water-Soluble Noncovalent Functionalized Graphene Sheets. *J Am Chem Soc*, 130:5857–5857.

Xu, Y., Z. Liu, X. Zhang, Y. Wang, J. Tian, Y. Huang et al. 2009. A Graphene Hybrid Material Covalently Functionalized with Porphyrin: Synthesis and Optical Limiting Property. *Adv Mater*, 21:1275–1279.

Xu, X., Q. Luo, W. Lv, Y. Dong, Y. Lin, Q. Yang et al. 2011. Functionalization of Graphene Sheets by Polyacetylene: Convenient Synthesis and Enhanced Emission. *Macromol Chem Phys*, 212:768–773.

Yang, Y., J. Wang, J. Zhang, J. Liu, X. Yang, and H. Zhao. 2009. Exfoliated Graphite Oxide Decorated by PDMAEMA Chains and Polymer Particles. *Langmuir*, 25:11808–11814.

Yoonessi, M., and J. R Gaier. 2010. Highly Conductive Multifunctional Graphene Polycarbonate Nanocomposite. *ACS Nano*, 4:7211–7220.

Zhao, X., Q. Zhang, and D. Chen. 2010. Enhanced Mechanical Properties of Graphene-Based Poly(vinyl alcohol) Composites. *Macromolecules*, 43:2357–2363.

12

Synthesis and Properties of Graphene-Based Nanocomposites in a Biodegradable Polymer

Gui Lin, Isao Noda, and James E. Mark*

CONTENTS

12.1 Introduction

Biodegradable polymers are becoming increasingly important as they can be degraded when placed in bioactive environments, such as bacteria, fungi, and algae, or by hydrolysis when placed in buffer solutions or seawater. They are therefore attractive substitutes for many synthetic materials because they can alleviate problems associated with solid-waste disposal (Gross and Kalra 2002; Mooney 2009; Ray and Bousmina 2005). As a specific example, worldwide

* Correspondence to: James E. Mark (markje@uc.edu)

consumption of such biodegradable polymers has increased from 14 million kilograms in 1996 to an estimated 68 million kilograms in 2001 (Gross and Kalra 2002). Of interest here are the poly(3-hydroxyalkanoate)s (PHAs), a type of biodegradable and biocompatible polymers that has attracted much research interest and industrial attention (Noda 2000; Noda et al. 2005; Philip et al. 2007; Phithakrotchanakoon et al. 2009; Satkowski et al. 2001; Tian et al. 2002; Venkitachalam et al. 2005). They are produced directly from renewable resources by microbes, which accumulate PHAs as carbon and energy storage materials under unbalanced growth conditions. Poly(3-hydroxybutyrate-*co*-3-hydroxyhexanoate) or PHBHx (trade name Nodax™) is one of the materials in this category of biodegradable semicrystalline aliphatic polyesters, since it provides a unique combination of properties, such as full anaerobic degradability, moisture resistance, barrier properties, shelf stability (under distribution and home storage conditions), processability on conventional equipment, and dyeability (Ray et al. 2002; Satkowski et al. 2001). Its potential applications include manufacture of storage bags, shopping bags, and slipcovers; coatings in automobiles; and furniture. This specific copolymer has the advantage of being biodegradable, but it is somewhat lacking in strength (Hassan et al. 2004, 2006; Noda et al. 2005; Venkitachalam et al. 2005; Zhang et al. 2007, 2008a,b). A number of techniques have been tried for improving its mechanical properties, including orientation induced by prestretching film samples (Hassan et al. 2004, 2006). One alternative, originally proposed by Toyota Corporation researchers (Usuki et al. 1993), involves incorporation of exfoliated layered silicates to yield polymer/layered-silicate nanocomposites. Another is the in situ generation of silica particles using sol-gel techniques (Hassan et al. 2004, 2006; Usuki et al. 1993; Yang et al. 2007; Zhang et al. 2007, Zhang et al. 2008b).

The focus here is on reinforcement of these copolymers by graphene, a flat monolayer of carbon atoms tightly packed into a two-dimensional honeycomb lattice, which has become one of the most exciting materials research topics in the last few years (Geim and Novoselov 2007; Stankovich et al. 2006). Outstanding properties of graphene include high thermal conductivity (Balandin et al. 2008), high Young's modulus, large surface areas (Stoller et al. 2008), and high mobility of charge carriers (Bolotin et al. 2008). These characteristics have already made graphene a promising, viable, and inexpensive filler substitute for carbon nanotubes in many applications such as hydrogen storage (Lee et al. 2010; Srinivas et al. 2009), solar cells (Hong et al. 2008; Valentini et al. 2010), and nanocomposites (Jang and Zhamu 2008; Liang et al. 2009; Park et al. 2008, 2009; Ramanathan et al. 2008; Salavagione et al. 2009; Stankovich et al. 2006; Yang et al. 2009).

The present study involves the structure–mechanical properties relationships of PHBHx reinforced by expanded graphite (EG) nanosheets, graphene oxide (GO), or reduced GO (RGO). The morphology and dispersion of EG, GO, and RGO in PHBHx were studied, as were their effects on nanocomposite crystalline structure, mechanical properties, and thermal stabilities.

12.2 Experimental Section

12.2.1 Raw Materials

The PHBHx (see Figure 12.1) required for the experiment, a copolymer consisting of 87.4 mol% 3-hydroxybutyrate and 12.6 mol% 3-hydroxyhexanoate (based on ^{13}C nuclear magnetic resonance [NMR] analysis), was provided by Procter & Gamble, Co., West Chester, Ohio (technology now transferred to Meredian, Inc., Georgia). It had a number-average molecular weight of 500,000–600,000 g·mol^{-1} and a polydispersity index M_w/M_n of 1.9–2.1 based on gel permeation chromatography (GPC) measurement relative to polystyrene standards with tetrahydrofuran (THF) as the solvent. The graphite flakes had an average size of 500 μm and were used to prepare EG; they were generously provided by Asbury Graphite Mills, Inc., Asbury, New Jersey. Chloroform, dimethylformamide (DMF), and THF used in the study were analytical grade and purchased from Sigma-Aldrich, Co., Michigan. The exfoliated clay30B was generously provided by Southern Clay Products, Inc. (Gonzales, Texas), and its average interlayer distance was about 18.6 Å. Multiwalled carbon nanotubes (MWCNTs) were purchased from Nanostructured & Amorphous Materials, Inc., Los Alamos, New Mexico. Concentrated sulfuric acid and concentrated nitric acid were obtained from Aldrich, Co., and all these chemicals were used as received without further purification.

12.2.2 Preparation of EG, GO, and RGO

The EG was prepared by the method described in the studies of Lin et al. (2008), Zhang et al. (2008a), and Zhou (2005). Briefly, graphite flakes were oxidized in a solution mixture of concentrated sulfuric acid and nitric acid (4:1 volume ratio) for 48 hours with vigorous stirring in an ice-water bath. The acid-treated graphite flakes were washed with enough deionized water until its pH value was in the range of 5–6 and then dried at

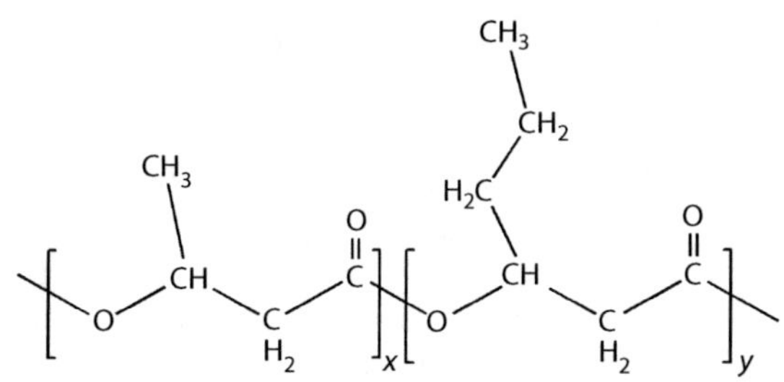

FIGURE 12.1
Molecular structure of PHBHx.

100°C under N_2 atmosphere to remove the remaining water. The dried graphite flakes were put in a silicon tube through which N_2 flowed, and the tube was heated at 1000°C for 60 seconds. The EG had over 350-fold C-direction expansion.

The GO was prepared according to modified Hummers method (Becerril et al. 2008) from EG. It is noted that 2.5 g of EG and 1.92 g of $NaNO_3$ were placed in a flask. Then, 190 mL of H_2SO_4 was added with stirring in an ice-water bath, and 22.5 g of $KMnO_4$ was slowly added over about 1 hour. Stirring was continued for 2 hours in the ice-water bath. After the mixture was stirred vigorously for 5 days at room temperature, 350 mL of 5 wt% H_2SO_4 aqueous solution was added over about 1 hour with stirring while the temperature was kept at 98°C. The resultant mixture was further stirred for 2 hours at 98°C. The temperature was reduced to 60°C, 7.5 mL of H_2O_2 (30 wt% aqueous solution) was added, and the mixture was stirred for 2 hours at room temperature. To remove the ions of the oxidant and any other inorganic impurities, the resultant mixture was purified by repeating the following procedure cycle until the pH value of the final product was nearly 6.0–7.0. The procedure cycle step was described as follows: The mixture was centrifuged at 6000 rpm, removal of the clear solution liquid, addition of 2 L of deionized water with vigorous stirring, and ultrasonication for 30 minutes at a power of 140 W. The resultant water dispersion was dried at 50°C under vacuum overnight. The final product yield was about 2.3 g.

The RGO was prepared from GO in DMF (1 mg/mL) after ultrasonic treatment over 30 minutes with the help of the surfactant poly(diallyldimethylaminonium chloride) [PDMA]. The GO was reduced by introducing N_2H_4 (2 μL per milligram of GO) and holding the mixture at room temperature for 24 hours. The brown solution of GO in DMF became black and the solution was stable in DMF for over 24 hours, permitting convenient mixing with PHBHx. There was a tiny deposit in the solution after 24 hours in room temperature.

12.2.3 Preparation of PHBHx/Nanosheets Nanocomposite

It is noted that PHBHx nanocomposites with differing contents (0–5 wt%) of EG, GO, RGO, clay30B, and MWCNT were prepared through solution mixing. Briefly, the desired amounts of EG, GO, RGO, clay30B, and MWCNT were first dissolved in a solvent (chloroform, THF, or DMF) and subjected to ultrasonic treatment (140 W for 10 minutes) to obtain stable solutions. In the case of RGO, its amount was calculated according to the amount of GO required for the reduction process, since RGO easily aggregates during the drying process and it is difficult to disperse RGO again into graphene in the solvents. Chloroform solutions of EG, GO, and RGO were stable for at least 48 hours. The desired amounts of PHBHx were also dissolved in chloroform at room temperature and then mixed with filler/chloroform suspensions.

The mixtures were again exposed to ultrasonic treatment for 10 minutes, and the solutions were continuously stirred vigorously for 2 hours. Further, the solutions were poured into Teflon™ molds to volatize the solvent at room temperature and then put into an oven at 100°C for 2 hours under vacuum to make certain that all solvent was gone. The PHBHx/EG, PHBHx/GO, PHBHx/RGO, PHBHx/clay30B, and PHBHx/MWCNT nanocomposites sheets were prepared by using a compression mold on a press at 145°C for 10 minutes, followed by cooling to 50°C. The final samples were ready for characterization, including extents of dispersion, mechanical properties, and biodegradability.

12.2.4 Characterization of Samples

Fourier transform infrared (FTIR) spectroscopy measurements for EG, GO, and RGO were conducted using a Digilab Excalibur FTIR equipment (Thermo Nicolet Nexus 670, Thermo Electron Corporation, Waltham, Massachusetts) in the mid-IR range (4000 to 500 cm^{-1}). The resolution was 4 cm^{-1}, and 16 scans per sample were analyzed using the Digilab Resolutions Software.

The x-ray diffraction (XRD) studies of original graphite, EG, RGO, PHBHx/EG, PHBHx/GO, and PHBHx/RGO sheets were carried out on an x-ray generator (a Siemens D500 diffractometer; Cu $K\alpha$ radiation, with 30 mA, 40 kV, and $\lambda = 1.5406$ Å). The 2θ was 2°–40°, with a step size of 0.05° per second.

The morphologies of the samples using broken fracture surfaces were characterized by scanning electron microscopy (SEM; Hitachi S-4100 SEM at 15 kV). Surfaces of the samples were coated with an alloy of Au and Pd. The dispersion of graphene-based fillers, clay, and MWCNT in the polymer matrix was characterized by transmission electron microscopy (TEM; JEM 1230 EX-II) using an ultrathin section under 60 kV.

The glass transition temperature (T_g), crystalline temperature (T_c), and melting temperature (T_m) were determined by differential scanning calorimetry (DSC) using a TA1501 thermal analyzer at 10°C/min in N_2 (20 mL/min) from–100°C to 200°C cycles. Results from a second DSC scan were used to ensure that the heat history of the sample played no role in our interpretation. Thermogravimetric analysis (TGA) was carried out on a TA Instruments' TGA 2050 thermogravimetric analyzer with nitrogen (30 mL/min) as purging gas. Tests were conducted from 40°C to 600°C at a heating rate of 20°C/min.

The mechanical properties were measured on a fully computerized INSTRON mechanical tester (model 1122, Acton, Massachusetts) using a sample of dimensions 30 × 5 × 1 mm^3 at a crosshead speed of 5 mm/min with the initial gauge length being 20 mm at 25°C. The values of modulus, tensile strength, elongation at break, and toughness were determined for each sample as an average of at least five tests.

12.3 Results and Discussion

12.3.1 Preparation of EG, GO, and RGO

The EG was prepared according to the methods stated in the literature starting with graphite (Jang and Zhamu 2008; Li 2007), in which the surfaces of graphite were oxidized by sulfuric acid and nitric acid at the edge and some of the acids were inserted into the layers of graphite. Ultrahigh temperatures aid the expansion of graphene sheets (see Figure 12.2). The morphology of EG after the high-temperature process clearly showed the layered structures (see Figure 12.3a). There was about 350-fold increase in the spaces between the layers. The FTIR results in Figure 12.4 show that when compared with pure graphite, there were new peaks at 1714 and 3300 cm^{-1}, which were assigned to carboxyl and hydroxyl groups, respectively, generated on the surface of graphite during the preparation process. The graphite crystalline structure was expanded, leading to the formation of graphite nanosheets after further treatment. However, the XRD results in Figure 12.5a also show the characteristic peak near 26.5° (d_{002}) of the graphite in EG, which means that EG still contains the crystalline structure of graphite, that is, the layers in EG are still made of several layers of graphene. Meanwhile, the introduction of carboxyl and hydroxyl groups greatly helped in dispersing EG in

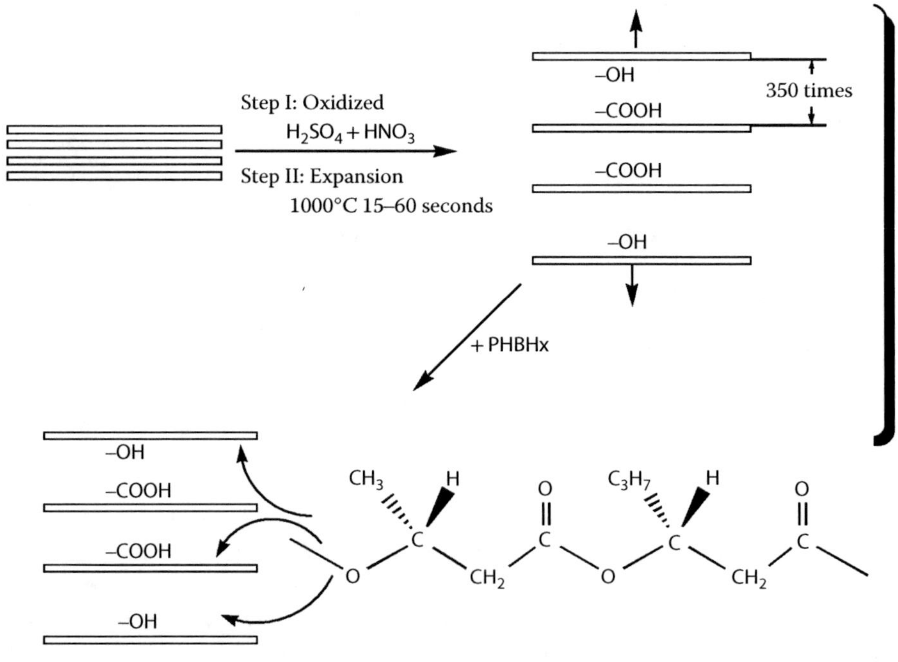

FIGURE 12.2
Scheme of preparation of PHBHx/EG nanocomposites, and their interactions.

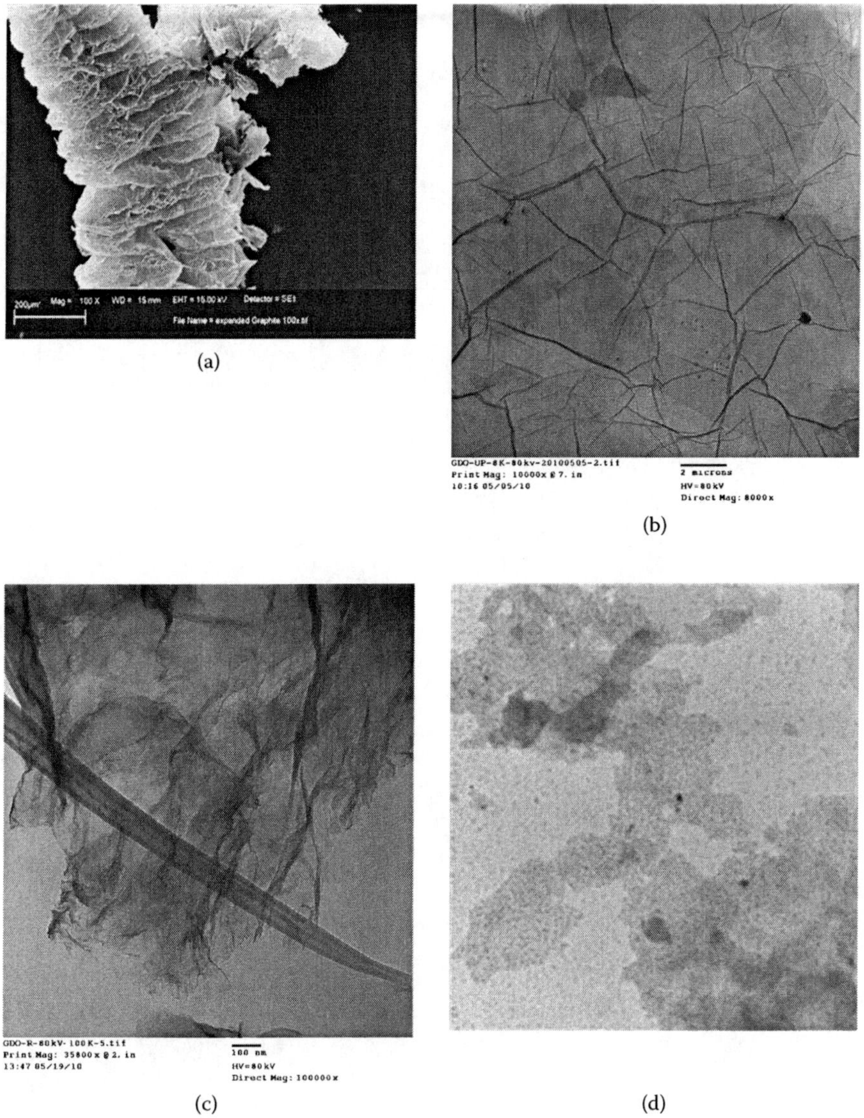

FIGURE 12.3
Figure showing the (a) SEM of expanded graphite, TEM images of (b) GO and (c) RGO, and (d) RGO with PDAM. (Courtesy of P. Potschke, Institute of Polymer Research, Dresden, Germany.)

polar solvents, such as chloroform (Tung et al. 2009). In fact, the EG/chloroform solutions were stable for 24 hours with only slight precipitations. It is noted that PHBHx was dispersible in chloroform as well, which obviously made chloroform a good solvent for the solution blending of PHBHx and EG in the preparation of desired PHBHx/EG nanocomposites.

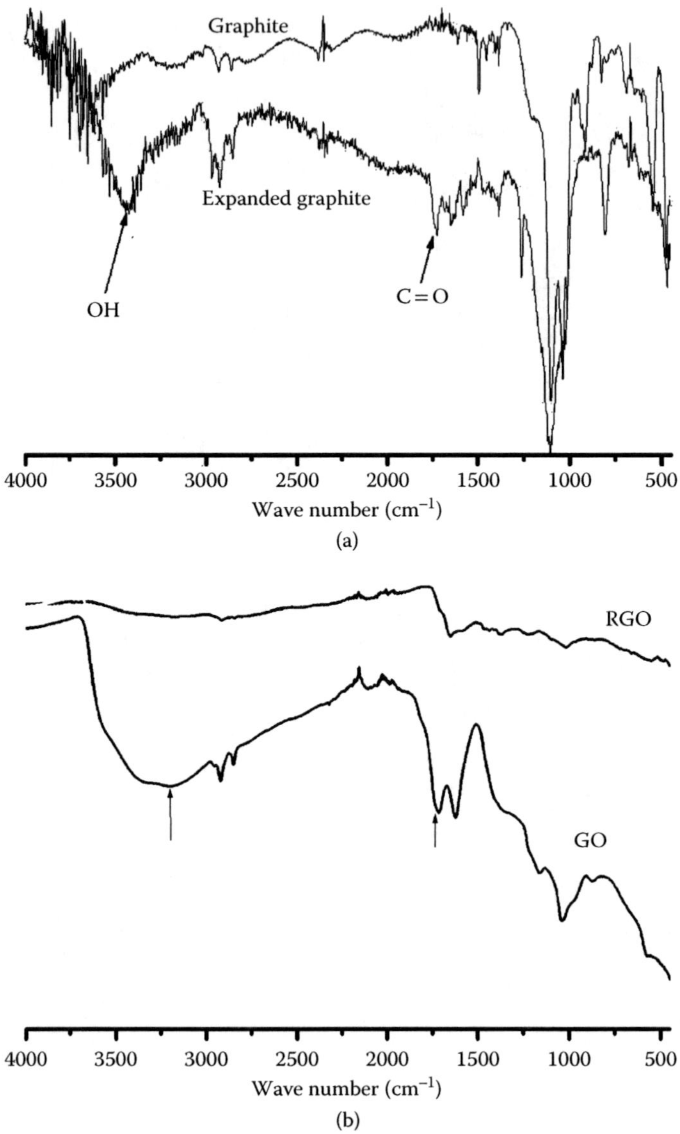

FIGURE 12.4
The FTIR results of (a) graphite and EG, and (b) GO and RGO.

The preparation of GO involves further oxidation of EG to expand any possible stacked sheets in the EG (McAllister et al. 2007; Stankovich et al. 2007). The TEM image in Figure 12.3b shows the presence of sheets of GO over 10 μm in length. The FTIR results showed that there were two extra peaks (1714 and 3351 cm^{-1}) indicating that numerous hydroxyl and carboxyl groups were generated during the process. Specifically, the GO contained a

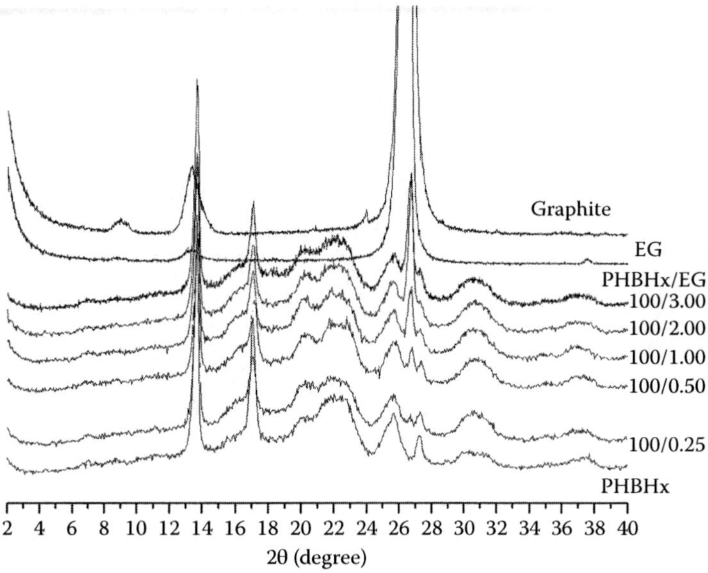

(a)

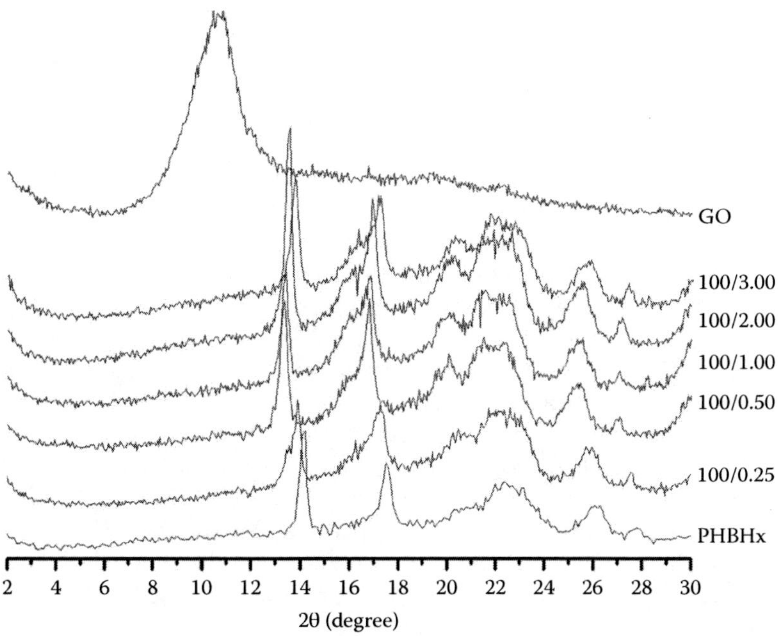

(b)

FIGURE 12.5
The XRD patterns of (a) EG, GO, PHBHx, and PHBHx/EG; (b) PHBHx/GO; and (c) PHBHx/RGO nanocomposites.

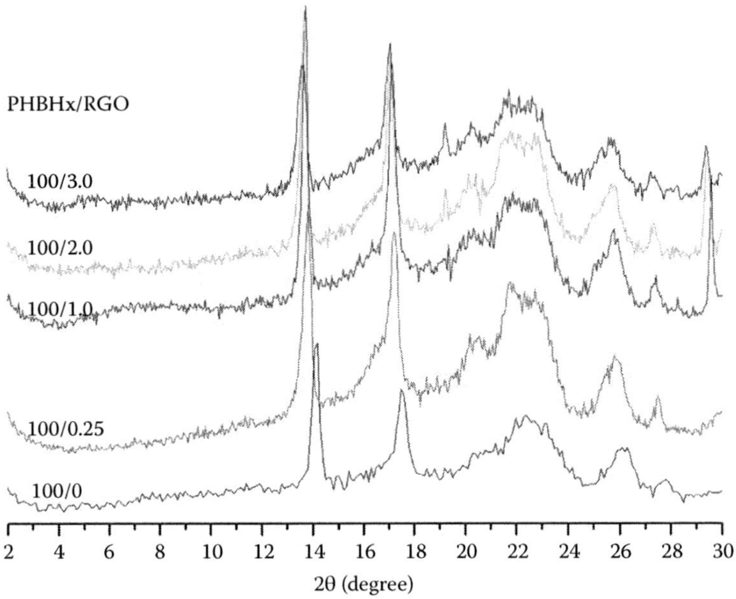

(c)

FIGURE 12.5
(Continued)

wide range of oxygen functional groups, both on the basal planes and at the edge of GO sheets, which greatly improved the compatibility of the polar biodegradable polymer PHBHx by interactions with GO. The TEM results presented in Figure 12.3 show that the GO exfoliated from EG formed GO sheets having an in-plane dimension of over 5 μm. The XRD result of pure GO in Figure 12.5 also shows that the characteristic peak of graphite near 26.5° has disappeared. The FTIR results in Figure 12.4 show the presence of new peaks near 3300 and 1714 cm^{-1} for hydroxyl and carbonyl groups, respectively. The stability of GO in pure water also shows that the presence of numerous oxygen-containing groups improved its dispersibility. However, the polar groups also limited its dispersibility in some organic solvents, such as chloroform and THF. Additional research work is currently underway on the surface modification of GO to improve its compatibility with several types of polymers to prepare nanocomposites (Zhang et al. 2010). It is noted that GO in DMF, consisting of single sheets of graphene with hydroxyl and carboxyl groups, facilitates the mixing of GO with biodegradable polyester polymers to prepare desired nanocomposites.

The preparation of RGO involved reduction of GO by N_2H_4 in DMF with the assistance of PDMA (Park et al. 2009). Reduction of GO first involves the formation of a hydrazone in a mechanism that is probably analogous to the formation of an imine. Successive deprotonations eventually result in the evolution of nitrogen, and this mechanism is supported by the evolution of nitrogen as the thermodynamic driving force. The TEM image of RGO in Figure 12.3d shows a single sheet of graphene formed by these procedures. However, graphene very likely deposits from solution without the assistance of surfactants or ionic polymers, since graphene sheets tend to form graphite again through static interactions. The ionized polymer PDMA helped in stabilizing the graphene and suppressing the tendency of RGO sheets in solution to precipitate as a result of electronic interactions. The FTIR results in Figure 12.4 also show the very weak peaks from the hydroxyl and carboxyl groups, which means most of the hydroxyl and carboxyl groups were reduced in the reaction. The stable solution of RGO in DMF also greatly helped in the preparation of RGO/polyester nanocomposites.

12.3.2 Dispersion of EG, GO, and RGO in PHBHx

The dispersion of EG in PHBHx, as judged by TEM images, indicated that EG was homogenously dispersed at the nanoscale level in the PHBHx matrix. The physical compatibility between EG and PHBHx enabled by interactions between the carboxyl and hydroxyl groups on EG and the polyester structures in PHBHx facilitated the formation of an intercalated or exfoliated state in PHBHx. The XRD results of PHBHx/EG nanocomposites (from compositions of 100/0.25 to 100/3.0) in Figure 12.7a also show the special peak of EG near 26.5° (d_{002}), which indicated that some graphite crystalline structures still exist in the nanocomposites. It also indicated that solution mixing with ultrasonic treatment helped to improve the fractures of EG into small layered sheets suitable for reinforcement. The unexpanded layers could not be further exfoliated to single pieces or sheets, even with the help of the ultrasonic treatment. However, the strong mechanical force during the mechanical hot melting mixing might have helped to improve the dispersion of GO in the polymer matrix. Zhang et al. (2008a) applied melt mixing of PHBHx with EG and obtained oriented graphite nanosheets in PHBHx, as shown in Figure 12.7b.

The dispersion of GO in PHBHx/DMF solution was very stable at room temperature, as shown in Figure 12.6. This is due to the fact that GO and PHBHx are both very stable in DMF, and the oxygen-containing groups

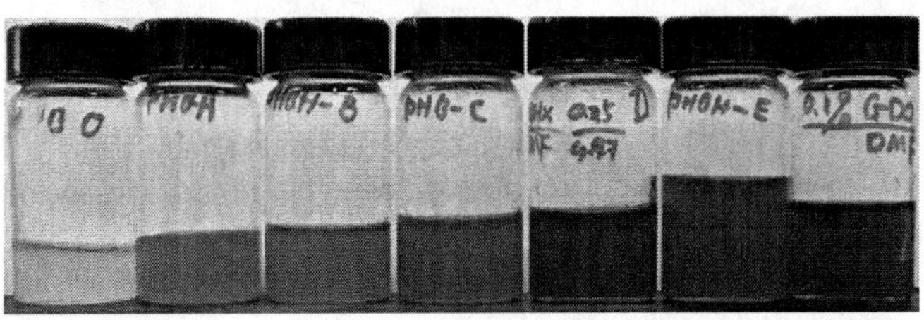

FIGURE 12.6
The dispersion of GO in PHBHx/DMF solution: From left to right, GO contents (in wt%) are 0, 0.25, 0.50, 1.00, 2.00, and 3.00, and GO in DMF solution is 0.1 wt%.

(hydroxyl and carboxyl groups) on the surface and at the edge of GO are likely to form hydrogen bonds with the polyester groups in PHBHx. The TEM of GO in PHBHx, which is shown in Figure 12.7a, shows a fine and homogenous dispersion of GO, although it is obviously not easy to detect single pieces of GO from the TEM. The dispersion mechanism is similar to that of clay30B, which is a layered silicate modified with quaternary alkyl-ammonium cation containing two hydroxyl ethyl groups in a biodegradable polyester matrix (see Figure 12.7c). Therefore, GO, being one atom in thickness, has considerable advantages over layered silicates or MWCNTs (Figure 12.7d) in reinforcing polymers, especially biodegradable polyesters.

12.3.3 Crystallinity of PHBHx and Its Thermal Characteristics

The DSC results for PHBHx/EG, PHBHx/GO, and PHBHx/RGO nanocomposites are showed in Figure 12.8 and in Table 12.1.

In case of PHBHx/EG, the glass transition temperature, T_g, decreased from –9.5°C in PHBHx to –16.4°C in PHBHx/EG (100/0.5) and T_c increased from 41.4°C in PHBHx to 59.9°C in PHBHx/EG (100/2.0). There are two crystalline melting temperatures, T_{m1} and T_{m2}, corresponding to the second melting point and the primary melting point of PHBHx. Both T_{m1} and T_{m2} increased with increasing EG content, and the secondary melting heat capacity of PHBHx, C_1, decreased while the primary melting heat capacity, C_2, increased in PHBHx/EG compared with the values for pure PHBHx. This suggested that the multilayered structure in EG allows more free volume for the motions of polymer chain segments in the glass transition status, and the results also suggested that the interaction between the EG and PHBHx chains may retard crystallization and melting. The strong interactions between the polar carboxyl and hydroxyl groups on the surfaces of EG and the polyester chains may be hydrogen bonds (Ramanathan et al. 2008). However, based on the XRD results for the PHBHx/EG and PHBHx/GO nanocomposites series shown in Figure 12.5, there are two characteristic crystallite peaks (13.9° and 17.4°) for the PHBHx matrix corresponding to the crystallites of PHBHx. It means that the introduction of EG did decrease the rate but did not affect the formation of PHBHx crystalline style (Yun et al. 2008). The PHBHx/EG is very similar with the case of PHBHx/MWCNT (see Figure 12.8a).

In the case of PHBHx/GO (see Figure 12.8b), the glass transition temperature, T_g, increased from 1.2°C in pure PHBHx to 5.4°C in PHBHx/GO (100/3.00) and the crystalline temperature, T_c, also slightly increased from 53.0°C in PHBHx to 56°C in PHBHx/GO (100/3.00). This may be the result of the greatly increased surface having rich oxygen-containing groups, which likely interact with the long main chains through hydrogen bonding and, therefore, retard the motions of polymer chain segments and their rearrangement into crystalline regions. But the low thermal conductivity of GO also leads to increased melting peaking area with increasing GO content.

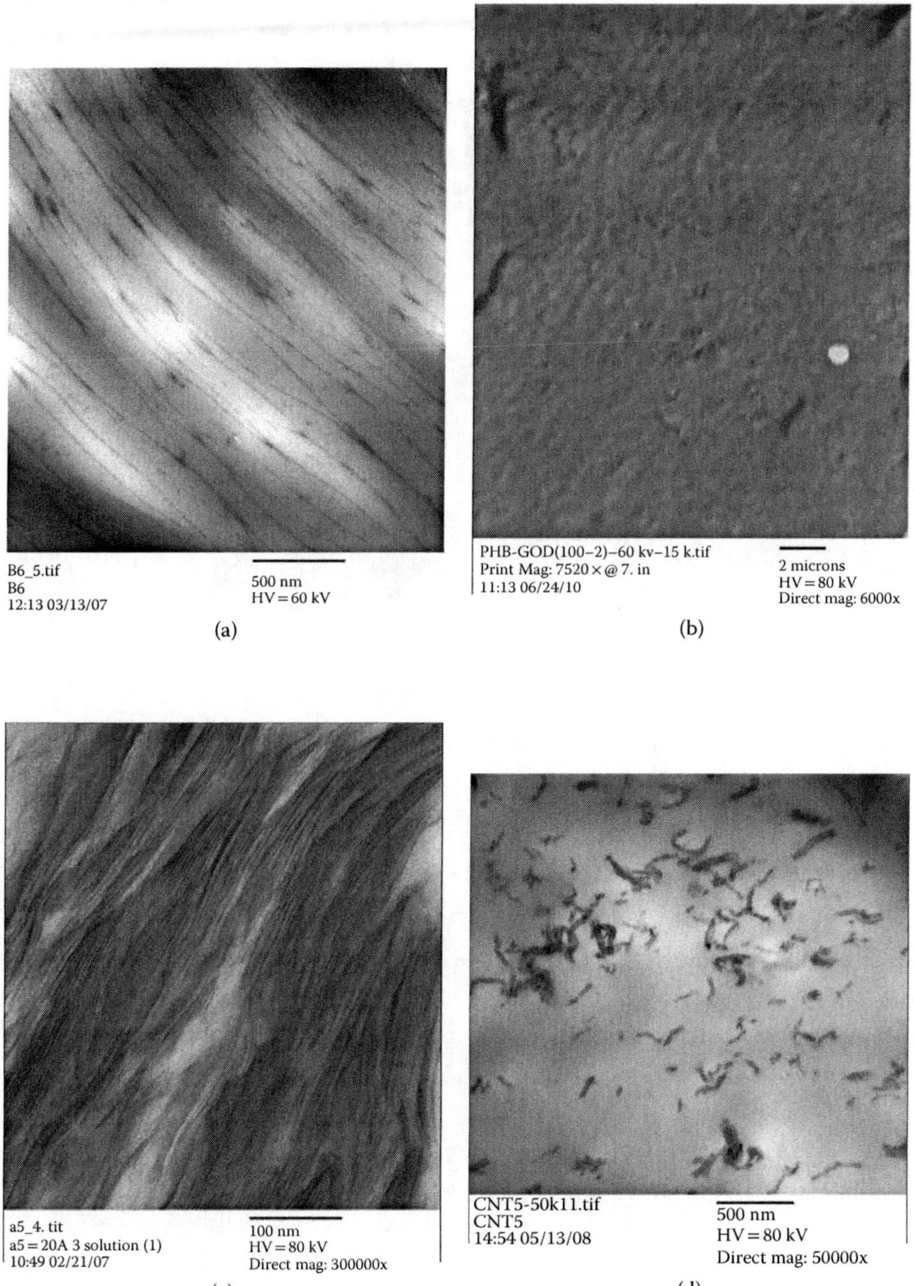

FIGURE 12.7
TEM images of (a) EG, (b) GO, (c) clay30B, and (d) MWCNT in the PHBHx matrix (all are 2 wt% based on PHBHx).

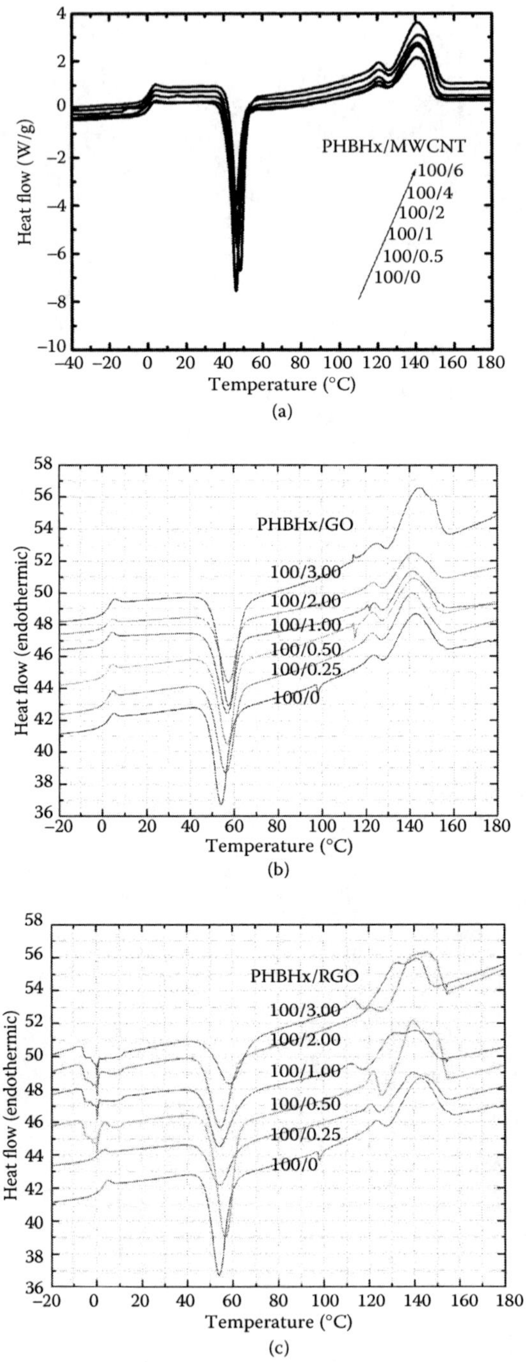

FIGURE 12.8
DSC results of (a) PHBHx/MWCNT, (b) PHBHx/GO, and (c) PHBHx/RGO nanocomposites.

TABLE 12.1

Thermal Characteristics of PHBHx/EG Nanocomposites

PHBHx/EG	T_g (°C)	T_c (°C)	T_{m1} (°C)	T_{m2} (°C)	C_1[a] (J/g·K)	C_2[b] (J/g·K)
100/0	−9.5	41.4	115.3	127.0	17.15	0.62
100/0.25	−12.7	44.5	118.8	131.8	2.57	3.12
100/0.5	−16.4	41.1	117.0	129.3	4.12	3.71
100/2.0	−10.1	59.9	111.2	130.4	5.26	7.37

[a] Secondary heat capacity of PHBHx.
[b] Primary heat capacity of PHBHx.

In case of PHBHx/RGO (see Figure 12.8c), T_g did not change when the ratio of PHBHx was 100/0.25. But specific glass transition occurred near 0°C when PHBHx/RGO was over 100/0.50. The crystalline temperature, T_c, also increased from 53.0°C in pure PHBHx to 60.3°C in PHBHx/RGO (100/3.00). The reduced hydroxyl and carboxyl groups lead to weak interaction through hydrogen bonding between RGO and PHBHx; therefore, introduction of GO does not affect the motion of polymer chain segments. But the total exfoliated layers greatly increased the surface area and retarded their rearrangement to form the PHBHx crystalline structure. The improved thermal conductivity from RGO may also lead to the lowering of melting peak compared with PHBHx/GO.

12.3.4 Thermal Stability

The thermal stabilities of the PHBHx/EG, PHBHx/GO, and PHBHx/RGO nanocomposites were gauged by TGA measurements (see Figure 12.9 and Table 12.2). The "onset temperature" (T_{onset}) is defined as the temperature at which 5 wt% of the samples was lost, and the "end-set temperature" (T_{end}) is defined as the temperature at which the final weight became stable. The TGA results in Figure 12.9a demonstrate that T_{onset} and T_{end} for PHBHx/EG decrease slightly with increasing EG content. Meanwhile, in the PHBHx GO series, both T_{onset} and T_{end} increase with increasing GO content. The T_{onset} increases from 275°C in PHBHx to nearly 300°C in PHBHx/GO (100/3.0; see Figure 12.9b). In the PHBHx/RGO series, there is some weight loss before 200°C, which might be due to the presence of surfactant residue after the preparation of RGO, although the nanocomposites were prepared though the dry and heating process. The T_{onset} for the thermal degradation of PHBHx decreased slightly and T_{end} also decreased slightly with increasing RGO content, compared with the values for pure PHBHx (see Figure 12.9c).

The final weight residue at thermal weight at T_{end} for PHBHx/EG, PHBHx/GO, and PHBHx/RGO is shown in Figure 12.9d. The results indicated that the weight residue in PHBHx/EG is almost the same as the ideal result; that is, the amount of EG introduced into the composite, and in the case of PHBHx/GO the weight residue at the end-set temperature was slightly more than this ideal amount when GO was over 100/1.00. The most interesting result is that in PHBHx/RGO, the weight residue was sevenfold larger than the ideal value, and the higher the amount of RGO introduced the more residue that remained.

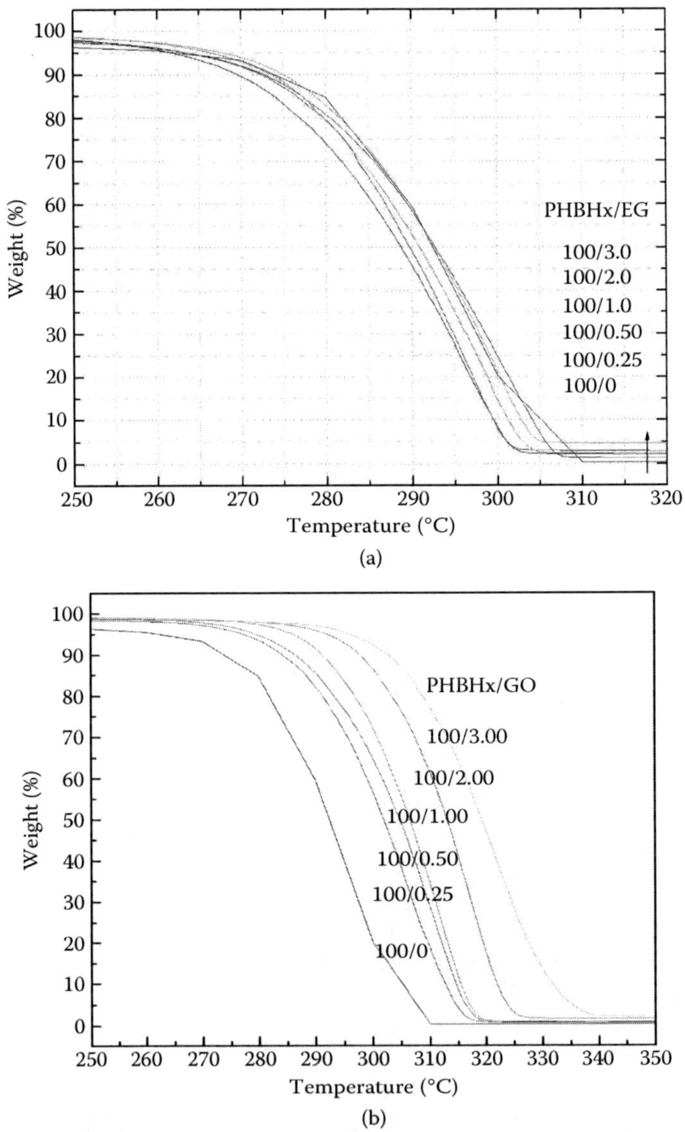

FIGURE 12.9
Thermal stability of (a) PHBHx/EG, (b) PHBHx/GO, (c) PHBHx/RGO, and (d) weight residues for the series nanocomposites.

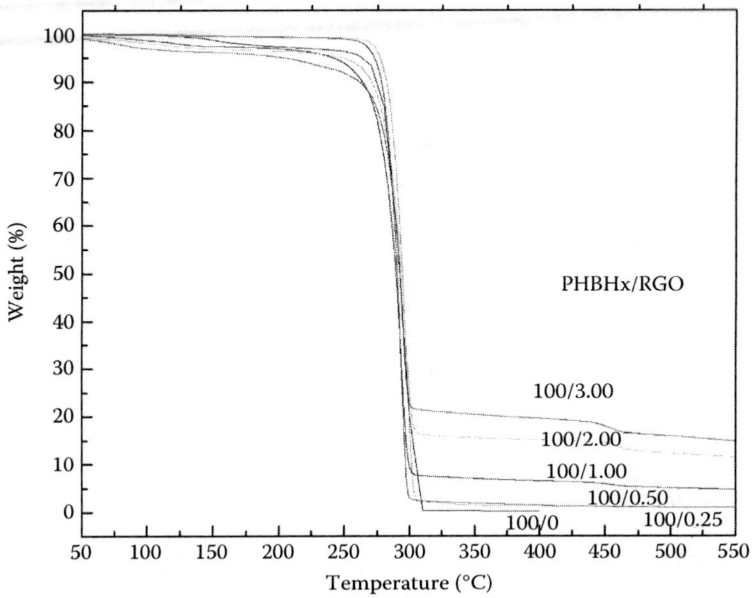

(c)

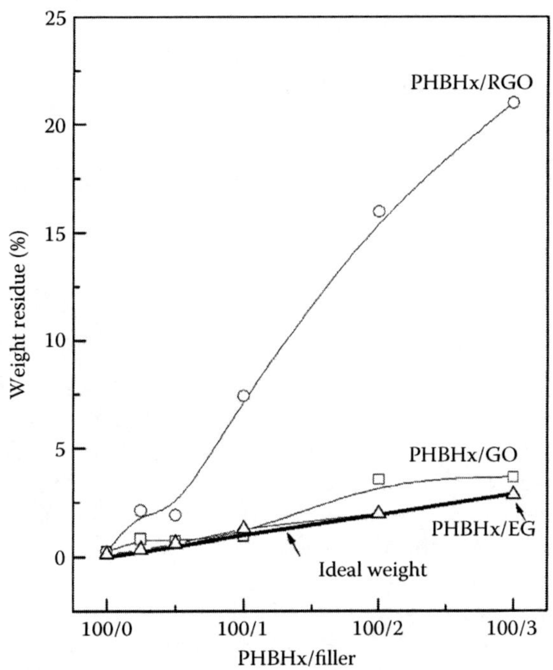

(d)

FIGURE 12.9
(*Continued*)

TABLE 12.2
Thermal Stability of PHBHx/EG Nanocomposites

PHBHx/EG	Loss (wt%)	Ideal Loss (wt%)	Loss on Graphene (wt%)	T_{onset} (°C)	T_{peak} (°C)	T_{end} (°C)	ΔT (°C)
100/0.00	99.9	100.0	−0.120	298.8	322.8	331.7	32.9
100/0.25	99.7	99.8	−0.081	298.1	318.2	326.1	27.9
100/0.50	99.4	99.5	−0.112	293.1	311.0	319.4	26.4
100/1.00	98.7	99.0	−0.280	295.3	314.1	321.7	26.5
100/2.00	98.0	98.0	0.001	294.1	317.3	324.0	30.0
100/3.00	97.2	97.1	0.073	289.4	311.7	320.6	31.1

The EG, made up of multiple layers, was expected to have improved thermal conductivity. In the case of PHBHx/GO, the homogenous dispersion of GO in PHBHx, the greatly increased surface with oxygen-containing groups would likely interact with the long main chains through hydrogen bonding, and the lower thermal conductivity of GO (compared with graphene), therefore, leads to more energy to destroy the polymer chains to thermal degradation.

In the case of PHBHx/RGO, the highest weight residue may result from the homogeneously dispersed RGO, which acts as the mass transfer barrier against the volatile pyrolized product in the polymer matrix, and the high thermal conductivity of RGO, which might help in heat transfer from the surface to the bulk material increasing the thermal degradation of PHBHx chains (Zhang et al. 2008a). However, the decrease in hydroxyl and carboxyl groups also affects the polymer chain degradation temperature (T_{onset} or T_{end}). Therefore, the combination of exfoliated GO and RGO could be applied as fire-resistant materials (Ávila et al. 2010).

12.3.5 Mechanical Properties

The reinforcement of PHBHx with EG, GO, and RGO is obviously a very interesting topic and is very important for further broadening the applications of this biodegradable polymer, such as in food packaging engineering, since the layered filler acts as the gas- or humidity-resistant material in the film and graphene is supposed to be the layered material with the highest gas resistance. The mechanical properties of PHBHx/EG nanocomposites are listed in Table 12.3.

The Young's modulus was greatly improved, specifically from 291.6 MPa for pure PHBHx to 573.7 MPa for PHBHx/EG (100/3.0), which is a much higher improvement than the predicted result from the Halpin–Tsai model (Yung et al. 2006) (see Figure 12.10). The improvement in Young' modulus is comparable to that obtained from layered silicates (Zhang et al. 2007).

TABLE 12.3
Mechanical Properties of PHBHx/EG Nanocomposites

PHBHx/EG	Young's Modulus from Tsai Model (MPa)	Young's Modulus (MPa)	Tensile Strength (MPa)	Toughness (kN·m)	Elongation at Break (%)
100/0	291.6	291.6	24.6	7.26	17.7
100/0.25	293.4	402.2	23.4	2.82	15.7
100/0.50	295.3	447.7	24.2	1.87	10.4
100/1.00	299.2	481.4	25.8	1.79	10.2
100/2.00	307.3	464.1	23.8	1.66	10.0
100/3.00	315.9	573.7	26.9	1.50	10.4

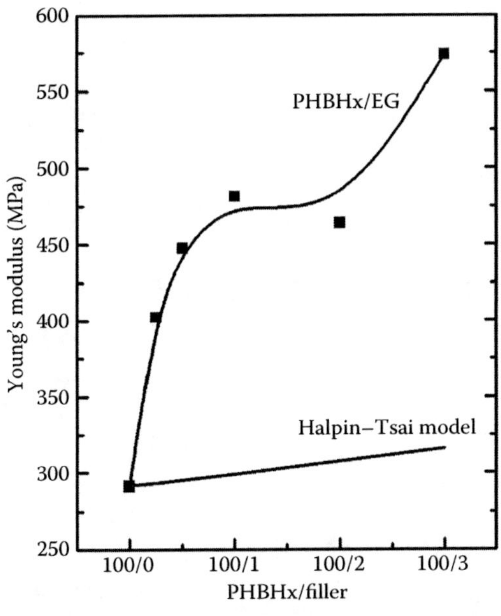

FIGURE 12.10
Mechanical properties of PHBHx/EG nanocomposites and prediction value based on the Halpin–Tsai model.

At the same time, the tensile strength, toughness, and elongation at break almost remained at the same level of pure PHBHx. The introduction of GO and RGO make the nanocomposites harder and therefore more brittle for measurement, because GO and RGO are layered fillers with one-atom thickness and huge surface areas to absorb polymer chains through physical and chemical interactions among the polar groups. A lower content of GO or RGO is more applicable for balancing strength and toughness in the nanocomposites.

12.3.6 Discussion

The incorporation of nanolayered fillers in PHBHx is a convenient way of improving the properties of this type of "green" materials and accelerating its application in commercial products. Previously, our group incorporated exfoliated layered silicates to PHBHx and achieved excellent properties (Zhang et al. 2007, 2008a). The incorporation of EG with numerous reactive groups at the edges and surfaces of the filler makes it easy to disperse it in solvents for further modifications aimed at improving its compatibility with polymers. The carboxyl, hydroxyl, and epoxy groups could be physically linked with groups on polymer chains through hydrogen bonding to generate novel interphases.

As in the case of graphene and GO, RGO is easily dispersed in thermoplastic polymers including PHBHx, and this greatly affects the crystallinity, mechanical properties, and thermal degradability of PHBHx. Compared with RGO, GO, which is further exfoliated from EG, with a higher content of polar groups greatly increased the interaction of PHBHx with GO, leading to higher thermal stability. However, RGO from GO greatly helps in the improvement of thermal stability of the nanocomposites. In view of application, when these nanocomposites are used as packing materials, additional properties including permeabilities to gases, such as oxygen and carbon dioxide, should also be considered. Therefore, it is necessary to study all properties of possible relevance. Although there have been a number of studies on the effect of graphene and its derivatives on permeabilities of some gases, there is none on the effect of these materials on the permeability of water vapor, which should also be considered. Decreasing permeability to water vapor is a critical issue in the development of some sustainable packaging materials (Arora and Padua 2010).

12.4 Concluding Remarks and Future Directions

It is noted that EG, GO, and RGO are prepared from graphite. The EG was made of multilayered graphene with graphite crystalline structure, with some carboxyl and hydroxyl groups. The GO was extensively exfoliated from EG, and single-layer graphene with oxygen-containing groups was achieved. The RGO was prepared from GO using N_2H_4 through the reduction of oxidation groups with the help of PDMA from aggregation. The biodegradable nanocomposites based on PHBHx and the graphene-based nanofillers (EG, GO, RGO, and MWCNT nanofillers) were prepared by solution mixing with ultrasonic treatment. The EG was homogenously dispersed with multilayered structure in PHBHx, even with a substantially high EG content. The EG greatly improved the mechanical properties especially Young's modulus from 283 MPa in pure PHBHx to 575 MPa in PHBHx/EG (100/3.0) without sacrificing toughness. Both GO and RGO in the PHBHx matrix were homogenously dispersed in a single-layered graphene. The improvement in mechanical properties of PHBHx especially Young's modulus from GO and RGO is even higher than that from EG. To achieve the optimal mechanical properties with proper toughness, GO and RGO must be kept in low levels; that is, less than the PHBHx/filler (100/3.0).

The effect of EG, GO, and RGO on glass transition temperature, T_g; crystalline temperature, T_c; and melting temperature, T_m, is slight when PHBHx/filler is less than 100/3.0, similar to that in the case of PHBHx/MWCNT. The thermal stability of PHBHx decreased with the incorporation of EG because of the introduction of oxygen. However, incorporation of GO greatly improved the thermal stability of PHBHx, leading to over 30°C increase in onset temperature in PHBHx/GO (100/3.0) compared with the onset temperature of pure PHBHx; this is because of the strong physical interaction between the polymer chains and the polar groups. Incorporation of RGO slightly affected the thermal degradation temperature of PHBHx nanocomposites, but the weight residue greatly increased over 20 wt% compared with that of pure PHBHx. The reason for the increase in residue is not known currently, but it is suggested that partial reduction of GO helps to increase thermal and weight residues. This makes graphene-based polymer nanocomposites have improved thermal properties, both in onset temperature and weight residue. More interesting study on the effects of graphene, GO, and RGO in biodegradable polymers is done in the direction of improvement of gas permeability and the natural degradability of nanocomposites using traditional degradation methods.

Acknowledgments

Gui Lin and James E. Mark are pleased to acknowledge the assistance provided by Michael Menard of the Department of Geology, University of Cincinnati, Ohio, on XRD measurements. The authors also thank Dr. Thuy D. Dang of Wright-Patterson Air Force Base in Dayton, Ohio, for generously providing some graphite nanosheets. The authors further thank Dr. Petra Potschke for allowing the use of the SEM image of expanded graphite. It is also a pleasure to acknowledge the financial support provided to James E. Mark by the National Science Foundation through the grant DMR-0803454 (Polymers Program, Division of Materials Research).

References

Arora, A., and G. Padua. (2010). "Review: Nanocomposites in food packaging." *Journal of Food Science* **75**(1): R43–R49.

Ávila, A. F., E. C. Dias, et al. (2010). "An investigation on graphene and nanoclay effects on hybrid nanocomposites post fire dynamic behavior." *Materials Research* **13**: 143–150.

Balandin, A., S. Ghosh, et al. (2008). "Superior thermal conductivity of single-layer graphene." *Nano Letters* **8**(3): 902–907.

Becerril, H. A., J. Mao, et al. (2008). "Evaluation of solution-processed reduced graphene oxide films as transparent conductors." *ACS Nano* **2**(3): 463–470.

Bolotin, K. I., K. J. Sikes, et al. (2008). "Ultrahigh electron mobility in suspended graphene." *Solid State Communications* **146**(9–10): 351–355.

Geim, A. K., and K. S. Novoselov. (2007). "The rise of graphene." *Nature Materials* **6**(3): 183–191.

Gross, R. A., and B. Kalra. (2002). "Biodegradable polymers for the environment." *Science* **297**(5582): 803–807.

Hassan, M. K., S. A. Abdel-Latif, et al. (2004). "Improvement in the mechanical properties of poly(3–hydroxybutyrate-co-3-hydroxyhexanoate) by preorientation." *Journal of Applied Polymer Science* **94**(6): 2257–2262.

Hassan, M. K., R. Abou-Hussein, et al. (2006). "Biodegradable copolymers of 3-hydroxybutyrate-co-3-hydroxyhexanoate (Nodax™), including recent improvements in their mechanical properties." *Molecular Crystals and Liquid Crystals* **447**(1): 23–44, 341–362.

Hong, W., Y. Xu, et al. (2008). "Transparent graphene/PEDOT-PSS composite films as counter electrodes of dye-sensitized solar cells." *Electrochemistry Communications* **10**(10): 1555–1558.

Jang, B., and A. Zhamu. (2008). "Processing of nanographene platelets (NGPs) and NGP nanocomposites: A review." *Journal of Materials Science* **43**(15): 5092–5101.

Lee, H., J. Ihm, et al. (2010). "Calcium-decorated graphene-based nanostructures for hydrogen storage." *Nano Letters* **10**(3): 793–798.

Li, Y. (2007). Processing of Nano Graphene Plates (NGPs) and NGP Nanocomposite. *Science in Engineering*. Dayton, Wright State University. Master.

Liang, J., Y. Xu, et al. (2009). "Infrared-triggered actuators from graphene-based nanocomposites." *The Journal of Physical Chemistry C* **113**(22): 9921–9927.

Lin, G., X. Zhang, et al. (2008). "Some nanocomposites based on a glycerol-derived alkyd resin and layered silicates." *Molecular Crystals And Liquid Crystals* **18**(1): 33–48.

McAllister, M. J., J.-L. Li, et al. (2007). "Single sheet functionalized graphene by oxidation and thermal expansion of graphite." *Chemistry of Materials* **19**(18): 4396–4404.

Mooney, B. P. (2009). "The second green revolution? Production of plant-based biodegradable plastics." *Biochemical Journal* **418**(2): 219–232.

Noda, I. (2000). Absorbent articles comprising biodegradable PHA copolymers. US Patent 6,160,199.

Noda, I., P. R. Green, et al. (2005). "Preparation and properties of a novel class of polyhydroxyalkanoate copolymers." *Biomacromolecules* **6**(2): 580–586.

Park, S., D. A. Dikin, et al. (2009). "Graphene oxide sheets chemically cross-linked by polyallylamine." *The Journal of Physical Chemistry C* **113**(36): 15801–15804.

Park, S., K.-S. Lee, et al. (2008). "Graphene oxide papers modified by divalent Ions; enhancing mechanical properties via chemical cross-linking." *ACS Nano* **2**(3): 572–578.

Philip, S., T. Keshavarz, et al. (2007). "Polyhydroxyalkanoates: Biodegradable polymers with a range of applications." *Journal of Chemical Technology & Biotechnology* **82**(3): 233–247.

Phithakrotchanakoon, C., Y. Rudeekit, et al. (2009). "Microbial degradation and physico-chemical alteration of polyhydroxyalkanoates by a thermophilic Streptomyces sp." *Biologia* **64**(2): 246–251.

Ramanathan, T., A. A. Abdala, et al. (2008). "Functionalized graphene sheets for polymer nanocomposites." *Nature Nanotechnology* **3**(6): 327–331.

Ray, S. S., and M. Bousmina. (2005). "Biodegradable polymers and their layered silicate nanocomposites: In greening the 21st century materials world." *Progress in Materials Science* **50**(8): 962–1079.

Ray, S. S., K. Yamada, et al. (2002). "Polylactide-layered silicate nanocomposite: A novel biodegradable material." *Nano Letters* **2**(10): 1093–1096.

Salavagione, H. J., M. A. Gomez, et al. (2009). "Polymeric modification of graphene through esterification of graphite oxide and poly(vinyl alcohol)." *Macromolecules* **42**(17): 6331–6334.

Satkowski, M. M., D. H. Melik, et al. (2001). *Biopolymers*. Weinheim, Wiley-VCH.

Srinivas, G., Y. Zhu, et al. (2009). "Synthesis of graphene-like nanosheets and their hydrogen adsorption capacity." *Carbon* **48**(3): 630–635.

Stankovich, S., D. A. Dikin, et al. (2006). "Graphene-based composite materials." *Nature* **442**(7100): 282–286.

Stankovich, S., D. A. Dikin, et al. (2007). "Synthesis of graphene-based nanosheets via chemical reduction of exfoliated graphite oxide." *Carbon* **45**(7): 1558–1565.

Stoller, M., S. Park, et al. (2008). "Graphene-based ultracapacitors." *Nano Letters* **8**(10): 3498–3502.

Tian, G., Q. Wu, et al. (2002). "Two-dimensional Fourier transform infrared spectroscopy study of biosynthesized poly(hydroxybutyrate-*co*-hydroxyhexanoate) and poly(hydroxybutyrate-*co*-hydroxyvalerate)." *Journal of Polymer Science Part B: Polymer Physics* **40**(7): 649–656.

Tung, V. C., M. J. Allen, et al. (2009). "High-throughput solution processing of large-scale graphene." *Nature Nanotechnology* **4**(1): 25–29.

Usuki, A., Y. Kojima, et al. (1993). "Synthesis of nylon 6-clay hybrid." *Journal of Materials Research* **8**(7): 1179–1184.

Valentini, L., M. Cardinali, et al. (2010). "Use of butylamine modified graphene sheets in polymer solar cells." *Journal of Materials Chemistry* **20**(5): 995–1000.

Venkitachalam, R., J. E. Mark, et al. (2005). "Drying biodegradable poly (hydroxybutyrate-co-3-hydroxyvalerate) gels in the elongated state to improve their mechanical properties." *Journal of Applied Polymer Science* **95**(6): 1519–1523.

Yang, K., X. Wang, et al. (2007). "Progress in nanocomposite of biodegradable polymer." *Journal of Industrial and Engineering Chemistry* **13**(4): 485–500.

Yang, Y., J. Wang, et al. (2009). "Exfoliated graphite oxide decorated by PDMAEMA chains and polymer particles." *Langmuir* **25**(19): 11808–11814.

Yun, S., G. Gadd, et al. (2008). "Mechanical properties of biodegradable polyhydroxyalkanoates/single wall carbon nanotube nanocomposite films." *Polymer Bulletin* **61**(2): 267–275.

Yung, K. C., J. Wang, et al. (2006). "Modeling Young's modulus of polymer-layered silicate nanocomposites using a modified Halpin—Tsai micromechanical model." *Journal of Reinforced Plastics and Composites* **25**(8): 847.

Zhang, X., G. Lin, et al. (2007). "Some novel layered-silicate nanocomposites based on a biodegradable hydroxybutyrate copolymer." *European Polymer Journal* **43**(8): 3128–3135.

Zhang, X., G. Lin, et al. (2008a). "Biodegradable nanocomposites based on the polyester poly(3-hydroxybutyrate- co-3-hydroxyhexanoate) and layered silicate or expanded graphite." *Journal of Macromolecular Science, Part A Pure and Applied Chemistry* **45**(6): 431–439.

Zhang, X., G. Lin, et al. (2008b). Novel biodegradable nanocomposites based on poly(3-hydroxybutyrate-co-3-hydrohexanote) and in-situ silica. *American Chemical Society Meeting*. Philadelphia, PA.

Zhang, Y., G. Lin, et al. (2010). "Chemical modification of graphene oxide by alkylamines fror improving dispersibility in chloroform." *Chemistry of Materials* Submitted.

Zhou, D. H. (2005). Evaluation of the correlation between properties and structure of polymer nanocomposites. *Department of Chemistry*. Cincinnati, University of Cincinnati. Master.

13

Effect of Modified Graphene on Properties of Polypropylene Nanocomposites

Olga Shepelev and Samuel Kenig*

CONTENTS

13.1 Introduction

Polymer nanocomposites based on nanoclays, carbon nanotubes, nanosilica, and polyhedral oligomeric silsesquioxane (POSS) have been used to enhance the mechanical, thermal, electrical, and barrier properties of polymers.

With the emergence of graphene nanoplatelets (GNPs), new potential routes have been opened to enhance polymer properties. GNPs are based on a two-dimensional (2D) nanoplatelet structure having a Young's modulus of 1 TPa and ultimate strength of 130 GPa. Single-layer graphene has very high electrical conductivity of up to 6000 S/cm. These properties in addition to very high surface area of more than 2000 m^2/g and gas impermeability indicate graphene's potential for improving mechanical, electrical, and gas

* Correspondence to: Samuel Kenig (shmkenig@bezeqint.net)

barrier properties of polymers (Novoselov et al. 2005, 2008; Stankovich et al. 2006, 2007 Zhang et al. 2005).

The successful use of graphene depends on the exfoliation of bulk graphite into individual sheets. Several chemical–mechanical routes have been developed to produce individual exfoliated graphene sheets, for example, mechanical exfoliation, chemical exfoliation, solvent-thermal reduction, chemical vapor deposition, or a combination of these methods. Each method has its own advantages and drawbacks related to the purity and the presence of defects (oxygen and functional groups on the surface) (Mukopadhyay and Gupta 2009).

The most common route to produce graphene involves the production of graphite oxide (GO) by oxidation chemistry followed by a reduction and mechanical exfoliation. This is the basis of Hummers and Offeman's (1958) process. GO is nonconductive, hydrophilic, and can readily swell and disperse in water. Two different approaches to convert GO to graphene are commonly used: thermal treatment (Chen et al. 2009; McAllister et al. 2007; Zhang et al. 2010) and chemical modification (Gomez-Navarro et al. 2008; Jin et al. 2009; Ramanathan et al. 2008; Stankovich et al. 2006, 2007). Recently, several new methods of graphene functionalization were reported. Functionalization proceeded by a covalent bonding of epoxy monomer to the surface of expanded graphite, through a coupling agent, such that the epoxy concentration reaches 4%, improves the dispersion of the modified graphite in epoxy resin (Miller et al. 2010). Sulfonated graphene was prepared from graphene oxide in four steps for final synthesis of hybrid material of graphene and poly(3,4-ethyldioxythiophene) with high conductivity, flexibility, and transparency (Xu et al. 2009).

Functionalized graphene sheets (FGS) demonstrate improved dispersibility in organic solvents and polymers. Isocyanate-modified GO showed better dispersion in thermoplastic polyurethane, the material that becomes electrically conductive at less than 0.5 wt% of FGS (Schniepp et al. 2006).

Polycarbonate-based composites with exfoliated FGS required a smaller amount of reinforcement for rigidity, electrical percolation, and decrease of gas permeability (Kim 2009). However, defects in sheet structure formed during oxidation and pyrolysis limited modulus and dimensional stability improvement. Graphene exhibited its potential in improving electrical conductivity of polymers—poly(ethylene terephthalate) (PET) (Zhang et al. 2010), poly(methyl metacrylate) (PMMA) (Ramanathan et al. 2008; Zheng and Wong 2003), polyester (Kim and Macosko 2008), and polystyrene (Stankovich et al. 2006).

Commercial low-cost GNPs have been produced by XG Sciences. The GNP is derived from thermal expansion of acid-intercalated graphite. The platelets are composed of several sheets of graphene with an overall thickness ranging from 1 nm to 15 nm and particle diameters of a few microns. Specific surface area is rather low—about 100 m^2/g. However, large particle aspect ratio prevents particle bucking (Kalaitzidou 2006; Kalaitzidou et al. 2008).

Nanocomposites fabricated using the latter GNP at various concentrations showed significant increase in flexural modulus and improved electrical and thermal conductivities (Drzal et al. 2009; Kalaitzidou et al. 2007b). GNP dispersed in a matrix material in amount of 15 wt% provided electrical conductivity in the range of 100 S/cm, intended for use in automotive friction plates and aircraft brake components (Song et al. 2009).

Generally, nanocomposites made of polypropylene (PP) reinforced with GNP nanoparticles have been prepared by melt mixing, polymer solution and coating are reported (Kalaitzidou et al. 2007a; Park et al. 2010).

Most of the production methods use acid, chemicals and organics, which are difficult to apply in mass production for the plastics technology. Therefore, the objective of this research work was to study the possible routes to modify the low-cost GNP for further compounding and conventional plastics processing with the aim to improve the performance of PP-based nanocomposites with respect to mechanical, thermal, and electrical properties.

13.2 Experimental

13.2.1 Materials

PP homopolymer was used throughout this work. The grade chosen had a melt flow index (MFI) of 13 produced by Carmel Olefins. GNP used was xGnP-05 obtained from XG Sciences. A few surface active agents (SAA) were used to disperse and steric stabilize the graphite particles like polyether polyols (Baycoll, BD 1110; hydroxyl number 112 mg KOH/g, Bayer), polyol of molecular weight (MW) 4000 (Polyol 4200, hydroxyl number 28 mg KOH/g, Bayer), polyol of MW 8000 (Polyol 8200, hydroxyl number 14 mg KOH/g, Bayer), octylphenol ethoxylate (Triton X-100, Sigma-Aldrich), poly(ethyleneoxide)/poly(propyleneoxide) (PEO/PPO) block copolymer (Pluronic 3100), and sodium alkyl sulfonate (Lutensit A-PS, BASF). In addition, fumed silica (Aerosil 200, Aerosil 380, R-974, Degussa) was used as co-fillers to improve the GNP dispersion. Glycerol monostearate (GMS) and ethylene bis stearamide (EBS) (Acrawax C) were used for preparation of GNP melt-blended masterbatches.

13.2.2 Modification of GNP

The necessity of surface treatment of nanographite was found to be mandatory in the early stages of the research to improve dispersion within the polymer matrix and improve adhesion, as also shown by Park et al. (2010). Furthermore, since the lower-cost GNP used is composed of stacked graphene, further exfoliation using milling, sonication, and microwaving has been studied.

Various techniques were applied aiming to achieve separation of strongly bound graphene sheets in GNP nanoparticles, to achieve stabilization of the nanoparticles, and avoid reaggregation during processing. Hence, thermally stable materials were chosen for modification to prevent degradation during processing of the PP.

Following surface treatments procedures were studied:

1. Incorporation of GNP into GMS or EBS in the molten state using melt mixing. The masterbatch route was chosen to prepare a 33% concentrate of GNP that was further ground to powder form.
2. Mixing of GNP with ethanol solutions of anionic or nonionic SAA, using the composition of (wt%) ethanol 90%, SAA 2%, and graphite 8%. Mixtures were subjected to treatment in a ball mill for 96 hours (rotation rate about 200 rpm), followed by sonication in ultrasonic bath at frequency of 55 kHz and input power of 70 W for 5 hours. The resulting viscous slurries were dried at 75°C–100°C and grinded to fine powder.

   For some compositions, half of the SAA was substituted by fumed silica (Aerosil 200, Aerosil 380, or hydrophobic Aerosil R-974).
3. Treatment of mixtures GNP/SAA in water solutions using microwave oven, followed with drying and grinding to fine powder.

13.2.3 Fabrication of PP-Based GNP Nanocomposites

Generally, properties of polymer nanocomposites depend on proper dispersion and distribution of nanofillers in the polymer matrix. The low bulk density of the high surface area graphene requires special attention for incorporation into polymers. Various methods have been investigated, like producing graphene nanocomposites through in situ intercalative polymerization or through blending with organic solvents followed by solvent removal (e.g., Janowska et al. 2010; Khan et al. 2010; Lotya et al. 2009; Tkalya et al. 2010). The most economically attractive and scalable method for dispersing nanoparticles into polymers is melt blending.

PP-based compositions with powdered 33% GNP masterbatches were compounded in a mixing cell (Brabender Plasticoder) at 200°C and 80 rpm for 10 minutes. The other PP compositions were compounded in a twin-screw extruder (L/d = 25) at 190°C and 220 rpm. The compositions contained both untreated GNP powder and modified GNP. Specimens were prepared by compression molding at 200°C.

13.2.4 Characterization

Thermomechanical properties of compositions were investigated by three-point bending or cantilever beam geometry, using Dynamical Mechanical Analyzer (DMA Q800, TA Instruments) according to ASTM D4065. Viscoelastic characteristics were evaluated by dynamic rheometry according to ASTM D4440 at 200°C using (AR 2000ex Dynamic Rheometer, TA Instruments).

Changes in transition temperatures were determined by differential scanning calorimetry according to ASTM D3418 (DSC Q100 device, TA Instruments). Dimensional distribution of pristine and treated GNP was examined by optical microscope at magnitude ×400 (Olympus BX51). The microstructures of modified GNP particles and PP-based compositions were characterized using scanning electron microscope (SEM).

The surface resistivity of PP-based compositions was determined using Keithely 8009 Resistivity Test Fixture device, while electrical resistivity of GNP was evaluated using Fluke 115 RMS Multimeter by applying them on nonconductive substrate.

13.3 Results and Discussion

13.3.1 Modification of GNP Particles

Initial experiments included compounding of untreated GNP and PP homopolymer at 200°C and 80 rpm. Films prepared from these compositions showed nonuniform distribution of the GNP in the polymer—visible points of the graphite filler were scattered in the transparent matrix. Addition of GMS resulted in some improvement in GNP distribution.

Consequently, a two-stage approach was investigated using a preblended masterbatch (33 wt% GNP) in GMS or EBS (dispersants differing in hydrophilic-lipophilic balance [HLB] and melting temperatures), followed by compounding with PP to the desired GNP content.

As can be seen in Table 13.1, a significant enhancement in mechanical and thermal properties, manifested in the storage modulus (SM) and crystallization temperature T_c of PP-based compositions, was obtained even at low loadings (0.5%–1%) of GNP nanoplatelets. Furthermore, an increase in melt viscosity and dynamic SM, especially for compositions with GMS-treated GNP, indicated improved distribution of nanoparticles in the polymer. It can be assumed that GMS characterized with lower melting point (MP) (63°C–68°C) and hydrophilic lipophilic balance (HLB) in the range 3–5 provided a certain degree of interaction with the functional groups located on the GNP particles resulting in better dispersion of the GNPs stacks and improved compatibility with the polymer matrix than EBS (MP about 143°C–146°C). Higher values of dynamic viscosity and dynamic modulus of compositions containing GMS-modified GNP confirm this assumption. Higher content of masterbatches resulted in reduction in SM of the PP nanocomposites due to the plasticizing effect of the masterbatch carriers, GMS and EBS, incorporated in large amounts (6% of carrier per 3% of GNP).

Hence, GNP modification was carried out at high concentrations of graphene nanoparticles. The steric stability of the exfoliated graphite nanoparticles was investigated through incorporation of nonionic and anionic

TABLE 13.1

PP-Based Compositions with GNP Melt Blended Masterbatches

Material	GNP Content (wt%)	Storage Modulus (MPa)		Dynamic Viscosity (PaxS) 200°C, Frequency		Dynamic Modulus (Pa) 200°C, Frequency		Tc [°C]
		25°C	70°C	6.3	31.5	6.3	31.5	
Neat PP, MFI = 13	0	1,123	614	816	476	2,142	8,715	115
PP composition with GNP/ GMS 33% masterbatch	0.5	1,416	685	1,500	844	4,300	16,240	120
	1	1,653	761	1,421	805	4,055	15,480	121
	3	1,314	620	—	—	—	—	124
PP composition with GNP/ EBS 33% masterbatch	0.5	1,435	630	1,060	915	2,490	1,980	120
	1	1,400	620	650	532	11,250	11,620	117
	3	1,270	740	—	—	—	—	119

Note: Prepared by Brabender at 200°C, 80 rpm, 10 minutes.

SAA. In addition, the effect of size reduction of the GNP agglomerates was investigated by milling the GNP in presence of ethanol and SAA solutions followed by ultrasonication, drying, and grinding to fine powder containing 80% of GNP and 20% of SAA.

X-ray diffraction (XRD) spectra showed that the main diffraction peak for both untreated and modified GNP nanoparticles were identical at 2-theta of 26.3°, indicating the existence of stacked graphene sheets. However, the performance of the polymer nanocomposites prepared with the modified GNP species was improved substantially—no visible agglomerates were observed and films molded from these compounds containing up to 1% GNP were almost transparent. Figure 13.1 demonstrates the effect of GNP loading on the mechanical properties of the PP nanocomposites. As can be noticed, an increase of up to 50% in SM took place at 1 wt%, while at 5–10 wt% loadings, SM was doubled. Best results, using the studied surfactants, were observed for GNP treated with sodium alkyl sulfonate and polyether polyol of MW 4000.

As discussed earlier, graphene addition results in improved electrical conductivity of polymers (Drzal et al. 2009; Schniepp et al. 2006) at very low concentration. For instance, polystyrene/graphene nanocompositions prepared by chemical modification in solution exhibited a percolation threshold at 0.1 vol% (Stankovich et al. 2006). In the case of the GNP (stacked graphene)/PP nanocomposites, only concentrations of 15 wt% exhibited a network formation (Kalaitzidou et al. 2010; Park et al. 2010; Song et al. 2009).

Since the electrical conductivity properties are of utmost importance in the case of GNP nanocomposites (Jing et al. 2000; Steurer et al. 2009), the effect of

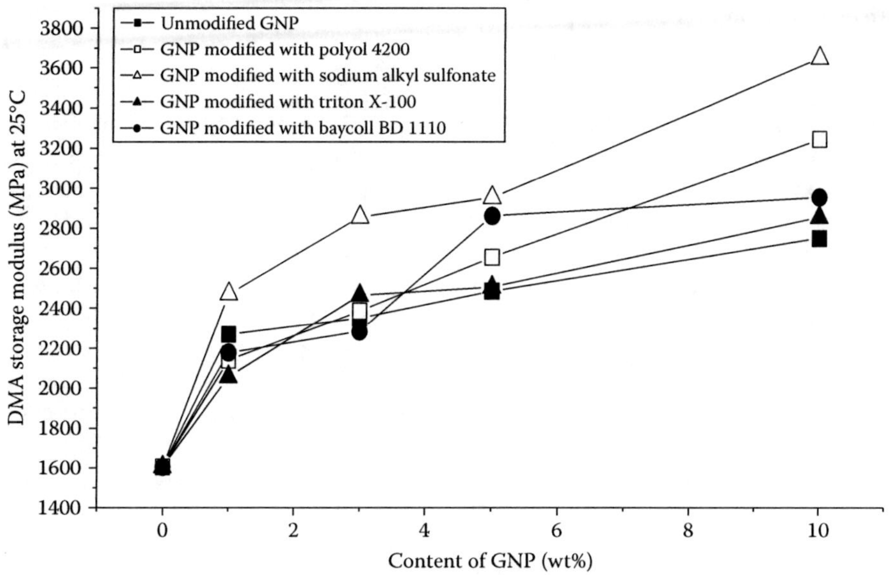

FIGURE 13.1
DMA storage modulus of PP-based compositions with GNP modified in SAA solutions.

TABLE 13.2
Electrical Resistivity of GNP Modified with SAA in Ratio 80: 20

Modification of GNP	Resistivity (Ohms)
Pristine GNP	1.5×10^3
Sodium alkyl sulfonate (ball mill + ultrasonics)	0.9×10^3
Sodium alkyl sulfonate (microwave)	5×10^3
Triton X-100 (ball mill + ultrasonics)	4×10^3
PEO/PPO copolymer (ball mill + ultrasonics)	0.2×10^3
Polyether polyol (ball mill + ultrasonics)	0.3×10^3
Polyether polyol Polyol 4200 (ball mill + ultrasonics)	0.5×10^3

SAA on the electrical properties of GNP was thoroughly studied. In the first step, the electrical resistivity of the GNP in ethanol solutions (after drying at 100°C) was evaluated.

As can be seen in Table 13.2, a decrease in electrical resistivity was achieved for GNP modified with most of the studied SAA when size reduction was achieved by ball mill/ultrasonication combination resulting in higher degree of exfoliation. It should be emphasized that GNP/sodium alkyl sulfonate aqueous mixture treated in microwave demonstrated an increase in electrical resistivity; this may possibly be due to agglomeration of GNP during treatment.

13.3.2 PP/GNP Nanocomposites

Materials prepared from PP pellets coated with GNP nanoparticles were processed by the method proposed by Drzal and coworkers (Kalaitzidou et al. 2010; Park et al. 2010). In the present work, the method was modified to include the addition of nonionic polyol as a SAA to improve the wettability of GNP nanoparticles. Thus, GNP nanoplatelets were first dispersed by ultrasonication for 1 hour in SAA/IPA solution. Then, PP pellets were added to the slurry and sonicated for another hour. Mixture was dried at 80°C–100°C. The material was divided into two groups. From one part, specimens for tests were prepared by compression molding at 200°C for 20 minutes of melting and 20 minutes under pressure. Molded plates exhibited macro-heterogeneous structure composed of polymer inclusions surrounded with a network of GNP. The second part of coated PP was processed at twin-screw compounder at 190°C and 220 rpm. Specimens for tests were prepared by compression molding at 200°C by the same procedure as the first part. For reference purposes, PP-based compound of identical composition—10 wt% of pristine GNP and 2.5% of polyol—was prepared using the twin-screw extruder by conventional technology. Electrical, mechanical, and thermal characteristics of compositions are represented in Table 13.3.

As can be observed, the electrical resistivity of materials prepared from the coated PP pellets dropped significantly at 3 wt% of GNP due to the controlled morphology leading to continuous path of the GNP particles arranged on the surface of PP granules. The use of the SAA improved the integrity of coating layer. An effect of SAA addition on the improved distribution of the GNP particles is manifested in higher values of dynamic viscosity and dynamic modulus for the composition containing 6% GNP/polyol combination. At 10 wt% of GNP, saturation took place, resistivity became close to that of pristine GNP.

When GNP-coated PP granules were further processed by conventional extrusion compounding, the electrical resistivity decreased at low GNP concentrations and was preserved only at higher values of 10 wt% of GNP. This is attributed to the fact that network of the graphene nanoparticles formed on the surface of PP was destroyed. Similar value of electrical resistivity was obtained for the composition with 10% of GNP prepared from dry blend using extrusion compounding.

13.3.3 Extrusion Compounded PP/Gnp Compositions, Morphology, and Properties

Electrical, thermomechanical, and viscoelastic properties of PP/GNP compositions prepared by conventional extrusion compounding are summarized in Table 13.4. Composites containing 10 wt% of pristine GNP exhibited some decrease in surface resistivity and significant rise in crystallization temperature and in SM compared to neat PP. Addition of the polyol SAA to the composition with untreated GNP sharply improved electrical conductivity,

TABLE 13.3
Properties of Compositions Prepared from PP Pellets Coated with GNP

	PP Pellets Coated with IPA/GNP Dispersion, Sonicated, Dried					Composition prepared by extrusion compounding
	Composition (wt%)					
PP pellets	100	100	94	94	90	87.5
GNP	1	3	6	6	10	10
Polyol 4200	0.26	0.8	—	1.5	2.5	2.5
Specimens Prepared by Compression Molding from Coated PP Granules at 200°C; Melting −20 Minutes + 20 Minutes at Pressure						
Surface resistivity (Ohms)	1×10^{14}	1.6×10^{11}	$2–7 \times 10^{4}$	3×10^{4}	2×10^{3}	—
Coated PP Granules Passed Twin-Screw Extruder, Specimens Prepared by Compression Molding at Regular Conditions						
Surface resistivity (Ohms)		0.4×10^{16}	7×10^{12}	1×10^{14}	9.6×10^{6}	6.2×10^{6}
Storage modulus (MPa)						
At °C	25	2,836	2,646	2,747	3,628	3,305
	70	1,432	1,425	1,393	1,867	1,723
Dynamic viscosity (PaS)						
At 200°C	6.3	1,245	1,296	1,579	2,454	3,038
Freq.	31.5	720	748	877	1,250	1,550
Dynamic modulus (PaS)						
At 200°C	6.3	3,426	3,625	4,601	8,923	10,860
Freq	31.5	13,570	14,210	17,050	27,410	33,790

storage modulus (SM) increased by 20%, and doubled the values of dynamic melt properties. This was attributed to the improved distribution of exfoliated GNP particles in polymer.

Incorporation of GNP modified with SAA only or in combinations with fumed silica brought further enhancement in crystallization temperature, mechanical properties, and thermal stability as could be concluded from the sharp increase in SM at ambient and elevated temperatures. Furthermore, significant increase in melt viscosity and strength, especially for compositions containing fumed silica, was observed due to the formation of smaller size platelets of the GNP. This effect was highly emphasized when fumed silica with high specific surface area (Aerosil 380) was used, resulting in reduction of surface resistivity of the nanocomposites.

When higher-MW polyols were used (Table 13.5), a decrease in surface resistivity took place at lower content of GNP (8 wt% of GNP). Another point of interest was that in the case of Polyol 8200—the effect of silica addition during the GNP treatment was noticeable only for SM and not in

TABLE 13.4

PP-Based Compositions with GNP Modified by Polyol 4200 and Silica Additives

Composition, Parts by Weight							
GNP content (wt%)		10	10	8	10	10	10
PP, MFI = 13		90	87.5	90	87.5	87.5	87.5
Polyol 4200		—	2.5	—	—	—	—
Pristine GNP		10	10	—	—	—	—
80% GNP + Polyol 4200		—	—	10	12.5	—	—
80% GNP + Polyol 4200 + A-200		—	—	—	—	12.5	—
80% GNP + Polyol 4200 + A-380		—	—	—	—	—	12.5
Properties of Compression Molded Specimens							
Tc (°C)		127.4	127.5	129.6	129.3	130.4	130.9
Surface resistivity (Ohms)		3.2×10^{13}	6.2×10^{6}	6.1×10^{14}	6.6×10^{7}	5×10^{6}	5.4×10^{5}
Cantilever DMA storage							
Modulus	25	2,751	3,305	3,290	4,090	3,800	3,868
(MPa), °C	70	1,548	1,723	1,737	2,215	2,071	2,001
Dynamic viscosity (PaS)							
200°C,	6.3	1,443	3,038	2,650	2,836	4,090	4,580
frequency	31.5	817	1,550	1,373	1,476	1,960	2,110
Dynamic modulus (Pa)							
200°C,	6.3	4,175	10,860	9,346	9,750	16,370	19,520
frequency	31.5	15,790	33,790	29,430	31,400	45,380	50,360

electrical properties improvement. This may be due to interference of the silica particles situated on the surface of the GNP, as it will be shown in section 13.3.4.

When the hydrophobic type of silica (Aerosil R-974) was used alone for GNP treatment, it was ineffective possibly due to insufficient wetting of nanoparticles in polar medium. Only the combination of the Polyol 8200/Aerosil R-974 resulted in a significant improvement in electrical properties. It was concluded that the effect of surface modification and surface area of the co-fillers on the properties of treated GNP deserves further attention.

Finally, two additional types of chemically active agents, aminosilane (AMEO) and octylphenol ethoxylate, were examined as a SAA for GNP.

TABLE 13.5
PP-Based Compositions with GNP Modified by Polyol 8200 and Silica Additives

Composition, Parts by Weight								
GNP content (wt%)		10	8	10	8	10	10	10
PP, MFI = 13		90	90	87.5	90	87.5	87.5	87.5
Polyol 8200		2.5	—	—	—	—	—	—
Pristine GNP		10	—	—	—	—	—	—
80% GNP + Polyol 8200		—	10	12.5	—	—	—	—
80% GNP + Polyol 8200 + A-200		—	—	—	10	12.5	—	—
80% GNP + Polyol 8200 + R-974		—	—	—	—	—	12.5	—
80% GNP + Aerosil R-974		—	—	—	—	—	—	12.5
Properties of Compression Molded Specimens								
Tc (°C)		127.9	128.4	128.8	129.8	130.7	130.5	
Surface resistivity (Ohms)		1.7×10^8	1.8×10^9	9×10^5	4×10^{13}	1×10^6	6.8×10^6	1×10^{11}
Storage modulus (MPa)								
At °C	25	3,143	3,301	3,274	3,787	3,664	4,061	3,626
	70	1,693	1,774	1,790	2,013	2,105	2,216	1,979
Dynamic viscosity (PaS)								
At 200°C	6.3	2,328	2,392	3,197	2,450	4,550	2,657	2,870
Freq.	31.5	1,219	1,289	1,601	1,257	2,184	1,326	1,411
Dynamic modulus (Pa)								
At 200°C	6.3	8,000	8,134	12,100	8,723	18,260	10,590	11,420
Freq.	31.5	25,900	27,300	35,890	27,070	50,590	30,700	32,250

They were added directly at the extrusion stage or applied during the GNP treatment, alone or in combination with fumed silica. Properties of PP-based compounds are given in Table 13.6.

As can be noticed, decrease in the electrical resistivity and improvement of mechanical properties were achieved when 10 wt% of GNP modified with both AMEO and octylphenol ethoxylate were combined with silica. The resulting higher values of resistivity indicate lower degree of GNP dispersion.

TABLE 13.6

PP-Based Compositions with GNP Modified by Aminosilane, Octylphenol Ethoxylate, and Silica Additives

Composition, Parts by Weight							
GNP content (wt%)		10	10	10	10	10	10
PP, MFI = 13		90	87.5	87.5	87.5	87.5	87.5
Aminosilane (AMEO)		—	2.5	—	—	—	—
Pristine GNP		10	10	—	—	10	—
80% GNP modified with AMEO		—	—	12.5	—	—	—
80% GNP + AMEO + A-200		—	—	—	12.5	—	—
Triton X-100		—	—	—	—	2.5	—
80% GNP + X-100 + A-200		—	—	—	—	—	12.5
Properties of Compression Molded Specimens							
Tc, °C, by DSC		127.4	126.6	128.5	130.8	126.6	129.5
Surface resistivity (Ohms)		3.2×10^{13}	1.9×10^{16}	2.5×10^{12}	1.4×10^{10}	1.9×10^{14}	1.3×10^{10}
Storage modulus (MPa)							
At °C	25	2,751	3,185	3,078	3,684	3,150	3,498
	70	1,548	1,621	1,671	2,086	1,565	1,830
Dynamic viscosity (PaS)							
At 200°C	6.3	1,443	1,587	2,009	3,783	1,238	2,498
Freq.	31.5	817	889	1,074	1,804	703	1,312
Dynamic modulus (Pa)							
At 200°C	6.3	4,175	4,671	6,406	15,210	3,575	8,785
Freq.	31.5	15,790	17,370	21,880	41,460	13,530	28,140

13.3.4 Microstructure of Modified GNP

The effect of SAA and silica on the morphology of GNP nanoparticles, treated in a ball mill and ultrasonicated in the presence of SAA solutions, was studied using optical microscopy and SEM.

The specimens for optical microscopy were prepared by deposition of the GNP powder on a film substrate; observations were conducted in transmittance mode.

Figure 13.2 images the pristine GNP and materials modified in solutions of polyol without and with addition of silica. In addition, specimens containing GNP, polyol, and silica mixed in ethanol, milled, and ultrasonicated were examined.

The image of pristine GNP exhibited a combination of small graphite particles and big black agglomerates. GNP modified with polyol was less

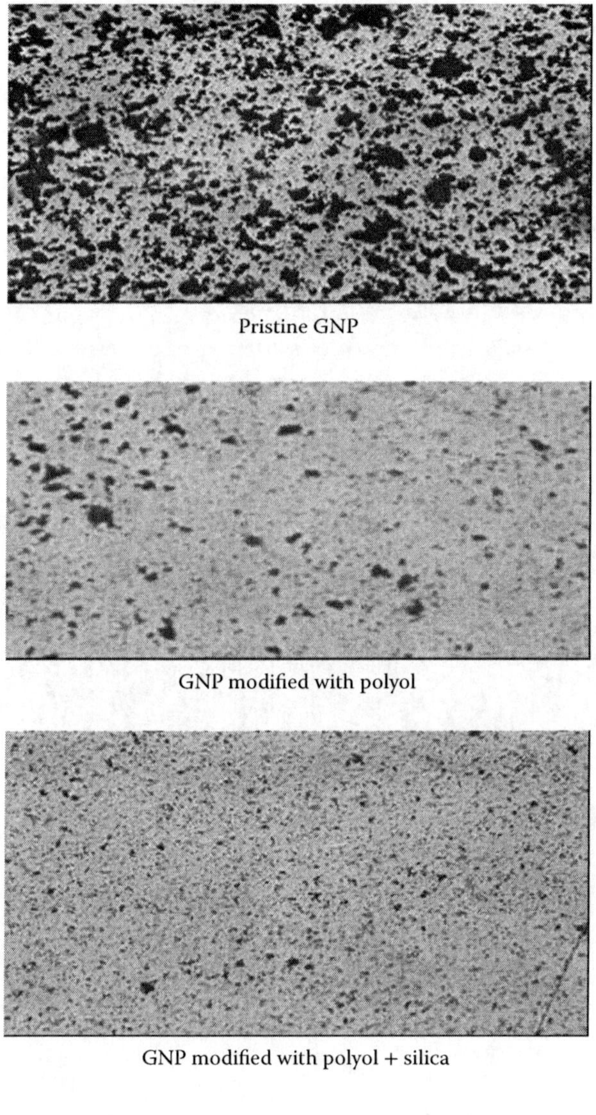

FIGURE 13.2
Optical microscope images (×400) of pristine GNP, polyol modified GNP, and polyol and silica modified GNP.

agglomerated and the particles' borders were less clear. Addition of silica resulted in disintegration of GNP to uniform small particles.

Figure 13.3 shows the SEM micrographs at magnification of ×50,000. As can be seen, the milling process followed by ultrasonication in the presence of the SAA polyol resulted in the disintegration of the GNP particles and formation of smaller fragments on the surface of GNP particles. The fumed silica (Aerosil 380) was found to decorate the surface of graphite particles and partly penetrated between the graphite layers promoting their separation. At magnification ×3000 (Figure 13.4), the surface of GNP particles, when treated in Polyol 4200, had the form of mini fish scales. It should be emphasized that the size of GNP particles changed depending on the composition and treatment process. The initial size of the pristine GNP decreased from 30–50 μm to 20–30 μm when modified with Polyol 4200, and only 10 μm when milled in the presence of silica (Figure 13.5).

The morphology of PP compositions prepared with 10 wt% of untreated and modified GNP particles was studied by SEM.

Figures 13.6 and 13.7 present the large particles of untreated GNP consisting of stacked graphene layers as well as smaller size needlelike particles distributed in the PP matrix. With GNP modified by Polyol 4200, a uniform microstructure was formed and sections of sharp edges of exfoliated graphite

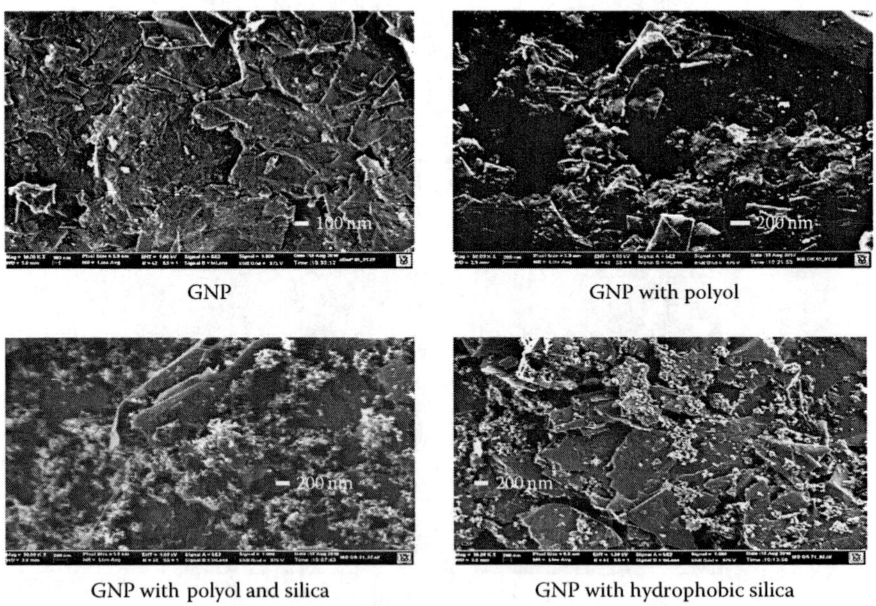

FIGURE 13.3
SEM images (×50,000) of pristine and modified GNP with polyol, polyol and silica, and hydrophobic silica.

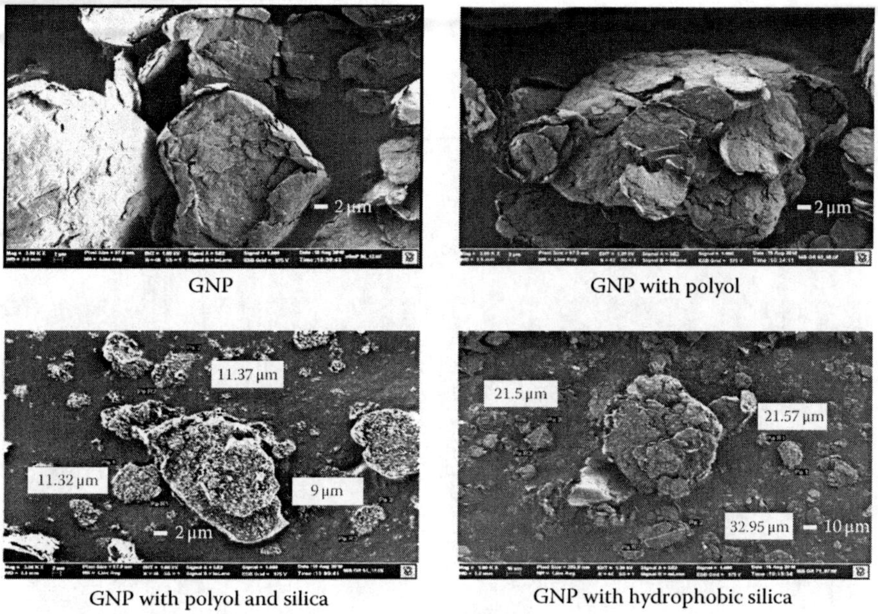

FIGURE 13.4
SEM images (×3000) of pristine and modified GNP with polyol, polyol and silica, and hydrophobic silica.

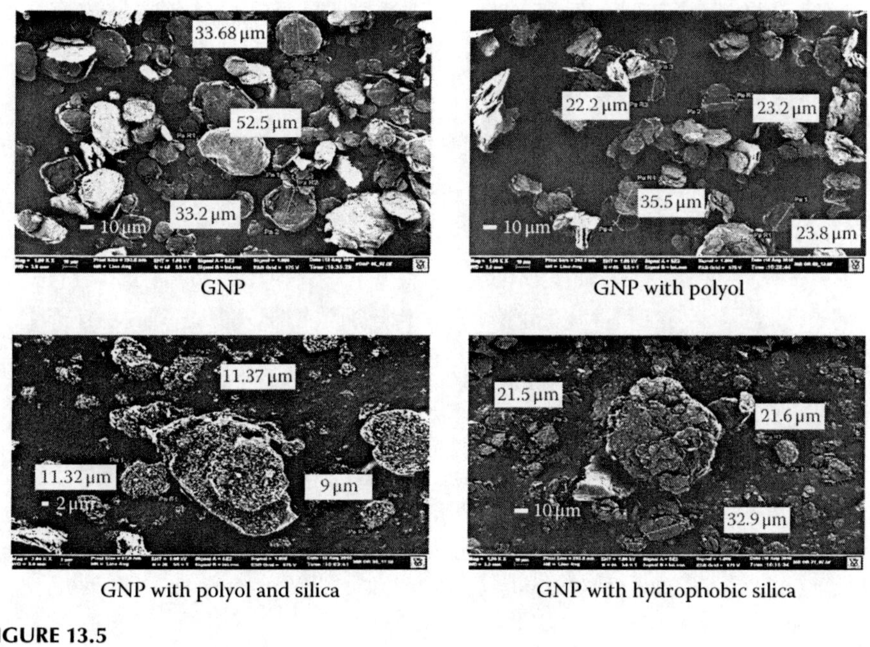

FIGURE 13.5
SEM images (×1000) of pristine and modified GNP indicating size reduction of GNP with modifications.

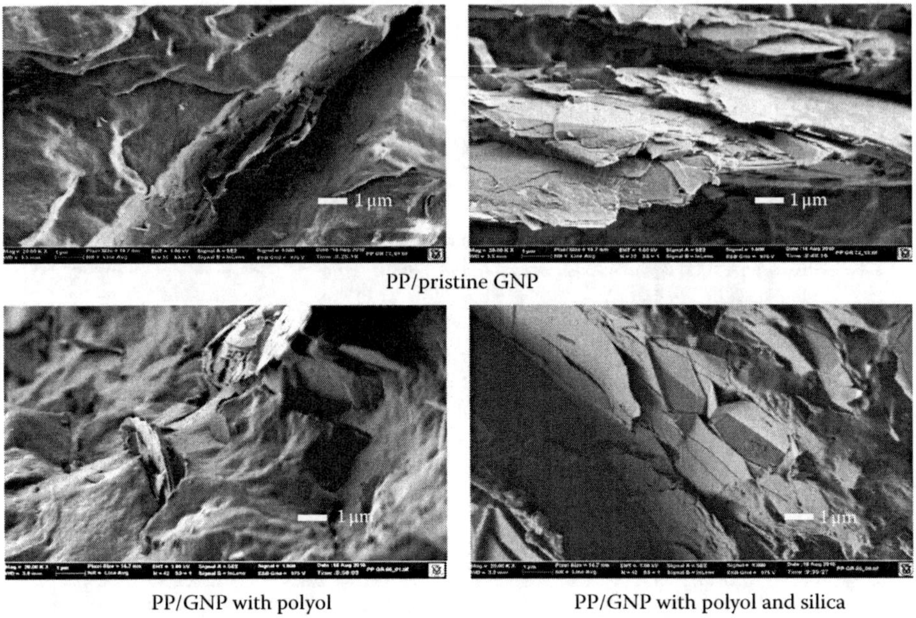

FIGURE 13.6
SEM images (×20,000) of PP/GNP nanocomposites containing pristine and modified GNP with polyol and polyol/silica combinations.

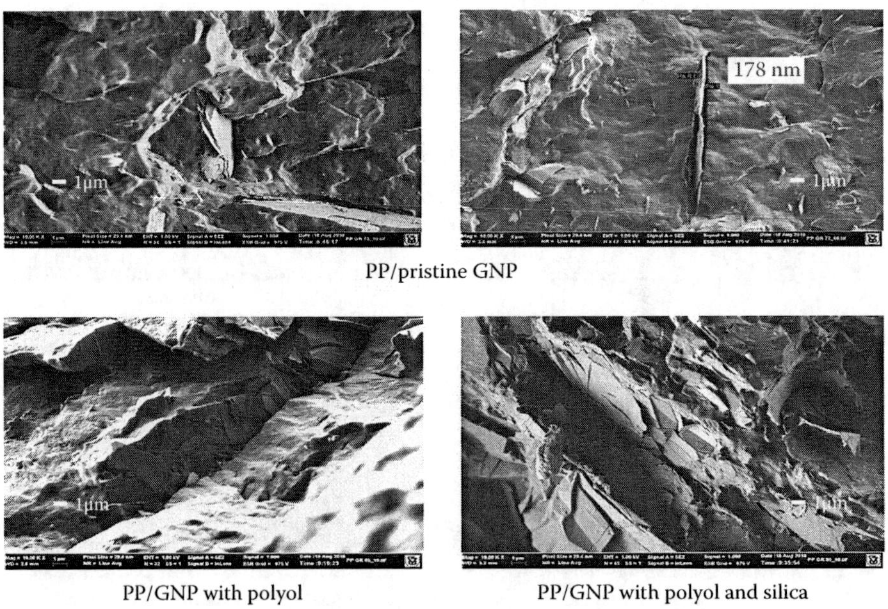

FIGURE 13.7
SEM images (×10,000) of PP/GNP nanocomposites containing pristine and modified GNP with polyol and polyol/silica combination.

particles are noticeable. PP compositions with GNP modified by SAA/silica combination are typified by network morphology. This is manifested in the low electrical resistivity of the latter compositions.

13.4 Conclusions and Outlook

The outstanding intrinsic properties of graphene, strength, stiffness, elongation, electrical and thermal conductivity, and potentially enormous surface area, present new routes for polymer nanocomposites. To exploit these properties in polymer nanocomposites, the exfoliation and dispersion of stacked graphene in the polymer matrix have to be ensured. Stabilization of the delaminated graphene layers to avoid restacking and agglomeration is of utmost importance.

Low-cost GNP derived from thermal expansion of acid-intercalated graphite results in large particle aspect ratio (100–1500) and high particle rigidity. However, the multilayer structure of GNP particles is detrimental to enhanced mechanical and electrical properties.

Surface modifications of GNP aimed to further delaminate the stacked graphene and better disperse the nanoparticles in PP matrix were studied using surfactants (nonionic or anionic), masterbatch preparation, ball milling, and ultrasonication.

In PP-based nanocomposites containing GNP, significant improvement in mechanical properties was observed already at low GNP loadings of 1%. Optical microscopy and SEM analysis indicated that changes in morphology and degree of dispersion took place when GNP particles were treated with polyether polyol. Furthermore, when the hard silica particles were added, an increased delamination was observed. The changes in GNP morphology were translated into the resulting properties of PP/GNP compositions leading to higher values of crystallization temperature, SM, thermal stability, increased melt viscosity, and improved electrical properties. When higher-MW polyol was used, a decrease in surface resistivity took place at lower content of GNP, and the effect of silica addition improved only the mechanical properties.

Finally, it was concluded that the suggested approach to modify low-cost GNP can yield polymer nanocomposites using conventional extrusion compounding. Further investigation regarding optimization of modification technology studying various types of SAAs, co-fillers, and their influence on the SAA should be pursued.

Future studies will include the effect of different sources of delaminated GNPs on the barrier and fire retardancy of PP and other plastics as well as polymer hybrid compositions containing carbon black and carbon nanotubes in addition to GNPs.

References

Chen, T., Zeng, B., Liu, J. L. et al. 2009. High through output exfoliation of graphene oxide from expanded graphite with assistance of strong oxidant in modified hummers method. *J. Physics: Conf. Ser.* 188(1): 2051.

Drzal, L. T., Fukushima, H., Do, I. et al. 2009. Polymer nanocomposites from exfoliated graphite (graphene) nanoplatelets. *Nanotech Conference*, May 3–7, Houston, TX.

Gomez-Navarro, C., Burghard, M., and Kern, K. 2008. Elastic properties of chemically derived single graphene sheets. *Nano Lett.* 8(7): 2045–9.

Hummers, W. S. Jr., and Offeman, R. E. 1958. Preparation of graphitic oxide. *J. Am. Chem. Soc.* 80(6): 1339.

Janowska, I., Chizari, K., Ersen, O. et al. 2010. Microwave synthesis of large few-layer graphene sheets in aqueous solution of ammonia. *Nano Res.* 3(2): 126–37.

Jin, Z., Lomeda, J. R., Price, B. K. et al. 2009. Mechanically assisted exfoliation of thermally converted graphene sheets. *Chem. Mater.* 21: 3045–7.

Jing, X., Zhao, W., and Lan, L. 2000. The effect of particle size on electric conducting percolation threshold in polymer/conducting particle composites. *J. Mater. Sci. Lett.* 19: 377–9.

Kalaitzidou, K. 2006. Exfoliated graphite nanoplatelets as reinforcement for multifunctional polypropylene nanocomposites. Ph.D. Thesis, Michigan State University, East Lansing, MI.

Kalaitzidou, K., Fukushima, H., Askeland, P. et al. 2008. The nucleating effect of exfoliated graphite nanoplatelets on the crystallization of polypropylene nanocomposites. *J. Mater. Sci.* 43: 2895–907.

Kalaitzidou, K., Fukushima, H., and Drzal, L. T. 2007a. A new compounding method for exfoliated graphite-polypropylene nanocomposites with enhanced flexural properties and lower percolation threshold. *Comp. Sci. Technol.* 67(10): 2045–51.

Kalaitzidou, K., Fukushima, H., and Drzal, L. T. 2007b. Mechanical properties and morphological characterization of exfoliated graphite-polypropylene nanocomposites. *Composites* 8(7): 1675–82.

Kalaitzidou, K., Fukushima, H., and Drzal, L. T. 2010. A route for polymer nanocomposites with engineered electrical conductivity and percolation threshold. *Materials* 3: 1089–103.

Khan, U., O'Neill, A., Lotya, M. et al. 2010. High-concentration solvent exfoliation of graphene. *Small* 6(7): 864–71.

Kim, H. 2009. Processing, morphology and properties of graphene reinforced polymer nanocomposites. Dissertation, University of Minnesota, St. Paul, MN.

Kim, H., and Macosko, C. W. 2008. Morphology and properties of polyester/exfoliated graphite nanocomposites. *Macromolecules* 41(9): 3317–27.

Lotya, M., Hernandez, Y., King, P. J. et al. 2009. Liquid phase production of graphene by exfoliation in surfactant/water Solutions. *J. Am. Chem. Soc.* 131(10): 3611–20.

McAllister, M. J, Li, J. L., Adamson, D. H. et al. 2007. Single sheet functionalized graphene by oxidation and thermal expansion of graphite. *Chem. Mater.* 19(18): 4396–404.

Miller, S. G., Bauer, J. L., Matyanski, M. L. et al. 2010. Characterization of epoxy functionalized graphite nanoparticles and the physical properties of epoxy matrix nanocomposites. *Comp. Sci. Technol.* 70: 1120–5.

Mukopadhyay, P., and Gupta, R. K. 2009. Trends in graphene and graphene-based nanocomposites. *Plastics Trends Net*, 9: 27.

Novoselov, K. S., Geim, A. K., Morozov, S. V. et al. 2005. Two-dimensional gas of massless dirac fermions in graphene. *Nature* 438(7065): 197–200.

Novoselov, K. S., Jiang, Z., Zhang, Y. et al. 2007. Room-temperature quantum hall effect in graphene. *Science* 315(5817): 1379.

Park, H. M., Kalaitzidou, K., Fukushima, H., and Drzal, L. T. 2010. Exfoliated graphite nanoplatelet (xGnP)/polypropylene nanocomposites. Michigan State University Publications, East Lansing, MI.

Ramanathan, T., Abdala, A. A., Stankovich, S. et al. 2008. Functionalized graphene sheets for polymer nanocomposites. *Nat. Nanotechnol.* 3: 327–31.

Schniepp, H. C., Li, J. L., McAllister, M. J. et al. 2006. Functionalized single graphene sheets derived from splitting graphite oxide. *J. Phys. Shem.* 110: 8535–9.

Song, L., Guo, J., and Zhamu, A. 2009. Highly conductive nano-scaled graphene plate nanocomposites. US Patent 7566410.

Stankovich, S., Dikin, D. A., Dommett, G. H. B. et al. 2006. Graphene-based composite materials. *Nature* 442(7100): 282–6.

Stankovich S., Dikin D. A., Piner, R. D. et al. 2007. Synthesis of graphene-based nanosheets via chemical reduction of exfoliated graphite oxide. *Carbon* 45: 1558–65.

Stankovich, S., Piner, R. D., Chen, X. et al. 2006. Stable aqueous dispersions of graphitic nanoplatelets via the reduction of exfoliated graphite oxide in the presence of poly (sodium4-styrenesulphonate). *J. Mater. Chem.* 16: 155–8.

Stankovich, S., Piner, R. D., Nguyen, S. T. et al. 2006. Synthesis and exfoliation of isocyanate treated graphene oxide nanoplatelets. *Carbon* 44: 3342–7.

Steurer, P., Wissert, R., Thomann, R. et al. 2009. Functionalized graphenes and thermoplastic nanocomposites based upon expanded graphite oxide. *Macromol. Rapid Comm.* 30: 316–27.

Tkalya, E., Ghislandi, M., Alekseev, A. et al. 2010. Latex-based concept for the preparation of graphene-based nanocomposites. *J. Mater. Chem.* 20: 3035–9.

Xu, Y., Wang, Y., Liang, J. et al. 2009. A hybrid material of graphene and poly (3,4-ethyl- dioxythiophene) with high conductivity, flexibility and transparency. *Nano Res.* 2(4): 343–8.

Zhang, H., Zheng, W., Yan, Q. et al. 2010. Electrically conductive polyethylene terephthalate/graphene nanocomposites prepared by melt compounding. *Polymer* 51: 1191–6.

Zhang, Y. B., Tan, Y. W., Stormer, H. L. et al. 2005. Experimental observation of the quantum Hall effect and Berry's phase in graphene. *Nature* 438(7065): 201–4.

Zheng, W., and Wong, S.-C. 2003. Electrical conductivity and dielectric properties of PMMA/expanded graphite composites. *Comp. Sci. Technol.* 63(2): 225–35.

14

Diffusion through Polymers Containing Platelike Nanomaterials

Matthew S. Thompson, Sushant Agarwal, Prithu Mukhopadhyay, and Rakesh K. Gupta*

CONTENTS

14.1 Introduction

Polymer nanocomposites (PNCs) formed by dispersing a few weight percent or less of nanometer-sized fillers, such as carbon nanotubes and montmorillonite (clay), in either thermoplastic or thermosetting polymers are now commonly available. Compared to "neat" polymers, PNCs tend to have higher tensile and flexural moduli, improved barrier properties, and enhanced flame resistance, among other properties. It is for this reason that they are of great research and commercial interest in many fields (Gupta et al. 2010). This chapter is concerned with the study of molecular diffusion of small molecules through PNCs. This situation is encountered in food packaging where PNCs help to prolong shelf life, by preventing oxygen ingress into foods such as ketchup, beer, and juice; by stopping carbon dioxide loss from carbonated beverages; or by preventing moisture

* Correspondence to: Rakesh K. Gupta (Rakesh.Gupta@mail.wvu.edu)

from reaching baked goods. In addition to food packaging, PNCs are utilized to develop more durable tennis balls, protective gloves, as well as coatings for tubing, fuels, and chemical tanks, and a pressure retention layer for tire applications (Goldberg et al. 2004). Moreover, they are applied in composite bridge decks in order to exclude moisture diffusion into fiber-reinforced polymers, since the presence of moisture can deteriorate the long-term durability and performance of composites (Rana et al. 2005; Ravindran and Gupta 2006; Shah et al. 2002). As a consequence, quantitatively predicting the reduction in diffusion is both fundamentally important and a practical necessity. Indeed, there is now substantial literature on this topic (Azeredo 2009; Beall 2001; Bharadwaj et al. 2002; Bhattacharya et al. 2011; Choudalakis and Gotsis 2009; Dunkerley and Schmidt 2010; Herrera-Alonso et al. 2009; Jacob et al. 2008; Lu and Mai 2007; Maji et al. 2010; Minelli et al. 2011; Petrovicova et al. 2000; Paul and Robeson 2008; Priolo et al. 2010; Sorrentino et al. 2005; Sun et al. 2008; Tortora et al. 2002; Takahashi et al. 2006; Xu et al. 2006; Yano et al. 1993).

Even in the absence of any filler–matrix interactions, diffusion through a heterogeneous, two-phase medium can be expected to depend on the properties of individual phases and the amount, shape, size, size distribution, and orientation of the dispersed phase. Early work on diffusion through arrays of spheres, cylinders, and ellipsoids has been summarized by Crank (1975). Barrer (1968) studied diffusion through two kinds of heterogeneous media: (1) Laminates in which layers of different properties are sandwiched together and (2) particulates in which discrete particles of one phase are dispersed in a continuum of another. Although the mass transfer resistance is often not explicitly defined in the literature, the series–parallel formulation to obtain the effective diffusivity is proposed for both the aforementioned situations by several authors. In addition, Fidelle and Kirk (1971) and Bell and Crank (1974), among others, numerically solved the two-dimensional Laplace equation over a domain appropriate to the second case and compared their results with the expressions predicted by series–parallel and parallel–series formulations. Not surprisingly, they found that the numerical results were bounded by the two analytical expressions. As far as filler shape is concerned, plates give the maximum reduction in diffusivity among shapes such as cylinders, spheres, and plates at a given loading level (Moggridge et al. 2003). Therefore, only plates are of interest in this chapter, and appropriate theories as well as representative data are presented on mass transport through polymers containing dispersed nanoclay and dispersed graphene platelets, although the latter material is the focus of this chapter. Available results testify to the effectiveness of nanomaterials, especially graphene, in limiting mass transfer even at low loading levels. We, however, begin by briefly describing the process of molecular diffusion through polymers, and we also give an account of the standard methods of measuring the diffusion coefficient for mass transfer through solids.

14.2 Diffusion through Polymers

When concentration gradients exist in a multicomponent system, there is a natural tendency to reduce and ultimately eliminate concentration differences by mass transfer. This is the process of diffusion, and mass transfer occurs by molecular means. The basic equation governing the rate of mass transfer of component A in a binary mixture of A and B is (in one dimension) given as follows (see, for example, the study by Kumar and Gupta [2003]):

$$J_{A,z} = -D_{AB} c \frac{dx_A}{dz} \tag{14.1}$$

This is often known as "Fick's first law." Here, $J_{A,z}$ is the flux of A in the z direction in units of moles per unit time per unit area, D_{AB} (assumed constant) is the diffusion coefficient or diffusivity, c is the molar concentration of the mixture, and x_A is the mole fraction of A. Clearly dx_A/dz is the concentration gradient, and diffusion occurs due to the presence of this quantity. Equation 14.1 governs diffusion through solids, and its form is similar to Fourier's law of heat conduction and Newton's law of viscosity.

If we consider the diffusion of species A through a region of infinitesimal size, then it is easy to carry out mass balance in three dimensions; by using Equation 14.1 to represent individual molar fluxes, it is easy to show (Kumar and Gupta 2003) that

$$\frac{\partial c_A}{\partial t} = D_{AB} \nabla^2 c_A \tag{14.2}$$

in which c_A is the molar concentration of species A, which equals cx_A. Equation 14.2 is often called "Fick's second law." The term ∇^2 is Laplace's operator ($\partial^2/\partial x^2 + \partial^2/\partial y^2 + \partial^2/\partial z^2$) in rectangular coordinates. Clearly, at steady state,

$$\nabla^2 c_A = 0 \tag{14.3}$$

In any given situation of practical interest, the concentration profile is obtained by solving either Equation 14.2 or Equation 14.3 (as appropriate), and it involves the diffusion coefficient in the transient diffusion case. The flux is subsequently computed with the help of Equation 14.1. Comparison with experimental data then yields the diffusivity. Because a new solution is generated each time a boundary condition is changed, a very large number of methods exist for the experimental determination of diffusivity. However, D (where we have dropped the subscripts A and B) is most commonly measured using either a sorption experiment or a permeation experiment. It is noted that if diffusivity is not constant but depends on mixture composition, the mass transfer situation is termed non-Fickian.

Diffusion of small molecules through solid, nonporous polymers is a three-step process. In the first step, the diffusing molecule has to dissolve in the polymer at the high-partial-pressure or high-concentration side. Then it has to diffuse as a solute to the low-partial-pressure side. In the third step, the solute desorbs or evaporates back to the gas or liquid phase. Thus, if we consider steady-state diffusion of a gas through a membrane of thickness L exposed to a partial pressure difference Δp, the mass or molar flux through the membrane is given by Equation 14.1 as follows:

$$J = DS\frac{(-\Delta p)}{L} \tag{14.4}$$

This is a permeation experiment, and S is solubility of the diffusing species in the polymer at pressure p. By definition, the concentration of the gas in the polymer is given by

$$c_A = Sp \tag{14.5}$$

in which S is commonly called the "Henry's law constant." For simple gases such as carbon dioxide and oxygen (i.e., gases above their critical temperatures) dissolved in rubbery polymers, S is independent of p.

It is evident from Equation 14.4 that a measurement of the steady-state flux alone does not yield D; instead, it gives the product $D{\cdot}S$, which is called "permeability." To obtain diffusivity, we have to know the value of solubility; else, we need to make one additional measurement, which is usually the "time lag," a quantity that is explained next in this section. We can use the method of separation of variables to solve Equation 14.2 for one-dimensional transient diffusion through a membrane that is initially solute-free, subject to the following boundary conditions (we have dropped the subscript A from c_A):

$$c(0,t) = c_1$$

$$c(L,t) = 0$$

$$c(0 < x < L, 0) = 0$$

The time-dependent concentration profile is as follows (Crank 1975; Daynes 1920):

$$c = c_1\left(1 - \frac{x}{L}\right) - \frac{2c_1}{\pi}\sum_{n=1}^{\infty}\frac{1}{n}\sin\frac{n\pi x}{L}\exp\left(-Dn^2\pi^2 t/L^2\right) \tag{14.6}$$

It is noted that as $t \to \infty$, the exponential terms vanish and we obtain the expected linear concentration profile. The flux of gas leaving the membrane is $-D(\partial c/\partial x)$ evaluated at $x = L$. When this term is integrated over time, we get $M(t)$, the amount of material that has actually passed through the membrane in time t:

$$\frac{M(t)}{Lc_1} = \frac{Dt}{L^2} - \frac{1}{6} - \frac{2}{\pi^2}\sum_{1}^{\infty}\frac{(-1)^n}{n^2}\exp\left(-Dn^2\pi^2/L^2\right) \tag{14.7}$$

This equation reduces to the equation of a straight line as $t \to \infty$, which is as follows:

$$M(t) = \frac{Dc_1}{L}t - \frac{Lc_1}{6} \tag{14.8}$$

Clearly, this straight line intersects the t axis at the point $t = L^2/6D$, and this is called the time lag. It allows us to obtain the diffusivity D from a plot of $M(t)$ versus time. It is noted that commercial instruments are available that use this principle, especially in the measurement of diffusion coefficients of carbon dioxide, oxygen, and moisture through polymer films as a function of temperature.

In a sorption experiment, a polymer film initially at equilibrium with a given partial pressure of vapor or having a fixed and known concentration c_0 of a diffusing species is suddenly exposed to a different constant partial pressure of the same vapor or is immersed into a liquid at a constant concentration. The diffusion coefficient is obtained from analyzing data on the gain or loss in mass of the polymer film with time. If the film's thickness is $2L$ and if the surfaces located at $x = \pm L$ instantly reach a constant concentration c_1, then the solution of Equation 14.2 is obtained by the method of Laplace transforms and after a very lengthy analysis one finds that (Crank 1975)

$$\frac{c - c_0}{c_1 - c_0} = 1 - \frac{4}{\pi}\sum_{n=0}^{\infty}\frac{(-1)^n}{2n+1}\exp\left[-D(2n+1)^2\pi^2 t/\left(4L^2\right)\right]\cos\frac{(2n+1)\pi x}{2L} \tag{14.9}$$

Equation 14.9 can be used to calculate the ratio of mass of vapor, $M(t)$, that has diffused into the film at any time to the equilibrium mass uptake, M_∞. At small values of t, this quantity is given by

$$\frac{M(t)}{M_\infty} = 2\left(\frac{Dt}{L^2}\right)^{1/2}\left[\pi^{-1/2} + 2\sum_{n=1}^{\infty}(-1)^n\, i\,erfc\,\frac{nL}{\sqrt{Dt}}\right] \tag{14.10}$$

which states that $M(t)/M_\infty$ is a unique function of (Dt/L^2). Here, $ierfc(x)$ is given by $\left(1/\sqrt{\pi}\right)e^{-x^2} - x\left(1 - erf(x)\right)$. On plotting Equation 14.10, we find that (Crank and Park 1968)

$$\frac{M(t)}{M_\infty} = 2\left(\frac{Dt}{L^2}\right)^{1/2} \tag{14.11}$$

until the point where $M(t)/M_\infty$ equals 0.5. This linear relationship on logarithmic coordinates facilitates the evaluation of D from experimental data.

During the final stages of a desorption experiment, however, the equivalent form of Equation 14.10 is as follows (Crank and Park 1968):

$$\frac{d}{dt}\{\ln[M(t)-M_\infty]\} = -\frac{\pi^2 D}{4L^2} \tag{14.12}$$

Further, a plot of $\ln[M(t) - M_\infty]$ versus time should approach a straight line whose slope is given by the right-hand side of Equation 14.12.

The permeation method works well for simple gases diffusing through polymers above the glass transition temperature, because in this situation both solubility and diffusivity are constant—independent of concentration—at a given temperature. In contrast to simple gases, organic vapors (at temperatures below the critical temperature) often tend to interact with polymers and cause them to swell. A result of this interaction is that even for polymers above the glass transition temperature, the diffusion coefficient becomes dependent on concentration and Henry's law is no longer obeyed. In this situation, the preferred method for measuring diffusivity is the sorption experiment. We also note that when diffusion takes place below the polymer glass transition temperature, negative deviations from Henry's law are often observed for the case of simple gases, whereas polymer swelling and non-Fickian behavior are observed for organic vapors.

Let us now examine diffusion through a simple polymer system such as vinyl ester (VE).

14.3 Moisture Diffusion through VE as a Polymer Matrix

It is noted that VE resins are amorphous polymers that are produced by the reaction of an epoxy resin with an unsaturated monocarboxylic acid. The result is a molecule having unsaturated ester groups at the ends, and these double bonds act as cross-linking sites with reactive molecules, such as styrene. Curing, using either styrene or methyl methacrylate, is generally done at room temperature using free radical initiators; samples are often postcured at high temperatures to complete the polymerization reactions. The resulting polymer has a glass transition temperature close to 100°C, and the polymer backbone contains (hydrophilic) hydroxyl groups. The VE polymer was chosen for study by Shah et al. (2002) because of its practical importance due to its use as the matrix polymer in fiber-reinforced thermosetting composites (Mallick 1993) and also because the absence of crystals removes a potential complicating factor from the data analysis. These authors conducted sorption experiments for the diffusion of moisture through neat VE films. Samples having a thickness of 0.2–0.6 mm and a dry mass ranging from about 120 to 400 mg were immersed in distilled water at room temperature. The samples

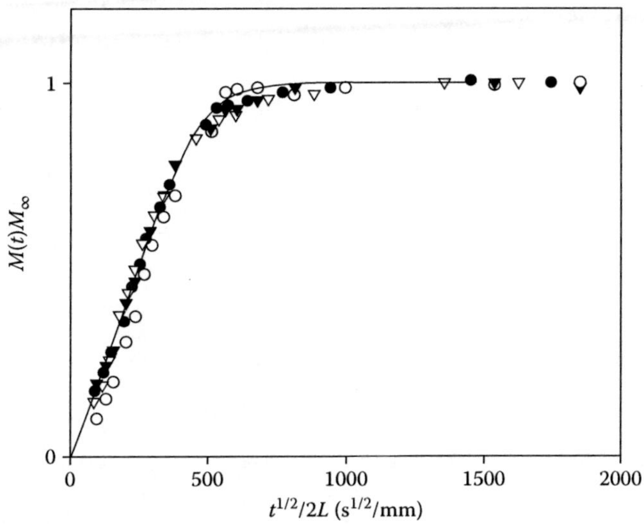

FIGURE 14.1
Water sorption curve of postcured neat VE resin at 25°C: Sample thicknesses are 0.529 mm for filled circles, 0.500 mm for unfilled circles and filled triangles, and 0.567 mm for unfilled triangles. Solid line represents Equation 14.10 with $D = 9.19 \times 10^{-7}$ mm^2/s. (From Shah et al. *Polym. Eng. Sci.*, 42, 1852–1863, 2002. With permission.)

were periodically removed, weighed, and reimmersed in water until equilibrium was attained and the sample weight became constant. A typical experiment lasted about 10 days, and Figure 14.1 shows the fractional increase in the mass of samples plotted versus $t^{1/2}/2L$. As expected, based on Equation 14.11, water uptake increases linearly at short times, but it levels off at large times as equilibrium is attained. Also, data from different runs superpose, and one can compute a (constant) diffusion coefficient; at 25°C, the diffusivity is 9.19×10^{-7} mm^2/s and the equilibrium moisture content is about 0.434%. The solid line in Figure 14.1 represents the prediction of the complete (time-dependent) solution to Equation 14.2, and this says that moisture transport through VE is Fickian in character.

Rana et al. (2005) carried out steady moisture permeation experiments and compared the results with those obtained using the aforementioned sorption experiment. For permeation experiments, the authors attached a VE film to the mouth of an aluminum dish containing a desiccant and put the assembly inside a constant-humidity chamber. As moisture diffused into the dish, the mass of the assembly increased progressively and ultimately the rate of mass increase became constant. This was determined by periodic weighing of the assembly using an electronic balance. Data were analyzed according to Equation 14.4 in conjunction with the measured values of solubility. It was found that the diffusion coefficient values obtained using the two methods were comparable to each other.

For the diffusion of moisture through neat VE, both solubility and diffusivity have been found to increase with increasing temperature (Chin et al. 1999) and results can be represented using the Arrhenius equation. However, one must be aware that the process of diffusion can alter polymer properties such as its glass transition temperature. In addition, polymer structure can change with increasing temperature; one can observe continued polymer curing and even hydrolysis in the case of VE. Also, polymers can swell, craze, or exhibit microcracking during use. The presence of these phenomena can complicate data analysis.

14.4 Theories of Diffusion through PNCs

The most popular theory explaining the observed diffusion behavior of small molecules through PNCs was originally proposed by Nielsen (1967), and this is based on the tortuous path concept that is illustrated in Figure 14.2. If one takes a two-dimensional view of the diffusion process, filler particles are positioned perpendicular to the direction of mass transfer. Filler particles act as impenetrable barriers that force diffusing molecules to go around them. As a consequence, the path length increases from d to d'. It is obvious that the additional distance traveled by the typical molecule is $(nh)/2$, where n is the number of clay platelets (or the number of clay platelet layers) encountered along the way and h is the width of each barrier. If the thickness of each platelet is t and if the spacing between adjacent platelets is small compared to the platelet width, then it is clear that the volume fraction of nanofiller, ϕ, is $(nt)/d$. As a result,

$$d' = d + \frac{nh}{2} = d + \frac{h}{2t}\phi d \tag{14.13}$$

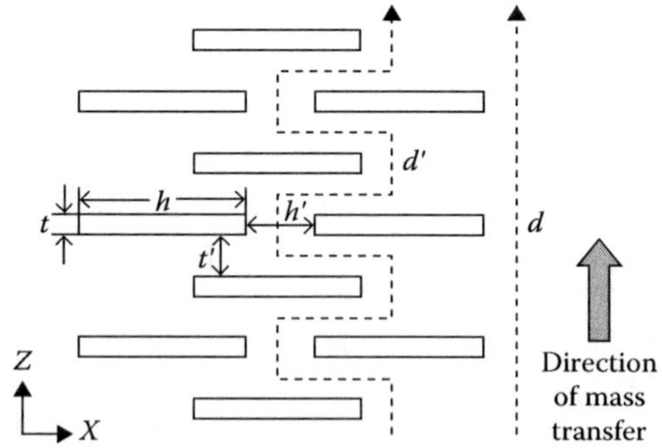

FIGURE 14.2
Two-dimensional idealized schematic diagram depicting the tortuous path concept.

Further, the tortuosity factor, τ, defined as the ratio of the distance that a molecule must travel to get through the film to the thickness of the film is as follows:

$$\tau = \frac{d'}{d} = 1 + \left(\frac{h}{2t}\right)\phi \tag{14.14}$$

Since the concentration difference across the film remains unchanged, the concentration gradient and, therefore, the mass flux each decrease by a factor of τ. Equivalently, one can say that

$$\frac{D_0}{D} = 1 + \left(\frac{h}{2t}\right)\phi \tag{14.15}$$

where D_0 is the diffusivity of the molecule in the neat polymer and D is the diffusivity through the nanocomposite. Since h/t is typically on the order of 10^2, a very significant reduction in diffusion coefficients is predicted even when ϕ is small. It is noted, though, that this relative reduction in diffusivity (and therefore permeability, provided that solubility is unaltered) is a purely geometric effect, since it is assumed that the diffusing molecules do not interact with the polymer or the filler platelets. As such, the change in diffusion coefficient should not depend on the applied concentration difference, the temperature of measurement, or the nature and size of the diffusing molecule. It should also not depend on the matrix polymer if filler dispersion remains unchanged. Although one often observes a large reduction in mass flux (Yano et al. 1993), the extent of reduction, even for the same nanocomposite, depends on whether the diffusing molecule is oxygen or water vapor (Chang et al. 2001). Also, the dependence on volume fraction is not the one predicted by Equation 14.15. Consequently, the value of filler aspect ratio, h/t, obtained from transmission electron microscopy (TEM) cannot be used in Equation 14.15 to quantitatively estimate the improvement in barrier properties. This mismatch between theory and data is usually ascribed to incomplete exfoliation of the nanoclay (Carrado 2003; Chang et al. 2001).

Over the years, Cussler and coworkers (Cussler et al. 1988; Eitzman et al. 1996; Falla et al. 1996; Moggridge et al. 2003) have studied the diffusion problem using flake-filled membranes and have developed an extensive theory for predicting the changes in diffusivity, again as a function of the loading level and aspect ratio of filler particles. Their basic equation is as follows:

$$\frac{D_0}{D} = 1 + \frac{(h/t)^2\phi^2}{4(1-\phi)} \tag{14.16}$$

Equation 14.16 shows that when compared to Equation 14.15, relative diffusivity depends much more strongly on both the aspect ratio and the filler volume fraction. This is because reduction in the area for diffusion is considered, in addition to the increase in path length. It is noted that Moggridge et al. (2003) assume that ϕ equals 0.2 or higher for Equation 14.16 to be valid, and this assumption is consistent with the tortuous path picture shown in Figure 14.2. It should, however, be recognized that the density of uncoated clay is more than twice that of the polymer matrix, and coated clay contains only about 60% of inorganic content by weight. As a consequence, 20 vol% inorganic material in the nanocomposite corresponds to an organoclay content of more than 50 wt%. This is an enormous amount of filler and it exceeds the normal loading level in a practical nanocomposite by one order of magnitude. Many other theories have also been developed over the years for various filler geometries using different approaches, but most yield equations that are similar in form to Equations 14.15 and 14.16 (Choudalakis and Gotsis 2009).

In the limit of low volume concentrations, the physical picture that one needs to work with is not the one shown in Figure 14.2 but rather the one depicted in Figure 14.3. Here, the filler volume fraction is $ht/(HT)$ where h and t are defined in Figure 14.2 and H and T are the spacing between neighboring clay layers in each direction. Clearly, the unit cell that repeats itself in two dimensions has an area of $2HT$. It is also seen that one can keep the filler volume fraction fixed even as one varies the layer spacing T; an increase in T must, however, be balanced by an appropriate reduction in H so that the product HT is kept unchanged. When T is small, there is no overlap between filler platelets in adjacent layers, and some of the diffusing molecules can go right through the film without encountering a single barrier. In this situation, the fraction of the total area that is available for one-dimensional mass transfer is $(1 - 2h/H)$. Thus, the predicted diffusivity for one-dimensional diffusion is (Statler and Gupta 2008) as follows:

$$D = D_0\left(1 - 2h/H\right) = D_0\left(1 - 2\phi T/t\right) \tag{14.17}$$

Equation 14.17 represents a lower bound on D since, in reality, diffusion is two-dimensional rather than one-dimensional in the situation being examined.

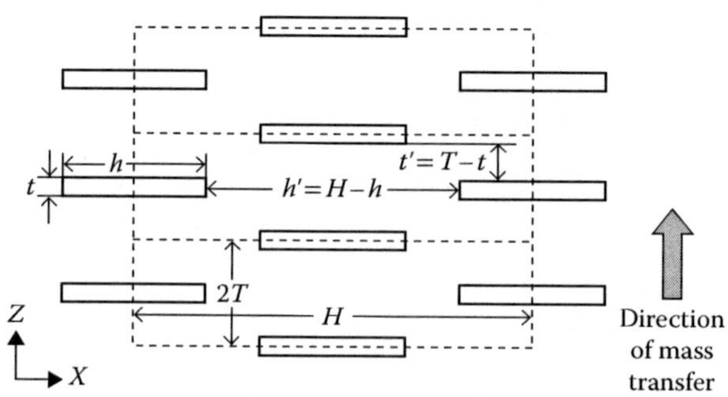

FIGURE 14.3
Two-dimensional schematic diagram of the effective area method with three repeating units.

This equation, which represents what we call the "effective area method," teaches us that for platelets of fixed dimensions h and t, the diffusivity, even at a low but fixed filler loading level, can be made to vary all the way from zero to almost D_0 simply by varying the layer spacing T. It also tells us that the diffusivity ought to remain unchanged if, at fixed values of t and T, the filler aspect ratio is changed by changing h; this is because h/H is constant at fixed ϕ. It is obvious that the predictions of Equation 14.17 are very different from those of Equations 14.15 and 14.16. As a practical matter, therefore, observed reductions in diffusion coefficient with the use of a few weight percentages of even a platelike nanofiller are likely to be quite modest if the filler is mixed in a random manner with the polymer. It must be noted also that in the foregoing, we have assumed the polymer is amorphous. If crystals are present, they must be taken into account as separate barriers to mass transfer.

14.5 Studies on Diffusion through Polymer/Clay Nanocomposites

The most common barrier material encountered in the PNCs literature is montmorillonite, a platelike layered silicate. It is inexpensive, since it is abundantly available, and it is mined in large quantities. It is a swelling clay comprising layers formed by the condensation of two silicon-oxygen tetrahedral sheets and one aluminum- or magnesium-oxygen-hydroxyl octahedral sheet; the layer thickness is 0.96 nm. The individual layers generally exist stacked together, but they can be separated by a variety of methods. However, since the clay is inherently hydrophilic, a surface treatment is needed when the clay must be dispersed into a hydrophobic polymer matrix. One generally replaces cations, such as sodium, that are found in the gallery spacing between silicate layers with an organic cation, and several such treated clays are commercially available. For use with VE, Cloisite 10A® has been found to be appropriate (Shah et al. 2002); Cloisite 10A is available from Southern Clay Products, Inc., Gonzales, Texas, and it is montmorillonite treated with benzyl (hydrogenated tallow alkyl) dimethyl quaternary ammonium chloride. The organic content of this coated clay is about 39 wt%, and it is expected to have lateral dimensions on the order of 180 nm.

Gupta and coworkers (Bhardwaj and Gupta 2005; Rana et al. 2005; Ravindran and Gupta 2006; Shah et al. 2002; Sridhar et al. 2006) have published many papers on moisture diffusion through VE–matrix nanocomposites. Bhardwaj and Gupta (2005) conducted moisture permeation experiments using VE films containing Cloisite 10A and examined the influence of variables such as temperature, moisture concentration gradient, and filler content on permeation. The equipment used for experiments was obtained from MOCON, Inc., Minneapolis, and Figure 14.4 shows a schematic diagram of the setup. The film of the material being tested is placed in the test cell, and a concentration difference is created across the sample by running dry nitrogen on one side and nitrogen containing water vapor on the other side. Water dissolves in the polymer on the high-concentration side, diffuses to the low-concentration side, and is carried away by dry nitrogen gas; the concentration of moisture in this nitrogen stream is detected by an infrared spectrometer. This information allows the calculation of a water vapor transmission rate (WVTR). Diffusivity values are then computed using Equation 14.8. To obtain 100% relative humidity (RH), sponges soaked in water are used instead of humid nitrogen gas.

The nanocomposite samples used by Bhardwaj and Gupta (2005) were made with the help of resin transfer molding, and the samples had dimensions of 10 cm × 10 cm and a thickness of about 1 mm. The matrix polymer was Derakane™ VE from Dow Chemical, Co., Midland, Michigan, and the maximum concentration of nanoclay used was 5 wt%. Figure 14.5 shows

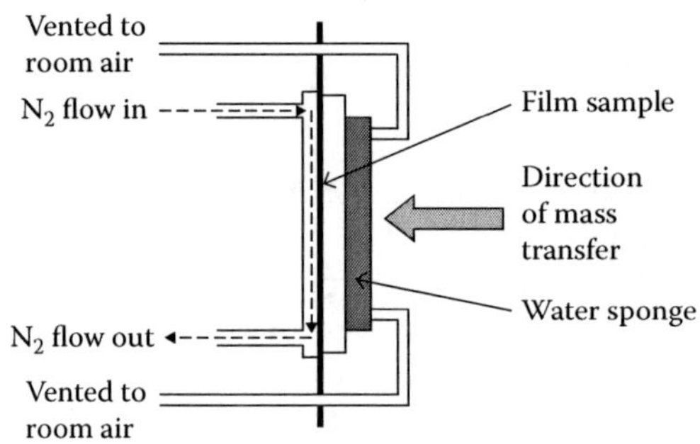

FIGURE 14.4
Schematic diagram of MOCON experimental setup for 100% RH.

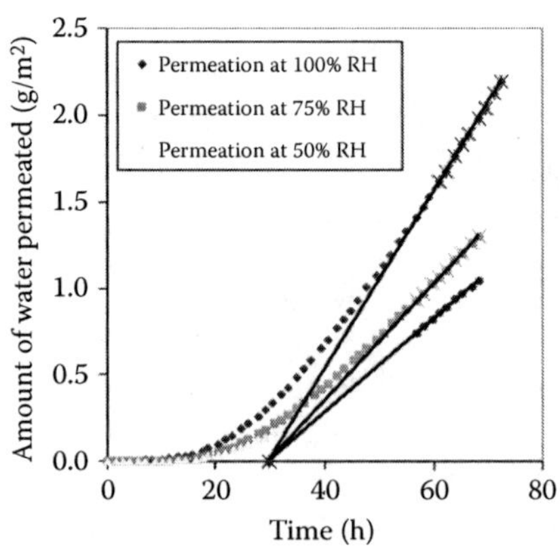

FIGURE 14.5
Time lag curves for neat VE at 37.8°C and three different RH values.

permeation data for moisture diffusion through the neat VE film at 37.8°C and three different relative humidity values. Similar results for a nanocomposite film containing 1 wt% organoclay are displayed in Figure 14.6. The calculated diffusion coefficients are independent of the applied concentration gradient, and their values, as a function of clay content, are listed in Table 14.1. The temperature variation of data for diffusion through neat polymer and polymer containing 5 wt% clay is shown in Table 14.2. The

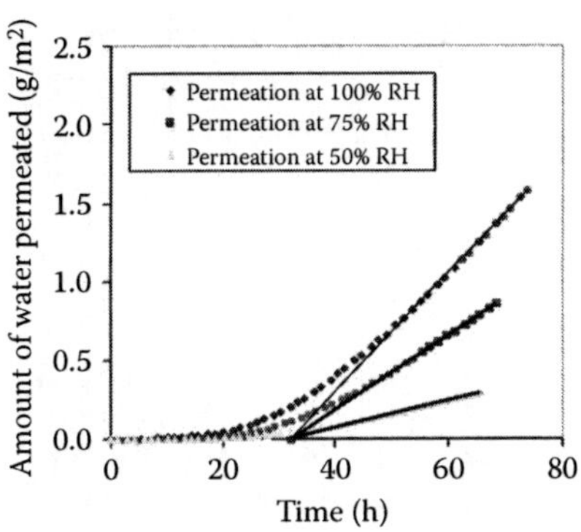

FIGURE 14.6
Time lag curves for 1 wt% clay/VE at 37.8°C and three different RH values.

TABLE 14.1
Diffusivity for Clay/VE Nanocomposites at 37.8°C

Clay Concentration (wt%)	Volume Fraction	Diffusivity × 10^8 (cm²/s)	D_0/D
0	0.0	1.56	1.0
1	0.006	1.42	1.10
2	0.011	1.25	1.25
3	0.017	1.12	1.39
4	0.023	1.07	1.45
5	0.029	0.99	1.57

TABLE 14.2
Variation of Diffusivity for Clay/VE Nanocomposites with Temperature

Temperature (°C)	Diffusivity × 10^8 of Neat VE (cm²/s)	Diffusivity × 10^8 of 5 wt% Clay/VE (cm²/s)	D_0/D
37.8	1.56	0.99	1.57
25.0	0.90	0.64	1.40
15.0	0.68	0.45	1.51

diffusivity of moisture through the neat resin is 0.90×10^{-8} cm²/s at 25°C, and this is in good agreement with the results of Shah et al. (2002). However, the extent of reduction in diffusion coefficient with added clay is less than what is predicted by either Equation 14.15 or Equation 14.16 even when experimentally determined clay aspect ratios are used. We speculate that this happens because most of the diffusing molecules do not follow a tortuous path since the volume fraction of the nanofiller is less than 0.2. It is noted that data in Table 14.2 are in good agreement with the Arrhenius equation, with the activation energy being independent of clay content. Further, within experimental error, it is observed that temperature does not influence the relative decrease in diffusivity obtained on adding clay. This allows us to conclude that these observations are a geometrical effect.

14.6 Literature on Graphene-Based PNCs

Similar to montmorillonite, graphene is a platelike material, but individual flakes of graphene are much thinner than nanoclay platelets. Consequently, it is of interest to us to determine how graphene acts as a mass transfer barrier when dispersed in plastics of interest. Previous chapters of this book have described the synthesis of graphene, as well as its different variations and their characterization. Also described in other chapters are techniques of incorporating and dispersing graphene into polymers and characterizing the resulting morphology. It is found, in general, that solution processing results in more uniform dispersion and distribution of a nanofiller as compared to melt compounding. In terms of graphene-based PNC (GPNC) properties, information on barrier properties is much more limited when compared to the literature on electrical, flow, mechanical, or thermal properties. It is also small compared to the data published on mass transport through clay-based PNCs. Much of the available results related to gas permeability have been summarized in Table 5 of the review article by Kim et al. (2010a). Fillers used include graphite oxide and thermally reduced graphene oxide, and the loading level is usually less than about 3 vol%. Polymers utilized include polycarbonate, polyolefins, polystyrene, rubbery polymers, and engineering thermoplastics. Gases examined are helium, hydrogen, nitrogen, and oxygen. Both sorption and permeation methods are used. In addition, several publications (Kim and Macosko 2008, 2009; Kim et al. 2010b) report the use of a constant-volume, variable-pressure permeation technique. Here, one side of a GPNC film is exposed to gas at atmospheric pressure, whereas the pressure is monitored in an evacuated chamber on the other side. Since defect-free graphene sheets are impermeable to gas molecules (Bunch et al. 2008), the permeability of any of the gases through the nanocomposite is always found to be lower than the corresponding value through the neat polymer and the effect increases with increasing concentration of the nanofiller. Indeed, the relative reduction can be as large as 99% (Potts et al. 2011). Further, the extent of permeability reduction for the diffusion of a specified gas through a given polymer is almost always superior to the corresponding value for the same polymer containing the same amount of organically modified nanoclay in place of graphene.

Kim and Macosko (2008) measured hydrogen permeability through 70-μm-thin melt-compounded films of poly(ethylene-2,6-naphthalate) containing either graphite flakes or functionalized graphite sheets (FGSs). Data were obtained at 35°C using a constant-volume, variable-pressure method. As expected, hydrogen permeability decreased progressively with increasing nanofiller content, and the nanocomposites made with FGSs were more effective in reducing permeability, possibly due to better exfoliation of FGSs as compared to graphite. These authors also showed that a version of the theory proposed by Cussler and coworkers (Lape et al. 2004) could fit both sets of data. However, the best-fit value of the FGS aspect ratio differed very significantly from the aspect ratio determined from TEM. Similarly, filler aspect ratio calculated from nitrogen permeation through polycarbonate/FGS nanocomposites differed from the aspect ratio calculated from helium permeation through the same nanocomposite (Kim and Macosko 2009). This suggests the existence of factors not accounted for in the tortuous path theory. Note that Kim et al. (2010b) used the same apparatus and observed as much as 90% decrease in nitrogen permeation through thermoplastic urethane sheets filled with 1.6 vol% isocyanate-treated graphite oxide (iGO). No attempt was made to determine if the reduction in permeability was the result of changes in diffusivity, changes in solubility, or changes in both quantities.

Whereas low-density polyethylene (LDPE) is a poor barrier against nonpolar hydrocarbon liquids and poly(vinyl alcohol) (PVA) is similarly a poor barrier against polar liquids, their mass transfer resistances against these same fluids can be improved by dispersing graphene nanosheets within them. Wang et al. (2011a) found that, whereas PVA was hydrophilic based on contact angle measurements, films of graphene/PVA nanocomposites became increasingly hydrophobic with increase in graphene content; graphene was synthesized in situ by the reduction of graphene oxide. When these films, containing up to 1 wt% graphene, were immersed in distilled water for 24 hours at 30°C, not only was the initial slope of the mass gain curves (in dimensional form) diminished but also the final moisture content was reduced with increasing amounts of added graphene. Thus, the incorporation of graphene within the polymer lowered its permeability, but this was more due to a reduction in solubility rather than a reduction in diffusivity. Wang et al. (2011b) similarly observed that there was a reduction, from 56% to 39%, in the absorption ratio of toluene in graphene/LDPE nanocomposites formulated using vinyl triethoxysilane as a coupling agent. Here, absorption ratio represents the percentage of solvent absorption by the nanocomposite within a certain time period at a specified temperature. The lower the absorption ratio the lower the solubility and the permeability.

A further example of the improvement in barrier properties by disproportionate reduction in solubility as compared to diffusivity is provided by Compton et al. (2010). These authors dispersed 1-nm-thick, 500-nm-long crumpled sheets of phenyl isocyanate–functionalized graphene in polystyrene at very low graphene concentrations. The graphene dispersed so well in the polymer matrix that it even reduced the free volume as determined by density measurements. This resulted in an unprecedented reduction in oxygen solubility in the nanocomposite relative to the neat polystyrene; the crumpling of the nanosheets was speculated to facilitate polymer–filler interactions and limit the formation of interstitial cavities. Thus, at a graphene content as low as 0.02 vol%, oxygen permeability, measured with a MOCON™ instrument, was reduced to 80% of the permeability of the unfilled polymer. It is noted that in this situation, one must not appeal to Equations 14.15 and 14.16 to explain the observed permeability results since the mechanism of improvement in barrier properties is a lowering of solubility and not an apparent reduction of diffusivity arising from an increase in the path length of diffusion.

The different variables that can be expected to influence the diffusivity of any molecule through a PNC include the amount of the nanofiller, its orientation relative to the direction of mass transfer, its polydispersity and aspect ratio, the temperature of measurement, the concentration gradient, any filler–matrix interactions, and the presence of coupling agents and impurities. However, only the influence of the amount of nanofiller seems to have been considered in the published literature. This paucity of information motivated us to explore the effects of some of the other variables and, in Section 14.7, we present our own data on moisture diffusion through graphene-filled polyvinyl acetate (PVAc) nanocomposites. The PVAc homopolymers are commercially available, low-cost polymers sold as emulsion, latex, or spray-dried solids. Due to their good adhesion properties, they are used in packaging and furniture applications, especially those involving porous surfaces such as paper and wood. Through formulations, they are also used in paints, coatings, adhesives, and textiles.

14.7 Studies on Diffusion through GPNCs

Thompson et al. (2012) surface treated commercially available graphenes with silanes to make them relatively hydrophobic for ease of dispersion in PVAc, a nonpolar, amorphous thermoplastic polymer. The PVAc sample was obtained from Scientific Polymer Products, Inc. (Ontario, New York), and it had a molecular weight of 170,000. Its glass transition temperature was measured to be 40°C using a differential scanning calorimeter. Two batches

(a) (b)

FIGURE 14.7
As-received exfoliated graphite nanoplatelets of lateral dimension 5 μm at (a) low and (b) high magnifications.

of graphite nanoplatelets (xGnP®) were purchased from XG Sciences, Inc. (Lansing, Michigan), with reported platelet diameters of 5 and 25 μm and a thickness of 7 nm. Figure 14.7 shows scanning electron microscopy (SEM) images of the as-received graphite nanoplatelets that should have had a platelet diameter of 5 μm. It is seen from Figure 14.7a that the powder assembles itself into accordion-like structures, and with increasing magnification the flaky, layered structure of xGnP can be seen in Figure 14.7b. The platelets appear agglomerated with agglomerates as large as 600 μm in size.

In order to make graphene compatible with the matrix polymer, the as-received nanoparticles were surface treated with 3-aminopropyltrimethoxysilane (APTMS) by mixing the graphene powder and an equal mass of APTMS in dimethylformamide (DMF) using an ultrasonicator and a magnetic stirrer and filtering the DMF. Figure 14.8 shows energy-dispersive x-ray spectroscopy (EDAX) results comparing the treated and untreated samples of 5-μm-diameter graphene. Surface oxygen atoms were replaced by silicon atoms; the added silanes react with surface hydroxyl groups to increase the hydrophobicity of the graphene platelets. Before use, the treated and untreated platelets were ultrasonicated in isopropyl alcohol (IPA) to separate the platelets, thereby reducing the overall thickness of the graphene stack. The IPA was then evaporated and the treated graphene was dried in an oven.

The PVAc-matrix PNCs containing up to 1.5 wt% graphene were prepared with the help of an internal mixer and then compression molded into thin sheets. The effects of nanofiller concentration, surface treatment, and platelet diameter on the dispersion of graphene in the PVAc matrix were examined using SEM. The GPNC samples were dried, cryogenically fractured, and coated with gold-palladium before SEM images were taken of the fracture surface; the results are shown in Figure 14.9. Graphene platelets appear as white horizontal lines, showing that the platelets lie in the plane of the nanocomposite sample and are, therefore, aligned normal to the direction of

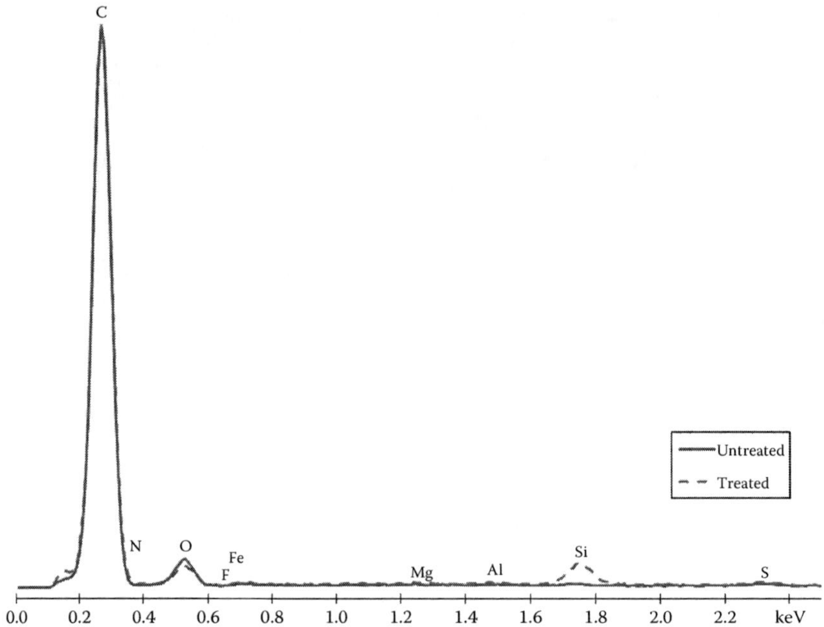

FIGURE 14.8
EDAX results showing replacement of surface oxygen atoms with silicon atoms from untreated graphene (solid line) to treated graphene (dotted line).

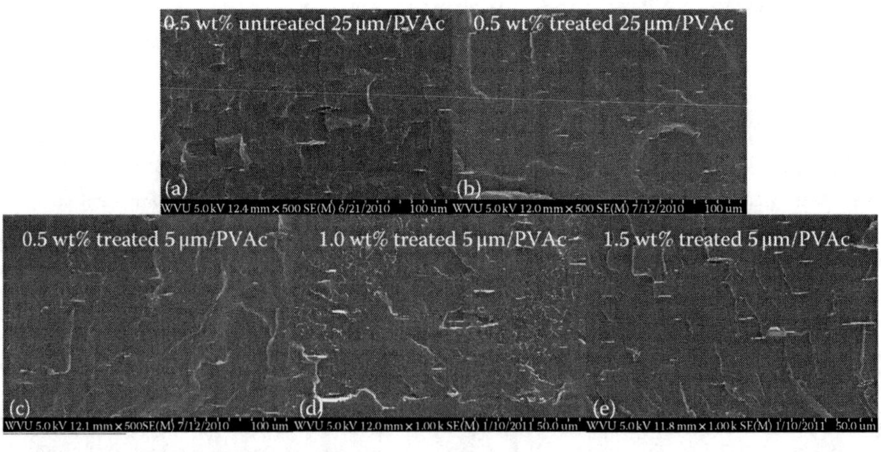

FIGURE 14.9
SEM results showing relative dispersion of graphene in PVAc containing (a) 0.5 wt% untreated 25-μm-diameter graphene, (b) 0.5 wt% treated 25-μm-diameter graphene, (c) 0.5 wt% treated 5-μm-diameter graphene, (d) 1.0 wt% treated 5-μm-diameter graphene, and (e) 1.5 wt% treated 5-μm-diameter graphene.

FIGURE 14.10
TEM results of 1.5 wt% 5-μm graphene in PVAc showing thickness of dispersed particles and relative distribution in matrix.

moisture diffusion. This morphology, in which platelets align parallel to one another, results from compression molding. The SEM images of GPNC pellets before being molded showed randomly oriented platelets.

As expected, Figure 14.9 shows that the number of platelets that are visible increases with increasing concentration of the filler. Further, fewer platelets can be seen in the treated sample, Figure 14.9b, as compared to the untreated sample, Figure 14.9a. This is most likely due to better dispersion and the reduction in thickness of platelet stacks. Average lateral particle sizes for the 25-μm-diameter sample, Figure 14.9b, are larger than those for the 5-μm-diameter sample, Figure 14.9c; however, these platelets do not appear to actually be 25 or 5 μm in diameter. This is possibly due to the breakage or fracture resulting from ultrasonication and stresses imposed during compounding. A typical TEM image that reveals the GPNC morphology is shown in Figure 14.10. Graphene stacks are visible, and these may be as thick as 15 nm. Also, the distribution is less homogeneous than what would be desired.

Sorption tests were performed by immersing 2.5 cm × 2.5 cm GPNC samples, having a thickness of 0.410–0.549 mm and a dry mass ranging from 310 to 407 mg, in deionized water at room temperature (about 23°C) and periodically removing, weighing, and reimmersing them. For the unfilled polymer, near-equilibrium state was reached in about 60 days, which is consistent with the study of Fornasiero et al. (2002). These authors noted that mass did increase further even after that point and, similarly, in this work sample mass still increased incrementally even after as many as 14 months. In addition, the timescale of the sorption tests increased progressively as more and more graphene was incorporated into the polymer. Experiments were conducted as a function of surface treatment, filler concentration (0.1, 0.2, 0.5, 1.0, and 1.5 wt%), and platelet size (5 and 25 μm).

Fickian diffusion theory, given by Equations 14.10 and 14.11, was used to analyze the sorption data, and this is reasonable for the PVAc/water system examined here (Kishimoto et al. 1960). Figure 14.11 shows the results for a typical sorption test for the unfilled polymer. Figure 14.11a shows short-timescale mass uptake, whereas Figure 14.11b shows the corresponding long-term data. All these data are replotted in Figure 14.11c in normalized form. Finally, Figure 14.11d displays normalized short-timescale mass uptake results according to Equation 14.11, and a diffusion coefficient of 1.29×10^{-6} cm^2/min can be calculated from the slope of the observed straight line, which is the correct order of magnitude based on results available in the literature (Kokes and Long 1953; Kishimoto et al. 1960). It is seen that, in each case, near-equilibrium state is reached fairly quickly and the final mass gain is about 7 wt% of the initial mass. It is noted from Figure 14.11 that on prolonged immersion, the samples absorb more water and sample weight continues to increase. Concomitantly, sample thickness also increases. This phenomenon is seen more prominently in the mass gain curves for filled samples since these are of longer duration, and we speculate that this effect

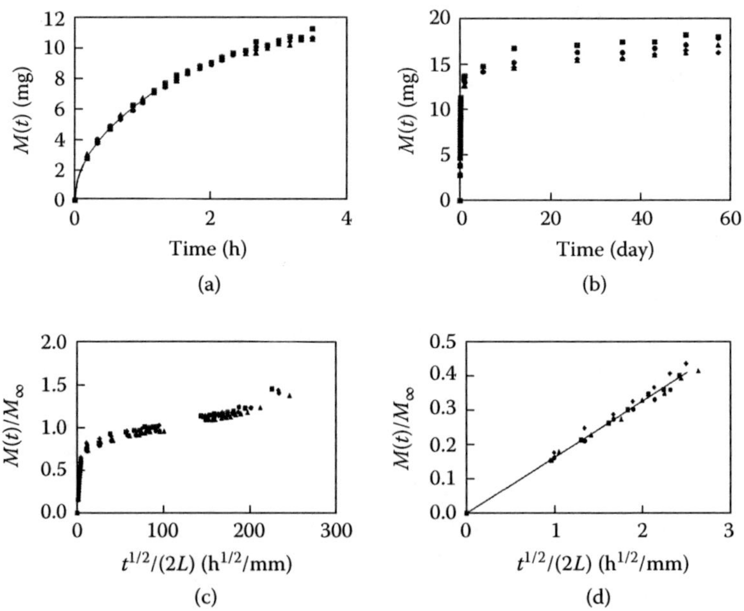

FIGURE 14.11
Sorption test results for water diffusion into unfilled PVAc for four replicate runs with solid line representing theoretical results using calculated diffusion coefficient. (a) Short-timescale mass uptake. (b) Long-timescale mass uptake. (c) Long-timescale normalized mass uptake. (d) Short-timescale normalized mass uptake from which D was calculated.

is unrelated to the diffusion process that is examined here. Figure 14.11c and d demonstrates that the initial fractional increase in mass varies linearly with $t^{1/2}/(2L)$, as expected. Given the selection of M_∞ used in this work to be described shortly, the final $M(t)/M_\infty$ values are greater than 1.

Mass gain versus time behavior is shown for the treated and untreated 25-μm graphene/PVAc samples in Figure 14.12a, and the effects of platelet size and surface treatment are shown in Figure 14.12b at a constant

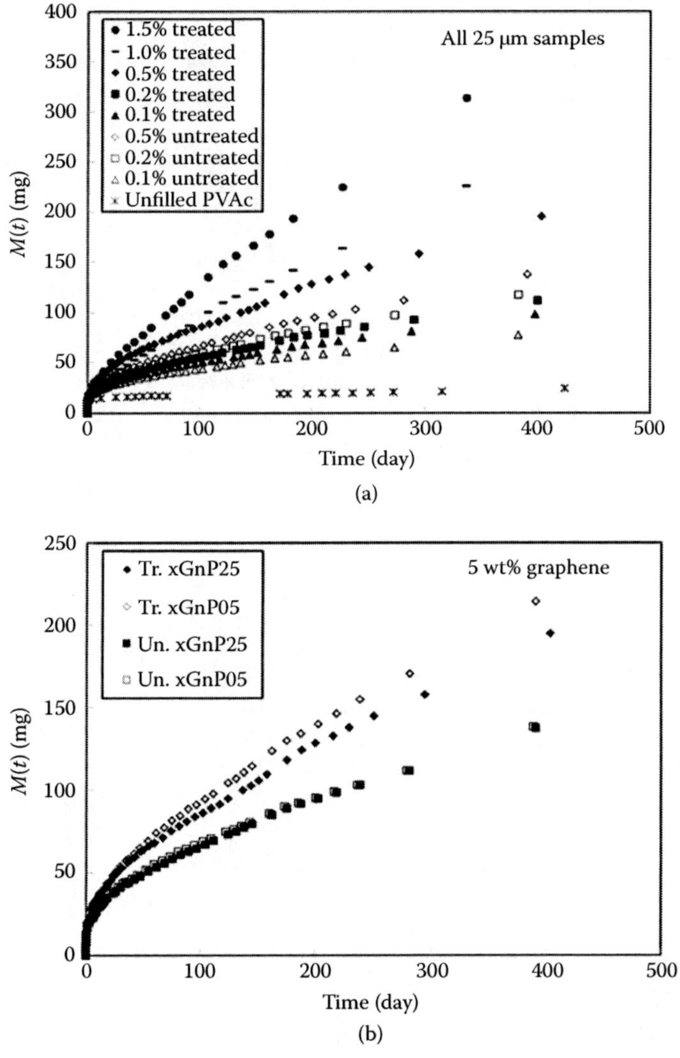

FIGURE 14.12
(a) Average mass uptake for treated and untreated 25-μm graphene in PVAc samples and (b) comparison of average mass uptakes for all 5 wt% treated and untreated graphene in PVAc samples.

filler concentration of 0.5 wt%. Each curve in Figure 14.12 is the average of four replicate runs. From Figure 14.12, it is found that increase in graphene concentration results in a very significant increase in the equilibrium moisture uptake; similar results were seen for the 5-μm graphene samples. This progressive increase in equilibrium moisture uptake with increasing graphene concentration is without doubt the result of the hydrophilic nature of the nanofiller. Indeed, when graphene powder was exposed to air of high humidity, it easily incorporated moisture and increased in mass by more than 30%. On average, treated samples were found to absorb more water on a long timescale, but it was noted that surface treatment decreases short-timescale water diffusion. Also, there is only a marginal difference between the results of the samples using different platelet sizes.

Values of M_∞ for the different situations considered here are listed in Table 14.3. Since true equilibrium was not reached even after 1 year of immersion, equilibrium was taken to have been attained at the point where the slope of the $M(t)$ curve begins to increase again. An increase in slope should not occur under normal Fickian diffusion, and it indicates the presence of another diffusion mechanism. It is noted that if we were to use values of M_∞ that are greater than the values shown in Table 14.3, the calculated diffusion coefficients would be smaller still. It is also noted that Mathew et al. (2011) have published moisture sorption data on cellulose nanowhisker-reinforced PVAc, and their sorption curves are very similar to those obtained by

TABLE 14.3

Average Equilibrium Mass Uptakes for Graphene/PVAc Samples

	Treated	
Platelet Size	**5 μm**	**25 μm**
Concentration (wt%)	M_∞ *(mg)*	M_∞ *(mg)*
PVAc	17.27 ± 0.80	
0.1	52.04 ± 8.08	46.68 ± 1.73
0.2	47.99 ± 3.81	48.26 ± 1.16
0.5	74.41 ± 4.89	78.35 ± 12.20
1.0	114.70 ± 5.19	100.84 ± 3.83
1.5	120.16 ± 8.23	135.43 ± 2.66
	Untreated	
Platelet Size	**5 μm**	**25 μm**
Concentration (wt%)	M_∞ *(mg)*	M_∞ *(mg)*
0.1	31.65 ± 0.91	34.16 ± 2.61
0.2	40.97 ± 10.97	46.26 ± 17.54
0.5	55.27 ± 7.02	53.76 ± 12.23

Thompson et al. (2012) in that the uptake of water was higher for the nanocomposites than for the neat matrix.

In Figure 14.13, short-timescale $M(t)/M_\infty$ versus $t^{1/2}/(2L)$ plots for moisture sorption through GPNCs containing the 25-μm graphene platelets, using M_∞ values given in Table 14.3, are shown. Results for treated and untreated samples are shown in Figure 14.13a and b, respectively. Again, each data set is the average of four replicate runs. It is seen that the initial slope of all the curves decreases with increasing graphene content. Corresponding values

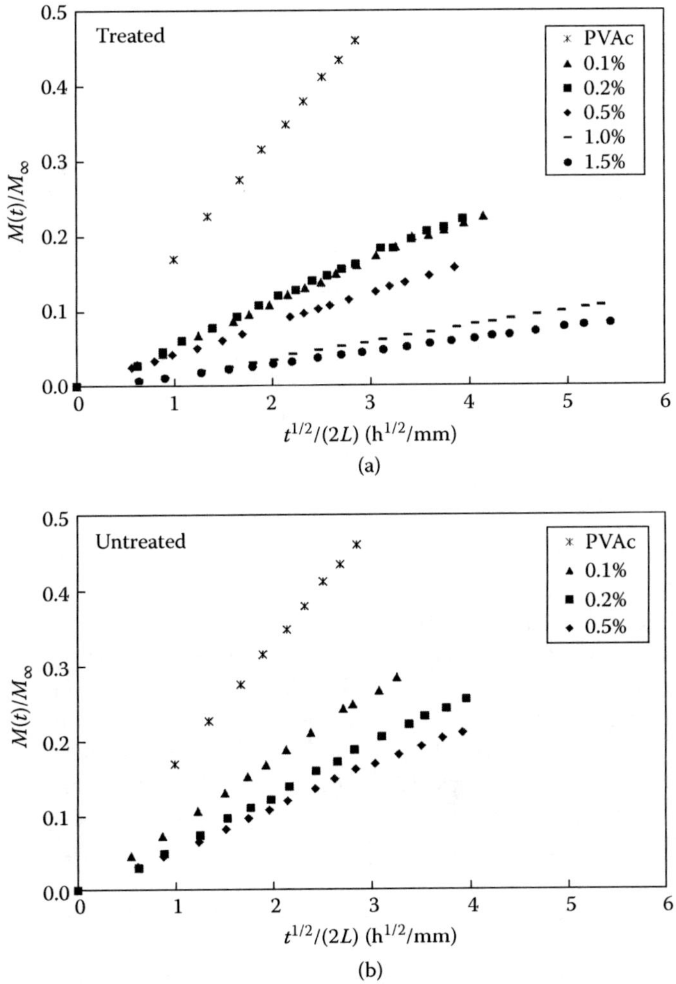

FIGURE 14.13
Short-timescale normalized mass gain plots from which diffusion coefficients were calculated for 25-μm graphene in PVAc samples: (a) Treated samples and (b) untreated samples.

TABLE 14.4

Average Diffusivities for Graphene/PVAc Samples

	Treated	
Platelet Size	**5 μm**	**25 μm**
Concentration (wt%)	$D \times 10^8$ *(cm²/min)*	$D \times 10^8$ *(cm²/min)*
PVAc	87.5 ± 8.1	
0.1	10.9 ± 3.1	10.1 ± 0.7
0.2	8.72 ± 0.38	10.7 ± 0.6
0.5	4.12 ± 0.54	5.48 ± 1.18
1.0	1.26 ± 0.21	1.28 ± 0.07
1.5	1.05 ± 0.15	0.777 ± 0.095
	Predicted $D \times 10^8$ (cm²/min) for 25 μm	
Concentration (wt%)	*Nielsen*	*Cussler et al.*
0.1	109	125
0.2	94.7	114
0.5	67.8	71.2
1.0	46.0	30.3
1.5	34.7	15.4
	Untreated	
Platelet Size	**5 μm**	**25 μm**
Concentration (wt%)	$D \times 10^8$ *(cm²/min)*	$D \times 10^8$ *(cm²/min)*
0.1	29.4 ± 3.0	25.2 ± 5.0
0.2	16.8 ± 6.9	14.7 ± 8.6
0.5	8.38 ± 2.09	10.1 ± 4.5

of the diffusion coefficient, D, are given in Table 14.4. It is seen that the diffusion coefficient decreases by almost two orders of magnitude when graphene content is increased to only 1.5 wt%. It is further found that the use of surface treatment yields smaller diffusion coefficients due to better nanofiller dispersion and distribution. These data show that graphene nanoplatelets form a highly effective moisture barrier, especially after surface treatment. Improvement in barrier properties occurs because the increase in M_∞ or solubility is overshadowed by a reduction in diffusivity. As a consequence, the permeability is predicted to decrease rapidly with increasing graphene content. Also presented in Table 14.4 are the predicted values of D for the treated 25-μm graphene samples using the theories of Nielsen and Cussler et al., Equations 14.15 and 14.16, respectively, where h was estimated to be about 10 μm and t was estimated to be about 15 nm from Figures 14.9 and 14.10. The measured diffusivities are much lower than those predicted by either of the two models, and this may be due to the fact that the filler concentrations used are outside the range of validity of these theories.

14.8 Conclusions and Future Outlook

Data available on the barrier properties of GPNCs are much more limited when compared with results on clay-based PNCs, but published information appears to indicate that quantitatively similar benefits may be obtained at much lower graphene loading levels in the former case. This tentative conclusion, however, needs to be subjected to rigorous testing and evaluation. It also must be determined if permeability reductions are due to diminished diffusion coefficients or lowered solubility of the diffusing molecule in the nanocomposite. In either case, the nanofiller contents are so low that they are outside the range of validity of most of the popular theories that rely on the concept of a tortuous diffusion path. Thus, there is a need for data over wide ranges of conditions of amount, orientation, size, and surface treatment (or extent of dispersion) of the filler, concentration gradient, temperature, and nature of the diffusing molecule using well-characterized GPNCs and using both permeation and sorption techniques. In addition, theories are required for quantitatively explaining results for diffusion through PNCs formulated with small amounts of nanofillers.

Acknowledgment

The TEM image shown in Figure 14.10 was obtained by courtesy of Professor S. C. Jana of the University of Akron, Akron, Ohio.

References

Azeredo, H. M. C. 2009. Nanocomposites for food packaging applications. *Food Res. Int.* 42: 1240–1253.

Barrer, R. M. 1968. Diffusion and permeation in heterogeneous media. In *Diffusion in Polymers*, ed. J. Crank and G. S. Park, Chapter 6. New York: Academic Press.

Beall, G. W. 2001. New conceptual model for interpreting nanocomposite behavior. In *Polymer-Clay Nanocomposites*, ed. T. J. Pinnavaia and G. W. Beall, 267–279. New York: Wiley.

Bell, G. E., and J. Crank. 1974. Influence of imbedded particles on steady-state diffusion. *J. Chem. Soc., Faraday Trans. 2: Mol. Chem. Phys.* 70: 1259–1273.

Bhardwaj, M., and R. K. Gupta. 2005. Water vapor diffusion through glass-fiber reinforced nanocomposites: A quantitative approach. *Soc. Plast. Eng. ANTEC*: 3332–3336.

Bharadwaj, R. K., A. R. Mehrabi, C. Hamilton, et al. 2002. Structure-property relationships in cross-linked polyester-clay nanocomposites. *Polymer* 43: 3699–3705.

Bhattacharya, M., S. Biswas and A. K. Bhowmick. 2011. Permeation characteristics and modeling of barrier properties of multifunctional rubber nanocomposites. *Polymer* 52: 1562–1576.

Bunch, J. S., S. S. Verbridge, J. S. Alden, et al. 2008. Impermeable atomic membranes from graphene sheets. *Nano Lett.* 8: 2458–2462.

Carrado, K. A. 2003. Polymer-clay nanocomposites. In *Advanced Polymeric Materials*, ed. G. O. Shonaike and S. G. Advani, 349–396. Boca Raton: CRC Press.

Chang, J. H., K. M. Park, D. H. Cho, H. S. Yang, and K. J. Ihn. 2001. Preparation and characterization of polyimide nanocomposites with different organo-montmorillonites. *Polym. Eng. Sci.* 41: 1514–1520.

Chin, J. W., T. Nguyen, and K. Aouadi. 1999. Sorption and diffusion of water, salt water, and concrete pore solution in composite matrices. *J. Appl. Polym. Sci.* 71: 483–492.

Choudalakis, G., and A. D. Gotsis. 2009. Permeability of polymer/clay nanocomposites: A review. *Euro. Polym. J.* 45: 967–984.

Compton, O. C., S. Kim, C. Pierre, J. M. Torkelson, and S. B. T. Nguyen. 2010. Crumpled graphene nanosheets as highly effective barrier property enhancers. *Adv. Mater.* 22: 4759–4763.

Crank, J. 1975. *Mathematics of Diffusion*, 2nd ed. Oxford: Clarendon Press.

Crank, J., and G. S. Park. 1968. Methods of measurement. In *Diffusion in Polymers*, ed. J. Crank and G. S. Park, 1–39. New York: Academic Press.

Cussler, E. L., S. E. Hughes, W. J. Ward, and R. Aris. 1988. Barrier membranes. *J. Membr. Sci.* 38: 161–174.

Daynes, H. A. 1920. The process of diffusion through a rubber membrane. *Proc. Roy. Soc. London A* 97: 286–307.

Dunkerley, E., and D. Schmidt. 2010. Effects of composition, orientation and temperature on the O_2 permeability of model polymer/clay nanocomposites. *Macromol.* 43: 10536–10544.

Eitzman, D. M., R. R. Melkote, and E. L. Cussler. 1996. Barrier membranes with tipped impermeable flakes. *AIChE J.* 42: 2–9.

Falla, W. R., M. Mulski, and E. L. Cussler. 1996. Estimating diffusion through flake-filled membranes. *J. Membr. Sci.* 119: 129–138.

Fidelle, T. P., and R. S. Kirk. 1971. A study of unidirectional versus tridirectional heat flux models and the effect of particle size on heat conduction in composite solids. *AIChE J.* 17: 1427–1434.

Fornasiero, F., M. M. Olaya, B. Esprester, V. Nguyen, and J. M. Prausnitz. 2002. Distribution coefficients and diffusivities in three polymers for nineteen aqueous nonvolatile solutes. *J. App. Polym. Sci.* 85: 2041–2052.

Goldberg, H. A., D. P. Karim, C. A. Feeney, and M. Farrell. 2004. Clay polymer nanocomposite barriers. *The 9th New Industrial Chemistry and Engineering (NIChE) Conference: Synthesis and Characterization of Customized Nanocomposites and Materials*, June 1–4. St. Pete Beach, FL.

Gupta, R. K., E. Kennel, and K. J. Kim, eds. 2010. *Polymer nanocomposites handbook*. Boca Raton: CRC Press.

Herrera-Alonso, J. M., E. Marand, J. C. Little, and S. S. Cox. 2009. Transport properties in polyurethane/clay nanocomposites as barrier materials: Effect of processing conditions. *J. Membr. Sci.* 337: 208–214.

Jacob, A., P. Kurian, and A. S. Aprem. 2008. Transport properties of natural rubber latex layered clay nanocomposites. *J. Appl. Polym. Sci.* 108: 2623–2629.

Kim, H., A. A. Abdala, and C. W. Macosko. 2010a. Graphene/polymer nanocomposites. *Macromol.* 43: 6515–6530.

Kim, H., and C. W. Macosko. 2008. Morphology and properties of polyester/exfoliated graphite nanocomposites. *Macromol.* 41: 3317–3327.

Kim, H., and C. W. Macosko. 2009. Processing-property relationships of polycarbonate/graphene composites. *Polymer* 50: 3797–3809.

Kim, H., Y. Miura, and C. W. Macosko. 2010b. Graphene/polyurethane nanocomposites for improved gas barrier and electrical conductivity. *Chem. Mater.* 22: 3441–3450.

Kishimoto, A., E. Maekawa, and H. Fujita. 1960. Diffusion coefficients for amorphous polymer and water systems. *Bull. Chem. Soc. Japan* 33: 988–992.

Kokes, R. J., and F. A. Long. 1953. Diffusion of organic vapors into polyvinyl acetate. *J. Am. Chem. Soc.* 75: 6142–6146.

Kumar, A., and R. K. Gupta. 2003. *Fundamentals of Polymer Engineering*, 2nd ed., Chapter 13. New York: Marcel Dekker.

Lape, N. K., E. E. Nuxoll, and E. L. Cussler. 2004. Polydisperse flakes in barrier films. *J. Membr. Sci.* 236: 29–37.

Lu, C., and Y.-W. Mai. 2007. Permeability modeling of polymer-layered silicate nanocomposites. *Comp. Sci. Tech.* 67: 2895–2902.

Maji, P. K., N. K. Das, and A. K. Bhowmick. 2010. Preparation and properties of polyurethane nanocomposites of novel architecture as advanced barrier materials. *Polymer* 51: 1100–1110.

Mallick, P. K. 1993. *Fiber-Reinforced Composites*, 2nd ed. New York: Marcel Dekker.

Mathew, A. P., G. Gong, N. Bjorngrim, D. Wixe, and K. Oksman. 2011. Moisture absorption behavior and its impact on the mechanical properties of cellulose whiskers-based polyvinylacetate nanocomposites. *Polym. Eng. Sci.* 51: 2136–2142.

Minelli, M., M. G. Baschetti, and F. Doghieri. 2011. A comprehensive model for mass transport properties in nanocomposites. *J. Membr. Sci.* 381: 10–20.

Moggridge, G. D., N. K. Lape, C. Yang, and E. L. Cussler. 2003. Barrier films using flakes and reactive additives. *Prog. Org. Coat.* 46: 231–240.

Nielsen, L. E. 1967. Models for the permeability of filled polymer systems. *J. Macromol. Sci. (Chem.)* A1: 929–942.

Paul, D. R., and L. M. Robeson. 2008. Polymer nanotechnology: Nanocomposites. *Polymer* 49: 3187–3204.

Petrovicova, E., R. Knight, L. S. Schadler, and T. E. Twardowski. 2000. Nylon 11/silica nanocomposite coatings applied by the HVOF process. II. Mechanical and barrier properties. *J. Appl. Polym. Sci.* 78: 2272–2289.

Potts, J. R., D. R. Dreyer, C. W. Bielawski, and R. S. Ruoff. 2011. Graphene-based polymer nanocomposites. *Polymer* 52: 5–25.

Priolo, M. A., D. Gamboa, K. M. Holder, and J. C. Grunlan. 2010. Super gas barrier of transparent polymer-clay multilayer ultrathin films. *Nano Lett.* 10: 4970–4974.

Rana, H. T., R. K. Gupta, H. V. S. GangaRao, and L. N. Sridhar. 2005. Measurement of moisture diffusivity through layered-silicate nanocomposites. *AIChE J.* 51: 3249–3256.

Ravindran, N., and R. K. Gupta. 2006. Dispersed nanoclay in GFRP as a barrier against environmental attack. *Soc. Plast. Eng. ANTEC*: 243–247.

Shah, A. P., R. K. Gupta, H. V. S. GangaRao, and C. E. Powell. 2002. Moisture diffusion through vinyl ester nanocomposites made with montmorillonite clay. *Polym. Eng. Sci.* 42: 1852–1863.

Sorrentino, A., M. Tortora, and V. Vittoria. 2005. Diffusion behavior in polymer-clay nanocomposites. *J. Polym. Sci. Part B: Polym. Phys.* 44: 265–274.

Sridhar, L. N., R. K. Gupta, and M. Bhardwaj. 2006. Barrier properties of polymer nanocomposites. *Ind. Eng. Chem. Res.* 45: 8282–8289.

Statler, D. L., Jr., and R. K. Gupta. 2008. Relating nanoclay dimensions and arrangement to barrier properties of polymer nanocomposites. *Soc. Plast. Eng. ANTEC*: 163–167.

Sun, L., W.-J. Boo, A. Clearfield, H.-J. Sue, and H. Q. Pham. 2008. Barrier properties of model epoxy nanocomposites. *J. Membr. Sci.* 318: 129–136.

Takahashi, S., H. A. Goldberg, C. A. Feeney, et al. 2006. Gas barrier properties of butyl rubber/vermiculite nanocomposite coatings. *Polymer* 47: 3083–3093.

Thompson, M. S., S. Agarwal, P. Mukhopadhyay, and R. K. Gupta. 2012. Barrier properties of graphene-based polymer nanocomposites. *Soc. Plast. Eng. ANTEC.*

Tortora, M., V. Vittoria, G. Galli, S. Ritrovati, and E. Chiellini. 2002. Transport properties of modified montmorillonite-poly(ε-caprolactone) nanocomposites. *Macromol. Mater. Eng.* 287: 243–249.

Wang, J., X. Wang, C. Xu, M. Zhang, and X. Shang. 2011a. Preparation of graphene/poly(vinyl alcohol) nanocomposites with enhanced mechanical properties and water resistance. *Polym. Int.* 60: 816–822.

Wang, J., C. Xu, H. Hu, et al. 2011b. Synthesis, mechanical, and barrier properties of LDPE/graphene nanocomposites using vinyl triethoxysilane as a coupling agent. *J. Nanopart. Res.* 13: 869–878.

Xu, B., Q. Zheng, Y. Song, and Y. Shangguan. 2006. Calculating barrier properties of polymer/clay nanocomposites: Effects of clay layers. *Polymer* 47: 2904–2910.

Yano, K., A. Usuki, A. Okada, T. Kurauchi, and O. Kamigaito. 1993. Synthesis and properties of polyimide-clay hybrid. *J. Polym. Sci. Part A: Polym. Chem.* 31: 2493–2498.

15

Using Functionalized Graphene Sheets to Restrain Chemically Driven Composite Aging

J. S. van der Wal*

CONTENTS

15.1 Introduction

15.1.1 Chemical Aging

With a growing demand for better materials in more demanding applications, polymer-based materials, combined with micro and nanomaterial, are often the best option. In many material applications, the transparency,

* Correspondence to: J. S. van der Wal (s.van.der.wal@composite-agency.com)

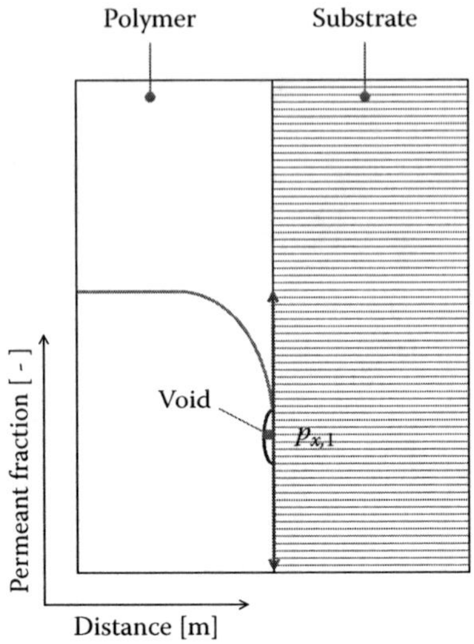

FIGURE 15.1
Permeant fraction in a composite configuration: a polymer on a substrate (e.g., a filler). The forthcoming interfacial swelling pressure ($p_{x,1}$) might cause void formation.

low weight, and corrosion resistance are key features. Nevertheless, polymer matrices are not as tight as they look. Solubility, and subsequent diffusion, of chemicals in polymers is a drawback.* On the first place, chemical solubility and diffusion might alter the integrity of the composite itself (see Figure 15.1).

Second, permeation might cause quality loss of packaged materials/goods or give rise to chemical emissions. Neglecting chemically driven component/interfacial degradation, chemical aging manifests itself as follows:

- Permeants can swell the polymer matrix. If the interface has insufficient potential to restrain the expansion, voids are formed (see Figure 15.1). This might result in an interfacial crack/delamination.
- Permeants may plasticize the polymer matrix (reduce the Young's modulus), which reduces the overall strength. Plasticizing also enhances the permeation rates of the structure. Especially for glassy polymers, this enhancement can be several orders of magnitude.

It is obvious that in more severe mechanical–chemical conditions, chemically driven aging is a more important issue than in mild or moderate conditions. Examples of demanding structural applications are high-pressure pipelines and windmill rotor blades operated in sea conditions. An example of a more or less stationary ambient condition is the storage of gasoline in a glass fiber-reinforced polyester tank.

* Apart from membrane applications where selective permeation is an essential property.

In this chapter, although only structural (construction) materials will be considered, we cannot ignore the importance of functional materials. For example, the long-term light yield of an organic light-emitting diode might decrease in humid conditions. Furthermore, the efficiency of polymer-based photovoltaics might reduce as a result of chemical interface interactions. Quantification of chemical aging will be even more important for the new class of structural–functional materials, such as self-repairing materials, materials with integrated lifetime monitoring systems, or car body parts with integrated solar cells.

15.1.2 Chemically Driven Expansion

The focus of this work is chemically driven expansion in polymer-based materials. For a composite material, this situation is schematically shown in Figure 15.2.

From thermal expansion experiments and modeling, the restraint of thermally driven composite expansion is widely known. Schapery (1968) modeled the thermal expansion of a composite, assuming that at equilibrium the thermal expansion of the fiber equals the thermal expansion of the matrix in the direction of the long fibers. In Figure 15.3, the result is depicted for a glass fiber–epoxy composite.

Although the interfacial mechanics of thermal stresses can be compared with chemical stresses, the calculation of effective chemically driven expansion is more complicated. Predictive models must include fracture mechanics as well as chemical mixture thermodynamics (chemical potentials).

The relevant fracture mechanics are described in Section 15.2. Subsequently, the chemical mixture thermodynamics are described in Section 15.3.

15.1.3 FGS-Reinforced Epoxy

Majority of publications on diffusion in fiber and nanofiller-reinforced materials focus on the increased path length due to the presence of the impermeable filler. Unfortunately, the barrier improvement is often quite limited,

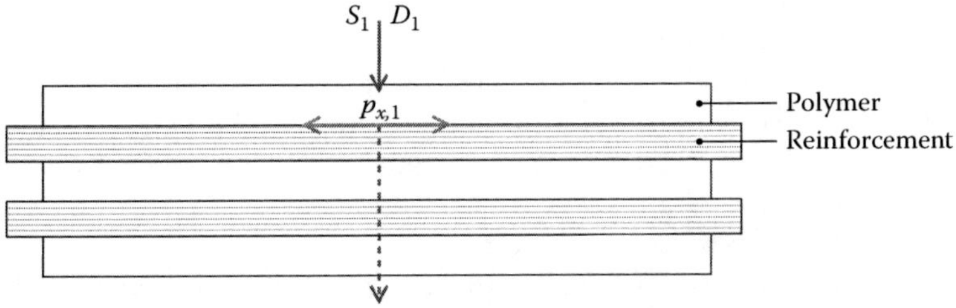

FIGURE 15.2
Chemical solubility (S_1), diffusivity (D_1), and forthcoming swelling pressure ($p_{x,1}$) on an interface.

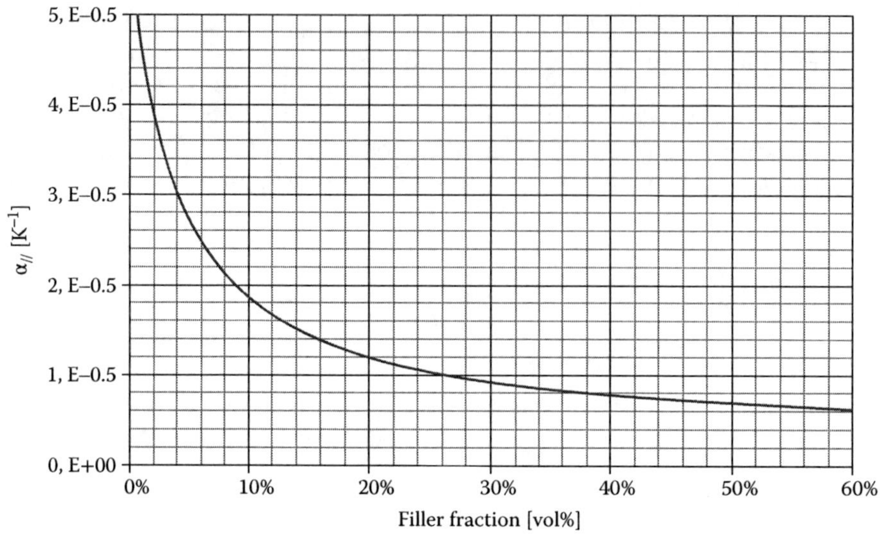

FIGURE 15.3
Effect of fiber volume on thermal expansion coefficient ($\alpha_{//}$) of a glass fiber–epoxy composite. Thermal expansion of glass fiber is $0.50{\cdot}10^{-5}$ K^{-1} and that of cast epoxy is $5.5{\cdot}10^{-5}$ K^{-1}.

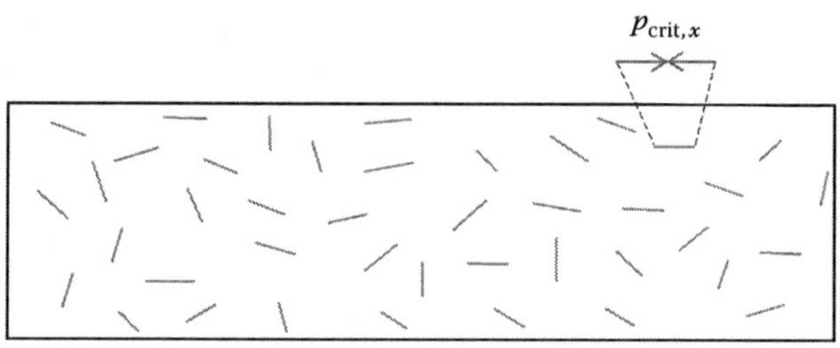

FIGURE 15.4
Restraint pressure ($p_{crit,x}$) of FGS filler in epoxy.

since swelling is not noticeably restrained due to insufficient filler aspect ratio, unidirectional filler orientation, or insufficient filler–matrix adhesion.

Indirectly, interesting work on this subject came from the field of mechanics. Rafiee et al. (2010) showed in their work on fracture and fatigue of functionalized graphene sheets (FGS) based epoxy composites that FGS with an impressive aspect ratio (dimensions: ~1 nm × ~4.4 μm × ~2.4 μm) and good adherence to epoxy can yield composite properties that are theoretically sufficient to prevent chemically driven interfacial fracture at chemical uptakes to around 15 volume percent.

Realize that compared to well-adhered glass fiber–epoxy interface, the surface-to-volume ratio of this FGS is easily a 1000-fold larger. Then, if the volume fraction of FGS would be a 100-fold lower, the effective surface will still be one order of magnitude larger (Figure 15.4).

15.2 Energy-Based Fracture Analysis

15.2.1 Introduction

If it is assumed that chemicals absorb and diffuse into the polymer matrix, the solubility of chemicals becomes stationary if the chemical potential in all environments and materials is equal. If chemical swelling of the polymer is restrained by a substrate, the resulting interfacial swelling pressure in the x direction can be estimated as follows (assuming isotropic polymer expansion):

$$p_{1,x} \approx \frac{1}{2}\left(\frac{\varepsilon_1}{3} \cdot E_{\mathrm{m}}\right) \qquad (15.1)$$

where ε_1 is the unrestrained volume fraction of a chemical in the polymer and E_{m} (N/m^2) is the Young's modulus of the polymer. For example, if the chemical uptake (ε_1) would be 0.15 and $E_{\mathrm{m}} = 3.3$ GPa, then $p_{1,x} = 83$ MPa.

Since chemical-driven expansion occurs in three dimensions, restraining filler substrates should ideally be mixed uniformly and oriented randomly in three dimensions. This is not as easy as it looks. For example, in nano- and micro-based composites, the volume fraction of fibers that can be mixed with the matrix is limited by the need for uniform dispersion. The maximum allowable volume fraction for uniform dispersion depends on the nature of the polymer matrix, the size, shape, and chemical composition of the filler material. For sized glass fibers, this mixing limit is in the order of 2–3 vol%.

For larger volume fractions of fibers, the matrix must be brought to the fibers (often the fibers are fixed and positioned in mats, like glass veils, chopped strand mats, or woven rovings). Examples of composite products produced in this manner are winded pipes, molded yachts, and molded windmill blades. Note that, apart from some exceptions, these structures are particularly strong in the direction of the fibers. Obviously, the thermal and swelling restraint potential is also more substantial in this direction.

15.2.2 Adhesive-Fracture Mechanics for a Polymer on a Substrate

In energy-based fracture mechanics, it is assumed that the initiation of fracture takes place if the resultant work input (δW, e.g., a mechanical or chemical expansion load) exceeds the required fracture energy ($\delta\Gamma$), taking into account the conversion in internal elastic energy ($\delta U_{\mathrm{elastic}}$).

The energy balance of this process looks as follows:

$$\partial G = \partial W - \partial U_{\mathrm{elastic}} - \partial\Gamma \qquad (15.2)$$

In case of equilibrium:

$$\delta G = 0 \qquad (15.3)$$

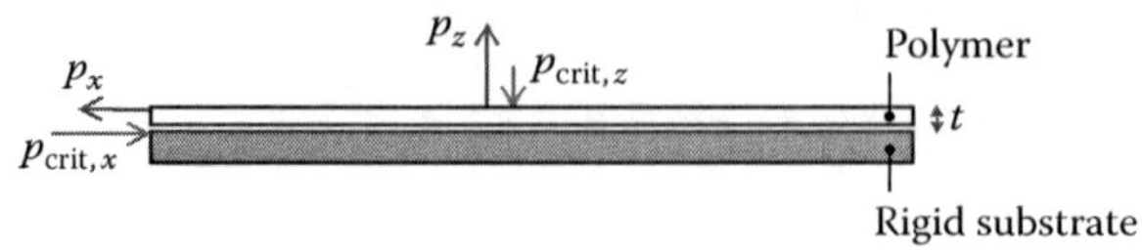

FIGURE 15.5
A 0° and 90° peel-off experiment.

As such:

$$\partial W = \partial U_{\text{elastic}} + \partial \Gamma \tag{15.4}$$

A solution of Equation 15.4 for the situation of a polymer on top of a substrate was derived by Anderson et al. (1974). If a mechanical load (p_x, see Figure 15.5) is applied to the polymer parallel to the substrate surface, the critical pressure for crack initiation becomes:

$$p_{\text{crit},x} = \sqrt{\frac{2 \cdot E_m \cdot \gamma}{t}} \tag{15.5}$$

where $p_{\text{crit},x}$ is the critical pressure in N/m^2, γ is the interfacial fracture energy (J/m^2), and t is the thickness (m) of polymer adhered to the substrate.

This solution is based on the assumption that the polymer is linearly elastic and that the substrate is mechanically rigid (no energy stored in the plate). Moreover, the friction force between the substrate and the polymer and the residual thermal stresses are neglected.

If the peeling load (p_z, see Figure 15.5) is applied perpendicularly to the substrate surface, the critical stress becomes:

$$p_{\text{crit},z} = \frac{\gamma}{t} \tag{15.6}$$

Although the previous formulae are fairly simple, they give a valuable basic insight in the conditions for crack initiation.*

15.2.3 Adhesive-Fracture Mechanics for Composites

For reinforcement filler in a polymer, the above-described principles for critical pressure still hold. Using basic mathematics, the polymer thickness (Equation 15.5) can be replaced by the radius of the fiber and the volume fraction of the polymer matrix. For a unidirectional continuous composite, this

* At least for the timescales relevant for processes regulated by chemical diffusion in tight polymers.

operation yields the following equation for the critical pressure, parallel to the fully embedded cylindrical fibers:

$$p_{\text{crit},x} = \sqrt{4\frac{\gamma}{r_\text{f}}\left(\frac{1}{\dfrac{1}{E_\text{f}}+\dfrac{\varepsilon_\text{m}}{E_{\text{m}\cdot(1-\varepsilon_\text{m})}}}\right)} \tag{15.7}$$

where r_f is the radius of the fiber, E_f is the axial Young's modulus of the fiber, and ε_m is the volume fraction of the matrix. The "4" in Equation 15.7 compares to the "2" in Equation 15.5 by the fact that the effective cylindrical surface is related to the cylindrical volume by $2/r_\text{f}$.

Obviously, a smaller volume fraction of polymer generates a higher critical pressure. However, lowering of the polymer volume fraction is limited because fibers need to be glued together effectively.

Under the following two conditions, Equation 15.7 also holds for discontinuous composites:

- The length of the filler (l) must exceed the critical reinforcement length (l_c).
- In formula:

$$l > l_\text{c} \tag{15.8}$$

The critical length for a fiber is calculated as follows:

$$l_\text{c} = \frac{2r_\text{f}\cdot\sigma_\text{f}}{p_{x,\text{crit}}} \tag{15.9}$$

where σ_f is the tensile strength of the fiber. Below this critical fiber length, the stresses are not completely transferred to the filler material.

- Adjustments must be made for filler orientation.
- For randomly oriented flakes and fibers, approximately 1/3 is oriented in the x direction. Hence, 1/3 is introduced in the square root of Equation 15.7.
- For fiber mats, approximately 1/2 is oriented in the x direction. Hence, a 1/2 is introduced in the square root of Equation 15.7.

The attentive reader may have noticed that friction stresses and residual thermal stresses have not been discussed.

Friction stresses are usually only significant after crack initiation. To assess the effective restraint pressure, primary concern is the condition before crack initiation.

Residual thermal stresses are important for the interfacial fracture energy of polymers operated (far) below the glass transition temperature. So, if the fracture energy is determined using a fiber pull-out or micro-bond experiment, residual thermal stress must be included. This ensures that the value for the interfacial fracture energy is exact, and not overestimated.

However, in the presence of chemicals, the effect of residual thermal stresses is reduced by the chemically driven plasticizing of the matrix. So, if the interfacial fracture energy has been determined with a proper fracture experiment, subsequent ignorance of the residual thermal stresses will yield a conservative (safe) estimation of the restraint potential in chemically oriented calculations.

Further discussion of friction stress, residual stresses, and traction in mechanically oriented experiments for determining interfacial fracture energy is beyond the scope of the work. The reader is made aware that Nairn (2000) did rigorous analytical work on this matter, especially with regard to the pull-out and micro-bond test.

Back to the defined equation for critical pressure (Equation 15.7), the presented equation can be simplified even further, if the following two conditions are obeyed:

- Filler with a much higher stiffness than the matrix
- A low fraction of filler in the matrix

These conditions then yield the bootstrap equation:

$$p_{\text{crit},x} = \sqrt{4\frac{\gamma}{r_\text{f}}\left(\frac{E_\text{m}\cdot(1-\varepsilon_\text{m})}{\varepsilon_\text{m}}\right)} \tag{15.10}$$

Note that the criticality condition (15.8) must hold!

Now, for low-volume randomly oriented long glass fibers and FGS in epoxy, the "1/3" is introduced. As such, the bootstrap expression becomes as follows:

$$p_{\text{crit},x} = \sqrt{\frac{4}{3}\frac{\gamma}{r_\text{f}}\left(\frac{E_\text{m}\cdot(1-\varepsilon_\text{m})}{\varepsilon_\text{m}}\right)} \tag{15.11}$$

where r_f is the radius of the glass fiber (in the order of micrometers) or the thickness of the FGS (in the order of nanometers).

Using Equation 15.11, in Figure 15.6, the critical pressure is depicted for a 2 vol% of glass fiber in an epoxy composite. The matrix modulus is 3.3 GPa. The radius of the fiber is 3.5 μm (at an average length of 350 μm). The interfacial fracture energy, y, is taken as 15 J/m^2.

The interfacial fracture energy of FGS in epoxy has yet not been determined by a proper pull-out, micro-bond, or resembling experiment. However, Rafiee et al. (2010) have found a tensile strength improvement of ~45% for a 0.06 vol% FGS-filled aerospace epoxy. When this result is calculated back to

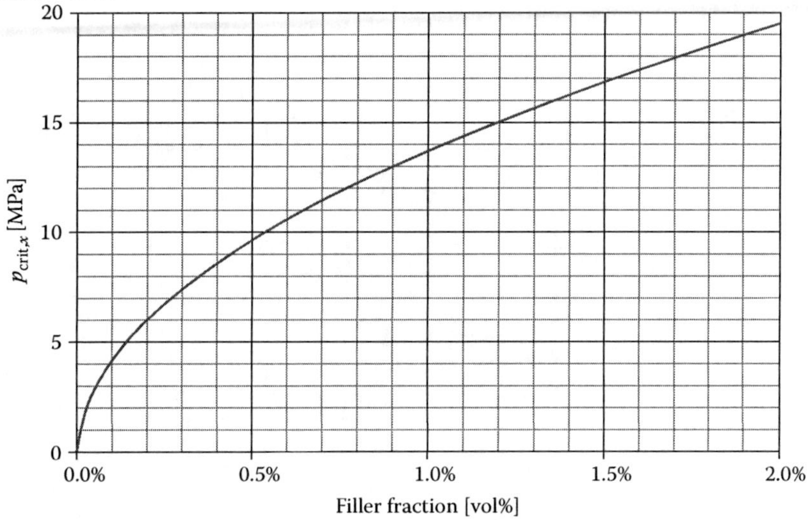

FIGURE 15.6
Average restraint pressure generated by randomly oriented glass fiber.

the DGEBA epoxy under evaluation, this gives an approximate value of 100 MPa for the interfacial shear strength. Using Equation 15.11, the interfacial fracture energy (γ) then becomes 4 J/m^2.

Below, the theoretical restraint pressure of FGS is plotted as a function of the volume fraction.

If one compares Figures 15.6 and 15.7, it is obvious that FGS generates a more than five times larger restraint pressure at a loading, which is almost a factor of 50 lower than the glass fiber filler. As such, the restraint potential of FGS with regard to chemically driven expansion is more than five times larger.

15.2.4 A Few Words on Crack Propagation

If the chemical pressure, as expressed by Equation 15.1, is larger than the critical restraint pressure, crack initiation will occur. Note that the effect of matrix plasticizing should be incorporated in this condition. This can be expressed as follows:

$$p_{1,x} > p_{crit,x} \tag{15.12}$$

In the first instance, this situation leads to the formation of voids that will release the local chemical stress. Subsequently, the initially restrained chemical concentration will increase locally.

Whether crack initiation leads to interfacial crack propagation is dependent on the magnitude of friction stresses, the magnitude and timescale of mechanical loads exerted on the composite, and chemical potentials within the composite.

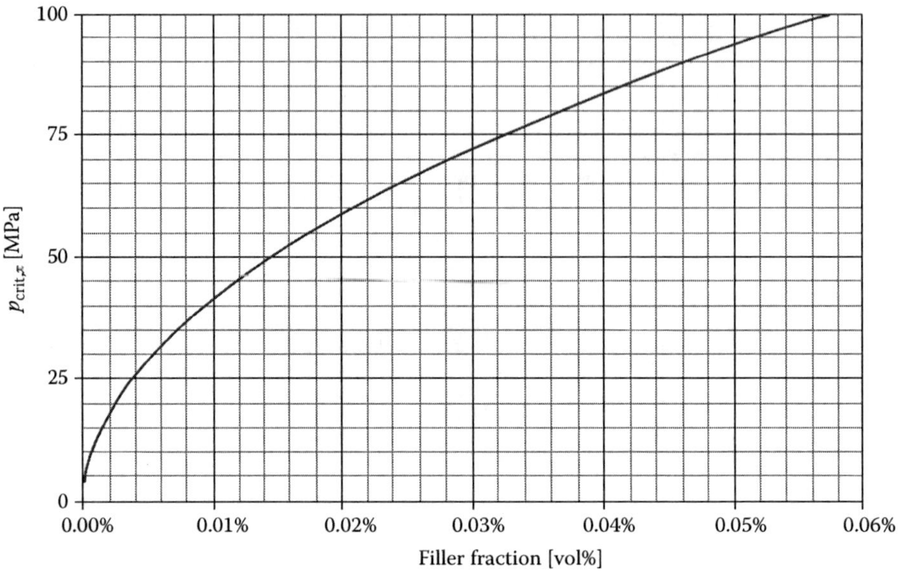

FIGURE 15.7
Average restraint pressure generated by FGS.

Either way, on the basis of Figures 15.6 and 15.7, one can conclude that FGS-based composites will be less susceptible to chemically driven crack initiation compared to the glass fiber composite considered. In Section 15.3, the effect of the presence of restraining fillers on the actual solvent concentration in the matrix will be described.

15.3 Chemical Potential of Solved Molecules

15.3.1 Introduction

In this section, the solubility (read: swelling) thermodynamics of methanol in unreinforced and reinforced epoxy resin in ambient conditions are described. Although many recent research programs have focused on nanocomposites, the fundamental works on solubility thermodynamics with regard to these materials are very limited. This is probably related to the issues described in Section 15.2.3. It is hoped that the discovery of graphene, and further development of its potential in functional and structural materials, will lead to a new élan in this area.

Therefore, we will reach back to a good and practical theory developed by Sanchez and Lacombe (1978) in the late 1970s. It is a lattice-based thermodynamical model, the principles of which were founded by Flory and Huggins in 1941.

Lattice-based chemical potentials are based on a lattice model description of the polymer. It is attractive to use this theory not only because of its good correlation with experimental thermodynamical results, but also because the baseline concepts such as free volume (which is in fact the compressed volume subtracted from the actual volume) appear in predictive models for diffusion and physical aging.

In this section, we will determine the chemical potential of a binary mixture at ambient temperature and pressure conditions using Sanchez–Lacombe-based chemical potentials. These environmental conditions enable incorporation of a swelling pressure from the cross-linked system and restraint stress generated by fillers, without getting too complex equations.

The ideal chemical potential of methanol in ambient conditions is given by

$$\mu_1 = \mu_1^0 + RT\ln\left(\frac{P_1}{P_1^{\text{vap}}}\right) \tag{15.13}$$

where μ_1^0 is the chemical potential in the reference state, R is the universal gas (8.314 J/mol), T is the temperature in Kelvin, P_1 is the methanol partial pressure, P_1^{vap} is the methanol vapor pressure at the given temperature and pressure. For saturated methanol, the partial pressure equals the vapor pressure. Hence:

$$\mu_1 = \mu_1^0 \tag{15.14}$$

The chemical potential equals the reference potential.

In the explanation below, only the parameters required for making a baseline calculation are described. Mind that a far more thorough explanation of the Sanchez–Lacombe theory would be required if we would deal with supercritical gases.

The basic chemical potential of methanol in epoxy resin reads as follows:

$$\mu_1' = \mu_1^0 + RT\ln(\varepsilon_1) + RT\left[(1-\varepsilon_1) + \chi_{12}(1-\varepsilon_1)^2\right] \tag{15.15}$$

Again the first term on the right side of the equal sign is the reference chemical potential. The second and third terms are related to the entropy of mixing, where ε_1 is the volume fraction of methanol in epoxy. The fourth term is for the heat of mixing of a polymer and solvent. This term includes the interaction parameter χ_{12} (-), which stands for the chemical compatibility between the permeant and the matrix. It is defined as follows:

$$\chi_{12} = \frac{\upsilon_1^* \cdot \Delta P}{RT} \tag{15.16}$$

where υ_1^* (m^3/mol) is the molar volume of methanol in the compressed state, and ΔP (N/m^2) is the difference between the characteristic pressures of methanol and epoxy. Note that the interaction term is inversely dependent on temperature. The compressed molar volume is more rigorously defined as

$$\upsilon_1^* = \frac{M_1}{\rho_1^*} \tag{15.17}$$

where M_1 is the molar mass and ρ_1^* is the compressed density. If one knows the molar volume of the chemical at the critical point (υ_1^c), the compressed molar volume can also be calculated as follows:

$$\upsilon_1^* = 0.286 \cdot \upsilon_1^c \tag{15.18}$$

For the sake of completeness, we mention that in Sanchez–Lacombe equation of state, the compressed volume is related to the number of lattice sites that a molecule can occupy (r_1). The characteristic volume of the molecule at the sites (υ_1^r) is related to the compressed molar volume as follows:

$$\upsilon_1^* = r_1 \cdot \upsilon_1^r \tag{15.19}$$

The term ΔP, in its most basic form, is defined as follows:

$$\Delta P = P_1^* + P_2^* - 2\zeta\sqrt{P_1^* P_2^*} \tag{15.20}$$

The factor ζ is a measure of the deviation of the mixing rule from the geometric mean. Usually, the factor deviates from 0.8 to 1.2 (±20% of geometric mean). In this work, it is assumed that ζ is 1.

P_1^* and P_2^* are the respective characteristic pressures of the permeant and the matrix in the compressed state (0 K). It is based on the heat of vaporization, ΔH (J/mol), as follows*:

$$P_1^* = \frac{\Delta H_1^{vap}}{\vartheta_1^*} \tag{15.21}$$

Referring back to the equilibrium condition (Equation 15.14). The thermodynamic equilibrium at constant pressure and temperature is defined as

$$\delta G = 0 \tag{15.22}$$

As such, the chemical potential of methanol in the environment equals the potential in the matrix:

$$\mu_1 = \mu_1' \tag{15.23}$$

* Note that this is related to a property known as the cohesive energy density (CED). However, the CED is usually based on the heat of vaporization and the molar volume at room conditions.

Using Equations 15.13 and 15.15, the methanol solubility in the matrix can be calculated. This also yields the unrestrained isotropic swelling potential in the *x*, *y*, and *z* directions ($\sim\varepsilon_1/3$).

15.3.2 Inclusion of Cross-Link

The advantage of using Sanchez–Lacombe-based chemical potentials is that additive forces, not only from swelling forces, but also from gravitational and electrical forces, can be added rather easily to the total expression for the chemical potential.

The chemical potential from cross-links can be calculated using the Phantom model (Mark 2007). For an isotropic expansion swelling pressure, the Phantom model can be expressed as follows:

$$\mu'_{cs} \approx \upsilon_1^* \cdot p_{cs} \cdot \left(\frac{{\varepsilon_1}^2}{3} + 2\varepsilon_1 \right) \tag{15.24}$$

where p_{cs} is the pressure from the cross-links. Although the Phantom model deals with the calculation of cross-link pressure, this will not be discussed here. Since, if all parameters in the expression for the chemical potential are known, the effective cross-link pressure can also be derived from an uptake experiment with an unfilled epoxy matrix (described in Section 15.4).

15.4 Experimental

15.4.1 Introduction

In this section, preliminary data are derived for a predictive analysis of the restraint potential of FGS in epoxy. As stated in Section 15.3, methanol is used as the permeant. Most commonly used (cross-linked) polymers are susceptible to the large solvation power of methanol (and ethanol). Reason is that methanol is relatively small (giving substantial penetration rates) and has an organic character combined with hydrogen-bonding capability.

In the experiment, plane sheets of samples (thickness of 3 mm) are exposed to the permeant, while the mass change of the sample is continuously recorded. For the interpretation of the mass change, we use a mass balance equation combined with the Maxwell–Stefan equation for diffusion. The reason for this approach is that chemical potentials (see Section 15.3) can be seamlessly incorporated. Mind that if the diffusion would only be driven by the concentration gradients, the more commonly used Fick's first law could have been used. Fickian and Maxwell–Stefan diffusivities, respectively, D_{Fick} and D_{MS}, are related as follows:

$$\frac{D_{\text{Fick}}}{D_{\text{MS}}} = \frac{\delta \ln(a_1)}{\delta \ln(\varepsilon_1)} \tag{15.25}$$

where a_1 is the chemical activity of the permeant in the matrix. More precisely,

$$a_1 = \varepsilon_1 \cdot e^{\alpha} \tag{15.26}$$

where the dimensionless alpha (α) is defined as:

$$\alpha = \left[(1-\varepsilon_1) + \chi_{12}(1-\varepsilon_1)^2\right] + \left(\frac{\upsilon_1^* \cdot p_{cs}}{RT}\right)\left[\frac{\varepsilon_1^{\,2}}{3} + 2\cdot\varepsilon_1\right] \tag{15.27}$$

From Equations 15.26 and 15.27, it is obvious that in very diluted situations (weak solvents), alpha becomes constant, and subsequently Equation 15.25 becomes unity.

Note that the diffusion coefficient is often not a constant, but a function of the chemical activity of the matrix plasticizing by the permeant. The reader is warned that this is not explicitly shown in Equations 15.25 through 15.27. Although thorough definition of the activity-dependent diffusion coefficient is beyond the scope of this chapter, it should be noted that the diffusivities can vary several orders of magnitudes as function of their chemical activities (especially in glassy polymers). As such, this phenomenon is important in many industrial applications. However, for our purposes, it is sufficient to only derive the weighted average diffusion coefficient as function of the total methanol activity range (from 0 to 1).

15.4.2 Methanol Uptake Experiments

The experiments are carried out with plane sheets of diglycidyl ether bisphenol A (DGEBA) epoxy cured with diamino diphenyl methane (DDM) and saturated methanol vapor at 40°C. This unfilled epoxy has a stiffness of 3.3 GPa at the given temperature and a high glass transition temperature (T_g) close to 190°C. A modified dynamic vapor sorption (DVS) setup is used to carry out the experiments.

As stated earlier, the first experiment is specifically useful for the calculation of the effective cross-link pressure of the epoxy network. Below, the result is depicted for unfilled epoxy (Figure 15.8).

From the graph, it becomes obvious that the equilibrium swelling is 15 vol% at an average diffusion coefficient of 3×10^{-13} m^2/s.

Using for methanol a characteristic pressure (P_1^*) of 32.6 MPa^{½} at a compressed molar volume (υ_1^*) of 3.3×10^{-5} m^3/mol and for the epoxy a characteristic pressure P_2^* of 24.5 MPa^{½}, the calculated cross-link pressure becomes 20 MPa.

We have performed a similar experiment, but now with 2 vol% of long glass fibers (see Figure 15.9).

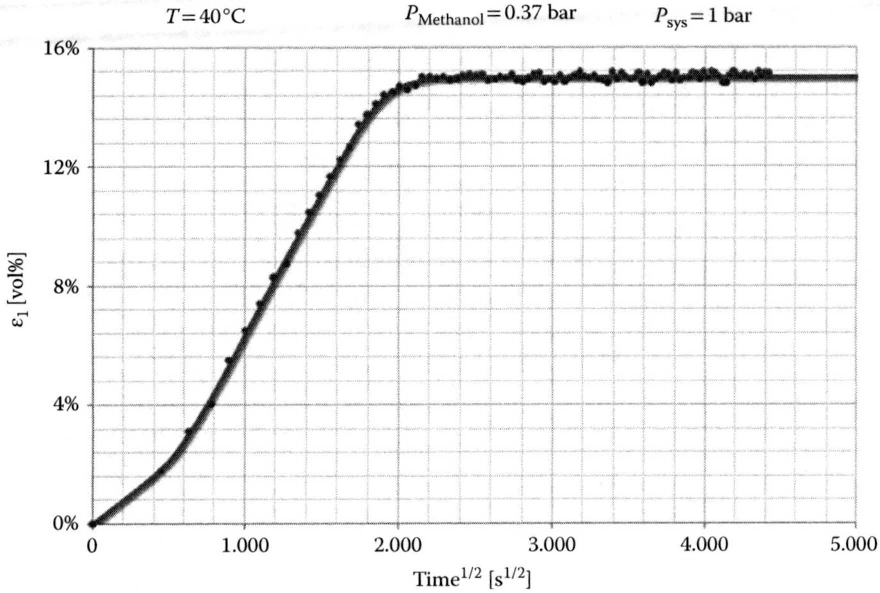

FIGURE 15.8
Saturated methanol uptake in unfilled DGEBA epoxy (plane sheets of 3 mm thickness).

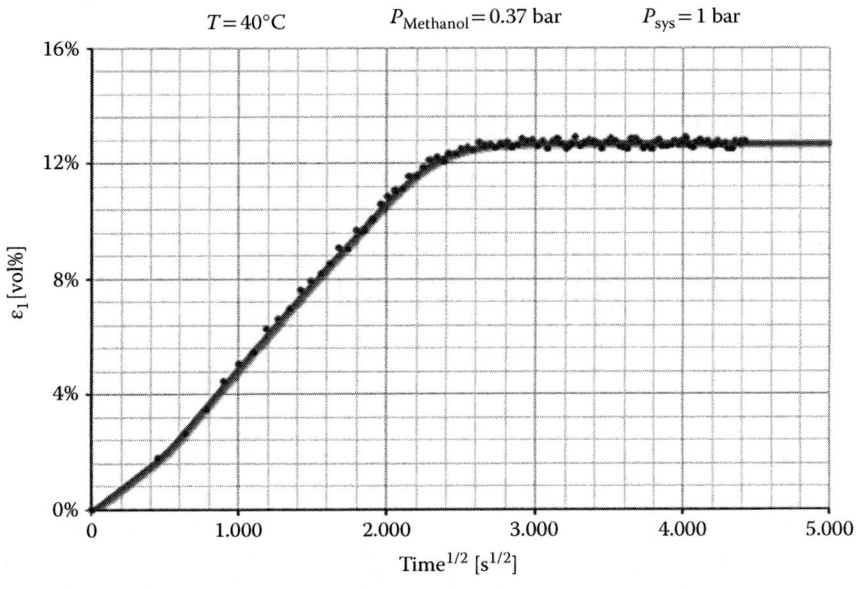

FIGURE 15.9
Liquid methanol uptake in 2 vol% glass Veil fiber-filled DGEBA epoxy (plane sheets of 3 mm thickness).

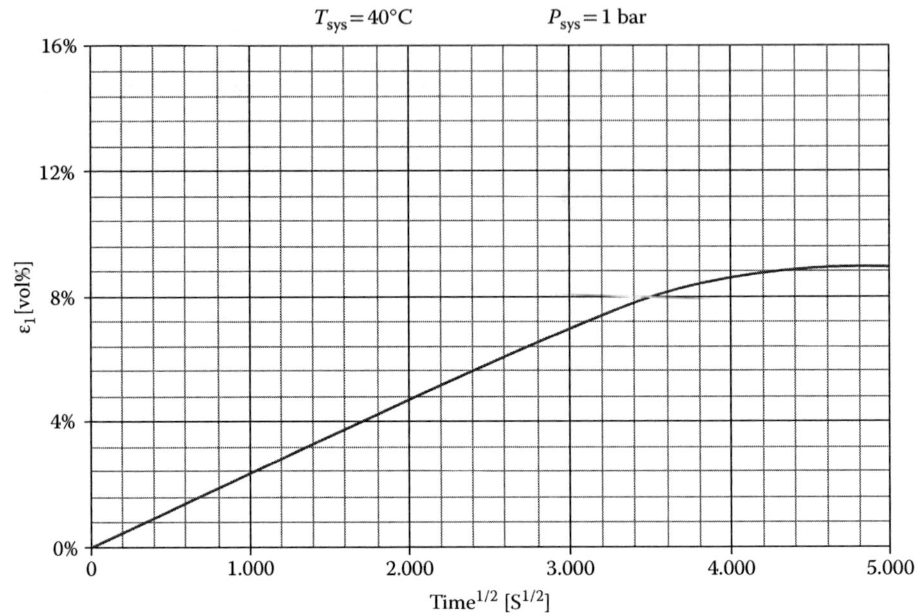

FIGURE 15.10
Simulation of liquid Methanol uptake in FGS/Glass Fibre filled DGEBA Epoxy (plane sheets of 3 millimeter thickness).

The determined weighted average diffusion coefficient is $2{\cdot}10^{-3}$ m^2/s.

It is obvious that also the diffusion rate decreases slightly due to the chemical restraint.

Now the uptake of methanol in DGEBA epoxy filled with 0.06 vol% of FGS/35 vol% Woven Roven Glass Fibre is simulated, using the previously determined parameters and the FGS restraint pressure of 100 MPa. This has been shown in Figure 15.10.

The equilibrium uptake now becomes 9 vol%. The reduction in uptake is 6 vol%! Moreover, the weighted average diffusion coefficient becomes $7{\cdot}10^{-14}$ m^2/s. This is four times lower. It can be concluded that the degree of plasticizing is much less; the FGS composite will remain quite strong in the presence of methanol.

15.5 Summary and Future Outlook

In this chapter, the potential of FGS-reinforced composite materials with regard to chemically driven expansion was demonstrated. It appears that the addition of FGS at proper loading levels might result in a better interfacial robustness, in a substantial decrease of chemically driven plasticizing (yielding a much stronger material) and in a lower permeation rate. This is not only limited to epoxies exposed to strong organic solvents, but might also be applicable to other polymers such as polyamide in the presence of moisture.

The subject of further scientific research is the incorporation of FGS in various thermoset and thermoplastic materials. It is expected that the chemical uptake experiments, and the interpretation methodology described in this chapter, will shine more light on FGS–matrix interfacial mechanisms. This could be a valuable contribution from the side of polymer chemistry to the present set of often mechanically oriented experiments and interpretation methodologies for the quantification of filler–matrix interfaces.

References

Anderson, G. P., De Vries, K. L., and Williams, M. L. 1974. The peel test in experimental adhesive-fracture mechanics. *Journal of Applied Polymer Science* 11: 11–15.

Mark, J. E. 2007. *Physical Properties of Polymers Handbook*. Springer-Verlag New York Inc, New York.

Nairn, J. A. 2000. Analytical fracture mechanics analysis of the pull-out test including the effects of friction and thermal stresses. *Advanced Composites Letters* 1: 1–10.

Rafiee, A. M., et al. 2010. Fracture and fatigue in graphene nanocomposites. *Small* 6(2): 179–183.

Sanchez, I. C., and Lacombe, H. L. 1978. Statistical thermodynamics of polymer solutions. *American Chemical Society* 11(6): 1145–1156.

Schapery, R. A. 1968. Thermal expansion coefficients of composite materials based on energy principles. *Journal of Composite Materials* 2(3): 380–404.

Additional Reading

Abdel-Magid, B. 2003. Flexure creep properties of e-glass reinforced polymers. *Composite Structures* 62: 247–253.

Anagnostopoulos, G., et al. 2008. Thermal stress development in fibrous composites. *Materials Letter* 62: 341–345.

Barber, A. H., et al. 2004. Interfacial fracture energy measurements for multi-walled carbon nanotubes pulled from a polymer matrix. *Composites Science and Technology* 64: 2283–2289.

Boyer, A. E., and Grollier, J. P. E. 2005. Modification of the glass transitions of polymers by high-pressure gas solubility. *Pure and Applied Chemistry* 77(3): 593–603.

Brovko, O., et al. 2002. Adhesion between differently treated fibres and a hybrid resin system. *Journal of Material Science* 21: 305–308.

Crank, J., Park, G. S. 1965. *The Mathematics of Diffusion*. Clarendon Press, Oxford.

Crank, J., Park, G. S. 1968. *Diffusion in Polymers*. Academic Press, London.
Einstein, A. 1905. *Investigation on the Theory of the Brownian Movement*. Dover Books, New York.
Gordon, J. E. 1968. *The New Science of Strong Materials*. Princeton University Press, Oxford.
Katsura, T., et al. 1985. Some properties of polypropylene filled with metal fibers. *Polymer Composites* 6(4): 282–295.
Nairn, J. A. 2001. Fracture mechanics analysis of the single-fiber pull-out test and the microbond test including the effects of friction and thermal stress. *Annual Technical Conference of the American Society for Composites*. Blacksburg VA.
Paipetis, A., and Galiotis, C. 2000. Modelling the stress-transfer efficiency of carbon-epoxy interfaces. *Proceedings of the Royal Society of London* 457: 1555–1577.
Paul, D. R., and Robeson, L. M. 2008. Polymer nanotechnology: Nanocomposites. *Polymer* 49: 3187–3204.
Soles, L. S. 1998. Contributions of the nanovoid structure to the moisture absorption properties of epoxy resins. *Journal of Polymer Science* 36: 3035–3048.
Struik, L. C. E. 1977. Physical ageing in plastics and other glassy materials. *Polymer Engineering and Science* 17(3): 165–173.
Strus, M. C., et al. 2009. Interfacial energy between carbon nanotubes and polymers measured from nanoscale peel tests in the atomic force microscope. *Composites Science and Technology* 69: 1580–1586.
Timoshenko, S. P., and Woinowksy-Krieger, S. 1959. *Theory of Plates and Shells*. McGraw-Hill, New York.
Wal, van der, J. S. 2006. Water resistance of glass and carbon reinforced epoxy composite. *Proceedings of 22nd European Symposium on Applied Thermodynamics* IVC-SEP, Denmark, 8–14.
Wesselingh, J. A., and Bollen, A. M. 1997. Multicomponent diffusivities from the free volume theory. *Transactions of the Institution of Chemical Engineers* 75A: 590–602.
Wesselingh, J. A., and Krishna, R. 2000. *Mass Transfer in Multicomponent Mixtures*. Delft University Press, Delft.
Xiaojing, X., et al. 2002. Mechanical properties and interfacial characteristics of carbon-nanotube-reinforced epoxy thin films. *Applied Physics Letters* 81: 15: 2833–2845.

16

*Graphene/Polymer Nanocomposites**

Hyunwoo Kim, Ahmed A. Abdala, and Christopher W. Macosko†

CONTENTS

* Reprinted from *Macromolecules, 43*, 6515–6530, 2010. With permission.
† Correspondence to: Christopher W. Macosko (macosko@umn.edu)

16.1 Introduction

Polymer nanocomposites based on carbon black (CB), carbon nanotubes (CNTs), and layered silicates are used to improve the mechanical, thermal, electrical, and gas barrier properties of polymers.[1–3] The discovery of graphene with its combination of extraordinary physical properties and ability to be dispersed in various polymer matrices has created a new class of polymer nanocomposites.

Graphene is an atomically thick, two-dimensional (2-D) sheet composed of sp^2 carbon atoms arranged in a honeycomb structure (Figure 16.1). It has been viewed as the building block of all other graphitic carbon allotropes of different dimensionality.[4] For example, graphite (three-dimensional [3-D] carbon allotrope) is made of graphene sheets stacked on top of each other and separated by 3.37 Å. The zero-dimensional (0-D) carbon allotrope, fullerenes (buckyballs), can be envisioned to be made by wrapping a section of graphene sheet. The one-dimensional (1-D) carbon allotropes, CNTs and nanoribbons, can be made by rolling and slicing graphene sheets, respectively. In reality, however, these carbon allotropes, with the exception of nanoribbons, are not synthesized from graphene. Graphite is a naturally occurring material with the first documented deposit[5] near Borrowdale, England, in 1555, but its first use may be dated back to 4000 years.[6] Single-walled CNT (SWCNT) was first synthesized in 1991[7] following the discovery of fullerene in 1985.[8] Although the first reported method for the production of graphene nanosheets can be traced back to 1970,[9] isolation of

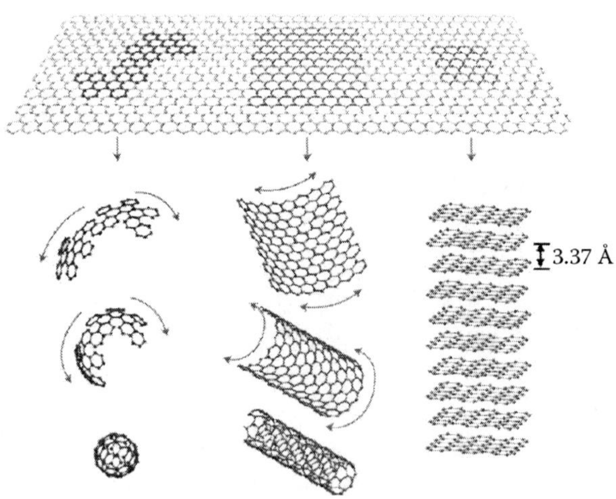

FIGURE 16.1
Graphene, the building block of all graphitic forms, can be wrapped to form the 0-D buckyballs, rolled to form the 1-D nanotubes, and stacked to form the 3-D graphite. (From Geim, A. K., and Novoselov, K. S. *Nat. Mater.*, 6, 183–191, 2007. With permission.)

free-standing single-layer graphene was first achieved in 2004 when graphene was separated from graphite using micromechanical cleavage.[10]

With Young's modulus of 1 TPa and ultimate strength of 130 GPa, single-layer graphene is the strongest material ever measured.[11] It has a thermal conductivity of 5000 W/m·fK, which corresponds to the upper bound of the highest values reported for SWCNT bundles.[12] Moreover, single-layer graphene has very high electrical conductivity, up to 6000 S/cm[13] and, unlike CNT, chirality is not a factor in its electrical conductivity. These properties in addition to extremely high surface area (theoretical limit: 2630 m^2/g) and gas impermeability[14] indicate graphene's great potential for improving mechanical, electrical, thermal, and gas barrier properties of polymers.

Owing to the great interest generated by the exceptional properties of graphene sheets and the discovery of methods for their production, researchers all over the world are studying graphene. This interest is clearly evident by the number of research publications. A simple search with *graphene* as a keyword using three of the most common databases, Institute for Scientific Information (ISI)-Web of Science, Sciencedirect, and SciFinder, as shown in Figure 16.2, indicates rapid growth of publications post 2005 with nearly 3000 publications in 2009. A similar trend in number of publications is also observed using *graphene composites* as a keyword.

In this chapter, we review this literature with a focus on graphene/polymer composites. We first review the different methods for the preparation of graphene sheets with an emphasis on the methods suitable for polymer composite applications. Then the methods to characterize graphene including number of layers, sheet size, and chemical modification are discussed. Dispersion routes

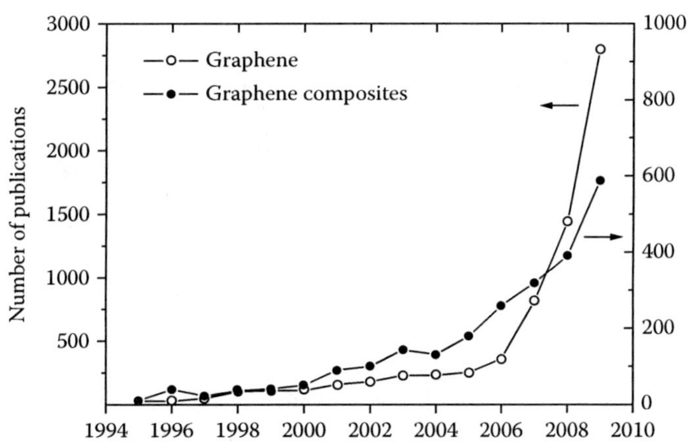

FIGURE 16.2
Number of publications returned using "graphene" and "graphene composites" as keywords in ISI-Web of Science, Sciencedirect, and SciFinder. Duplicates were removed.

for graphene into polymers and the resultant polymer/graphene composite properties are also reviewed. We conclude with challenges for future growth of this exciting new class of nanocomposites.

16.2 Bottom-Up Graphene

In bottom-up processes, graphene is synthesized by a variety of methods such as chemical vapor deposition (CVD),[15–21] arc-discharge,[22,23] epitaxial growth on SiC,[24–30] chemical conversion,[31–33] reduction of CO,[34] unzipping CNTs,[35–37] and self-assembly of surfactants.[38] CVD and epitaxial growth often produce tiny amounts of large-sized, defect-free graphene sheets. They may be more attractive than the mechanical cleavage method[10] for the production of graphene sheets for fundamental studies and electronic applications, but are not a suitable source for polymer nanocomposites that require a large amount of graphene sheets preferably with modified surface structure. The nature, average size, and thickness of the graphene sheets produced by different bottom-up methods, as well as the advantages and disadvantages of each method are summarized in Table 16.1.

TABLE 16.1

Bottom-up Processes for Graphene Production

	Typical Dimension				
Method	**Thickness**	**Lateral**	**Advantage**	**Disadvantage**	**References**
Confined self-assembly	Single layer	100s of nm	Thickness control	Existence of defects	38
CVD	Few layers	Very large (cm)	Large size; high quality	Small production scale	15–21
Arc-discharge	Single-, bi-, and few layers	Few 100 nm to a few μm	Can produce ~10 g/h of graphene	Low yield of graphene; carbonaceous impurities	22, 23
Epitaxial growth on SiC	Few layers	Up to cm size	Very large area of pure graphene	Very small scale	24–30
Unzipping of carbon nanotubes	Multiple layers	Few μm long nanoribbons	Size controlled by selection of the starting nanotubes	Expensive starting material; oxidized graphene	35–37
Reduction of CO	Multiple layers	Sub-μm	Unoxidized sheets	Contamination with α-Al_2O_3 and α-Al_2S	34

16.3 Top-Down Graphene

In top-down processes, graphene or modified graphene sheets are produced by separation/exfoliation of graphite or graphite derivatives (such as graphite oxide [GO] and graphite fluoride[39]). In general, these methods are suitable for large-scale production required for polymer composite applications. Starting from graphite or its derivatives offers significant economic advantages over the bottom-up methods; graphite is a commodity material with current annual global production of over 1.1 million tons at $825/t in 2008.[40] Therefore, the top-down approaches will be discussed in more detail. Figure 16.3 shows a block diagram, which summarizes the different routes reported for the production of graphene or modified graphene starting from graphite or GO.

Alkali metal[41] or acid[42,43]-intercalated graphite can be expanded upon heat treatment to produce a thicker (~100 nm) form of 2-D carbon known as expanded graphite (EG), which is commonly used as a filler for polymer composites. However, the layered structure of graphite is still maintained in EG.[44] Recently, a thinner form (~10 nm) of EG known as graphite nanoplatelets (GNPs) was produced by either thermal expansion of fluorinated graphite intercalation compounds[45] or microwave radiation of acid-intercalated graphite followed by pulverization using ball milling or ultrasonication.[46,47]

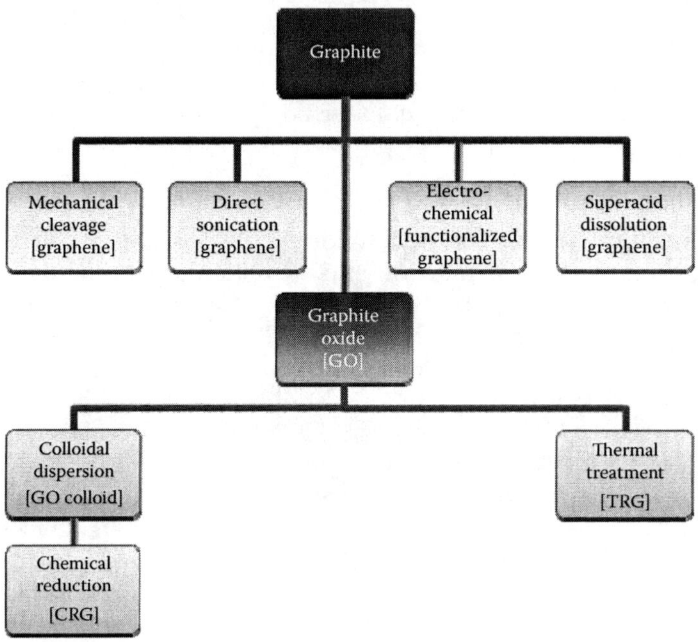

FIGURE 16.3
Top-down methods for the production of graphene and modified graphene starting from graphite or via graphite oxide (GO).

Because the large diameter and rigidity of graphite flakes are preserved in this process, even without complete exfoliation, GNP can improve electrical conductivity and mechanical properties of polymers at substantially smaller loadings than graphite or EG.[47,48] Properties of GNP-reinforced polymers will be mentioned for comparison in Section 16.6, but since the focus of this review is on single- or a few-layer graphene materials, GNP will not be discussed further.

16.3.1 Direct Exfoliation of Graphite

Micromechanical cleavage of graphite gave birth to the interest in graphene.[10] It can produce large-sized, high-quality sheets but in very limited quantities, which makes it suitable only for fundamental studies or electronic applications.[10] However, recently, graphite has also been directly exfoliated to single- and multiple-layer graphene via sonication in the presence of polyvinylpyrrolidone[49] or *N*-methyl-pyrrolidone,[50] electrochemical functionalization of graphite assisted with ionic liquids,[51] and through dissolution in super acids.[52] The direct sonication method has potential to be scaled up to produce large quantities of single- and multiple-layer graphene or functionalized graphene that can be used for composite applications. However, separation of the exfoliated graphene sheets from the bulk graphite could be a challenge. On the other hand, dissolution of graphite in chlorosulfonic acid[52] has potential for a large-scale production, but the hazardous nature of the hydrosulfonic and the cost of its removal may limit this potential. Electrochemical exfoliation methods produce graphene sheets functionalized with imidazolium groups that assist dispersion in aprotic solvents.[51]

16.3.2 Graphite Oxide

Currently, the most promising methods for large-scale production of graphene are based on the exfoliation and reduction of GO. GO was first prepared over 150 years ago by Brodie.[53] It is also produced using different variations of Staudenmaier[54] or Hummers[55] methods in which graphite is oxidized using strong oxidants such as $KMnO_4$, $KClO_3$, and $NaNO_2$ in the presence of nitric acid or its mixture with sulfuric acid. Analogous to graphite that is composed of stacks of graphene sheets, GO is composed of graphene oxide sheets stacked with an interlayer spacing between 6 and 10 Å depending on the water content.[56] The structure of graphene oxide has been the subject of theoretical[57–60] and experimental studies.[61–72] The Lerf–Klinowski model[66,67] is believed to be the most likely description of GO structure. The model describes GO as built of pristine aromatic "islands" separated from each other by aliphatic regions containing epoxide and hydroxyl groups, and double bonds as shown in Figure 16.4a. Recently, Gao et al.[61] studied the structure of GO using ^{13}C NMR. They proposed that GO contains ketones, six-membered lactol rings, and tertiary alcohol (Figure 16.4b) in addition to

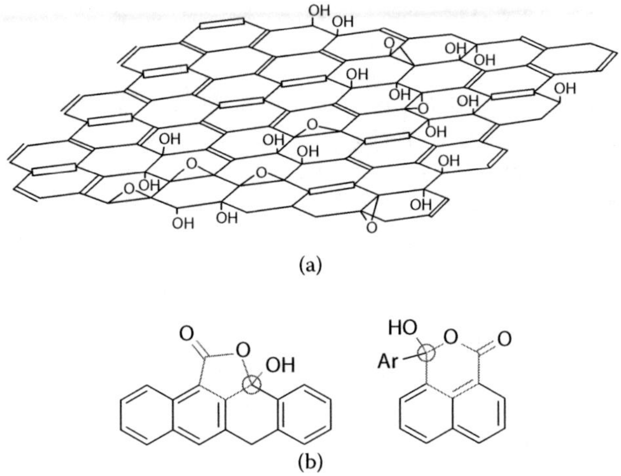

FIGURE 16.4
Structure of GO (a) consisting of aromatic islands separated by aliphatic regions containing oxygen-bonded carbons as described by the Lerf-Klinowski model (Image adapted and modified from He, H. et al., *Chem. Phys. Lett.*, 287, 53–56, 1998) and (b) ketone groups in six and five-membered ring as proposed by Gao et al.[61] The actual oxygen content of GO corresponds to C/O ratio of ~2, which is higher than what is shown in (a). This is due to the difficulty in representing bonds on the other side of the sheet and because a majority of C atoms on the edges (not shown) are O substituted.

epoxide and hydroxyl groups. GO has an approximate C/O/H atomic ratio of 2:1:0.8.[53,73] During oxidation, graphene oxide sheets undergo unzipping resulting in size reduction compared to the parent graphite flake size.[74] For more details about GO, we refer the reader to the extensive review of GO preparation, structure, and reactivity by Dreyer et al.[75]

Exfoliation of GO to produce chemically modified graphene sheets provides different routes for large-scale production of functionalized graphene sheets. Although GO can be readily dispersed in water[76] and in organic solvents after chemical modification, graphene oxide is electrically insulating and thermally unstable. Therefore, at least partial reduction of graphene oxide is necessary to restore electrical conductivity. A number of different methods currently exist for the exfoliation and reduction of GO to produce chemically modified graphene. The term "chemically modified" is chosen because complete reduction of graphene oxide to graphene has not yet been observed. These methods are described in Sections 16.3.3 and 16.3.4. For more details about these routes, we refer the reader to Park and Ruoff's recent review.[56]

16.3.3 Chemical Reduction of GO

In these methods, a stable colloidal dispersion of GO is produced followed by chemical reduction of the exfoliated graphene oxide sheets. Stable colloids of graphene oxide can be obtained using solvents such as water, alcohol, and other protic solvents combined with either sonication or long stirring. Alternatively, GO can be exfoliated in polar aprotic solvents by reacting with organic compounds such as isocyanate[77] and octadecylamine[78] or treating with surfactants.[76,78,79] Although these suspensions can be used for the production of GO/polymer composites, the low electrical conductivity and poor thermal stability of graphene oxide are significant drawbacks.

Colloidal graphene oxide or the organically treated version can be chemically reduced producing chemically reduced graphene (CRG) using hydrazine,[77,79,80] dimethylhydrazine,[81] sodium borohydride followed by hydrazine,[82] hydroquinone,[83] and UV-irradiated TiO_2.[84] Stankovich et al. proposed the following mechanism for the reduction of graphene oxide using hydrazine[77]:

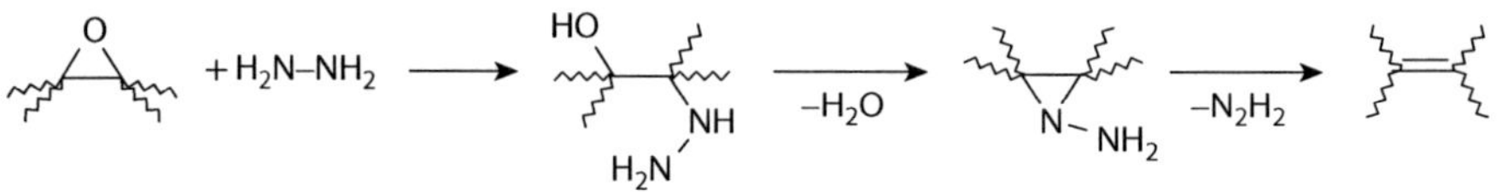

The reduction of graphene oxide restores electrical conductivity. However, significant oxygen content remains: C/O ~ 10/1.[76]

Although chemical reduction of graphene oxide provides an efficient route for the production of CRG, the hazardous nature and cost of the chemicals used in reduction may limit its application. An alternative chemical reduction is the dehydration of the hydroxyl groups on graphene oxide in water[85,86] at high pressure and temperature 120°C–200°C (Figure 16.5a). Aluminum powder appears to catalyze this process in an acidic condition.[87]

16.3.4 Thermal Exfoliation and Reduction

Thermally reduced graphene (TRG) oxide can be produced by rapid heating of dry GO under inert gas and high temperature.[90–92] Heating GO in an inert environment at 1000°C for 30 seconds leads to the reduction and exfoliation of GO, producing TRG sheets. Exfoliation takes place when the pressure generated by the gas (CO_2) evolved due to the decomposition of the epoxy and hydroxyl sites of GO exceeds van der Waals forces holding the graphene oxide sheets together. About 30% weight loss is associated with the decomposition of the oxygen groups and evaporation of water.[89] The exfoliation leads to volume expansion of 100–300 times producing very low-bulk density TRG sheets (Figure 16.5d). Owing to the structural defects caused by the loss of CO_2, these sheets are highly wrinkled as shown in Figure 16.5e.[89,90] As high as 80% of the TRG sheets are single layers with an average size of about 500 nm independent of the starting GO size.[90] The advantage of the thermal reduction methods is the ability to produce chemically modified graphene sheets without the need for dispersion in a solvent. TRG has C/O ratio of about 10:1 compared to 2:1 for GO.[90] This ratio has been increased up to 660:1 through

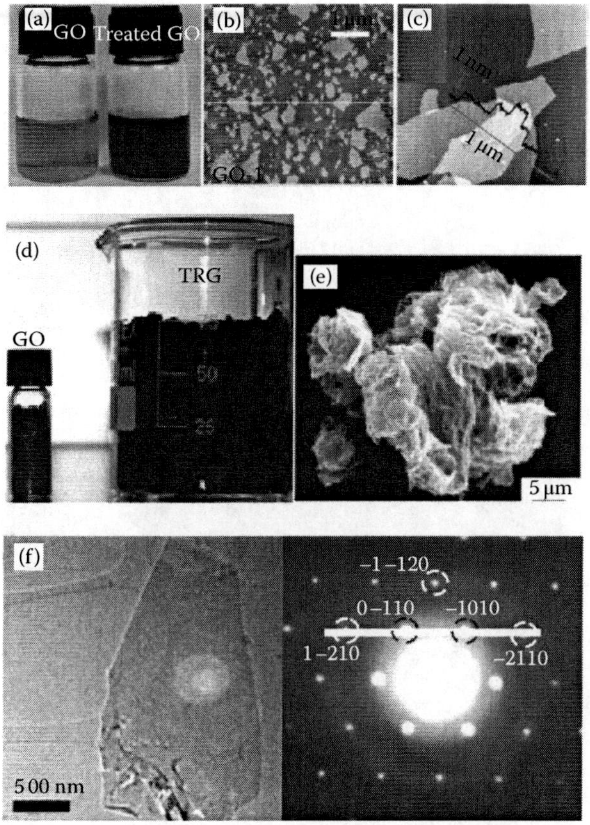

FIGURE 16.5
(a) GO suspension (0.5 mg/mL) before and after hydrothermal treatment at 180°C;[85] (b) and (c) noncontact mode AFM images of GO on mica;[81,88] (d) 0.5 g of GO expands to 75 mL of TRG upon rapid heating to ~1000°C; (e) SEM of TRG suggests a structure-like crumpled sheets of paper;[89] (f) TEM image of a single-layer graphene with an electron diffraction pattern with Miller–Bravais indices.[50] (All images were reproduced with permission.)

heat treatment at higher temperature (1500°C) or for longer time.[93] TRG sheets have high surface area, 1700 m^2/g, as measured in methylene blue and can be well dispersed in organic solvents such as *N,N*-dimethylformamide (DMF) and tetrahydrofuran. The thermal reduction also leads to restoration of the electrical conductivity with reported electrical conductivity of a compacted film with a density of 0.3 g/cm^3 ranging between 10 and 20 S/cm,[89] compared to 6000 S/cm for defect-free single graphene sheets.[13]

The nature, average size, and thickness of the graphene sheets produced by different top-down methods as well as the advantages and disadvantages of each method are summarized in Table 16.2. As indicated in the earlier discussion and in Table 16.2, the most promising routes to the preparation of graphene for polymer nanocomposites start from GO. Thus, in the rest of this review, we will concentrate on these routes.

TABLE 16.2

Top-down Processes for the Production of Graphene, Chemically Reduced (CRG) and Thermally Reduced Graphene Oxide (TRG)

Method	Typical Dimension: Thickness	Typical Dimension: Lateral	Advantage	Disadvantage	References
Directly from Graphite					
Micromechanical exfoliation	Few layers	μm to cm	Large-sized and unmodified graphene sheets	Very small scale production	10
Direct sonication of graphite	Single and multiple layers	μm or sub-μm	Unmodified graphene; inexpensive	Low yield; separation	49,50
Electrochemical exfoliation/functionalization of graphite	Single and few layers	500–700 nm	Single-step functionalization and exfoliation; high electrical conductivity of the functionalized graphene	Cost of ionic liquids	51
Superacid dissolution of graphite	Mostly single layer	300–900 nm	Unmodified graphene; scalable	Use of hazardous chlorosulfonic acid; cost of acid removal	52
From Graphite Derivatives (Graphite Oxide [GO] or Graphite Fluoride)					
Li alkylation of graphite fluoride	Single layer	μm	Large size; functionalized sheets; no oxygen functionality	Cost of the starting material; restacking after annealing	39
Chemical reduction of colloidal GO in water	Single and multiple-layer CRG	μm or sub-μm size	Large sheet size; some routes use only water	Some of these methods use hazardous chemicals; only dispersed in hydrophilic polymers	77, 80, 82–87
Chemical reduction of organically treated GO	Mostly single-layer CRG	Few 100 nm to a μm	Colloidal stability in organic solvents; better exfoliation	Low thermal stability; in situ chemical reduction degrades some polymers	77–79
Thermal exfoliation/reduction of GO	Single- and few-layer TRG	~500 nm	One-step exfoliation/reduction; short heating time; dry basis	High heating temperature; smaller sheet size compared to chemically reduced sheets	92, 89

16.4 Characterization of Graphene

It is important to verify that the synthesis methods described earlier do in fact produce graphene single sheets. Moreover, the size of these sheets and attached functional groups are important for dispersion in polymers. In this section, we briefly review the techniques best suited for the characterization of graphene sheets.

16.4.1 Layer Number and Size

X-ray diffraction is used to demonstrate that graphite has been intercalated. For example, the sharp reflection at $2\theta = 26.3°$ (CuKα radiation, x-ray wavelength = 0.154 nm) in graphite shifts to 14.1–14.9° in GO.[64] However, x-ray diffraction disappears as the sheets of GO exfoliate into single sheets.[89,90]

Although indirect, surface area has been used as an indicator of exfoliation. Since theoretically the specific surface area is inversely proportional to the thickness of disk-like particles (~2/density/thickness), well-exfoliated sheets will have higher surface area. Surface area can be determined by N_2 or methylene blue adsorption.[89,90] However, Schniepp et al.[89] noted that N_2 adsorption measurements were highly dependent on the compressibility of TRG. Although methylene blue adsorption is done on exfoliated sheets in solution and thus avoids compressibility issues, surface topology and chemistry may influence the area occupied by each methylene blue molecule.

Atomic force microscopy (AFM) imaging provides more reliable measures of sheet dimensions. Contact or tapping mode AFM can be used to probe surface topology, defects, and bending properties.[89,94] Lateral size and layer thickness of particles lying on substrates can be determined from the steps in height scans, as illustrated in Figure 16.5b and c. Folded[94] or wrinkled sheets[72,89] as well as adsorbed solvents or moisture[77,94] can complicate measurements. Scanning electron microscopy (SEM) can give qualitative insight into the 3-D structure of graphene sheets, as illustrated in Figure 16.5e.[90]

Figure 16.5f shows a transmission electron microscopy (TEM) image of a single graphene sheet on a holey carbon-covered copper grid.[50] In addition to size determination by TEM imaging, electron diffraction patterns can clearly differentiate single- from bilayer sheets.[96] High-resolution TEM (HR-TEM) can identify atomic bonds on functionalized sheets (C–OH vs. C–O–C) and atomistic defects.[72] HR-TEM has also confirmed the existence of aliphatic islands containing oxygen-bonded carbons as proposed in Figure 16.4a.

Size and morphology of platelets can be estimated indirectly in dilute solution. Viscosity measurements of suspensions in the dilute limit do not have orientation problems and can give particle aspect ratio. Pasquali and coworkers[97] have used intrinsic viscosity measurements to evaluate the average length of surfactant-stabilized SWCNT in water assuming rigid rods, but, as yet, intrinsic viscosity has not been applied to characterize graphene. Static light scattering from dilute GO in water can give fractal dimensions d_f of GO. One study showed nearly flat sheets ($d_f = 2.15$),[98] while another reported GO as crumpled membranes (2.54).[99] However, the results were sensitive to solvent polarity: GO collapsed to a compact structure upon addition of acetone leading to higher fractal dimension.[98,99]

16.4.2 Identifying Chemical Modification

As indicated in Table 16.2, the routes most suitable for producing graphene sheets in large quantities start from GO and thus all have some remaining oxygen. The overall degree of oxidation can be quantified by standard elemental analysis. X-ray photoelectron spectroscopy (XPS) can quantify the amount of oxygen on the surface and also indentify the types of carbon oxygen bonds. Chemical shifts in XPS C_{1s} spectra can be evidence for the existence of C–O, C=O, or O–C=O on GO and its derivatives,[76,77,100] but are limited in quantifying their relative amounts. Infrared absorption[101–103] has similar limitations. ^{13}C NMR may be the most direct method to distinguish oxygen functional groups.[61,64,66] However, since the NMR-active ^{13}C allotrope occurs naturally at only ~1%, signal-to-noise ratio is low. By CVD of ^{13}C-labeled graphene on nickel substrates and subsequent oxidation, Ruoff and coworkers[63] enhanced the signal and identified chemical groups and their connectivity.

Raman spectroscopy can quantify the transformation of sp^3-hybridized carbons back to sp^2 on reduction of GO[77,104] and the presence of disordered stacking in graphite samples.[105] The transformation of sp^3 to sp^2 restores electrical conductivity, thus conductivity is also a valuable qualitative measure of the conversion of graphene oxide to graphene.[77,100,106,107]

16.5 Dispersion of Graphene into Polymers

The properties of polymer nanocomposites depend strongly on how well they are dispersed. Much of nanocomposite research with CNTs has focused on finding better methods for dispersing nanotubes into polymers.[2] Surface functionalization via fluorination,[108] acid modification,[108] and radical addition[110] improve solubility of CNT in solvents and polymers. However, disentangling the bundles during dispersion into polymers cannot be done easily and sonication often shortens the tubes. The synthesis of graphene from graphene oxide leaves some epoxide and hydroxyl groups; these greatly facilitate functionalization.[73,78,79,81] Since graphene oxide and CRG are flat sheets, entangled bundles are not an issue. However, restacking of the flat sheets, especially after chemical reduction, can significantly reduce their effectiveness. Restacking can be prevented by either use of surfactants that can stabilize the reduced particle suspensions[76] or blending with polymers prior to chemical reduction.[81]

GO readily exfoliates in water or other protic solvents via hydrogen bonding interaction.[98,99] Nanocomposites have been created with GO and water-soluble polymers such as poly(ethylene oxide) (PEO)[111] or poly(vinyl alcohol) (PVA).[112] Using GO after chemical modification with isocyanate or amine, composites have also been produced in aprotic solvents with hydrophobic polymers such as polystyrene (PS),[81] polyurethane (PU)[112,114], or poly(methyl methacrylate) (PMMA).[115] As discussed in Section 16.3.3, electrical conductivity can be restored via chemical reduction of the graphene oxide. This can also be done in situ in the presence of a polymer. For example, Stankovich et al. added sulfonated PS and then reduced graphene oxide with hydrazine hydrate.[76] Without the sulfonated PS, the reduced sheets rapidly aggregated. However, depending on the polymer type and the reducing agent, this in situ reduction technique may result in polymer degradation.

Composites of CRG and TRG have been made with a number of polymers via blending with organic solvents followed by solvent removal. Unlike most chemically modified GO that retain some layered structure from GO, thermal expansion of GO (TRG) leads to nearly complete exfoliation.[89,90] Therefore, dispersion of TRG can be easier while stacked layers of CRG have to be exfoliated by applying mechanical stress and via inter-gallery polymer diffusion in solvents.[113] Due to its wrinkled structure, TRG may experience less restacking after solvent removal than the flatter CRG sheets.

Graphene composites can be produced via in situ intercalative polymerization of monomers. Successful polymerization of PVA,[116] PMMA,[117] epoxy[118], and poly(arylene disulfide)[119] with graphene oxide, or silicone foams[120] and PU[113,121] with TRG has been reported. Especially for poly(arylene disulfide), graphene oxide was used as an oxidation agent, which converts thiol salts to disulfide. However, so far monomers have only been polymerized in solvents. The high viscosity of even dilute dispersion of graphene makes bulk-phase polymerization difficult. If functional groups on the chemically modified graphene are reactive with the monomer, grafting of polymer chains onto graphene surfaces can occur. Chain grafting has been demonstrated with the polymerization of poly(2-(dimethylamino)ethyl methacrylate)[122] and PVA,[123] and with PU formation.[113]

The most economically attractive and scalable method for dispersing nanoparticles into polymers is melt blending. However, due to thermal instability of most chemically modified graphene, use of melt blending for graphene has so far been limited to a few studies with the thermally stable TRG. Successful melt compounding of TRG into elastomers[113,124] and glassy polymers[125,126] has been reported. In the few direct comparisons between solvent and melt blending,[113] solvent blending produces better dispersion (Tables 16.3 through 16.5). Another challenge for melt compounding is the low bulk density of graphene such as TRG, which makes feeding into melt mixers difficult. Torkelson and coworkers[127] have attempted to bypass all graphene synthesis steps by exfoliating graphite directly into polypropylene (PP) using the very high stresses generated in the solid

TABLE 16.3

Electrically Conductive Graphene/Polymer Nanocomposites

Polymer	Graphene Type	Processing	Electrical Percolation Threshold (vol%)	References
Chemically Modified				
PS	iGO[a]	Solvent blending + hydrazine	0.1	81
Vinyl chloride/ vinyl acetate copolymer	GO	Solvent blending + hydrazine	0.15	128
PS	Electrochemically exfoliated graphite	Solvent blending + ionic liquid	0.13–0.37	51
Epoxy	Partially reduced GO	In situ polymerization at 250°C	0.52	129
PMMA	GO (17 mol% O)	In situ polymerization	(1.3)[b]	117
PMMA	GO (12 mol% O)	In situ polymerization	(1.6)	130
Thermally Exfoliated				
TPU	TRG ~800 m²/g 4 mol% oxygen	Solvent blending	0.3	113
		In situ polymerization	0.4	
		Melt compounding	0.8	
PEN		Melt compounding	0.5	126
PC		Melt compounding	0.6	125
Natural rubber	>600 m²/g 5 mol% oxygen	Melt or solvent blending + vulcanization	(0.8, solvent blend) (>2.0, melt blend)	124
PS-PI-PS		Melt or solvent blending	(0.6)	
PDMS		Oligomer blending + polymerization	(0.6)	
TPU	Apparent specific volume: 410 cm³/g 5 mol% oxygen	Solvent blending	(1.0)	131
TPU		Solvent blending	(1.0)	132
TPU		In situ polymerization	(1.6)	121
PVDF	—	Solvent blending	(1.6)	133
SAN	600–950 m²/g up to 14 mol% oxygen	Preblending using solvents, followed by melt compounding	(1.9)	91
PC			(1.3)	
PP			(2.0)	
PA6			(3.8)	

PS-PI-PS, poly(styrene-co-isoprene-co-styrene) triblock copolymer; PDMS, polydimethylsiloxane; SAN, poly(styrene-ran-acrylonitrile); PA6, polyamide 6.

[a] Isocyanate-treated GO.

[b] Values in parenthesis are percolation volume fraction converted from the weight fraction reported using the density of the polymer and graphite (2.28 g/cm³).

TABLE 16.4

Mechanical Properties of Graphene/Polymer Nanocomposites

Polymer	Reinforcements	Processing	E_{matrix} (MPa)	Graphene Concentration (vol%)	Modulus Increase (%)	Tensile Strength Increase (%)	Ultimate Strain Increase (%)	References
PVA	GO	Solvent	2100	2.5	128	70	32	134
PVA	GO	Solvent	2130	(0.49)[a]	62	76	−70	135
PMMA	GO	In situ polymerization	520[b]	(1.7)	54[b]	N/A		117
PCL	GO	Solvent	340	(2.4)	108	36	−90	136
PCL	GO	Solvent	260	(0.46)	50	N/A		137
Epoxy	TRG	In situ polymerization	2850	(0.05)	31	40	N/A	138
PEN	TRG	Melt	2350	2.4	57	N/A		126
PC	TRG	Melt	2080	1.3	25	N/A		125
PMMA	TRG	Solvent	2100	(0.005, 0.5)	33, 80	N/A		139
PVDF	TRG	Solvent	1280	(3.1)	92	N/A		133
SAN	TRG	Solvent + melt	2350	(2.3)	34	N/A	−58	91
PC			1480	(2.5)	52		−98	
PP			980	(1.9)	43		−99	
PA6			1650	(2.4)	32		−94	
Natural rubber	TRG	Solvent/melt	1.3	(1.2)	750	N/A		124
PDMS		In situ polymerization	0.6	(2.2)	1100			
Styrene-butadiene rubber		–	10	(0.8)	390			
TPU	TRG	Solvent	458	(1.5)	43	−23	−15	132

(Continued)

TABLE 16.4 (*Continued*)

Mechanical Properties of Graphene/Polymer Nanocomposites

Polymer	Reinforcements	Processing	E_{matrix} (MPa)	Graphene Concentration (vol%)	Modulus Increase (%)	Tensile Strength Increase (%)	Ultimate Strain Increase (%)	References
Silicone foam	TRG	In situ polymerization	250[c]	(0.12)	200[c]	N/A		120
PVA	Acid functionalized TRG	Solvent	660	(0.4)	35	N/A		115
PMMA	Amine-treated, acid-functionalized TRG		2120	(0.3)	70			
TPU	TRG	Melt	6.1–7.1	1.6	250	N/A		113
		Solvent		1.6	680			
		In situ polymerization		1.5	210			
	iGO	Solvent		1.6	490–900			
PS	PS-functionalized, chemically reduced GO	Solvent	1450	(0.4)	57	N/A		140
TPU	Chemically reduced sulfonated-graphene	Solvent	9.8	(0.5)	120	75	N/A	114
TPU	GO	Solvent	6	(2.4)	900	−19	−60	141
PAN	Exfoliation of alkali intercalated graphite	Electrospinning, solvent	2450	(2.1)	100	N/A		142

[a] Values in parenthesis are percolation volume fraction estimated from the weight fraction reported using the density of the polymer and graphite (2.28 g/cm^3).

[b] Measured at 60°C.

[c] Compressive modulus, normalized by relative polymer density.

TABLE 16.5
Gas Permeability of Graphene/Polymer Nanocomposites

Polymer	Filler	Processing	Permeant	Relative Reduction (%)	Graphene loading (vol%)	References
PEN	TRG	Melt	Hydrogen	44	1.8	126
PC	TRG	Melt	Helium	32	1.6	125
			Nitrogen	39	1.6	
TPU	TRG	Melt	Nitrogen	52	1.6	113
		Solvent	Nitrogen	81	1.6	
		In situ polymerization	Nitrogen	71	1.5	
	iGO	Solvent	Nitrogen	94–99	1.6	
	GO	In situ polymerization	Nitrogen	62	1.5	
Natural rubber	TRG	Melt/solvent/oligomer polymerization	Air	60	1.7	124
PS-PI-PS				~80	2.2	
PDMS				~80	2.2	

state shear pulverization process. Their x-ray diffraction and TEM data indicate, however, that the resulting composite is primarily small stacks of graphite.

16.5.1 Quantifying Dispersion

Most studies have relied on improved physical properties to demonstrate that their graphene is well dispersed in the polymer matrix. These indirect methods will be reviewed in Section 16.5.2. However, TEM can give direct images of dispersion and has been widely used to visualize layered silicates in polymers.[143,144] Graphite can be imaged in TEM without staining[91,113,126] although it has a lower atomic number contrast with polymers than layered silicates. The smaller thickness of isolated graphene sheets (~0.34 nm) makes them more difficult to be resolved in TEM micrographs. Moreover, functionalization and defects formed during thermal exfoliation distort flat graphene into highly wrinkled sheets. In an extensive study of TEM images of layered silicates/polymer nanocomposites, Fornes and Paul[144] discuss other difficulties in using TEM to quantify the dispersion of thin plate-like fillers.

Despite these difficulties, there has been at least one attempt to quantify graphene dispersion embedded in polymers directly with TEM. Figure 16.6 highlights the difference between graphite and TRG dispersed in poly(ethylene naphthalate) (PEN).[126] The wrinkled nature of TRG can be seen in some of the sheets in Figure 16.6b. A measure of dispersion is the aspect ratio of the particles, A_f = (sheet length)/thickness. Using similar micrographs, sheet lengths were evaluated by drawing straight contour lines, as shown in Figure 16.6b. Thickness was determined from full-width at half maximum of a linear intensity profile crossing each particle. Values for ~100 sheets are plotted in Figure 16.6c. Despite the possibility of missing thinner layers, the distribution of these aspect ratios is a useful measure of TRG dispersion.

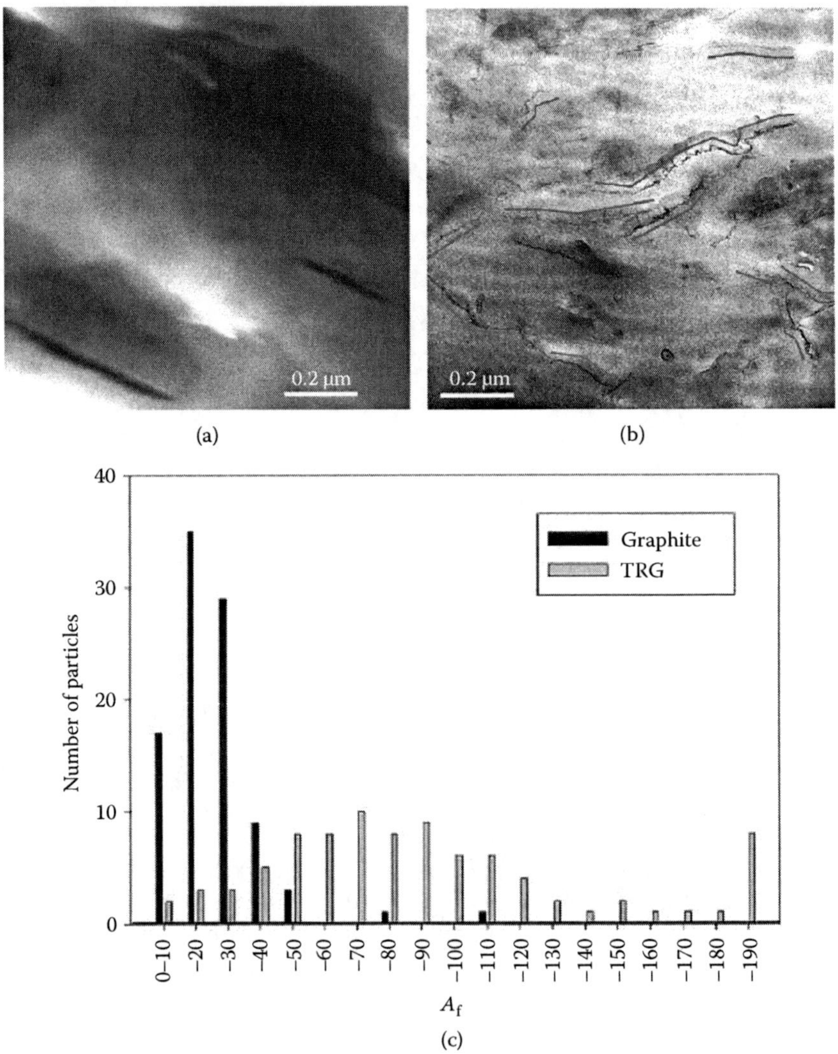

FIGURE 16.6
Comparison of the dispersion of graphite and TRG at 3 wt% in PEN by TEM: (a) graphite and (b) TRG. Solid lines indicate lengths used to evaluate aspect ratio, A_f. (c) Aspect ratio distribution from the measurements on several micrographs. Average A_f = 21 (graphite) and 88 (TRG). (From Wakabayashi, K. et al. *Macromolecules*, *41*, 1905–1908, 2008. With permission.)

16.5.2 Rheology

Rheology can be an effective tool for quantifying nanocomposite dispersion.[145] It averages over many particles and is also useful in its own right for predicting processability. An experiment that can yield A_f is the onset of network formation. Figure 16.7 shows the shear storage modulus, G', from small strain oscillatory shear versus frequency tests for a series of concentrations of graphite and TRG dispersed in PEN at 290°C.[126] At about 5 wt% graphite and 1 wt% TRG, G' becomes independent of frequency at low frequency, a signature of solid-like network formation. In this rigidity percolation regime, the concentration ϕ dependence of the elasticity of particle suspensions can be described by power law scaling[145,146]

$$G' \propto (\phi - \phi_{\text{perc}})^{\nu} \tag{16.1}$$

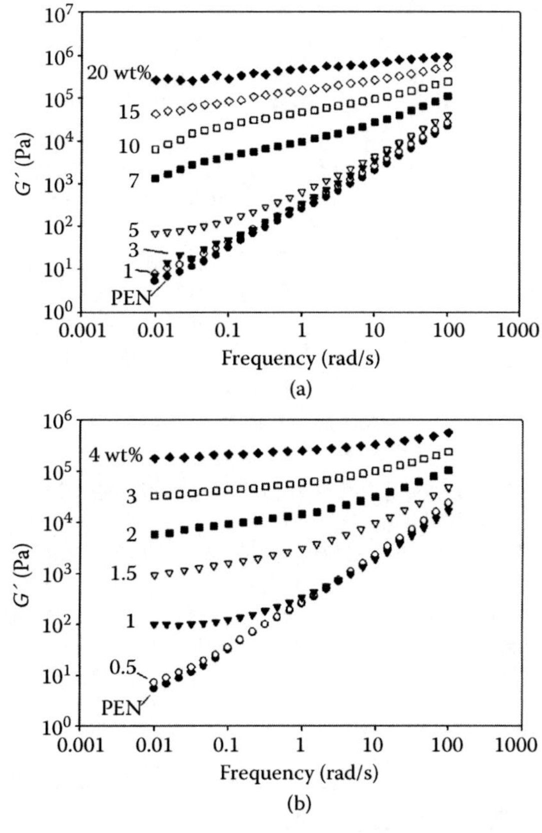

FIGURE 16.7
Dynamic frequency swe\eps of PEN containing (a) graphite and (b) TRG at 290°C. (Adapted from Kim, H., and Macosko, C. W., *Polymer*, 50, 3797–3809, 2009.)

where ϕ_{perc} is the percolation threshold and ν is a power-law exponent. ϕ_{perc} of graphene can be determined experimentally from a plot of low-frequency G' versus $\phi - \phi_{perc}$. Compared with spheres, particles with shape anisotropy (disks and rods) percolate at smaller volume fraction.[147,148] In extreme oblate/prolate limits, the percolation threshold is inversely proportional to the particle aspect ratio. With a few simplifying assumptions (randomly oriented, monodispersed, and disk-shaped particles), Ren et al.[149] identified a proportionality between A_f and ϕ_{perc}:

$$A_f = \frac{3\phi_{sphere}}{2\phi_{perc}} \tag{16.2}$$

where $\phi_{sphere} = 0.29$, the onset of percolation of interpenetrating, randomly packed spheres.[150] Using the data of Figure 16.7 to determine ϕ_{perc}, Equation 16.2 gives $A_f = 18$ for graphite, in good agreement with the TEM results in Figure 16.6c. For TRG, $A_f = 160$, considerably higher than the value determined from the electron micrographs. At least part of this difference may be due to TEM missing the thinnest sheets as discussed previously. The percolation threshold for electrical conductivity (see Table 16.3) can also be used to quantify dispersion through A_f. From the electrical percolation values, $A_f \sim 13$ for graphite in PEN and ~100 for TRG. The higher values of A_f (160) calculated using G' data may indicate that polymer chains can bridge between particles causing network percolation at a lower concentration than conductivity percolation.[151]

However, there are several limitations in using rheological measurements to quantify graphene dispersion. Rheological responses are sensitive to the orientation of anisotropic particles.[151,152] Typical analyses assume random particle orientation, which can be far from the fact since processing and even loading into a rheometer induce particle alignment, especially for high aspect ratio particles concentrated in viscous polymer matrices. For example, 1 wt% TRG was dispersed in a polycarbonate (PC) melt matrix of viscosity 4000 Pa·s.[126] Its elastic shear modulus showed no sign of reaching equilibrium even 10 hours after loading the sample into the rheometer at 230°C. Brownian randomization was extremely slow under typical experimental conditions.[153] Similar temporal changes in melt viscoelasticity and also electrical conductivity through annealing were reported from layered silicates,[153] CNT,[154] and carbon nanofiber (CNF)[155] composites. Although meaningful results can be obtained if time after loading is held constant, sufficient thermal stability of the matrix is required for prolonged rheological tests.

16.6 Properties of Graphene/Polymer Nanocomposites

As reviewed in the previous sections, exfoliated carbon sheets obtained from graphene oxide via either chemical reduction or rapid pyrolysis can be dispersed in polymers to modify their physical properties. In this section, we present electrical, thermal, mechanical, and gas barrier properties of graphene/polymer nanocomposites.

16.6.1 Electrical Conductivity

Graphene sheets can provide percolated pathways for electron transfer making the composites electrically conductive. Similar benefits can be achieved with other conductive carbon fillers such as CB, CNF, and EG. However, graphene enables the insulator to conductor transition at significantly lower loading,[91] comparable to electrical percolation thresholds for CNTs. When particles possess the same aspect ratio, theory[147] predicts that prolates (rod-like) percolate at one-half the volume fraction of oblates (disk-like). However, since real composite morphology is rather complex (particle interaction, flexibility, and entanglement), this does not apply in reality: rod-like CNT may not necessarily percolate at lower concentration than disk-like graphene. Particle orientation also plays an important role: the percolation threshold becomes greater as particles are aligned parallel.[156,157]

Production of electrically conductive polyolefin,[91] vinyl[51,81,91,116,128,133] and acrylic[117,130] polymers, polyester,[91,125,126] polyamide (PA),[91] PU,[113,131,132] epoxy,[129] natural and synthetic rubbers[124] with graphene has been reported. These materials can be used, for example, for electromagnetic shielding,[129,158] anti-static coating, and conductive paints.[159] Table 16.3 shows the wide range of electrically conductive polymer/graphene composites, their production methods, and minimum filler volume fraction for electrical conduction. When electrical percolation thresholds were provided only in weight fraction, they were converted to volume fraction (values in parenthesis in Tables 16.3 through 16.5) using density of the polymer[160] and graphite, 2.28 g/cm^3.[126]

The lowest electrical percolation threshold was 0.1 vol% reported by Stankovich et al.[81] for PS solvent blended with isocyanate-treated GO (iGO)[94] followed by solution-phase reduction with dimethylhydrazine. This onset of percolation is comparable with those of SWCNT or multiwalled carbon nanotubes (MWCNTs).[2] Also, electrochemical exfoliation of graphite in ionic liquids is reported to produce graphitic layers that can percolate at 0.13–0.37 vol% in PS.[51]

As mentioned in Sections 16.3 and 16.5, TRG[89] retains high conductivity without additional reduction steps and can be melt processed due to its thermal stability. However, for thermoplastic polyurethane (TPU),[113] films cast from a DMF solution still showed better dispersion of these carbon sheets and required lower loading for electrical conduction (0.3 vol%) than films compression molded after melt compounding (0.8 vol%). The dispersion state of TRG and electrical conductivity of the composites may also vary depending on thermal exfoliation conditions and compatibility with the polymer matrix. Steurer et al.[91] found percolation thresholds of TRG ranged from 1.3 to 3.8 vol% for different polymers even though processed using similar procedures (also see Table 16.3). Kim and Macosko[125] reported that electrical percolation of TRG takes place in PC at ~0.6 vol%, notably lower than Steurer. This difference may be due to the fact that Steurer et al. used relatively milder conditions (~600°C) for the rapid pyrolysis of GO. Orientation of anisotropic graphitic disks[125] also plays an important role, which can be induced by flow during processing: compression-molded TRG/PC composites are less conductive than samples annealed for long time.

It should be noted that Drzal and coworkers report that the electrical percolation of PP can be achieved at as low as 0.1–0.3 vol% with GNP derived from thermal expansion of acid-intercalated graphite.[47,48] GNP is not graphene; it still maintains the multilayer structure of graphite and relatively low specific surface area (~100 m^2/g).[161] However, its large particle aspect ratio (100–1500) and high rigidity of stacked sheets prevent particle buckling, which might reduce the amount of particles necessary for electrical percolation. Surprisingly, for GNP in linear low-density polyethylene (PE), a polymer with very similar polarity to PP, electrical percolation was reported to be significantly higher, that is, 12–15 wt%.[162] In the PP study, polymer pellets were coated with GNP from solvents prior to blending.[47] This strategy may break particle aggregates and improve the dispersion quality.

16.6.2 Thermal Conductivity

Superior thermal transport properties of graphene dispersion have potential for thermal management in miniaturized electronic devices,[118,163,164] for thermal pastes,[165] and for heat-actuated, shape-memory polymers.[114] It has been reported that 2-D, platelet-like GNP can improve thermal conductivity more effectively than 1-D, rod-like CNT or CNF.[166–169] However, unlike the exponential increase in electrical conductivity, thermal conductivity enhancement by the carbon nanofillers is not as dramatic,[168] but even lower than expected from effective medium theory.[170] This is partly due to smaller contrast in thermal conductivity between polymers (0.1–1 W/m·K) and graphitic carbons (MWCNT[171] 3000 and graphene[12] 5000 W/m·K) compared to the contrast in electrical conductivity (graphene[13] 6000 and nonconducting polymers 10^{-18}–10^{-13} S/cm). Moreover, since thermal energy is transferred mainly in the form of lattice vibration (phonons), poor coupling in vibration modes at the filler–polymer and filler–filler[172] interfaces will impart significant thermal resistance (Kapitza resistance).[173] Huxtable et al. estimated that the thermal resistance at CNT–polymer interface is equivalent to that of 20-nm-thick polymer layers.[170] Matching surface chemistries[174] or covalent interface coupling[175] can be pursued to minimize the interfacial phonon scattering. However, excessive functionalization also tends to reduce the intrinsic thermal conductivity of carbon materials.[176]

Improving thermal conductivity using GNP was demonstrated for epoxy,[169,177,178] PP,[161] PE,[179] PA,[179] and paraffin wax.[180] For epoxy, up to 30-fold increase in thermal conductivity was attained by incorporating 33[177] or 25 vol%[169] of a few nanometer thick graphite platelets via solution polymerization. However, thermal conductivity increase for PP, PE, and PA composites via melt compounding with similar amount of GNP was less significant (less than 15-fold increase).[161,179] Degree of exfoliation, orientation, and interfacial interaction have an influence on the thermal transport in composites. Haddon and coworkers controlled A_f of GNP through choosing different thermal expansion temperatures and reported higher thermal conductivity increase for higher A_f of reinforcing platelets.[169] When the platelets are oriented by extrusion or solvent casting, conductivity was higher in the direction of graphite alignment than perpendicular implying macroscopic anisotropy.[167,177] Silane modification of EG[178] was used to enhance the conductivity gain for an epoxy; covalent bonding between graphite surface and polymer matrix may reduce the acoustic phonon scattering at the interface. For graphene oxide, strong interaction of surface –OH and –O– groups with epoxy resins enabled fourfold increase in thermal conductivity at 5 wt%,[118] which can be compared with the performance of SWCNT. Also, TRG increased thermal conductivity of silicone foams[120] by 6% at 0.25 wt% although part of the conductivity increase may be attributed to the foam density increase (higher solid content) by graphene addition.

16.6.3 Mechanical Properties

Defect-free graphene is the stiffest material ($E \sim 1$ TPa) ever reported in nature and also has superior intrinsic strength, ~130 GPa.[11] Despite some structural distortion, the measured elastic modulus of CRG sheets is still as high as 0.25 TPa.[181] Advantages of graphene in mechanical reinforcement over existing carbon fillers such as CB, EG, and SWCNT have also been discussed.[91,124,138,139]

Table 16.4 provides a summary of mechanical properties of graphene-reinforced polymer nanocomposites. Although modulus increase with graphene dispersion is evident for all polymers, it is more pronounced for elastomeric matrices. This is due to greater stiffness contrast between the reinforcement and the matrix. This difference is more clearly illustrated in Figure 16.8, where we have plotted relative modulus E gains normalized by filler volume fraction, $(E/E_{matrix} - 1)/\phi$ versus neat matrix modulus, E_{matrix}. For glassy polymers, the most surprising results were found from PMMA (33% improvement at only 0.01 wt%)[139] and epoxy (31% increase at 0.1 wt%)[138] reinforced with TRG. Ramanathan et al.[139] and Rafiee et al.[138] attributed this to the strong hydrogen bonding interaction of oxygen-functionalized TRG and the mechanical interlocking at the wrinkled surface that may restrict segmental mobility of polymer chains near TRG surfaces. Still, the increase in PMMA modulus with only 0.01 wt% of TRG is unrealistically high. This enhancement even exceeds the Voigt upper bound prediction (the limit of perfect alignment and infinite particle aspect ratio A_f). To achieve the reported modulus, in addition to being perfectly aligned and infinitely long, the TRG sheets would need to increase the modulus of PMMA to 1 TPa for at least 5 nm on either side of each sheet.

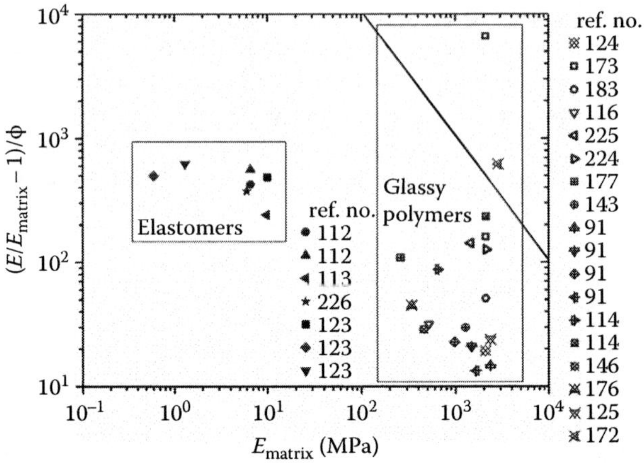

FIGURE 16.8
Relative modulus increase for polymer/graphene nanocomposites normalized by volume fraction of graphene as a function of matrix stiffness. The symbols represent solvent-blended, melt-blended, and in situ-polymerized composites and composites produced via preblend formation in solvent, followed by melt blending.[91] The diagonal solid line represents the Voigt limit, infinitely long sheets all aligned in the tensile direction.

Kim and coworkers[113] compared the stiffening performance of different types of functionalized graphene, dispersed via different blending routes in TPU elastomers. Despite different particle sizes, iGO and TRG gave nearly comparable tensile modulus increase (Table 16.4). This result indicates that stiffness increase is determined by the persistence length of the flexible graphene layers rather than A_f of fully extended sheets. For TRG composites, solvent-processed (directly cast from the solution) composites have higher stiffness than melt-processed ones. The lower modulus of melt-processed composites may be due to reaggregation of particles under extensional flow during melt compounding. However, unlike the significant modulus gains for TPU (three- to eightfold increase with 3 wt%), TRG only modestly improved the tensile stiffness for rigid glassy polymers, PEN[126] and PC[125]; <50% increase at 3 wt%. Defects in the sheet structure formed by oxidation and pyrolysis possibly reduce the effective in-plane stiffness of TRG, which may account for the relatively small increase in tensile properties observed for glassy polymers.[126] Mori and Tanaka's micromechanical theory[182,183] was used to estimate the transverse tensile stiffness E_{11} of polymers (Poisson's ratio, $\nu = 0.4$) filled with perfectly aligned, 1 vol% $A_f = 200$ ellipsoids (e.g., graphene). These calculations are plotted in Figure 16.9 as a function of modulus ratio between pure reinforcement and matrix, $E_{ellipsoid}/E_{matrix}$. Reduction in graphene stiffness from 1000 GPa (modulus of defect-free) to 100 GPa diminishes modulus of rigid glassy polymers ($E_{matrix} = 3$ GPa), while it hardly affects that of elastomers ($E_{matrix} = 10$ MPa). These predictions also

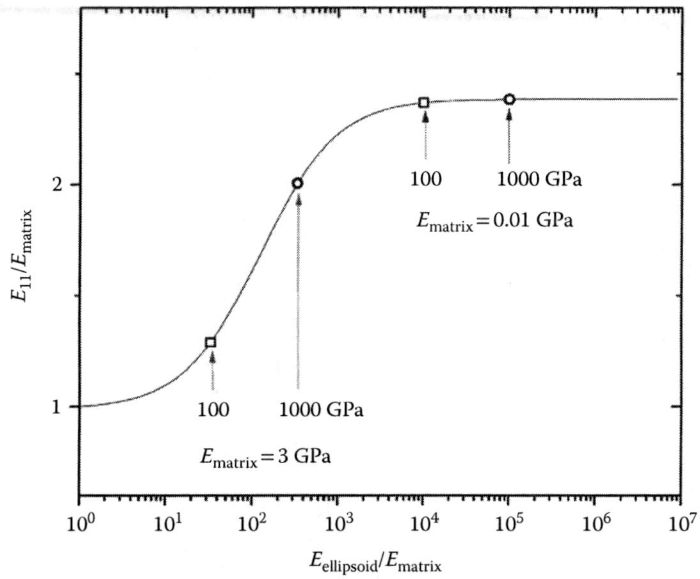

FIGURE 16.9
Relative increase in tensile modulus predicted by the theory of Mori and Tanaka[182,183] of elastomeric ($E_{matrix} = 0.01$ GPa) and glassy (3 GPa) polymer composites reinforced with ellipsoids (graphene), 1 vol% $A_f = 200$. Reducing ellipsoid stiffness $E_{ellipsoid}$ from 1000 (circles) to 100 GPa (squares) barely affects modulus enhancement for elastomers, but almost eliminates it for glassy polymers.

indicate that the composite stiffness cannot be improved further in the limit of extremely large stiffness contrast ($E_{ellipsoid}/E_{matrix} > 10^4$).

In addition to stiffness increase, improved tensile strength has been reported. Elongation to break typically decreases severely with the addition of rigid fillers. While elongation of most polymers decreased (negative signs in Table 16.4) with the addition of graphene, the decrease is less than one would expect, for example, for a similar amount of CB and certainly for the amount of CB needed to reach the same electrical conductivity. A nanoindentation study also revealed increased hardness of TRG-reinforced PVA and PMMA.[115]

16.6.4 Thermo-Calorimetric Transitions

Significant increases in glass transition temperature T_g[117,123,139] and altered crystallization kinetics[133] of polymers by functionalized graphene dispersion have been observed. Ramanathan et al. report that TRG increased T_g of oxygen-containing polymers such as PMMA (by 30°C at 0.05 wt%) and poly(acrylonitrile) (PAN) (by 46°C at 1 wt%) by hindering segmental motions of polymer chains via mechanical interlocking and hydrogen bonding with surface oxygen functionalities.[139] For graphene oxide sheets esterified with PVA chains, PVA T_g elevation can be attributed to reduced chain mobility by covalent cross-linking.[123]

Graphene can nucleate crystallization of polymers. For polycaprolactone (PCL) and PVA, melt crystallization takes place at higher temperature and degree of crystallinity becomes greater after isothermal crystallization in the presence of graphene oxide[136,137] or TRG.[115,184] This is suggestive of strong interfacial interaction between polar polymer chains and functionalized graphene surface. TRG also induces new crystalline phases for poly(vinylidene fluoride) (PVDF).[133] Contradictory trends (suppressed crystallization by graphene addition) have also been reported. As a result of covalent grafting on GO, semicrystalline PVA[123] becomes completely amorphous.

16.6.5 Dimensional Stability

Graphite has a negative coefficient of thermal expansion (-1.5×10^{-6}/°C) in basal plane near room temperature.[185] Thermal expansion along the thickness direction is also far smaller (2.7×10^{-5}/°C)[186] than that of typical polymeric materials. Therefore, graphene can prevent dimensional changes of polymers when incorporated and oriented appropriately. Kalaitzidou et al. demonstrated that GNP suppresses thermal expansion of PP as effectively as other carbon fillers such as CB and CNF, especially in the direction of platelet alignment.[161] Graphene oxide also reduced thermal expansion of epoxy (~30% reduction at 5 wt% incorporation).[118] However, Kim and Macosko found TRG is only marginally better or even less effective than graphite in improving dimensional stability of glassy polymers despite the higher aspect ratio.[125,126] They attributed this lower reinforcement efficiency of TRG to its wrinkled structure and flexibility.

16.6.6 Gas Permeation

Defect-free graphene sheets are impermeable to all gas molecules.[14] Combining graphene with polymeric hosts makes possible large-scale barrier membranes with mechanical integrity, a significant advantage over 1-D SWCNT, MWCNT, or CNF. Kalaitzidou et al. reported that GNP can reduce oxygen permeability more efficiently than 0-D (CB), 1-D (CNF), or even other 2-D fillers such as organically modified montmorillonite (MMT) at similar loading.[161] Gas permeation data of graphene/polymer nanocomposites in literature are summarized in Table 16.5. Kim and coworkers[113,125,126] compared gas permeation through polymeric membranes filled with different types of graphitic reinforcements. Figure 16.10 shows some of their results for TPU. Thermally or chemically treated GO layers decreased permeability of TPU more than organically modified

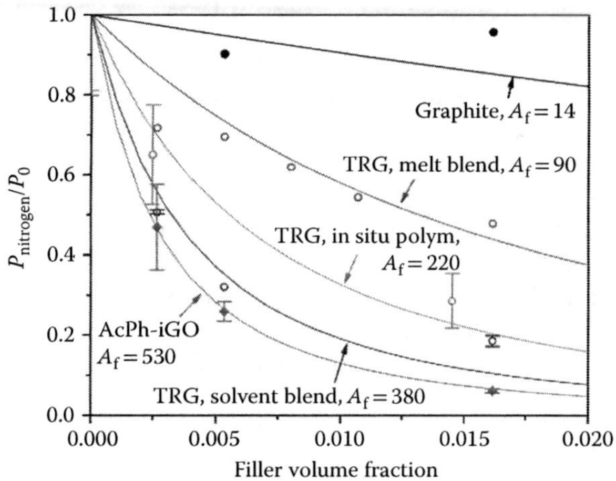

FIGURE 16.10
N_2 permeability of thermoplastic polyurethane (TPU) composites normalized by permeability P_0 of neat TPU. (Reproduced from Hirata, M. et al., *Carbon*, 42, 2929–2937, 2004.) Gas permeation of TRG/TPU composites produced via melt, solvent mixing, and in situ polymerization are compared. AcPh-iGO represents acetylphenyl isocyanate-treated GO. Solid curves are predictions based on Lape et al.[188]

MMT layers at similar loadings. Osman et al. reported a 25% decrease in oxygen permeation at 2.1 vol% MMT.[187] Especially, incorporation of iGO led to ~90% reduction in nitrogen permeability at 3 wt% loading, which theoretically can be attained by perfectly aligned, impermeable platelets with A_f of ~500 according to Lape and coworkers' model[188] for gas permeation of filled membranes. Also, solvent-based processing resulted in more reduction in permeation than melt-based one implying better dispersion of graphene platelets in agreement with review of dispersion methods in Section 16.5, and electrical and mechanical properties in Sections 16.6.1 and 16.6.3.

16.6.7 Thermal Stability

Improved thermal stability of host polymers is another benefit expected from graphene-based reinforcements. Thermal degradation temperature, characterized by the maximum weight loss rate in thermogravimetry, shifts up by 10°C–100°C at <10 wt% graphene for PS,[51] PVA,[123,134] PMMA,[189] and silicone foams.[120] Decomposition of graphene composites is substantially slower than neat polymers, which is attributed to restricted chain mobility[51] of polymers near the graphene surface. During combustion, inflammable anisotropic nanoparticles form a jammed network of char layers that retards transport of the decomposition products.[190,191] This suggests application of graphene/polymer nanocomposites for flame retardation.

16.6.8 Synergy with Other Carbon Nanofillers

It is worth mentioning that a combination of carbon fillers with different dimensionality may lead to synergistic physical property gains for polymers. Epoxy resins containing a binary mixture of GNP and SWCNT in 3:1 weight ratio have higher thermal conductivity than those reinforced with either individual filler.[168] Yu and colleagues explained this synergistic effect by bridging interactions between GNP and SWCNT that can reduce the interfacial resistance for thermal conduction. Also, mechanical properties such as hardness and stiffness of PVA can be enhanced more dramatically when graphene is dispersed with nanodiamonds.[184] 0-D nanodiamond particles at the interface may prevent clustering of graphene sheets.

16.7 Summary and Future Perspectives

16.7.1 Summary

Graphene is a multifunctional reinforcement that can improve electrical, thermal, mechanical, and gas barrier properties of polymers at extremely small loading. Graphene can provide a combination of the benefits achieved with layered silicates (gas permeation barrier and stiffness) and CNTs (electrical and thermal conductivity). These exfoliated carbon sheets can be produced via either bottom-up or top-down approaches. Bottom-up strategies, including CVD[15–21] and epitaxial growth from SiC,[24–30] produce large-sized graphene sheets with less defects, but in very limited quantity. Methods starting from chemically modified graphene precursors, particularly GO, followed by chemical (CRG)[77–79] or thermal reduction (TRG)[89,90] are better suited for large-scale production of reinforcements for the polymer composites.

Both CRG and TRG are electrically conductive but still have residual oxygen groups on the surface. These enable dispersion into solvents directly or with chemical functionalization,[56] which then leads to facile blending with soluble polymers. Alternatively, composites can be processed by polymerizing monomers containing graphene. No studies yet have shown an advantage of in situ polymerization over solvent blending, and in some cases, functional groups on the graphene may have interfered with the property improvements.[113] Melt compounding generally results in poorer dispersion but can be done more economically in large scale using melt extrusion. However, melt compounding is not an option for some chemically modified graphene materials that are prone to thermal degradation.

Dispersion of graphene in polymers can be characterized with TEM and by measuring solid properties. Linear viscoelastic measurements on composite melts[125,126] are also a useful tool to quantify dispersion. In the vicinity of particle percolation, nanocomposite melts start to develop a yield stress as well as a shear thinning behavior, which can be probed by nonlinear viscoelastic tests.[145] As discussed earlier, dilute viscosity measurements can also be considered provided that graphene forms a stable dispersion.[97]

The onset concentration for electrical percolation is a good indicator for graphene dispersion since it is inversely proportional to the aspect ratio of randomly oriented disks. Graphene oriented parallel to the surface of polymer films can reduce gas permeation of polymeric membranes by as much as 90% at only 3 wt%.[113] Stiffness improvement by graphene is more pronounced for elastomers than glassy polymers due to the higher stiffness contrast between the matrix and the reinforcement. Extraordinarily high increase in modulus or glass transition temperature for PMMA, PAN, and epoxy by TRG[138,139] has been attributed to hydrogen bonding interaction with oxygen-modified graphene surfaces, but we question some of these results in this review.

16.7.2 Challenges

This versatility of graphene/polymer nanocomposites indicates their potential application in automotive, aerospace, electronics, and packaging. However, there are several challenges. The first is cost. Micromechanically produced mono- and bilayer graphene sheets are sold for 0.5–3 British pounds per μm^2 supported on SiO_2 substrate.[192] Although graphene can be derived more economically via GO, it will not see practical applications until commercial-scale production of graphene is available at costs below CNT. Currently, the pricing of commercial, research grade CNT ranges from 1 to 2 USD/g (MWCNT)[193] to hundreds of USD/g (SWCNT).[194] The average cost of GO estimated by the authors is 20–30 USD/kg, which is still higher than that of GNP (~10 USD/kg).[179] Moreover, GO synthesis produces acid wastes and the oxidation reaction involves potential risks for explosion. The thermal or chemical treatments required to exfoliate GO will add to the cost. There are currently two commercial sources for TRG: Vorbeck[195] and Angstron materials.[196] Although there is no announced price for graphene, we estimate a target price of 50 USD/kg. Scale-up of graphene production (e.g., continuous fluidized bed reactor for TRG[93]) and recycling acid wastes would reduce the production cost of graphene materials.

Another barrier to practical use is difficult handling of graphene sheets in processing. Extremely low bulk density of high-surface-area graphene layers such as TRG makes processing into polymers in commercial scale challenging (e.g., difficult feeding into melt compounders). Densification for easy handling and transport may be necessary. Lightness of graphene sheets may also lead to human intake during handling. Biotoxicity of carbon nanomaterials including CNT to human has been frequently reported.[197,198] Prior to use, biocompatibility and safety of these 2-D nanocarbons must be tested, although their large lateral dimension and lack of entanglement make toxicity less likely than with CNT.

To date, the most commonly researched routes to graphene are via GO due to scalability. Even though these methods start from the same material, structure and surface characteristics may differ significantly depending on how the GO is exfoliated and reduced. Oxidation of graphite, and thermal or chemical reduction lead to irreversible deformation of graphitic carbons, which can modify the micromechanical and electrical transport properties. However, mechanisms for defect formation, and their influence on nanocomposite physical properties have not been fully understood yet. For instance, highly wrinkled structure and atomistic perforations in TRG may make the sheets less stiff and more permeable but may inhibit restacking. Moreover, there have been no direct comparisons of CRG and TRG for their dispersibility and interaction with polymers. Synthetic routes to graphene layers avoiding the structural deformation or unnecessary functionalization, and posttreatments that can restore graphitic planar domains need to be sought to maximize property increase for polymeric materials. For instance, thermal curing is reported to increase carbon-to-oxygen ratio in TRG.[61,93,199] More use of molecular modeling will help to guide experimental strategies.[57,59,74,200,201]

Even at the same loading, nanocomposite properties strongly depend on spatial distribution of the anisotropic inclusions. As a result of high shear or elongational flow fields, melt processing tends to yield highly aligned morphology of platelet-like fillers than solvent mixing.[113] Morphology and properties of the composites are sensitive to postshaping such as stretching or injection molding, which induces orientation and migration of particles.[125,202–204] The influence of shear deformation on viscosity and electrical conductivity of CNT/polymer melts has been extensively studied.[205,206] These properties are based on particle connectivity and thus generally benefit from random orientation of anisotropic inclusions.[151,207] On the other hand, particle orientation typically improves mechanical and gas permeation properties.[208,209] Since no single composite morphology will attain all desired properties simultaneously, it is important to engineer the spatial arrangement of dispersed nanoparticles appropriately for target properties. Current strategies and challenges to control composite morphology are well discussed in the review by Vaia and Maquire.[210]

Adhesion between graphene and polymers is also an important factor to be considered in designing composite materials. Superior mechanics of carbon nanofillers can be fully realized when they are strongly bonded to matrix polymers.[211,212] Improved adhesion may prevent interfacial scattering of phonons and yield higher composite thermal conductivity as well as minimize interfacial free volume, which affects gas permeation. Shenogina et al. found that interfacial thermal conductance can be correlated with the wetting properties.[174] Covalently attaching alkyl chains on graphene edges can improve thermal conduction across the graphene–hydrocarbon interface.[213] TRG appears to bind strongly with some oxygen-containing polymers as substantiated by highly enhanced stiffness and T_g of the matrix.[138,139] However, such an appreciable T_g increase was never observed by us for PC and PEN.

16.7.3 Future Application

As reviewed in Section 16.6.7, the excellent thermal stability of graphene-based nanocomposites can be used for producing flame retardant materials. Networks of CNT,[190] CNF,[214] and EG[215] are known to retard flammability of PMMA and also PU foams. Its aromatic and 2-D nature makes graphene an ideal alternative for the flame-retarding additives. Also, high electrical or thermal conductivity of graphene/polymer nanocomposites may satisfy conditions for heat- or electricity-activated shape memory,[114,216] static charge dissipation, and electromagnetic wave reflective materials.[158,217] Especially near the percolation threshold, resistivity of composites can vary dramatically upon temperature change,[218] solvent attack,[219] and external strain.[220] This on–off phenomena in electrical conductivity[158] by external stimuli can be used for electrical switching and strain/solvent sensing. Graphene nanocomposites have potential as photo- or electromechanical actuators since mechanical response has been induced by infrared radiation[221] or electric potential[222] in CNT composites.

The transmittance of visible light is attenuated by ~2.3% as it penetrates through a single layer of graphene.[223] If thin layers of electrically conductive graphene can be deposited on glass or polymer surfaces, flexible displays,[224] thin-film transistors,[225] photovoltaic,[226] and liquid crystal devices[227] will be made possible. TRG-conductive inks are already being offered commercially.[159] Recently, homogeneous coating of a few-layer-thick graphene has been demonstrated via direct transfer of graphene films from CVD metal substrates[224] or from vacuum filtration membranes[225] as well as by conventional spin coating.[226,228]

Acknowledgments

Financial support from the Abu Dhabi-Minnesota Institute for Research Excellence (ADMIRE), a partnership between the Petroleum Institute of Abu Dhabi, and the Department of Chemical Engineering and Materials Science of the University of Minnesota is acknowledged. The authors also appreciate helpful input from Prof. K. Andre Mkhoyan at the University of Minnesota, and Prof. Matteo Pasquali and Natnael Behabtu at Rice University.

References

1. Huang, J.-C., Carbon black filled conducting polymers and polymer blends. *Adv. Polym. Technol.* **2002**, *21*, 299–313.
2. Moniruzzaman, M., and Winey, K. I., Polymer nanocomposites containing carbon nanotubes. *Macromolecules* **2006**, *39*, 5194–5205.
3. Okamoto, M., *Polymer/clay nanocomposites*, Vol. 8. Stevenson Ranch, CA: American Scientific Publishers, **2004**.
4. Geim, A. K., and Novoselov, K. S., The rise of graphene. *Nat. Mater.* **2007**, *6*, 183–191.
5. Lax, A., and Maxwell, R., The seathwaite connection: Some aspects of the graphite mine and its owners. *National Trust Annual Archaeological Review* **1998–1999**, 18–23.
6. Weinberg, S. S., *The stone age in the agean*, 10th ed., Vol. 1. Cambridge: Cambridge University Press, **2007**.
7. Iijima, S., Helical microtubules of graphitic carbon. *Nature* **1991**, *354*, 56–58.
8. Kroto, H. W., Heath, J. R., O'Brien, S. C., Curl, R. F., and Smalley, R. E., C60: Buckminsterfullerene. *Nature* **1985**, *318*, 162–163.
9. Eizenberg, M., and Blakely, J. M., Carbon monolayer phase condensation on Ni(111). *Surf. Sci.* **1970**, *82*, 228–236.
10. Novoselov, K. S., Geim, A. K., Morozov, S. V., Jiang, D., Zhang, Y., Dubonos, S. V., Grigorieva, I. V., and Firsov, A. A., Electric field effect in atomically thin carbon films. *Science* **2004**, *306*, 666–669.
11. Lee, C., Wei, X., Kysar, J. W., and Hone, J., Measurement of the elastic properties and intrinsic strength of monolayer graphene. *Science* **2008**, *321*, 385–388.
12. Balandin, A. A., Ghosh, S., Bao, W., Calizo, I., Teweldebrhan, D., Miao, F., and Lau, C. N., Superior thermal conductivity of single-layer graphene. *Nano Lett.* **2008**, *8*, 902–907.
13. Du, X., Skachko, I., Barker, A., and Andrei, E. Y., Approaching ballistic transport in suspended graphene. *Nat. Nanotechnol.* **2008**, *3*, 491–495.
14. Bunch, J. S., Verbridge, S. S., Alden, J. S., van der Zande, A. M., Parpia, J. M., Craighead, H. G., and McEuen, P. L., Impermeable atomic membranes from graphene sheets. *Nano Lett.* **2008**, *8*, 2458–2462.
15. Wang, X., You, H., Liu, F., Li, M., Wan, L., Li, S., Li, Q., Xu, Y., Tian, R., Yu, Z., Xiang, D., and Cheng, J., Large-scale synthesis of few-layered graphene using CVD. *Chem. Vapor Deposit.* **2009**, *15*, 53–56.
16. Wang, Y., Chen, X., Zhong, Y., Zhu, F., and Loh, K. P., Large area, continuous, few-layered graphene as anodes in organic photovoltaic devices. *Appl. Phys. Lett.* **2009**, *95*, 063302/1–063302/3.
17. Xianbao, W., Haijun, Y., Fangming, L., Mingjian, L., Li, W., Shaoqing, L., Qin, L., Yang, X., Rong, T., Ziyong, Y., Dong, X., and Jing, C., Large-scale synthesis of few-layered graphene using CVD. *Chem. Vapor Deposit.* **2009**, *15*, 53–56.
18. Dervishi, E., Li, Z., Watanabe, F., Biswas, A., Xu, Y., Biris Alexandru, R., Saini, V., and Biris Alexandru, S., Large-scale graphene production by RF-CCVD method. *Chem. Commun.* **2009**, *27*, 4061–4063.
19. Li, X., Cai, W., An, J., Kim, S., Nah, J., Yang, D., Piner, R., Velamakanni, A., Jung, I., Tutuc, E., Banerjee, S. K., Colombo, L., and Ruoff, R., S., Large-area synthesis of high-quality and uniform graphene films on copper foils. *Science* **2009**, *324*, 1312–1314.

20. Chong-an, D., Dacheng, W., Gui, Y., Yunqi, L., Yunlong, G., and Daoben, Z., Patterned graphene as source/drain electrodes for bottom-contact organic field-effect transistors. *Adv. Mater.* **2008**, *20,* 3289–3293.
21. Chae, S. J., Gu¨nes, F., Kim, K. K., Kim, E. S., Han, G. H., Kim, S. M., Shin, H.-J., Yoon, S.-M., Choi, J.-Y., Park, M. H., Yang, C. W., Pribat, D., and Lee, Y. H., Synthesis of large-area graphene layers on poly-nickel substrate by chemical vapor deposition: Wrinkle formation. *Adv. Mater.* **2009**, *21,* 2328–2333.
22. Li, N., Wang, Z., Zhao, K., Shi, Z., Gu, Z., and Xu, S., Large scale synthesis of n-doped multi-layered graphene sheets by simple arc-discharge method. *Carbon* **2009**, *48,* 255–259.
23. Karmakar, S., Kulkarni, N. V., Nawale, A. B., Lalla, N. P., Mishra, R., Sathe, V. G., Bhoraskar, S. V., and Das, A. K., A novel approach towards selective bulk synthesis of few-layer graphenes in an electric arc. *J. Phys. D: Appl. Phys.* **2009**, *42,* 115201/1–115201/14.
24. Rollings, E., Gweon, G.-H., Zhou, S. Y., Mun, B. S., McChesney, J. L., Hussain, B. S., Fedorov, A. V., First, P. N., de Heer, W. A., and Lanzar, A., Synthesis and characterization of atomically thin graphite films on a silicon carbide substrate. *J. Phys. Chem. Solids* **2006**, *67,* 21722177.
25. de Heer, W. A., Berger, C., Wu, X., First, P. N., Conrad, E. H., Li, X., Li, T., Sprinkle, M., Hass, J., Sadowski, M. L., Potemski, M., and Martinez, G., Epitaxial graphene. *Solid State Commun.* **2007**, *143,* 92–100.
26. Alexander, M., and Oleg, P., Density functional study of graphene overlayers on sic. *Physica Status Solidi B* **2008**, *245,* 1425–1435.
27. Ni, Z. H., Chen, W., Fan, X. F., Kuo, J. L., Yu, T., Wee, A. T. S., and Shen, Z. X., Raman spectroscopy of epitaxial graphene on a SiC substrate. *Phys. Rev. B: Condens. Matter Mater. Phys.* **2008**, *77,* 115416/1–115416/6.
28. Sutter, P. W., Flege, J.-I., and Sutter, E. A., Epitaxial graphene on ruthenium. *Nat. Mater.* **2008**, *7,* 406–411.
29. Seyller, T., Bostwick, A., Emtsev, K. V., Horn, K., Ley, L., McChesney, J. L., Ohta, T., Riley, J. D., Rotenberg, E., and Speck, F., Epitaxial graphene: A new material. *Physica Status Solidi B* **2008**, *245,* 1436–1446.
30. Sprinkle, M., Soukiassian, P., de Heer, W. A., Berger, C., and Conrad, E. H., Epitaxial graphene: The material for graphene electronics. *Physica Status Solidi (RRL)—Rapid Res. Lett.* **2009**, *3,* A91–A94.
31. Yannick, C., and Wim, K., Growing graphene sheets from reactions with methyl radicals: A quantum chemical study. *ChemPhysChem* **2006**, *7,* 1770–1778.
32. Yang, X., Dou, X., Rouhanipour, A., Zhi, L., Rader Hans, J., and Mullen, K., Two-dimensional graphene nanoribbons. *J. Am. Chem. Soc.* **2008**, *130,* 4216–7.
33. Zhi, L., and Muellen, K., A bottom-up approach from molecular nanographenes to unconventional carbon materials. *J. Mater. Chem.* **2008**, *18,* 1472–1484.
34. Kim, C.-D., Min, B.-K., and Jung, W.-S., Preparation of graphene sheets by the reduction of carbon monoxide. *Carbon* **2009**, *47,* 1610–1612.
35. Kosynkin, D. V., Higginbotham, A. L., Sinitskii, A., Lomeda, J. R., Dimiev, A., Price, B. K., and Tour, J. M., Longitudinal unzipping of carbon nanotubes to form graphene nanoribbons. *Nature* **2009**, *458,* 872–876.
36. Hirsch, A., Unzipping carbon nanotubes: A peeling method for the formation of graphene nanoribbons. *Angew. Chem. Int. Ed.* **2009**, *48,* 6594–6596.

37. Jiao, L., Zhang, L., Wang, X., Diankov, G., and Dai, H., Narrow graphene nanoribbons from carbon nanotubes. *Nature* **2009**, *458*, 877–880.
38. Zhang, W., Cui, J., Tao, C.-a., Wu, Y., Li, Z., Ma, L., Wen, Y., and Li, G., A strategy for producing pure single-layer graphene sheets based on a confined self-assembly approach. *Angew. Chem. Int. Ed.* **2009**, *48*, 5864–5868.
39. Worsley, K. A., Ramesh, P., Mandal, S. K., Niyogi, S., Itkis, M. E., and Haddon, R. C., Soluble graphene derived from graphite fluoride. *Chemi. Phys. Lett.* **2007**, *445*, 51–56.
40. Kelly, T. D., and Matos, G. R. Historical Statistics for Mineral and Material Commodities in the United States: U.S. Geological Survey Data Series 140. Reston, VA: US Department of the Interior, US Geological Survey, **2005**.
41. Viculis, L. M., Mack, J. J., Mayer, O. M., Hahn, H. T., and Kaner, R. B., Intercalation and exfoliation routes to graphite nanoplatelets. *J. Mater. Chem.* **2005**, *15*, 974–978.
42. Carr, K. E., Intercalation and oxidation effects on graphite of a mixture of sulfuric and nitric acids. *Carbon* **1970**, *8*, 155–166.
43. Chen, G., Wu, D., Weng, W., and Wu, C., Exfoliation of graphite flake and its nanocomposites. *Carbon* **2003**, *41*, 619–621.
44. Celzard, A., Mareche, J. F., and Furdin, G., Surface area of compressed expanded graphite. *Carbon* **2002**, *40*, 2713–2718.
45. Lee, J. H., Shin, D. W., Makotchenko, V. G., Nazarov, A. S., Fedorov, V. E., Kim, Y. H., Choi, J.-Y., Kim, J. M., and Yoo, J.-B., One-step exfoliation synthesis of easily soluble graphite and transparent conducting graphene sheets. *Adv. Mater.* **2009**, *21*, 4383–4387.
46. Fukushima, H. Graphite nanoreinforcements in polymer nanocomposites. Ph.D. Thesis, Michigan State University, **2003**.
47. Kalaitzidou, K., Fukushima, H., and Drzal, L. T., A new compounding method for exfoliated graphite-polypropylene nanocomposites with enhanced flexural properties and lower percolation threshold. *Compos. Sci. Technol.* **2007**, *67*, 2045–2051.
48. Kalaitzidou, K., Fukushima, H., Askeland, P., and Drzal, L. T., The nucleating effect of exfoliated graphite nanoplatelets and their influence on the crystal structure and electrical conductivity of polypropylene nanocomposites. *J. Mater. Sci.* **2008**, *43*, 2895–2907.
49. Bourlinos, A. B., Georgakilas, V., Zboril, R., Steriotis, T. A., and Stubos, A., Liquid-phase exfoliation of graphite towards solubilized graphenes. *Small* **2009**, *5*, 1841–1845.
50. Hernandez, Y., Nicolosi, V., Lotya, M., Blighe, F. M., Sun, Z., De, S., McGovern, I. T., Holland, B., Byrne, M., Gun'Ko, Y. K., Boland, J. J., Niraj, P., Duesberg, G., Krishnamurthy, S., Goodhue, R., Hutchison, J., Scardaci, V., Ferrari, A. C., and Coleman, J. N., High-yield production of graphene by liquid-phase exfoliation of graphite. *Nat. Nanotechnol.* **2008**, *3*, 563–568.
51. Liu, N., Luo, F., Wu, H., Liu, Y., Zhang, C., and Chen, J., One-step ionic-liquid-assisted electrochemical synthesis of ionic-liquid-functionalized graphene sheets directly from graphite. *Adv. Funct. Mater.* **2008**, *18*, 1518–1525.
52. Behabtu, N., Lomeda, J. R., Green, M. J., Higginbotham, A. L., Sinitskii, A., Kosynkin, D. V., Tsentalovich, D., Parra-Vasquez, A. N. G., Schmidt, J., Kesselman, E., Cohen, Y., Talmon, Y., Tour, J. M., and Pasquali, M., Spontaneous high-concentration dispersions and liquid crystals of graphene. *Nat. Nanotechnol.* **2010**, *5*, 406–411.

53. Brodie, B. C., On the atomic weight of graphite. *Philos. Trans. R. Soc. Lond.* **1859**, *149*, 249–259.
54. Staudenmaier, L., Method for the preparation of graphitic acid. *Ber. Dtsch. Chem. Ges.* **1898**, *31*, 1481–87.
55. Hummers, W. S., Jr., and Offeman, R. E., Preparation of graphitic oxide. *J. Am. Chem. Soc.* **1958**, *80*, 1339.
56. Park, S., and Ruoff, R. S., Chemical methods for the production of graphenes. *Nat. Nanotechnol.* **2009**, *4*, 217–224.
57. Boukhvalov, D. W., and Katsnelson, M. I., Modeling of graphite oxide. *J. Am. Chem. Soc.* **2008**, *130*, 10697–10701.
58. Lahaye, R. J. W. E., Jeong, H. K., Park, C. Y., and Lee, Y. H., Density functional theory study of graphite oxide for different oxidation levels. *Phys. Rev. B: Condens. Matter Mater. Phys.* **2009**, *79*, 125435/1–125435/8.
59. Paci, J. T., Belytschko, T., and Schatz, G. C., Computational studies of the structure, behavior upon heating, and mechanical properties of graphite oxide. *J. Phys. Chem. C* **2007**, *111*, 18099–18111.
60. Zhang, W. H., Carravetta, V., Li, Z. Y., Luo, Y., and Yang, J. L., Oxidation states of graphene: Insights from computational spectroscopy. *J. Chem. Phys.* **2009**, *131*, 244505/1–244505/6.
61. Gao, W., Alemany, L. B., Ci, L., and Ajayan, P. M., New insights into the structure and reduction of graphite oxide. *Nat. Chem.* **2009**, *1*, 403–408, S403/1–S403/20.
62. Jeong, H.-K., Lee, Y. P., Lahaye, R. J. W. E., Park, M.-H., An, K. H., Kim, I. J., Yang, C.-W., Park, C. Y., Ruoff, R. S., and Lee, Y. H., Evidence of graphitic AB stacking order of graphite oxides. *J. Am. Chem. Soc.* **2008**, *130*, 1362–1366.
63. Cai, W., Piner, R. D., Stadermann, F. J., Park, S., Shaibat, M. A., Ishii, Y., Yang, D., Velamakanni, A., An, S. J., Stoller, M., An, J., Chen, D., and Ruoff, R. S., Synthesis and solid-state NMR structural characterization of 13c-labeled graphite oxide. *Science* **2008**, *321*, 1815–1817.
64. Szabo, T., Berkesi, O., Forgo, P., Josepovits, K., Sanakis, Y., Petridis, D., and Dekany, I., Evolution of surface functional groups in a series of progressively oxidized graphite oxides. *Chem. Mater.* **2006**, *18*, 2740–2749.
65. Hristea, G., and Panaitescu, C., Structural aspects about graphite oxides. *Revue Roumaine de Chimie* **2002**, *46*, 1107–1111.
66. Lerf, A., He, H., Forster, M., and Klinowski, J., Structure of graphite oxide revisited. *J. Phys. Chem. B* **1998**, *102*, 4477–4482.
67. He, H., Klinowski, J., Forster, M., and Lerf, A., A new structural model for graphite oxide. *Chem. Phys. Lett.* **1998**, *287*, 53–56.
68. Mermoux, M., Chabre, Y., and Rousseau, A., FTIR and ^{13}C NMR study of graphite oxide. *Carbon* **1991**, *29*, 469–474.
69. Nakajima, T., Mabuchi, A., and Hagiwara, R., A new structure model of graphite oxide. *Carbon* **1988**, *26*, 357–361.
70. Scholz, W., and Boehm, H. P., Graphite oxide. Vi. Structure of graphite oxide. *Zeitschrift fuer Anorganische und Allgemeine Chemie* **1969**, *369*, 327–340.
71. Clauss, A., Plass, R., Boehm, H. P., and Hofmann, U., Investigations of the structure of graphite oxide. *Zeitschrift fuer Anorganische und Allgemeine Chemie* **1957**, *291*, 205–220.
72. Mkhoyan, K. A., Contryman, A. W., Silcox, J., Stewart, D. A., Eda, G., Mattevi, C., Miller, S., and Chhowalla, M., Atomic and electronic structure of graphene-oxide. *Nano Lett.* **2009**, *9*, 1058–1063.

73. Herrera-Alonso, M., Abdala, A. A., McAllister, M. J., Aksay, I. A., and Prud'homme, R. K., Intercalation and stitching of graphite oxide with diaminoalkanes. *Langmuir* **2007**, *23*, 10644–10649.
74. Li, J.-L., Kudin, K. N., McAllister, M. J., Prud'homme, R. K., Aksay, I. A., and Car, R., Oxygen-driven unzipping of graphitic materials. *Phys. Rev. Lett.* **2006**, *96*, 176101/1–176101/4.
75. Dreyer, D. R., Park, S., Bielawski, C. W., and Ruoff, R. S., The chemistry of graphene oxide. *Chem. Soc. Rev.* **2010**, *39*, 228–240.
76. Stankovich, S., Piner, R. D., Chen, X., Wu, N., Nguyen, S. T., and Ruoff, R. S., Stable aqueous dispersions of graphitic nanoplatelets via the reduction of exfoliated graphite oxide in the presence of poly(sodium 4-styrenesulfonate). *J. Mater. Chem.* **2006**, *16*, 155–158.
77. Stankovich, S., Dikin, D. A., Piner, R. D., Kohlhaas, K. A., Kleinhammes, A., Jia, Y., Wu, Y., Nguyen, S. T., and Ruoff, R. S., Synthesis of graphene-based nanosheets via chemical reduction of exfoliated graphite oxide. *Carbon* **2007**, *45*, 1558–1565.
78. Wang, G., Shen, X., Wang, B., Yao, J., and Park, J., Synthesis and characterisation of hydrophilic and organophilic graphene nanosheets. *Carbon* **2009**, *47*, 1359–1364.
79. Lomeda, J. R., Doyle, C. D., Kosynkin, D. V., Hwang, W.-F., and Tour, J. M., Diazonium functionalization of surfactant-wrapped chemically converted graphene sheets. *J. Am. Chem. Soc.* **2008**, *130*, 16201–16206.
80. Wang, H., Robinson, J. T., Li, X., and Dai, H., Solvothermal reduction of chemically exfoliated graphene sheets. *J. Am. Chem. Soc.* **2009**, *131*, 9910–9911.
81. Stankovich, S., Dikin, D. A., Dommett, G. H. B., Kohlhaas, K. M., Zimney, E. J., Stach, E. A., Piner, R. D., Nguyen, S. T., and Ruoff, R. S., Graphene-based composite materials. *Nature* **2006**, *442*, 282–286.
82. Si, Y., and Samulski, E. T., Synthesis of water soluble graphene. *Nano Lett.* **2008**, *8*, 1679–1682.
83. Wang, G., Yang, J., Park, J., Gou, X., Wang, B., Liu, H., and Yao, J., Facile synthesis and characterization of graphene nanosheets. *J. Phys. Chem. C* **2008**, *112*, 8192–8195.
84. Williams, G., Seger, B., and Kamat, P. V., TiO_2-graphene nanocomposites. UV-assisted photocatalytic reduction of graphene oxide. *ACS Nano* **2008**, *2*, 1487–1491.
85. Zhou, Y., Bao, Q., Tang, L. A. L., Zhong, Y., and Loh, K. P., Hydrothermal dehydration for the "green" reduction of exfoliated graphene oxide to graphene and demonstration of tunable optical limiting properties. *Chem. Mater.* **2009**, *21*, 2950–2956.
86. Nethravathi, C., and Rajamathi, M., Chemically modified graphene sheets produced by the solvothermal reduction of colloidal dispersions of graphite oxide. *Carbon* **2008**, *46*, 1994–1998.
87. Fan, Z., Wang, K., Wei, T., Yan, J., Song, L., and Shao, B., An environmentally friendly and efficient route for the reduction of graphene oxide by aluminum powder. *Carbon* **2010**, *48*, 1670–1692.
88. Prud'homme, R. K., Aksay, I. A., Adamson, D., and Abdala, A. Thermally exfoliated graphite oxide and its use in nanocomposites. U.S. Patent 20070092432, April 26, **2007**.

89. Schniepp, H. C., Li, J.-L., McAllister, M. J., Sai, H., Herrera-Alonso, M., Adamson, D. H., Prud'homme, R. K., Car, R., Saville, D. A., and Aksay, I. A., Functionalized single graphene sheets derived from splitting graphite oxide. *J. Phys. Chem. B* **2006**, *110*, 8535–8539.
90. McAllister, M. J., Li, J.-L., Adamson, D. H., Schniepp, H. C., Abdala, A. A., Liu, J., Herrera-Alonso, M., Milius, D. L., Car, R., Prud'homme, R. K., and Aksay, I. A., Single sheet functionalized graphene by oxidation and thermal expansion of graphite. *Chem. Mater.* **2007**, *19*, 4396–4404.
91. Steurer, P., Wissert, R., Thomann, R., and Muelhaupt, R., Functionalized graphenes and thermoplastic nanocomposites based upon expanded graphite oxide. *Macromol. Rapid Commun.* **2009**, *30*, 316–327.
92. Aksay, I. A., Milius, D. L., Korkut, S., and Prud'homme, R. K. Functionalized graphene sheets having high carbon to oxygen ratios for polymer composite. W.O. Patent 2009/134492 A2, November 5, **2009**.
93. Schniepp, H. C., Kudin, K. N., Li, J.-L., Prud'homme, R. K., Car, R., Saville, D. A., and Aksay, I. A., Bending properties of single functionalized graphene sheets probed by atomic force microscopy. *ACS Nano* **2008**, *2*, 2577–2584.
94. Stankovich, S., Piner, R. D., Nguyen, S. T., and Ruoff, R. S., Synthesis and exfoliation of isocyanate-treated graphene oxide nanoplatelets. *Carbon* **2006**, *44*, 3342–3347.
95. Meyer, J. C., Geim, A. K., Katsnelson, M. I., Novoselov, K. S., Obergfell, D., Roth, S., Girit, C., and Zettl, A., On the roughness of single- and bi-layer graphene membranes. *Solid State Commun.* **2007**, *143*, 101–109.
96. Parra-Vasquez, A. N. G., Stepanek, I., Davis, V. A., Moore, V. C., Haroz, E. H., Shaver, J., Hauge, R. H., Smalley, R. E., and Pasquali, M., Simple length determination of single-walled carbon nanotubes by viscosity measurements in dilute suspensions. *Macromolecules* **2007**, *40*, 4043–4047.
97. Spector, M. S., Naranjo, E., Chiruvolu, S., and Zasadzinski, J. A., Conformations of a tethered membrane: Crumpling in graphite oxide? *Phys. Rev. Lett.* **1994**, *73*, 2867–2870.
98. Wen, X., Garland, C. W., Hwa, T., Kardar, M., Kokufuta, E., Li, Y., Orkisz, M., and Tanaka, T., Crumpled and collapsed conformations in graphite oxide membranes. *Nature* **1992**, *355*, 426–428.
99. Shin, H.-J., Kim, K. K., Benayad, A., Yoon, S.-M., Park, H. K., Jung, I.-S., Jin, M. H., Jeong, H.-K., Kim, J. M., Choi, J.-Y., and Lee, Y. H., Efficient reduction of graphite oxide by sodium borohydride and its effect on electrical conductance. *Adv. Funct. Mater.* **2009**, *19*, 1987–1992.
100. Hontoria-Lucas, C., Lopez-Peinado, A. J., Lopez-Gonzalez, J. d. D., Rojas-Cervantes, M. L., and Martin-Aranda, R. M., Study of oxygen-containing groups in a series of graphite oxides: Physical and chemical characterization. *Carbon* **1995**, *33*, 1585–1592.
101. Szabo, T., Berkesi, O., and Dekany, I., Drift study of deuterium-exchanged graphite oxide. *Carbon* **2005**, *43*, 3186–3189.
102. Titelman, G. I., Gelman, V., Bron, S., Khalfin, R. L., Cohen, Y., and Bianco-Peled, H., Characteristics and microstructure of aqueous colloidal dispersions of graphite oxide. *Carbon* **2005**, *43*, 641–649.
103. Kudin, K. N., Ozbas, B., Schniepp, H. C., Prud'homme, R. K., Aksay, I. A., and Car, R., Raman spectra of graphite oxide and functionalized graphene sheets. *Nano Lett.* **2008**, *8*, 36–41.

104. Cancado, L. G., Takai, K., Enoki, T., Endo, M., Kim, Y. A., Mizusaki, H., Speziali, N. L., Jorio, A., and Pimenta, M. A., Measuring the degree of stacking order in graphite by Raman spectroscopy. *Carbon* **2008**, *46*, 272–275.
105. Jung, I., Dikin, D., Park, S., Cai, W., Mielke, S. L., and Ruoff, R. S., Effect of water vapor on electrical properties of individual reduced graphene oxide sheets. *J. Phys. Chem. C* **2008**, *112*, 20264–20268.
106. Jung, I., Dikin, D. A., Piner, R. D., and Ruoff, R. S., Tunable electrical conductivity of individual graphene oxide sheets reduced at "low" temperatures. *Nano Lett.* **2008**, *8*, 4283–4287.
107. Mickelson, E. T., Huffman, C. B., Rinzler, A. G., Smalley, R. E., Hauge, R. H., and Margrave, J. L., Fluorination of single-wall carbon nanotubes. *Chem. Phys. Lett.* **1998**, *296*, 188–194.
108. Chen, J., Hamon, M. A., Hu, H., Chen, Y., Rao, A. M., Eklund, P. C., and Haddon, R. C., Solution properties of single-walled carbon nanotubes. *Science* **1998**, *282*, 95–98.
109. Bahr, J. L., Yang, J., Kosynkin, D. V., Bronikowski, M. J., Smalley, R. E., and Tour, J. M., Functionalization of carbon nanotubes by electrochemical reduction of aryl diazonium salts: A bucky paper electrode. *J. Am. Chem. Soc.* **2001**, *123*, 6536–6542.
110. Matsuo, Y., Tahara, K., and Sugie, Y., Structure and thermal properties of poly(ethylene oxide)-intercalated graphite oxide. *Carbon* **1997**, *35*, 113–120.
111. Hirata, M., Gotou, T., Horiuchi, S., Fujiwara, M., and Ohba, M., Thin-film particles of graphite oxide 1: High-yield synthesis and flexibility of the particles. *Carbon* **2004**, *42*, 2929–2937.
112. Kim, H., Miura, Y., and Macosko, C. W., Graphene/polyurethane nanocomposites for improved gas barrier and electrical conductivity. *Chem Mater.* **2010**, *22*, 3441–3450.
113. Liang, J., Xu, Y., Huang, Y., Zhang, L., Wang, Y., Ma, Y., Li, F., Guo, T., and Chen, Y., Infrared-triggered actuators from graphene-based nanocomposites. *J. Phys. Chem. C* **2009**, *113*, 9921–9927.
114. Das, B., Prasad, K. E., Ramamurty, U., and Rao, C. N. R., Nano-indentation studies on polymer matrix composites reinforced by few-layer graphene. *Nanotechnology* **2009**, *20*, 125705/1–125705/5.
115. Liu, P., Gong, K., Xiao, P., and Xiao, M., Preparation and characterization of poly(vinyl acetate)-intercalated graphite oxide nanocomposite. *J. Mater. Chem.* **2000**, *10*, 933–935.
116. Jang, J. Y., Kim, M. S., Jeong, H. M., and Shin, C. M., Graphite oxide/poly(methyl methacrylate) nanocomposites prepared by a novel method utilizing macroazoinitiator. *Compos. Sci. Technol.* **2009**, *69*, 186–191.
117. Wang, S., Tambraparni, M., Qiu, J., Tipton, J., and Dean, D., Thermal expansion of graphene composites. *Macromolecules* **2009**, *42*, 5251–5255.
118. Du, X. S., Xiao, M., Meng, Y. Z., and Hay, A. S., Direct synthesis of poly(arylenedisulfide)/carbon nanosheet composites via the oxidation with graphite oxide. *Carbon* **2005**, *43*, 195–213.
119. Verdejo, R., Barroso-Bujans, F., Angel Rodriguez-Perez, M., de Saja, J. A., and Angel Lopez-Manchado, M., Functionalized graphene sheet filled silicone foam nanocomposites. *J. Mater. Chem.* **2008**, *18*, 2221–2226.
120. Lee, Y. R., Raghu, A. V., Jeong, H. M., and Kim, B. K., Properties of waterborne polyurethane/functionalized graphene sheet nanocomposites prepared by an in situ method. *Macromol. Chem. Phys.* **2009**, *210*, 1247–1254.

121. Yang, Y., Wang, J., Zhang, J., Liu, J., Yang, X., and Zhao, H., Exfoliated graphite oxide decorated by pdmaema chains and polymer particles. *Langmuir* **2009**, *25*, 11808–11814.
122. Salavagione, H. J., Gomez, M. A., and Martinez, G., Polymeric modification of graphene through esterification of graphite oxide and poly(vinyl alcohol). *Macromolecules* **2009**, *42*, 6331–6334.
123. Prud'homme, R. K., Ozbas, B., Aksay, I. A., Register, R. A., and Adamson, D. H. Functional graphene-rubber nanocomposites. W.O. Patent 2008045778 A1, **2008**.
124. Kim, H., and Macosko, C. W., Processing-property relationships of polycarbonate/graphene composites. *Polymer* **2009**, *50*, 3797–3809.
125. Kim, H., and Macosko, C. W., Morphology and properties of polyester/exfoliated graphite nanocomposites. *Macromolecules* **2008**, *41*, 3317–3327.
126. Wakabayashi, K., Pierre, C., Dikin, D. A., Ruoff, R. S., Ramanathan, T., Brinson, L. C., and Torkelson, J. M., Polymer-graphite nanocomposites: Effective dispersion and major property enhancement via solid-state shear pulverization. *Macromolecules* **2008**, *41*, 1905–1908.
127. Drummy, L. F., Wang, Y. C., Schoenmakers, R., May, K., Jackson, M., Koerner, H., Farmer, B. L., Mauryama, B., and Vaia, R. A., Morphology of layered silicate-(nanoclay-) polymer nanocomposites by electron tomography and small-angle x-ray scattering. *Macromolecules* **2008**, *41*, 2135–2143.
128. Fornes, T. D., and Paul, D. R., Modeling properties of nylon 6/clay nanocomposites using composite theories. *Polymer* **2003**, *44*, 4993–5013.
129. Vermant, J., Ceccia, S., Dolgovskij, M. K., Maffettone, P. L., and Macosko, C. W., Quantifying dispersion of layered nanocomposites via melt rheology. *J. Rheol.* **2007**, *51*, 429–450.
130. Zosel, A., Rheological properties of dispersed systems at low shear stress. *Rheol. Acta* **1982**, *21*, 72–80.
131. Garboczi, E. J., Snyder, K. A., Douglas, J. F., and Thorpe, M. F., Geometrical percolation threshold of overlapping ellipsoids. *Phys. Rev. E: Stat. Phys. Plasmas Fluids Relat. Interdiscip. Topics* **1995**, *52*, 819–28.
132. Celzard, A., McRae, E., Deleuze, C., Dufort, M., Furdin, G., and Mareche, J. F., Critical concentration in percolating systems containing a high-aspect-ratio filler. *Phys. Rev. B: Condens. Matter* **1996**, *53*, 6209–14.
133. Ren, J., Silva, A. S., and Krishnamoorti, R., Linear viscoelasticity of disordered polystyrene-polyisoprene block copolymer based layered-silicate nanocomposites. *Macromolecules* **2000**, *33*, 3739–3746.
134. Shante, V. K. S., and Kirkpatrick, S., An introduction to percolation theory. *Adv. Phys.* **1971**, *20*, 325.
135. Du, F., Scogna, R. C., Zhou, W., Brand, S., Fischer, J. E., and Winey, K. I., Nanotube networks in polymer nanocomposites: Rheology and electrical conductivity. *Macromolecules* **2004**, *37*, 9048–9055.
136. Krishnamoorti, R., and Giannelis, E. P., Rheology of end-tethered polymer layered silicate nanocomposites. *Macromolecules* **1997**, *30*, 4097–4102.
137. Ren, J., Casanueva, B. F., Mitchell, C. A., and Krishnamoorti, R., Disorientation kinetics of aligned polymer layered silicate nanocomposites. *Macromolecules* **2003**, *36*, 4188–4194.
138. Alig, I., Skipa, T., Lellinger, D., and Poetschke, P., Destruction and formation of a carbon nanotube network in polymer melts: Rheology and conductivity spectroscopy. *Polymer* **2008**, *49*, 3524–3532.

139. Cipriano, B. H., Kota, A. K., Gershon, A. L., Laskowski, C. J., Kashiwagi, T., Bruck, H. A., and Raghavan, S. R., Conductivity enhancement of carbon nanotube and nanofiber-based polymer nanocomposites by melt annealing. *Polymer* **2008**, *49*, 4846–4851.
140. Munson-McGee, S. H., Estimation of the critical concentration in an anisotropic percolation network. *Phys. Rev. B* **1991**, *43*, 3331–3336.
141. Balberg, I., Binenbaum, N., and Wagner, N., Percolation thresholds in the three-dimensional sticks system. *Phys. Rev. Lett.* **1984**, *52*, 1465–1468.
142. Wei, T., Luo, G., Fan, Z., Zheng, C., Yan, J., Yao, C., Li, W., and Zhang, C., Preparation of graphene nanosheet/polymer composites using in situ reduction-extractive dispersion. *Carbon* **2009**, *47*, 2296–2299.
143. Ansari, S., and Giannelis, E. P., Functionalized graphene sheet-poly(vinylidene fluoride) conductive nanocomposites. *J. Polym. Sci. B: Polym. Phys.* **2009**, *47*, 888–897.
144. Wang, W.-P., and Pan, C.-Y., Preparation and characterization of poly(methyl methacrylate)-intercalated graphite oxide/poly(methyl methacrylate) nanocomposite. *Polym. Eng. Sci.* **2004**, *44*, 2335–2339.
145. Raghu, A. V., Lee, Y. R., Jeong, H. M., and Shin, C. M., Preparation and physical properties of waterborne polyurethane/functionalized graphene sheet nanocomposites. *Macromol. Chem. Phys.* **2008**, *209*, 2487–2493.
146. Nguyen, D. A., Lee, Y. R., Raghu, A. V., Jeong, H. M., Shin, C. M., and Kim, B. K., Morphological and physical properties of a thermoplastic polyurethane reinforced with functionalized graphene sheet. *Polym. Int.* **2009**, *58*, 412–417.
147. Liang, J., Wang, Y., Huang, Y., Ma, Y., Liu, Z., Cai, J., Zhang, C., Gao, H., and Chen, Y., Electromagnetic interference shielding of graphene/epoxy composites. *Carbon* **2009**, *47*, 922–925.
148. Chung, D. D. L., Electrical applications of carbon materials. *J. Mater. Sci.* **2004**, *39*, 2645–2661.
149. Lettow, J. S., Aksay, I. A., Korkut, S., and Chiang, K. S. Printed electronics. U.S. Patent 20070092432 A1, **2009**.
150. Brandrup, J., Immergut, E. H., Grulke, E. A., Abe, A., and Bloch, D. R. *Polymer handbook*, 4th ed. New York: John Wiley & Sons, **2005**.
151. Kalaitzidou, K., Fukushima, H., and Drzal, L. T., Multifunctional polypropylene composites produced by incorporation of exfoliated graphite nanoplatelets. *Carbon* **2007**, *45*, 1446–1452.
152. Kim, S., Do, I., and Drzal, L. T., Multifunctional xgnp/lldpe nanocomposites prepared by solution compounding using various screw rotating systems. *Macromol. Mater. Eng.* **2009**, *294*, 196–205.
153. Prasher, R. S., Chang, J.-Y., Sauciuc, I., Narasimhan, S., Chau, D., G., C., Myers, A., Prstic, S., and Hu, C., Nano and micro technology-based next-generation package-level cooling solutions. *Intel Technol. J.* **2005**, *9*, 285–296.
154. Biercuk, M. J., Llaguno, M. C., Radosavljevic, M., Hyun, J. K., Johnson, A. T., and Fischer, J. E., Carbon nanotube composites for thermal management. *Appl. Phys. Lett.* **2002**, *80*, 2767–2769.
155. Leong, C.-K., and Chung, D. D. L., Carbon black dispersions as thermal pastes that surpass solder in providing high thermal contact conductance. *Carbon* **2003**, *41*, 2459–2469.
156. Xie, S. H., Liu, Y. Y., and Li, J. Y., Comparison of the effective conductivity between composites reinforced by graphene nanosheets and carbon nanotubes. *Appl. Phys. Lett.* **2008**, *92*, 243121/1–243121/3.

157. Ghose, S., Working, D. C., Connell, J. W., Smith, J. G., Jr., Watson, K. A., Delozier, D. M., Sun, Y. P., and Lin, Y., Thermal conductivity of ultem/carbon nanofiller blends. *High Perform. Polym.* **2006**, *18*, 961–977.
158. Yu, A., Ramesh, P., Sun, X., Bekyarova, E., Itkis, M. E., and Haddon, R. C., Enhanced thermal conductivity in a hybrid graphite nanoplatelet - carbon nanotube filler for epoxy composites. *Adv. Mater.* **2008**, *20*, 4740–4744.
159. Yu, A., Ramesh, P., Itkis, M. E., Bekyarova, E., and Haddon, R. C., Graphite nanoplatelet-epoxy composite thermal interface materials. *J. Phys. Chem. C* **2007**, *111*, 7565–7569.
160. Huxtable, S. T., Cahill, D. G., Shenogin, S., Xue, L., Ozisik, R., Barone, P., Usrey, M., Strano, M. S., Siddons, G., Shim, M., and Keblinski, P., Interfacial heat flow in carbon nanotube suspensions. *Nat. Mater.* **2003**, *2*, 731–734.
161. Kim, P., Shi, L., Majumdar, A., and McEuen, P. L., Thermal transport measurements of individual multiwalled nanotubes. *Phys. Rev. Lett.* **2001**, *87*, 215502/1–215502/4.
162. Zhong, H., and Lukes, J. R., Interfacial thermal resistance between carbon nanotubes: Molecular dynamics simulations and analytical thermal modeling. *Phys. Rev. B: Condens. Matter Mater. Phys.* **2006**, *74*, 125403/1–125403/10.
163. Pollack, G. L., Kapitza resistance. *Rev. Mod. Phys.* **1969**, *41*, 48–81.
164. Shenogina, N., Godawat, R., Keblinski, P., and Garde, S., How wetting and adhesion affect thermal conductance of a range of hydrophobic to hydrophilic aqueous interfaces. *Phys. Rev. Lett.* **2009**, *102*, 156101/1–156101/4.
165. Shenogin, S., Bodapati, A., Xue, L., Ozisik, R., and Keblinski, P., Effect of chemical functionalization on thermal transport of carbon nanotube composites. *Appl. Phys. Lett.* **2004**, *85*, 2229–2231.
166. Liu, C. H., and Fan, S. S., Effects of chemical modifications on the thermal conductivity of carbon nanotube composites. *Appl. Phys. Lett.* **2005**, *86*, 123106/1–123106/3.
167. Veca, L. M., Meziani, M. J., Wang, W., Wang, X., Lu, F., Zhang, P., Lin, Y., Fee, R., Connell, J. W., and Sun, Y.-P., Carbon nanosheets for polymeric nanocomposites with high thermal conductivity. *Adv. Mater.* **2009**, *21*, 2088–2092.
168. Ganguli, S., Roy, A. K., and Anderson, D. P., Improved thermal conductivity for chemically functionalized exfoliated graphite/epoxy composites. *Carbon* **2008**, *46*, 806–817.
169. Fukushima, H., Drzal, L. T., Rook, B. P., and Rich, M. J., Thermal conductivity of exfoliated graphite nanocomposites. *J. Therm. Anal. Calorim.* **2006**, *85*, 235–238.
170. Kim, S., and Drzal, L. T., High latent heat storage and high thermal conductive phase change materials using exfoliated graphite nanoplatelets. *Sol. Energ. Mater. Sol. Cells* **2009**, *93*, 136–142.
171. Gomez-Navarro, C., Burghard, M., and Kern, K., Elastic properties of chemically derived single graphene sheets. *Nano Lett.* **2008**, *8*, 2045–2049.
172. Rafiee, M. A., Rafiee, J., Wang, Z., Song, H., Yu, Z.-Z., and Koratkar, N., Enhanced mechanical properties of nanocomposites at low graphene content. *ACS Nano* **2009**, *3*, 3884–3890.
173. Ramanathan, T., Abdala, A. A., Stankovich, S., Dikin, D. A., Herrera-Alonso, M., Piner, R. D., Adamson, D. H., Schniepp, H. C., Chen, X., Ruoff, R. S., Nguyen, S. T., Aksay, I. A., Prud'Homme, R. K., and Brinson, L. C., Functionalized graphene sheets for polymer nanocomposites. *Nat. Nanotechnol.* **2008**, *3*, 327–331.

174. Mori, T., and Tanaka, K., Average stress in matrix and average elastic energy of materials with misfitting inclusions. *Acta Metallurg.* **1973**, *21*, 571–574.
175. Tandon, G. P., and Weng, G. J., The effect of aspect ratio of inclusions on the elastic properties of unidirectionally aligned composites. *Polym. Compos.* **1984**, *5*, 327–333.
176. Kai, W., Hirota, Y., Hua, L., and Inoue, Y., Thermal and mechanical properties of a poly(e-caprolactone)/graphite oxide composite. *J. Appl. Polym. Sci.* **2008**, *107*, 1395–1400.
177. Cai, D., and Song, M., A simple route to enhance the interface between graphite oxide nanoplatelets and a semi-crystalline polymer for stress transfer. *Nanotechnology* **2009**, *20*, 315708/1–315708/6.
178. Prasad, K. E., Das, B., Maitra, U., Ramamurty, U., and Rao, C. N. R., Extraordinary synergy in the mechanical properties of polymer matrix composites reinforced with 2 nanocarbons. *Proc. Natl. Acad. Sci. USA* **2009**, *106*, 13186–13189, S13186/1–S13186/3.
179. Kelly, B. T., Thermal expansion coefficient of graphite parallel to the basal planes. *Carbon* **1972**, *10*, 429–433.
180. Nelson, J. B., and Riley, D. P., The thermal expansion of graphite from 15 °C to 800 °C: Part I. Experimental. *Proc. Phys. Soc.* **1945**, *57*, 477.
181. Osman, M. A., Mittal, V., Morbidelli, M., and Suter, U. W., Polyurethane adhesive nanocomposites as gas permeation barrier. *Macromolecules* **2003**, *36*, 9851–9858.
182. Lape, N. K., Nuxoll, E. E., and Cussler, E. L., Polydisperse flakes in barrier films. *J. Membr. Sci.* **2004**, *236*, 29–37.
183. Xu, Y., Hong, W., Bai, H., Li, C., and Shi, G., Strong and ductile poly(vinyl alcohol)/graphene oxide composite films with a layered structure. *Carbon* **2009**, *47*, 3538–3543.
184. Villar-Rodil, S., Paredes, J. I., Martinez-Alonso, A., and Tascon, J. M. D., Preparation of graphene dispersions and graphene-polymer composites in organic media. *J. Mater. Chem.* **2009**, *19*, 3591–3593.
185. Kashiwagi, T., Du, F., Douglas, J. F., Winey, K. I., Harris, R. H., and Shields, J. R., Nanoparticle networks reduce the flammability of polymer nanocomposites. *Nat. Mater.* **2005**, *4*, 928–933.
186. Kashiwagi, T., Mu, M., Winey, K., Cipriano, B., Raghavan, S. R., Pack, S., Rafailovich, M., Yang, Y., Grulke, E., Shields, J., Harris, R., and Douglas, J., Relation between the viscoelastic and flammability properties of polymer nanocomposites. *Polymer* **2008**, *49*, 4358–4368.
187. Graphene-Industries Raw graphene, Devices and Membranes for Sale. http://grapheneindustries.com/?Sample+Catalog, Accessed on June 14, 2010.
188. Baytubes, www.baytubes.com, Accessed on June 14, 2010.
189. SouthWest NanoTechnologies, www.swentnano.com/, Accessed on June 14, 2010.
190. Vorbeck Materials, www.vorbeck.com, Accessed on June 14, 2010.
191. Angstron Materials, www.angstronmaterials.com, Accessed on June 14, 2010.
192. Magrez, A., Kasas, S., Salicio, V., Pasquier, N., Seo, J. W., Celio, M., Catsicas, S., Schwaller, B., and Forro, L., Cellular toxicity of carbon-based nanomaterials. *Nano Lett.* **2006**, *6*, 1121–1125.
193. Panessa-Warren, B. J., Warren, J. B., Wong, S. S., and Misewich, J. A., Biological cellular response to carbon nanoparticle toxicity. *J. Phys.: Condens. Matter* **2006**, *18*, S2185–S2201.

194. Yang, D., Velamakanni, A., Bozoklu, G., Park, S., Stoller, M., Piner, R. D., Stankovich, S., Jung, I., Field, D. A., Ventrice, C. A., Jr., and Ruoff, R. S., Chemical analysis of graphene oxide films after heat and chemical treatments by x-ray photoelectron and micro-Raman spectroscopy. *Carbon* **2009**, *47*, 145–152.
195. Li, Z., Zhang, W., Luo, Y., Yang, J., and Hou, J. G., How graphene is cut upon oxidation? *J. Am. Chem. Soc.* **2009**, *131*, 6320–6321.
196. Kim, M. C., Hwang, G. S., and Ruoff, R. S., Epoxide reduction with hydrazine on graphene: A first principles study. *J. Chem. Phys.* **2009**, *131*, 064704/1–064704/5.
197. Thostenson, E. T., and Chou, T.-W., Aligned multi-walled carbon nanotube-reinforced composites: Processing and mechanical characterization. *J. Phys. D: Appl. Phys.* **2002**, *35*, L77–L80.
198. Haggenmueller, R., Gommans, H. H., Rinzler, A. G., Fischer, J. E., and Winey, K. I., Aligned single-wall carbon nanotubes in composites by melt processing methods. *Chem. Phys. Lett.* **2000**, *330*, 219–225.
199. Hong, C.-M., Kim, J., and Jana, S. C., Shear-induced migration of conductive fillers in injection molding. *Polym. Eng. Sci.* **2004**, *44*, 2101–2109.
200. Kharchenko, S. B., Douglas, J. F., Obrzut, J., Grulke, E. A., and Migler, K. B., Flow-induced properties of nanotube-filled polymer materials. *Nat. Mater.* **2004**, *3*, 564–568.
201. Obrzut, J., Douglas, J. F., Kharchenko, S. B., and Migler, K. B., Shear-induced conductor-insulator transition in melt-mixed polypropylene-carbon nanotube dispersions. *Phys. Rev. B: Condens. Matter Mater. Phys.* **2007**, *76*, 195420/1–195420/9.
202. Du, F., Fischer, J. E., and Winey, K. I., Effect of nanotube alignment on percolation conductivity in carbon nanotube/polymer composites. *Phys. Rev. B: Condens. Matter* **2005**, *72*, 121404/1–121404/4.
203. Hull, D., and Clyne, T. W., *An introduction to composite materials*, 2nd ed. London: Cambridge University Press, **1996**.
204. Bharadwaj, R. K., Modeling the barrier properties of polymer-layered silicate nanocomposites. *Macromolecules* **2001**, *34*, 9189–9192.
205. Vaia, R. A., and Maguire, J. F., Polymer nanocomposites with prescribed morphology: Going beyond nanoparticle-filled polymers. *Chem. Mater.* **2007**, *19*, 2736–2751.
206. Mamedov, A. A., Kotov, N. A., Prato, M., Guldi, D. M., Wicksted, J. P., and Hirsch, A., Molecular design of strong single-wall carbon nanotube/polyelectrolyte multilayer composites. *Nat. Mater.* **2002**, *1*, 190–194.
207. Velasco-Santos, C., Martinez-Hernandez, A. L., Fisher, F. T., Ruoff, R., and Castano, V. M., Improvement of thermal and mechanical properties of carbon nanotube composites through chemical functionalization. *Chem. Mater.* **2003**, *15*, 4470–4475.
208. Konatham, D., and Striolo, A., Thermal boundary resistance at the graphene-oil interface. *Appl. Phys. Lett.* **2009**, *95*, 163105/1–163105/3.
209. Zammarano, M., Kramer, R. H., Harris, R., Jr., Ohlemiller, T. J., Shields, J. R., Rahatekar, S. S., Lacerda, S., and Gilman, J. W., Flammability reduction of flexible polyurethane foams via carbon nanofiber network formation. *Polym. Adv. Technol.* **2008**, *19*, 588–595.
210. Thirumal, M., Khastgir, D., Singha, N. K., Manjunath, B. S., and Naik, Y. P., Effect of expandable graphite on the properties of intumescent flame-retardant polyurethane foam. *J. Appl. Polym. Sci.* **2008**, *110*, 2586–2594.
211. Koerner, H., Price, G., Pearce, N. A., Alexander, M., and Vaia, R. A., Remotely actuated polymer nanocomposites-stress-recovery of carbon-nanotube-filled thermoplastic elastomers. *Nat. Mater.* **2004**, *3*, 115–120.

212. Al-Saleh, M. H., and Sundararaj, U., Electromagnetic interference (EMI) shielding effectiveness of PP/PS polymer blends containing high structure carbon black. *Macromol. Mater. Eng.* **2008**, *293*, 789.
213. Mather, P. J., and Thomas, K. M., Carbon black/high density polyethylene conducting composite materials. Part II. The relationship between the positive temperature coefficient and the volume resistivity. *J. Mater. Sci.* **1997**, *32*, 1711–1715.
214. Kobashi, K., Villmow, T., Andres, T., Haeussler, L., and Poetschke, P., Investigation of liquid sensing mechanism of poly(lactic acid)/multi-walled carbon nanotube composite films. *Smart Mater. Struct.* **2009**, *18*, 035008/1–035008/15.
215. Wang, X., Fu, X., and Chung, D. D. L., Strain sensing using carbon fiber. *J. Mater. Res.* **1999**, *14*, 790–802.
216. Ahir, S. V., and Terentjev, E. M., Photomechanical actuation in polymer-nanotube composites. *Nat. Mater.* **2005**, *4*, 491–495.
217. Baughman, R. H., Cui, C., Zakhidov, A. A., Iqbal, Z., Barisci, J. N., Spinks, G. M., Wallace, G. G., Mazzoldi, A., De Rossi, d., Rinzler, A. G., Jaschinski, O., Roth, S., and Kertesz, M., Carbon nanotube actuators. *Science* **1999**, *284*, 1340–1344.
218. Nair, R. R., Blake, P., Grigorenko, A. N., Novoselov, K. S., Booth, T. J., Stauber, T., Peres, N. M. R., and Geim, A. K., Fine structure constant defines visual transparency of graphene. *Science* **2008**, *320*, 1308.
219. Kim, K. S., Zhao, Y., Jang, H., Lee, S. Y., Kim, J. M., Kim, K. S., Ahn, J.-H., Kim, P., Choi, J.-Y., and Hong, B. H., Large-scale pattern growth of graphene films for stretchable transparent electrodes. *Nature* **2009**, *457*, 706–710.
220. Eda, G., Fanchini, G., and Chhowalla, M., Large-area ultrathin films of reduced graphene oxide as a transparent and flexible electronic material. *Nat. Nanotechnol.* **2008**, *3*, 270–274.
221. Tung, V. C., Chen, L.-M., Allen, M. J., Wassei, J. K., Nelson, K., Kaner, R. B., and Yang, Y., Low-temperature solution processing of graphene-carbon nanotube hybrid materials for high-performance transparent conductors. *Nano Lett.* **2009**, *9*, 1949–1955.
222. Blake, P., Brimicombe, P. D., Nair, R. R., Booth, T. J., Jiang, D., Schedin, F., Ponomarenko, L. A., Morozov, S. V., Gleeson, H. F., Hill, E. W., Geim, A. K., and Novoselov, K. S., Graphene-based liquid crystal device. *Nano Lett.* **2008**, *8*, 1704–1708.
223. Becerril, H. A., Mao, J., Liu, Z., Stoltenberg, R. M., Bao, Z., and Chen, Y., Evaluation of solution-processed reduced graphene oxide films as transparent conductors. *ACS Nano* **2008**, *2*, 463–470.
224. Liang, J., Huang, Y., Zhang, L., Wang, Y., Ma, Y., Guo, T., and Chen, Y., Molecular-level dispersion of graphene into poly(vinyl alcohol) and effective reinforcement of their nanocomposites. *Adv. Funct. Mater.* **2009**, *19*, 2297–2302.
225. Fang, M., Wang, K., Lu, H., Yang, Y., and Nutt, S., Covalent polymer functionalization of graphene nanosheets and mechanical properties of composites. *J. Mater. Chem.* **2009**, *19*, 7098–7105.
226. Cai, D., Yusoh, K., and Song, M., The mechanical properties and morphology of a graphite oxide nanoplatelet/polyurethane composite. *Nanotechnology* **2009**, *20*, 085712/1–085712/5.
227. Mack, J. J., Viculis, L. M., Ali, A., Luoh, R., Yang, G., Hahn, H. T., Ko, F. K., and Kaner, R. B., Graphite nanoplatelet reinforcement of electrospun polyacrylonitrile nanofibers. *Adv. Mater.* **2005**, *17*, 77–80.
228. Zhang, L., Liang, J., Huang, Y., Ma, Y., Wang, Y., and Chen, Y., Size-controlled synthesis of graphene oxide sheets on a large scale using chemical exfoliation. *Carbon* **2009**, *47*, 3365–3368.

17

Highly Filled Graphite–Polymer Composites: Synthesis, Processing, and Characterization

Sadhan C. Jana* **and Ling Du**

CONTENTS

17.1 Introduction

Tetravalent carbon has several ramifications encompassing both good and bad effects on human life. As hydrocarbons, it is the primary source of fossil energy and thus constitutes the fabric of our daily life. It forms the backbone of various dietary nutritional elements such as carbohydrates, fat, and proteins. Lightweight plastic components used in automobiles, aircraft, and space craft and the bulk of rubber materials used in tires are derived from tetravalent carbon compounds primarily of hydrogen. Grocery bags and synthetic clothing materials are made of polymeric tetravalent carbon compounds to satiate the ever increasing demands of consumer industry. The oxides of carbon such as gaseous carbon monoxide and carbon dioxide are also tetravalent, but appear as the villain in almost all policy talks on sustainable environment. The environmental impacts of most chemical processes are now evaluated and measured in terms of carbon footprints.

* Correspondence to: Sadhan C. Jana (janas@uakron.edu)

This chapter is about one form of elemental carbon and its solid functionalized compounds with oxygen, nitrogen, and hydrogen that hold great promise in an array of high technology applications, such as in electronics, sensors, devices, and fuel cells. This chapter specifically deals with graphite nanoparticles and their compounds with an array of host polymers. Superior electrical and thermal conductivity of various forms of graphite particles, such as exfoliated graphite, expanded graphite (EG), and even single graphite platelets of 1 nm in thickness have made this field very appealing to a broader audience including scientists, engineers, and even policy makers. The 2010 Nobel Prize in Physics given for identification of superior optical, electrical, and thermal properties of single graphite platelets (Geim and Novoselov 2007; Novoselov et al. 2004, 2005) caught the attention of many researchers not earlier involved in this field and consequently research involving graphene and in general graphite grew exponentially fast. This chapter presents a review of existing literature on polymer–graphite compounds both at low loading (less than 5 wt%) and high loading (greater than 20 wt%). The underlying principles of attaining desired electrical and thermal conductivity and mechanical properties are discussed.

The chapter is divided as follows. First, we present brief discussions on various forms of carbon nanomaterials. The electrical properties of polymer composites of various forms of graphitic nanomaterials are then presented in dilute regime with graphite loading less than 5 wt%. In this regard, the intrinsic electrical conductivity of graphite nanoplatelets is first discussed followed by a discussion on percolation at graphite content lower than 5% by volume. However, a major emphasis in this chapter is placed on preparation and properties of electrically conductive polymer compounds with volumetric electrical conductivity of the compounds reaching at least 10 S/cm. Such compounds will be useful in preparation of bipolar plates of proton exchange membrane (PEM) fuel cells and usually requires graphite loading exceeding 20% by volume. In this context, the limitations of graphite in providing large through-plane conductivity are discussed and the means of improving such conductivities using synergistic combinations of several nanofillers are reviewed.

17.2 Various Forms of Carbon Nanomaterials Including Graphite

In order to understand the basic structure and properties of graphite, let us first review the structure and properties of the various forms of carbon family. Carbon is the lightest element in Group IV in the periodic table with some unique properties. At ambient conditions, sp^2-bonded graphite (Figure 17.1a) (Yoshimura and Chang 1998) is the ground state phase of carbon. At higher temperatures and pressures, sp^3-bonded cubic diamond (Figure 17.1b)

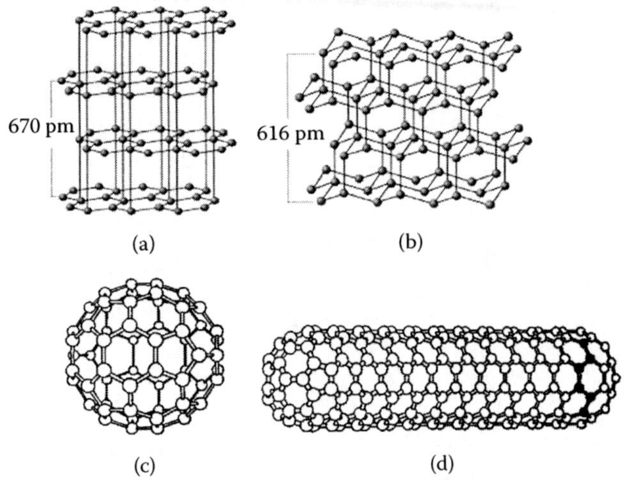

FIGURE 17.1
Carbon nanomaterials: (a) Graphite, (b) diamond, (c) fullerene, and (d) carbon nanotubes.

(Yoshimura and Chang 1998) is a stable form while other regions show stability ranges for hexagonal diamond, hexagonal carbines, and liquid carbon. In addition to the bulk phases, much attention has recently been focused on the small carbon clusters since the discovery of fullerenes by Kroto et al. in 1985 and of carbon nanotubes (CNTs) by Iijima in 1991. The physical reason behind these nanostructures is energy minimization from the high-energy state of a graphene layer. The graphene layer of finite size has many edge atoms carrying dangling bonds, thus giving rise to the high-energy states. The total energy of a small number (30–100) of carbon atoms is reduced by eliminating the dangling bonds, even at the expense of increasing the strain energy, thereby promoting the formation of closed cage fullerene molecules, the most stable of which are C_{60} (Figure 17.1c). Under slightly different growth conditions, CNTs (Figure 17.1d) are formed. A CNT can be seen as a cylinder rolled from a graphene sheet, capped at both ends by hemispheres of fullerenes. Although the high curvature of the graphene sheets increases the total energy, CNT is more stable than single sheet of graphene as the energy is substantially lowered due to elimination of dangling bonds at the edges.

Many commercial carbon materials, such as carbon fibers and glassy carbon, contain carbon networks that can be approximated by graphene ribbons. Studies on edge states might help the tailoring of commercial carbons to possess certain desired properties, such as the properties of porous carbons that have a very high density of edge states. Graphite edge sites are more reactive than the bulk sites because of the availability of dangling bonds. Thus, carbon clusters have been investigated in terms of their potential for the uptake of active species, which could be of interest for many applications such as carbon batteries.

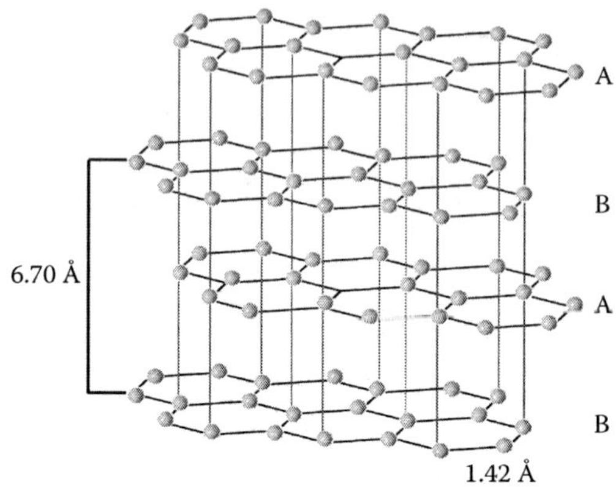

FIGURE 17.2
Graphite lattice showing the ABAB stacking of carbon layers. (Adapted from Zabel, H., and Solin, S. A., *Graphite Intercalation Compounds I—Structure and Dynamics*, Springer-Verlag, Berlin, 1990.)

The pure graphite structure shown in Figure 17.2 consists of hexagonal net planes of carbon stacked along the c-axis in a staggered array usually denoted as ABAB … (Zabel and Solin 1990). While this is the normal graphite stacking, ABCA …, ABAC …, and so on, are possible sequences and the commonly observed stacking faults in graphite are a reflection of these discrete stacking alternatives. Single graphite crystals can be both natural and synthetic with various degrees of perfection.

Graphite is available in natural form or can be synthesized. Natural graphite (NG) flakes have been combined with polymeric binders to obtain electrically conductive composites with the intent of using these composites as inert electrodes (Krohn et al. 1988). These composites offered electrical percolation at 5–20 wt% loading of NG flakes, and as high as 70–80 wt% flakes were used to make inert electrodes. The high loading of NG flakes required for attaining electrical percolation in composites with polymeric binders obviated further development. This was followed by research efforts on the separation of individual graphite layers from the flakes.

One area of research involved the use of graphite intercalation compounds, such as sulfuric acid (Hennig 1951), mixture of nitric and sulfuric acid (Dedov 2001), or other charge donor/acceptors (Zaghib et al. 2000) to penetrate into the interstices between two adjacent graphene layers, followed by rapid heating at high temperature to obtain EG particles. EG particles exhibit much higher surface area per unit weight; for example, an increase from 1.2 m^2/g before expansion to 43 m^2/g after expansion by heating the intercalated compound at 1000°C (Vieira et al. 2006) indicates opening of galleries of adjacent graphite layers. A representative scanning electron microscope (SEM) image of EG is presented in Figure 17.3 showing partial delamination of adjacent graphite layers.

Research publications now claim good handle on large-scale production of EG particles using combinations of thermal expansion of

FIGURE 17.3
Scanning electron microscope (SEM) image of expanded graphite (EG) particle. (From Du, L., "Highly conductive epoxy/graphite polymer composite bipolar plates in proton exchange membrane (PEM) fuel cells," PhD Thesis, University of Akron, OH, 2008. With permission.)

graphite-intercalated compounds and ultrasonication (Choi et al. 2011). Although the EG particles have been considered as nanoparticles in a number of research publications, the expectation of single platelet delamination has not been fully realized. A recent study showed that single- and double-layer graphene nanoplatelets can be easily obtained from re-exfoliation of EG particles using sonication and centrifugation in organic solvent (Dhakate et al. 2011). Figure 17.4 presents a schematic of the process.

A parallel research effort ensued in converting graphite into graphite oxide (GO) and graphene oxide in several research laboratories across the globe. A recent review article presents comprehensive coverage of various chemical reactions associated with the synthesis of GO and subsequent functionalization (Dreyer et al. 2010). The first study on synthesis of GO was reported in 1859 by Brodie, where graphite flakes were oxidized by a mixture of potassium chlorate and fuming nitric acid, which was later perfected by Staudenmaier in 1898. However, the method of Hummers and Offeman (1958), based on the use of a mixture of potassium permanganate and concentrated sulfuric acid as the oxidizing medium, has become more popular among researchers for the synthesis of GO. This method is known as Hummers method in literature and often uses naturally occurring graphite flakes after purification (Wissler 2006).

Several structural models of GO exists in literature, including the lattice models (Hofmann and Holst 1939; Ruess 1946) and the amorphous model

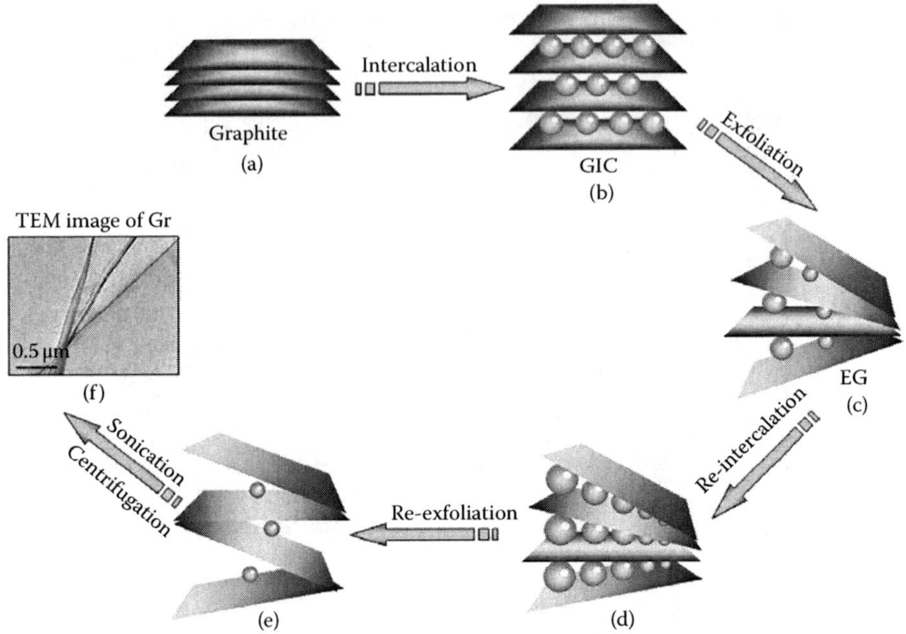

FIGURE 17.4
Schemes for the production of single and double layer graphite nanoplatelets. (From Dhakate, S. R., Chauhan, N., Sharma, S., Tawale, J. S. S., Sahare, P. D., and Mathur, R. B., *Carbon*, 49, 1946, 2011. With permission.)

(Lerf et al. 1998). These models took into account the C:H:O ratios determined from the analysis of composition, reactivity of GO, x-ray diffraction (XRD), and nuclear magnetic resonance (NMR) and accommodated the existence of polar functional groups, such as epoxide, carbonyl, ether, carboxylic acid, and alcoholic –OH, in the structures of GO particles. The epoxide and alcoholic –OH functional groups are predominant among these.

GO particles are electrical insulators due to loss of π bonds in oxidation reactions. Therefore, GO particles must be converted back into graphitic structures, preferably into single platelet graphene, for applications requiring high electrical conductivity. This is achieved by chemical reduction, electrochemical reduction, or by thermal means, such as annealing (Dreyer et al. 2010). The methods of chemical reduction generally involve dispersion of GO particles in aqueous medium and reactions with a strong reductant, such as hydrazine monohydrate (Stankovich et al. 2007) or sodium borohydride (Shin et al. 2009). Readers are referred to Dreyer et al. (2010) for a summary of other reductants used in chemical reduction of GO. The quality of graphene materials obtained from reduction process is monitored from measurement of sheet resistance, and the quantity of surface functional groups is measured from x-ray photoelectron spectra (Dreyer et al. 2010; Geim and Novoselov 2007; Warner et al. 2009).

Thermal reduction of GO particles into graphene involves rapid heating of the materials to approximately 1000°C. The gas, usually carbon dioxide, emanating inside the particles due to thermal decomposition of the organic functional groups develop high pressure as high as 130 MPa (McAllister et al. 2007; Schniepp et al. 2006). Consequently, graphite platelets become exfoliated with the mass loss of approximately 30 wt% (Schniepp et al. 2006).

The different organic functional groups of graphene oxide particles can be used to plant other functional groups, such as amines on the surfaces of GO particles. These can be later used to obtain covalently bonded ester and amide molecules such that functionalized GO particles can be easily dispersed in hosts of polymers.

The values of electrical and thermal conductivity are presented in Tables 17.1 and 17.2. It is seen from Table 17.1 that through-plane and in-plane electrical conductivity of EG differs by three orders of magnitude. The thermal conductivity of bulk graphite is almost half than that of single layer graphene (Table 17.2). Some representative mechanical properties of graphene are listed in Table 17.3.

TABLE 17.1

Electrical Conductivity of NG and EG

Material	Electrical Conductivity (S/cm)	Source
NG	In plane: 2,500, through plane: ~10	Zheng et al. (2004) and Nam and Lee (2007)
EG	In plane: 10,000, through plane: ~10	Zheng et al. (2004) and Nam and Lee (2007)

TABLE 17.2

Thermal Conductivity of Graphene

Material	Thermal Conductivity (W/m/K)	Source
Single layer graphene	3080–5030	Balandin et al. (2008)
Bulk graphite (in plane)	2000	Klemens (2000, 2001)

TABLE 17.3

Representative Mechanical Properties of Graphene

Properties	Value	Source
Young's modulus	~1000 GPa	Lee et al. (2008)
Fracture strength	~125 GPa	Lee et al. (2008)
Elastic modulus	~0.25 TPa	Gomez-Navarro et al. (2008)
Specific surface area (calculated value)	~2630 m^2/g	Stoller et al. (2008)

17.3 Composites of EG

The research on polymer composites of EG is mostly driven by the need for obtaining high electrical conductivity. Applications involving lightweight antistatic materials, electromagnetic shielding materials, conductive films, conductive coatings, and phase-transfer catalysis materials all benefit from development of electrically conductive compounds of polymers preferably with smaller fractions of conductive particles. The usual conductive fillers are metal powder, carbon black (CB), CNT and carbon nanofiber, and graphite. Many factors influence the conductivity of composites, for example, the properties of conductive particles, such as the intrinsic conductivity; particle size distribution and particle aspect ratio; and the structure characterized in terms of the surface area, particle concentration, shear and extensional flows involved in compounding and processing steps, and intrinsic conductivity and crystallinity of the host polymers.

The conductive composites are characterized by a percolation threshold or a critical value at which the conductivity starts to increase dramatically as a function of filler content. However, the mechanical properties decrease drastically at high filler contents, typically at greater than 15 wt%. Therefore, much attention has been centered on the preparation of conducting composites with the objective of achieving a low percolation threshold (Blunk et al. 2006; Chen et al. 2002; Hong and Weil 2007; Lafront et al. 2007; Nam and Lee 2007; Nikam and Reddy 2005; Novak and Krupa 2004; Qu and Wong 2007; Ramirez-Garcia et al. 2004; Schniepp et al. 2006; Stankovich et al. 2006a, 2006b; Staschewski and Mao 1999; Tawfik et al. 2007; Wang and Northwood 2006, 2007a, 2007b, 2008; Wang and Pan 2004a, 2004b; Wu and Shaw 2005; Xiao et al. 2006).

EG is more efficient than CB particles in producing electrically conductive polymer composites. These particles have large specific surface area, a unique layered nanostructure, high adsorption capacity, and large occluded volume. The naturally occurring graphite has strong affinity for nonpolar molecules. During the oxidation process, however, the surface becomes polar as functional groups such as –OH and –COOH are formed at the surface of the graphite particles. Consequently, EG particles can absorb polar molecules. Moreover, it has excellent thermal conductivity of about 600 W/m/k and relatively high electrical conductivity of the order of 10^5 S/cm at room temperature (Novak and Krupa 2004; Ramirez-Garcia et al. 2004; Staschewski and Mao 1999).

The design and manufacture of lightweight polymer composites with high electrical conductivity have been a research focus in recent years. Zheng et al. (2004) investigated EG/polymer composites prepared by in situ polymerization. Poly(styrene-co-acrylonitrile) (PSAN)/EG composite sheets with very low in-plane resistivity ~8.5×10^{-3} Ω·cm and through-plane resistivity ~1.2×10^{-2} Ω·cm have been prepared. The EG was made by the oxidation of NG flakes, followed by the thermal expansion at 600°C. A closer inspection of the optical microscopy (OM) images of PSAN/EG composites was conducted by Zheng et al. (2004) at higher magnification. The OM images showed that all the graphite layers were well-connected. They observed that EG formed a network throughout the entire composite, which was coined as graphite-interpenetrating network structure. The network structure established from OM was later confirmed by SEM. A sample specimen was etched with tetrahydrofuran to remove PSAN and to expose the graphite phase. The images showed the honeycomb structure of graphite particles along with the diamond-shaped pores. A lightweight, highly conductive polymer composite based on PSAN was obtained by mixing 11 wt% EG (Zheng et al. 2004).

The oxidation and expansion conditions for EG manufacture are critical and must be carefully controlled in order to avoid excessive oxidation. The honeycomb structure is produced at low expansion ratio (~30) of EG. The in situ polymerization of styrene and acrylonitrile monomers inside the diamond-shaped pores of the EG produced a PSAN/EG composite with a well-connected graphite network in the compound. XRD and transmission electron microscopy results confirmed that the graphite network possesses the same crystal structure as that of NG. Both the in-plane and through-thickness electrical conductivities of the PSAN/EG composite were found to be very high. Especially, the through-thickness electrical conductivity of the composite was slightly higher than that of NG, as shown in Table 17.4. Once again, the higher electrical conductivity was attributed to the graphite network formed in the composites.

Wang and Pan (2004a, 2004b) prepared polystyrene (PS)/EG composite by cationic grafting polymerization. The electrical conductivity of the composite increased from 10^{-16} S/cm for unfilled PS to 10^{-2} S/cm with the inclusion of 13.4 wt% EG. The percolation threshold was found to be approximately 2.6 wt% EG. The conductivity of PS/EG composite was found to be higher than that of the NG powder/polymer composite at the same graphite loading of 13.4 wt%. Poor electrical conductivity in composites of NG was attributed to dense structures and poor dispersion of NG in the polymer.

Graphite is frequently used as electrically conductive filler for manufacturing high-performance antistatic and/or electrically conductive articles found in casting and coating applications. Improvements in mechanical

TABLE 17.4

Resistivity of NG, Compressed EG, and PSAN/EG Composites at Room Temperature

Direction of Measurement	Resistivity of NG (Ω·cm)	Resistivity of EG (Ω·cm)	Resistivity of Composites (Ω·cm)
Through plane	1.2×10^{-1}	1.1×10^{-1}	1.2×10^{-2}
In plane	4.0×10^{-5}	1.0×10^{-5}	8.5×10^{-3}

Source: Zheng, G. et al. *Carbon* 42:2839–2847, 2004. With permission.

properties, as well as thermal and dimensional stability are also observed. A few basic polymeric matrices have been used for casting –acrylics, polyesters, polyurethanes (PUs), and epoxies. Novak and Krupa (2004) studied electrical conductivity of epoxy resins filled with two grades of synthetic graphite (SG) and reported percolation threshold concentration at 22 vol% of the filler. However, the impact strength and the strength of adhesive joint decreased with the incorporation of SG particles in epoxy.

Ramirez-Garcia et al. (2004) carried out research on charge percolation and electroanalytical performance of carbon composite microelectrodes and reported percolation conductivity of SG-filled epoxy at 52 vol%. The composite exhibited a bulk conductivity of 0.15 S/cm, although the electrical conductivity increased dramatically to 96 S/cm when the SG loading was increased to 67 vol%.

Chen et al. (2002) investigated electrical conductivity of polypropylene (PP)/EG nanocomposites. They prepared maleic anhydride (MA) grafted polypropylene (MA-g-PP)/EG nanocomposites via solution intercalation. The results indicated that the MA-g-PP molecules infiltrated the interplanar spaces of EG particles due to porous feature and polar interaction between the MA group of MA-g-PP and the –OH, –COOH groups on the graphite layers. It was argued that MA-g-PP macromolecules supported and fixed the EG/MA-g-PP compound networks and maintained the high aspect ratio of the graphite sheets in the nanocomposites. Thus, a conductive network spanning the entire specimen could form at a very low EG concentration of about 0.67 vol%, thus causing the system to "percolate."

Qu and Wong (2007) investigated piezoresistive behavior of EG-reinforced maleated PP composites. The percolation threshold appeared to be very low, that is, ~0.4 wt%, and the conductivity increased to 10^{-8} S/cm. The electrical resistivity decreased dramatically with applied pressure on the PP/EG composite at a filler loading of 7.5 wt%. The resistivity decreased rapidly with compressive stress of less than 10 MPa and reached a plateau value at a higher compressive stress from 10 to 30 MPa. The resistivity increased slightly when the compressive stress increased beyond 30 MPa. These researchers believed that the observed piezoresistive phenomenon was caused by the formation of conductive paths at lower compressive stress and the degeneration of such networks at higher compressive stress. The network degeneration originated from reorientation of the graphite platelets in the matrix and their rupture due to the significant shear forces at large compressive stress.

EG particles and bifunctional epoxy molecules were separately modified by 3-isocyanatopropyltriethoxysilane, and the resultant materials were subjected to sol-gel reactions to obtain epoxy–EG composites (Chiang and Hsu 2010). These composites showed reduction of toxic gas release upon thermal decomposition. Similarly, polyethylene oxide methyl ether was successfully grafted onto EG particles via reactions of toluene diisocyanate with the hydroxyl and carboxylic acid groups on graphite particles and with the hydroxyl groups on polyethylene oxide methyl ether molecules (Wang and Pan 2004a, 2004b).

Ethylene vinyl acetate rubber mixed in solution with 4 wt% EG upon curing showed 35% increase in tensile strength and 150% increase in secant modulus at 100% elongation although the tensile strength decreased at 8 wt% loading of EG (George and Bhowmick 2008). Xie and Qu (2001) reported that EG used together can augment the performance of nonhalogen flame retardant materials and simultaneously can improve the mechanical properties of some polyolefin compounds.

Melt compounded composites of EG and high-density polyethylene (HDPE) showed electrical conductivity of 1×10^{-5} S/cm at approximately 15 wt% EG loading, while the electrical conductivity increased to approximately 0.13 S/cm at 50 wt% EG loading (Li et al. 2007). PSAN/EG composites with approximately 11 wt% EG were prepared by intercalating EG particles in styrene and acrylonitrile and carrying out in situ polymerization at 80°C. Such in situ polymerization preserved the honeycomb structures of EG particles and increased electrical conductivity (Zheng et al. 2004). An electrically conductive nanocomposite of poly(ether sulfone) was prepared by sonication of EG in solution in dichloromethane. The composite showed electrical percolation at 3 wt% filler loading and with 5 wt% filler exhibited volumetric conductivity of 10^{-2} S/cm (Ramanujam et al. 2010). Methyl methacrylate was polymerized in the pores of EG using an azo initiator first by soaking the monomer with EG and second by heating the particles at 60°C–80°C (Wang et al. 2006). The graphite particles with pores containing the polymer were then compression molded at 180°C to obtain composite specimens. It was found that approximately 7 wt% EG was needed for volume electrical conductivity of approximately 4 S/cm. Such higher conductivity was made possible due to retention of continuous honeycomb structures of the graphite platelets in the composite, as shown in Figure 17.5.

Compounds of polylactic acid with EG particles were prepared in twin screw extruder (Kim and Jeong 2010). The resultant composites showed electrical percolation at 3–5 wt% EG loading, and thermal degradation was delayed by almost 10 K with the addition of 0.5 wt% EG. The interfacial

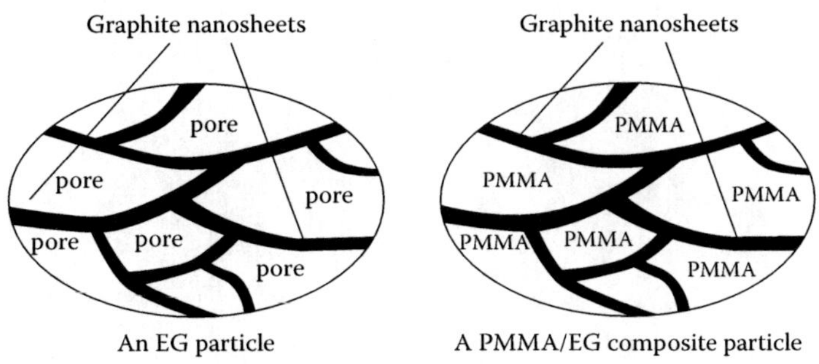

FIGURE 17.5
The use of honeycomb morphology used in absorption of monomer followed by polymerization. (a) EG showing pores. (b) Pores filled with polymethyl methacrylate (PMMA) polymer. (From Wang, W-P., Liu, Y., Li, X-X., and You, Y-Z., *J. Appl. Polym. Sci.*, 100, 1427, 2006. With permission.)

interactions between EG particles and the host polymer PP was improved in a recent study (Mirzazadeh et al. 2011) by considering maleated polypropylene (PP-g-MAH) and maleated ethylene propylene diene monomer (EPDM) rubber as compatibilizers. It was observed that maleated EPDM encapsulated EG particles and shielded them from forming conductive networks, while increased electrical conductivity was observed with the use of PP-g-MAH. A representative set of research reports on improvement of electrical conductivity are highlighted in Table 17.5.

TABLE 17.5

Electrical Conductivity of Representative Graphite–Polymer Composites

System	Compound Preparation Method	Percolation Fraction	Comments	Source
Ultrahigh molecular weight polyethylene/graphene nanosheet	Dispersion in water/ethanol by ultrasound followed by compression molding	0.7% by volume	Two-dimensional networks	Xu and Gao (2010)
PS/graphene nanosheets	Aqueous dispersion of PS latex and graphene followed by freeze drying	Less than 1 wt%	Graphene remained at the interfaces of polymer domains	Tkalya et al. (2010)
PMMA/graphene nanosheets	Solution mixing in acetonitrile/methanol mixture	Approximately 5 wt%	Homogeneous mixture	Jang et al. (2009)
PS/graphene nanosheet	Dispersed in solution in DMF	Less than 0.1% by volume		Liu et al. (2008)
PU/graphite nanosheet	In situ polymerization of PU	>2 wt%		Wang et al. (2011)
Ethylene–vinyl acetate/EG	Melt mixing	~7 vol%	Compression molded plaques. Particle aspect ratio varied between 20 and 250 with length 5 μm	Tlili et al. (2010)
PP/EG with maleated PP as compatibilizer	Melt mixing	~3 vol%		Mirzazadeh et al. (2011)

In summary, the following lessons are learned from existing literature on composites of polymers with low (<10 wt%) loading of EG:

1. The polar functional groups residing at the edges of EG can be easily exploited for covalent bond formation with polymer chains. Chemical reactions of functional groups such as isocyanates, amines, acid chlorides, and carboxylic acid promote such covalent bond formation. These reactions are usually carried out in solutions where improved dispersion of EG particles can be easily achieved with sonication. The polymer composites provided better electrical and thermal properties than composites of NG.
2. The porous structures of EG can absorb mixtures of monomers and thermal initiators into the pores (Wang et al. 2006). The monomer can be polymerized later raising the temperature.
3. Electrical percolation tends to occur usually at 3–5 wt% loading if EG is dispersed well. Lower than 1 wt% EG gives electrical percolation when the honeycomb structures are retained (Chen et al. 2002). On the other hand, much higher loading, as high as 15 wt%, is needed for percolation if particles are not properly dispersed (Ramanujam et al. 2010).

17.4 Highly Conductive EG–Polymer Composites

17.4.1 Technical Problems and Opportunities

As presented above, EG composites of an array of polymers show electrical percolation with less than 5 wt% EG. The electrical percolation conductivity ranges from 10^{-8} to 10^{-3} S/cm in most polymeric systems. However, such values of conductivity are well below the specifications of many interesting applications requiring volume conductivity greater than 10 S/cm. Examples include bipolar plates for PEM fuel cells and inert electrodes for batteries. If EG or graphene particles are used in such examples, it is imperative that the filler loading has to exceed 10 wt% and most such electrically conductive composites may require loading as high as 60 wt%. These limits of filler loading bring new scientific questions. For example, achieving nanoscale dispersion of graphite platelets in the host polymer is no longer a key objective. Instead, ease of processing of the compounds and manufacturing of the articles at such high graphite loading become determining factors. In addition, the anisotropic electrical conductivity of EG, for example, much higher in-plane versus through-plane conductivity (Table 17.1), creates a bottleneck on the final electrical conductivity of the composites. Specifically, it is conceivable that graphite platelets, due to high concentrations, will tend to orient more along the major direction of the finished articles, for example, along the flow direction in injection-molded or profile-extruded articles.

Consequently, the through-plane electrical conductivity of these articles may be inferior. Therefore, two important issues for such composites are inferior through-plane conductivity and adverse effects of high viscosity on processing. Some secondary effects must also be considered; for example, graphite particles may migrate from the surface of injection-molded articles due to shear-induced migration (Gadala-Maria and Acrivos 1980; Hong et al. 2004; Leighton and Acrivos 1987; Phillips et al. 1992) and reduce surface conductivity. In case of low viscosity resins, such as when epoxy and unsaturated polyesters are used, the resin may drain out from the regions of high particle concentration, thus inducing inhomogeneity in cured articles.

17.4.2 Highly Filled Thermoset Systems

First a brief review of existing peer-reviewed literature on highly filled thermoset polymeric systems is presented. This is followed by discussion on electrical, thermal, and mechanical properties of epoxy-filled expanded and SG systems prepared in our laboratory.

Du and Jana (2007) developed highly filled graphite–epoxy composite materials for potential application as bipolar plates in PEM fuel cells. The composites were prepared by solution intercalation mixing of EG and CB particles with epoxy resin and curatives, followed by compression molding and curing. Electrical conductivity, thermal and mechanical properties, and hygrothermal characteristics were determined as function of carbon-filler content. It was found that a synergistic combination of EG and CB particles provided desired in-plane and through-plane conductivities. These composites showed high glass transition temperatures (T_g ~ 180°C), high thermal degradation temperatures (~415°C), in-plane conductivity of 200–500 S/cm with 50 wt% carbon fillers, in addition to offering high values of flexural modulus, flexural strength, and impact strength, respectively, 2×10^4 MPa, 72 MPa, and 173 J/m. The presence of carbon fillers helped reduce water uptake from 4 to 5 wt% for unfilled epoxy resins to 1–2 wt%. In addition, morphological, electrical, mechanical, and thermal properties remain unchanged on exposure to boiling water and acid reflux. Du and Jana (2008) investigated the hygrothermal effects on mechanical, thermal, and electrical properties of highly conductive graphite-based epoxy composites described in Ref. [73]. The composite specimens were immersed in boiling water, aqueous sulfuric acid solution, and aqueous solution of hydrogen peroxide. The water uptake, changes in surface appearance and dimensions, glass transition behavior and thermal stability, and electrical and mechanical properties were evaluated. The water uptake at short time increased linearly with the square root of time as in linear Fickian diffusion. The presence of graphite significantly reduced both the rate and extent of water uptake. No discernible changes in specimen dimensions, surface appearance, and morphology of the composites were observed. The electrical conductivity and mechanical properties remained almost unchanged. The wet specimens showed slight reduction of T_g due to plasticization of epoxy networks by absorbed water, while the re-dried specimens showed small increase of T_g. The composites maintained high electrical conductivity of about 300–500 S/cm and good mechanical properties and showed thermal stability up to 350°C.

The synergistic combination of EG and CB was later used in the work of Hui et al. (Chen et al. 2009) for the development of highly conductive composites of modified phenolic resins. These composites showed electrical conductivity close to 90 S/cm, flexural strength ~40 MPa, and thermal stability up to 420°C.

Tjong (2011) provided a review of literature on manufacturing electrically conductive polymeric composites suitable for bipolar plates of fuel cells. Apart from the use of synergistic combinations of EG and CB described above (Chen et al. 2009; Du and Jana 2007, 2008), Tjong (2011) also discussed the use of combinations of SG, EG, graphene nanoplatelets, and CNTs in achieving the dual goals of high electrical conductivity and strong mechanical properties.

17.4.3 Epoxy–Graphite Composites for Bipolar Plates

We now discuss how EG can be used to solve an important problem in fuel cell industry. Du and Jana (2007, 2008) developed electrically conductive epoxy compounds with a target of at least 100 S/cm of volume electrical conductivity. Such compounds were intended to replace graphite bipolar plates in PEM fuel cells. Bipolar plates are a major component in the PEM fuel cells, in terms of weight, volume, cost, and functions. Metals have been considered as bipolar plates due to high electrical and thermal conductivity, strong mechanical properties, and good gas impermeability, but they suffer from corrosion. Consequently, major focus is given in alternative materials based on polymers.

Polymer composite bipolar plates have chemical and electrochemical stability, good gas impermeability, lightweight, and low cost. However, polymer composite bipolar plates encounter the problem of low electrical and thermal conductivity. High filler loading is needed to obtain the high electrical and thermal conductivity, but results in poor mechanical properties.

Du and Jana (2007, 2008) considered synthetic or EG as the major conductive filler and small fractions of electrically conductive CB as a synergistic combination to obtain higher values of both in-plane and through-plane electrical conductivities. The intrinsic conductivity of graphite is very high, about 10^5 S/cm (Table 17.1). The compounds were prepared using a low shear solution mixing method to keep the special layered network structure of EG intact in the polymer composites such that electron conductive paths of the EG are also maintained. These authors used epoxy mixtures of bisphenol A type epoxy (Epon 826) and aliphatic epoxy, polypropylene glycol diglycidyl ether (Araldite DY3601) in various weight ratios and cured the epoxy mixture with diaminodiphenyl sulfone to obtain materials with higher ductility and strength than with Epon 826 epoxy alone. The use of Araldite DY3601 added flexibility to the cross-linked epoxy network as reflected from the plot of storage modulus versus temperature in Figure 17.6 and flexural stress versus strain in Figure 17.7. Table 17.6 lists flexural strength and toughness of cured epoxy mixture at various weight ratios of Epon 826 and Araldite DY3601. It is seen that the epoxy system with 10 wt% Araldite DY3601 has higher

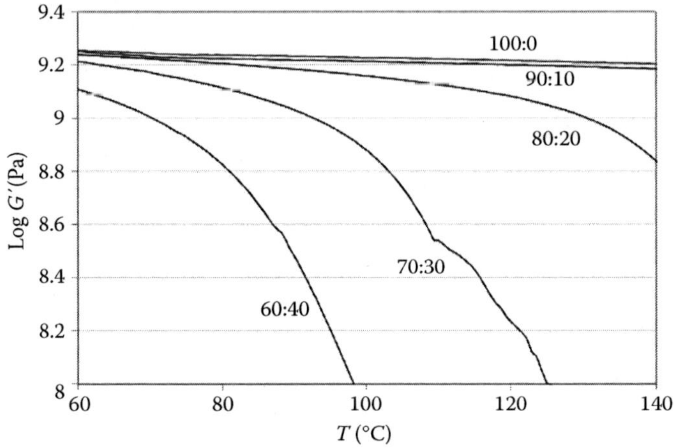

FIGURE 17.6
Storage modulus as a function of temperature for epoxy resins with different weight ratios of Epon 826 and Araldite DY3601. (From Du, L., "Highly conductive epoxy/graphite polymer composite bipolar plates in proton exchange membrane (PEM) fuel cells," PhD Thesis, University of Akron, OH, 2008. With permission.)

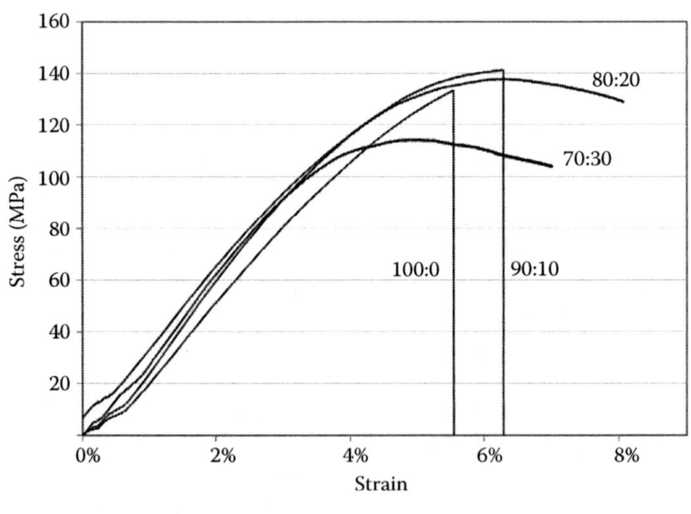

FIGURE 17.7
Flexural stress versus strain for epoxy resins with different weight ratios of Epon 826 and Araldite DY3601. (From Du, L., "Highly conductive epoxy/graphite polymer composite bipolar plates in proton exchange membrane (PEM) fuel cells," PhD Thesis, University of Akron, OH, 2008. With permission.)

TABLE 17.6

Flexural Strength and Toughness of Cured Epoxy Resin Systems

Epoxy Resins	Flexural Strength (MPa)	Toughness
EP100	136	3.91
EP90	141	5.58
EP80	138	7.69
EP70	114	4.79

Source: Du, L., PhD Thesis, University of Akron, OH, 2008.

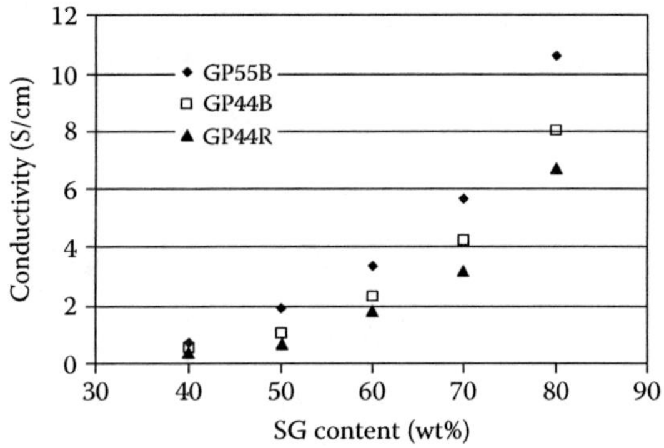

FIGURE 17.8
In-plane electrical conductivity of 90:10 epoxy mixture with synthetic graphite (SG) obtained from GrafTech (Cleveland, OH). (From Du, L., "Highly conductive epoxy/graphite polymer composite bipolar plates in proton exchange membrane (PEM) fuel cells," PhD Thesis, University of Akron, OH, 2008. With permission.)

toughness than cured Epon 826. This formulation was used by Du (2008) for the development of graphite-filled compounds using both synthetic and EG.

It is observed in Figure 17.8 that the conductivity of the composites increased with the increase of graphite content for all three SG grades, that is, GP55B, GP44B, and GP44R. Also, GP55B gave the highest conductivity at the same filler loadings compared to GP44B and GP44R. However, the overall conductivity of the composites is low, only about 10 S/cm even at the filler loading as high as 80 wt%. This conductivity is still much lower than the Department of Energy (DOE) target value of 100 S/cm for these materials to qualify for bipolar plates. It was observed that the resultant materials were very brittle. The combinations of SG and CB did not improve conductivity. The SEM image in Figure 17.9a indicates that the particle size of the synthetic was large, of the order of 75 μm with very dense structures. These particles were possibly coated by a thin layer of epoxy resin, resulting in very low electrical conductivity of the composites.

In comparison, the EG exhibited porous structures, low density, and large surface area due to layered and network structures (Figure 17.9b). It was

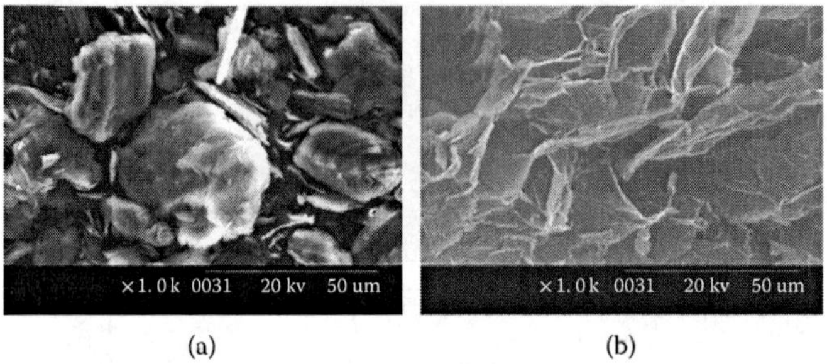

FIGURE 17.9
Comparison of SEM images of SG GP55B (a) and EG (b). (From Du, L., "Highly conductive epoxy/graphite polymer composite bipolar plates in proton exchange membrane (PEM) fuel cells," PhD Thesis, University of Akron, OH, 2008. With permission.)

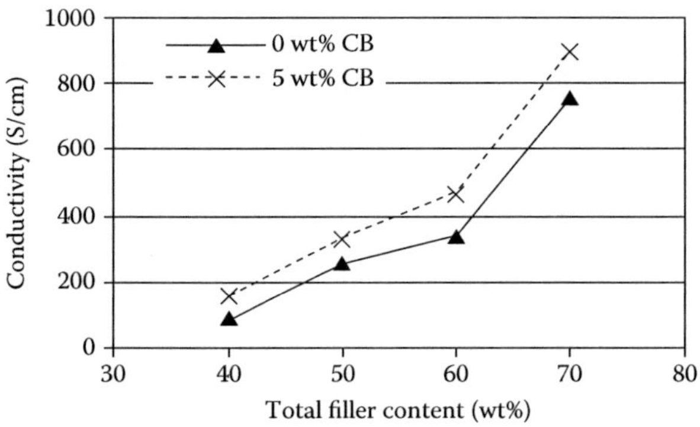

FIGURE 17.10
Electrical conductivity of composites filled with EG and EG combined with 5 wt% of CB. (From Du, L., "Highly conductive epoxy/graphite polymer composite bipolar plates in proton exchange membrane (PEM) fuel cells," PhD Thesis, University of Akron, OH, 2008. With permission.)

expected that the layer and network structures, if kept in the composites, would result in high electrical conductivity of the composites. The solution mixing method with low shear force was used to prepare the composites. Thus, the layered and network structures of the EG were maintained in the composites. These layered and network structures formed large fractions of electron-conducting paths in the composites, resulting in high electrical conductivity. The electrical conductivity of the EP90/EG/CB composites was measured by conventional four-point probe conductivity measurement device, and the data are shown in Figure 17.10. It is observed that the electrical conductivity increased with the increase of the total filler content for both composites filled with EG and composites filled with EG combined with 5 wt% CB. It is also seen that incorporation of 5 wt% of CB increased the electrical conductivity. For both composites, the electrical conductivity is higher than the DOE target of 100 S/cm (Robberg and Trapp 2003) at the total filler loading as low as 40 wt%. For SG-filled composites, the conductivity was only 15 S/cm at the filler loading as high as 80 wt%. Therefore, EG is much more effective as conductive filler compared to SG. Composites with combined fillers of EG and small amount of CB have higher conductivity compared to single filler EG-filled composites.

From the above results, we can see that the electrical conductivity increased when small amounts of CB was added to the systems for both EG- and SG-filled composites at the same filler loading content. There is a positive synergistic effect between graphite and CB. The optimization of CB content was discussed below. The electrical conductivity of epoxy composites filled with 60 wt% total filler content of EG and CB was measured and presented in Figure 17.11. It is seen that the in-plane electrical conductivity increased with the increase of the CB content up to 5 wt% and then decreased with further increase of the CB content. Composites with 5 wt% CB showed the highest electrical conductivity. This is because there is a positive synergistic effect on electrical conductivity between EG and CB. However, the intrinsic conductivity of CB is much lower than that of EG, thus incorporation of too much CB into the composites decreases the electrical conductivity. In this study, 5 wt% of CB was selected as the minor conductive filler loading.

The in-plane electrical conductivity was largely influenced by the total filler content as well as the filler type and the degree of filler dispersion in the polymer matrix. In Figure 17.12, CB content was fixed at 5 wt% and EG content was increased up to 85 wt%. It is seen that the conductivity of epoxy composites increased with the increase of EG content. The conductivity increased monotonically with the increase of filler content up to 600 S/cm for 75 wt% and then jumped sharply to above 1000 S/cm at 85 wt% EG. The very high electrical conductivity of larger than 1000 S/cm at 85 wt% can be

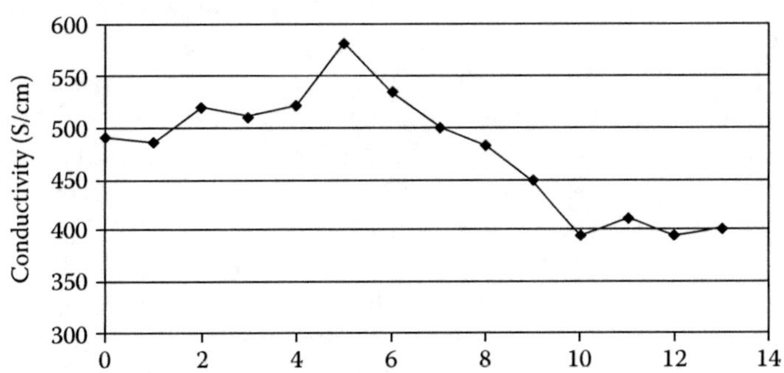

FIGURE 17.11
In-plane electrical conductivity of EG-filled epoxy composites as a function of CB content. The total filler content was 60 wt%. (From Du, L., "Highly conductive epoxy/graphite polymer composite bipolar plates in proton exchange membrane (PEM) fuel cells," PhD Thesis, University of Akron, OH, 2008. With permission.)

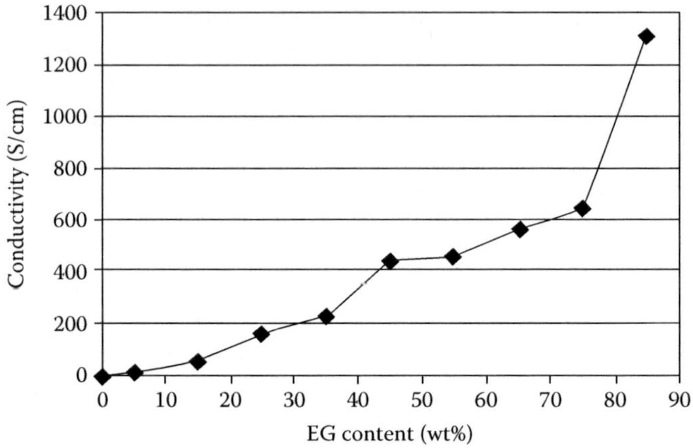

FIGURE 17.12
In-plane electrical conductivity of the EP90 composites as a function of EG content. The amount of CB is 5 wt%. (From Du, L., "Highly conductive epoxy/graphite polymer composite bipolar plates in proton exchange membrane (PEM) fuel cells," PhD Thesis, University of Akron, OH, 2008. With permission.)

attributed to physical contacts between the filler particles, which established large numbers of electron conducting pathways. One can argue that EG content as high as 85 wt% should be used to capitalize on the very high values of electrical conductivity. However, the mechanical properties were found to be poor when the content of EG was greater than 65 wt%. It was found that compression-molded laminates with filler content greater than 70 wt% were fragile and in some case can be fractured easily with the application of small load by hand. In view of this, we focused in this study on composites containing 35–65 wt% EG and a maximum of 5 wt% CB.

Next, in-plane electrical conductivity of EP90 composites without CB was measured and compared with that of EP90 composites with 5 wt% CB, as shown in Figure 17.10. It is seen that composites with 5 wt% CB provided higher conductivities over those without CB at the same total filler loading. This indicates that CB particles effectively formed additional conductive networks in the composites. This is counterintuitive as the intrinsic conductivity of EG is 10^5 S/cm, much higher than the value 10–100 S/cm of CB, and the presence of CB particles should have lowered the overall conductivity instead of increasing it. This can be interpreted in terms of a positive synergistic effect between EG and CB on electrical conductivity. Clingerman et al. (2002) also noticed positive synergistic effects using a combination of CB, SG, and carbon fiber.

The synergistic effect can be explained on the basis of additional conductive pathways formed in the presence of CB particles. Note that the intrinsic conductivity of EG is around 10^5 S/cm and that of CB is between 10 and 100 S/cm. Nevertheless, the macroscopic resistance of polymer composites depends on two contributions (Donnet et al. 1993): (1) the resistance of aggregates, $R_a = \rho_i/d$, where ρ_i is the resistivity of the filler and d is the diameter of the contact area and (2) the resistance of the inter-aggregate space R_e, which results from the tunneling resistance. An expression of R_e can be written as $R_e = \rho_t/A$, where ρ_t is the tunneling resistivity and A is the contact surface area of the particles. The total resistance of the composite, R, is the sum of all aggregate resistances $R_{a,i}$ and inter-aggregate space resistances $R_{e,i}$ and given as

$$R = \Sigma\, R_{a,i} + \Sigma\, R_{e,i} \tag{17.1}$$

Figure 17.13 schematically shows how various resistances contribute to the total resistance of the composites. The resistance of EG particles is very small owing to high value of intrinsic conductivity. In this case, therefore, the total resistance comes primarily from the resistance of the inter-aggregate space, R_e (Figure 17.13a). Accordingly, CB particles, present in the composites, provided additional connections between EG layers, thus reducing the value of R_e, and hence the total resistance R (Figure 17.13b). Note also that CB can efficiently impart electrical conductivity with a minimum loading because of highly branched and hollow structures, high surface area, and small aggregate sizes.

An evidence of the structural arrangement of EG and CB particles included in the resistivity model in Figure 17.13 can be obtained from the SEM images presented in Figure 17.14. These SEM images clearly show that CB particles uniformly distributed in epoxy/EG/CB composites, and that many CB aggregates were present between graphite layers (Figure 17.14b), thus imparting both high in-plane and through-plane conductivities between the adjacent graphite layers. In this manner, R_e was substantially reduced, resulting in a lower total resistance or higher total conductivity of the composites.

Bipolar plates must possess good mechanical properties to support thin membranes and electrodes and to withstand high clamping forces for the

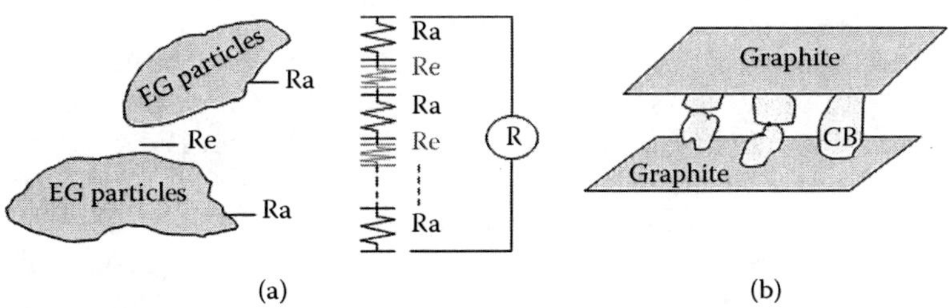

FIGURE 17.13
Schematic illustration of (a) how various resistances act in series and define the overall resistance, R; (b) CB particles placed between two graphite platelets to help reduce the value of R_e. (From Du, L., "Highly conductive epoxy/graphite polymer composite bipolar plates in proton exchange membrane (PEM) fuel cells," PhD Thesis, University of Akron, OH, 2008. With permission.)

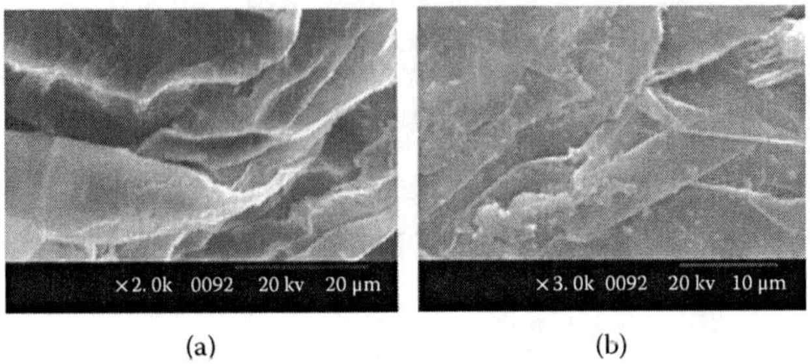

FIGURE 17.14
SEM images of (a) EP90/EG/CB (50/50/0) and (b) EP90/EG/CB (50/45/5). The granular images inside circles in (b) indicate CB particles. (From Du, L., "Highly conductive epoxy/graphite polymer composite bipolar plates in proton exchange membrane (PEM) fuel cells," PhD Thesis, University of Akron, OH, 2008. With permission)

stack assembly. However, high loading of conductive fillers required to reach high electrical conductivity may also introduce voids and defects in the resulting composites, especially if the filler loading is higher than a critical volume concentration, known in additive industry as the critical pigment volume concentration (Blunk et al. 2006; Donnet et al. 1993; Kawamoto 1982). In view of this, it is difficult to obtain simultaneously high electrical conductivity and sufficient mechanical strength from the same materials. In this study, a solution intercalation method was used to mix the epoxy and fillers with low shear force, which to some extent preserved the layered and network structures of EG particles in the composites. Also, low viscosity allowed epoxy resins and the curing agent to penetrate into the EG gallery and fill the micro- and macropores of the EG particles. In addition, the large pressure applied during epoxy curing led to the reduction of voids and defects. Furthermore, synergistic combination of EG and CB allowed substantial reduction of total filler loading to reach high electrical conductivity. Thus, several measures were taken in composite preparation method to eliminate possible sources of voids and defects.

Table 17.7 lists the mechanical properties of the composites. Impact strengths of the samples, 173 and 144 J/m for composites epoxy/EG/CB (50/50/0) and epoxy/EG/CB (50/45/5), respectively, significantly exceed the DOE target value of 40.5 J/m (Clulow et al. 2002; Robberg and Trapp 2003). The flexural strengths of epoxy composites without CB are also higher than the DOE target value of 59 MPa (Clulow et al. 2002; Robberg and Trapp 2003), although the flexural strengths decreased with the addition of 5 wt% CB. The tensile strengths of epoxy composites are lower than the DOE target value of 41 MPa except in the case of 50:50:0 ratio of epoxy/EG/CB.

It was seen earlier that a synergistic effect of combination of EG and CB on the electrical conductivity was observed because of the additional conductive bridges formed by CB between graphite layers. However, from Table 17.7,

TABLE 17.7

Mechanical Properties of Composites

Resin/EG/CB (by weight)	Flexural Modulus (MPa)	Flexural Strength (MPa)	Tensile Strength (MPa)	Impact Strength (J/m)
60/40/0	1.77×10^4	61	40	
60/35/5	0.75×10^4	40	29	
50/50/0	2.08×10^4	72	45	173
50/45/5	1.49×10^4	44	26	144
40/60/0	2.66×10^4	82	31	
40/55/5	1.76×10^4	56	25	
Target value (Clulow et al. 2002; Robberg and Trapp 2003)		59	41	40.5

we see that incorporation of 5 wt% of CB into the composites reduced the mechanical properties compared with single filler EG composites at the same total filler loadings. This is probably due to the disturbance of interfacial adhesion between EG particles and epoxy in the presence of CB particles. Note that small amounts of reactive functional groups such as carboxyl and epoxide groups on the surface of EG might have readily reacted with the epoxy matrix and developed strong adhesion between the interface of EG particles and epoxy matrix. This in turn resulted in stronger mechanical properties. The CB particles, on the other hand, did not contain much reactive functional groups on the surface, and the surface area was much smaller than that of EG particles. Thus, incorporation of CB into epoxy/EG composites generated poor adhesion between the interface of CB and polymer matrix and, thereby, resulted in inferior mechanical properties compared with those without CB.

The data regarding mechanical properties presented in Table 17.7 compare well with the properties of commercial bipolar plates or research reported in literature. For example, composites of polyphenylene sulfide (PPS) and polyethylene terephthalate with graphite and carbon nanofibers provided in-plane and through-plane conductivity of 230–271 and 18–25 S/cm, respectively (Huang et al. 2005). In addition, PPS composites provide flexural strengths of 96 MPa and tensile strength of 58 MPa (Huang et al. 2005). Therefore, the polymer composites developed in this study have the advantages of higher electrical conductivity and better mechanical properties over the commercial bipolar plates.

The flexural modulus and flexural strength of the epoxy composites as function of filler content are presented in Figures 17.15 and 17.16, respectively. We see that both flexural modulus and flexural strength are higher

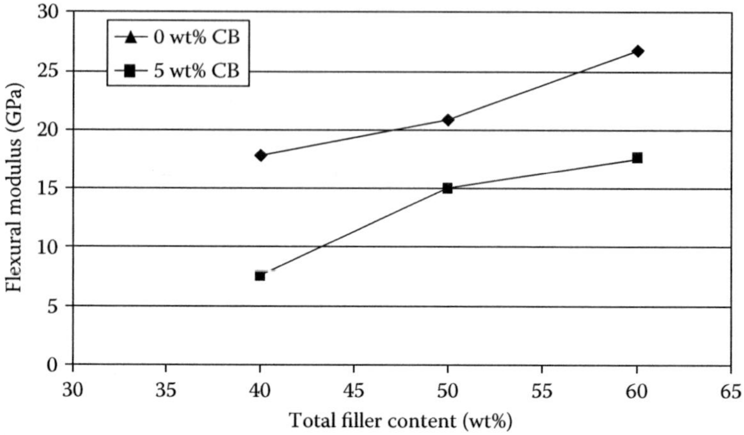

FIGURE 17.15
Flexural modulus of epoxy composites filled with EG or with combined EG and CB fillers. (From Du, L., "Highly conductive epoxy/graphite polymer composite bipolar plates in proton exchange membrane (PEM) fuel cells," PhD Thesis, University of Akron, OH, 2008. With permission.)

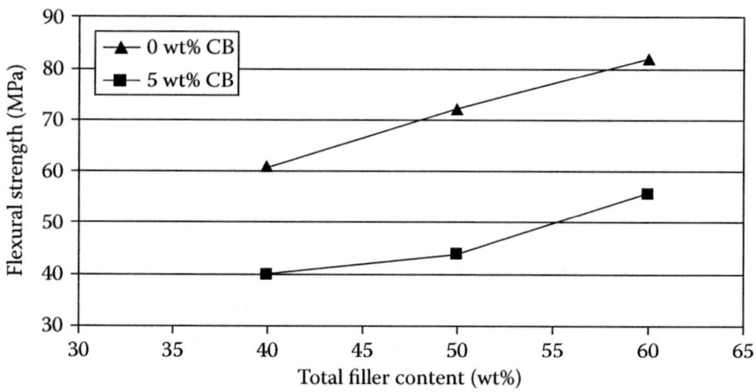

FIGURE 17.16
Flexural strength of epoxy composites filled with EG or filled with combined EG and CB fillers. (From Du, L., "Highly conductive epoxy/graphite polymer composite bipolar plates in proton exchange membrane (PEM) fuel cells," PhD Thesis, University of Akron, OH, 2008. With permission.)

for epoxy composites filled with the single filler EG than those of epoxy composites filled with combined EG and CB. It is also seen that both flexural modulus and flexural strength increased with the increase of total filler content. However, the flexural strain decreased with the increase of the total filler content. This shows that the materials became brittle at higher filler loadings.

17.5 Graphene-Based Polymer Composites

Graphene sheets—one-atom-thick two-dimensional layers of sp^2-bonded carbon—are predicted to have a range of unusual properties. Their thermal conductivity and mechanical stiffness may rival the remarkable in-plane values for graphite (3000 W/m/K and 1060 GPa, respectively); their fracture strength should be comparable to that of CNTs for similar types of defects. Recent studies have shown that individual graphene sheets have extraordinary electronic transport properties (Geim and Novoselov 2007; Novoselov et al. 2004, 2005).

One possible route to harnessing these properties would be to incorporate graphene sheets in composite materials, although several challenges must be overcome. For example, graphene sheets should be produced in commercial quantities and homogeneously distributed in polymer matrices. Graphite, inexpensive and available in large quantity, unfortunately does not readily exfoliate to yield individual graphene sheets. A general approach for the preparation of graphene–polymer composites entails complete exfoliation of graphite and molecular-level dispersion of individual, chemically modified graphene sheets within the polymer hosts.

Stankovich et al. (2006a) successfully developed fully exfoliated GO nanoplatelets. A number of chemically modified GOs were prepared by treating GO with organic isocyanates. The isocyanate treatment reduces the hydrophilic character of graphene oxide sheets by forming amide and carbamate ester bonds to the carboxyl and hydroxyl groups of GO, respectively. As a result, such isocyanate-derivatized GOs readily form stable dispersions in polar aprotic solvents such as *N,N*-dimethylformamide (DMF), consisting of completely exfoliated, functionalized individual graphene oxide sheets with thickness ~1 nm as determined by atomic force microscopy (AFM). These dispersions of isocyanate-derivatized GO allow graphene oxide sheets to be intimately mixed with many organic polymers, facilitating synthesis of graphene–polymer composites. Coupled with the possibility that GO can be chemically reduced, nanocomposites of polymer and isocyanate-derivatized GO were expected to be electrically conductive. Schniepp et al. (2006) used thermal exfoliation to produce graphene sheets. The dried GO is rapidly heated (>2000°C/min) to 1050°C and then split into individual sheets through the evolution of CO_2.

Stankovich et al. (2006a) developed electrically conductive graphene–polymer nanocomposites by solution-phase mixing of the exfoliated phenyl isocyanate-treated GO sheets with PS, followed by chemical reduction of GO. These composites feature individual graphene sheets well dispersed in the polymer matrix. The presence of the polymer in solution during the reduction step was the key to prevent agglomeration of the grapheme sheets. As the reduction proceeds, the sheets become coated with the polymer and remain individually dispersed. This PS/graphene composite exhibits a percolation threshold of 0.1 vol% for room temperature electrical conductivity, the lowest reported value for any carbon-based composite except for those involving CNTs; at only 1 vol% graphite loading, this composite has a conductivity of 0.1 S/cm, which is sufficient for many electrical applications. This bottom-up chemical approach of tuning the graphene sheet properties provides a path to a broad new class of graphene-based materials and their use in a variety of applications. An important factor that determines the final properties of the composites is the adhesion between the interface of filler and the polymer matrix. Therefore, knowing the surface chemistry of the filler phase and the polymer phase is very important.

Potts et al. (2011) observed about 15°C increase in glass transition temperature in polymethyl methacrylate (PMMA) in the presence of 0.05 wt% chemically modified graphene. The monomer methyl methacrylate was polymerized in solution in DMF with graphene particles predispersed by ultrasound. The elastic modulus increased by 28% with the addition of 1 wt% of graphene. These improvements were attributed to exfoliated particles of graphene as revealed from XRD. In similar effort, Fang et al. (2010) were able to produce single layer graphene nanocomposites of in situ polymerized PS. These composites also exhibited grafted PS chains and an increase of glass transition temperature by 9°C and 18°C, respectively, for low and high degree of grafting density. PS–graphene composites were prepared by dispersing chemically modified graphene oxide in PS latex (Tkalya et al. 2010). These composites showed percolation threshold for conduction below 1 wt% of graphene, and a maximum conductivity of about 0.15 S/cm was obtained with 1.6–2 wt% nanofiller.

The presence of exfoliated graphene particles also increased the microwave absorbing capacity of polyethylene oxide (PEO) (Bai et al. 2011). Chemically reduced graphene (~1 wt%) was dispersed in aqueous solution of PEO to fully exfoliated state of monolayers with thickness of ~1 nm and length of ~1.5 μm. Such nanoscale dispersion of graphene in PEO was attributed to hydrogen bonding. Once dispersed, agglomeration of graphene was prevented by hydrogen-bonded PEO chains. Chemically reduced graphene oxide nanosheets were also useful in bringing more than 200% increase in tensile strength and storage modulus of in situ polymerized elastomeric PUs at a small loading of only 2 wt% (Wang et al. 2011). The electrical conductivity increased from $\sim 1 \times 10^{-11}$ to $\sim 2.0 \times 10^{-8}$ S/cm. Gong et al. (2010) recently studied Raman spectra of single sheet of graphene placed on a film of PMMA and then subjected the film to tensile deformation. They showed that shift in Raman spectra could be easily recorded and translated into calculation of states of stress at the filler–polymer interface. They also showed that micromechanics models based on continuum theory were valid to describe load transfer at nanoscale. Xu and Gao (2010) prepared polyamide 6/graphene nanocomposites by in situ polymerization of caprolactam in the presence of thermally reduced graphene oxide. These authors presented evidence of formation of two-dimensional brush-like morphology of polymer chains grafted on the surfaces of graphene. The melt-spun fibers of the composite showed 210% increase in tensile strength and 240% increase in Young's modulus with only 0.1 wt% graphene. Composites of GO with polycarbonate, high impact PS, and acrylonitrile–butadiene–styrene terpolymers were prepared by dispersing the filler particles in polymer solutions. The resultant composites offered much better flame retardance than pristine polymers (Sengupta et al. 2011).

An NMR study result described by Schniepp et al. (2006) has shown that GO as used in the thermal exfoliation process contains aromatic regions randomly interspersed with oxidized aliphatic six-membered rings. The oxidized rings contain C–O–C (epoxide) and C–OH groups, while the sheets terminate with C–OH and –COOH groups. Some of these functional groups are retained in the thermally exfoliated nanoplatelets of graphene stacks. Thus, when used as a nanofiller in a polymer, for example, PMMA matrix, these functionalized graphene stacks offer comparable or better thermal, mechanical, and electrical property enhancements than single walled CNTs. In contrast to pristine graphite, the graphene-derived sheets in GO (graphene oxide sheets) are heavily oxidized, bearing hydroxyl and epoxide functional groups on their basal planes, in addition to carbonyl and carboxyl groups located at the sheet edges. The presence of these functional groups makes graphene oxide sheets strongly hydrophilic, which allows GO to readily swell and disperse in water. A mild ultrasonic treatment of GO in water results in its exfoliation to form stable aqueous dispersions that consist almost entirely of 1-nm-thick sheets. In fact, at present, exfoliation of GO is the only way to produce stable suspensions of quasi-two-dimensional carbon sheets, making this a strategic starting point for large-scale synthesis of graphene sheets.

17.6 General Conclusions and Future Outlook

The following general concluding remarks can be made about the field of research on graphite–polymer composites:

- The field of research on graphenes has reached some maturity in identification of chemical functionalities and characterization of various physical forms of graphite particles, such as NG, SG, EG, GO, and finally chemically modified GO or graphenes. Interested readers are referred to three excellent review articles on the subject (Dreyer et al. 2010; Kuilla et al. 2010; Sengupta et al. 2011). It is our observation that researchers interchangeably use the words "graphene" and "graphene oxide" to indicate chemically, thermally, or electrochemically reduced graphene oxide, although they are not the same materials in terms of chemical functionality and particle morphology.

We also observed that some researchers use the term "exfoliated" graphite to really mean EG. In true sense, exfoliated graphite should refer to single platelet sheet of graphene, whether fully oxidized or partially reduced following chemistry as prescribed in Hummers method (Hummers and Offeman 1958). Therefore, an array of composite materials has been reported in literature with wide variation of percolation volume fractions and percentage improvements in mechanical and thermal transport properties. Only a handful of researchers uses companion electron microscope or AFM images to infer the state of graphite dispersion. A majority of authors delve directly into reporting significant improvements in properties.

- It is evident that the field of research involving chemically modified graphenes is expanding, while the use of NG and EG is waning. In graphene research, Hummers method has become a standard for the preparation of graphene oxides. Several investigators used such graphene oxides and exploited the carboxylic acid groups or alcoholic hydroxyls for further functionalization with polymerizable monomers or oligomers. Some researchers even were able to grow grafted polymer chains perpendicular to the surface of graphenes. Only in these cases, do we see large enhancements in glass transition of the polymers.
- Only a handful of studies considered melt blending of polymers and EG particles due to apparent difficulty in feeding high volume, low density particles. On the other hand, a large majority of studies considered the use of ultrasound for delamination and dispersion of chemically modified GO particles suspended in a polar solvent. The same state of dispersion is captured by in situ polymerization of the monomer or by dissolving a premade polymer.
- It is observed that retaining honeycomb structures of EG particles is a necessary condition for achieving high values of electrical conductivity. This also predetermines the methods of polymerization, which will be amenable to retaining such structures. Both condensation polymerization (epoxy, phenolic esters, etc.) and in situ free-radical chain polymerization (PS, PSAN, PMMA, etc.) have been successful.
- It is anticipated that the next 5 to 10 years will see a lot of growth on single platelet electronics and manufacturing in conjunction with flexible polymeric films used on flexible displays and photovoltaics. We also anticipate that graphene research may not permeate heavily in consumer industries. Companion processing research addressing the problems in rheology and structure formation by crystallization at or near graphene particle/polymer interface is still lacking. In addition, the cost of graphene is relatively higher than CNTs.

References

Bai, X., Zhai, Y., and Zhang, Y. 2011. Green approach to prepare graphene-based composites with high microwave absorption capacity. *J. Phys. Chem. C* 115:11673–11677.

Balandin, A. A., Ghosh, S., Bao, W., Calizo, I., Teweldebrhan, D., Miao, F., and Lau, C. N. 2008. Superior Thermal Conductivity of Single-Layer Graphene. *Nano Lett.* 8:902.

Blunk, R., Zhong, F., and Owens, J. 2006. Automotive composite fuel cell bipolar plates: Hydrogen permeation concerns. *J. Power Sources* 159:533–542.

Brodie, B. C. 1859. *Philos. Trans. R. Soc. London* 149:249–259.

Chen, H., Liu, H-B., Li, J-X., Yang, L., and He, Y-D. 2009. Characteristics and preparation of polymer/graphite composite bipolar plate for PEM fuel cells. *J. Compos. Mater.* 43:755–767.

Chen, X-M., Shen, J-W., and Huang, W-Y. 2002. Novel electrically conductive polypropylene/graphite nanocomposites. *J. Mater. Sci. Lett.* 21:213–214.

Chiang, C-L., and Hsu, S-W. 2010. Synthesis, characterization and thermal properties of novel epoxy/expandable graphite composites. *Polym. Int.* 59:119–126.

Choi, S. Y., Mamak, M., Cordola, E., and Stadler, U. 2011. Large scale production of high aspect ratio graphite nanoplatelets with tunable oxygen functionality. *J. Mater. Chem.* 21:5142–5147.

Clingerman, M. L., Weber, E. H., King, J. A., and Schulz, K. H. 2002. Synergistic effects of carbon fillers in electrically conductive nylon 6,6 and polycarbonate based resins. *Polym. Compos.* 23:911–924.

Clulow, J. G., Zappitelli, F. E., Carlstrom, C. M., Zemsky, J. I. L., Busick, D. N., and Wilson, M. S. 2002. In *Fuel Cell Technology: Opportunities and Challenges, Topical Conference Proceedings, 2002 AIChE Spring National Meeting*, New Orleans, LA, March 10–14, 2002, pp. 417–425.

Dedov, A. V. 2001. Sorption properties of thermally expanded graphite. *Chem. Tech. Fuels Oils* 37:134–135.

Dhakate, S. R., Chauhan, N., Sharma, S., Tawale, J. S. S., Sahare, P. D., and Mathur, R. B. 2011. An approach to produce single and double layer graphene from re-exfoliation of expanded graphite. *Carbon* 49:1946–1954.

Donnet, J-B., Bansal, R. C., and Wang, M-J. (Eds.). 1993. *Carbon Black, Science and Technology*. New York: CRC Press.

Dreyer, D. R., Park, S., Bielawski, C. W., and Ruoff, R. S. 2010. The chemistry of graphene oxide. *Chem. Soc. Rev.* 39:228–240.

Du, L. 2008. Highly conductive epoxy/graphite polymer composite bipolar plates in proton exchange membrane (PEM) fuel cells. Ph.D. Thesis, University of Akron, OH.

Du, L., and Jana, S. C. 2007. Highly conductive epoxy/graphite composites for bipolar plates in proton exchange membrane fuel cells. *J. Power Sources* 172:734–741.

Du, L., and Jana, S. C. 2008. Hygrothermal effects on properties of highly conductive epoxy/graphite com posites for applications as bipolar plates. *J. Power Sources* 182:223–229.

Fang, M., Wang, K. G., Lu, H. B., Yang, Y. L., and Nutt, S. 2010. Single-layer graphene nanosheets with controlled grafting of polymer chains. *J. Mater. Chem.* 20:1982–1992.

Gadala-Maria, F., and Acrivos, A. 1980. Shear-induced structure in a concentrated suspension of solid spheres. *J. Rheol.* 24:799.

Geim, A. K., and Novoselov, K. S. 2007. The rise of graphene. *Nat. Mater.* 6:183–191.

George, J. J., and Bhowmick, A. K. 2008. Ethylene vinyl acetate/expanded graphite nanocomposites by solution intercalation: preparation, characterization and properties. *J. Mater. Sci.* 43:702–708.

Gomez-Navarro, C., Burghard, M., and Kern, K. 2008. Elastic Properties of Chemically Derived Single Graphene Sheets. *Nano Lett.* 8:2045.

Gong, L., Kinloch, I. A., Young, R. J., Riaz, I., Jalil, R., and Novoselov, K. S. 2010. Interfacial Stress Transfer in a Graphene Monolayer Nanocomposite. *Adv. Mater.* 22:2694–2697.

Hennig, G. 1951. The properties of interstitial compounds of graphite. I. The electronic structure of graphite bisulfate. *J. Chem. Phys.* 19:922–929.

Hofmann, U., and Holst, R. 1939. The acid nature and methylation of graphitic oxide. *Ber. Dtsch. Chem. Ges. B* 72:754–771.

Hong, C. M., Kim, J., and Jana, S. C. 2004. Shear-induced migration of conductive fillers in injection molding. *Polym. Eng. Sci.* 44:2101–2109.

Hong, S.-T., and Weil, K. S. 2007. Niobium-clad 304L stainless steel PEMFC bipolar plate material. *J. Power Sources* 168:408–417.

Huang, J., Baird, D. G., and McGrath, J. E. 2005. Development of fuel cell bipolar plates from graphite filled wet-lay thermoplastic composite materials. *J. Power Sources* 150:110–119.

Hummers, W. S., and Offeman, R. E. 1958. Preparation of graphitic oxide. *J. Am. Chem. Soc.* 80:1339.

Iijima, S. 1991. Helical microtubules of graphitic carbon. *Nature* 354:56–58.

Jang, J. Y., Kim, M. S., Jeong, H. M., and Shin, C. M. 2009. Graphite oxide/poly(methyl methacrylate) nanocomposites prepared by a novel method utilizing macroazoinitiator. *Compos. Sci. Technol.* 69:186–191.

Kawamoto, H., 1982. Rheology of carbon-polymer composites. In Sichel, E. K. (Ed.), *Carbon Black-Polymer Composites*. New York: Marcel Dekker.

Kim, I-H., and Jeong, Y. G. 2010. Polylactide/exfoliated graphite nanocomposites with enhanced thermal stability, mechanical modulus, and electrical conductivity. *J. Polym. Sci. B Polym. Phys.* 48:850–858.

Klemens, P. G. 2000. Theory of the a-plane thermal conductivity of graphite. *J. Wide Bandgap Mater.* 7:332.

Klemens, P. G. 2001. Theory of thermal conduction in thin ceramic films. *Int. J. Thermophysics* 22:265.

Krohn, H., Beck, F., and Zahn, A. 1988. *Die Ang. Makro. Chem.* 164:143–159.

Kroto, H. W., Heath, J. R., O'Brien, S. C., Curl, R. F., and Smalley, R. E. 1985. C60: buckminsterfullerene. *Nature* 318:162–163.

Kuilla, T., Bhadra, S., Yao, D., Kim, N. H., Bose, S., and Lee, J. H. 2010. Recent advances in graphene based polymer composites. *Prog. Polym. Sci.* 35:1350–1375.

Lafront, A-M., Ghali, E., and Morales, A. T. 2007. Corrosion behavior of two bipolar plate materials in simulated PEMFC environment by electrochemical noise technique. *Electrochim. Acta* 52:5076–5085.

Lee, C., Wei, X. D., Kysar, J. W., and Hone, J. 2008. Measurement of the Elastic Properties and Intrinsic Strength of Monolayer Graphene. *Science* 321:385.

Lerf, A., He, H., Forster, M., and Klinowski, J. 1998. Structure of Graphite Oxide Revisited. *J. Phys. Chem. B* 102:4477–4482.

Li, Y., Wu, D., and Chen, G. 2007. Preparation and characterization of high-density polyethylene/expanded graphite conducting masterbatch. *J. Appl. Polym. Sci.* 106:3119–3124

Liu, N., Luo, F., Wu, H. X., Liu, Y. H., Zhang, C., and Chen, J. 2008. One-step ionic-liquid-assisted electrochemical synthesis of ionic-liquid-functionalized graphene sheets directly from graphite. *Adv. Funct. Mater.* 18:1518–1525.

McAllister, M. J., Li, J-L., Adamson, D. H., Schniepp, H. C., Abdala, A. A., Liu, J., Herrera-Alonso, M., et al. 2007. Single Sheet Functionalized Graphene by Oxidation and Thermal Expansion of Graphite. *Chem. Mater.* 19:4396–4404.

Mirzazadeh, H., Katbab, A. A., and Hrymak, A. N. 2011. The role of interfacial compatibilization upon the microstructure and electrical conductivity threshold in polypropylene/expanded graphite nanocomposites. *Polym. Adv. Technol.* 22:863–869.

Nam, D-G., and Lee, H-C. 2007. Thermal nitridation of chromium electroplated AISI316L stainless steel for polymer electrolyte membrane fuel cell bipolar plate. *J. Power Sources* 170:268–274.

Nikam, V. V., and Reddy, R. G. 2005. Corrosion studies of a copper-beryllium alloy in a simulated polymer electrolyte membrane fuel cell environment. *J. Power Sources* 152:146–155.

Novak, I., and Krupa, I. 2004. Electro-conductive resins filled with graphite for casting applications. *Eur. Polym. J.* 40:1417–1422.

Novoselov, K. S., Geim, A. K., Morozov, S. V., Jiang, D., Katsnelson, M. I., Grigorieva, I. V., Dubonos, S. V., and Firsov, A. A. 2005. Two-dimensional gas of massless Dirac fermions in graphene. *Nature* 438:197.

Novoselov, K. S., Geim, A. K., Morozov, S. V., Jiang, D., Zhang, Y., Dubonos, S. V., Grigorieva, I. V., et al. 2004. Electric Field Effect in Atomically Thin Carbon Films. *Science* 306:666–9.

Phillips, R. J., Armstrong, R. C., and Brown, R. A. 1992. A constitutive equation for concentrated suspensions that accounts for shear-induced particle migration. *Phys. Fluids A* 4:30.

Potts, J. R., Lee, S. H., Alam, T. M., An, J., Stoller, M. D., Piner, R. D., and Ruoff, R. S. 2011. Thermomechanical properties of chemically modified graphene/poly(methyl methacrylate) composites made by in situ polymerization. *Carbon* 49:2615–2623.

Qu, S., and Wong, S-C. 2007. Piezoresistive behavior of polymer reinforced by expanded graphite. *Compos. Sci. Technol.* 67:231–237.

Ramanujam, B. T. S., Mahale, R. Y., and Radhakrishnan, S. 2010. Polyethersulfone-expanded graphite nanocomposites: Charge transport and impedance characteristics. *Compos. Sci. Technol.* 70:2111–2116.

Ramirez-Garcia, S., Alegret, S., Céspedes, F., and Forster, R. J. 2004. Carbon composite microelectrodes: charge percolation and electroanalytical performance. *Anal. Chem.* 76:503–512.

Robberg, K., and Trapp, V. 2003. Bipolar plate materials and flow field design. In Vielstich, W., Gasteiger, H. A., and Lamm, A. (Eds.), *Handbook of Fuel Cells—Fundamentals, Technology and Applications, Vol. 3: Fuel Cell Technology and Applications*. New York: Wiley & Sons, pp. 308–314.

Ruess, G. 1946. *Monatsh. Chem.* 76:381–417.

Schniepp, H. C., Li, J-L., McAllister, M. J., Sai, H., Herrera-Alonso, M., Adamson, D. H., Prud'homme, R. K., Car, R., Saville, D. A., and Aksay, I. A. 2006. Functionalized

Single Graphene Sheets Derived from Splitting Graphite Oxide. *J. Phys. Chem. B* 110:8535–8539.

Sengupta, R., Bhattacharya, M., Bandyopadhyay, S., and Bhowmick, A. K. 2011. A review on the mechanical and electrical properties of graphite and modified graphite reinforced polymer composites. *Prog. Polym. Sci.* 36:638–670.

Shin, H-J., Sim, K., Benayad, K. A., Yoon, S-M., Park, H. K., Jung, I-S., Jin, M., et al. 2009. Efficient reduction of graphite oxide by sodium borohydride and its effect on electrical conductance. *Adv. Funct. Mater.* 19:1987–1992.

Stankovich, S., Dikin, D. A., Dommett, G. H. B., Kohlhaas, K. M., Zimney, E. J., Stach, E. A., Piner, R. D., Nguyen, S. T., and Ruoff, R. S. 2006a. Graphene-based composite materials. *Nature* 442:282–286.

Stankovich, S., Dikin, D. A., Piner, R. D., Kohlhaas, K. A., Kleinhammes, A., Jia, Y., Wu, Y., Nguyen, S. T., and Ruoff, R. S. 2007. Synthesis of graphene-based nanosheets via chemical reduction of exfoliated graphite oxide. *Carbon* 45:1558–1565.

Stankovich, S., Piner, R. D., Nguyen, S. T., and Ruoff, R. S. 2006b. Synthesis and exfoliation of isocyanate-treated graphene oxide nanoplatelets. *Carbon* 44:3342–3347.

Staschewski, D., and Mao, Z-Q. 1999. Hydrogen-air PEMFC operation with extraordinarily low gas pressures and internal humidification—conception and experimental prototype stack. *Int. J. Hydrogen Energy* 24:543–548.

Staudenmaier, L. 1898. Preparation of graphitic acid. *Ber. Dtsch. Chem. Ges.* 31:1481–1487.

Stoller, M. D., Park, S. J., Zhu, Y. W., An, J. H., and Ruoff, R. S. 2008. Graphene-based ultracapacitors. *Nano Lett.* 8:3498.

Tawfik, H., Hung, Y., Hung, Y., and Mahajan, D. 2007. Metal bipolar plates for PEM fuel cell—A review. *J. Power Sources* 163:755–767.

Tjong, S. C. 2011. Polymer nanocomposite bipolar plates reinforced with carbon nanotubes and graphite nanosheets. *Energy Environ. Sci.* 4:605–626.

Tkalya, E., Ghislandi, M., Alekseev, A., Koning, C., and Loos, J. 2010. Latex-based concept for the preparation of graphene-based polymer nanocomposites. *J. Mater. Chem.* 20:3035–3039.

Tlili, R., Boudenne, A., Cecen, V., Ibos, L., Krupa, I., and Candau, Y. 2010. Thermophysical and electrical properties of nanocomposites based on ethylene-vinylacetate copolymer (EVA) filled with expanded and Unexpanded graphite. *Int. J. Thermophys.* 31:936–948.

Vieira, F., Cisneros, I., Rosa, N. G., Trindade, G. M., and Mohallem, N. D. S. 2006. Synthesis of a carbon nanotube monolith with controlled macroscopic shape. *Carbon* 44:2587–2592.

Wang, X., Hu, Y., Song, L., Yang, H., Xing, W., and Lu, H. 2011. In situ polymerization of graphene nanosheets and polyurethane with enhanced mechanical and thermal properties. *J. Mater. Chem.* 21:4222–4227.

Wang, W-P., Liu, Y., Li, X-X., and You, Y-Z. 2006. Synthesis and characteristics of poly(methyl methacrylate)/expanded graphite nanocomposites. *J. Appl. Polym. Sci.* 100:1427–1431.

Wang, Y., and Northwood, D. O. 2006. An investigation on metallic bipolar plate corrosion in simulated anode and cathode environments of PEM fuel cells using potential-pH diagrams. *Int. J. Electrochem. Sci* 1:447–455.

Wang, Y., and Northwood, D. O. 2007a. An investigation of the electrochemical properties of PVD TiN-coated SS410 in simulated PEM fuel cell environments. *Int. J. Hydrogen Energy* 32:895–902.

Wang, Y., and Northwood, D. O. 2007b. An investigation into TiN-coated 316L stainless steel as a bipolar plate material for PEM fuel cells. *J. Power Sources* 165:293–298.

Wang, Y., and Northwood, D. O. 2008. An investigation into the effects of a nano-thick gold interlayer on polypyrrole coatings on 316L stainless steel for the bipolar plates of PEM fuel cells. *J. Power Sources* 175:40–48.

Wang, W., and Pan, C-Y. 2004a. Preparation and characterization of polystyrene/graphite composite prepared by cationic grafting polymerization. *Polymer* 45:3987–3995.

Wang, W-P., and Pan, C-Y. 2004b. Synthesis and characterizations of poly(ethylene oxide) methyl ether grafted on the expanded graphite with isocyanate groups. *Eur. Polym. J.* 40:543–548.

Warner, J. H., Rümmeli, M. H., Ge, L., Gemming, T., Montanari, B. N., Harrison, M. Büchner, B., and Briggs, G. A. D. 2009. Structural transformations in graphene studied with high spatial and temporal resolution. *Nat. Nanotechnol.* 4:500–504.

Wissler, M. 2006. Graphite and carbon powders for electrochemical applications. *J. Power Sources* 156:142–150.

Wu, M., and Shaw, L. L. 2005. A novel concept of carbon-filled polymer blends for applications in PEM fuel cell bipolar plates. *Int. J. Hydrogen Energy* 30:373–380.

Xiao, M., Lu, Y., Wang, S. J., Zhao, Y. F., and Meng, Y. Z. 2006. Poly(arylene disulfide)/graphite nanosheets composites as bipolar plates for polymer electrolyte membrane fuel cells. *J. Power Sources* 160:165–174.

Xie, R., and Qu, B. 2001. Expandable graphite systems for halogen-free flame-retarding of polyolefins. I. Flammability characterization and synergistic effect. *J. Appl. Polym. Sci.* 80:1181–1189.

Xu, Z., and Gao, C. 2010. In situ Polymerization Approach to Graphene-Reinforced Nylon-6 Composites. *Macromolecules* 43:6716–6723.

Yoshimura, S., and Chang, R. P. H. 1998. *Supercarbon—Synthesis, Properties and Applications*. Berlin: Springer-Verlag.

Zabel, H., and Solin, S. A. 1990. *Graphite Intercalation Compounds I—Structure and Dynamics*. Berlin: Springer-Verlag.

Zaghib, K., Nadeau, G., and Kinoshita, K. J. 2000. Effect of graphite particle size on irreversible capacity loss. *Electrochem. Soc.* 147:2110–2115.

Zheng, G., Wu, J., Wang, W., and Pan, C. 2004. Characterizations of expanded graphite/polymer composites prepared by in situ polymerization. *Carbon* 42:2839–2847.

Index

D

E

H

T

U

V